Extrinsic and Intrinsic Approaches to Self-Healing Polymers and Polymer Composites

Extrinsic and Intrinsic Approaches to Self-Healing Polymers and Polymer Composites

Ming Qiu Zhang and Min Zhi Rong

Materials Science Institute, Zhongshan University
Guangzhou, China

WILEY

This edition first published 2022

Registered Office
John Wiley & Sons, Inc., 111 River Street, Hoboken, NJ 07030, USA

Editorial Office
111 River Street, Hoboken, NJ 07030, USA

For details of our global editorial offices, customer services, and more information about Wiley products visit us at www.wiley.com.

Wiley also publishes its books in a variety of electronic formats and by print-on-demand. Some content that appears in standard print versions of this book may not be available in other formats.

Library of Congress Cataloging-in-Publication Data applied for

[Hardback]: 9781119629955

Cover Design: Wiley
Cover Images: courtesy of Ming Qiu Zhang. Min Zhi Rong, Dong Yu Zhu and Nan Nan Xia

Set in 9.5/12.5pt STIXTwoText by Straive, Pondicherry, India

10 9 8 7 6 5 4 3 2 1

Contents

Preface

The global study of self-healing polymers and polymer composites started roughly at the beginning of this century based on the consensus that it represents a next-generation technology, which would greatly improve the performance of products, including, but not limited to, their reliability and durability. Two decades have past since then. When summarizing the present status, we find there has not yet been a landmark example of a commercial application. It turns out that our prediction of the development of the newly emerging innovation made in our book *Self-Healing Polymers and Polymer Composites*, which was published by Wiley in 2011, was too optimistic. Nowadays, scientists are still working hard in both theoretical and applied research. In this context, an updated book is necessary to reflect the latest achievements, which fits with the vigor and vitality of the study activities.

Being involved in the early stage of the research, we witnessed the evolution track of the movement. In general, the strategies of self-healing polymers and polymer composites are classified into two categories: extrinsic and intrinsic. The former makes use of the healing vessels (e.g. microcapsules, micropipelines, and vascular networks) embedded in the target materials to be repaired, which would be broken upon damaging the materials, releasing fluidic healing agent to the cracked sites and rebinding the cracks via chemical/physical interactions across the interface. In contrast, the latter operates by means of reversible intra- and/or intermacromolecular interactions, and no additional healing agent is required. A few groups investigated intrinsic self-healing of polymers as early as the 1970s from the angle of polymer physics. Entanglement of molecular chains acted as the main mechanism of wound healing. In 2001, Professor Scott White and his coworkers at the University of Illinois at Urbana-Champaign published a paper about self-healing in terms of healing capsules. Afterwards, extrinsic self-healing attracted plenty of research interest worldwide. In the meantime, intrinsic self-healing via chemical reversible covalent and noncovalent interactions grew so fast that it was ahead of extrinsic self-healing once again with respect to the materials species and properties to be restored. Nevertheless, chemistry has made a greater contribution this time, as more and more chemists have become involved, focusing on molecular design and synthesis as well as functionalities recovery. Intrinsic self-healing ability has been a standard feature of materials in some cases. In addition, the development of intrinsic self-healing has created other new accompanying techniques, such as reprocessing, reshaping and recycling, topology rearrangement, and processing difficult-to-process materials of traditionally nonreworkable thermosetting polymers. The new possibilities go beyond the scope of classic polymer engineering and enrich the measures of material diversification.

The style of writing and organization of our abovementioned book is followed herein, so that this book will provide the reader with an overall view of the ongoing research, and more importantly, inspiration for the development of novel materials and functionalities.

Chapter 1 serves as an introduction to the field of self-healing polymers and polymer composites. In addition to the general scope, the achievements made by different laboratories are carefully reviewed. Personal viewpoints of the authors are addressed showing the challenges, trends, and future directions. In Chapters 2–8, the self-healing strategies proposed by the authors' group are introduced, which are structured in accordance to the chemical reactions responsible for crack healing. Chapter 2 is dedicated to the healing reaction governed by addition polymerization and the self-healing systems based microcapsules and plastic tubes as well, which are able to work without manual intervention. High healing efficiency, determined by static fracture, impact and fatigue tests, is acquired for epoxy composites. To improve healing speed and widen windows of processing and operation of self-healing thermosetting materials, a few fast hardeners are introduced to work with epoxy monomer forming new groups of healing agent, as discussed in Chapter 3. Taking advantage of cationic polymerization, not only repair of cracks is possible at and below room temperature like the epoxy-mercaptan pair, but also crack healing is greatly accelerated. Besides, redox cationic polymerization is used for healing of thermoplastics. Chapter 4 describes the self-healing polymeric materials for advanced engineering applications, driven by anionic polymerization. The healing system consists of epoxy-loaded microcapsules and imidazole latent hardener. The latter can be well pre-dissolved in an uncured composites' matrix, leading to homogenous distribution of the reagent on a molecular scale. The major concern of Chapter 5 lies in usage of small molecule monomers as healing agent instead of epoxy monomer. Accordingly, nucleophilic addition, ring-opening reaction, atom transfer radical polymerization, and free radical polymerization prove effective in rebinding the cracked planes. Unlike the extrinsic self-healing approaches surveyed in Chapters 2–5, Chapter 6 deals with design, synthesis and characterization of intrinsic self-healing epoxy, in which thermally reversible Diels–Alder bonds account for crack healing via chain reconnection. In addition to the remendability, the cured version of the novel epoxy has similar mechanical performance to conventional epoxy. In Chapter 7, the intrinsic self-healing via reversible C—ON bonds is analyzed. Because of the synchronous fission/radical recombination of C—ON bonds, the polymers containing this type of reversible covalent bond can be self-healed in a one-step fashion, which prevents material distortion during healing as for those containing Diels–Alder bonds. Lastly, Chapter 8 demonstrates the application of reversible S—S bonds in self-healing polymers. The disulfide bonds in the tailor-made polymers as well as the commercial silicone elastomer and vulcanized rubber can be triggered under the stimuli of heating and sunlight, offering satisfactory healing efficiency.

It is our intention to emphasize integration of existing techniques and/or inventing novel synthetic approaches for application-oriented material design and fabrication. Having gone through the book, readers would have a comprehensive knowledge of the field, while new researchers might have an idea of the framework for creating new materials or new applications. Readers from both academic and industrial communities will be provided with a grasp of the achievements to date and an insight into future developments. In addition, graduate students may be able to combine theories learnt in the classroom with practical research and development of materials. These are the goals of this book.

We would like to acknowledge support from the Natural Science Foundation of China (Grants 52 033 011, 51 773 229, 51 673 219, 51 333 008, and 51 873 235). We would also like to thank the team at John Wiley & Sons for their assistance throughout the publication process. In addition, we hope that the publisher is successful with this new book.

Guangzhou, February 2021

Ming Qiu Zhang
Min Zhi Rong

1

Basics of Self-Healing – State of the Art

CHAPTER MENU

1.1 Background

Polymers and polymer composites have been widely used in important engineering fields including aerospace, marine, automotive, surface transport, and sports equipment because of their advantages including light weight, good processibility, and chemical stability in any atmospheric conditions. However, long-term durability and reliability of polymeric materials are still problematic when they are used for structural applications [1, 2]. This is particularly true when impact resistance is concerned, which is a critical aspect of vehicle design. Their inability to undergo plastic deformation results in energy adsorption via the creation of defects and damage. Besides, exposure to a harsh environment would easily lead to degradation of polymeric components. Comparatively, microcracking or hidden damage is one of the fatal deteriorations generated either during manufacturing or in service as a result of mechanical stress or cyclic thermal fatigue (Figure 1.1). Its propagation and coalescence would bring about catastrophic failure of the materials and hence significantly shorten the lifetime of the structures.

Extrinsic and Intrinsic Approaches to Self-Healing Polymers and Polymer Composites, First Edition. Ming Qiu Zhang and Min Zhi Rong.

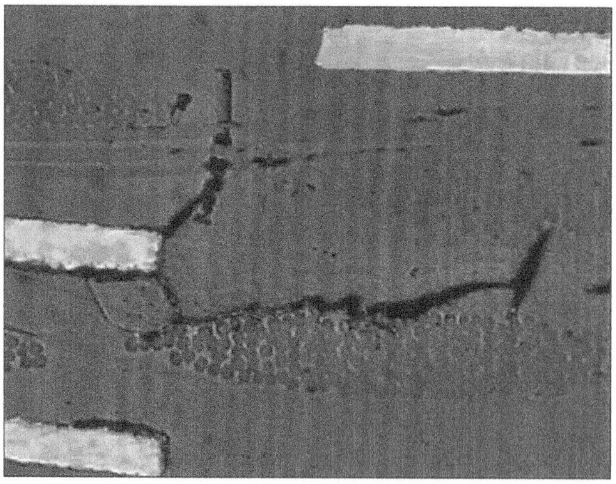

Figure 1.1 Cohesive failures beside plated through holes on copper clad laminates, which used to be produced by thermal stress.

1.1.1 Adhesive Bonding for Healing Thermosetting Materials

Damaged composites or composite structures should be repaired when significant structural degradation is detected [3]. The routine repair procedures for thermosetting composites are shown in Scheme 1.1. Some damage to composites is obvious and easily assessed but in some cases the damage may first appear quite small, although the real damage is much greater. Impact damage to a fiber can appear as a small dent on the reinforced composite surface but the underlying damage can be much more extensive. The decision to repair or scrap is determined by considering the extent of repair needed to replace the original structural performance of the composite [4]. Other considerations are the repair costs, the position and accessibility of the damage, and the availability of suitable repair materials. Easy repairs are usually small or do not affect the structural integrity of the component. Complex repairs are needed when the damage is extensive and the original structural performance of the component needs to be replaced. The best choice of materials would be to use the original fibers, fabrics, and matrix resin. Any alternative would need careful consideration of the service environment of the repaired composite, i.e. hot, wet, and mechanical performance. The proposed repair scheme should meet all the original design requirements for the structure. Some repairs need specialist equipment of a workshop and some form of improvised repair is needed to return the component to a suitable repair workshop. A temporary repair, usually in the form of a patch, can be fixed to the component. Usually a "belt and braces" approach is taken to ensure safety until the component can be repaired at a later date.

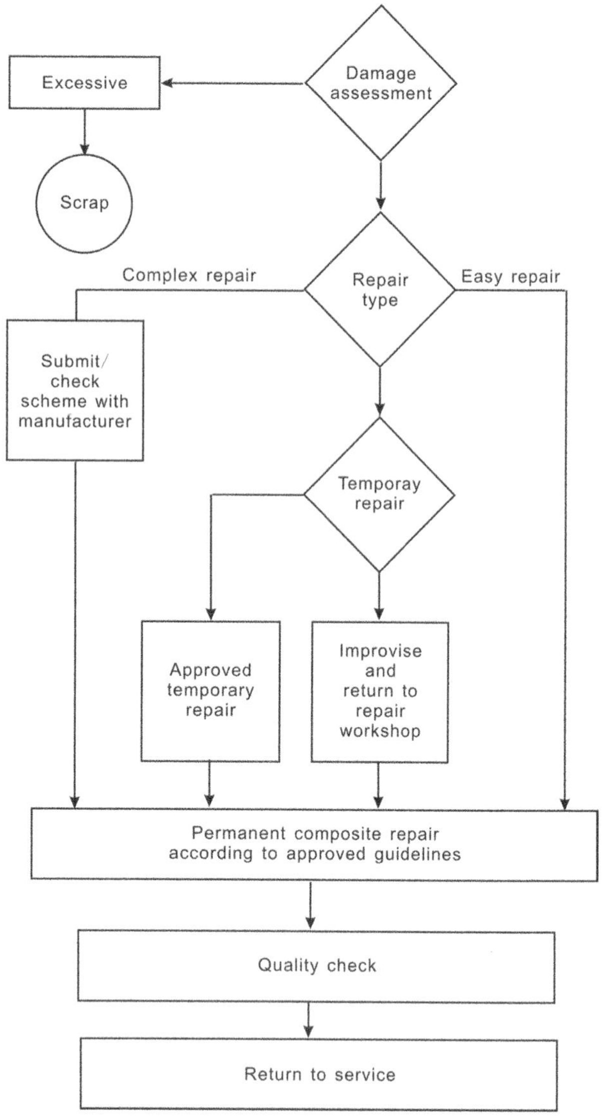

Scheme 1.1 Flow chart of the key stages for thermosetting composite repair.

Most damage to fiber reinforced polymer composites is a result of low velocity (and sometimes high velocity) impact [5]. In metals the energy is dissipated through elastic and plastic deformations, and a good deal of structural integrity is retained. In polymer composites the damage is usually more extensive than that seen on the surface. Typical damage is summarized in the following. (i) Delamination following impact on a monolithic laminate. (ii) Laminate splitting, which does not extend through the full length of the part. Its influence on the mechanical performance depends on the length of split relative to the component thickness. (iii) Heat damage, a local fracture with separation of surface plies. Its effect on the mechanical performance depends on the thickness of the part.

(iv) Dents in a sandwich structure. (v) Puncture damage in a sandwich structure. (vi) Bolt hole damage, which could be elongation of the hole causing laminate splitting, or damage to the upper plies.

Patch repair, a main technique based on adhesive bonding, involves covering or replacing of the damaged portions with a new material [5–7]. It restores the load path weakened or removed by damage or cracking, ideally without significantly changing the original load distribution. Reinforcements or doublers are used to replace lost strength or stiffness, correct design errors, or to improve performance [8]. Since the main purpose of composite repair is to fully support applied loads and to transmit applied stresses across the repaired area, the patch repair materials must overlap, and be adequately bonded to the plies of the original laminate. In this case, the thickness of the original laminate is made up of filler plies and the repair materials are bonded to the surface of the laminate. The advantages of this approach include the fact that it is quick and simple to do, and there is minimum preparation, while the repaired laminate is thicker and heavier than the original and very careful surface preparation is needed for good adhesion. The degree of property recovery is a function of bonding between the patch and the original material, the presence/orientation of reinforcing fibers and patch thickness [6, 9–12].

In addition to patch repair, there are two similar techniques: (i) taper sanded or scarf repair; and (ii) step sanded repair. For the first one, an area around the hole is sanded to expose a section of each ply in the laminate. Sometimes one filler ply is added to produce a flatter surface. Taper is usually in the region of 30–60:1. Comparatively, the repaired version is only marginally thicker than the original. Each repair ply overlaps the ply that it is repairing, so a straighter and stronger load path is obtained. The freshly exposed surfaces help to achieve tight bonds at the interface. With respect to the second technique, the laminate is sanded down so that a flat band of each layer is exposed, producing a stepped finish. Typical steps are 25–50 mm per layer. Nevertheless, it is worth noting that the method needs high skill and is difficult to do.

To conduct bonded external patch repair for structural components, equipment and ancillaries must be employed (Figure 1.2). The vacuum bag is suited to components with thin sections and large sandwich structures. It involves the placing and sealing of a flexible bag over a composite lay-up and evacuating all the air under the bag. The removal of air forces the bag down onto the lay-up with consolidation pressure of 1 atm. The completed assembly, with vacuum still applied, is placed inside an oven with good air circulation, and the composite is produced after a relatively short cycle cure.

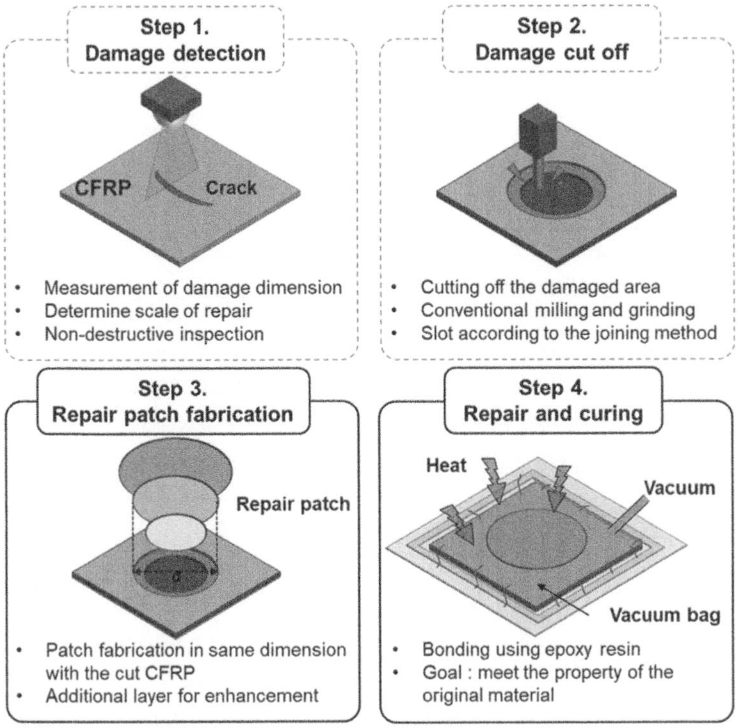

Figure 1.2 Typical repair processes of thermosetting composites represented by carbon fiber reinforcement polymer (CFRP). *Source:* Reprinted from Kim et al. [4]. Copyright 2019, with permission from Elsevier.

1.1.2 Fusion Bonding for Healing Thermoplastic Materials

In general, the aforementioned thermosetting adhesive bonding is not directly transferable to thermoplastic polymer-based composite. Fusion bonding, or welding, a long-established technology in the thermoplastic industry, offers an effective way for rejoining fractured surfaces with thermal flowability [13]. Although welding may induce residual stresses if performed without adequate control, it eliminates the stress concentrations created by holes required for mechanical fasteners and so does thermosetting adhesive bonding. In addition, welding reduces processing times and surface preparation requirements [14]. However, the high content of carbon fiber reinforcement in the composites, resulting in high thermal and electrical conductivity, imposes difficulties such as uneven heating, delamination and distortion of the laminates. These problems become more difficult when bonding large components [15]. In addition, as fiber volume fraction increases, the amount of resin available to melt and reconsolidate into a fused joint is reduced and this can affect the welding quality [16].

Fusion bonding techniques can be classified according to the technology used for introducing heat [13, 16] (Scheme 1.2), namely bulk heating (co-consolidation, hot melt adhesives, and dual resin bonding), frictional heating (spin welding, vibration welding, and ultrasonic welding), electromagnetic heating (induction welding, microwave heating, dielectric heating, and resistance welding), and two-stage techniques (hot plate welding, hot gas welding, and radiant welding).

Bulk heating techniques such as autoclaving, compression molding or diaphragm forming are available for performing co-consolidation [17]. Co-consolidation is an ideal joining method as no weight is added to the final structure, no foreign material is introduced at the bond line, essentially no surface preparation is required, and the bond strength is potentially equal to that of the parent laminate. However, the entire part is brought to the melt temperature, and this generally implies the need for complex tooling to maintain pressure on the entire part and to prevent de-consolidation. Hot melt thermoplastic adhesive films may be inserted at the bond line to improve filling of parts mismatch. Inserting of an amorphous polymer interlayer proved to reduce the scatter of strength [18], which widens the processing window. The dual resin bonding, or amorphous bonding, involves co-molding an amorphous thermoplastic film to a semi-crystalline thermoplastic matrix laminate prior to bonding [19]. During the joining step, the amorphous polyetherimide (PEI) film can be fused at a temperature above its glass transition; below the melting temperature of the semi-crystalline polyether ether ketone (PEEK) polymer; avoiding any deterioration of the bonded structure [20].

Spin welding and vibration welding have been extensively used in the plastics industry but are less appropriate to joining thermoplastic composites as the motion of the substrates relative to one another may cause deterioration

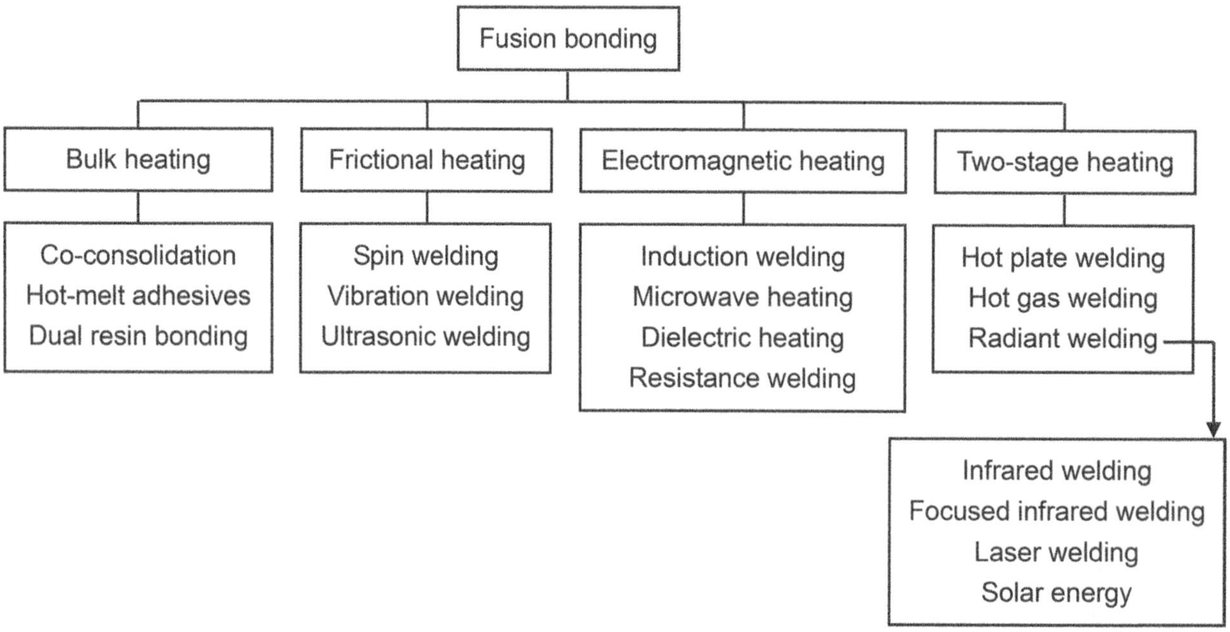

Scheme 1.2 Fusion bonding techniques. *Source:* Reprinted from Ageorges et al. [13]. Copyright 2001, with permission from Elsevier.

of the microstructure, such as fiber breaking. The process was however investigated for joining carbon fiber/PEEK [16] and glass fiber/polypropylene (PP) [21] systems. Microwave and dielectric welding are available for joining thermoplastics [22] but the fact that heating occurs volumetrically and that multilayer composites are excellent shields in the microwave range [23] make these techniques unsuitable for the welding of thermoplastic composites particularly when they are reinforced by carbon fibers.

Ultrasonic welding [24], induction welding [25], and resistance welding [26] are the three most promising fusion bonding techniques. Only the welding interface is brought to the melt temperature, and the impact on the rest of the structure is minimized. Welding times are very short. Large-scale welding may be performed through sequential or scanning approaches, and on-line monitoring of the consolidation is possible.

In two-stage techniques the heating device needs to be removed from between the substrate surfaces between the stages of heating and forging. This aspect involves limitations on size of the component since the whole welding surface must be heated in a single step [27]. Heating times are normally long as they rely on the low thermal conduction of heat through the polymer. Between the heating and forging steps, surface temperature drops and the region experiencing the maximum temperature is located below the skin of the laminate. The high pressure required to consolidate the bond line may cause warpage/flow in the higher temperature inner region [22].

1.1.3 Bioinspired Self-Healing

Although the repair strategies discussed in the previous sections have demonstrated their capability of recovering the load-bearing property of polymers and polymer composites, the complicated procedures typically represented by Figure 1.2 show that they are time consuming and cost ineffective, not including the losses resulting from malfunction of the components. The possible solution of this problem lies in early elimination of cracks, so that no macroscopic damage would eventually occur. As the cracks deep inside the materials are difficult to be perceived and to repair, it would be better if the materials have the ability of self-healing like biological systems.

In fact, self-healing is almost universal in nature. Most structures can repair themselves, after undergoing nonfatal trauma or injury [28–33]. Exceptions are teeth and cartilage, which do not possess any significant vascularity. It is also true that brains cannot self-repair; however, other parts of the brain take up the lost functions.

When an injury causes a blood vessel wall to break, for example, platelets are activated. They change shape from round to spiny, stick to the broken vessel wall and each other, and begin to plug the break. They also interact with other blood proteins to form fibrin. Fibrin strands form a net that entraps more platelets and blood cells, producing a clot that plugs the break [34] (Figure 1.3). The blood clotting is factually a protective mechanism that prevents excessive blood from being lost after an injury and prevents bacteria from getting into the wound. Normal clotting takes place within five minutes. For healing of a broken bone, the following processes are conducted in an autonomic way: internal bleeding forming a fibrin clot, development of unorganized fiber mesh, calcification of fibrous cartilage, and conversion of calcification into fibrous bone and lamellar bone. Clearly, the natural healing in living bodies depends on rapid transportation of repair substance to the injured part and reconstruction of the tissues.

Having been inspired by these findings, continuous efforts are now being made to mimic natural materials and to integrate self-healing capability into polymers and polymer composites. A series of healing concepts that offer the ability to restore the mechanical performance of materials have been proposed and successfully applied. The healing mechanisms and methods involved in nature, such as bleeding, blood cells, and blood flow vascular network, were simulated in the form of microcapsules [36], hollow fibers [37],

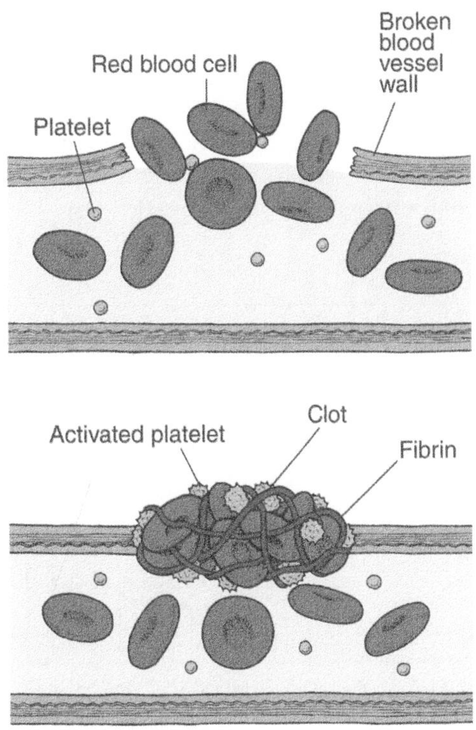

Figure 1.3 Blood clotting in an injured vessel. *Source:* Reprinted from Merck Sharp & Dohme Corp. [35]. Copyright 2020, Merck Sharp & Dohme Corp., a subsidiary of Merck & Co., Inc., Kenilworth, NJ.

nanoparticles [38], and interconnected microchannels [39]. The timescale for realization of self-healing within engineered structures is considerably reduced by a comprehensive exploration and study of the many examples of how the natural world undertakes the process. Biomimicry of the complex integrated microstructures and micromechanisms found in biological organisms offers considerable scope for the improvement in the design of future multifunctional materials [29]. The progress has opened an era of new intelligent materials.

Among the important achievements, the approach using microencapsulation of fluidic healing agent developed by White et al. in 2001 [36] marks a milestone. Since then, the number of scientific publications on self-healing polymers and polymer composites has been significantly increasing (Figure 1.4). The year 2007 was a turning point, meaning an "international race" started, as reflected by both the quantities of published papers and the affiliations of the authors. As of 2015, the number of related reports has multiplied as more and more researchers from different disciplines have become involved. The driving forces may have come from, for example, the rapid consumption of unrenewable crude oil, ecological concerns, advanced application requirements, and the miniaturization and integration of products. A biennial international forum, the International Conference on Self-Healing Materials, has also been established. The first one was held at Noordwijk aan Zee, the Netherlands in 2007. Innovative measures and revolutionary knowledge of the related mechanisms are constantly emerging. It is interesting that the results from different groups complement each other. As a result, the knowledge framework of self-healing polymeric materials is gradually being perfected like a jigsaw puzzle.

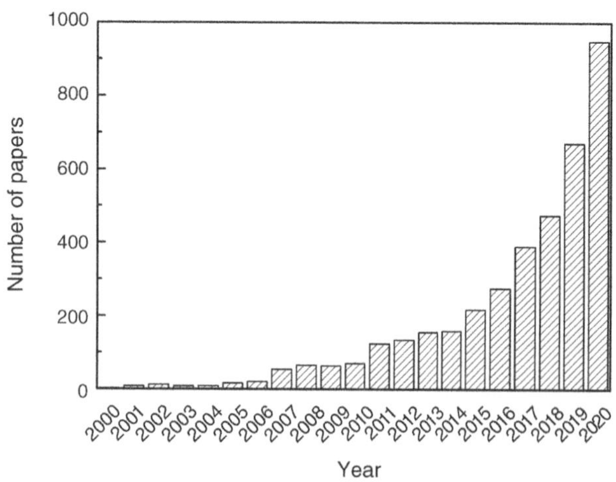

Figure 1.4 Number of research papers on self-healing polymers and polymer composites published between 2000 and 2020 according to the statistics of ISI Web of Knowledge.

One thing that needs to be mentioned is that self-healing has been developed in many laboratories in the world, while a universally accepted definition of this issue is yet to be made. We tend to include a wider scope that considers the following two types according to the ways of healing [40]:

- autonomic (without any intervention);
- non-autonomic (needs intervention/external triggering).

Since this monograph is devoted to the development of new healing chemistry, the focus of the review of the state-of-the-art in the subsequent sections of this chapter will be on the origination of healing capability and the ways of substance supply and energy supply. Accordingly, self-healing polymers and polymer composites will be classified into two categories [41] for the convenience of discussion: (i) intrinsic ones that are able to heal cracks by the polymers themselves without the need of additional healing agent; and (ii) extrinsic in which healing agent has to be intentionally pre-embedded. The prototypes worked out by our group are not analyzed in this chapter but are discussed in Chapters 2–8.

Finally, as viewed from the ultimate outcomes of repair, self-healing would lead to: (i) full-scale restoration; and (ii) functionality restoration. The former recovers the materials to their previous status quo, while the latter recovers the principal function of the materials. A self-healing anticorrosion coating that operates relying on the embedded inhibitors is a typical example of the latter effect. The authors of this book are more interested in the former.

1.2 Intrinsic Self-Healing

The so-called intrinsic self-healing polymers and polymer composites are based on specific molecular structures and performance of the polymers and polymeric matrices that enable crack healing through inter- and intramacromolecular disconnection and reconnection. Autonomic healing without manual intervention is not available in all cases for the time being. As viewed from the predominant molecular mechanisms involved in the healing processes, the reported achievements mainly consist of three modes: (i) physical interactions; (ii) reversible covalent chemistry; and (iii) supramolecular interactions.

The self-healing based on physical interactions started with crack healing in thermoplastic polymers in the late 1970s. Thermal activation induced macromolecular chain entanglements across the cracked interface, which represents the main healing mechanism. Wool et al. systematically studied the theory involved [42, 43]. They pointed out that the healing process goes through five phases: (i) surface rearrangement, which affects initial diffusion function and topological feature; (ii) surface approach, related to healing patterns; (iii) wetting; (iv) diffusion, the main factor that controls recovery of mechanical properties; and (v) randomization, ensuring disappearance of the cracking interface. In addition, Kim and Wool [44] proposed a microscopic model for the last two phases on the basis of the reptation model that describes longitudinal chain diffusion responsible for crack healing.

On the other hand, thermomechanical healing proves to be valid for some specific polymers, such as ionomers. An ionomer is a copolymer that comprises repeat units of both electrically neutral repeating units and a fraction of ionized units (usually no more than 15%). With a rise in temperature, the polymer exhibits an order-to-disorder transition as a result of which the ionic clusters, although persisting, lose order and strength. As the temperature is further increased, the semi-crystalline polymer matrix melts, even though the disordered clusters remain and continue to provide increased melt strength [45]. These transformations play a vital role in the self-healing process immediately after impact. When the thermal energy of impact dissipates, the aforementioned reversibility ensures rapid solidification while reordering of the ionic clusters and physical cross-links follows more slowly. The works by Fall [46] and Kalista et al. [47–50] have shown the unique self-healing response in poly(ethylene-*co*-methacrylic acid), which is beyond the aforementioned external thermal activation induced melting [51, 52]. These films are able to heal upon ballistic puncture and sawing damage. This occurs through a heat generating frictional process, which heats the polymer to the viscoelastic melt state and provides the ability to rebond and repair damage. In contrast, a low speed friction event fails to produce sufficient thermal energy favorable to healing. Varley and van der Zwaag conducted a careful investigation on the mechanism involved in the ballistic stimulated self-healing process [45].

Hereinafter, self-healing driven by reversible covalent chemistry and supramolecular interactions is discussed.

1.2.1 Self-Healing Based on Reversible Covalent Chemistry

Reversible polymers share one property in common – reversibility – either in the polymerization process or in the cross-linking process [53, 54]. Such a feature offers versatile possibilities of repeated healing on a molecular scale. So far, intrinsic self-healing polymers can be classified into two groups in view of the properties to be recovered. One deals with restoration of mechanical properties, while the other concerns regaining non-structural functional properties (such as electrical conductivity, transparency, corrosion resistance, and superhydrophobicity) [55]. In this section, a few typical intrinsic self-healing polymers belonging to the former group are quoted according to the type of reversible covalent reaction involved (Table 1.1).

Reversible covalent reactions include general reversible covalent reactions (e.g. reversible addition, reversible condensation, and reversible redox) and dynamic reversible covalent reactions (e.g. reversible exchange and reversible fission/recombination) (Figure 1.5). The key difference between these two groups of reversible covalent reactions lies in the fact that the types of starting reactants of the forward and backward reactions are different in the case of general reversible covalent reactions. In contrast, the forward and reverse reactions of a dynamic reversible covalent reaction occur under the stimulus of a single triggering factor, without completed dissociation of the reversible covalent bonds. As defined by the characteristics of the reversible reactions, the corresponding healing processes are quite different. That is, the healing driven by general reversible covalent chemistry is conducted in a two-step fashion (e.g. at two temperatures), while that by dynamic reversible covalent chemistry proceeds in a one-step fashion (e.g. at a single temperature). The latter not only simplifies the healing process, but also avoids completely losing the load-bearing capacity and integrity of the materials due to disconnection of the reversible bonds during crack healing.

1.2.1.1 Healing Based on General Reversible Covalent Reactions

Thermally Reversible Cycloaddition

Many reported thermally reversible self-healing polymers are bonded by Diels–Alder (DA) linkages. The healing principle lies in disconnection of DA bonds of a polymer through retro-DA reaction at higher temperature followed by reconnection of the disconnected DA bonds via DA reaction at lower temperature. On account of the randomness of the molecular chains at the crack interface, the broken parts can be sewn up by the successive retro-DA and DA reactions. In 2002, Chen et al. first applied DA cycloaddition between multi-furan and multi-maleimide monomers to construct a thermally remendable cross-linked polymer (Scheme 1.3) [56]. The damaged specimen was treated at 150 °C and then cooled down to room temperature, recovering about 50% fracture toughness, which proved the feasibility of solid state reversible covalent reaction for crack rehabilitation. No catalyst was needed for the entire healing process. To improve solubility of the monomers, the authors synthesized two bismaleimide (BMI) monomers with lower melting points in a later work [57]. Higher healing efficiency was observed for the cross-linked polymer owing to the higher mobility of the dangling chains.

Table 1.1 Typical intrinsic self-healing polymers toward mechanical properties restoration based on reversible covalent chemistry.

Polymer[a]	T_g[b] (°C)	Mechanical strength[c–j]	Failure strain (%)	Healing condition	Mechanism	Healing efficiency (%)	Ref.
DA network	N/A	~150[a]	N/A	150 °C	DA cycloaddition	50% recovery of fracture load	[56]
DA network	N/A	~60[a]	N/A	115 °C, 30 min; 40 °C, 6 h	DA cycloaddition	80% recovery of fracture load	[57]
PEA	−30 (DMA)	25.9[b]	1077	RT, 53 d	DA cycloaddition	Average 45% recovery of strain at break	[58]
PFMES	7.8 (DSC)	1.53 ± 0.3[b]	458 ± 30	RT, 10 d	DA cycloaddition	73.7% of the toughness calculated from the area underneath stress-strain curve	[59]
Epoxy-anhydride	93.5–136.5 (DSC)	8.51–29.1[c]	N/A	110–126 °C, 20 min; 80 °C, 72 h	DA cycloaddition	65.9–96% recovery of critical stress of DCDC test	[60]
PU	N/A	25–46.5[b]	250–450	120–130 °C, 1.5–5 min; 55–60 °C, 24 h	DA cycloaddition	>80% recovery of tensile strength	[61–63]
PU	~60 (DMA)	22[b]	1385	NIR, 0.75 W cm^{-2}, 60 s	DA cycloaddition	96% recovery of tensile strength	[64]
PA	4 (DSC)	~0.25[b]	>300	90 °C, 7 h	DA cycloaddition	85% recovery of tensile strength	[65]
PMMA	N/A	44.3[d]	N/A	>280 nm, 10 min	Cinnamate dimerization	0.1~7.8% recovery of flexural strength	[66]
PU	−7.1, 41.2 (DSC)	2.26[b]	141	254 nm, 1 min; 365 nm, 90 min	Coumarin dimerization	70.2% recovery of tensile strength	[67]
Tetra-coumarins	74.5, 76.3 (DSC)	21.1–25.1[f]	N/A	254 nm, 5.41 J cm^{-2}; 365 nm, 16.66 J cm^{-2}	Coumarin dimerization	90% recovery of Vickers hardness	[68]
Bis-thymines	59.8 (DSC)	18.1[f]	N/A	<240 nm, 10 J cm^{-2}; 302 nm, 12.5 J cm^{-2}	Thymine dimerization	95.6% recovery of Vickers hardness	[69]
Dienes-thiol	−3.5 to −16 (DSC)	1.5–4[b]	N/A	RT, 3 d, 85% humidity	Boronic ester hydrolysis/formation	Complete recovery of tensile strength	[70]
PDMS	65 (DSC)	9.64 ± 0.28[b]	9.72 ± 0.19	70 °C, 12 h, water	Boroxine hydrolysis/formation	Complete recovery of tensile strength	[71]
PPG	14.6	3.9[b]	292	35% humidity, 48 h	Iminoboronate hydrolysis/formation	93% recovery of strain at break	[72]
PPG-PAA	−55 (DSC)	1.7–12.7[b]	182–659	Water, 1 min; RT, 6~18 h	Boroxine hydrolysis/formation	87.3–99% recovery of tensile strength	[73]
P(DA-co-BA)	N/A	4.6 ± 0.2[b]	300 ± 30	RT, 3 d, 75% humidity	Boroxine hydrolysis/formation	96.1% recovery of toughness	[74]
SBR	~−5 (DMA)	1.94 ± 0.18[b]	262 ± 18	80 °C, 24 h	Boronic ester exchange	80% recovery of tensile strength	[75]
PU	41 (DSC)	11	~3	70 °C, overnight	Iminoboronate exchange	Complete recovery of tensile strength	[76]
Epoxy-thiol	−35 (DMA)	~0.5[b]	~70	60 °C, 60 min	Disulfide-thiol exchange	Almost 100% recovery of tensile strength	[77, 78]

Epoxy-amine	−35.5 (DMA)	0.23^b	113	25 °C, 24 h, air	Disulfide exchange	91% recovery of tensile strength	[79]
PUU	−50.39 (DSC)	0.81 ± 0.05^b	3100 ± 50	RT, 24 h	Disulfide exchange	97% recovery of strain at break	[80]
PDS	−30 (DMA)	$\sim 12^b$	50–160	320–390 nm, 2000 mW cm^{-2}, 5 min	Disulfide exchange	Almost 100% recovery of tensile strength	[81]
VPB	N/A	$\sim 3^b$	~400	110 °C, 12 h	Disulfide exchange	75% recovery of tensile strength	[82]
VCR	N/A	$\sim 9^b$	~650	120 °C, 5 h	Disulfide exchange	92.0% recovery of apparent shear strength	[83]
PU	29.8 (DMA)	9.66^b	550	Sunlight, 4 h	Disulfide exchange	92.2% recovery of tensile strength	[84]
PHMA	20 (DSC)	7.57 ± 0.82^b	276 ± 17	120 °C, 70 kPa, 24 h	Disulfide exchange	85% recovery of tensile strength	[85]
PU	−8.2 (DMA)	6.8^b	9.23	RT, 2 h	Disulfide exchange	75% recovery of tensile strength	[86]
PU	−50; 12.5–62.5 (DMA)	$16.0\sim25.0^b$	1400–2000	70 °C, 24 h	Disulfide exchange	85.0–89.4% recovery of tensile strength	[87]
Epoxy-anhydride	113 (DMA)	5.24^g	0.5	160 °C, 1 h	Disulfide exchange	95% recovery of shear strength	[88]
PU	−26.9	1.31^b	380	0.6 W cm^{-2}, 457 nm blue laser, 35 °C, 30 min	Diselenide exchange	84% recovery of tensile strength	[89]
PU	−39.59 (DSC)	0.161^b	~140	RT, 12 h	Ditelluride exchange	93.8% recovery of tensile strength	[90]
PA	−21.2 (DMA)	0.19^b	270	25 °C, 24 h, air	Schiff base exchange	98.1% recovery of tensile strength	[91]
Epoxy-amine	19 (DSC)	$27–33^b$	~60	90 °C, 3 h	Schiff base exchange	86.3% recovery of tensile strength	[92]
Epoxy-vanillin	75.7 (DMA)	20.5^b	12	100 °C, 12 h	Schiff base exchange	93.2% recovery of tensile strength	[93]
PS	125 (DMA)	99.4^c	N/A	130 °C, 2.5 h, Ar	Alkoxyamine	75.7% recovery of critical stress of DCDC test	[94]
Epoxy-amine	52.2 (DMA)	0.38^e	N/A	90 °C, 1.5 h, Ar	Alkoxyamine	62.2% recovery of impact strength	[95]
PU	−60 (DMA)	0.2^e	N/A	15–25 °C, 48 h, air	Alkoxyamine	Over 90% recovery of impact strength	[96]
PU	20.3 (DMA)	6.78^b	175.2	80 °C, 2.5 h, Ar	Alkoxyamine	70.7% recovery of tensile strength	[97]
Epoxy-thiol	12 (DMA)	$\sim 0.25^b$	~55	25 °C, 24 h, air	Alkoxyamine	65.6% recovery of tensile strength	[98]
Acrylate-thiol	27 (DMA)	0.42^b	266	80 °C, 4 h, air	Alkoxyamine	Almost 100% recovery of tensile strength	[99]
PU/SBS/MWCNTs	N/A	6.63^b	~300	100 °C, 24 h	Alkoxyamine	81.4% recovery of tensile strength	[100]
PU	−7 (DMA)	$\sim 25^b$	~900	100 °C, 24 h	Alkoxyamine	90.1% recovery of tensile strength	[101]
PBA	−50	0.065 ± 0.011^b	N/A	Swollen in acetonitrile, 330 nm UV, 4 h	Trithiocarbonate	Almost 100% recovery of tensile strength	[102]

(Continued)

Table 1.1 (Continued)

Polymer[a]	T_g[b] (°C)	Mechanical strength[c–j]	Failure strain (%)	Healing condition	Mechanism	Healing efficiency (%)	Ref.
PU	−34	0.392 ± 0.196[b]	202 ± 46	Visible light, RT, 24 h	Thiuram disulfide	Average 97.4% recovery of tensile strength	[103]
PU	N/A	~0.08[b]	~500	RT, 24 h, air	Diarylbibenzofuranone unit	98% recovery of strain at break	[104]
PU	−58 (DSC)	~0.8[b]	~800	50 °C, 24 h	Diarylbibenzofuranone unit	Almost 100% recovery of tensile strength	[105]
PBD	N/A	~0.6[b]	~0.9	5–22 °C, 10–30 kPa, 15 min–24 h	Olefin metathesis	Restoring most of the strength	[106]
PCO	N/A	2.85 ± 0.38[b]	345 ± 80	50 °C, 16 h	Boronic ester exchange	Average 94.6% recovery of tensile strength	[107]
PLA	50–57 (DSC)	48–60[b]	4.5–5.5	$Sn(Oct)_2$, 140 °C, 30 min, 4 MPa	Transesterification	Maximum 67% recovery of strain at break and 102% recovery of tensile strength	[108]
Epoxy-acid	~20 (DMA)	~0.6[b]	~0.3	$Zn(Ac)_2$, 160 °C, 2 h	Transesterification	Almost 100% recovery of tensile strength	[109]
PBZ	144 (DMA)	2.88[h]	N/A	$Zn(Ac)_2$, 140 °C, 1 h	Transesterification	99% recovery of storage modulus	[110]
ENR/CABt	−11.3 (DMA)	4.5b	400	150 °C, 3 h	Transesterification	96% recovery of tensile strength	[111]
Epoxy-anhydride	70.3 (DMA)	46.5 ± 2.3[b]	1.95 ± 0.10	150 °C, 1 h	Transesterification	Almost 100% recovery of shear strength	[112]
PU	−52 (DSC)	0.93 ± 0.06[b]	301 ± 12	37 °C, 12 h	Urea bond	Average 87% recovery of strain at break	[113]
PU	−72.9 (DMA)	14.8[b]	1200	RT, 130 h	Oxime-urethane	92% recovery of tensile strength	[114]

[a] CABt, citric acid-modified bentonite; PAA, polyacrylic acid; PBA, poly(n-butyl acrylate); PBZ, polybenzoxazine; PCO, polycyclooctene; PDS, polydisulfide; P(DA-co-BA), poly(dopamine acrylamide-*co*-n-butyl acrylate); PDS, polydisulfide; PFMES, poly(2,5-furandimethylene succinate); PHMA, poly(hexyl methacrylate); PMMA, poly(methyl methacrylate); PPG, polypropylene glycol; PUU, polyurethane urea; SBR, styrene-butadiene rubber; Sn(Oct)2, stannous(II) octoate; VCR, vulcanized chloroprene rubber; VPB, vulcanized polybutadiene; $Zn(Ac)_2$, zinc(II) acetate.
[b] DMA, dynamic mechanical analysis; DSC, dfferential scanning calorimetry.
[c] Compact tension facture load, N.
[d] Tensile stress, MPa.
[e] Critical stress measured by double cleavage drilled compression (DCDC) test, MPa.
[f] Flexural strength, MPa.
[g] Impact strength, kJ m^{-2}.
[h] Vickers hardness, Hv.
[i] Shear strength, MPa.
[j] Storage modulus, GPa.

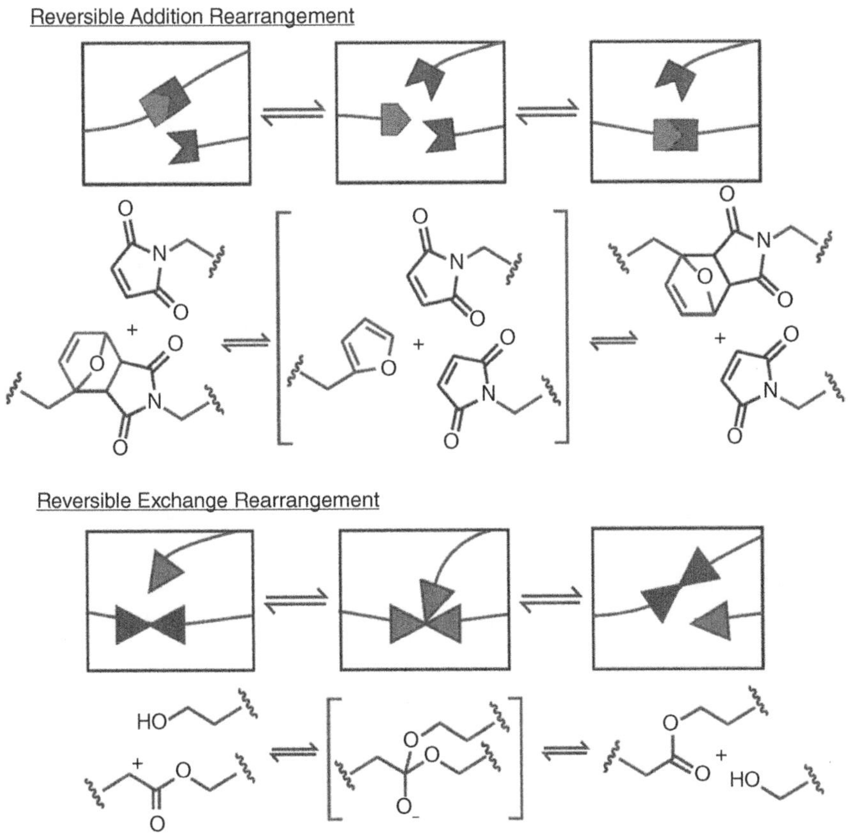

Figure 1.5 Connectivity transition difference for the reversible addition (top) and reversible exchange (bottom) cross-linking reactions. Based on these differences in connectivity, it is clear that both the equilibrium state of these two reactions as well as the dynamic exchange, i.e. the kinetics, between the various states are important. *Source:* Reprinted from Kloxin and Bowman [54]. Copyright 2013, with permission from the Royal Society of Chemistry.

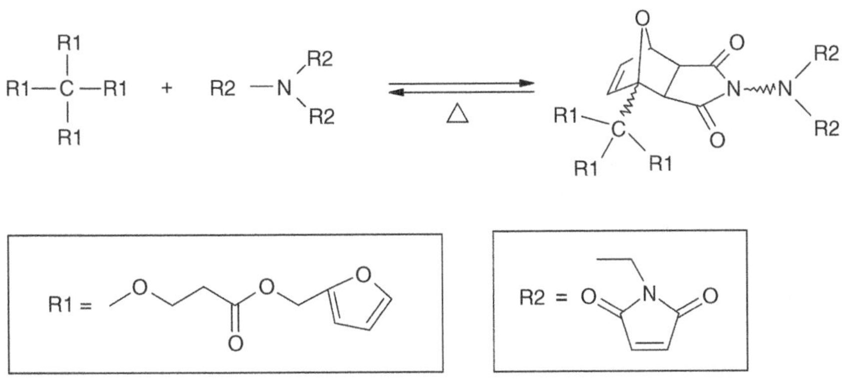

Scheme 1.3 Thermally reversible cross-linking based on Diels–Alder reaction. *Source:* Chen et al. [56]/Elsevier.

Subsequently, composite panels were prepared by sandwiching the DA monomers between carbon fiber fabric layers [115]. Microcracks that were induced on the resin-rich surface of the composites disappeared after treatment at 180 °C for one hour as a result of resistance heating. The experiments demonstrated the self-healing ability of the remendable polymers in both bulk form and fiber composites. Owing to the fast kinetic rate of the reaction of the polymer constituents, which requires a fast injection of the healable resin into the carbon fiber perform, a modified resin transfer mold (RTM) technique was developed accordingly [116]. Besides, Plaisted et al. proposed the concept of multifunctional composites [117], in which the cross-linked polymer from multi-furan and multi-maleimide [56] serves as the matrix, arrays of straight copper wires and copper coils as the conductive electromagnetic scattering elements, and synthetic fibers as the reinforcement. The scattering elements provide controlled electromagnetic response for tasks such as tuning the dielectric constant and filtering radio frequency radiation. Internal damage, in the form of polymer matrix cracking, may be repaired when heat is applied through the metallic wires. In the meantime, Kwok and Hahn carefully studied the influential factors when using carbon fibers as a resistive heating network for activating the self-healing function [118]. Various electrode methods were tested in an attempt to reduce the contact resistance that caused localized heating around the electrode. In addition, Park et al. confirmed multiple healing and shape memory effects of the composites after electrical resistive heating [119].

Afterwards, Liu and co-workers prepared a self-healable DA bond cross-linked epoxy and high toughness polyamide [41, 120–122], respectively. When the scratched epoxy surface was thermally treated at 120 °C for 20 minutes and at 50 °C for 24 hours, the fissure completely disappeared. However, the cracks on the polyamide film were only partly healed even when the healing time was as long as five days due to the low mobility of polyamide chains. The result showed that the healing behavior was diffusion-controlled [123].

Kavitha and Singha applied click chemistry to make poly(methyl acrylate) (PMA) bearing reactive furfuryl functionality, which was then reacted with BMI to form cross-linked poly(furfuryl methacrylate)-bismaleimide (PFM-BMI) [124]. DA bonds acted as the cross-linking sites, so that the polymer can be de-cross-linked through the retro-DA reaction. Accordingly, a complete notch (knife-cut) recovery to regain the original structure was detected by scanning electron microscopy (SEM) on PFM-BMI film after treatment at 120 °C for four hours.

Murphy et al. simplified the dual monomer route for the synthesis of remendable polymers from DA adducts by replacing the furan-maleimide pair with a single monomer that contained a dicyclopentadiene core unit [125]. The new single-component remendable polymer system utilized the dicyclopentadiene moiety as both diene and dienophile in the thermally reversible DA cycloaddition reaction. Additionally, the Staudinger cross-linking of the dienophilic DA dimer adduct double bond of the growing polymer chains was found to be the key to the strength of the materials. Fracture tests showed that the healing treatment of 120 °C for 20 hours in an argon atmosphere led to an average 46% healing efficiency. Similarly, thermally remendable composites with a dicyclopentadiene-based polymer as the matrix and graphite fibers as the reinforcements were also fabricated [126]. Since the graphite fibers can act as electrical conductors to provide the necessary heat to the polymer, resistance heating was used to stimulate the healing. It was found that microcracks can be healed within minutes at temperatures ranging from 70 to 100 °C.

Syrett et al. used DA chemistry to synthesize polymerization initiators and a dimethacrylic cross-linker that leads to cleavage and reformation required by self-healing [127]. The linker exhibited cleavage properties with 50% reformation occurring upon the reheating cycle. Linear and star methyl methacrylate polymers bearing DA adducts within their macromolecular backbone prepared via living radical polymerization were also preliminarily evaluated to check their ability to cleave and reform under external thermal stimuli.

Cheng et al. [128] extended DA chemistry to organic/inorganic hybrid materials with polyphosphazenes as matrices. Crack healing was performed by DA de-cross-linking at 120 °C for two hours and re-cross-linking at 60 °C for 3~5 days. The authors suggested that appropriate content of the pendant furan moieties and density of DA cross-links were important for achieving better healing effect. Xu et al. [129] reported a stiff and transparent polyhedral oligomeric silsesquioxane-based nanocomposite with crack-healing ability due to the thermally reversible reaction of furan and maleimide moieties. Lee et al. [130] prepared graphene nanoplate (GNP)/polyurethane (PU) nanocomposites via bulk in-situ DA reaction between GNP and furfuryl side chains of PU. The thermally self-healing capability was confirmed by the disappearance of scratch at 110 °C through DA reaction at the interface. Moreover, DA cross-linked bulk polymers with crack healability, such as triple-shape memory poly(*p*-dioxanone)-poly(tetramethylene oxide) (PPDO-PTMEG) co-network [131] and novolac epoxy [132], were also synthesized. These works only assessed the crack-healing ability qualitatively.

In the works by Du et al. [61–63], linear and cross-linked PUs containing DA bonds were developed. The thermally remendable PU displayed tensile strengths ranging from about 25 to 46.5 MPa, and the corresponding healing efficiency could reach more than 80% for the first time. Multiple healing was allowed with gradual decrease in efficiency. Due to the poor heat resistance of PU, thermal treatment at elevated temperature had to result in dissociation of hydrogen bonds and deterioration of mechanical properties [133], which would no doubt affect the healing efficiency.

Thermal reversibility can also be imparted to rubbers. Chen and Jiao used diglycidyl dicyclopentadienedicarboxylic acid ester (DGDCA) as a cross-linking monomer in alkyl aluminum system catalyzed copolymerization with epoxide monomers such as epichlorohydrin (ECH) [134].

The resultant polyether thermoplastic elastomer can be molded into sheets at 215 °C. When the opposite parts of a broken specimen were joined together, the crack became invisible under certain conditions. It means that de-cross-linking of the cross-linked copolymer occurred at elevated temperature, leading to melting and plastic flow. When the material was cooled down, the DA cycloaddition between the side group cyclopentadiene rings reconstructed the elastic cross-linking networks.

Watanabe and Yoshie developed network polymers with recyclability using telechelic prepolymers with reversible reactivity [135]. On the basis of the research, they produced an elastomer from bisfuranic terminated poly(ethylene adipate) (PEA) and tris-maleimide through DA reaction at 60 °C [136]. The resultant polymer, PEAF2M3, possessed a glass transition temperature (T_g) of −34 °C, and can be de-cross-linked by retro-DA reaction at 145 °C for 20 minutes. When a film sample was cut into two pieces and the cut surfaces were kept in contact with each other at 60 °C, rejoining of the cut pieces was observed. This mending was believed to be induced by the following three mechanisms. (i) The reversible cross-linking reaction bridged the cut surfaces. At the cut front, the weak DA adducts were selectively dissociated sacrificially to release the stress to protect the chemical structure of the prepolymer and the linker against the scission or degradation. (ii) Exchange of the maleimide (furan) group in a DA adduct with that in another DA adduct. (iii) Entanglement of dangling chains.

Generally, a fast dynamic exchange rate not only means quick healing of polymer materials, but also rapid creep. To overcome this limitation, Konkolewicz et al. [65] explored the cross-linked acrylate polymer containing both multiple hydrogen bond (weak but fast) and DA linkage (strong but slow), which showed different self-healing behaviors and mechanisms depending on temperature. At room temperature, the exchange of dynamic hydrogen bond allowed for recovering 50% of strain at break within seven hours, while at elevated temperature (e.g. 90 °C), the combination of physical hydrogen bond and DA linkage could recover the majority of its initial mechanical property (90% of strain at break).

Although most of the DA healing requires heating assistance, several systems that possess self-healing ability at room temperature have been developed. Early in 2009, Reutenauer et al. proposed such a DA cross-linked polymer despite the fact that quantitative characterization of healing efficiency was not available [137]. It involved fulvenes as the dienes and bis(dicyanofumarates) or bis(tricyanoethylenecarboxylates) as the dienophiles, which was different from the conventional DA chemistry between furan and maleimide. Repairing of the elastomer film was demonstrated by lapping followed by pressing to ensure microscopic contact, and the overlapping area was unable to separate by elongating. Then, Yoshie et al. [58] reported a DA cross-linked polyester, which allowed for autonomous mending of cracks at room temperature based on DA reaction between anthracene and maleimide. The average recovery due to healing at room temperature for 53 days was 17% and 45% in terms of tensile strength and elongation at break, respectively. By increasing the healing temperature to 100 °C, a better recovery was observed after seven days. The healing efficiencies reached 45% and 79%, respectively. Compared with the DA system from the furan-maleimide pair, the DA adduct between anthracene and maleimide is a potential replacement for designing self-healing polymers with high thermal stability, despite the fact that the healing time was obviously longer and the healing efficiency needed to be improved. Similarly, the authors also developed bio-based polymers from bis(hydroxymethyl)furan, which could undergo room temperature self-healing at bulk or with the aid of BMI solutions [59]. Because the kinetic rate and mobility of macromolecular chain segments decrease with decreasing temperature, however, the DA reaction rate has to be suppressed at ambient temperature. Consequently, healing efficiencies of the above polymers were not promising.

Recently, taking advantage of the photothermal effect of polydopamine particles (PDAPs) and the reversible properties of DA linkages, PU/PDAP composites with near-infrared (NIR) triggered shape memory and self-healing properties were fabricated by Xia et al. [64] Typically, the composites (1 wt% PDAPs with 220 nm diameter) were deformed at 65 °C and then cut, followed by irradiation with NIR light (0.75 W cm^{-2}) for 60 seconds to achieve 96% recovery of mechanical strength. The crack was immediately closed in the presence of NIR light due to its shape memory function and healed after the crack closure through DA reaction. The healing efficiency of the broken sample with heating assistance (110 °C, 1 hours; 80 °C, 24 hours) was only 76%.

On the whole, the DA reaction driven self-healing has to be carried out in a two-step fashion as a result of successive retro-DA and DA reactions. During the first stage of healing (i.e. retro-DA reaction/de-cross-linking), creep deformation or even collapse of materials with DA cross-linked linkages often occurs [138, 139] due to molecular cleavage. Evidently, it is a challenge for structural application, where distortion of end-use products is not allowed even if self-healing is proceeding.

In fact, the low viscosity and high mobility of DA cross-linked networks at elevated temperature are ideal for application of self-healing coatings, where load bearing capacity is not the requisite property. Wouters et al. [140] synthesized methacrylate- and epoxy-based DA cross-linked coatings, which proved to be excellent for scratch healing even if small amounts of coating were scraped off. In comparison with methacrylate-based copolymers obtained from linear precursors bearing furan moieties and BMI, the epoxy networks from DA adduct monomer had much lower initial flow temperature (below 95 °C) and ultimate viscosity (see Chapter 3, Section 3.1 for details). Kötteritzsch et al. [141] described several DA pendant chains containing methacrylate-based copolymers for the development of a one-component coating, which exhibited diverse repair abilities as demonstrated by atomic force microscopy (AFM) and SEM measurements. Various methacrylate comonomers (e.g. MMA, butyl methacrylate [BMA] and lauryl methacrylate [LMA]) with different chain lengths were used to tune flexibility and mobility of the final copolymers. Due to higher molecular mobility and sub-ambient T_g, the polymer with LMA as co-monomer was able to remend scratches within two minutes at 160 °C or within a much longer time (four hours) at the retro-DA temperature (110 °C). Scheltjens et al. [142, 143] reported a DA cross-linked epoxy coating (thickness = 200 μm) prepared by solution casting, which allowed for flowing above 90 °C. The manual scratches can be completely remended at 130 °C within two minutes. Moreover, Amato et al. [144] used DA linkage-containing soybean-based PU as a two-component automotive topcoat, which displayed remendability of abrasion above 95 °C.

Photoreversible Cycloaddition

In addition to the thermally initiated [4+2] cycloaddition, photoinitiated [2+2] and [4+4] cycloaddition can also be used for the photochemical healing of polymers, as reversion of the cycloaddition to C═C bonds can readily take place in the solid state [145, 146]. Self-healing based on photoreversible reaction is attractive because the use of light is clean, cost-effective, and easily accessible. It is especially suitable for the repair of specific injured regions where the thermal effect is unavailable. In general, ultraviolet (UV) light having shorter wavelength ($\lambda \leq 280$ nm) and that at longer wavelength ($\lambda \geq 280$ nm) are employed for successive photocleavage and photodimerization of the photoresponsive reversible bonds. Due to fast decay of incident light in bulk polymer, however, the cracks deep inside products could not be healed by this method. Therefore, the most promising application of photo-stimulated self-healing lies in coatings and films.

Chung et al. [66] proposed the first example of photochemical crack healing of a rigid and transparent polymer based on cinnamate [2+2] photocycloaddition reaction. The material was first cross-linked by UV irradiation at $\lambda > 280$ nm. When it was subjected to impact load, cleavage of the resultant cyclobutane ring occurred due to its low bond strength. Re-irradiation with UV light at $\lambda > 280$ nm allowed recovery of the cross-linked networks within 10 minutes, but the healing efficiency in terms of flexural strength was low. Froimowicz et al. [147] prepared a polyglycerol dendrimer grafted with anthracene groups, which showed photoreversible cross-linking by exposing to 366 and 254 nm UV light. The cross-linked film with an artificial scratch was first irradiated with UV light at 254 nm to regenerate the macromolecular building blocks, and then kept in darkness for overnight to refill the damaged region. Finally, the cross-linked film was recovered by anthracene [4+4] cycloaddition under 366 nm UV light. Obviously, the material is somewhat difficult to be used in practice since the de-cross-linking procedure led to complete disintegration of the networks. More recently, UV-triggered self-healing cross-linked polyphosphazenes were explored by Hu et al. [148]. They introduced photoresponsive ethyl 4-aminocinnamate to the main chains of linear polyphosphazenes as pendant groups, which were then exposed to UV irradiation at 365 nm to obtain cross-linked networks. The retro-[2+2] cycloaddition reaction under 254 nm UV light and the [2+2] cycloaddition reaction under 365 nm UV light resulted in self-healing of surface cutting of the elastomer as revealed by UV–visible spectroscopy and SEM. Polyphosphazenes are known for their unique backbone consisting of alternating phosphorus and nitrogen, and their attractive properties such as heat resistance and flame retardancy. The photo-self-healability undoubtedly adds to their attractiveness.

In the works by Ling et al. [67, 149, 150], photocross-linked PUs containing coumarin in the main chain and side chain, respectively, were synthesized. By taking advantage of the photoreversibility of coumarin, repeatedly self-healing as characterized by restoration of mechanical strength under UV illumination at room temperature was observed. The influence of the soft and hard segments as well as the coumarin moiety content were discussed. It was found that the rubbery domains originating from phase separation were necessary for improving the efficiency of photo-remending. Subsequently, a coumarin-modified tetrafunctional monomer was successfully synthesized and polymerized using irradiation with 365 nm UV light by Saito et al. [68] The self-healing capability of the resultant polymers was first investigated in terms of scratch healing. Meanwhile, two pieces of polymers were partially irradiated with 254 nm UV light to promote the formation of free coumarin units on the surface, and then reconnected by irradiation with 365 nm UV light to achieve recovery of Vickers hardness over 90%. More recently, Abdallh et al. reported a dynamic linear polymer formed by reversible cycloaddition reaction of a thymine unit-containing functional monomer [69]. Irradiation at wavelengths lower than 240 nm could lead to depolymerization of the linear polymer and generation of the oligomeric units, resulting in a decrease in T_g value and improvement of chain mobility. Consequently, a 30–40 μm wide scratch disappeared after being exposed to <240 nm UV (10 J cm^{-2}). Then, repolymerization at 302 nm led to most recovery of the molecular weight, T_g value and Vickers hardness.

In summary, although photochemically triggered cycloaddition has been successfully applied to self-healing polymeric materials, the high energy UV radiation used results in irreversible side reactions and decreases the reproducibility of the self-healability. In-depth fundamental research needs to be carried out in the future to solve these problems.

Hydrolysis-Bonding Equilibrium

Hydrolysis-bonding equilibrium of condensation reaction (e.g. borate ester linkage formation) has been exploited for self-healing hydrogels as they contain a large amount of water. Self-healing hydrogels with borate ester bonds [151–153], for example, have enabled autonomous crack healing without external intervention, despite relatively low mechanical strength. The underlying mechanism was attributed to the fact that borate esters were under hydrolysis equilibrium with boronic acid and diol constituents, providing free boronic acid and diol groups on the fracture surfaces for the formation of borate ester bonds across the damage interface [154].

Cash et al. [70] developed a self-healable boronic ester network (Figure 1.6), which could completely recover the tensile strength of bisected specimens by healing at room temperature and 85% humidity for three days. The healing was proved to be dependent on water content, which shifted the equilibrium of boronic ester toward dissociation to generate more free boronic acid and diol moieties, followed by reformation of boronic ester linkages across the damage interface for remending. Meanwhile, Lai et al. [71] demonstrated that the boroxine/boronic acid equilibrium can be readily shifted by the addition/removal of Lewis bases or water. They provided a new mechanism of self-healing of boroxine-containing polydimethylsiloxane (PDMS) networks, which had both strong and stiff mechanical properties (Young's modulus = 182 ± 15.8 MPa; compressive modulus = 142 ± 9.8 MPa). The self-healing was carried out by cutting the specimen into two pieces, which were then wetted with water and healed at 70 °C for 12 hours. Complete recovery of mechanical strength was detected after drying.

As proposed by Raquez et al., iminoboronate-based boroxine derivatives were introduced to obtain ambient humidity-sensitive polymers with self-healing ability [155]. The Lewis acid–base interactions between the boron atom of boroxine and the nitrogen atom of the imine bond can inhibit the reverse hydrolysis reaction of boroxine to a certain extent. Moreover, due to the improvement of thermodynamic stability, the imine bonds were hard to hydrolyze. As a result, the sample obtained from 2-formylbenzeneboronic acid and oligo-(propylene glycol) bis(2-aminopropyl ether) (400 g mol^{-1}) could recover 93% of strain at break (i.e. 272%), when the material was exposed to 35% relative humidity (RH) for 48 hours. Subsequently, the authors further developed a water-responsive self-healing epoxy based on the same mechanism, achieving 80% recovery in modulus and tensile strength upon water contact [72]. Meanwhile, nitrogen-coordinated boroxines were also exploited for the fabrication of self-healing and recyclable polymers, whose tensile strengths could be tailored from 1.7 to 12.7 MPa [73]. It was demonstrated that the secondary amine ligands accelerated the formation and dissociation of nitrogen-coordinated boroxines at room temperature. The resultant

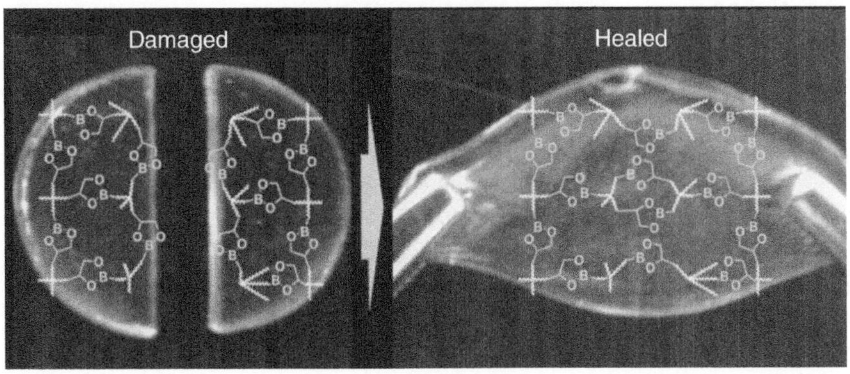

Figure 1.6 Cross-linked polymers constructed with dynamic-covalent boronic esters were synthesized via photoinitiated radical thiol-ene click chemistry. The reversible bonds of the boronic esters were shown to shift their exchange equilibrium at room temperature when exposed to water. The facile reversibility of the networks allowed intrinsic self-healing under ambient conditions. *Source:* Cash et al. [70], UNP 01, p. 1/with permission of American Chemical Society.

material was first bisected and then the damaged ends were dipped into water for one minute to open the nitrogen-coordinated boroxines, followed by contacting and then removal of water to completely heal the damage within 18 hours. The reversibility of nitrogen-coordinated boroxines and hydrogen-bonding interactions accounted for the healing process.

Since the above neutral trigonal boronic ester prefers to shift toward hydrolysis in the presence of water, this would undoubtedly decrease the cross-link density of the polymers. To complete the healing effect, water should be removed by drying to shift the equilibrium toward bond formation. Generally, tetrahedral boronic ester is more stable against hydrolysis; the bond formation constant is 10^2–10^4 times higher than that of the neutral trigonal boronic ester [156–158]. Taking this into consideration, Yoshie et al. [74] developed a moisture-triggered autonomous self-healing polymer by introducing tetrahedral boronic ester bonds. The result could always maintain a certain number of cross-links to ensure mechanical strength even under highly humid condition (e.g. 90% RH) for over 10 days. Furthermore, two separate pieces of the material were almost fully healed after three days under 75% RH.

Deng et al. synthesized an acylhydrazone cross-linked polymer gel, which displayed self-healing ability based on the reversible breaking and regeneration of acylhydrazone bonds in the presence of acetic acid [159, 160]. The free acylhydrazine and aldehyde moieties produced from reversible equilibrium could diffuse across the crack interface and reform acylhydrazone bonds, resulting in the healing of two separated plates. Almost identical results were obtained subsequently by Apostolides et al. [161] using tetrahydrazide-containing monomer and dialdehyde oligomer as building blocks. In fact, multiple reversible reactions (i.e. formation reaction and exchange reaction of C=N bonds) may be simultaneously involved in the healing system containing C=N bonds, which are rather complicated and sometimes difficult to distinguish.

1.2.1.2 Healing Based on Dynamic Reversible Covalent Reactions

Exchange Reaction of Disulfide Bonds

Canadell et al. [78] reported self-healing behavior of a commercially available epoxy elastomer containing disulfide bonds (20 wt%) for the first time. The tensile strength of the broken specimens could be fully restored by treatment at 60 °C for one hour. It was believed that disulfide exchange accounted for the self-healability. Subsequently, Rekondo et al. [80] developed a catalyst-free room temperature self-healing polyurethane urea (PUU) elastomer based on aromatic disulfide metathesis. The repaired sample showed 97% recovery of original strain at break, despite the fact that almost 51% of healing efficiency was contributed by hydrogen bonding.

Otsuka et al. [85] proposed a facile strategy to prepare poly(hexyl methacrylate) networks that contain thermally exchangeable bis(2,2,6,6-tetramethylpiperidin-1-yl)disulfide cross-linkers. Typically, the cut sample was brought into contact in a mold and then healed at 120 °C under 70 kPa by a paper clip, achieving 85% and 92% recoveries of strength and strain at break, respectively. Kim et al. [86] reported a transparent and easily processable thermoplastic polyurethane (TPU) containing aromatic disulfide bonds, which simultaneously showed high tensile strength and toughness (6.8 MPa and 26.9 MJ m^{-3}, respectively), as well as excellent self-healing ability at room temperature. After the TPU film was cut into half and rejoined, the mechanical properties could achieve about 75% healing efficiency within two hours. Moreover, a highly transparent (~100% transparency) self-healing TPU based on dynamic disulfide bonds was also achieved by Lai et al. [87]. About 85.0–89.4% of tensile strength could be recovered after healing at 70 °C for 24 hours. Recently, Zhang et al. [88] designed a disulfide bond-containing anhydride hardener (1,4,5-oxadithiepane-2,7-dione, DSAA) for constructing self-healing epoxy resins. For the resin containing 22.7 wt% DSAA, the healed sample can recover 96% of the original lap shear strength (i.e. 5.05 MPa) by heating at 160 °C for one hour.

Besides thermal activation, photoirradiation can also initiate disulfide metathesis in aliphatic disulfide-containing polyester [162]. Michal et al. developed a semi-crystalline polymer containing disulfide bonds [81], which was self-healable under UV light, despite the fact that the power intensity of the lamp had to be as high as 2000 mW cm^{-2} for partial melting of the sample to liberate the frozen disulfide bonds.

The disulfide bond is one of the most widely used dynamic linkages in self-healing polymers, which combines both stimulus diversity (e.g. heat, UV light and sunshine) and the flexibility of reaction mechanism (e.g. base-catalyzed disulfide exchange, metal ion-catalyzed disulfide exchange, thiol-disulfide exchange, and radical-mediated exchange). Nevertheless, the self-healing polymers based on disulfide bonds are facing some challenges, including relatively lower mechanical strength, compatibility and dispersion of catalysts, easy oxidation of thiol, and UV light induced aging.

Exchange Reaction of C=N Bonds

In recent years, dynamic hydrogels with self-healability based on imine and acylhydrazone bonds have been designed and prepared for biological applications [163]. A hydrogel made from Schiff base-containing chitosan [164], for example, could undergo a dynamic exchange reaction between the Schiff base bond and residual aldehyde and amine reactants under multiple biochemical stimuli such as pH, amino acids. and vitamin B6 derivatives. Self-healing capability was demonstrated in terms of qualitative visual inspection of crack closure and quantitative rheological analysis of rapid recovery of elastic modulus. By further combination with ferroferric oxide [165], self-healing could even be driven by an external magnetic field.

Wei et al. [166] synthesized a polysaccharide-based hydrogel containing both acylhydrazone and Schiff base bonds with different pH response ranges. Generally, acylhydrazone bonds take part in reversible reactions only in slightly acidic conditions (pH = 4.0–6.0). Introduction of more reactive imine linkages can endow self-healing ability to the hydrogel under physiological conditions. As a result, healing of the fragmented hydrogel was successfully conducted by immersion in phosphate buffer saline (pH = 7) at physiological temperature for 48 hours, achieving 95% recovery of compression strength.

Deng et al. [167] developed a multi-responsive dynamic polymer hydrogel containing both acylhydrazone and disulfide bonds, which exhibited self-healing property through diverse mechanisms adapting to different chemical stimuli (Scheme 1.4). The exchange reaction of acylhydrazone and base-triggered disulfide exchange was responsible for the self-healing in acidic (pH = 3 and 6) and basic conditions (pH = 9), respectively. Since acylhydrazone and disulfide bonds were dynamically locked in neutral conditions (pH = 7), the hydrogel lost its self-healability in this case. Nevertheless, acylhydrazone exchange could be activated in the presence of aniline accelerator at pH = 7, which enabled similar healing manner [168, 169]. In view of the fact that multi-stimulus responsive polymer gels possess unique properties and potential applications, Zhang et al. [170] developed a dual-responsive gel from benzhydrazide-containing polytriazole and disulfide-containing dialdehyde. When using aniline as catalyst, both acylhydrazone exchange and disulfide exchange reactions were triggered, leading to maximum healing efficiency of 85% in terms of compressive stress at break.

Polymer hydrogels and organogels can provide sufficient mobility for the exchange reaction of C=N bonds, so that healing could be carried out at ambient temperature. However, the gel-like materials usually have poor mechanical strength and limited application range.

Kuhl et al. [171] applied the mechanism of acylhydrazone bond exchange to crack healing of methylacrylate-based copolymer solid. When the 1:1 mixture of triethylene glycol methyl ether methacrylate and 2-hydroxyethyl methacrylate was copolymerized with 10 mol% acylhydrazone cross-linker in the presence of the initiator azobisisobutyronitrile (AIBN), a bulk polymer (T_g = 45 °C) was yielded, which was able to remend scratches up to 1 cm long through treatment at 100 °C for 64 hours. Mo et al. [92] constructed a series of dynamic epoxy-imine networks via post cross-linking between linear polyimine with residual secondary amines and biphenyl A type epoxy. The resultant network with 51.1% gel content exhibited outstanding mechanical strength and self-healing ability, where the tensile strength could recover 86.3% of the original strength by healing at 90 °C for three hours. As proposed by Song et al. [93], a biomass-based dynamic polymer cross-linked by multiple hydrogen bonds and imine bonds was achieved from epoxy soybean oil and vanillin. With increase of the imine bond content, the as-prepared polymers showed controllable mechanical strength from softness (tensile strength = 0.61 MPa, strain at break = >2800%) to stiffness (tensile strength = 20.50 MPa, strain at break = 12%). When the stiff sample was cut and re-spliced, followed by heat treating at 100 °C for 12 hours, 93.2% of the original tensile strength could be regained.

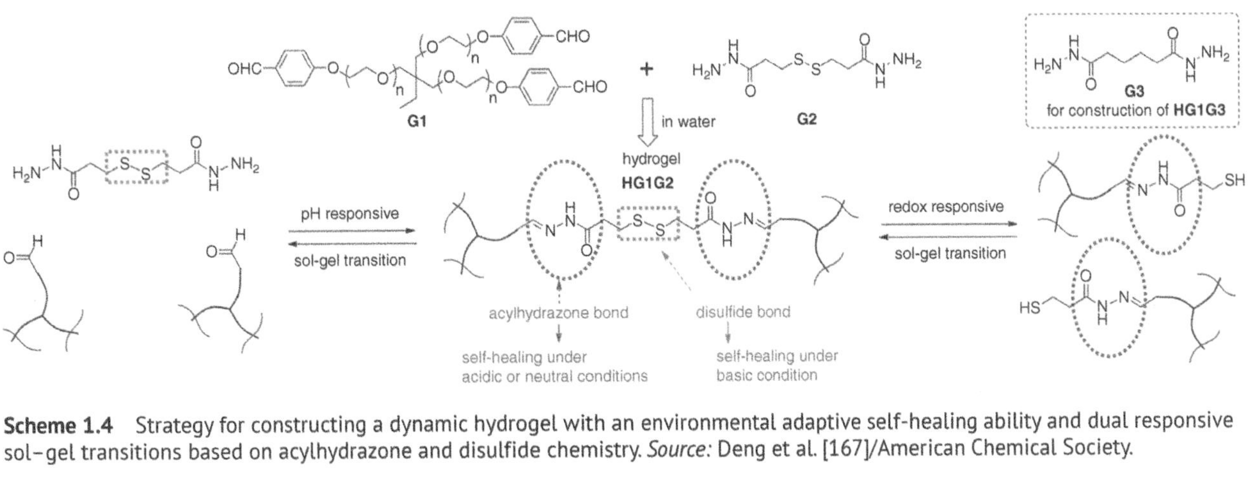

Scheme 1.4 Strategy for constructing a dynamic hydrogel with an environmental adaptive self-healing ability and dual responsive sol–gel transitions based on acylhydrazone and disulfide chemistry. *Source:* Deng et al. [167]/American Chemical Society.

When aryl groups are attached to both nitrogen and carbon atoms of the Schiff base bond, dynamic reversible exchange of the aromatic Schiff base bonds can be conducted at room temperature free of catalyst, as reported by Lei et al. [91]. Accordingly, polyacrylate with aromatic aldehyde groups reacted with 4,4′-diaminodiphenyl methane, producing self-healing networks (T_g = −21.2 °C) cross-linked by aromatic Schiff base bonds. As Schiff base polymers are known for their diverse properties and stimulus-responsivities, triggering the dynamic exchange of the Schiff base bonds in a similar way would arouse a few novel functionalities in addition to self-healability.

Transesterification

Proposed by Leibler's group [172–174], dynamic covalent interaction (i.e. transesterification) was first introduced into epoxy networks to achieve malleability (Figure 1.7). For instance, when the hard networks of epoxy-anhydride containing 10% zinc acetate (serving as transesterification catalyst) were ground into powders, they were able to be remolded at 240 °C for three minutes [172]. The regained bulk sample had almost the same mechanical properties and insolubility as the original version. Subsequently, a large number of research studies used transesterification to realize reprocessing and recycling [175–177].

Transesterification of ester bonds in covalent polymers is an effective method for healing but a considerable amount of catalyst must be added. Brutman et al. [108] prepared bioderived polylactide (PLA) vitrimers with self-healing ability, which were synthesized from hydroxyl-terminated star-shaped PLA and cross-linked by methylene diphenyl diisocyanate (MDI). Stannous(II) octoate ($Sn(Oct)_2$) served as the catalyst for both cross-linking and transesterification. Eventually, many of the materials exhibited full recovery of tensile strength due to transesterification assisted healing at 140 °C. Although $Sn(Oct)_2$-catalyzed transesterification allowed the PLA vitrimers to present shorter mean relaxation time at moderate temperature, the catalyst is highly toxic and sensitive to oxygen and water. Subsequently, Fu et al. [110] developed a self-healing polybenzoxazine from bisphenol F derived benzoxazine monomer and succinic anhydride, which behaved as a thermoplastic polymer above 140 °C in the presence of zinc(II) acetate ($Zn(Ac)_2$). The multiple self-healing ability of the dynamic network was evaluated by a three-point bending method, during which the sample was first subjected to impact breakage and then healed at 140 °C for one hour. The result showed that healing efficiencies were 99, 92, and 89% estimated from the recovery of storage modulus for three healing cycles, respectively.

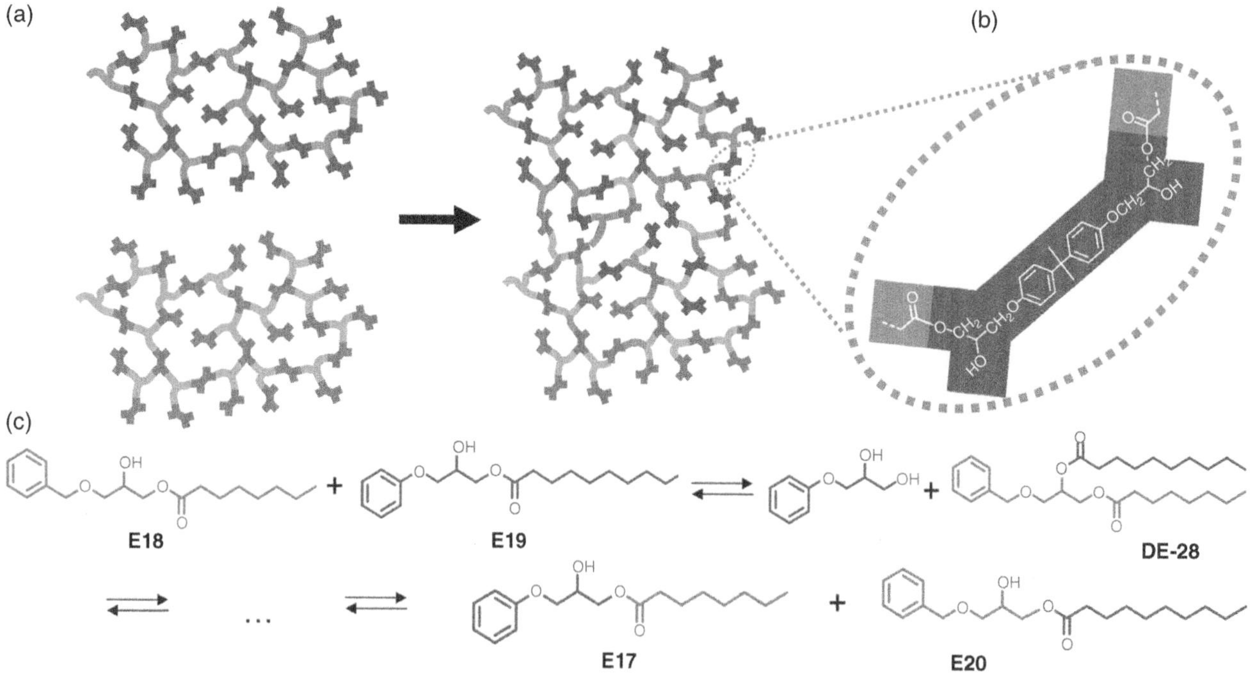

Figure 1.7 (a) Scheme illustrating the welding of two pieces of epoxy resin by transesterification exchange reactions. (b) Structure of the epoxy–acid network used in the study. The epoxy resin DGEBA is linked through β-hydroxy esters to dicarboxylic and tricarboxylic acids. (c) Model molecules E18 and E19 investigated to mimic β-hydroxy ester links. Numerous transesterification products appear when the mixture is heated, such as diester DE28 and β-hydroxy esters E17 and E20. *Source:* Reprinted from Capelot et al. [174]. Copyright 2012, with permission from American Chemical Society.

As reported by Norvez and Leibler et al. [178], the epoxidized natural rubber (ENR) network cross-linked by dodecanedioic acid in the presence of $Zn(Ac)_2$ could relax stress at high temperatures and exhibited vitrimer-like properties. Subsequently, Tang et al. [179] synthesized an ENR vitrimer using carboxyl group-functionalized carbon nanodot (CD) as cross-linker, which could be recycled and welded through $Zn(Ac)_2$ catalyzed exchange reactions of β-hydroxyl ester linkages on the ENR-CD interphase. More recently, citric acid-modified bentonite (CABt) was utilized as cross-linker to achieve a new ENR composite without $Zn(Ac)_2$ by Xu et al. [111]. Interestingly, even in the absence of $Zn(Ac)_2$, the tensile strength and elongation at break of ENR composite (20 wt% CABt) could recover to 4.3 MPa and 380% by healing at 150 °C for three hours, corresponding to about 96 and 94% of the original values, respectively.

As some special cases, Altuna et al. [109] produced a thermosetting bio-based polymer by cross-linking epoxidized soybean oil (ESO) with citric acid (CA) under the catalysis of the protons derived by the CA dissociation in water. Lap-shear test at 160 °C for two hours demonstrated the self-healing performance as a result of thermally activated transesterification of β-hydroxyester groups created in the course of polymerization. To overcome the disadvantages of catalysts (e.g. toxicity and poor miscibility), two hyperbranched epoxy (HBE) prepolymers were synthesized and then reacted with succinic anhydride to obtain a catalyst-free epoxy vitrimer system [112]. Since the abundant hydroxyl groups in the HBE structure served as both reacting moieties and catalysts, the produced cross-linked materials exhibited a fast relaxation above 120 °C. As a result, the sample welded at 150 °C for one hour could completely restore the shear strength (~47.3 MPa).

Miscellaneous

To realize heat-/photostimulation of covalent self-healing polymer, Matyjaszewski et al. [102, 180] prepared several dynamic gels (from poly(n-butyl acrylate) [PBA], polystyrene [PS], and poly(methyl methacrylate) [PMMA]) cross-linked by trithiocarbonate (TTC) unit. Healing of the damaged PBA swollen in acetonitrile was completed in N_2 protection as a result of network reorganization based on UV driven dynamic chain transfer reaction. Additionally, macroscopic fusion of shredded samples was obtained by UV irradiation under similar circumstances. Meanwhile, healing of PMMA and PS gels was conducted through heat-triggered network reshuffling in the presence of AIBN and CuBr/*N,N,N′,N″,N″*-pentamethyldiethylenetriamine (PMDETA) complex as initiators, respectively. Nevertheless, there is a distance between these findings and practical application as solvent swelling and inert gas protection serve as the prerequisites. On the other hand, Amamoto et al. [103] demonstrated that thiuram disulfide (TDS) moieties could undergo exchange reaction in air under visible light through two possible mechanisms: radical transfer and crossover. Therefore, cross-linked PU containing TDS in the main chain was able to self-heal under the stimulation of visible light through radical reshuffling of TDS units. Since TTC and TDS units are generally employed as chain transfer agents of reversible addition–fragmentation chain-transfer (RAFT) polymerization, the self-healing systems based on TDS and TTC provide a reference for the synthesis of self-healing polymers through RAFT polymerization.

Compared with disulfide bonds, diselenide bonds have relatively lower bond energy ($172\,kJ\,mol^{-1}$), so that they could be involved in a reshuffling reaction under milder conditions. In the work by Xu et al. [84] and Ji et al. [181, 182], the diselenide exchange reaction was enabled through radical mechanism under both visible-light irradiation and heating without catalyst. Accordingly, linear PU containing diselenide bonds with visible-light-induced self-healing ability was prepared. A significant enhancement of healing efficiency could be achieved by applying pressure, but simultaneously leading to great decrease of Young's modulus (by more than 20%). By using 0.6 W blue laser (457 nm) instead of a table lamp, the healing time was greatly shortened from 48 hours to only 30 minutes, achieving 84% recovery of breaking stress. Although visible light is a convenient and inexpensive means of stimulation, the dynamic linkages included in the visible-light-induced self-healing polymer must be highly active, which would lead to structural instability as a result of continuous stress relaxation similar to room temperature intrinsic self-healing polymers. Similarly, ditelluride bonds with low bond energy ($126\,kJ\,mol^{-1}$) were also utilized to prepare self-healing systems at room temperature without external intervention. According to the work of Lang et al. [90], a novel ditelluride compound bis(2-hydroxyethyl) ditelluride was synthesized and used as the chain extender to fabricate a series of ditelluride-containing linear PUs. The results showed that only the appropriate proportion of hard and soft segments could make the resultant polymer possess the appropriate chain mobility and dynamic bond content, leading to satisfactory healing efficiency. The healing mechanism was attributed to dynamic exchange of ditelluride bonds in the main chain and hydrogen bonds among macromolecular chains. Nevertheless, the mechanical strength is poor, only 0.1–0.3 MPa.

Zheng and McCarthy [183] prepared a siloxane cross-linked polymer containing reactive tetramethylammonium dimethylsilanolate end groups. The "living" anionic species are stable to water, oxygen, and carbon dioxide. Anionic equilibration of the networks allowed for self-healing of cut sample at 90 °C. Fracture toughness measurements indicated that the critical strain energy release rate, G_{IC}, of the healed sample was $70 \pm 12\,J\,m^{-2}$, even surpassing the original value ($61 \pm 14\,J\,m^{-2}$). To overcome the shortcomings of ref. [183] (i.e. elevated temperature and long exchange time), Schmolke et al. [184] regulated the dynamic characteristics and mechanical strength of cross-linked PDMS by varying contents of the ionic initiator (i.e. tetramethylammonium hydroxide) and cross-linker (i.e. bis(heptamethylcyclotetrasiloxanyl) ethane). Consequently, stress relaxation and self-healing at room temperature through siloxane equilibration were detected. Besides, their results also revealed that there is a strong dependence of the stress relaxation rate on the concentration of "living" anion ends in the dynamic network.

As the reversible units (i.e. "living anionic species") are not purposely introduced into the above polymers, and anionic ring-opening polymerization of cyclosiloxanes is the most commonly used method of preparing polysiloxanes, all the polysiloxanes synthesized by anionic equilibrium polymerization should have the potential to achieve self-healing.

Lu and Guan [106] explored olefin metathesis aided self-healing of polymers through dynamic exchange of C=C bonds. In their model material, the second-generation Grubbs ruthenium catalyst was incorporated into cross-linked polybutadiene. Increase of catalyst concentration, pressure and temperature was beneficial for promoting reshuffling of the cross-linking networks. Comparatively, the catalyst-free cross-linked polybutadiene showed minimal healability. Imato et al. [104] investigated the feasibility of self-healing based on exchange of C—C bonds. To verify their idea, a chemical gel was made from polypropylene glycol (PPG)-based PU with diarylbibenzofuranone (DABBF) as the reversible cross-linkage. The enormous steric hindrance near the center carbon–carbon bond of DABBF reduced C—C bond energy, so that a dynamic equilibrium between carbon free radical and carbon–carbon bond can be established at room temperature. Tensile tests indicated that 98% recovery of strain at break was obtained by healing at room temperature for 24 hours with the assistance of dimethyl formamide for wetting the fractured surfaces in advance. For overcoming the shortcomings of gels, which used to suffer from solvent volatilization and low mechanical property, the authors [105] prepared a cross-linked PU elastomer containing DABBF bonds, which was healable in the bulk state without the help of solvent. Almost complete restoration of tensile properties was achieved after healing at 50 °C for 12 hours due to network rearrangement originating from exchange of the DABBF linkages. It is worth noting that although stable and strong C=C and C—C bonds were included in the above dynamic polymers toward synthesis of strong self-healing polymers [104–106], tensile strengths of the resultants were not high enough (<0.8 MPa [105]) for structural application.

To couple reversible covalent polymers with both high strength and toughness, Zhang et al. [185] designed a cross-linked PU containing aromatic pinacol as a novel reversible C—C bond provider (Figure 1.8), of which tensile strength and failure strain were tunable from 27.3 MPa to as high as 115.2 MPa and from 324 to 1501%, respectively. Autonomous healing capabilities of the resulting polymer were firstly verified by AFM pull-off curves, which indicated the adhesive energy of polymer at 100 °C (i.e. $6.4\,mJ\,m^{-2}$) was about eight times higher than that at 25 °C (i.e. $0.78\,mJ\,m^{-2}$). Meanwhile, the overlapped AFM pull-off curves collected during the repeated tests were indicative of reversibility of the self-bonding ability in the same place. Moreover, the penetrating crack could also be healed at 100 °C by taking advantage of the dynamic equilibrium of pinacol.

Poly(hydroxy urethane)s (PHUs) that could undergo network reconfiguration through nucleophilic addition of free hydroxyl groups to the carbamate bonds (transcarbamoylation) in the absence of catalyst were described by Fortman et al. [186]. The ground sample could be reprocessed into bulk material at 160 °C for eight hours under 4 MPa, achieving 74% recovery of tensile stress and 69% recovery of strain at break. Liu et al. [187] reported novel dynamic reversible poly(oxime urethane)s obtained from multifunctional oximes and hexamethylene diisocyanate (HDI). Based on the network reconfiguration via oxime-enabled transcarbamoylation, the cross-linked polymer showed catalyst-free self-healing capability at 120 °C. Subsequently, a covalently cross-linked PU elastomer with triple dynamic bonds (i.e. oxime-urethane, copper-oxime-urethane coordination complex, and hydrogen bond) were prepared by Zhang et al. [114]. Both experiments and density functional theory (DFT) calculations demonstrated that the copper ions could significantly facilitate the transcarbamoylation reactions. As a result, the samples can be rejoined under ambient conditions leading to the regain of 92% of tensile strength after 130 hours.

Cheng et al. [113] studied a dynamic covalent polymer containing hindered urea bonds, which could reversibly dissociate into isocyanate and amine under mild conditions without catalyst. The results showed that there was a direct relationship between dynamic characteristics of urea bonds and the polymer repair behavior. The sample containing hindered urea bonds with large equilibrium constant K_{eq} ($7.9 \times 10^5\,M^{-1}$) and reasonable reverse rate constant k_{-1} ($0.042\,h^{-1}$) was healed at 37 °C for 12 hours, recovering 87% of its initial strain. Cromwell et al. [107] developed two kinds of cross-linked covalent polymers

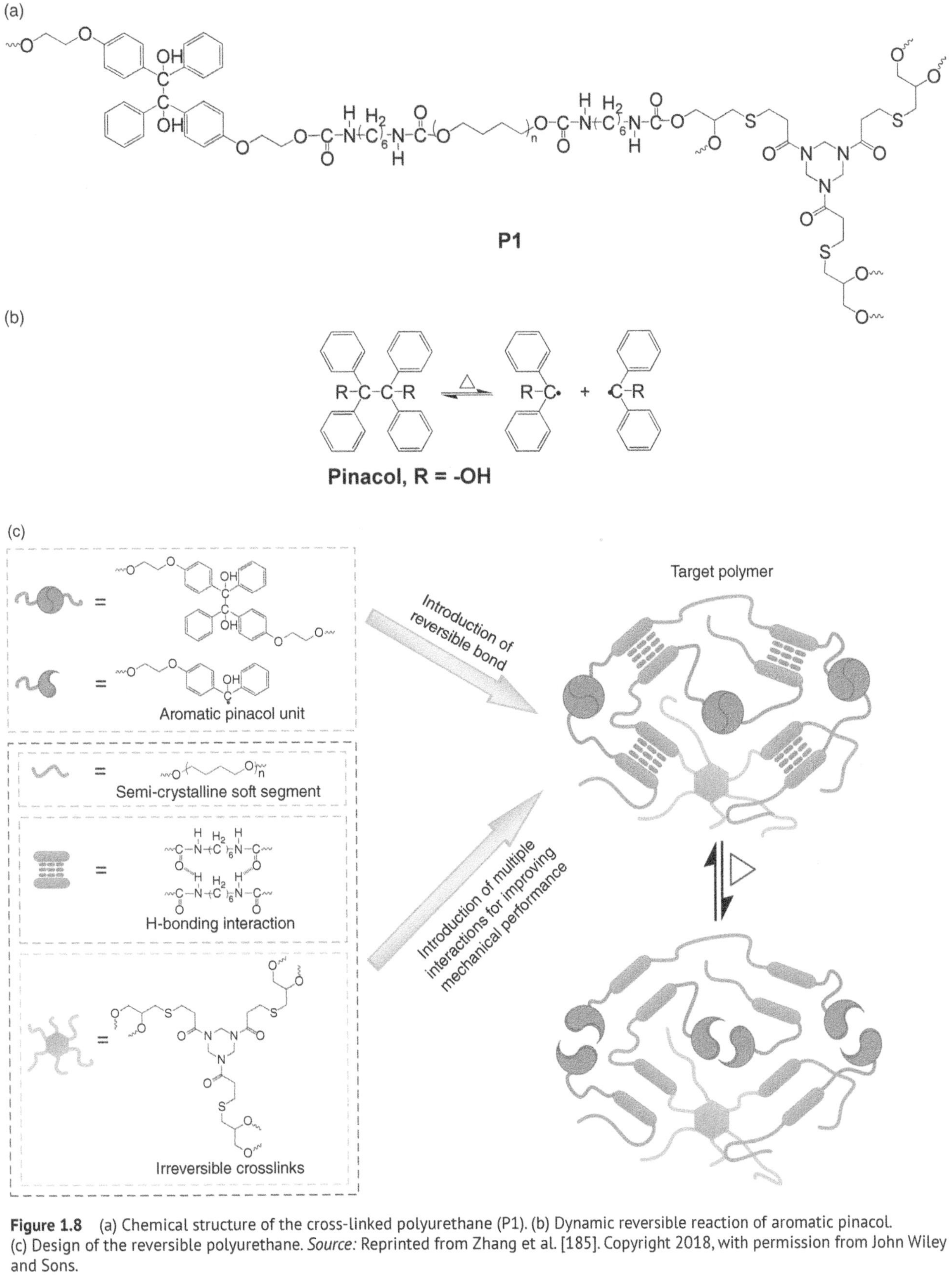

Figure 1.8 (a) Chemical structure of the cross-linked polyurethane (P1). (b) Dynamic reversible reaction of aromatic pinacol. (c) Design of the reversible polyurethane. *Source:* Reprinted from Zhang et al. [185]. Copyright 2018, with permission from John Wiley and Sons.

containing dynamic boronic ester linkages with different kinetic exchange rates. As a result, completely different efficiencies of self-healing due to boronic ester transesterification were measured. Only the sample bearing the cross-linker with fast exchange rate was able to effectively heal at 50 °C, achieving almost the same strength as the original one. It was thus demonstrated that small molecule kinetics can be correlated with dynamic properties of bulk polymer, which would in turn be used to tune the performance of the latter. Recently, a similar strategy was applied by the authors to adjust the dynamic properties of silicone ether bond cross-linked polystyrene copolymers [188]. Afterwards, Chen et al. [75] designed a self-healing diene-rubber by incorporating boronic ester linkages into the network, which was synthesized through a thermally initiated thiol-ene click reaction between a novel dithiol-containing boronic ester cross-linker and common styrene-butadiene rubber. As a consequence, the samples could recover their original mechanical properties (~2 MPa) to a larger extent after healing at 80 °C for 24 hours.

As compared with boronic ester, boroxine can enhance the cross-linking density and is more suitable for preparation of self-healing materials with robust mechanical properties. Delpierre et al., for example, exploited the iminoboronate-based boroxine chemistry for designing a non-isocyanate-based PU network with high stiffness (Young's modulus = 551 MPa, tensile strength = 11 MPa) and outstanding healing property [76]. By combination of dynamic exchange reactions of imine bonds and boroxines at 70 °C, complete recovery of the mechanical properties can be observed.

Aza-Michael addition between an amine and α, β-unsaturated double bond was also explored as dynamic reversible chemistry for self-healing materials [154, 189]. Baruah et al. [189] synthesized a Knoevenagel condensation product (TPMD) from terepthalaldehyde and diethyl malonate, which contained four carboxylate functionalities and two highly activated double bonds. Then, TPMD was cross-linked with oligo-amines to achieve a set of dynamic networks. The crack on the resultant material surface was able to be healed at 50 °C within one hour by thermal reversibility of the aza-Michael addition between the secondary amine and activated double bond. Subsequently, oligo-amine (pK_a = ~10–11) was replaced by carbonyl hydrazide with lower pK_a (~3–4) for understanding the influence of pK_a value of the donor on the dynamic reversible character of the Michael adduct [154]. The results showed that the decrease of pK_a value of the donor could retard the rate of β-elimination and enhance the stability of the corresponding hydrazide Michael adducts. The healing ability was verified as in the following. Two hard samples were first softened by dipping in tetrahydrofuran:water (95:5, vol:vol) mixture for three hours, and then overlapped and pressed at room temperature for 24 hours to regain 88% of the ultimate tensile strength after softening (1.7 MPa).

It is interesting to note that conformational change is a useful measure of self-healing, as sol–gel transformation or directional fluidization of azobenzene compounds can be induced by the conformational changes under light stimulation [190, 191]. Watanabe et al. [190] prepared a photo-healable ion gel by self-assembly of tetra-arm diblock copolymer grafted with azobenzene groups in a hydrophobic ionic liquid. The dumbbell sample was cut into two pieces, and then irradiated by UV light at 38 °C for 50 hours to initiate the gel-to-sol transition, followed by visible-light irradiation for 30 minutes to heal the gel by the sol-to-gel transition. Consequently, the tensile strength can recover 80% (i.e. 0.06 MPa) of the original value.

In the works of Zhou et al. [192] and Xu et al. [193], photo-switching of azopolymers from solids to isotropic liquids was carried out by reversible cis–trans photoisomerization of azobenzene moieties, which provides a strategy for designing healable and processable hard coatings. Typically, a scratch on a trans azopolymer film (T_g = 68 °C) was locally irradiated with UV light (365 nm at 67 mW cm^{-2}) for 10~20 minutes to convert the surrounding polymer into the cis azopolymer at liquid state (T_g = 48 °C). As a result, the scratch could be healed due to the flow of the azopolymer, followed by irradiating with visible light (530 nm at 5 mW cm^{-2}) for 10 minutes to transform back to the solid state.

1.2.2 Self-Healing Based on Supramolecular Interactions

Intrinsic self-repair without the aid of healing agent or external stimuli requires high chain mobility and therefore seems to be in direct contradiction with the fixation needed to form covalent bonding in polymers [194]. Nevertheless, the healing strategies via supramolecular interactions that develop rapidly in recent years have shown capabilities comparable with those based on covalent bond reconnection (Table 1.2).

1.2.2.1 Coordination Bonds

A coordination bond (also known as a dative bond) is a special type of covalent bond described as a covalent bond between two atoms in which the two shared electrons are from the same atom. Burnworth et al. reported a metallo-supramolecular elastomer that was capable of healing a crack by using photothermal conversion of UV light [223]. The metallo-supramolecular elastomer was made from 2,6-bis (1′-methylbenzimidazolyl) pyridine (MeBip) ligands end-capped amorphous poly(ethylene-*co*-butylene) and zinc

Table 1.2 Typical intrinsic self-healing polymers toward mechanical properties restoration based on supramolecular interactions.

Polymer[a]	T_g (°C)	Mechanical strength[b,c]	Failure strain (%)	Healing condition	Mechanism	Healing efficiency (%)	Ref.
PU/CNTs	<−50 (DSC)	22.8[b]	1076	NIR, 4.2 mW mm^{-2}, 30 min	Zn^{2+}-terpyridine	Complete recovery of toughness	[195]
PDMS	<−90 (DSC)	0.23 ± 0.05[b]	1860 ± 80	0 °C, 72 h	Fe^{3+}-pyridyl, Fe^{3+}- carboxamido	90% recovery of fracture strain	[196]
PDMS	55.7 (DSC)	9.146 ± 0.45[c]	3.39 ± 0.39	80 °C, 4 h	Zn^{2+}-carboxylate	Complete recovery of flexural strength	[197]
PBz	N/A	~0.8[b]	~275	RT, 24 h	Fe^{3+}-O-N, H-bonding	61% recovery of toughness	[198]
PU	N/A	2.5[b]	~1700	25 °C, 36 h	Al^{3+}-urea, H-bonding	90% recovery of tensile strength	[199]
PU	−63.3 (DMA)	~2.5[b]	~2900	25 °C, artificial seawater	Fe^{3+}-DOPA	82.7% recovery of tensile strength	[200]
P(DA-co-BA)	12–13 (DSC)	3.5–3.8[b]	260–310	24 h, seawater	Ca^{2+}-DOPA, Mg^{2+}-DOPA	58–79% recovery of toughness	[201]
PMMA-b-P(MTEA-co-BA)	−27, 20, 103 (DSC)	~2[b]	~570	RT, 6 h	Ag^{+}-S	80% recovery of toughness	[202]
POA	−5.7 to −3.0 (DSC)	3.9–5.2[b]	178–231	RT, 1–1.5 h	His-Zn^{2+}, His-Cu^{2+}, His-Fe^{3+}	Complete recovery of tensile property	[203]
BIIR	−67 (DSC)	9[b]	1000	RT, 192 h	Ionic imidazolium bromide	96% recovery of tensile strain	[204]
NR/ZDMA	−59.2 (DMA)	0.63[b]	340	RT, 5 min	Zn^{2+} ion cluster	95% recovery of tensile strength	[205]
NR/MAA/ZnO	4 (DSC)	0.64–1.89[b]	435–478	RT, 20 min	Zn^{2+} ion cluster	>96% recovery of tensile strength	[206]
PAA/PAH	N/A	3.2[b]	500	2.5 M NaCl, 6 h	Carboxyl-amino	Complete recovery of tensile property	[207]
PA-b-PMMA-b-PA	2–4	4.38–6.07[b]	3.16–8.08	60 °C, 24 h	H-bonding	70–90% recovery of tensile strength	[208]
ENR/t-CNs	−25	4.04 ± 0.18[b]	198 ± 12	RT, 24 h	H-bonding	86% recovery of tensile strength	[209]
PBA-b-PS	−42 (DSC), 74 (DMA)	1.40[b]	600	45 °C, 18 h	Upy quadruple H-bonding	90% recovery of tensile strength	[210]
PTMEG/PCL	−25.5 (DSC)	54.38 ± 3.8[b]	24.04 ± 2.27	40 °C, 48 h	Upy quadruple H-bonding	87% recovery of tensile strength	[211]

(Continued)

Table 1.2 (Continued)

Polymer[a]	T_g (°C)	Mechanical strength[b,c]	Failure strain (%)	Healing condition	Mechanism	Healing efficiency (%)	Ref.
PU	20.2 (DMA)	48.5 ± 6.3[b]	25.8 ± 0.6	100 °C, 24 h	Upy quadruple H-bonding	90.7% recovery of tensile strength	[212]
PB/TMPMP	10~15 (DSC)	1.78–2.52[b]	158–322	RT, 12 h	Ionic H-bonding	35–90% recovery of tensile strength	[213]
CT-ENR/C-CNC	~−10 (DMA)	0.85[b]	~550	RT, 15 s	H-bonding	Complete recovery of toughness	[214]
Copolyester	N/A	3[b]	~90	35 °C, 2 min	H-bonding	Complete recovery of ultimate strength	[215]
p(MMA/nBA)	−25 (DSC)	9.74 ± 0.72[b]	620 ± 62	RT, 14 h	van der Waals forces	83.8% recovery of tensile strain	[216]
PI/PU	~−20 (DMA)	~4[b]	170	100 °C, 4 h	π–π stacking, H-bonding	91% recovery of tensile strain	[217]
PI/PAm/CNC	N/A	~2.5[b]	20	85 °C, 30 min	π–π stacking	Complete recovery of tensile modulus	[218]
Crosslinked HEA	−20 (DSC)	5[b]	~800	RT, 24 h	β-CD/phenolphthalein	65% recovery of tensile strength	[219]
PHEMA-SWCNTs	16.9 (DSC)	0.13 ± 0.02[b]	553 ± 8.31	RT, 20% humidity, 5 min	β-CD/adamantane	53.8% recovery of tensile strength	[220]
PEI-SWCNTs	N/A	4[b]	0.07	RT, 50–60% humidity	β-CD/adamantane	>90% recovery of tensile strength	[221]
Epoxy/GNs	87.7 (DMA)	20.8[b]	~25	120 °C, 30 min	β-CD/azobenzene	79.2% recovery of tensile strength	[222]

[a] BIIR, bromobutyl rubber; C-CNC, carboxyl cellulose nanocrystal; CT, chitosan; ENR, epoxidized natural rubber; GN, graphene nanosheet; HEA, hydroxyethyl acrylate; His, histidine; NR, natural rubber; PAA, polyacrylic acid; PA-b-PMMA-b-PA, poly(acrylate amide)-*b*-poly(methyl methacrylate)-poly(acrylate amide); PAm, polyamide; PB, polybutadiene; PBA-b-PS, poly(n-butyl acrylate)-*b*-polystyrene; PBz, polybenzoxazine; PCL, poly(ε-caprolactone); P(DA-co-BA), poly(dopamine acrylamide-*co*-n-butyl acrylate); PEI, polyethylenimine; PHEMA, poly(2-hydroxyethyl methacrylate); PI, polyimide; PMMA-b-P(MTEA-co-BA), poly(methyl methacrylate)-*b*-poly(methyl thioethyl acrylate-*co*-butyl acrylate); p(MMA/nBA), poly(methyl methacrylate)/n-butyl acrylate; POA, poly(oleic acid); PPG, polypropylene glycol; PTMEG, poly(tetramethylene ether) glycol; t-CN, tunicate cellulose nanocrystal; TMPMP, trimethylolpropane tris(3-mercaptopropionate); UPy, 2-ureido-4-pyrimidinone; ZDMA, zinc dimethacrylate.

[b] Tensile strength, MPa.

[c] Flexural strength, MPa.

ions (Zn^{2+}) or lanthanum ions (La^{3+}). Relatively, the samples derived from La^{3+} displayed higher toughness and healing efficiency due to more dynamic nature of La-MeBip complexes as compared with Zn-MeBip complexes. Bode et al. [224] developed a series of self-healing cross-linked metallo-supramolecular copolymers based on the well-investigated iron(II)-bis-terpyridine system. The polymers with low T_g (before the cross-linking) were able to heal large and wide scratches (10–35 μm in width and 180–250 μm in length) at 100 °C within 82 hours. However, the mechanical strengths of these self-healing metallopolymers were rather weak.

Zheng et al. [195] synthesized a novel Zn^{2+}-terpyridine coordinated metallo-supramolecular carbon nanotube (CNT)/PU self-healing nanocomposite with excellent mechanical properties (tensile strength = 22.8 MPa, elongation at break = 1076%). The dynamic nature of the Zn^{2+}-terpyridine interaction endowed the nanocomposite with rapid and efficient self-healing capabilities in the presence of NIR light, heat, or solvent, leading to almost complete recovery of toughness (141 MJ m^{-3}).

Li et al. [196] reported a PDMS network cross-linked by the coordination complexes between 2,6-pyridinedicarboxamide ligands and iron ions (Fe^{3+}), showing the potential application for self-healing artificial muscle actuators. The ligand can simultaneously form one strong (i.e. pyridyl) and two weak (i.e. carboxamido) metal–ligand bonds (Figure 1.9). The weaker bonds could readily break and reform, favorable for energy dissipation during stretching and self-healing upon damage. Meanwhile, the Fe^{3+} ions were still maintained near the ligands by the stronger interactions, which allowed for rapid bond re-formation. The samples healed at −20 and 0 °C for 72 hours could recover 68 and 90% of the fracture strain, respectively. The authors also designed a rigid and healable PDMS polymer cross-linked by abundant Zn^{2+}–carboxylate interactions [197]. The resultant dynamic PDMS had a high Young's modulus (~480 MPa) at 25 °C, but could be healed and reprocessed upon heating. When two separate pieces were brought back into contact and healed at 80 °C for four hours, the maximal tensile strength (8.92 MPa) was almost completely recovered.

Marine mussels are a fascinating species that can anchor themselves to the turbulent intertidal zone by secreting byssal adhesion protein to form an adhesive plaque. This class of proteins contain lots of 3,4-dihydroxyphenylalanine (DOPA) residues, which can form robust complexes with Fe^{3+} ions. The catechol-Fe^{3+} coordination number has been proved to be adjustable in the order of mono-, bis-, and tris-catechol complex when the pH is increased from about 4 to 9 [225, 226]. Therefore, pH-responsive biomimetic self-healing polymers can be designed on the basis of the catechol-Fe^{3+} coordination system. In the work of Dana and Guler et al. [227], an amphiphilic polypeptide (DopaK-PA) containing a dopamine catechol structure was synthesized and then cross-linked with Fe^{3+}. The thixotropy test was used to analyze the self-repairing property of the resulting material (DopaK-PA/Fe(III)). When the cross-linked materials suffered high shear load far beyond the linear viscoelastic behavior (1000%), both covalent and non-covalent bonds were broken, and then the non-covalent bonds were able to rapidly recover after the load was removed. The results show that the storage modulus of DopaK-PA/Fe(III) could recover 77.6% of the original value within one hour, while only 7.1% was achieved in the case of irreversible cross-linked reference. Xia et al. [200] synthesized hyperbranched polyurethane (HBPU) with functional catechol and hydrophilic carboxyl end groups, which was crosslinked by Fe^{3+} at pH = 9. The catechol-Fe^{3+} coordinate bond became dynamically reversible in the presence of water, which provided the polymers with both self-healing and recycling ability in seawater. Recycling of the polymer was performed by soaking the fragments in water of pH = 4 for 24 hours to dissociate the catechol-Fe^{3+} coordinate bonds, and then transferring the mass into seawater for another 24 hours to rebuild tris-coordinated catechol-Fe^{3+} cross-linking bonds. The wetted polymer fragments were finally pressed under 6 MPa at room temperature for

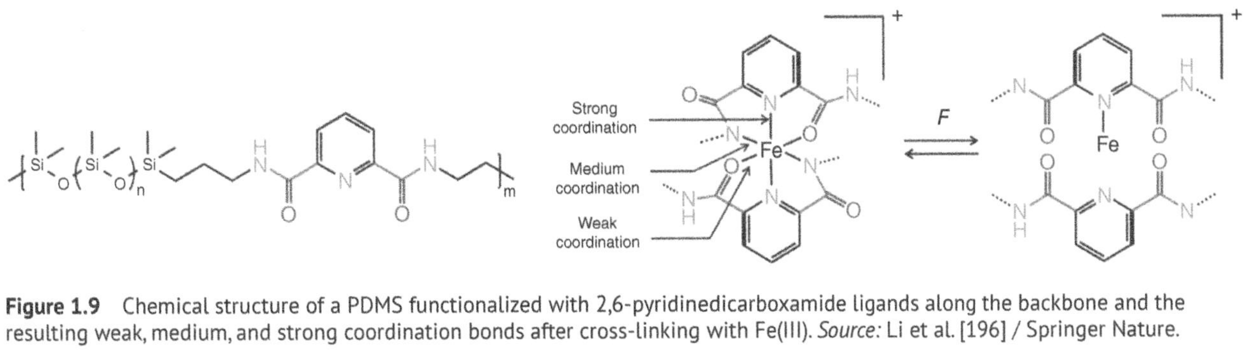

Figure 1.9 Chemical structure of a PDMS functionalized with 2,6-pyridinedicarboxamide ligands along the backbone and the resulting weak, medium, and strong coordination bonds after cross-linking with Fe(III). *Source:* Li et al. [196] / Springer Nature.

48 hours to obtain a fresh dynamic polymer network, leading to recovery of tensile strength and self-healing ability. To decrease pH sensitivity and increase adaptability to different aquatic environments of the underwater reversible lipophilic bulk polymer, the same groups of authors developed HBPU bearing hydrophilic quaternary ammonium salts and dynamic reversible catechol-boron ion (B^{3+}) cross-linkages [228]. The electron-withdrawing effect of quaternary ammonium cations from catechol-B^{3+} hindered hydrolysis of the boronic ester bonds at lower pH. Consequently, recycling in the same way as mentioned above [200] showed that the polymer can regain its mechanical properties in the water of pH of both 7 and 9. Moreover, Yoshie synthesized catechol-functionalized polymers through copolymerization of dopamine acrylamide and n-butyl acrylate, followed by cross-linking with calcium ions (Ca^{2+}) or magnesium ions (Mg^{2+}) in the presence of triethylamine [201]. When the Ca^{2+} or Mg^{2+} content was increased, the Young's modulus and tensile strength were raised accordingly. To reveal the healing mechanism, P–Ca^{2+} 20% and P–Mg^{2+} 20% films (where P stands for metallopolymer) were cut into two pieces and allowed to rejoin for over 24 hours in air, water, and seawater, respectively. Consequently, P–Ca^{2+} 20% achieved 78% toughness recovery in seawater, but only 25% in air and water. P–Mg^{2+} 20% gained 59%, 22% and 17% recovery of toughness in seawater, air, and water, respectively. Both hydrogen bonding and coordination contributed to the high healing efficiencies in seawater.

As a new type of phenolic resin, polybenzoxazine (PBz) resin not only has excellent heat resistance and flame retardancy, but also overcomes the shortcomings of traditional phenolic resin in releasing small molecules during curing. Arslan et al. [198] proposed a novel approach for self-healing of PBz thermosets based on hydrogen bond between nitrogen atoms and phenolic hydroxyl groups, as well as Fe(III)–O(phenolic)–N(tertiary amine) metal–ligand interactions. The resultant PBz film was prepared by using bis(3-aminopropyl)-terminated PDMS, formaldehyde and bisphenol A, followed by ring opening polymerization at 100–120 °C in the presence of $FeCl_3$. Consequently, 61% of fracture toughness could be regained through healing at room temperature for 24 hours.

Recently, a series of biomass-based self-healing elastomers were designed and fabricated from histidine (His) moieties-grafted poly(oleic acid) (POA) and transition-metal ion (Zn^{2+}, Cu^{2+}, and Fe^{3+}) [203]. Since the order of the stability constants of metal ion-His complexes increases in the order of Zn^{2+}, Cu^{2+}, and Fe^{3+}, the tensile strength of the corresponding metal ion cross-linked POA-His films increases from 3.9 to 5.2 MPa, while the fracture strain decreases from 231 to 178%. To test the self-healing ability, two separate pieces of the resultant films were pushed together and then healed at room temperature for 1–1.5 hours, which could completely restore the original tensile properties. Moreover, Wu et al. reported a self-healing polyurea elastomer cross-linked by hydrogen bond and aluminum ion (Al^{3+})-nitrogen atom (urea group) coordination bond [199]. The as-prepared elastomer exhibits a tensile strength of 2.6 MPa and failure strain of ~1700%, as well as a high self-healing efficiency of 90%. In the work of Zhang et al. [202], self-healable thioether-based acrylate block polymers were fabricated by RAFT polymerization and silver ion (Ag^{+})-thioether coordination. The resultant specimens were first cut to a depth of 70–90% of the initial thickness and left for six hours at room temperature. As a result, the dynamic nature of the silver–thioether bonds enabled 80% recovery of breaking strain and toughness.

1.2.2.2 Ionic Associations

Das et al. [204] reported that commercial bromobutyl rubber (BIIR) can be converted into a physical cross-linking version with extraordinary self-healing properties by transforming the bromine functionalities of BIIR into ionic imidazolium bromide groups. The reversibility of ionic associations facilitated the cut sample to regain 57.8 and 96% of its original tensile strength and failure strain after 192 hours of healing at room temperature, respectively. Later, Chen et al. [205] designed a self-healable rubber through controlled peroxide-induced vulcanization of natural rubber (NR) in the presence of zinc dimethacrylate (ZDMA), which allowed for the graft-polymerization of ZDMA onto NR chains. The powerful electrostatic interaction of Zn^{2+} ion pairs in the polymerized ZDMA promoted the generation of ionic multiplets, which further constituted ion clusters. Due to the breaking and rebuilding of ionic cross-links, the specimen healed for 20 minutes at ambient temperature got a recovery of more than 95% of the original strength (~0.63 MPa). Afterwards, the authors also conducted in-situ reaction of methacrylic acid (MAA) with the excessive zinc oxide (ZnO) in NR [206]. Except for participating in the formation of ion cross-links, the residual ZnO could tailor the mechanical properties of NR by changing the MAA/ZnO molar ratios. When the molar ratio was adjusted from 2:1 to 2:1.4, tensile strength of NR increased to 1.89 MPa, which was almost 3 times that of the former. The healed samples almost fully regained their original strength after healing at room temperature for 20 minutes.

Reisch et al. [207] prepared a self-healing polyelectrolyte by mixing polyacrylic acid (PAA) and polyallylamine hydrochloride (PAH) in NaCl solution. The healing ability and healing speed of the resultant polyelectrolyte were proportional to the concentration of NaCl solution, and all of the original tensile strength (i.e. 3.2 MPa) could be restored within 20 minutes by dropping 2.5 M NaCl onto the fracture surfaces. By means of the carbon-13 nuclear magnetic resonance relaxation test, the influence of ion concentration on the mobility of molecular segments was evaluated. The results indicated that the greater ion concentration led to higher molecular segment mobility and diffusion capacity, and eventually caused better healing.

1.2.2.3 Hydrogen Bonds

Chino and Ashiura synthesized thermally reversible cross-linking rubber having similar mechanical properties to vulcanized rubber by using supramolecular hydrogen bonding networks [123]. The molecules can be rearranged in the heated state, while the network can be reformed when cooled down. Leibler's group further demonstrated complete restoration of tensile properties of a thermoplastic elastomer by simply bringing together fractured surfaces [229, 230]. The system was characterized by indefinitely repeatable healing with thermoplastic behavior in the melt and presumably with low toxicity. Dimer and trimer fatty acids from vegetable oils were reacted with diethylene triamine and subsequently with urea. This reaction yielded a transparent glassy material with a T_g of 28 °C. A supramolecular network was formed in the mixture of a large number of di- and trifunctional building blocks with various strongly hydrogen bonding urea and amide molecules. Crystallization was prevented by the presence of a multitude of different species, which warranted rubbery properties at higher temperatures. The temperature-dependent strength of the hydrogen-bonding units resulted in a strong decrease of the viscosity of the material at high temperatures, thus the material can easily be melt-processed. While adhesion was almost immediately sufficient for the pieces to sustain their own weight, the bonding gradually increased over time to reach the initial strength after only three hours. The key mechanism responsible for the self-healing was related to the dynamics and the density of the strongly associating hydrogen-bonding groups. When the elastomer failed by rupture, these hydrogen bonds, rather than covalent bonds, were broken, leading to a high density of non-associated groups at the interface. The non-associated hydrogen bonds remained in their free state for a significant time, waiting to find a new partner when brought into contact with another freshly cut surface. Relatively fast recombination of a high density of groups at the surface gave rise to a unique repair mechanism that did not rely on reptation of complete polymer chains to recover mechanical strength.

The above synthetic pathways relied on grafting hydrogen-bonding moieties onto previously synthesized appropriate backbones. The group of Leibler developed a facile one-pot synthesis of supramolecular plastics and self-healing rubbers [231, 232]. By combining and controlling the backbone building, with the grafting of the functional groups, a large variety of supramolecular materials were obtained from the same one-pot reaction by merely varying the stoichiometry of the reactants.

Harreld et al. synthesized a self-healing non-covalently cross-linked organosiloxane-polypeptide block copolymer [233]. The key issue lay in the coupling of an aminopropyl-functionalized PDMS with protected polypeptide. The polypeptide segments not only provided toughness to the material, but also allowed for rehealing by allowing reforming of the cross-links (by virtue of ionic and/or hydrogen bonding) after a disruptive stress incidence.

Chen and Guan synthesized a triblock copolymer composed of a rigid PMMA middle block and two poly(acrylate amide) soft blocks carrying multivalent, pervasive hydrogen bonds [208]. The mechanical properties of the ABA block copolymers (4.38–6.07 MPa) were tunable by simply adjusting the chain length of the soft blocks. Healing was carried out by bringing together two separated halves of the specimen at 60 °C for 24 hours, which led to significant regain of tensile strength (70–90%).

For improving mechanical properties of self-healable ENR-based elastomers, tunicate cellulose nanocrystals (t-CNs) were utilized as reinforcing fillers to obtain a high strength ENR via latex mixing [209]. The resultant ENR-based nanocomposite with 20 phr t-CNs (i.e. ENR/t-CNs-20) displayed a tensile strength of 4.04 MPa, which was almost four times that of ENR. Due to the dynamic hydrogen bonds between oxygenous groups of ENR and hydroxyl groups on the t-CN surface, together with molecular interdiffusion, ENR/t-CNs-20 achieved a healing efficiency of 86%, while neat ENR only regained 48% of its tensile strength. Subsequently, the authors also developed a kind of self-healing thermoplastic vulcanizate (TPV) composed of PLA, ENR and Fe_3O_4, which were prepared by a Haake rheochord in the presence of dicumyl peroxide (DCP) as vulcanized agent [234]. The self-healing mechanism was proposed as in the following. (i) The shape memory effect led to the physical contact of damaged surfaces. (ii) The desorption–absorption of ENR/Fe_3O_4-bound rubber promoted interdiffusion of ENR chains, resulting in the healing of the ENR phase. (iii) Since PLA chains had been grafted onto the ENR chains induced by DCP, the PLA segments at the damaged surfaces could rearrange and entangle again with the aid of ENR chains. Consequently, the partial damaged sample P7E3F1-0.5DCP (70 wt% PCL, 30 wt% ENR, 10 phr Fe_3O_4, and 0.5 phr DCP) exhibited the highest healing efficiency of 82.6% by healing at 70 °C for three hours.

2-Ureido-4-pyrimidinone (UPy) motifs can form dimers with well-defined quadruple hydrogen bonding, which display high thermodynamic stability and rapid kinetic reversibility [235]. Hentschel at el. [210] designed a kind of poly(n-butyl acrylate)-*b*-polystyrene (PBA-b-PS) diblock copolymers end-functionalized with UPy groups, whose dimerization led to the formation of supramolecular elastomers with dynamic and self-healing properties. By taking advantage of the association/dissociation of hydrogen-bonding junctions, a significant recovery of tensile strength (~90%) and elongation at break (~75%) was observed for the resultant elastomer (molar ratio of styrene to butyl methacrylate = 37:63) after healing at 45 °C for 18 hours. Subsequently, Yoshie et al. [236] reported a self-healing and tough ABA-type triblock copolymer bearing UPy side chains, whose A blocks formed hard domains with dense UPy junctions that prevented macroscopic deformation, while the B blocks formed a softer matrix with sparse UPy cross-links that serve as sacrificial bonds (Figure 1.10). As a result, the resultant copolymer film exhibited high toughness (~62 MJ m^{-3}), self-healing and excellent self-recovery. A scratch on the elastomer almost disappeared within three hours at 50 °C.

In the work of Yang et al. [211], a physical cross-linking network was successfully prepared via self-complementary quadruple hydrogen bonding through UPy-telechelic poly(tetramethylene ether) glycol (PTMEG) and four-arm star-shaped poly(ε-caprolactone) (PCL). Typically, the resultant network with 80 wt% PTMEG (54.38 MPa) was partially damaged with a razor blade (24.04 MPa) and then healed at 40 °C for 48 hours (32.74 MPa). Zhang et al. [237] developed a series of self-healing gelatin hydrogels cross-linked by

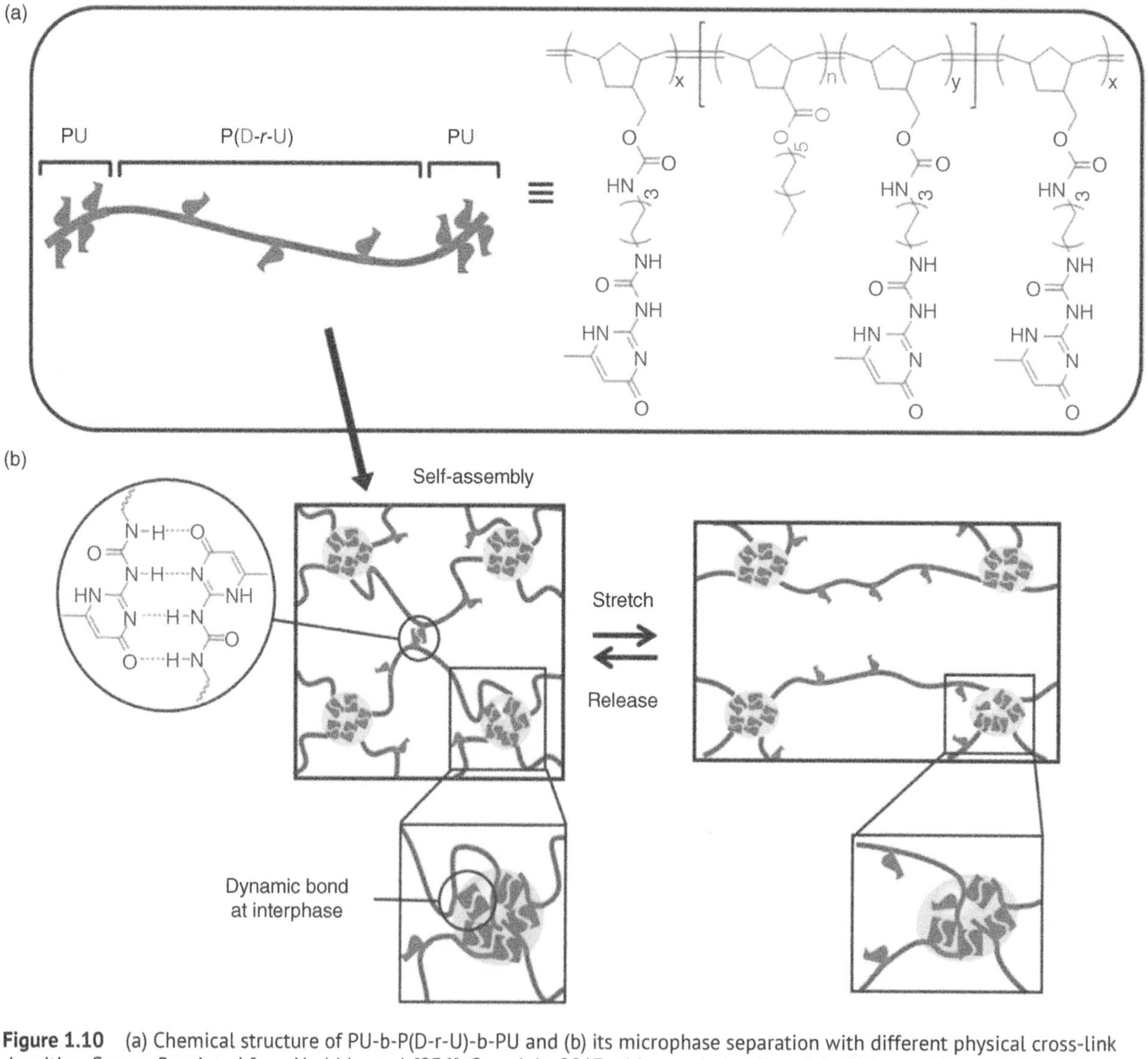

Figure 1.10 (a) Chemical structure of PU-b-P(D-r-U)-b-PU and (b) its microphase separation with different physical cross-link densities. *Source:* Reprinted from Yoshida et al. [236]. Copyright 2017, with permission from John Wiley and Sons.

combining multiple hydrogen bonds from UPy side chains, and ionic coordination of Fe^{3+} and carboxyl group in the gelatin. As a result, a crack on the specimen surface could be completely healed within five hours. Recently, transparent and healable elastomers featuring supertoughness (386 MJ m^{-3}) and high tensile strength (48.5 MPa) were achieved by Song et al. [212]. A diol bearing UPy side chain was first prepared and used as chain extender, leading to PUs with hierarchical hydrogen bonds formed with urethane, urea, and UPy groups. Consequently, the cut sample (e.g. PT-HM-U20) was able to recover 90.7% of tensile strength by treating at 100 °C for 24 hours.

Cao et al. [214] synthesized a self-healing nanocomposite elastomer from carboxyl cellulose nanocrystals (CNCs) and chitosan-decorated ENR latex, which exhibited a rapid self-healing speed within 15 seconds to completely recover its strength (~0.85 MPa) and toughness (~3 MJ m^{-3}) in the first healing cycle based on multiple hydrogen bonding. By further adding electrically conductive CNTs, self-healing flexible strain sensor can be fabricated through layer-by-layer self-assembly. Later, a humidity-responsive and self-healing copolyester was also fabricated through emulsion polymerization of poly (ethylene glycol) (PEG), PTMEG, ethylene glycol, phthalic acid, and citric acid, followed by cross-linking with tri-arm aziridine [215]. The cutting–healing process demonstrated that 100% of ultimate strength (~3 MPa) and 88% of stretchability was regained by healing at 35 °C for two minutes. The resulting hygroscopic soft robot was also capable of shaping and reshaping into arbitrary 3D geometric structures benefitting from the hydrogen bonding and covalent bonding interpenetrating network.

In contrast to most self-healing polymers based on hydrogen bonds, which are elastomers, Yanagisawa et al. developed a rigid poly(ether-thioureas) cross-linked by dense hydrogen bonds [238]. The room temperature healing was enabled by the following key factors. The thioureas anomalously formed zigzag hydrogen-bonded arrays that did not induce unfavorable crystallization. The relatively short polymer chains permitted greater segmental motions.

An ionic hydrogen bond is a special type of hydrogen bond with potential proton transfer between organic cations (e.g. primary or secondary amine, quaternary ammonium) and anions (e.g. carboxylate and sulfate). Because of its high stability and facile accessibility, this type of reversible interaction has been utilized for constructing self-healing supramolecular polymers. Mecerreyes et al. developed a simple and convenient method for ionic supramolecular networks by mixing cationic Priamine 1074 and anionic citrate [239]. The cut pieces of cylinder-shaped sample were reconnected for 30 minutes, and could be stretched to twice the size again without breaking. In the work of Wang et al. [213], carboxylic acid and amine groups were first introduced into polybutadiene oligomers through thiol-ene click reaction. Afterwards, the precursors were self-assembled through reversible ionic hydrogen bonds via the acid–base reaction, and then further covalently cross-linked by trimethylolpropane tris(3-mercaptopropionate) (TMPMP) to obtain a self-healing and shape memory rubber. The specimen with lower covalent crosslinking density (2 wt% TMPMP) could be healed at ambient temperature to regain more than 90% of the original strength (1.78 MPa) within 12 hours, while the specimen with 6 wt% TMPMP (2.52 MPa) could only achieve a healing efficiency of 35% after 12 hours.

1.2.2.4 Other Intermolecular Forces

Unlike the supramolecular hydrogen bonded self-healing systems, a healable supramolecular polymer blend that assembled via π–π stacking interactions was reported by Burattini et al. [240, 241] They used two relatively low molecular weight polymers: a polydiimide that is capable of adopting a chain-folded conformation, thereby generating π-electron-deficient "tweezer-type" receptor units; and a linear polysiloxane that features complementary π-electron-rich pyrene end groups. The blend of these two polymers cast from solution afforded a thermodynamically stable material capable of autonomous healing when exposed to above ambient temperatures. The complementary π-electron-rich and -poor receptors exhibited rapid and reversible complexation behavior in solution, and healable characteristics in the solid state in response to temperature. A mechanism was proposed for this thermoreversible healing behavior that involved disruption of the intermolecular π–π stacking cross-links as the temperature of the supramolecular film was increased. The low T_g polysiloxane component can then flow and as the temperature of the blend was decreased, π–π stacking interactions drove formation of a new network and so led to good damage-recovery characteristics of the two-component blend. Later, a healable supramolecular polymer blend comprising a chain-folding polyimide and a telechelic PU with pyrenyl end groups was fabricated [217]. By taking advantage of aromatic π–π stacking between the π-electron-deficient diimide units and the π-electron-rich pyrenyl groups, the fractured sample could regain more than 95, 91, and 77% of the tensile modulus, elongation at break, and modulus of toughness, respectively, of the pristine material after healing at 100 °C for 240 minutes. Moreover, similar healable supramolecular polymer nanocomposites with varying weight percentages of CNCs were prepared via solvent casting followed by compression molding [218]. The nanocomposite with 7.5 wt% CNCs displayed a tensile modulus (149 MPa) of about 20 times that of the original supramolecular polymer (8 MPa), which could be fully regained at 85 °C within 30 minutes.

In view of the need for additional design and synthesis of intrinsic self-healing materials, it is difficult to use commercial raw materials directly. Thereby, Urban et al. [216] developed a self-healing poly(methyl methacrylate)/n-butyl acrylate [p(MMA/nBA)] system based on van der Waals force, which was expected to become the prototype of the next generation of self-healing materials. Compared with hydrogen bonding, the weaker van der Waals force exists more widely in polymers. The results showed that the copolymers with 45/55 to 50/50 MMA/nBA molar ratios can be self-healed at ambient temperature within 14 hours. In fact, Young's modulus and T_g increased with increasing MMA/nBA molar ratios, but the copolymer with lower MMA content (e.g. 40/60 MMA/nBA) exhibited no significant healing under the same condition. The alternative sequence of methyl methacrylate and butyl acrylate units in the polymer chain should be responsible for the efficient healing via the key-and-lock interactions stabilized by van der Waals forces.

1.2.2.5 Host–Guest Inclusion

As one of the important supramolecular interactions, host–guest inclusion plays a significant role in the design and synthesis of self-healing materials. Due to its high selectivity and dynamic equilibrium, the corresponding self-healing materials possess advantages such as no need for external stimuli and high healing rate [242].

Cyclodextrins (CDs) are a series of cyclic oligosaccharides, which usually consist of 6, 7, or 8 glucopyranose units linked by α-1,4-glycosidic bonds and are denoted as α-CD, β-CD, and γ-CD, respectively [243]. Alcohol hydroxyl groups on the rim of CDs are beneficial for chemical modification, and the hydrophobic cavities of CDs can form host–guest complexes with various types of guest molecules. In 2011, Nakahata et al. reported a kind of self-healing supramolecular hydrogel using β-CD-modified poly(acrylic acid) (PAA) and ferrocene (Fc)-containing PAA as host polymer and guest polymer, respectively [244]. Compared with the oxidized state of the Fc group (Fc^+), β-CD showed a high affinity for the reduced state of the Fc group due to its hydrophobic nature. Therefore, the redox stimuli induced sol–gel phase transition in the supramolecular hydrogel could regulate re-adhesion between the fractured surfaces. To demonstrate the healing effect, the hydrogel was cut in half, and the cut surfaces were coated with oxidant solution (e.g. NaClO) to induce oxidation of the Fc group. As a result, healing was not observed after 24 hours. Re-adhesion was observed after spreading reductant solution (e.g. glutathione [GSH]) onto the oxidized cut surface. Subsequently, the authors developed two kinds of polyacrylamide hydrogels cross-linked by host–guest complexation of cyclodextrins (α-CD and β-CD) and hydrocarbon-group (n-butyl [Bu], adamantane [Ad]) moieties [245]. Due to the formation of inclusion complexes between free CDs and Ad units on the cut surfaces, the ruptured β-CD-Ad hydrogel can completely recover its adhesive strength after rejoining in wet conditions for 24 hours. More recently, stimuli-responsive coloring hydrogel and self-healing xerogel using phenolphthalein unit as guest molecule were prepared by the same research group [219]. It is known that the phenolphthalein unit turns purple under basic conditions in aqueous solutions, while it becomes colorless when it forms a complex with β-CD. Taking advantage of this feature, β-CDs and phenolphthalein units were incorporated into the side chains of polyacrylamide or hydroxyethyl acrylate (HEA). The obtained polyacrylamide hydrogel could undergo color change upon thermal stimuli or Joule heating due to the association–disassociation equilibrium of the β-CD/phenolphthalein inclusion complex. For the HEA xerogel with lower T_g exhibited self-healing properties under dry conditions, 65% of the initial strength (~5 MPa) was restored after 24 hours at room temperature.

In the work of Yu et al. [246], Fe_3O_4-doped self-healing host–guest gels based on β-CD and imidazole unit were obtained through magnetically induced frontal polymerization (MIFP) of β-CD, *N*-vinylimidazole (VI), 2-hydroxypropyl acrylate (HPA) and methylenebisacrylamide. The resultant supramolecular gels (CD/VI (w/w) = 2/10) showed a healing efficiency of 95% at ambient temperature as estimated from recovery of tensile modulus (0.282 MPa). Moreover, the Fe_3O_4-doped supramolecular gels exhibited accelerated self-healing rate under the applied magnetic field, which was attributed to the synergy of host–guest healing and magnetocaloric effect.

Later, Zhang's group developed a series of conductive elastomers with autonomic self-healing properties by combination of single-walled carbon nanotubes (SWCNTs) and β-CD/Ad complexation [220, 221, 242]. First, the surface of the SWCNTs was decorated by β-CD through π–π interactions between SWCNTs and pyrene-modified β-CD (Py-β-CD). Then, Py-β-CD/SWCNTs reacted with adamantine modified polymer matrix (e.g. poly(2-hydroxyethyl methacrylate) [PHEMA], polyethylenimine [PEI]) via host–guest interactions to obtain the self-healing conductive composites. The PHEMA-SWCNTs composite with 10 wt% SWCNTs has a tensile stress of 0.13 MPa and a conductivity of $5.55\,S\,m^{-1}$, which could regain 53.8 and 94.6%, respectively, by healing at room temperature and 20% RH for five minutes. Moreover, the PEI-SWCNTs composite could also be rapidly healed to over 90% of the original electrical ($0.15\,S\,m^{-1}$) and mechanical properties (~4 MPa, 0.07%) with the aid of water (50–60% RH).

In addition, other guest molecules such as azo [247], α-bromonaphthalene [247], and cholic acid [248] were also utilized for preparing self-healing supramolecular hydrogels. It is worth noting that although these gels possess fast healing rate and good repairing efficiency, they usually suffer from poor mechanical strength.

The addition of nanofillers into polymer matrices is a practically useful approach to improve mechanical properties of polymer materials, but generally causes a decrease in healing efficiency due to the restriction of molecular chain movement. For resolving the contradiction between mechanical robustness and healability, a strategy was proposed by Xiang et al. to homogeneously disperse reduced graphene oxide (RGO) nanosheets within self-assembled PAA/PEI film, in which the interactions between RGO and polymer are based on reversible dissociated/associated host–guest interaction of β-CD and Fc [249]. Consequently, the as-prepared self-healing films displayed Young's modulus of 17.2 GPa and hardness of 1.00 GPa. A cut ~70 μm wide completely disappeared after being immersed in 70 mM H_2O_2 solution for 10 minutes, followed by immersing in aqueous GSH solution (70 mM) for 30 minutes to restore the original storage modulus (117.5 MPa in water). Similarly, the host–guest interaction between β-CD and azobenzene was combined with graphene nanosheets to fabricate self-healing epoxy with enhanced strength [222]. It was found that the films exhibited excellent tensile strength (20.8 MPa) and healing efficiency (79.2%) after healing at 120 °C for 30 minutes. Furthermore, photothermally triggered self-healing could also be achieved to regain 59.4% of tensile strength after NIR (2 W) radiation for 10 minutes.

Since the synthesis of crown ether host compounds is more complex, and the selection range of appropriate guest molecules (e.g. secondary ammonium) is relative narrow, there have been few self-healing materials based on the host–guest interaction of crown ether so far. Huang et al. first designed two non-covalently cross-linked polymer gels by mixing a pendant dibenzo-24-crown-8 (DB24C8) group-containing PMMA and two bisammonium cross-linkers with different end-group sizes (i.e. phenyl and cyclohexyl) [250]. The supramolecular gel based on DB24C8 and phenyl-end-capped bisammonium could form immediately after mixing of the reactants, while that from DB24C8 and bisammonium with terminal cyclohexyl units took more than two months to achieve the reaction equilibrium due to the larger size of the cyclohexyl groups. Although both the resultant polymer gels could heal cracks within 20 minutes, the healing mechanisms were quite different. The self-healing of the former was caused by reversible host–guest interactions, while the latter relied on electrostatic and hydrogen-bonding interactions between bisammonium cross-linker and DB24C8. Later, the same authors proposed a linear supramolecular polymer assembled from host–guest complexation between bis(benzo-21-crown-7) and 1,2,3-triazole group-containing bis(dialkylammonium salt), which was then crosslinked by Pd^{2+} through coordination with 1,2,3-triazole ligand [251]. The resultant free-standing cross-linked supramolecular gel exhibited rapidly crack healing capability.

Chen et al. described a supramolecular polymer gel formed between DB24C8-based bis(crown ether) cross-linker and a styrene acrylate copolymer containing dibenzylammonium salt (DBAS) moieties [252]. Accordingly, crack healing capability was endowed to the resulting gel. In the work of Ikeda et al. [253], DB24C8 and DBAS were incorporated into the side chains of glycidyl triazole polymer (GTP). Self-healable supramolecular organogel was obtained by mixing the two kinds of functionalized GTP to form physical cross-links of the inclusion complex. Two pieces of the gel can be connected again immediately after attaching.

At present, all the crown ether-based systems are limited to secondary ammonium guests, and the crown ether-based interactions are only built up in organic solvents, which greatly restricts their practical applications.

1.3 Extrinsic Self-Healing

In the case of extrinsic self-healing, the matrix resin itself lacks the intrinsic self-healing mechanism. Healing agent has to be stored in some media and embedded into the materials in advance. As soon as the cracks destroy the fragile reservoirs, the healing agent would be released into the crack planes due to capillary effect and heal the cracks. In accordance with the types of container, there are two main modes of repair activity: (i) self-healing in terms of healant loaded pipelines; and (ii) self-healing in terms of healant loaded microcapsules. With the advantages of crack triggered delivery of healing agent, manual intervention (e.g. heating that used to be applied for intrinsic self-healing) might be no longer necessary.

1.3.1 Self-Healing in Terms of Healant Loaded Pipelines

1.3.1.1 Hollow Tubes and Fibers

The core issue of this technique lies in filling the brittle-walled vessels with polymerizable medium, which should be fluid at least at the healing temperature. Subsequent polymerization of the chemicals flowing to the damaged area plays the role of crack elimination. Dry first identified the potential applicability of hollow glass tubes [254–257]. A similar approach was adopted by Motuku et al. [258] and Zhao et al. [259]. The hollow glass capillaries have diameters (on millimeter scale) much larger than those of the reinforcing fibers in composites, so they have to act as initiation for composites failure [37]. Instead, Bleay et al. employed hollow glass fiber (with an external diameter of 15 μm and an internal diameter of 5 μm) to minimize the detrimental effect associated with large diameter fibers [37]. Complete filling of the tiny tubes with healing agent was achieved by vacuum assisted capillary action filling technique.

Accordingly, the following three types of healing system were developed (Figure 1.11) [254–269]. (i) Single-part adhesive. All hollow pipettes contained only one kind of resin-like epoxy particles (that can be flowable upon heating and then cured by the residual hardener) or cyanoacrylate (that can be consolidated under the induction of air). (ii) Two-part adhesive. In general, epoxy and its curing agent were used in this case. Neighboring hollow tubes were filled by them. (iii) Two-part adhesive. One component was incorporated into hollow tubes and the other in microcapsules.

With the aid of the pre-embedded healing system in hollow pipettes, Motuku et al. studied the healing ability of glass fiber/unsaturated polyester composites subjected to low velocity impact [258, 264]. The species of healing agent, characteristic parameters of the hollow pipes (amount, type of tubing materials, and spatial distribution), composite panel thickness, and impact energy level were found to be critical to the healing efficiency. Meanwhile, Bleay et al. proved that the epoxy-based composites reinforced by hollow glass fibers containing solvent diluted two-part epoxy became repairable as assessed by compression after impact test [37]. Trask et al. [268] considered the placement of self-healing hollow glass fiber layers within both glass fiber/epoxy and carbon fiber/epoxy composite laminates to mitigate damage and restore mechanical strength. The hollow fibers were bespoke

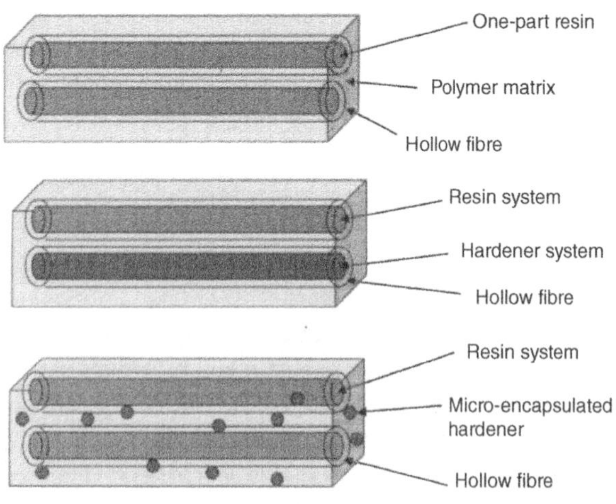

Figure 1.11 Schematic diagram of repair concept for polymer matrix composites using pre-embedded hollow tubes. *Source:* Reprinted from Bleay et al. [37]. Copyright 2001, with permission from Elsevier.

with diameters between 30 μm and 100 μm and were approximately 50% hollow. The study revealed that after the laminates were subjected to quasi-static impact damage, a significant fraction of flexural strength can be restored by the self-repairing effect of a healing resin stored within hollow fibers. More details of such a healing system can be found elsewhere [265–267, 269–271]. For example, Pang et al. added UV fluorescent dye to the healing resin within the hollow fibers so that bleeding of the repair substance in the composites can be visualized [265]. Williams et al. demonstrated the importance of optimizing the location and amount of healing resin filled hollow glass fibers to provide an effective self-healing function within carbon fiber reinforced epoxy laminates [270]. Kousourakis and Mouritz carefully assessed the influence of hollow glass fibers (external diameter = 170, 320, 430, and 680 μm, wall thickness = 0.6–1.5 μm) on mechanical properties of continuous carbon fibers/epoxy composites because ply waviness and resin-rich zones are changed [271]. Their results indicate that tension and compression properties of the composites are not significantly affected when the hollow fibers are aligned parallel to the load-bearing plies. However, the static strength properties are reduced at relatively large fiber sizes (above ~200 μm) when aligned normal to the load-bearing plies.

Earlier studies used to pay attention to the composites with restorative systems containing identical microtube diameters. Gumula and Szatkowski carried out comparative studies with multiple diameter structures [272]. They proved that a self-healing system containing two types of glass capillaries with diameters of 100 and 50 μm was more efficient (68% recovery of bending strength and Young's modulus after third regeneration) than a regeneration system containing capillaries with the same diameter (44% recovery of bending strength and 44% recovery of Young's modulus after third regeneration).

Instead of hollow glass tubes and glass fibers, hollow carbon fiber [273], hollow polymer tube [274, 275], and hollow metallic tube [276] were also proposed as optional micropipeline configurations. In particular, compartmented alginate fibers containing *ortho*-dichlorobenzene (DCB) as healing agent were embedded in PMMA and used as the model composite system [277]. The compartmented fibers offered a high potential for multiple locally distributed release of healing agent into the damaged polymer matrix in fiber reinforced polymer composites. After crack opening of a compartment the fiber still possessed healing capability because of its specific morphology – the inner volume of the fiber was divided into independent compartments filled with a healing liquid. Thus, the healing agent could be released independently from different compartments (i.e. in predesigned doses) responding to the crack density in a surrounding polymer matrix. Meanwhile,

Neisiany et al. [278] manufactured polyacrylonitrile shelled nanofibers separately containing the dual components of a self-healing epoxy system (comprising a low viscosity epoxy resin and its amine-based curing agent) via coaxial electrospinning. The nanofiber layers were then incorporated between carbon fiber fabric sheets during layup followed by vacuum-assisted resin transfer molding to fabricate self-healing carbon fiber composites. Mechanical analysis demonstrated an 11% improvement in tensile strength and a 4% gain in impact energy absorption compared with the control composite without nanofibers, which forms a striking contrast to conventional systems with healant loaded micropipelines that used to the reduce mechanical properties of composites. Three-point bending tests further affirmed the room temperature healing characteristics of the nanofiber containing composites, with a 96% recovery in flexural strength observed 24 hours after the initial bending fracture.

1.3.1.2 Three-Dimensional Microvascular Networks

In general, extrinsic self-healing composites are hard to use for repeated healing, because rupture of the embedded healant loaded containers would lead to depletion of the healing agent after the first damage. To overcome this difficulty, Toohey et al. proposed a self-healing system consisting of a 3D microvascular network capable of autonomously repairing repeated damage events [39]. Their work mimicked the architecture of human skin. When a cut in the skin triggers blood flow from the capillary network in the dermal layer to the wound site, a clot would rapidly form, which serves as a matrix through which cells and growth factors migrate as healing ensues. Owing to the vascular nature of this supply system, minor damage to the same area can be healed repeatedly. The 3D microvascular networks were fabricated by deposition of fugitive ink (a mixture of Vaseline/microcrystalline wax [60/40 by weight]) in terms of direct-write assembly [279] through a cylindrical nozzle. Then, the yielded multilayer scaffold was infiltrated with epoxy resin. When the resin was consolidated, a structural matrix was obtained. With the help of heating and light vacuum, the fugitive ink was removed, and 3D microvascular networks were created. By inserting a syringe tip into an open channel at one end of the microvascular networks, fluidic polymerizable healing agent was injected into the networks.

By means of a four-point bending configuration monitored with an acoustic emission sensor, the above approach proved to be feasible. The authors imagined extending this approach further to integrate pumps, valves, and internal reservoirs, as well as to introduce new functionalities, including self-diagnosis or self-cooling, through the circulation of molecular signals, coolants, or other species [39].

In the meantime, Hansen et al. made advances in direct-write assembly – dual ink deposition and vertical ink writing [280]. They independently deposited two fugitive organic inks to construct a 3D interpenetrating microvascular network to enable repeated, autonomous healing of mechanical damage in a biomimetic coating/substrate architecture. A healing efficiency of ~50% was retained after 30 cycles of healing, which compared favorably with the self-healing microcapsule and single-network geometries.

On the other hand, the reliability of a self-healing vascular network for engineering applications was evaluated by identifying the critical failure modes via a failure modes, effects and criticality analysis (FMECA) [281]. This technique has also been applied to the human circulatory system to gain insight and understanding. Plant vasculature was mimicked to propose a segregated network to address the risk of fluid leakage. The approach could allow a network to be segregated into six separate paths with a system mass penalty of only approximately 25%. Fluid flow interconnections that mimic the anastomosis of animal vasculatures can be used within a segregated network to balance the risk of failure by leakage and blockage. These biomimetic approaches defined a design space that considered the existing published literature in the context of system reliability. Hamilton et al. studied the issue at a different angle in terms of experiments [282]. They explored the interaction of cracks with a vascular system of microchannels (ca. 200 μm) embedded in a bulk epoxy structural material. Mechanical recovery from multiple cycles of damage was observed owing to the continuous supply of healing agents, and despite the intersection of crack damage with individual microchannels and subsequent disruption of the vascular network.

As an extension of the hollow glass fiber approach, vascular networks were also developed by Bond's group at the University of Bristol for providing a replenishable and repeated self-healing function [274, 275]. They first worked out glass fiber/epoxy laminates with a vascular sandwich core, giving a relatively thin-skinned configuration [274]. The healing networks consisted of tubing bonded into the midplane of the core and vertical risers supplying the skin–core bond region. The distribution of the vertical risers can be varied by design. In another work [275], a vascular sandwich structure with horizontal supply channels and vertical riser channels in a closed-cell foam core was prepared. Such a network had negligible effect on the baseline static mechanical properties of the composite panel. Rupture of the vessels released the healing fluid, filling the void that formed as a result of impact damage on the sample, as characterized by a complete recovery of compressive stress at failure after impact damage. Later, the researchers in Bristol proposed to construct vascular networks in composites following a biomimetic concept of "ray cell" structure [283, 284], analogous to vasculature in hardwoods. In addition to the analytical assessment that proved the eligibility of such plantae vascules for creating self-healing functionality [283], a prototype of this system was introduced within a carbon fiber reinforced epoxy polymer composite laminate [284, 285]. The vascules were formed by a "lost-wax" type process using low melting temperature solder wire (0.25 mm diameter, tin [Sn] 60%, lead [Pb] 40%) positioned at key interfaces between plies to form parallel channels within the laminated composite before autoclave processing. The solder wire was selected such that it could survive exposure to the manufacturer's recommended cure cycle for the composite resin system and was subsequently removed under the application of a post-cure process of 190 °C for two hours whilst held under vacuum. Oncc rcmoved, open channels or vascules were created. The results of this conceptual study showed that damage morphology is influenced by vascule orientation and that a 10 J low-velocity impact damage event is sufficient to breach the vasculature; a prerequisite for any subsequent self-healing function [284]. The residual compressive strength after a 10 J impact was found to decrease to 70% of undamaged strength when vasculature was aligned parallel to the local host ply and a value of 63% when aligned transverse. Such a vasculature might offer a self-healing function with minimum mass penalty, without initiating premature failure within a composite structure.

The abovementioned "lost-wax" type method for creating microchannels belongs to the techniques based on removable solid cores/preforms. Nylon fibers [286] and high melting temperature rigid steel wires [287] have also effectively served as the removable cores for producing hollow microchannels in composites. Besides, micromachining (including precision machining, computer numerical controlled machining, electric discharge machining, and laser machining [288–290]) was applied to create microchannels of a few hundred micrometers. Considering that the non-removable and removable cores as well as laser processing are only able to create simple and straight microchannels, vaporization of sacrificial components was developed [291], enabling production of 3D interconnected microchannel networks (Figure 1.12). The approach involved weaving of fabric preform using a combination of

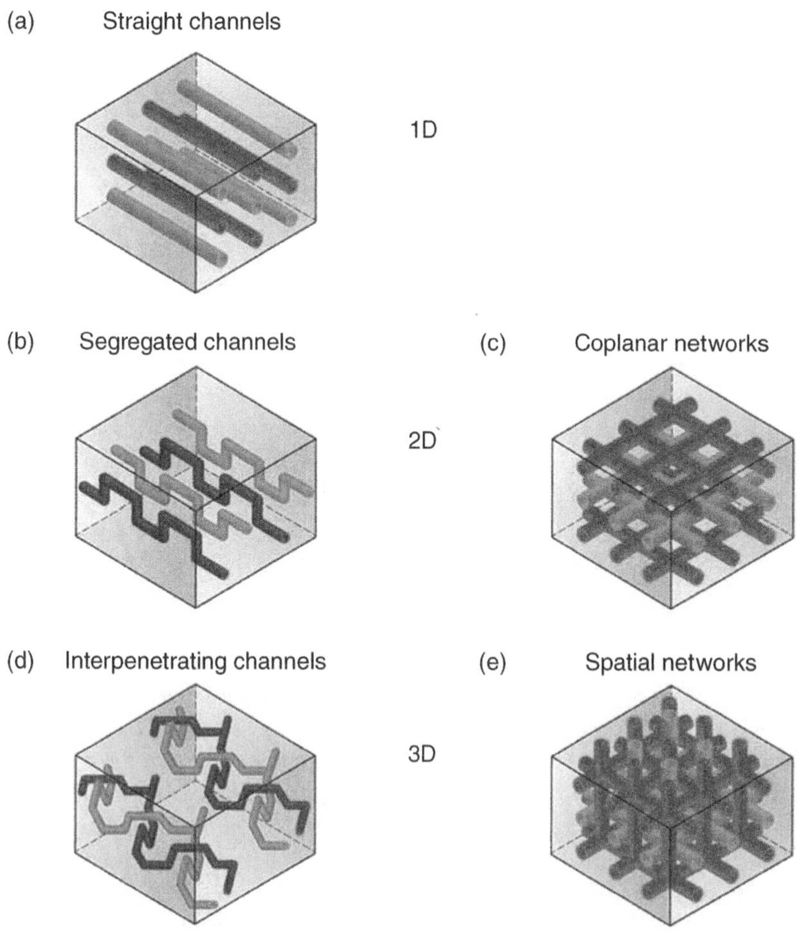

Figure 1.12 Vascular hierarchy. (a) 1D straight channels; (b) 2D segregated channels; (c) 2D coplanar interconnected networks; (d) 3D interpenetrating channels; and (e) 3D spatial interconnected networks. *Source:* Reprinted from Patrick et al. [291]. Copyright 2017, with permission from Elsevier.

glass or carbon fibers and catalyst impregnated polylactic acid sacrificial fibers [292]. When the composites were impregnated with matrix resin and cured, the sacrificial fibers were vaporized at elevated temperature and removed by application of modest vacuum, leaving behind the 3D microchannel network that had negligible effect on mechanical properties of the composites [293] but remarkable recovery ability [294].

1.3.2 Self-Healing in Terms of Healant Loaded Microcapsules

The principle of this approach resembles the aforementioned pipelines but the containers for storing healing agent are replaced by fragile microcapsules. The technique of microencapsulation has been rapidly developed since its emergence in the 1950s and mass production of microcapsules can be easily industrialized, so self-healing composites might thus be used in practice accordingly.

So far, five types of microencapsulated healing agent systems have been proposed (Figure 1.13). The single-capsule system consists of only one type of encapsulated healing agent. The healing agent can be a reactive chemical, suspension, solvent, or a low-melting point metal. Following release of the healing agent, reaction with the latent functional groups of the matrix, polymerization induced by the surrounding moisture environment or light, chain entanglement across the fractured surfaces, or conductive

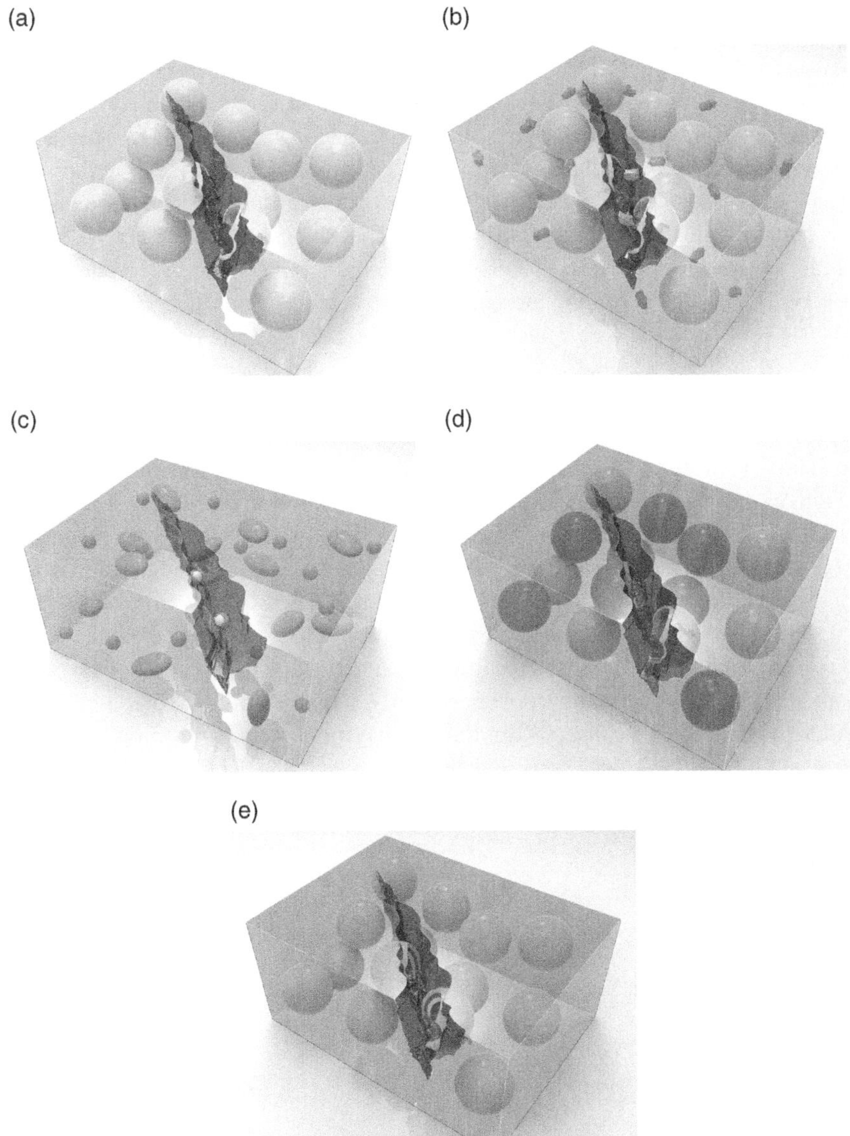

Figure 1.13 Typical extrinsic self-healing systems based on microcapsules. (a) Single capsules; (b) capsules/dispersed catalyst; (c) phase separated droplets/capsules; (d) double capsules; and (e) all-in-one microcapsules.

bridging would occur. The capsules/dispersed catalyst system is based on monomer-loaded capsules and catalyst dispersed (or dissolved) in the matrix. Soon after breakage of the microcapsules due to crack propagation in the matrix, the released monomer is polymerized under the action of the neighbor catalyst. In the phase separated droplets/capsule system, at least one healing component undergoes phase separation, whereas the remaining component is encapsulated. The two fluids react with each other upon release from capsules. The double-capsule system is often employed if two or more reactive liquid healing agents require compartmentalization. Furthermore, the self-healing material has to be subjected to stringent processing conditions such as heating and shear mixing. The all-in-one microcapsule system is entirely self-contained. Both the healing agent (monomer) and any required catalyst (initiator) are either held in the core and shell wall of the same capsule, and isolated from each other by layers (this generates a multilayer microcapsule), or encapsulated in separate smaller spheres that are stored within a larger sphere (this generates a capsule-in-capsule). When the capsules are ruptured, the released healing agent comes into contact with the catalyst, and is then polymerized in the absence of additional chemicals.

1.3.2.1 Methods of Microencapsulation

In the case of extrinsic self-healing based on healing capsules, the healing agents are microencapsulated for shielding against influences from the environment (before being compounded with the matrix) and matrix (during and after compounding) so that they can remain inactive for extended periods of time. Although a variety of alternative microencapsulation techniques are available, they are not necessarily suited for self-healing, and the encapsulating method is very often only suited for specific types of core material. To determine which microencapsulation technique and wall matrix should be adopted and developed, the following factors need to be considered: key properties of the core material, capsule size, permeability of the shell, encapsulation efficiency, resistance of the capsules to compounding with matrix, difference in the toughness properties and interfacial interactions between the microcapsule and polymer matrix, and processing parameters of the self-healing composite system. The microencapsulation technique must be tailored accordingly to satisfy the diverse requirements of a specific self-healing material as mentioned above.

In-Situ Polymerization

In-situ polymerization was used as an encapsulation method in the 1980s and progressed into industrial applications in the 1990s. It has many advantages such as controllable microcapsule size and shell thickness, simplicity of procedure, low costs, and ease of industrialization. However, compared with other encapsulation methods, in-situ polymerization requires longer reaction times.

During in-situ polymerization, oil emulsions in water or water emulsions in oil are first produced under vigorous agitation or sonication of a biphasic liquid. The core chemical acts as the dispersed phase. The monomers and initiators used for constructing the capsule wall are dissolved in either the dispersed or continuous phase. The polymer synthesized from the monomers is insoluble in the emulsion, so polymerization often takes place on the surface of the core material droplets or the resulting polymer accumulates onto the droplet surface, generating microcapsules with the desired core material (Figure 1.14).

In fact, most capsules used in self-healing materials are prepared by in-situ or interfacial polymerization in an oil-in-water emulsion system. The first self-healing microcapsule containing dicyclopentadiene (DCPD) was produced by in-situ polymerization with poly(urea-formaldehyde) (PUF) as wall material [295]. This study became one of the main references adopted by the following researchers for synthesizing different micro- and nanocapsules for self-healing polymers. For example, Yin et al. prepared PUF capsules containing epoxy monomers [296]. Suryanarayana et al. fabricated linseed oil-loaded PUF microcapsules [297],

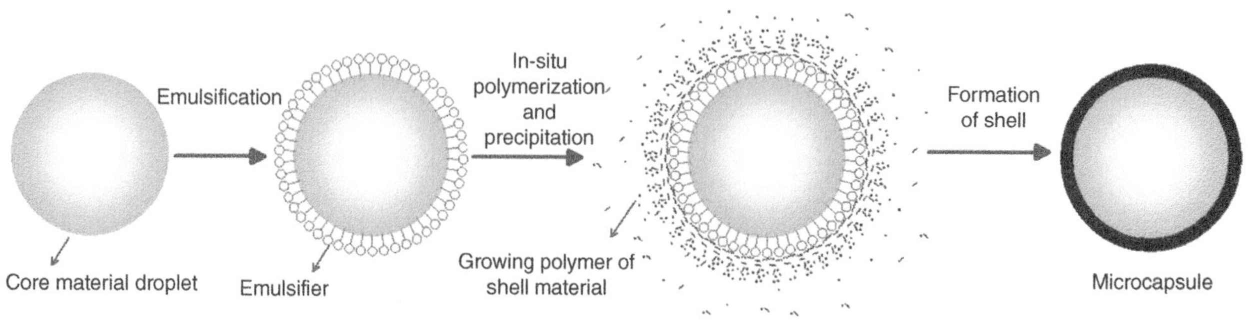

Figure 1.14 Schematic drawing of the formation of microcapsules by in-situ polymerization.

and Gragert et al. encapsulated two reactive components for the azide/alkyne click reaction (healing reaction) via in-situ polymerization of urea-formaldehyde [298]. Meanwhile, Cai et al. [299] modified the surface of PUF-walled capsules containing epoxy monomers using fluorinated silane to improve the hydrophobicity of the microcapsules in epoxy matrix and enhance the interfacial interaction.

Although PUF is frequently used, it has low weather resistance, poor sealing property, and high brittleness, which thwart attempted applications in practice. To avoid the shortcomings, Yuan et al. replaced the urea component with melamine to fabricate poly(melamine-formaldehyde) (PMF) capsules containing diglycidyl tetrahydro-*o*-phthalate (DTP, a low-viscosity and high-activity epoxy monomer) and polythiol (highly active low-temperature hardener for epoxy materials) [300]. The epoxy composites containing the microcapsules offered satisfactory autonomous restoration capabilities even after exposure at 250 °C for 24 hours, as assessed by the fracture mechanics and fatigue tests [301]. As PMF is more expensive than PUF, preparation of a copolymer of them is a feasible strategy to obtain both comprehensive properties and cost effectiveness. Liu et al. synthesized poly(melamine-urea-formaldehyde) (PMUF) microcapsules containing ethylidene norbornene (ENB)-based self-healing agents [302]. ENB is a potential self-healing agent candidate because of its rapid polymerization and lower cost Grubbs catalyst loading requirement. The PMUF microcapsules exhibited superior properties over the UF counterparts for self-healing composites. Moreover, the synthesis of PMUF microcapsules is much simpler than those prepared with UF.

For adapting to the requirement of high temperature application, other wall materials were introduced. Heat-resistant poly(phenylene oxide) (PPO) microcapsules filled with diglycidyl ether of bisphenol A (DGEBA) epoxy were made by in-situ polymerization [303]. The PPO resin was easily synthesized via oxidative polymerization of 2,6-dimethylphenol (DMP) at 30–50 °C in the presence of a small amount of a copper-amine complex (copper-ethylenediamine [Cu-EDA]) catalyst. The microcapsules remained stable up to 258 °C, indicating that they are resistant to thermal processes or operation environment under relatively high temperatures.

Interfacial Polymerization

Interfacial polymerization is similar to in-situ polymerization; however, the reactants involved in the synthesis of the shell wall are obtained from both the dispersed and continuous phases. In general, the core material is first emulsified or dispersed in the continuous phase containing one type of the pre-dissolved monomer precursor for the shell material. Then, another monomer that is soluble in the dispersed phase is added. Polymerization immediately occurs at or near the aqueous–organic interface, generating microcapsules with the droplets as the core material. Polymers may form from condensation reactions, namely between amines and aldehydes, acid chlorides, or isocyanates. The technique has the following advantages: (i) fast encapsulation, (ii) high encapsulation efficiency, (iii) mild reaction conditions, and (iv) flexible requirements relating to the purity and amount of reactants. However, the resulting microcapsules generally contain some unreacted shell monomers, which can react with the core material, potentially deactivating the core material.

For developing self-healing acrylic bone cement, PU-walled microcapsules containing medical cyanoacrylate tissue adhesive were prepared by interfacial polymerization of toluene-2,4-diisocyanate (TDI)-based PU prepolymer and 1,4-butanediol (BD) to encapsulate 2-octylcyanoacrylate (OCA) [304]. The capsule size and shell thickness both decreased with increasing agitation rates. However, a consistent wall thickness-to-diameter ratio of 0.01–0.02 was observed regardless of the agitation speed.

By means of interfacial polymerization of TDI prepolymer and BD [305], PU microcapsules can further include highly active isophorone diisocyanate (IPDI) as the core material, which is a potential catalyst-free healing agent for use in humid or wet environments. Du Prez and his group developed a one-pot approach for the synthesis of polyurea microcapsules with a high content of an encapsulated reactive liquid tri-isocyanate core and a shell bearing hydrophobic groups [306]. By the choice of the hydrophobic agents comprising hexamethyldisilazane (HMDS), alkylamine, fluorinated aromatic amine and perfluoride amine compounds, both the core content and stability of the encapsulated core as well as the shell morphology, varying from smooth, rough to wrinkled exterior surface, could be tuned.

Besides in-situ polymerization, interfacial polymerization can also be used for encapsulating epoxy healing agent. Liu et al. prepared epoxy microcapsules for self-healing coating by interfacial polymerization of epoxy resin droplets with EDA [307]. The encapsulation was achieved by reaction between the epoxy (which is water insoluble) and EDA (which is readily dissolved in water) at the interface of the epoxy droplets. When these microcapsules are embedded within the epoxy matrix to obtain self-healing epoxy composites, the inherent compatibility between the capsule shell and matrix affords efficient interfacial bonding.

In ref. [308] a modified encapsulation procedure that combined interfacial polymerization of PU and in-situ polymerization of UF was used to prepare liquid-filled healing microcapsules with two distinct shell walls. The commercial PU prepolymer (Desmodur L 75) was dissolved in the ethyl phenylacetate (EPA) core liquid prior to addition to the aqueous phase during in-situ polymerization of UF. The method yielded capsules with improved stability at high temperatures when compared with that of standard UF microcapsules, while retaining excellent interfacial bonding.

Pickering Emulsion Templating

Encapsulation via Pickering emulsion templating includes two steps: preparation of a stable Pickering emulsion (i.e. solid particle-stabilized emulsion) and immobilization of the colloidal particles adsorbed onto the droplets of the core material. To form a stable Pickering emulsion, the colloidal particles are first assembled at the interface through mechanical action. Then, the colloidal particles are immobilized by physical or chemical cross-linking to form a stable and compact shell structure. In comparison with conventional molecular surfactant-stabilized emulsions, a Pickering emulsion generally affords highly stable emulsions (inhibition of droplet coalescence), a lower degree of foaming, low toxicity, negligible pollution, easy separation, reusability, and with low costs, thus making the resulting emulsions popular candidates for templating microcapsule formation. Moreover, the resulting microcapsules can be imparted with distinct functionalities [309].

Li et al. described an approach for the preparation of solvent-based microcapsules via Pickering emulsion templated interfacial atom transfer radical polymerization (PETI-ATRP) [310]. The process involved the electrostatic deposition of a polyanionic ATRP initiator onto cationic nanoparticle surfaces. The resulting modified nanoparticles were good Pickering emulsifiers. Subsequently, the microcapsules were synthesized by PETI-ATRP of *N,N′*-methylene bisacrylamide to form the shell wall. The method allowed encapsulation of core solvents (xylene, hexadecane, and perfluoroheptane) of different solubility properties, and the microcapsule wall forming chemistry afforded the use of different vinyl monomers, demonstrating the versatility of this method.

Pickering emulsion is a good alternative method for encapsulating chemicals that are difficult to encapsulate via traditional emulsion systems (such as amine used for curing epoxy). A typical example is shown in ref. [311], which developed a method for the preparation of microcapsules containing reactive amine for self-healing polymers. The encapsulation was performed by interfacial polymerization of an isocyanate and an amine stabilized by an inverse Pickering emulsion. Hydrophobically modified nanoclay was used to form a Pickering emulsion at the polar–nonpolar interface, enabling excellent droplet stabilization. This method relies on the physical stabilization of emulsion droplets for wall formation and rapid condensation reactions between isocyanate and amine to form a stable shell wall.

The Pickering emulsion method can also be conveniently employed to prepare specific structured microcapsules. For example, EPA-loaded microcapsules with polyurea/PMF shells were fabricated using a silica nanoparticle-stabilized oil-in-water emulsion template and subsequent interfacial reaction and in-situ polymerization [312]. SiO_2 nanoparticles adsorbed at the interface between oil and water to stabilize the emulsions. The oil droplets containing EPA, IPDI, and TDI-terminated poly(propylene glycol) (PPG-TDI) subsequently reacted with melamine-formaldehyde (MF) prepolymer dissolved in the water phase. The interfacial reaction between pre-MF and IPDI produced an inner polyurea wall. Similarly, Patchan et al. [313] proposed a simple synthesis for the production of Pickering microcapsules containing octadecyltrimethoxysilane with outstanding strength, solvent resistance and barrier properties via interfacial polycondensation, except that silica nanopowder (10–20 nm diameter) replaced the conventional surfactant or hydrocolloid emulsifier. Compared with pure polyurea, the silica-polyurea composite improves the shelf life of the octadecyltrimethoxysilane by 10 times.

Because Pickering emulsion-based microcapsule preparation methods afford the isolation of the core material from the external continuous phase, the core loss is reduced, and core loading can be maximized. A wide variety of colloidal particles as the shell material can be employed, including micro- and nanoparticles, inorganic particles, and polymer microspheres. Shell permeability can be easily controlled by adjusting the size of the voids between particles. Meanwhile, the shell mechanical property can be regulated by controlling the particle size, particle array formation, and immobilization method.

Miniemulsion Polymerization

Miniemulsions are a special class of emulsions that are stabilized by surfactants (including co-stabilizer against Ostwald ripening) and high-energy homogenization. Miniemulsions usually yield stable and narrowly distributed droplets with a size ranging from 50 to 500 nm [314]. As polymers can be synthesized in the dispersed phase of a miniemulsion system, at the interface of the droplets, or in the continuous phase, miniemulsion droplets can be used as templates for the formation of a micro-/nanocapsule shell [314, 315]. This approach is termed miniemulsion polymerization encapsulation, suitable for encapsulating various materials that can be either hydrophobic or hydrophilic, and either a solid or a liquid [316].

Van den Dungen and Klumperman reported the synthesis of liquid-filled nanocapsules for self-healing using miniemulsion techniques [317]. The well-defined nanocapsules (150–350 nm) containing pentaerythritol tetrakis(3-mercaptopropionate) (PETMP) as a core material featured a styrene–maleic anhydride copolymer (PSMA) shell. The surfactant poly(styrene-maleic anhydride)-*block*-polystyrene (PSMA-b-PS) block copolymer stabilized the nanocapsules via reaction with formaldehyde. The extent of stabilization was greater when a PS block of an adequate length was used. The authors also synthesized nanocapsules containing 1,6-hexanediol di-(*endo*, *exo*-norborn-2-ene-5-carboxylate) (dinorbornene) through miniemulsion polymerization of styrene (60–150 nm). Dinorbornene also acted as a co-stabilizer owing to its high hydrophobicity and high boiling point. These core–shell nanocapsules present potential in self-healing coatings based on the thiol-ene chemistry [318, 319].

Moreover, miniemulsion polymerization can produce materials capable of satisfying both requirements of encapsulation and functionality. For example, functional polymer nanocapsules (hydrodynamic diameter ~300 nm) incorporating liquid healing agent DCPD in the core were successfully synthesized by free radical polymerization of the miniemulsions [317]. Polymerization between two monomers partitioned in the two liquid phases of the miniemulsion yielded nanocapsules with various functional groups such as sulfonate, amine, carboxylic acid, and PEG in the shell. As the range of functional groups (pendant groups) offered by vinyl monomers is very large, the compatibility requirement between the capsules and polymer to be repaired can be satisfied. The approach is not limited to encapsulation of core materials capable for ring-opening metathesis polymerization (ROMP) but can be extended to encapsulation of other reactants of polyaddition and polycondensation.

Zhang et al. directly prepared microcapsules containing pure polyamine through the integration of microfluidic emulsion and interfacial polymerization [320]. This integration made full use of the advantages but avoided the shortcomings of the two traditional encapsulation techniques involved, and thus required no stable emulsion and was not selective for the physicochemical properties of polyamines.

Solvent Evaporation/Solvent Extraction

In solvent evaporation-assisted microencapsulation, the polymer shell is dissolved in a water-immiscible volatile organic solvent, i.e. dichloromethane and chloroform, whereby the core material is also dissolved or dispersed. Next, the solution is added dropwise to an aqueous solution containing a suitable stabilizer (i.e. poly(vinyl alcohol) [PVA] and polyvinylpyrrolidone [PVP]), under stirring to form small polymeric droplets containing the core material. As the stirring proceeds, the droplets gradually harden to generate the corresponding polymer microcapsules. The hardening process is accomplished by removal of the solvent from the droplets through evaporation at elevated temperatures or reduced pressures. For encapsulation via solvent extraction, the polymer solution or suspension containing the healing agent is poured into a third liquid under stirring. Formation of the microencapsulated healing agent occurs because the solvent is extracted by the third liquid. The latter liquid acts as a precipitant for the polymer and is miscible with the solvent in which the healing agent is dissolved or dispersed [321–324].

Compared with other encapsulation techniques, these two processes offer many advantages, namely simple principles, ease of operation, and mild condition requirements. Furthermore, the processes are applicable to a wide variety of core materials, expensive equipment is not required, and the processes can be conveniently applied on an industrial scale. However, because the shell material needs to be predissolved in the organic solvent, this implies that the prepared microcapsules would be accordingly intolerant to the selected organic solvent, hence limiting the applicability of the methods.

Emulsion solvent evaporation affords the preparation of thermoplastic polymer-walled microcapsules with sizes ranging from the nanometer to the millimeter scale in the absence of polymerization. For instance, micrometer-sized PS-protected Grubbs catalyst particles (~1.5 μm) were synthesized in this manner [325]. Most of the reported encapsulated healing agent capsules feature sizes ranging from 10 to 100 μm. Previously reported Grubbs catalyst encapsulation methods produced stable particles that were larger than 50 μm [326]. It was believed that this size range was only ideal for healing large cracks and not for healing micrometer- and submicrometer-sized cracks [325]. To restore rather small cracks (i.e. micrometer-sized), smaller healing agents are required. In addition, to prepare highly stable microcapsules containing Grubbs catalyst, PMMA was added to the PS-walled catalyst particles using the emulsion solvent evaporation technique, and the resulting particles were subsequently coated with silica. The composite particles containing PS, PMMA, and silica shells were approximately 0.4 μm in diameter.

The most important advantages of the solvent extraction technique are simplicity and rapidity, but this method is more suitable for encapsulation of a solid core than liquid; therefore, it is usually not a main approach for producing self-healing microcapsules. Instead, this method serves as an auxiliary tool for preparation of complex microcapsules. For example, Xiao et al. [327] and Zhu et al. [328] employed solvent extraction as an intermediate step for the preparation of microcapsules.

Sol–Gel Reaction

The sol–gel reaction is an important strategy for preparing inorganic micro- and nanocapsules under relatively low temperatures and mild conditions. Basically, it involves the formation of an oxide network through polycondensation of a molecular precursor in a liquid. In the sol process, the inorganic shell material precursor is first dissolved in water to form a low-viscosity solution so that the shell reactant can be well mixed at the molecular level prior to gelling; the shell can be modified by other elements or functional groups. The sol–gel process is relatively time consuming. Also, micropores and microcracks may form on the resulting capsule shell after drying if the latter process is not well controlled.

In the area of self-healing materials, sol–gel chemistry has been mostly used to prepare silica capsules because silica is a potentially good diffusion barrier to maintain the chemical activity of the healing agent. In general, silica-based sol–gel processes involve the hydrolysis and condensation of an orthosilicate such as tetraethyl orthosilicate (TEOS). There are several silica-based condensation chemistries available [178–180], owing to the extensive studies carried out in silica chemistry. Using silica-based sol–gel reactions, aqueous, oil, and solid particles can all be encapsulated within the silica shell.

The most typical sol–gel reaction is based on Stöber chemistry. For realizing processable microcapsules via downstream processing of thermoplastic polymer matrices, direct encapsulation of a sustainable solvent system in silica microcapsules via sol–gel chemistry was developed [329]. The silica microcapsules were obtained by hydrolysis and condensation of TEOS, using an aqueous glycerol microemulsion as a template, via two different routes. The encapsulated solvent system, aqueous glycerol, has solvent welding capabilities for thermoplastic resins such as PVA [330]. Thus, silica microcapsules filled with aqueous glycerol solvent would be a potential candidate for self-healing systems of thermoplastic resins such as PVA. In this encapsulation process, pre-hydrolysis of TEOS was necessary to obtain isolated free-flowing silica microcapsules. The silica microcapsules had a higher micro-compression performance over PUF-based organic microcapsules. Similarly, the sol–gel process was applied to form silica microcapsules containing photopolymerizable trimethylolpropane triacrylate [331], meeting the technical requirements of space applications – low and high temperatures: −120 °C (dark side) to +250 °C (solar side); UV radiation: 200–400 nm; and low pressure: 10^{-4} Pa.

In addition, fluoride-catalyzed silica condensation chemistry proved to be feasible to coat both DCPD-loaded PUF capsules and polymer-protected Grubbs catalyst with silica, creating a protective and functional layer [332]. The importance of the silica shell was highlighted from the comparison studies examining the dispersion of the coated capsules and non-coated capsules in an epoxy matrix. The non-coated capsules could not be incorporated into epoxy at concentrations higher than 3 wt% and drying resulted in significant particle aggregation. In contrast, centrifugation and freeze drying of the aqueous solution of coated capsules resulted in a free-flowing powder of individual capsules. The powder could be incorporated into the epoxy matrix at significantly higher concentrations. Compared with the Stöber chemistry, the fluoride-catalyzed silica sol–gel process is conducted in a neutral aqueous medium, thereby minimizing the extent of catalyst deactivation and loss of the healing agent while building a non-porous shell. Larger capsules in micrometer or submicrometer size can be obtained. Nevertheless, this process is more costly and time consuming.

1.3.2.2 Healing Chemistries

The chemical reaction of the healing agent at the damage site is responsible for re-bonding the fractured surface and acts as the last step of the self-healing process based on microcapsules [42]. The extent of the reaction determines the efficiency of the healing. Many chemical reactions have been explored during the rapid development of self-healing polymeric materials. The most suitable healing chemistry can be chosen according to the type of material that requires repairs, the environmental conditions, and the user requirements.

Ring-Opening Metathesis Polymerization

The most known and extensively explored healing chemistry is ROMP of DCPD, assisted with Grubbs catalyst. In 2001, White et al. developed the first microcapsule-based self-healing material based on this chemistry [36]. The self-healing material consisted of an epoxy matrix with embedded DCPD microcapsules and Grubbs catalyst particles. Upon damage of the composite, the microcapsule shell ruptured, releasing DCPD into the newly created cavities through capillary action, and DCPD underwent ROMP when it came into contact with the pre-embedded, dispersed Grubbs catalyst (Scheme 1.5). The pioneering work was later improved to achieve higher healing efficiencies, lower synthesis costs, and higher diversity of the matrices and composites. For example, the healing system of Grubbs catalyst/DCPD capsules was incorporated into PMMA-based bone cement [332, 333]. Self-healing properties were imparted to epoxy vinyl ester [334] and polystyrene-*block*-polybutadiene-*block*-polystyrene (PS-b-PBD-b-PS) [335] using the same healing system. The healing system was applied to prepare self-healing glass fiber-reinforced epoxy composite [336].

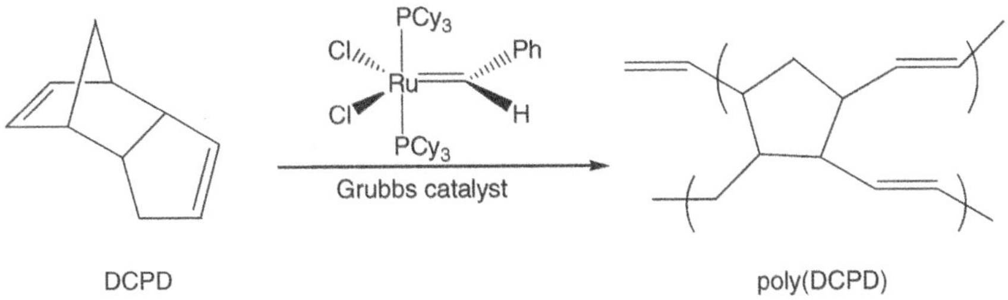

Scheme 1.5 Ring-opening metathesis polymerization of DCPD.

At the early investigation stages of ROMP-based self-healing systems [36], the first-generation Grubbs catalyst served as the polymerization initiator. However, the catalyst presents limitations because of its chemical and thermal instability that is typically encountered during processing and curing of epoxy resins [337]. To expand the scope of this particular healing chemistry to more challenging application conditions, comparison studies of three types of Grubbs catalysts (first-generation (benzylidene-bis(tricyclohexylphosphine) dichlororuthenium), second-generation (benzylidene[1,3-bis(2,4,6-trimethylphenyl)-2-imidazolidinylidene] dichloro (tricyclohexylphosphine) ruthenium), and Hoveyda–Grubbs second-generation (1,3-bis(2,4,6-trimethylphenyl)-2-imidazo lidinylidene) dichloro (*o*-isopropoxyphenylmethylene) ruthenium)) were conducted for use in self-healing polymers [337, 338]. Owing to the thermal stability and functional groups, the second-generation Grubbs catalyst emerged as the most versatile catalyst, especially for high-temperature processes and applications involving healing chemistries aimed at improving self-healing performance via non-covalent interactions. By choosing the appropriate curing cycle, the catalytic activity of the Hoveyda–Grubbs second-generation catalyst in ROMP reactions could be controlled [339]. More specifically, the catalyst remained intact during the preparation of the epoxy matrix via curing. Thus, the catalytic activity of the catalyst was preserved, and the catalyst efficiently catalyzed the polymerization of the healing monomer.

Besides DCPD, ENB has been examined as a potential candidate for the polymerizable monomer of the two-component healing agent via the ROMP mechanism, owing to its rapid polymerization and lower Grubbs catalyst loading requirements [302, 340–345]. Furthermore, ENB has a low viscosity that would afford complete infiltration in the cracked parts to achieve a high healing efficiency. In this context, application of ENB-loaded microcapsules and Grubbs catalyst might achieve metathesis product formation in the cracked epoxy material at low temperatures [344, 345]. However, because the resulting polymer possesses a linear molecular structure, with physical properties that may be unsuitable to effectively heal microcracks, the incorporation of either norbornene-based ROMP crosslinking agents [346] or DCPD [342, 343] was examined to improve the adhesive properties of the resulting polymer. Yang and Lee further demonstrated that methyl 5-norbornene-2-carboxylate (MNC) can successfully act as the adhesion promoter of ENB [347]. The adhesion strength increased remarkably with MNC loadings of up to 10 wt % compared with ENB alone, so that a significant improvement in healing efficiency was observed for the ENB/MNC mixture.

Majchrzak et al. reported an autonomous self-healing process at ambient temperature based on ROMP of mono- and difunctional norbornene dicarboximides [348]. A mixture of mono- and difunctional norbornene dicarboximides was encapsulated as healing agents and Grubbs ruthenium complexes served as the initiator. The authors claimed that the healing chemistry offered several advantages: (i) excellent level of control over the crosslinking density and T_g, in accordance with the material properties of the matrices; (ii) the monomers produce fewer unpleasant odors; (iii) the dicarboximide groups possess good adhesion properties; and (iv) other functional groups, such as hydroxyls, can be easily introduced into the monomer to enhance adhesion to a wide range of matrices. The healing experiments indicated that at least 10% healing was achieved after 30 minutes, and a maximum healing efficiency of 33% was observed between 11 hours and 24 hours. In contrast, a lower healing efficiency of 27% was achieved in the control specimen containing microencapsulated DCPD.

Polycondensation

In earlier attempts to remend cracks in vinyl ester matrices, siloxane polycondensation was used as the healing chemistry [349]. A typical system consists of phase-separated droplets containing hydroxyl end-functionalized polydimethylsiloxane (HOPDMS) and its cross-linker polydiethoxysiloxane (PDES) and the PU microcapsules containing catalyst, DBTL. When the composite was subjected to mechanical damage, the phase-separated healing agent and catalyst escaped from the broken microcapsules, then mixed and imbibed the entire crack plane. Polycondensation between HOPDMS and PDES proceeded (Scheme 1.6). Organotin catalysts are highly desirable for curing PDMS-based systems because side reactions are limited even under an open air atmosphere [350, 351].

The approach possesses some advantages: (i) the healing chemistry remains stable in a water or air environmen;, (ii) the chemistry is stable at elevated temperatures (>100 °C), enabling healing in high-temperature thermoset systems; and (iii) the components are widely available and comparatively low cost.

However, the phase-separated healing agent and microencapsulated catalyst system have some limitations. As the PDMS healing agent is in direct contact with the matrix, it is thus susceptible to matrix-initiated reactions. For example, the amine-curing agent remaining in the epoxy matrix would catalyze PDMS-based polycondensation. To overcome this disadvantage, a dual-capsule system was developed [352]. PDMS was filled into PUF microcapsules, whereas the catalyst solution of dimethyldineodecanoate tin (DMDNT) or Si[OSn $(n\text{-}C_4H_9)_2OOCCH_3]_4$ (TKAS) was loaded into the PU microcapsules. The two types of capsules were then dispersed in an epoxy–amine matrix to form an epoxy-based self-healing coating. The healing experiments showed that when DMDNT catalyst was used, a healing temperature of 50 °C was required. In contrast, in the presence of TKAS catalyst, autonomous corrosion protection was realized in an ambient environment.

Unlike catalyst-initiated siloxane polycondensation, catalyst-free hydrolysis and polycondensation of hydrolysable organic silane was also used for self-healing. To prepare self-healing corrosion-resistant polymer coating, POTS, a commercially available hydrolyzable organic silane, was selected and microencapsulated as the healing agent [353]. POTS is hydrolyzable under a wet environment, forming a silane-based film that is hydrophobic. Such a specific wetting property will enable repulsion of an aqueous electrolyte solution from the metal, thereby providing further corrosion protection to the metal substrate. As exemplified, the synthesized POTS microcapsules were incorporated into epoxy resin and coated on a steel substrate to form a corrosion-resistant organic coating. The self-healing coating had good corrosion resistance, which was attributed to an autonomous self-healing mechanism of the hydrolysis of POTS in water following release upon scratching. Nevertheless, the healing system is only effective in thin films with access of environmental water.

Free Radical Polymerization

To impart self-healing properties to PMMA bone cement (or simply bone cement), five types of peroxide initiators (i.e. benzoyl peroxide [BPO], lauroyl peroxide, methyl ethyl ketone peroxide, *tert*-butyl peroxide, and *tert*-butyl peroxybenzoate) were compared for radical polymerization-based self-healing [354]. This is because the proposed healing chemistry is identical to the curing chemistry of the target material. BPO performed the best among the five peroxide initiators. Accordingly, it was dissolved in phenyl acetate (9.9 wt%) and encapsulated to produce epoxy vinyl ester composites containing 10 wt% of the microcapsules. A healing efficiency of ~80% at ambient temperature was reflected by the preliminary tests, whereby a mixture of acrylic monomers and tertiary amine activator was injected into the crack plane after initial fracture. In a later report by White et al. [355], a self-healing bone cement was formulated using medical grade bone cement in which healing microcapsules (~100 μm) were dispersed. The healing approach based on dual-capsule free radical reaction chemistry was presented.

Epoxy can be self-healed through free radical polymerization of styrene [356]. Compared with the epoxy monomer, styrene has a considerably lower viscosity. Two types of microcapsules, styrene/cobaltous naphthenate (accelerant)-loaded PMF capsules and BPO/toluene solution-loaded PMF capsules, were prepared and embedded in the epoxy matrix. Microscopic observations indicated that fracture of the composites resulted in rupture of the capsules. The core substances (i.e. healing agents) were delivered to the cracked planes and free radical polymerization of styrene initiated by BPO occurred at room temperature. The cracks were mended, as characterized by the observed recovery of impact strength of the material. Here,

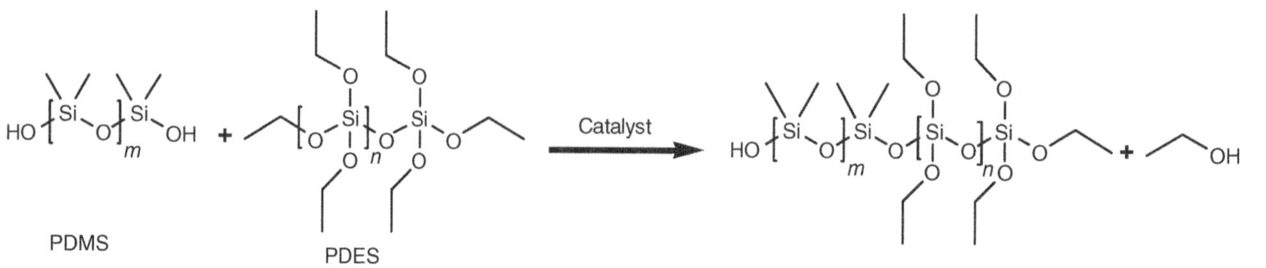

Scheme 1.6 Schematic illustration of polycondensation between HOPDMS and PDES.

PS served as a binder to mend the cracks in the epoxy matrix. On account of (i) the high volatility of styrene, (ii) the poor adhesion of PS to the matrix epoxy, and (iii) the lower mechanical properties of PS than that of matrix epoxy, full restoration was not available for the system.

Self-healing of polymers under environmental exposure is always desired. In addition to trace water in air, oxygen is an interesting auxiliary agent to initiate healing reactions. Based on the spontaneous oxidation of unsaturated aliphatic acid in air, a self-healing epoxy coating containing PUF-encapsulated linseed oil was developed for the fabrication of corrosion-resistant steel panels [297]. Materials with self-healing paint containing linseed oil microcapsules displayed no corrosion features at the scribed lines. The mechanism for the superior corrosion resistance involves the initial release of the linseed oil from the ruptured microcapsules and subsequent filling in the cracks, followed by formation of a film by oxidative polymerization in the presence of oxygen, consequently inhibiting the ingress of moisture and oxygen.

Meanwhile, tung oil was chosen as an oxidative healing agent that was encapsulated within the PUF shell by in-situ polymerization [357]. The prepared microcapsules containing tung oil were then incorporated into paint formulations. Corrosion resistance tests of the composite coating showed that the tung oil released from the broken microcapsules upon scratching was polymerized in the presence of air, creating a barrier between the substrate and air, thereby extending the coating lifespan against attacks from corrosive media. On the other hand, epoxy ester made from epoxy resin and fatty acid was encapsulated by PUF through in-situ polymerization [358]. One of the notable features of epoxy ester is that it can be cured at room temperature through auto-oxidation without additional cross-linkers, which ensures that the epoxy ester resin is utilized as the healing agent to repair the cracks without additional heating or catalyst. The coatings containing epoxy ester-loaded PUF microcapsules exhibited satisfactory self-healing property and sufficient corrosion resistance recovery.

Photo-induced reactions are attractive because light is readily available, and its consumption is clean and cheap. The first example of a capsule-type photo-induced self-healing system based on free radical polymerization was described in ref. [359]. Methacryloxypropyl-terminated polydimethylsiloxane (MAT-PDMS) was employed as a healing agent, and benzoin isobutyl ether (BIE) was used as a photoinitiator. The mixture of MAT-PDMS and BIE was encapsulated within the PUF shell and then embedded into a coating matrix to obtain a self-healing protective coating. Upon damage-induced cracking, the microcapsules were ruptured, resulting in release of the healing agent and photoinitiator into the crack plane. Afterwards, self-healing of the coating was achieved by sunlight-triggered free radical polymerization.

On the other hand, thiol-ene chemistry is a robust technique for the development of highly cross-linked polymer networks [318, 319, 360]. The mechanism of photo-induced thiol-ene chemistry proceeds via a step-growth free radical polymerization. Owing to the delayed onset of gelation, increasing amounts of uniform polymer networks form and reduce stress within the polymer network. Moreover, thiol-ene chemistry is insensitive to oxygen, unlike (meth) acrylate-based traditional free radical polymerization. It was adopted for synthesizing autonomous self-healing waterborne coatings [361], where the thiol and ene compounds were separately encapsulated to form a dual-capsule system and embedded in various coating formulations. Upon crack formation with a razor blade, the capsules ruptured, and the healing agent diffused into the crack via capillary action, initiating the thiol-ene healing process to restore the coating and supply continuing protection of the substrate. The selected thiol and ene compounds were PETMP (thiol), and 1,6-hexanediol diacrylate (diacrylate) and dinorbornene, respectively. PETMP and diacrylate were incorporated into either PUF micro- or nanocapsules, whereas dinorbornene was incorporated into PS nanocapsules. Although the healing reaction did not require an activator or a catalyst, the network formation was accelerated by UV irradiation and radical photoinitiator 2,2-dimethoxy-2-phenylacetophenone (DMPA). It appeared that both monomer combinations (PETMP-diacrylate and PETMP-dinorbornene), containing ~1 wt% photoinitiator with respect to PETMP, allowed for rapid network formation at ambient temperature upon UV exposure of the solution. The study successfully demonstrates the application of thiol-ene chemistry as a healing mechanism for the development of self-healing coatings based on micro- or nanocapsules. However, the healing efficiency of the fabricated coatings was not provided. Besides, healing in the presence of UV intervention instead of sunlight would limit its usage in practice.

Addition Reaction

Epoxy–amine curing chemistry is particularly attractive for self-healing epoxy composites because it has been widely used in the fabrication of epoxy materials. During this process, the healed region is chemically and mechanically compatible with the host matrix that possesses the same chemistry. Although synthesis of capsules containing a liquid amine is difficult when compared with the encapsulation of epoxy, there were some breakthroughs in this area. Various amines have been successfully incorporated into microcapsules, owing to the development of diverse microencapsulation techniques [311, 362–366]. However, sometimes the healing ability of the amine capsules is not as good as expected, due to the high viscosity of the core substance [362].

In an attempt to impart self-healing abilities to high-performance polymers, thermally stable epoxy-loaded PPO capsules were incorporated into thermosetting 4,4-bismale imido-diphenylmethane (BDM)/diallylbisphenol A (BA) resin [367]. The released core material (epoxy) containing EDA complexes of copper(I) (Cu-EDA) from the microcapsules can be polymerized in the presence of various catalytic groups, such as –OH and amine groups, on the matrix or capsule shell under controlled temperatures to mend the cracks.

In fact, the reaction between BMIs and amines (or thiols) is also applicable to the self-healing of epoxy materials. For example, it was found that thiols and maleimides react readily in the presence of tertiary amines in the epoxy material [368]. This shows that the multifunctional thiols and difunctional maleimides forming the two-component healing agent can react to fill the voids of the cracked area. Moreover, conjugation reactions between the maleimide and residual amines in the epoxy material ensured chemical bonding between the newly formed network and the original materials during crack healing. Healing efficiency evaluated though manual injection of the healing agent revealed a maximum value up to 121% for EPON 828 epoxy material.

On the other hand, isocyanate-based adhesives have been used in diverse applications. The isocyanate chemical group is extremely reactive and can form chemical bonds with any chemical group that contains an active hydrogen atom. Moreover, isocyanates are very soluble in many solvents, owing to their low molecular weight, and can easily wet and penetrate porous structures to form strong mechanical interlocks [369]. Isocyanate has been considered a candidate in catalyst-free one-part healing agent systems, owing to its reactivity with moisture. For example, HDI-loaded PU microcapsules were manufactured via interfacial polymerization of commercial MDI prepolymer and 1,4-butanediol in an oil-in-water emulsion [370]. Then, the synthesized microcapsules were dispersed into epoxy resin to fabricate a self-healing anticorrosion coating on steel substrates. Microscopic observation of the scratched area of the coated panels showed that the cracks were filled with the product formed between HDI released from the ruptured microcapsules and water. The cracks were thus sealed and healed autonomously, thereby delaying the diffusion of salt ions.

In addition to the above nucleophilic addition reaction, cycloaddition reaction has also been employed as an effective healing chemistry for extrinsic self-healing. Palmese et al. developed a combined healing system comprising extrinsic healing via microcapsules and intrinsic healing via reversible DA bonds, thereby exploiting the unique advantages that each individual system offers in a single self-healing system [371]. The study showed that a furan-functionalized epoxy–amine thermoset network was repaired in situ in the presence of a multimaleimide solution at room temperature, and significant load recovery was repeatedly achieved in a given location. The healing mechanisms involved a combination of solvent-induced mechanical interlocking and DA bond formation to restore the cracks [372, 373]. The structure and concentration of maleimide influenced the degree of DA bonding in a crack, whereas the type of solvent influenced both operating healing mechanisms. Moreover, in short-term healing (<1 day), the structure of the healing agent influenced the healing efficiency of the thermoset. However, in long-term healing (20 days), the structure of the healing agents had negligible effects on the healing efficiency for the range of multimaleimide structures studied [374]. Based on the above investigation, phenyl acetate–maleimide microcapsules with PUF walls were prepared and incorporated into a furan functional epoxy, generating an autonomous self-healing material. The fabricated specimens containing 10 wt% capsules showed a 75% strength recovery 24 hours after the first fracture at ambient temperature. The capsule healing system based on DA chemistry also allowed for repeated healing cycles [371].

Another microcapsule-type protective coating, which had repeated self-healing capabilities using the combined extrinsic and intrinsic healing strategy, was developed in ref. [375]. Cinnamide moiety-containing polydimethylsiloxane (CA-PDMS) was synthesized and used as a healing agent. Under photoirradiation, CA-PDMS generated a viscoelastic substance that had intrinsic self-healing capabilities. By using this behavior, CA-PDMS was encapsulated into PUF resin and incorporated into a coating matrix. When the CA-PDMS-based self-healing coating was damaged, the healing agent was released from the ruptured microcapsules and filled the damaged regions, generating viscoelastic materials by photocycloaddition reaction that contributed to the subsequent multiple healing cycles of the coating via hydrogen-bond interactions.

For the preparation of a self-healing system involving a liquid polymer as a reactive healing agent, Cu(I)-catalyzed azide/alkyne cycloaddition click chemistry was employed [376, 377]. This afforded the preparation of self-healing PIB composites based on encapsulation [298, 378]. Owing to the immiscibility of the liquids, azido-telechelic three-arm star polyisobutylene and trivalent alkynes were separately encapsulated into PUF microcapsules, and subsequently embedded into high-molecular weight PIB. The catalyst $CuBr(PPh_3)_3$ was dispersed in the polymer matrix. Self-healing of the obtained PIB composite containing 10 wt% polymer azide-filled microcapsules and 5 wt% alkyne-filled microcapsules was performed for five days. A tensile modulus recovery of 91% at 25 °C and 107% at 60 °C was attained as a result of cycloaddition cross-linking between the multivalent liquid azide and trivalent alkyne.

1.4 Insights for Future Work

Owing to their smart nature, self-healing materials might solve technical problems that are hard to be solved or simply cannot be solved by traditional approaches. They have opened broad prospects for key engineering communities and cutting-edge technologies. For example, aviation and automobile industries treat self-healing materials as an advanced means of upgrade. Materials capable of stopping possible crack formation and extending fatigue longevity are certainly attractive for every aspect of modern industries. However, the structural materials applied in practice have to withstand extremes of temperature, pressure, and acceleration, as well as unexpected impacts. There is still a long way to prove the feasibility of the self-help concept. As concluded by Trask et al. [29], most of the self-healing work to date has been bioinspired and not biomimetic, although this is slowly changing. According to their opinion, biomimetic self-healing must mimic self-healing to a higher degree, but the distinction is very difficult to define because in practice exact replication of natural systems is very challenging.

Many research groups worldwide have explored self-healing fiber-reinforced polymer composites in hopes of achieving autonomous structural recovery to hinder damage formation in its early stages before the integrity of the structure is compromised [379]. The composite structures with high safety factors would account for this defect sensitivity to ensure they will remain load bearing throughout their service life. Although both intrinsic and extrinsic approaches have shown their abilities in this aspect, the performance of the composites needs to be further improved. As shown in Figure 1.15, it would be rather difficult for the intrinsic self-healing systems to reach the strength of commercial ones. Meanwhile, the healing efficiencies of extrinsic self-healing composites are mostly lower than the intrinsic ones. In reality, incorporating an extrinsic additional functionality into an already highly architectured structure may also be detrimental to the manufacturing process.

Despite the fact that the technological achievements in self-healing polymers and polymer composites are far from satisfactory, the new opportunities found during scientific

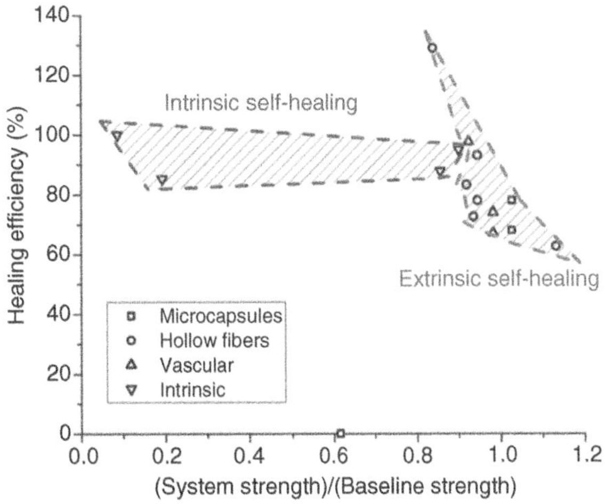

Figure 1.15 Performance map of self-healing structural composites. Maximum healing efficiency is shown as a function of system strength normalized by the baseline strength (i.e. without healing functionality). Data are divided into intrinsic or extrinsic, and further subdivided into microcapsules, hollow fibers, vascular, and intrinsic. For the case of intrinsic systems, these reference materials simply employ a commercially available thermosetting polymer matrix. *Source:* Reprinted from Cohades et al. [379]. Copyright 2018, with permission from John Wiley and Sons.

research and development have demonstrated that it is feasible to either invent new polymers with inherent crack repair capability or integrate existing materials with novel healing systems. Interdisciplinary studies based on close collaboration among scientists and engineers are prerequisites for overcoming the difficulties. The related works and outcomes have broadened the application possibilities of polymeric materials. Also, the extended service life of components made from these intelligent materials would contribute to reduce waste disposal. It is undoubtedly important for building up a sustainable society.

Application-oriented studies of self-healing materials and devices have increased rapidly in the past decade. Most cases involve the integration of existing sciences, rather than the creation of new healing chemistry, which may promote upgrading of many more traditional materials in practice. The healing of wide-open cracks, for example, has attracted tremendous research interest. White et al. proposed a large damage volume recovery strategy through two-stage polymer chemistry based on a vascular approach [380, 381]. When damage occurred, the vasculatures were ruptured and polymerizable liquids were then released that rapidly formed a shape-conforming gel as a semisolid scaffold. Polymerization followed forming a solid structural polymer with robust mechanical properties. In this strategy, the gelling solutions can fill damaged spans from 9.0 to 11.2 mm. In addition, Li's group proposed and validated a biomimetic two-step close-and-heal scheme [382–384], which proved to heal structural-length scale crack repeatedly by using heating induced melting of the embedded thermoplastic particles as the healing agent and confined shape memory effect of the matrix polymer. This scheme mimics the self-healing of human skin; that is, for a wide cut, the wound needs to be stitched together (close) and new cells gradually grow (heal). For enabling inherent repeated self-healing of wider wounds, Fan et al. [100] coupled an external stress-free two-way shape memory effect with intrinsic self-healing driven by dynamic reversible covalent bonds in a single polymer (crosslinked PU, Figure 1.16). That is, a wider crack was allowed to be repeatedly recovered by the successive functioning of the two-way shape memory effect (that closed the crack in advance) and intrinsic self-healing (that chemically re-bonded the cracked surfaces). The core issue lies in the fact that the homolysis temperature of the included reversible C—ON bond (i.e. self-healing temperature of the material) was tuned to be higher than the triggering temperature of the two-way shape memory effect.

The research progress clearly indicates that the benefit that self-healing brings to materials is nowadays no longer limited to mechanical performance, but is gradually expanding to include, for example, electrical, optical and physicochemical properties [55]. From a practical perspective, functionality restoration that recovers the principal function of materials might be more cost effective than full-scale restoration that recovers the materials to their previous status quo. A few new disciplines are being derived following this trend. Skin-inspired electronics [385], for example, which allows for advanced health monitoring, disease detection, medical therapies, and human–machine interfacing, incorporate skin-like properties including self-healing for long-term durability under harsh mechanical conditions. These demands have fueled the development of a new generation of electronic materials, primarily composed of polymers and polymer composites with both high electrical performance and skin-like properties. Additionally, self-healing wearable electronics, self-healing energy storage devices, and self-healing supercapacitors are also receiving growing attention [386]. Not only intrinsic but also extrinsic self-healing has contributed to the underlying healing paradigm. For example, the dual microcapsules system containing polyamine as a multirole trigger and epoxy monomer dyed with a pH indicator

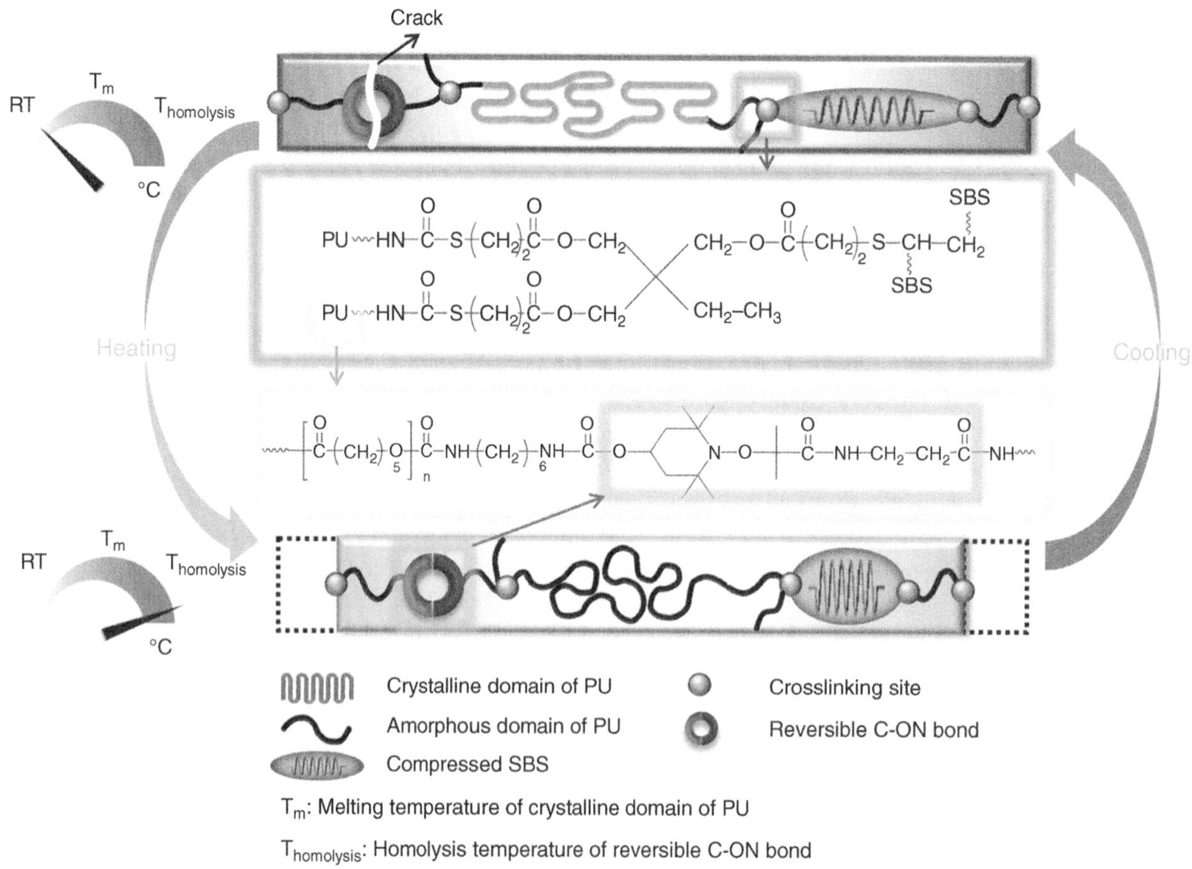

Figure 1.16 Schematic drawing of a cross-linked polyurethane containing a reversible C—ON bond for self-healing of a wide crack. *Source:* Reprinted from Fan et al. [100]. Copyright 2018, with permission from American Chemical Society.

was able to self-warn microscopic "subcutaneous" damage and macroscopic surface damage by a conspicuous red color and conduct self-healing as well [387].

From a long-term point of view, the synthesis of new polymers accompanied by intrinsic self-healing function through molecular design would be a reasonable objective. Recent studies have shown the prospects for this trend. Many more reports on intrinsic self-healing than on extrinsic self-healing are appearing. Nevertheless, the automatic triggering mechanism for intrinsic self-healing remains open. In most cases, intrinsic self-healing of polymers cannot take effect upon cracking like the extrinsic self-healing materials by making use of the crack response of healing capsules. This implies that mechanochemical exploration in this area needs to continue.

Piermattei et al. [387] designed latent catalysts that combine a metal with polymer-functionalized ligands, producing a long chain with a single coordination complex in the center. In these latent catalysts, selective scission of a coordination bond, which is weak compared with the covalent bonds in the polymer, results in either a catalytically active ligand or a catalytically active metal site (Figure 1.17). They found that silver(I) complexes of polymer-functionalized N-heterocyclic carbenes catalyzed a transesterification reaction when exposed to ultrasound in solution. Furthermore, ultrasonic activation of a ruthenium biscarbene complex with appended polymer chains led to catalysis of olefin metathesis reactions. In each case,

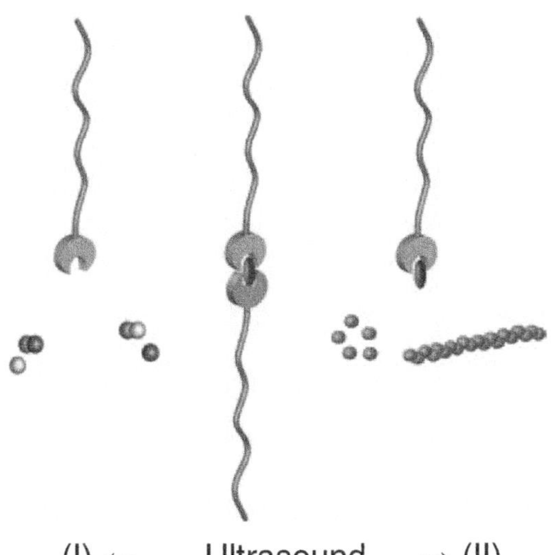

Figure 1.17 Cartoon depicting the ultrasound-induced mechanochemical scission of a metal complex with polymeric ligands, resulting in the activation of a catalyst. In (I), the active ligand catalyzes transesterification; and in (II) the metal center catalyzes polymerization. *Source:* Reprinted from Piermattei et al. [387]. Copyright 2009, with permission from Springer Nature.

the catalytic activity originating from ligand dissociation was brought about by transfer of mechanical forces from the polymeric substituents to the coordination bond. Such mechanochemical catalyst activation has potential applications not only in transduction and amplification of mechanical signals, but also in mechanically initiated polymerizations for self-healing materials by restoring the original molecular weight of the target polymer [388]. Similarly, Kersey et al. suggested inclusion of reversible interactions in a material for engineering repair without an external stimulus [389]. They described a family of hybrid polymer gels in which covalent cross-links created a permanent, stiff scaffold onto which reversible metal–ligand coordinative cross-links were added. In this way, the reversible component would relieve stress from the permanent network, break preferentially under significant load, and would then reform once that stress was removed to restore strength to the material. Takaffoli et al. [390] focused on mechanochemical responses in viscoelastic elastomers and developed a theoretical model to characterize the coupling between viscoelasticity and chemical reactions of mechanochemically responsive elastomers. Li et al. [391] observed the interfacially selective mechanochemical activation of a maleimide-anthracene mechanophore anchored PMA grafted SiO_2 nanoparticles system in sonication generated elongational flow. The model system is valid for designing mechanochemically self-healing composite materials.

These works show the possibility that conventional mechanical force might automatically trigger the start of the intrinsic repair process. The challenges lie in the fact that the strength of the mechanosensitive bonds should be high enough to provide the polymeric materials with adequate load bearing capability (which is of particular importance for structural materials).

Lenhardt et al. provided a possible solution for mechanochemistry-aided intrinsic self-healing from a different angle [392]. They found a polymer molecule that was so springy it snapped back from stretching to be much smaller than it was before. In this context, self-healing polymeric materials can be developed in which the macromolecules near cracked portions would be able to perform new chemistry to build bridges across the cracks.

In our latest work in this area [393, 394], our group synthesized a PU with its soft segments and hard segments cross-linked by DOPA-Fe^{3+} and His-Zn^{2+} coordination bonds, respectively, for self-blocking and self-healing of early minor damage in polymers. Upon loading, the weaker His-Zn^{2+} bonds in the cross-linked PU had to be preferentially dissociated, and the imidazole rings of the unbound His ligands raised the microenvironmental pH (Figure 1.18). Accordingly, the adjacent bis-DOPA-Fe^{3+} bonds changed to tris-coordination as driven by the

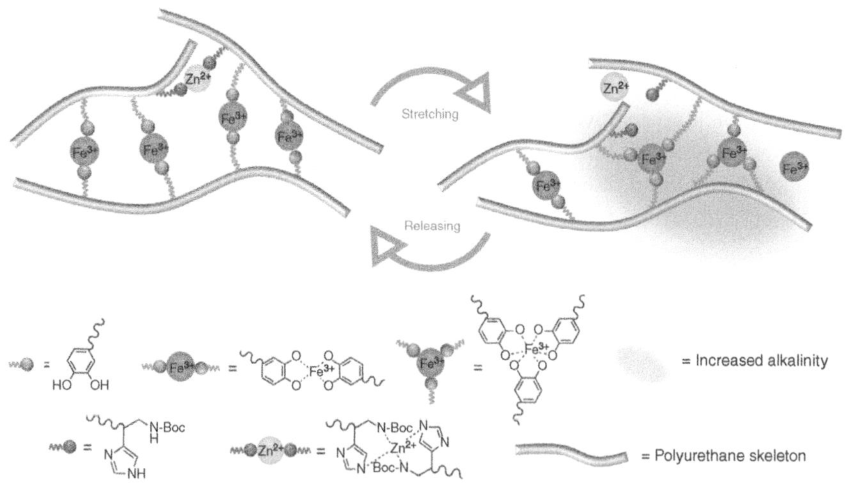

Figure 1.18 Schematic drawing of the mechanochemically initiated reversible cascading variation of the metal–ligand complexations.

alkalinity, which increased the amount of chains to be connected to the DOPA-Fe^{3+} cross-linkages, i.e. cross-link density. Owing to the temporarily increased cross-link density, the PU is strengthened so that the microvoids were blocked. When the applied force was removed, the minor damage was closed along with contraction of the deformed PU network. In the meantime, the imidazole rings of the unbound His ligands attached to the soft segments easily re-associated with the neighboring free Zn^{2+}. As a result, the minor damage originating from broken His-Zn^{2+} bonds was restored, and the microenvironmental pH was recovered by the reduced amount of free imidazole rings. Eventually, the tris-DOPA-Fe^{3+} bonds transformed from bis-DOPA-Fe^{3+} bonds also returned to their original status. Such mechanochemically initiated cascading variation of metal–ligand coordination bonds proved to proceed repeatedly free of manual intervention and catalyst. Besides, the static force offered by conventional universal tester, rather than ultrasonic, was sufficient for initiating the molecular-level internal equilibrium prevention mechanisms.

In conclusion, many proof-of-principle experiments, prototype formulations and structures for introduction of self-healing mechanism into various types of polymers and polymer composites have been reported, but very few are devoted to examples of commercial application. The laboratory- or small-scale studies used to neglect manufacturability, cost, aging, durability, stability, and validation of the self-healing technology. As a result, this would lead to a missed opportunity for justifying further investment in the realization of the technology. This may explain why self-healing functionalities are hard to find in industrialized products. Investigations on a larger scale that consider application scenarios are required. Only then will it be possible to talk about the availability of self-healing technology for daily use.

References

1 Riefsnider, K.L., Schulte, K., and Duke, J.C. (1983). Long term failure behavior of composite materials. *ASTM Special Technical Publication* 813: 136–159.

2 Cardon, A.H. and Verchery, G. (1991). *Durability of Polymer Based Composite Systems for Structural Applications*. Springer.

3 Dorworth, L. and Gardiner, G. (2007). Repair of composite structures – a review. *Journal of Advanced Materials* 39: 3–13.

4 Kim, H.-J., Kim, H.-S., Lee, G.-Y. et al. (2019). Three-dimensional carbon fiber composite printer for CFRP repair. *Composites Part B: Engineering* 174: 106945.

5 Paul, J. and Jones, R. (1992). Repair of impact damaged composites. *Engineering Fracture Mechanics* 41: 127–141.

6 Soutis, C., Duan, D.-M., and Goutas, P. (1999). Compressive behaviour of CFRP laminates repaired with adhesively bonded external patches. *Composite Structures* 45: 289–301.

7 Zhang, H., Motipalli, J., Lam, Y.C., and Baker, A. (1998). Experimental and finite element analyses on the post-buckling behaviour of repaired composite panels. *Composites Part A: Applied Science and Manufacturing* 29: 1463–1471.

8 Baker, A.A., Rose, L.R.F., and Jones, R. (2002). *Advances in the Bonded Composite Repairs of Metallic Aircraft Structure*. Amsterdam: Elsevier Science Ltd.

9 Zimmerman, K.B. and Liu, D. (1996). An experimental investigation of composite repair. *Experimental Mechanics* 36: 142–147.

10 Chotard, T.J., Pasquiet, J., and Benzeggagh, M.L. (2001). Residual performance of scarf patch-repaired pultruded shapes initially impact damaged. *Composite Structures* 53: 317–331.

11 Hosur, M.V., Vaidya, U.K., Myers, D., and Jeelani, S. (2003). Studies on the repair of ballistic impact damaged S2-glass/vinyl ester laminates. *Composite Structures* 61: 281–290.

12 Tse, P.-C., Lau, K.-J., and Wong, W.-H. (2002). Stress and failure analysis of woven composite plates with adhesive patch-reinforced circular hole. *Composites Part B: Engineering* 33: 57–65.

13 Ageorges, C., Ye, L., and Hou, M. (2001). Advances in fusion bonding techniques for joining thermoplastic matrix composites: a review. *Composites Part A: Applied Science and Manufacturing* 32: 839–857.

14 Beevers, A. (1991). Welding: the way ahead for thermoplastics? *Engineering* 231: ACE11–ACE12.

15 Davies, P., Courty, C., Xanthopoulos, N., and Mathieu, H.-J. (1991). Surface treatments for adhesive bonding of carbon fibre/poly(etherether ketone) composites. *Journal of Materials Science Letters* 10: 335–338.

16 Schwartz, M.M. (1994). *Joining of Composite Matrix Materials*, 35–88. Materials Park, OH: ASM International.

17 Todd, S.M. (1990). Joining thermoplastic composites. In: *Proceedings of the 22nd International SAMPE Technical Conference*, 383–392. Boston, MA (6–8 November 1990): SAMPE.

18 Don, R.C., Bastien, L., Jakobsen, T.B., and Gillepsie, J.W. Jr. (1990). *Fusion bonding of thermoplastic composites by resistance heating*. *SAMPE Journal* 26: 59–66.

19 Cogswell, F.N., Meakin, P.J., Smiley, A.J. et al. (1989). Thermoplastic interlayer bonding for aromatic polymer composites. In: *Proceedings of the 34th International SAMPE Symposium*, 2315–2325. Reno, NV (8–11 May 1989): SAMPE.

20 Smiley, A.J., Halbritter, A., Cogswell, F.N., and Meakin, P.J. (1991). Dual bonding of thermoplastic composite structures. *Polymer Engineering and Science* 31: 526–532.

21 Rudolf, R., Neitzel, M., and Strohfuss, W. (1997). Interaction of variables in vibration welding of continuous fiber reinforced thermoplastics. In: *Proceedings of the 29th International SAMPE Technical Conference*, 462–474. Orlando, FL (28 October–1 November 1997): SAMPE.

22 Benatar, A. and Gutowski, T.G. (1986). Methods for fusion bonding thermoplastic composites. *SAMPE Quarterly* 18 (1): 35–42.

23 Lee, W.I. and Springer, G.S. (1984). Interaction of electromagnetic radiation with organic matrix composites. *Journal of Composite Materials* 18: 357–386.

24 Brent, S.A., Johnson, D.P., and Johnson, B.A. (1990). Variable interactions in ultrasonic welding of thermoplastic composites. *SAMPE Quarterly* 21 (2): 36–41.

25 Ahmed, T.J., Stavrov, D., Bersee, H.E.N., and Beukers, A. (2006). Induction welding of thermoplastic composites - an overview. *Composites Part A: Applied Science and Manufacturing* 37: 1638–1651.

26 Stavrov, D. and Bersee, H.E.N. (2005). Resistance welding of thermoplastic composites-an overview. *Composites Part A: Applied Science and Manufacturing* 36: 39–54.

27 Fernie, J.A., Threadgill, P.L., and Watson, M.N. (1991). Progress in joining advanced materials. *Welding and Metal Fabrication* 5: 179–184.

28 Meyers, M.A., Chen, P.-Y., Lin, A.Y.-M., and Seki, Y. (2008). Biological materials: structure and mechanical properties. *Progress in Materials Science* 53: 1–206.

29 Trask, R.S., Williams, H.R., and Bond, I.P. (2007). Self-healing polymer composites: mimicking nature to enhance performance. *Bioinspiration & Biomimetics* 2: 1–9.

30 Hastings, G.W. and Mahmud, E.A. (1993). Intelligent orthopaedic materials. *Journal of Intelligent Material Systems and Structures* 4: 452–457.

31 Martin, P. (1997). Wound healing-aiming for perfect skin regeneration. *Science* 276: 75–81.

32 Caplan, A.I. (1988). Bone development, cell and molecular biology of vertebrate hard tissues. *Ciba Foundation Symposium* 136: 3–21.

33 Albert, S.F. (1981). Electrical stimulation of bone repair. *Clinics in Podiatric Medicine and Surgery* 8: 923–935.

34 David Darling (n.d.). Blood clotting. http://www.daviddarling.info/encyclopedia/B/blood_clotting.html (accessed 19 February 2020).

35 Merck Sharp & Dohme Corp. (2020). MSD manual consumer version. http://www.merckmanuals.com/home (accessed 19 February 2020).

36 White, S.R., Sottos, N.R., Geubelle, P.H. et al. (2001). Autonomic healing of polymer composites. *Nature* 409: 794–797.

37 Bleay, S.M., Loader, C.B., Hawyes, V.J. et al. (2001). A smart repair system for polymer matrix composites. *Composites Part A: Applied Science and Manufacturing* 32: 1767–1776.

38 Lee, J.Y., Buxton, G.A., and Balazs, A.C. (2004). Using nanoparticles to create self-healing composites. *Journal of Chemical Physics* 121: 5531–5540.

39 Toohey, K.S., Sottos, N.R., Lewis, J.A. et al. (2007). Self-healing materials with microvascular networks. *Nature Materials* 6: 581–585.

40 Ghosh, S.K. (ed.) (2009). *Self-Healing Materials: Fundamentals, Design Strategies, and Applications*. Weinheim: Wiley-VCH.

41 Yuan, Y.C., Yin, T., Rong, M.Z., and Zhang, M.Q. (2008). Self-healing in polymers and polymer composites. Concepts, realization and outlook: a review. *Express Polymer Letters* 2: 238–250.

42 Wool, R.P. and O'Connor, K.M. (1981). A theory of crack healing in polymers. *Journal of Applied Physics* 52: 5953–5963.

43 Wool, R.P., Yuan, B.L., and McGarel, O.J. (1989). Welding of polymer interfaces. *Polymer Engineering & Science* 29: 1340–1367.

44 Kim, Y.H. and Wool, R.P. (1983). A theory of healing at a polymer-polymer interface. *Macromolecules* 16: 1115–1120.

45 Varley, R.J. and van der Zwaag, S. (2008). Towards an understanding of thermally activated self-healing of an ionomer system during ballistic penetration. *Acta Materialia* 56: 5737–5750.

46 Fall, R. (2001). Puncture reversal of ethylene ionomers – mechanistic studies. Master of Science thesis. Virginia Polytechnic Institute and State University, Blacksburg.

47 Kalista, S., Ward, T.C., and Oyetunji, Z. (2007). Self-healing of poly(ethylene-co-methacrylic acid) copolymers following projectile puncture. *Mechanics of Advanced Materials and Structures* 14: 391–397.

48 Kalista, S., Ward, T.C., Kalista, S.J., and Ward, T.C. (2007). Thermal characteristics of the self-healing response in poly(ethylene-co-methacrylic acid) copolymers. *Journal of the Royal Society Interface* 4: 405–411.

49 Kalista, S. (2003). Self-healing of thermoplastic poly(ethylene-co-methacrylic acid) copolymers following projectile puncture. Master of Science thesis. Virginia Polytechnic Institute and State University, Blacksburg.

50 Kalister, S.J. (2009). *Self-healing ionomers*. In: *Self-Healing Materials: Fundamentals, Design Strategies, and Applications* (ed. S.K. Ghosh), 73–100. Weinheim: Wiley-VCH.

51 Meure, S., Wu, D.Y., and Furman, S.A. (2009). Polyethylene-co-methacrylic acid healing agents for mendable epoxy resins. *Acta Materialia* 57: 4312–4320.

52 Meure, S., Furman, S., and Khor, S. (2010). Poly[ethylene-co-(methacrylic acid)] healing agents for mendable carbon fiber laminates. *Macromolecular Materials and Engineering* 295: 420–424.

53 Bergman, S.D. and Wudl, F. (2008). Mendable polymers. *Journal of Materials Chemistry* 18: 41–62.

54 Kloxin, C.J. and Bowman, C.N. (2013). Covalent adaptable networks: smart, reconfigurable and responsive network systems. *Chemical Society Reviews* 42: 7161–7173.

55 Zhu, M., Rong, M.Z., and Zhang, M.Q. (2014). Self-healing polymeric materials towardsnon-structural recovery of functional properties. *Polymer International* 63: 1741–1749.

56 Chen, X.X., Dam, M.A., Ono, K. et al. (2002). A thermally re-mendable cross-linked polymeric material. *Science* 295: 1698–1702.

57 Chen, X.X., Wudl, F., Mal, A.K. et al. (2003). New thermally remendable highly cross-linked polymeric materials. *Macromolecules* 36: 1802–1807.

58 Yoshie, N., Saito, S., and Oya, N. (2011). A thermally-stable self-mending polymer networked by Diels-Alder cycloaddition. *Polymer* 52: 6074–6079.

59 Zeng, C., Seino, H., Ren, J. et al. (2013). Bio-based furan polymers with self-healing ability. *Macromolecules* 46: 1794–1802.

60 Tian, Q., Rong, M.Z., Zhang, M.Q., and Yuan, Y.C. (2010). Optimization of thermal remendability of epoxy via blending. *Polymer* 51: 1779–1785.

61 Du, P., Liu, X., Zheng, Z. et al. (2013). Synthesis and characterization of linear self-healing polyurethane based on thermally reversible Diels-Alder reaction. *RSC Advances* 3: 15475–15482.

62 Du, P., Wu, M., Liu, X. et al. (2014). Diels-Alder-based crosslinked self-healing polyurethane/urea from polymeric methylene diphenyl diisocyanate. *Journal of Applied Polymer Science* 131: 40234/1–40234/7.

63 Du, P., Wu, M., Liu, X. et al. (2014). Synthesis of linear polyurethane bearing pendant furan and cross-linked healable polyurethane containing Diels-Alder bonds. *New Journal of Chemistry* 38: 770–776.

64 Yang, L., Lu, X., Wang, Z., and Xia, H. (2018). Diels–Alder dynamic crosslinked polyurethane/polydopamine composites with NIR triggered self-healing function. *Polymer Chemistry* 9: 2166–2172.

65 Zhang, B., Digby, Z.A., Flum, J.A. et al. (2015). Self-healing, malleable and creep limiting materials using both supramolecular and reversible covalent linkages. *Polymer Chemistry* 6: 7368–7372.

66 Chung, C.-M., Roh, Y.-S., Cho, S.-Y., and Kim, J.-G. (2004). Crack healing in polymeric materials via photochemical [2+2] cycloaddition. *Chemistry of Materials* 16: 3982–3984.

67 Ling, J., Rong, M.Z., and Zhang, M.Q. (2011). Coumarin imparts repeated photochemicalremendability to polyurethane. *Journal of Materials Chemistry* 21: 18373–18380.

68 Abdallh, M., Yoshikawa, C., Hearn, M.T. et al. (2019). Photoreversible smart polymers based on $2\pi+2\pi$ cycloaddition reactions: nanofilms to self-healing films. *Macromolecules* 52: 2446–2455.

69 Abdallh, M., He, P., Hearn, M.T. et al. (2019). Light-switchable self-healing dynamic linear polymers: reversible cycloaddition reactions of thymine-containing units. *ChemPlusChem* 84: 333–337.

70 Cash, J.J., Kubo, T., Bapat, A.P., and Sumerlin, B.S. (2015). Room-temperature self-healing polymers based on dynamic-covalent boronic esters. *Macromolecules* 48: 2098–2106.

71 Lai, J.C., Mei, J.F., Jia, X.Y. et al. (2016). A stiff and healable polymer based on dynamic-covalent boroxine bonds. *Advanced Materials* 28: 8277–8282.

72 Yuan, D., Delpierre, S., Ke, K. et al. (2019). Biomimetic water-responsive self-healing epoxy with tunable properties. *ACS Applied Materials & Interfaces* 11: 17853–17862.

73 Bao, C., Jiang, Y.J., Zhang, H. et al. (2018). Room-temperature self-healing and recyclable tough polymer composites using nitrogen-coordinated boroxines. *Advanced Functional Materials* 28: 1800560/1–1800560/10.

74 Kim, C., Ejima, H., and Yoshie, N. (2018). Polymers with autonomous self-healing ability and remarkable reprocessability under ambient humidity conditions. *Journal of Materials Chemistry A* 6: 19643–19652.

75 Chen, Y., Tang, Z., Zhang, X. et al. (2018). Covalently cross-linked elastomers with self-healing and malleable abilities enabled by boronic ester bonds. *ACS Applied Materials & Interfaces* 10: 24224–24231.

76 Delpierre, S., Willocq, B., Manini, G. et al. (2019). Simple approach for a self-healable and stiff polymer network from iminoboronate-based boroxine chemistry. *Chemistry of Materials* 31 (10): 3736–3744.

77 Pepels, M., Filot, I., Klumperman, B., and Goossens, H. (2013). Self-healing systems based on disulfide-thiol exchange reactions. *Polymer Chemistry* 4: 4955–4965.

78 Canadell, J., Goossens, H., and Klumperman, B. (2011). Self-healing materials based on disulfide links. *Macromolecules* 44: 2536–2541.

79 Lei, Z.Q., Xiang, H.P., Yuan, Y.J. et al. (2014). Room-temperature self-healable and remoldable cross-linked polymer based on the dynamic exchange of disulfide bonds. *Chemistry of Materials* 26: 2038–2046.

80 Rekondo, A., Martin, R., de Luzuriaga, A.R. et al. (2014). Catalyst-free room-temperature self-healing elastomers based on aromatic disulfide metathesis. *Materials Horizons* 1: 237–240.

81 Michal, B.T., Jaye, C.A., Spencer, E.J., and Rowan, S.J. (2013). Inherently photohealable and thermal shape-memory polydisulfide networks. *ACS Macro Letters* 2: 694–699.

82 Xiang, H.P., Qian, H.J., Lu, Z.Y. et al. (2015). Crack healing and reclaiming of vulcanized rubber by triggering the rearrangement of inherent sulfur crosslinked networks. *Green Chemistry* 17: 4315–4325.

83 Xiang, H.P., Rong, M.Z., and Zhang, M.Q. (2016). Self-healing, reshaping, and recycling of vulcanized chloroprene rubber: a case study of multitask cyclic utilization of cross-linked polymer. *ACS Sustainable Chemistry & Engineering* 4: 2715–2724.

84 Xu, W.M., Rong, M.Z., and Zhang, M.Q. (2016). Sunlight driven self-healing, reshaping and recycling of a robust, transparent and yellowing-resistant polymer. *Journal of Materials Chemistry A* 4: 10683–10690.

85 Takahashi, A., Goseki, R., Ito, K., and Otsuka, H. (2017). Thermally healable and reprocessable bis(hindered amino)disulfide-cross-linked polymethacrylate networks. *ACS Macro Letters* 6: 1280–1284.

86 Kim, S.M., Jeon, H., Shin, S.H. et al. (2018). Superior toughness and fast self-healing at room temperature engineered by transparent elastomers. *Advanced Materials* 30: 1705145/1–1705145/8.

87 Lai, Y., Kuang, X., Zhu, P. et al. (2018). Colorless, transparent, robust, and fast scratch-self-healing elastomers via a phase-locked dynamic bonds design. *Advanced Materials* 30: 1802556/1–1802556/8.

88 Zhang, Y., Yuan, L., Liang, G., and Gu, A. (2018). Developing reversible self-healing and malleable epoxy resins with high performance and fast recycling through building cross-linked network with new disulfide-containing hardener. *Industrial & Engineering Chemistry Research* 57: 12397–12406.

89 Ji, S., Cao, W., Yu, Y., and Xu, H. (2015). Visible-light-induced self-healing diselenide-containing polyurethane elastomer. *Advanced Materials* 27: 7740–7745.

90 Liu, J., Ma, X., Tong, Y., and Lang, M. (2018). Self-healing polyurethane based on ditelluride bonds. *Applied Surface Science* 455: 318–325.

91 Lei, Z.Q., Xie, P., Rong, M.Z., and Zhang, M.Q. (2015). Catalyst-free dynamic exchange of aromatic Schiff base bonds and its application to self-healing and remolding of crosslinked polymers. *Journal of Materials Chemistry A* 3: 19662–19668.

92 Mo, R., Hu, J., Huang, H. et al. (2019). Tunable, self-healing and corrosion inhibiting dynamic epoxy–polyimine network built by post-crosslinking. *Journal of Materials Chemistry A* 7: 3031–3038.

93 Song, F., Li, Z., Jia, P. et al. (2019). Tunable "soft and stiff", self-healing, recyclable, thermadapt shape memory biomass polymers based on multiple hydrogen bonds and dynamic imine bonds. *Journal of Materials Chemistry A* 7: 13400–13410.

94 Yuan, C., Rong, M.Z., Zhang, M.Q. et al. (2011). Self-healing of polymers via synchronous covalent bond fission/radical recombination. Chemistry of Materials 23: 5076–5081.

95 Yuan, C., Zhang, M.Q., and Rong, M.Z. (2014). Application of alkoxyamine in self-healing of epoxy. Journal of Materials Chemistry A 2: 6558–6566.

96 Zhang, Z.P., Rong, M.Z., Zhang, M.Q., and Yuan, C. (2013). Alkoxyamine with reduced homolysis temperature and its application in repeated autonomous self-healing of stiff polymers. *Polymer Chemistry* 4: 4648–4654.

97 Yuan, C., Rong, M.Z., and Zhang, M.Q. (2014). Self-healing polyurethane elastomer with thermally reversible alkoxyamines as crosslinkages. *Polymer* 55: 1782–1791.

98 Zhang, Z.P., Rong, M.Z., and Zhang, M.Q. (2014). Room temperature self-healable epoxy elastomer with reversible alkoxyamines as crosslinkages. *Polymer* 55: 3936–3943.

99 Zhang, Z.P., Lu, Y., Rong, M.Z., and Zhang, M.Q. (2016). A thermally remendable and reprocessable crosslinked methyl methacrylate polymer based on oxygen insensitive dynamic reversible CON bonds. *RSC Advances* 6: 6350–6357.

100 Fan, L.F., Rong, M.Z., Zhang, M.Q., and Chen, X.D. (2018). Repeated intrinsic self-healing of wider cracks in polymer via dynamic reversible covalent bonding molecularly combined with a two-way shape memory effect. *ACS Applied Materials & Interfaces* 10: 38538–38546.

101 Fan, L.F., Rong, M.Z., Zhang, M.Q., and Chen, X.D. (2018). Dynamic reversible bonds enable external stress-free two-way shape memory effect of a polymer network and the interrelated intrinsic self-healability of wider crack and recyclability. *Journal of Materials Chemistry A* 6: 16053–16063.

102 Amamoto, Y., Kamada, J., Otsuka, H. et al. (2011). Repeatable photoinduced self-healing of covalently cross-linked polymers through reshuffling of trithiocarbonate units. *Angewandte Chemie* 123: 1698–1701.

103 Amamoto, Y., Otsuka, H., Takahara, A., and Matyjaszewski, K. (2012). Self-healing of covalently cross-linked polymers by reshuffling thiuram disulfide moieties in air under visible light. *Advanced Materials* 24: 3975–3980.

104 Imato, K., Nishihara, M., Kanehara, T. et al. (2012). Self-healing of chemical gels cross-linked by diarylbibenzofuranone-based trigger-free dynamic covalent bonds at room temperature. *Angewandte Chemie International Edition* 51: 1138–1142.

105 Imato, K., Takahara, A., and Otsuka, H. (2015). Self-healing of a cross-linked polymer with dynamic covalent linkages at mild temperature and evaluation at macroscopic and molecular levels. *Macromolecules* 48: 5632–5639.

106 Lu, Y.X. and Guan, Z. (2012). Olefin metathesis for effective polymer healing via dynamic exchange of strong carbon-carbon double bonds. *Journal of the American Chemical Society* 134: 14226–14231.

107 Cromwell, O.R., Chung, J., and Guan, Z. (2015). Malleable and self-healing covalent polymer networks through tunable dynamic boronic ester bonds. *Journal of the American Chemical Society* 137: 6492–6495.

108 Brutman, J.P., Delgado, P.A., and Hillmyer, M.A. (2014). Polylactide vitrimers. *ACS Macro Letters* 3: 607–610.

109 Altuna, F.I., Pettarin, V., and Williams, R.J.J. (2013). Self-healable polymer networks based on the cross-linking of epoxidised soybean oil by an aqueous citric acid solution. *Green Chemistry* 15: 3360–3366.

110 Fu, F., Huang, M., Zhang, W. et al. (2018). Thermally assisted self-healing behavior of anhydride modified polybenzoxazines based on transesterification. *Scientific Reports* 8: 10325/1–10325/6.

111 Xu, C., Cui, R., Fu, L., and Lin, B. (2018). Recyclable and heat-healable epoxidized natural rubber/bentonite composites. *Composites Science and Technology* 167: 421–430.

112 Han, J., Liu, T., Hao, C. et al. (2018). A catalyst-free epoxy vitrimer system based on multifunctional hyperbranched polymer. *Macromolecules* 51: 6789–6799.

113 Ying, H., Zhang, Y., and Cheng, J. (2014). Dynamic urea bond for the design of reversible and self-healing polymers. *Nature Communications* 5: 3218/1–3218/9.

114 Zhang, L., Liu, Z., Wu, X. et al. (2019). A highly efficient self-healing elastomer with unprecedented mechanical properties. *Advanced Materials* 31: 1901402/1–1901402/8.

115 Wang, Y., Bolanos, E., Wudl, F. et al. (2007). Self-healing polymers and composites based on thermal activation. In: *Proceedings of the SPIE: Smart Structures and Materials + Nondestructive Evaluation and Health Monitoring*, vol. 6526, 65261I. San Diego, CA (18–22 March 2007): SPIE.

116 Ghezzo, F., Smith, D.R., Starr, T.N. et al. (2010). Development and characterization of healable carbon fiber composites with a reversibly cross linked polymer. *Journal of Composite Materials* 44: 1587–1603.

117 Plaisted, T.A., Amirkhizi, A.V., Arbelaez, D. et al. (2003). Self-healing structural composites with electromagnetic functionality. In: *Proceedings of the SPIE: Smart Structures and Materials*, vol. 5054, 372–381. San Diego, CA (2–6 March 2003): SPIE.

118 Kwok, N. and Hahn, H.T. (2007). Resistance heating for self-healing composites. *Journal of Composite Materials* 41: 1635–1654.

119 Park, J.S., Darlington, T., Starr, A.F. et al. (2010). Multiple healing effect of thermally activated self-healing composites based on Diels–Alder reaction. *Composites Science and Technology* 70: 2154–2159.

120 Aldridge, M., Shankar, C., Zhen, C. et al. (2010). Combined experimental and simulation study of the

cure kinetics of DCPD. *Journal of Composite Materials* 44: 2605–2618.

121 Liu, Y.L. and Hsieh, C.Y. (2006). Crosslinked epoxy materials exhibiting thermal remendability and removability from multifunctional maleimide and furan compounds. *Journal of Polymer Science Part A: Polymer Chemistry* 44: 905–913.

122 Liu, Y.L. and Chen, Y.W. (2007). Thermally reversible cross-linked polyamides with high toughness and self-repairing ability from maleimide- and furan-functionalized aromatic polyamides. *Macromolecular Chemistry and Physics* 208: 224–232.

123 Chino, K. and Ashiura, M. (2001). Themoreversible cross-linking rubber using supramolecular hydrogen-bonding networks. *Macromolecules* 34: 9201–9204.

124 Kavitha, A.A. and Singha, N.K. (2009). "Click chemistry" in tailor-made polymethacrylates bearing reactive furfuryl functionality: a new class of self-healing polymeric material. *ACS Applied Materials & Interfaces* 7: 1427–1436.

125 Murphy, E.B., Bolanos, E., Shaffner-Hamann, C. et al. (2008). Synthesis and characterization of a single-component thermally remendable polymer network: Staudinger and Stille revisited. *Macromolecules* 41: 5203–5209.

126 Park, J.S., Takahashi, K., Guo, Z. et al. (2008). Towards development of a self-healing composite using a mendable polymer and resistive heating. *Journal of Composite Materials* 42: 2869–2881.

127 Syrett, J.A., Mantovani, G., Barton, W.R.S. et al. (2010). Self-healing polymers prepared via living radical polymerization. *Polymer Chemistry* 1: 102–106.

128 Cheng, X., Peng, C., Zhang, D. et al. (2013). A facile method for the preparation of thermally remendable cross-linked polyphosphazenes. *Journal of Polymer Science Part A: Polymer Chemistry* 51: 1205–1214.

129 Xu, Z., Zhao, Y., Wang, X., and Lin, T. (2013). A thermally healable polyhedral oligomeric silsesquioxane (POSS) nanocomposite based on Diels-Alder chemistry. *Chemical Communications* 49: 6755–6757.

130 Oh, C.R., Lee, S.H., Park, J.H., and Lee, D.S. (2019). Thermally self-healing graphene-nanoplate/polyurethane nanocomposites via Diels–Alder reaction through a one-shot process. *Nanomaterials* 9: 434/1–434/12.

131 Zhang, J., Niu, Y., Huang, C. et al. (2012). Self-healable and recyclable triple-shape PPDO-PTMEG co-network constructed through thermoreversible Diels-Alder reaction. *Polymer Chemistry* 3: 1390–1393.

132 Li, J., Zhang, G., Deng, L. et al. (2015). Thermally reversible andself-healing novolac epoxy resins based on Diels-Alder chemistry. *Journal of Applied Polymer Science* 132: 42167/1–42167/7.

133 Petrovíc, Z.S. and Ferguson, J. (1991). Polyurethane elastomers. *Progress in Polymer Science* 16: 695–836.

134 Chen, X.N. and Jiao, S.K. (1999). Synthesis of thermally reversible covalent crosslinked polyether rubber by ring-opening polymerization. *Acta Polymerica Sinica* 5: 564–570. (in Chinese).

135 Watanabe, M. and Yoshie, N. (2006). Synthesis and properties of readily recyclable polymers from bisfuranic terminated poly(ethylene adipate) and multi-maleimide linkers. *Polymer* 47: 4946–4952.

136 Yoshie, N., Watanabe, M., Araki, H., and Ishida, K. (2010). Thermo-responsive mending of polymers crosslinked by thermally reversible covalent bond: polymers from bisfuranic terminated poly(ethylene adipate) and tris-maleimide. *Polymer Degradation and Stability* 95: 826–829.

137 Reutenauer, P., Buhler, E., Boul, P.J. et al. (2009). Room temperature dynamic polymers based on Diels–Alder chemistry. *Chemistry - A European Journal* 15: 1893–1900.

138 Bose, R.K., Kötteritzsch, J., Garcia, S.J. et al. (2014). A rheological and spectroscopic study on the kinetics of self-healing in a single-component Diels-Alder copolymer and its underlying chemical reaction. *Journal of Polymer Science Part A: Polymer Chemistry* 52: 1669–1675.

139 Adzima, B.J., Aguirre, H.A., Kloxin, C.J. et al. (2008). Rheological and chemical analysis of reverse gelation in a covalently cross-linked Diels-Alder polymer network. *Macromolecules* 41: 9112–9117.

140 Wouters, M., Craenmehr, E., Tempelaars, K. et al. (2009). Preparation and properties of a novel remendable coating concept. *Progress in Organic Coatings* 64: 156–162.

141 Kötteritzsch, J., Stumpf, S., Hoeppener, S. et al. (2013). One-component intrinsic self-healing coatings based on reversible crosslinking by Diels-Alder cycloadditions. *Macromolecular Chemistry and Physics* 214: 1636–1649.

142 Scheltjens, G., Diaz, M.M., Brancart, J. et al. (2013). A self-healing polymer network based on reversible covalent bonding. *Reactive and Functional Polymers* 73: 413–420.

143 Scheltjens, G., Brancart, J., de Graeve, I. et al. (2011). Self-healing property characterization of reversible thermoset coatings. *Journal of Thermal Analysis and Calorimetry* 105: 805–809.

144 Amato, D.N., Strange, G.A., Swanson, J.P. et al. (2014). Synthesis and evaluation of thermally-responsive coatings based upon Diels-Alder chemistry and renewable materials. *Polymer Chemistry* 5: 69–76.

145 Cardenas-Daw, C., Kroeger, A., Schaertl, W. et al. (2012). Reversible photocycloadditions, a powerful tool for tailoring (nano)materials. *Macromolecular Chemistry and Physics* 213: 144–156.

146 Oya, N., Sukarsaatmadja, P., Ishida, K., and Yoshie, N. (2012). Photoinduced mendable network polymer from poly(butylene adipate) end-functionalized with cinnamoyl groups. *Polymer Journal* 44: 724–729.

147 Froimowicz, P., Frey, H., and Landfester, K. (2011). Towards the generation of self-healing materials by means of a reversible photo-induced approach. *Macromolecular Rapid Communications* 32: 468–473.

148 Hu, L., Cheng, X., and Zhang, A. (2015). A facile method to prepare UV light-triggered self-healing polyphosphazenes. *Journal of Materials Science* 50: 2239–2246.

149 Ling, J., Rong, M.Z., and Zhang, M.Q. (2012). Photo-stimulated self-healing polyurethane containing dihydroxyl coumarin derivatives. *Polymer* 53: 2691–2698.

150 Ling, J., Rong, M.Z., and Zhang, M.Q. (2014). Effect of molecular weight of PEG soft segments on photo-stimulated self-healing performance of coumarin functionalized polyurethanes. *Chinese Journal of Polymer Science* 32: 1286–1297.

151 Xu, J., Yang, D., Li, W. et al. (2011). Phenylboronate-diol crosslinked polymer gels with reversible sol-gel transition. *Polymer* 52: 4268–4276.

152 Deng, C.C., Brooks, W.L.A., Abboud, K.A., and Sumerlin, B.S. (2015). Boronic acid-based hydrogels undergo self-healing at neutral and acidic pH. *ACS Macro Letters* 4: 220–224.

153 He, L., Fullenkamp, D.E., Rivera, J.G., and Messersmith, P.B. (2011). pH responsive self-healing hydrogels formed by boronate-catechol complexation. *Chemical Communications* 47: 7497–7499.

154 Debnath, S., Ujjwal, R.R., and Ojha, U. (2018). Self-healable and recyclable dynamic covalent networks based on room temperature exchangeable hydrazide Michael adduct linkages. *Macromolecules* 51: 9961–9973.

155 Delpierre, S., Willocq, B., De Winter, J. et al. (2017). Dynamic iminoboronate-based boroxine chemistry for the design of ambient humidity-sensitive self-healing polymers. *Chemistry – A European Journal* 23: 6730–6735.

156 Bosch, L.I., Fyles, T.M., and James, T.D. (2004). Binary and ternary phenylboronic acid complexes with saccharides and Lewis bases. *Tetrahedron* 60: 11175–11190.

157 Furikado, Y., Nagahata, T., Okamoto, T. et al. (2014). Universal reaction mechanism of boronic acids with diols in aqueous solution: kinetics and the basic concept of a conditional formation constant. *Chemistry – A European Journal* 20: 13194–13202.

158 Fujita, N., Shinkai, S., and James, T.D. (2008). Boronic acids in molecular self-assembly. *Chemistry – An Asian Journal* 3: 1076–1091.

159 Deng, G., Tang, C., Li, F. et al. (2010). Covalent cross-linked polymer gels with reversible sol-gel transition and self-healing properties. *Macromolecules* 43: 1191–1194.

160 Liu, F., Li, F., Deng, G. et al. (2012). Rheological images of dynamic covalent polymer networks and mechanisms behind mechanical and self-healing properties. *Macromolecules* 45: 1636–1645.

161 Apostolides, D.E., Patrickios, C.S., Leontidis, E. et al. (2014). Synthesis and characterization of reversible and self-healable networks based on acylhydrazone groups. *Polymer International* 63: 1558–1565.

162 Otsuka, H., Nagano, S., Kobashi, Y. et al. (2010). A dynamic covalent polymer driven by disulfide metathesis under photoirradiation. *Chemical Communications* 46: 1150–1152.

163 Wei, Z., Yang, J.H., Zhou, J. et al. (2014). Self-healing gels based on constitutional dynamic chemistry and their potential applications. *Chemical Society Reviews* 43: 8114–8131.

164 Zhang, Y., Tao, L., Li, S., and Wei, Y. (2011). Synthesis of multiresponsive and dynamic chitosan-based hydrogels for controlled release of bioactive molecules. *Biomacromolecules* 12: 2894–2901.

165 Zhang, Y., Yang, B., Zhang, X. et al. (2012). A magnetic self-healing hydrogel. *Chemical Communications* 48: 9305–9307.

166 Wei, Z., Yang, J.H., Liu, Z.Q. et al. (2015). Novel biocompatible polysaccharide-based self-healing hydrogel. *Advanced Functional Materials* 25: 1352–1359.

167 Deng, G., Li, F., Yu, H. et al. (2012). Dynamic hydrogels with an environmental adaptive self-healing ability and dual responsive sol-gel transitions. *ACS Macro Letters* 1: 275–279.

168 Bhat, V.T., Caniard, A.M., Luksch, T. et al. (2010). Nucleophilic catalysis of acylhydrazone equilibration for protein-directed dynamic covalent chemistry. *Nature Chemistry* 2: 490–497.

169 Dirksen, A., Dirksen, S., Hackeng, T.M., and Dawson, P.E. (2006). Nucleophilic catalysis of hydrazone formation and transimination: implications for dynamic covalent chemistry. *Journal of the American Chemical Society* 128: 15602–15603.

170 Zhang, P., Deng, F., Peng, Y. et al. (2014). Redox- and pH-responsive polymer gels with reversible sol-gel transitions and self-healing properties. *RSC Advances* 4: 47361–47367.

171 Kuhl, N., Bode, S., Bose, R.K. et al. (2015). Acylhydrazones as reversible covalent crosslinkers for self-healing polymers. *Advanced Functional Materials* 25: 3295–3301.

172 Montarnal, D., Capelot, M., Tournilhac, F., and Leibler, L. (2011). Silica-like malleable materials from permanent organic networks. *Science* 334: 965–968.

173 Capelot, M., Unterlass, M.M., Tournilhac, F., and Leibler, L. (2012). Catalytic control of the vitrimer glass transition. *ACS Macro Letters* 1: 789–792.

174 Capelot, M., Montarnal, D., Tournilhac, F., and Leibler, L. (2012). Metal-catalyzed transesterification for healing and assembling of thermosets. *Journal of the American Chemical Society* 134: 7664–7667.

175 Yu, K., Taynton, P., Zhang, W. et al. (2014). Reprocessing and recycling of thermosetting polymers based on bond exchange reactions. *RSC Advances* 4: 10108–10117.

176 Yu, K., Taynton, P., Zhang, W. et al. (2014). Influence of stoichiometry on the glass transition and bond exchange reactions in epoxy thermoset polymers. *RSC Advances* 4: 48682–48690.

177 Li, Y., Rios, O., Keum, J.K. et al. (2016). Photo-responsive liquid crystalline epoxy networks with shape memory behavior and dynamic ester bonds. *ACS Applied Materials & Interfaces* 8: 15750–15757.

178 Imbernon, L., Norvez, S., and Leibler, L. (2016). Stress relaxation and self-adhesion of rubbers with exchangeable links. *Macromolecules* 49: 2172–2178.

179 Tang, Z., Liu, Y., Guo, B., and Zhang, L. (2017). Malleable, mechanically strong, and adaptive elastomers enabled by interfacial exchangeable bonds. *Macromolecules* 50: 7584–7592.

180 Nicolay, R., Kamada, J., van Wassen, A., and Matyjaszewski, K. (2010). Responsive gels based on a dynamic covalent trithiocarbonate cross-linker. *Macromolecules* 43: 4355–4361.

181 Ji, S., Cao, W., Yu, Y., and Xu, H. (2014). Dynamic diselenide bonds: exchange reaction induced by visible light without catalysis. *Angewandte Chemie International Edition* 53: 6781–6785.

182 Ji, S., Xia, J., and Xu, H. (2015). Dynamic chemistry of selenium: se-N and se–se dynamic covalent bonds in polymeric systems. *ACS Macro Letters* 5: 78–82.

183 Zheng, P. and McCarthy, T.J. (2012). A surprise from 1954: siloxane equilibration is a simple, robust, and obvious polymer self-healing mechanism. *Journal of the American Chemical Society* 134: 2024–2027.

184 Schmolke, W., Perner, N., and Seiffert, S. (2015). Dynamically cross-linked polydimethylsiloxane networks with ambient-temperature self-healing. *Macromolecules* 48: 8781–8788.

185 Zhang, Z.P., Rong, M.Z., and Zhang, M.Q. (2018). Mechanically robust, self-healable, and highly stretchable "living" crosslinked polyurethane based on a reversible C-C bond. *Advanced Functional Materials* 2: 1706050/1–1706050/8.

186 Fortman, D.J., Brutman, J.P., Cramer, C.J. et al. (2015). Mechanically-activated, catalyst-free polyhydroxyurethane vitrimers. *Journal of the American Chemical Society* 137: 14019–14022.

187 Liu, W.X., Zhang, C., Zhang, H. et al. (2017). Oxime-based and catalyst-free dynamic covalent polyurethanes. *Journal of the American Chemical Society* 139: 8678–8684.

188 Nishimura, Y., Chung, J., Muradyan, H., and Guan, Z. (2017). Silyl ether as a robust and thermally stable dynamic covalent motif for malleable polymer design. *Journal of the American Chemical Society* 139: 14881–14884.

189 Baruah, R., Kumar, A., Ujjwal, R.R. et al. (2016). Recyclable thermosets based on dynamic amidation and aza-Michael addition chemistry. *Macromolecules* 49: 7814–7824.

190 Ma, X., Usui, R., Kitazawa, Y. et al. (2015). Photo-healable ion gel with improved mechanical properties using a tetra-arm diblock copolymer containing azobenzene groups. *Polymer* 78: 42–50.

191 Christogianni, P., Moniruzzaman, M., and Kister, G. (2015). Light-triggered enhancement of mechanical properties and healing effect in azobenzene-based polymer films. *Polymer* 77: 272–277.

192 Zhou, H., Xue, C., Weis, P. et al. (2017). Photoswitching of glass transition temperatures of azobenzene-containing polymers induces reversible solid-to-liquid transitions. *Nature Chemistry* 9: 145–151.

193 Xu, W.C., Sun, S., and Wu, S. (2019). Photoinduced reversible solid-to-liquid transitions using photoswitchable materials. *Angewandte Chemie International Edition* 58: 9712–9740.

194 Wietor, J.L. and Sijbesma, R.P. (2008). A self-healing elastomer. *Angewandte Chemie International Edition* 47: 8161–8163.

195 Zheng, Q., Ma, Z., and Gong, S. (2016). Multi-stimuli-responsive self-healing metallo-supramolecular polymer nanocomposites. *Journal of Materials Chemistry A* 4: 3324–3334.

196 Li, C.H., Wang, C., Keplinger, C. et al. (2016). A highly stretchable autonomous self-healing elastomer. *Nature Chemistry* 8: 618–624.

197 Lai, J.C., Li, L., Wang, D.P. et al. (2018). A rigid and healable polymer cross-linked by weak but abundant Zn(II)-carboxylate interactions. *Nature Communications* 9: 2725/1–2725/9.

198 Arslan, M., Kiskan, B., and Yagci, Y. (2018). Benzoxazine-based thermoset with autonomous self-healing and shape recovery. *Macromolecules* 51: 10095–10103.

199 Wu, X., Wang, J., Huang, J., and Yang, S. (2019). Robust, stretchable, and self-healable supramolecular elastomers synergistically cross-linked by hydrogen bonds and coordination bonds. *ACS Applied Materials & Interfaces* 11: 7387–7396.

200 Xia, N.N., Xiong, X.M., Wang, J. et al. (2016). A seawater triggered dynamic coordinate bond and its application for underwater self-healing and reclaiming of lipophilic polymer. *Chemical Science* 7: 2736–2742.

201 Li, J., Ejima, H., and Yoshie, N. (2016). Seawater-assisted self-healing of catechol polymers via hydrogen bonding and coordination interactions. *ACS Applied Materials & Interfaces* 8: 19047–19053.

202 Zhang, J., Li, M., Cheng, L., and Li, T. (2018). Self-healable and tough thermoplastic materials from metal-thioether block polymers. *Macromolecular Chemistry and Physics* 219: 1700430/1–1700430/8.

203 Guo, W., Wang, X., Lu, X. et al. (2019). Plant oil and amino acid-derived elastomers with rapid room temperature self-healing ability. *Journal of Materials Chemistry A* 7: 21927–21933.

204 Das, A., Sallat, A., Böhme, F. et al. (2015). Ionic modification turns commercial rubber into a self-healing material. *ACS Applied Materials & Interfaces* 7: 20623–20630.

205 Xu, C., Cao, L., Lin, B. et al. (2016). Design of self-healing supramolecular rubbers by introducing ionic cross-links into natural rubber via a controlled vulcanization. *ACS Applied Materials & Interfaces* 8: 17728–17737.

206 Xu, C., Cao, L., Huang, X. et al. (2017). Self-healing natural rubber with tailorable mechanical properties based on ionic supramolecular hybrid network. *ACS Applied Materials & Interfaces* 9: 29363–29373.

207 Reisch, A., Roger, E., Phoeung, T. et al. (2014). On the benefits of rubbing salt in the cut: self-healing of saloplastic PAA/PAH compact polyelectrolyte complexes. *Advanced Materials* 26: 2547–2551.

208 Chen, Y. and Guan, Z. (2014). Multivalent hydrogen bonding block copolymers self-assemble into strong and tough self-healing materials. *Chemical Communications* 50: 10868–10870.

209 Cao, L., Yuan, D., Xu, C., and Chen, Y. (2017). Biobased, self-healable, high strength rubber with tunicate cellulose nanocrystals. *Nanoscale* 9: 15696–15706.

210 Hentschel, J., Kushner, A.M., Ziller, J., and Guan, Z. (2012). Self-healing supramolecular block copolymers. *Angewandte Chemie International Edition* 51: 10561–10565.

211 Wei, M., Zhan, M., Yu, D. et al. (2015). Novel poly(tetramethylene ether) glycol and poly(ε-caprolactone) based dynamic network via quadruple hydrogen bonding with triple-shape effect and self-healing capacity. *ACS Applied Materials & Interfaces* 7: 2585–2596.

212 Song, Y., Liu, Y., Qi, T., and Li, G.L. (2018). Towards dynamic but supertough healable polymers through biomimetic hierarchical hydrogen-bonding interactions. *Angewandte Chemie International Edition* 57: 13838–13842.

213 Wang, D., Guo, J., Zhang, H. et al. (2015). Intelligent rubber with tailored properties for self-healing and shape memory. *Journal of Materials Chemistry A* 3: 12864–12872.

214 Cao, J., Lu, C., Zhuang, J. et al. (2017). Multiple hydrogen bonding enables the self-healing of sensors for human–machine interactions. *Angewandte Chemie International Edition 56*: 8795–8800.

215 Cao, J., Zhou, C., Su, G. et al. (2019). Arbitrarily 3D configurable hygroscopic robots with a covalent-noncovalent interpenetrating network and self-healing ability. *Advanced Materials* 31: 1900042/1–1900042/8.

216 Urban, M.W., Davydovich, D., Yang, Y. et al. (2018). Key-and-lock commodity self-healing copolymers. *Science* 362: 220–225.

217 Burattini, S., Greenland, B.W., Merino, D.H. et al. (2010). A healable supramolecular polymer blend based on aromatic π-π stacking and hydrogen-bonding interactions. *Journal of the American Chemical Society* 132: 12051–12058.

218 Fox, J., Wie, J.J., Greenland, B.W. et al. (2012). High-strength, healable, supramolecular polymer nanocomposites. *Journal of the American Chemical Society* 134: 5362–5368.

219 Takashima, Y., Yonekura, K., Koyanagi, K. et al. (2017). Multifunctional stimuli-responsive supramolecular materials with stretching, coloring, and self-healing properties functionalized via host–guest interactions. *Macromolecules* 50: 4144–4150.

220 Guo, K., Zhang, D.L., Zhang, X.M. et al. (2015). Conductive elastomers with autonomic self-healing properties. *Angewandte Chemie International Edition* 54: 12127–12133.

221 Zhang, D.L., Ju, X., Li, L.H. et al. (2015). An efficient multiple healing conductive composite via host–guest inclusion. *Chemical Communications* 51: 6377–6380.

222 Hu, Z., Zhang, D., Lu, F. et al. (2018). Multistimuli-responsive intrinsic self-healing epoxy resin constructed by host–guest interactions. *Macromolecules* 51: 5294–5303.

223 Burnworth, M., Tang, L., Kumpfer, J.R. et al. (2011). Optically healable supramolecular polymers. *Nature* 472: 334–337.

224 Bode, S., Zedler, L., Schacher, F.H. et al. (2013). Self-healing polymer coatings based on crosslinked metallosupramolecular copolymers. *Advanced Materials* 25: 1634–1638.

225 Vaccaro, E. and Waite, J.H. (2001). Yield and post-yield behavior of mussel byssal thread: a self-healing biomolecular material. *Biomacromolecules* 2: 906–911.

226 Zeng, H., Hwang, D.S., Israelachvili, J.N., and Waite, J.H. (2010). Strong reversible Fe3+−mediated bridging between dopa-containing protein films in water. *Proceedings of the National Academy of Sciences* 107: 12850–12853.

227 Ceylan, H., Urel, M., Erkal, T.S. et al. (2013). Mussel inspired dynamic cross-linking of self-healing peptide nanofiber network. *Advanced Functional Materials* 23: 2081–2090.

228 Xia, N.N., Rong, M.Z., and Zhang, M.Q. (2016). Stabilization of catechol-boronic ester bonds for underwater self-healing and recycling of lipophilic bulk polymer in wider pH range. *Journal of Materials Chemistry A* 4: 14122–14131.

229 Mynar, J.L. and Aida, T. (2008). The gift of healing. *Nature* 451: 895–896.

230 Cordier, P., Tournilhac, F., Soulié-Ziakovic, C., and Leibler, L. (2008). Self-healing and thermoreversible rubber from supramolecular assembly. *Nature* 451: 977–980.

231 Montarnal, D., Cordier, P., Soulie-Ziakovic, C. et al. (2008). Synthesis of self-healing supramolecular rubbers from fatty acid derivatives, diethylene triamine, and urea. *Journal of Polymer Science Part A: Polymer Chemistry* 46: 7925–7936.

232 Montarnal, D., Tournilhac, F., Hidalgo, M. et al. (2009). Versatile one-pot synthesis of supramolecular plastics and self-healing rubbers. *Journal of the American Chemical Society* 131: 7966–7967.

233 Harreld, J.H., Wong, M.S., Hansma, P.K. et al. (2004). Self-healing organosiloxane materials containing reversible and energy-dispersive crosslinking domains. US Patent 6783709.

234 Huang, J., Cao, L., Yuan, D., and Chen, Y. (2018). Design of novel self-healing thermoplastic vulcanizates utilizing thermal/magnetic/light-triggered shape memory effects. *ACS Applied Materials & Interfaces* 10: 40996–41002.

235 Sçntjens, S.H.M., Sijbesma, R.P., van Genderen, M.H.P., and Meijer, E.W. (2000). Stability and lifetime of quadruply hydrogen bonded 2-ureido-4[1*H*]-pyrimidinone dimers. *Journal of the American Chemical Society* 122: 7487–7493.

236 Yoshida, S., Ejima, H., and Yoshie, N. (2017). Tough elastomers with superior self-recoverability induced by bioinspired multiphase design. *Advanced Functional Materials* 27: 1701670/1–1701670/9.

237 Zhang, G., Lv, L., Deng, Y., and Wang, C. (2017). Self-healing gelatin hydrogels cross-linked by combining multiple hydrogen bonding and ionic coordination. *Macromolecular Rapid Communications* 38: 1700018/1–1700018/9.

238 Yanagisawa, Y., Nan, Y., Okuro, K., and Aida, T. (2018). Mechanically robust, readily repairable polymers via tailored noncovalent cross-linking. *Science* 359: 72–76.

239 Aboudzadeh, A., Fernandez, M., Muñoz, M.E. et al. (2014). Ionic supramolecular networks fully based on chemicals coming from renewable sources. *Macromolecular Rapid Communications* 35: 460–465.

240 Burattini, S., Colquhoun, H.M., Greenland, B.W., and Hayes, W. (2009). A novel self-healing supramolecular polymer system. *Faraday Discussions* 143: 251–264.

241 Burattini, S., Colquhoun, H.M., Fox, J.D. et al. (2009). A self-repairing, supramolecular polymer system: healability as a consequence of donor–acceptor–stacking interactions. *Chemical Communications* 6717–6719.

242 Guo, K., Lin, M.S., Feng, J.F. et al. (2017). The deeply understanding of the self-healing mechanism for self-healing behavior of supramolecular materials based on cyclodextrin–guest interactions. *Macromolecular Chemistry and Physics* 218: 1600593/1–1600593/7.

243 Yang, X., Yu, H., Wang, L. et al. (2015). Self-healing polymer materials constructed by macrocycle-based host–guest interactions. *Soft Matter* 11: 1242–1252.

244 Nakahata, M., Takashima, Y., Yamaguchi, H., and Harada, A. (2011). Redox-responsive self-healing materials formed from host-guest polymers. *Nature Communications* 2: 511/1–511/6.

245 Kakuta, T., Takashima, Y., Nakahata, M. et al. (2013). Preorganized hydrogel: self-healing properties of supramolecular hydrogels formed by polymerization of host-guest-monomers that contain cyclodextrins and hydrophobic guest groups. *Advanced Materials* 25: 2849–2853.

246 Yu, C., Wang, C.F., and Chen, S. (2014). Robust self-healing host-guest gels from magnetocaloric radical polymerization. *Advanced Functional Materials* 24: 1235–1242.

247 Chen, H., Ma, X., Wu, S., and Tian, H. (2014). A rapidly self-healing supramolecular polymer hydrogel with photostimulated room-temperature phosphorescence responsiveness. *Angewandte Chemie International Edition* 53: 14149–14152.

248 Jia, Y.G. and Zhu, X.X. (2014). Self-healing supramolecular hydrogel made of polymers bearing cholic acid and β-cyclodextrin pendants. *Chemistry of Materials* 27: 387–393.

249 Xiang, Z., Zhang, L., Li, Y. et al. (2017). Reduced graphene oxide-reinforced polymeric films with excellent mechanical robustness and rapid and highly efficient healing properties. *ACS Nano* 11: 7134–7141.

250 Zhang, M., Xu, D., Yan, X. et al. (2012). Self-healing supramolecular gels formed by crown ether based host–guest interactions. *Angewandte Chemie International Edition* 51: 7011–7015.

251 Yan, X., Xu, D., Chen, J. et al. (2013). A self-healing supramolecular polymer gel with stimuli-responsiveness constructed by crown ether based molecular recognition. *Polymer Chemistry* 4: 3312–3322.

252 Zeng, F., Han, Y., Yan, Z.-C. et al. (2013). Supramolecular polymer gel with multi stimuli responsive, self-healing and erasable properties generated by host–guest interactions. *Polymer* 54: 6929–6935.

253 Liu, D., Wang, D., Wang, M. et al. (2013). Supramolecular organogel based on crown ether and secondary ammoniumion functionalized glycidyl triazole polymers. *Macromolecules* 46: 4617–4625.

254 Dry, C. (1992). Passive tunable fibers and matrices. *International Journal of Modern Physics B* 6: 2763–2771.

255 Dry, C. (1994). Matrix cracking repair and filling using active and passive modes for smart timed release of chemicals from fibers into cement matrices. *Smart Materials and Structures* 3: 118–123.

256 Dry, C. and McMillan, W. (1996). Three-part methylmethacrylate adhesive system as an internal delivery system for smart responsive concrete. *Smart Materials and Structures* 5: 297–300.

257 Dry, C. (1996). Procedures developed for self-repair of polymer matrix composite materials. *Composite Structures* 35: 263–269.

258 Motuku, M., Vaidya, U.K., and Janowski, G.M. (1999). Parametric studies on self-repairing approaches for resin infused composites subjected to low velocity impact. *Smart Materials and Structures* 8: 623–638.

259 Zhao, X.P., Zhou, B.L., Luo, C.R. et al. (1996). A model of intelligent material with self-repair function. *Chinese Journal of Materials Research* 10: 101–104. (in Chinese).

260 Dry, C. and Sottos, N.R. (1993). Passive smart self-repair in polymer matrix composites. In: *Proceedings of the SPIE: Smart Structures and Materials*, vol. 1916, 438–444. Albuquerque, NM (1–4 February 1993): SPIE.

261 Dry, C. and McMillan, W. (1996). Crack and damage assessment in concrete and polymer matrices using liquids released internally from hollow optical fibers. In: *Proceedings of the SPIE: Smart Structures and Materials*, vol. 2718, 448–451. San Diego, CA (25–29 February 1996): SPIE.

262 Dry, C. and Haven, P. (1998). Smart-fiber-reinforced matrix compositions. US Patent 5803963.

263 Dry, C., Line, S., and Winona, M. (2006). Self-repairing reinforced matrix materials. US Patent 7022179 B1.

264 Motuku, M., Janowski, G.M., Vaidya, U.K. et al. (1999). Low velocity impact characterization of unreinforced vinyl ester 411-350 and 411-C50 resin systems. *Polymers and Polymer Composites* 7: 383–407.

265 Pang, W.C. and Bond, I.P. (2005). A hollow fibre reinforced polymer composite encompassing self-healing and enhanced damage visibility. *Composites Science and Technology* 65: 1791–1799.

266 Pang, W.C. and Bond, I.P. (2005). Bleeding composites' damage detection and self-repair using a biomimetic approach. *Composites Part A: Applied Science and Manufacturing* 36: 183–188.

267 Trask, R.S. and Bond, I.P. (2006). Biomimetic self-healing of advanced composite structures using hollow glass fibres. *Smart Materials and Structures* 15: 704–710.

268 Trask, R.S., Williams, G.J., and Bond, I.P. (2007). Bioinspired self-healing of advanced composite structures using hollow glass fibres. *Journal of the Royal Society Interface* 4: 363–371.

269 Williams, G., Trask, R.S., and Bond, I.P. (2007). A self-healing carbon fibre reinforced polymer for aerospace applications. *Composites Part A: Applied Science and Manufacturing* 38: 1525–1532.

270 Williams, G.J., Bond, I.P., and Trask, R.S. (2009). Compression after impact assessment of self-healing CFRP. *Composites Part A: Applied Science and Manufacturing* 40: 1399–1406.

271 Kousourakis, A. and Mouritz, A.P. (2010). The effect of self-healing hollow fibres on the mechanical properties of polymer composites. *Smart Materials and Structures* 19: 085021/1–085021/9.

272 Gumula, T. and Szatkowski, P. (2016). Regeneration efficiency of composites containing two-sized capillaries. *Polymer Composites* 37: 1223–1230.

273 Kling, S. and Czigany, T. (2013). Analysis of applicability of the hollow carbon fibres for self-repairing composites. *Materials Science Forum* 729: 246–251.

274 Williams, H.R., Trask, R.S., and Bond, I.P. (2008). Self-healing sandwich panels: restoration of compressive strength after impact. *Composites Science and Technology* 68: 3171–3177.

275 Williams, H.R., Trask, R.S., and Bond, I.P. (2007). Self-healing composite sandwich structures. *Smart Materials and Structures* 16: 1198–1207.

276 Hemrick, J.G., Lara-Curzio, E., Loveland, E.R. et al. (2011). Woven graphite fiber structures for use in ultra-light weight heat exchangers. *Carbon* 49: 4820–4829.

277 Prajer, M., Wu, X., Garcia, S.J., and van der Zwaag, S. (2015). Direct and indirect observation of multiple local healing events in successively loaded fibre reinforced polymer model composites using healing agent-filled compartmented fibres. *Composites Science and Technology* 106: 127–133.

278 Neisiany, R.E., Lee, J.K.Y., Khorasani, S.N., and Ramakrishna, S. (2017). Self-healing and interfacially toughened carbon fibre-epoxy composites based on electrospun core-shell nanofibres. *Journal of Applied Polymer Science* 134: 44956/1–44956/10.

279 Therriault, D., Shepherd, R.F., White, S.R., and Lewis, J.A. (2005). Fugitive inks for direct-write assembly of three-dimensional microvascular networks. *Advanced Materials* 17: 394–399.

280 Hansen, C.J., Wu, W., Toohey, K.S. et al. (2009). Self-healing materials with interpenetrating microvascular networks. *Advanced Materials* 21: 4143–4147.

281 Williams, H.R., Trask, R.S., Knights, A.C. et al. (2008). Biomimetic reliability strategies for self-healing vascular networks in engineering materials. *Journal of the Royal Society Interface* 5: 735–747.

282 Hamilton, A.R., Sottos, N.R., and White, S.R. (2010). Self-healing of internal damage in synthetic vascular materials. *Advanced Materials* 22: 5159–5163.

283 Huang, C.-Y., Trask, R.S., and Bond, I.P. (2009). Analytical study of vascular networks for composite laminates. In: *Proceedings of the Seventeenth International Conference on Composite Materials (ICCM-17)*, D2:2. Edinburgh, UK (27–31 July 2009): IOM Communication Ltd.

284 Trask, R.S. and Bond, I.P. (2010). Bioinspired engineering study of plantae vascules for self-healing composite structures. *Journal of the Royal Society Interface* 7: 921–931.

285 Trask, R.S. and Bond, I.P. (2009). Biomimicry of plantae vascules in the development of self-healing composite structures. In: *Proceedings of the Seventeenth International Conference on Composite Materials (ICCM-17)*, B5:1. Edinburgh, UK (27–31 July 2009): IOM Communication Ltd.

286 Hamilton, A.R., Sottos, N.R., and White, S.R. (2009). Local strain concentrations in a microvascular network. *Experimental Mechanics* 50: 255–263.

287 Norris, C.J., Bond, I.P., and Trask, R.S. (2011). Interactions between propagating cracks and bioinspired self-healing vascules embedded in glass fibre reinforced composites. *Composites Science and Technology* 71: 847–853.

288 Hung, J.-C., Chang, D.-H., and Chuang, Y. (2012). The fabrication of high-aspect-ratio microflow channels on metallic bipolar plates using die-sinking micro-electrical discharge machining. *Journal of Power Sources* 198: 158–163.

289 Dixit, T. and Ghosh, I. (2015). Review of micro- and mini-channel heat sinks and heat exchangers for single phase fluids. *Renewable and Sustainable Energy Reviews* 41: 1298–1311.

290 Bremus, E., Gillner, A., Hellrung, D. et al. (1999). Laser processing for manufacturing microfluidic devices. In: *Proceedings of the Third International Conference on Microreaction Technology*, 80–89. Frankfurt, Germany (18–21 April 1999): Springer.

291 Patrick, J.F., Krull, B.P., Garg, M. et al. (2017). Robust sacrificial polymer templates for 3D interconnected microvasculature in fiber-reinforced composites. *Composites Part A: Applied Science and Manufacturing* 100: 361–370.

292 Esser-Kahn, A.P., Thakre, P.R., Dong, H. et al. (2011). Three-dimensional microvascular fiber-reinforced composites. *Advanced Materials* 23: 3654–3658.

293 Coppola, A.M., Thakre, P.R., Sottos, N.R., and White, S.R. (2014). Tensile properties and damage evolution in vascular 3D woven glass/epoxy composites. *Composites Part A: Applied Science and Manufacturing* 59: 9–17.

294 Patrick, J.F., Hart, K.R., Krull, B.P. et al. (2014). Continuous self-healing life cycle in vascularized structural composites. *Advanced Materials* 26: 4302–4308.

295 Brown, E.N., Kessler, M.R., Sottos, N.R., and White, S.R. (2003). In situ poly(urea-formaldehyde) microencapsulation of dicyclopentadiene. *Journal of Microencapsulation* 20: 719–730.

296 Yin, T., Rong, M.Z., Zhang, M.Q., and Yang, G.C. (2007). Self-healing epoxy composites– preparation and effect of the healant consisting of microencapsulated epoxy and latent curing agent. *Composites Science and Technology* 67: 201–212.

297 Suryanarayana, C., Rao, K.C., and Kumar, D. (2008). Preparation and characterization of microcapsules containing linseed oil and its use in self-healing coatings. *Progress in Organic Coatings* 63: 72–78.

298 Gragert, M., Schunack, M., and Binder, W.H. (2011). Azide/alkyne-"click"-reactions of encapsulated reagents: toward self-healing materials. *Macromolecular Rapid Communications* 32: 419–425.

299 Cai, X.L., Fu, D.T., and Qu, A.L. (2016). Effects of fluorinated silane compound on the repeated self-healing properties of nanocapsules. *Journal of Polymer Research* 23: 18/1–18/8.

300 Yuan, Y.C., Rong, M.Z., and Zhang, M.Q. (2008). Preparation and characterization of microencapsulated polythiol. *Polymer* 49: 2531–2541.

301 Yuan, Y.C., Ye, X.J., Rong, M.Z. et al. (2011). Self-healing epoxy composite with heat-resistant healant. *ACS Applied Materials & Interfaces* 3: 4487–4495.

302 Liu, X., Sheng, X., Lee, J.K., and Kessler, M.R. (2009). Synthesis and characterization of melamine-urea-formaldehyde microcapsules containing ENB-based self-healing agents. *Macromolecular Materials and Engineering* 294: 389–395.

303 Yuan, L., Gu, A., Nutt, S. et al. (2013). Novel polyphenylene oxide microcapsules filled with epoxy resins. *Polymers for Advanced Technologies* 24: 81–89.

304 Brochu, A.B., Chyan, W.J., and Reichert, W.M. (2012). Microencapsulation of 2-octylcyanoacrylate tissue adhesive for self-healing acrylic bone cement. *Journal of Biomedical Materials Research Part B: Applied Biomaterials* 100: 1764–1772.

305 Yang, J.L., Sottos, N.R., Keller, M.W. et al. (2008). Microencapsulation of isocyanates for self-healing polymers. *Macromolecules* 41: 9650–9655.

306 Nguyen, L.T.T., Hillewaere, X.K.D., Teixeira, R.F.A. et al. (2015). Efficient microencapsulation of a liquid isocyanate with in situ shell functionalization. *Polymer Chemistry* 6: 1159–1170.

307 Liu, X., Zhang, H., Wang, J. et al. (2012). Preparation of epoxy microcapsule based self-healing coatings and their behavior. *Surface and Coatings Technology* 206: 4976–4980.

308 Caruso, M.M., Blaiszik, B.J., Jin, H. et al. (2010). Robust, double-walled microcapsules for self-healing polymeric materials. *ACS Applied Materials & Interfaces* 2: 1195–1199.

309 Aveyard, R., Binks, B.P., and Clint, J.H. (2003). Emulsions stabilised solely by colloidal particles. *Advances in Colloid and Interface Science* 100–102: 503–546.

310 Li, J., Hitchcock, A.P., and Stover, H.D. (2010). Pickering emulsion templated inter-facial atom transfer radical polymerization for microencapsulation. *Langmuir* 26: 17926–17935.

311 McIlroy, D.A., Blaiszik, B.J., Caruso, M.M. et al. (2010). Microencapsulation of a reactive liquid-phase amine for self-healing epoxy composites. *Macromolecules* 43: 1855–1859.

312 Yang, Y., Wei, Z.J., Wang, C.Y., and Tong, Z. (2013). Versatile fabrication of nanocomposite microcapsules with controlled shell thickness and low permeability. *ACS Applied Materials & Interfaces* 5: 2495–2502.

313 Patchan, M.W., Fuller, B.W., Baird, L.M. et al. (2015). Robust composite-shell microcapsules via Pickering emulsification. *ACS Applied Materials & Interfaces* 7: 7315–7323.

314 Crespy, D., Musyanovych, A., and Landfester, K. (2006). Synthesis of polymer particles and nanocapsules stabilized with PEO/PPO containing polymerizable surfactants in miniemulsion. *Colloid and Polymer Science* 284: 780–787.

315 Ugelstad, J., El-Aasser, M.S., and Vanderhoff, J.W. (1973). Emulsion polymerization: initiation of polymerization in monomer droplets. *Journal of Polymer Science Part C: Polymer Letters* 11: 503–513.

316 Fickert, J., Makowski, M., Kappl, M. et al. (2012). Efficient encapsulation of self-healing agents in polymer nanocontainers functionalized by orthogonal reactions. *Macromolecules* 45: 6324–6332.

317 van Den Dungen, E.T.A. and Klumperman, B. (2010). Synthesis of liquid-filled nanocapsules via the miniemulsion technique. *Journal of Polymer Science Part A: Polymer Chemistry* 48: 5215–5230.

318 van Den Dungen, E.T.A. and Klumperman, B. (2010). Self-healing composition. EP2208611A1.

319 Fickert, J., Rupper, P., Graf, R. et al. (2012). Design and characterization of functionalized silica nanocontainers for self-healing materials. *Journal of Materials Chemistry* 22: 2286–2291.

320 Zhang, H., Zhang, X., Bao, C.L. et al. (2018). Direct microencapsulation of pure polyamine by integrating microfluidic emulsion and interfacial polymerization for practical self-healing materials. *Journal of Materials Chemistry A* 6: 24092–24099.

321 Yamakawa, L., Tsushima, Y., Machida, R., and Watanabe, S. (1992). Preparation of neurotensin analogue-containing poly(dl-lactic acid) micro-spheres formed by oil-in-water solvent evaporation. *Journal of Pharmaceutical Sciences* 81: 899–903.

322 Hausberger, A.G. and Deluca, P.P. (1995). Characterization of biodegradable poly(d,l-lactide-co-glycolide) polymers and microspheres. *Journal of Pharmaceutical and Biomedical Analysis* 13: 747–760.

323 Yang, Y.Y., Chia, H.H., and Chung, T.S. (2000). Effect of preparation temperature on the characteristics and release profiles of PLGA microspheres containing protein fabricated by double-emulsion solvent. *Journal of Controlled Release* 69: 81–96.

324 Jeffery, H., Davis, S.S., and O'Hagan, D.T. (1993). The preparation and characterization of poly(lactide-co-glycolide) microparticles: II. The entrapment of a model protein using a (water-in-oil)-in-water emulsion solvent evaporation technique. *Pharmaceutical Research* 10: 362–368.

325 Jackson, A.C., Bartelt, J.A., Marczewski, K. et al. (2011). Silica-protected micron and sub-micron capsules and particles for self-healing at the microscale. *Macromolecular Rapid Communications* 32: 82–87.

326 Rule, J., Brown, E.N., Sottos, N.R. et al. (2005). Wax-protected catalyst microspheres for efficient self-healing materials. *Advanced Materials* 72: 205–208.

327 Xiao, D.S., Yuan, Y.C., Rong, M.Z., and Zhang, M.Q. (2009). A facile strategy for preparing self-healing polymer composites by incorporation of cationic catalyst-loaded vegetable fibers. *Advanced Functional Materials* 19: 2289–2296.

328 Zhu, D.Y., Rong, M.Z., and Zhang, M.Q. (2013). Preparation and characterization of multilayered microcapsule-like microreactor for self-healing polymers. *Polymer* 54: 4227–4236.

329 Brandrup, J., Immergut, E.H., and Grulke, E.A. (1999). *Polymer Handbook*, 2336. New York: John Wiley & Sons, Inc.

330 Yang, Z.X., Hollar, J., He, X.D., and Shi, X.M. (2011). A self-healing cementitious composite using oil core/silica gel shell microcapsules. *Cement and Concrete Composites* 33: 506–512.

331 Ballout, W., Perichaud, A., Caserta, L. et al. (2016). Encapsulation methods for photo-polymerisable self-healing formulations. *Journal of Microencapsulation* 33: 331–343.

332 Biggs, P., Jones, I.I.L., and Lewis, G. (2009). An autonomically-healed PMMA bone cement: influence of the crystal size of Grubbs' catalyst on fracture toughness and polymerisation rate. *International Journal of Biomaterials* 2: 494–504.

333 Lewis, G., Wellborn, B., Jones, L., and Biggs, P. (2009). A room-temperature autonomically-healing PMMA bone cement: influence of composition on fatigue crack propagation rate. *Journal of Applied Biomaterials & Biomechanics* 7: 90–96.

334 Wilson, G.O., Moore, J.S., White, S.R. et al. (2008). Autonomic healing of epoxy vinyl esters via ring opening metathesis polymerization. *Advanced Functional Materials* 18: 44–52.

335 Chipara, M.D., Chipara, M., Shansky, E., and Zaleski, J.M. (2009). Self-healing of high elasticity block copolymers. *Polymers for Advanced Technologies* 20: 427–431.

336 Moll, J.L., White, S.R., and Sottos, N.R. (2010). A self-sealing fiber-reinforced composite. *Journal of Composite Materials* 44: 2573–2585.

337 Wilson, G.O., Caruso, M.M., Reimer, N.T. et al. (2008). Evaluation of ruthenium catalysts for ring-opening metathesis polymerization-based self-healing applications. *Chemistry of Materials* 20: 3288–3297.

338 Wilson, G.O., Porter, K.A., Weissman, H. et al. (2009). Stability of second generation Grubbs' alkylidenes to primary amines: formation of novel ruthenium-amine complexes. *Advanced Synthesis & Catalysis* 351: 1817–1825.

339 Guadagno, L., Longo, P., Raimondo, M. et al. (2011). Use of Hoveyda–Grubbs' second generation catalyst in self-healing epoxy mixtures. *Composites Part B: Engineering* 42: 296–301.

340 Liu, X., Lee, J.K., Yoon, S.H., and Kessler, M.R. (2006). Characterization of diene monomers as healing agents for autonomic damage repair. *Journal of Applied Polymer Science* 101: 1266–1272.

341 Larin, G.E. (2006). Rheokinetics of ring-opening metathesis polymerization of norbornene-based monomers intended for self-healing applications. *Polymer Engineering and Science* 46: 1804–1811.

342 Lee, J.K., Hong, S.J., and Liu, X. (2004). Characterization of dicyclopentadiene and 5-ethylidene-2-norbornene as self-healing agents for polymer composite and its microcapsules. *Macromolecular Research* 12: 478–483.

343 Lee, J.K., Liu, X., Yoon, S.H., and Kessler, M.R. (2007). Thermal analysis of ring-opening metathesis polymerized healing agents. *Journal of Polymer Science Part B: Polymer Physics* 45: 1771–1780.

344 Guadagno, L., Longo, P., Raimondo, M. et al. (2009). A composite material which is self-repairing even at low temperature. International Patent WO2009/113025.

345 Guadagno, L., Raimondo, M., Iannuzzo, G., and Russo, S. (2010). Self-healing structures in aerospace applications. In: *Proceedings of the Fifth International Conference on Times of Polymers and Composites*, 267–269. Ischia, Italy: AIP.

346 Sheng, X., Lee, J.K., and Kessler, M.R. (2009). Influence of cross-link density on the properties of ROMP thermosets. *Polymer* 50: 1264–1269.

347 Yang, G. and Lee, J.K. (2016). Evaluation of 5-ethylidene-2-norbornene with an adhesion promoter for self-healing applications. *Journal of Polymer Science Part B: Polymer Physics* 54: 1170–1179.

348 Majchrzak, M., Hine, P.J., and Khosravi, E. (2012). An autonomous self-healing system based on ROMP of norbornene dicarboximide monomers. *Polymer* 53: 5251–5257.

349 Cho, S.H., Andersson, H.M., White, S.R. et al. (2006). Polydimethylsiloxane-based self-healing materials. *Advanced Materials* 18: 997–1000.

350 Shah, G.B. (1998). Effect of length of ligand in organotin compounds on their catalytic activity for the polycondensation of silicone. *Journal of Applied Polymer Science* 70: 2235–2239.

351 van der Weij, F.W. The action of tin compounds in condensation-type RTV silicone rubbers. *Macromolecular Chemistry and Physics* 181: 2541–2548.

352 Cho, S.H., White, S.R., and Braun, P.V. (2009). Self-healing polymer coatings. *Advanced Materials* 21: 645–649.

353 Huang, M.X., Zhang, H., and Yang, J.L. (2012). Synthesis of organic silane microcapsules for self-healing corrosion resistant polymer coatings. *Corrosion Science* 65: 561–566.

354 Wilson, G.O., Henderson, J.W., Caruso, M.M. et al. (2010). Evaluation of peroxide initiators for radical polymerization-based self-healing applications. *Journal of Polymer Science Part A: Polymer Chemistry* 48: 2698–2708.

355 White, S.R., Gladman, A.S., Celestine, A.D. et al. (2013). Self-healing PMMA biomaterials. Concepts, realization and outlook: a review. In: *Proceedings of the Fourth International Conference on Self-Healing Materials (ICSHM 2013)*, 616. Ghent, Belgium (16–20 June 2013): Magnel Laboratory for Concrete Research.

356 Zhang, C.Y., Jiang, X.B., Rong, M.Z., and Zhang, M.Q. (2014). Free radical polymerization aided self-healing. *Journal of Intelligent Material Systems and Structures* 25: 31–39.

357 Samadzadeh, M., Hatami, B.S., Peikari, M. et al. (2011). Tung oil-an autonomous repairing agent for self-healing epoxy coatings. *Progress in Organic Coatings* 70: 383–387.

358 Zhang, C., Wang, H.R., and Zhou, Q.X. (2018). Preparation and characterization of microcapsules based self-healing coatings containing epoxy ester as healing agent. *Progress in Organic Coatings* 125: 403–410.

359 Song, Y.K., Jo, Y.H., Lim, Y.J. et al. (2013). Sunlight-induced self-healing of a microcapsule-type protective coating. *ACS Applied Materials & Interfaces* 5: 1378–1384.

360 Zhong, N. and Post, W. (2015). Self-repair of structural and functional composites with intrinsically self-healing polymer matrices: a review. *Composites Part A: Applied Science and Manufacturing* 69: 226–239.

361 van den Dungen, E.T.A. (2009). Self-healing coatings based on thiol–ene chemistry. PhD thesis. University of Stellenbosch.

362 Jin, H.H., Mangun, C.L., Stradley, D.S. et al. (2012). Self-healing thermoset using encapsulated epoxy–amine healing chemistry. *Polymer* 53: 581–587.

363 Zhang, H., Wang, P., and Yang, J. (2014). Self-healing epoxy via epoxy–amine chemistry in dual hollow glass bubbles. *Composites Science and Technology* 94: 23–29.

364 Yang, J.L. and Zhang, H. (2012). Etching of hollow glass beads loaded with amines for self-healing composites. In: *Proceedings of the 8th Asian–Australasian Conference on Composites Materials*, 284. Kuala Lumpur, Malaysia (6–8 November 2012): AACM.

365 Utada, A.S., Lorenceau, E., Link, D.R. et al. (2005). Monodisperse double emulsions generated from a microcapillary device. *Science* 308: 537–541.

366 Sun, G. and Zhang, Z. (2002). Mechanical strength of microcapsules made of different wall materials. *International Journal of Pharmaceutics* 242: 307–311.

367 Lin, C., Yuan, L., Gu, A.J. et al. (2013). High performance self-healing bismaleimide/diallylbisphenol a/poly(phenylene oxide) microcapsules composites with low temperature processability. *Polymer Composites* 34: 335–342.

368 Billiet, S., van Camp, W., Hillewaere, X.K.D. et al. (2012). Development of optimized autonomous self-healing systems for epoxy materials based on maleimide chemistry. *Polymer* 53: 2320–2326.

369 Schollenberger, C. (1990). *Polyurethane and isocyanate-based structural adhesives*. In: *Handbook of Adhesives* (ed. I. Skeist), 359–380. New York: Van Nostrand-Reinhold.

370 Huang, M.X. and Yang, J.L. (2011). Facile microencapsulation of HDI for self-healing anticorrosion coatings. *Journal of Materials Chemistry* 21: 11123–11130.

371 Palmese, G.R., Pratama, P.A., Peterson, A.M., and Sharifi, M. (2013). Capsule-based self-healing epoxy systems using Diels–Alder reactions. In: *Proceedings of the Fourth International Conference on Self-Healing Materials (ICSHM 2013)*, 720. Ghent, Belgium (16–20 June 2013): Magnel Laboratory for Concrete Research.

372 Peterson, A.M., Jensen, R.E., and Palmese, G.R. (2010). Room-temperature healing of a thermosetting polymer using the Diels-Alder reaction. *ACS Applied Materials & Interfaces* 2: 1141–1149.

373 Pratama, P.A., Peterson, A.M., and Palmese, G.R. (2013). The role of maleimide structure in the healing of furan-functionalized epoxy–amine thermosets. *Polymer Chemistry* 4: 5000–5006.

374 Peterson, A.M. (2008). Remendable polymeric materials using reversible covalent bonds. PhD thesis. Drexel University, Philadelphia.

375 Song, Y.K. and Chung, C.M. (2013). Repeatable self-healing of a microcapsule-type protective coating. *Polymer Chemistry* 4: 4940–4947.

376 Binder, W.H. and Sachsenhofer, R. (2008). 'Click' chemistry in polymer and material science: an update. *Macromolecular Rapid Communications* 29: 952–981.

377 Binder, W.H. and Sachsenhofer, R. (2007). 'Click' chemistry in polymer and mate-rials science. *Macromolecular Rapid Communications* 28: 15–54.

378 Schunack, M., Gragert, M., Döhler, D. et al. (2012). Low-temperature Cu(I)-catalyzed "click" reactions for self-healing polymers. *Macromolecular Chemistry and Physics* 213: 205–214.

379 Cohades, A., Branfoot, C., Rae, S. et al. (2018). Progress in self-healing fiber-reinforced polymer composites. *Advanced Materials Interfaces* 5: 1800177/1–1800177/20.

380 White, S.R., Moore, J.S., Sottos, N.R. et al. (2014). Restoration of large damage volumes in polymers. *Science* 344: 620–623.

381 Krull, B.P., Gergely, R.C.R., Cruz, W.A.S. et al. (2016). Strategies for volumetric recovery of large scale damage in polymers. *Advanced Functional Materials* 26: 4561–4569.

382 Li, G. and Uppu, N. (2010). Shape memory polymer based self-healing syntactic foam: 3-D confined thermomechanical characterization. *Composites Science and Technology* 70: 1419–1427.

383 Zhang, P. and Li, G. (2016). Advances in healing-on-demand polymers and polymer composites. *Progress in Polymer Science* 57: 32–63.

384 Li, G. and Meng, H. (ed.) (2015). *Recent Advances in Smart Self-Healing Polymers and Composites.* Cambridge: Woodhead Publishing.

385 Wang, S., Oh, J.Y., Xu, J. et al. (2018). Skin-inspired electronics: an emerging paradigm. *Accounts of Chemical Research* 51: 1033–1045.

386 Li, M.-X., Wu, H.-P., Zhu, Y.-L., and Wang, C. (2019). Mechanically adaptive electronic polymers: introduction, design and applications. *Acta Polymerica Sinica* 50: 247–260. (in Chinese).

387 Piermattei, A., Karthikeyan, S., and Sijbesma, R.P. (2009). Activating catalysts with mechanical force. *Nature Chemistry* 1: 133–137.

388 Paulusse, J.M.J. and Sijbesma, R.P. (2004). Reversible mechanochemistry of a PdII coordination polymer. *Angewandte Chemie International Edition* 43: 4460–4462.

389 Kersey, F.R., Loveless, D.M., and Craig, S.L. (2007). A hybrid polymer gel with controlled rates of cross-link rupture and self-repair. *Journal of the Royal Society Interface* 4: 373–380.

390 Takaffoli, M., Zhang, T., Parks, D., and Zhao, X. (2016). Mechanochemically responsive viscoelastic elastomers. *Journal of Applied Mechanics* 83: 071007/1–071007/10.

391 Li, J., Shiraki, T., Hu, B. et al. (2014). Mechanophore activation at heterointerfaces. *Journal of the American Chemical Society* 136: 15925–15928.

392 Lenhardt, J.M., Ong, M.T., Choe, R. et al. (2010). Trapping a diradical transition state by mechanochemical polymer extension. *Science* 329: 1057–1060.

393 Li, M.X., Rong, M.Z., and Zhang, M.Q. (2021). Reversible mechanochemistry enabled autonomous sustaining of robustness of polymers - an example of next generation self-healing strategy. *Chinese Journal of Polymer Science* 39: 545–553.

394 Li, M.X., Rong, M.Z., and Zhang, M.Q. (2021). Enhancement of mechanochemical self-blocking and self-healing of early minor damages in polymers through hyperbranched architecture. *Express Polymer Letters* 15: 781–790.

2

Extrinsic Self-Healing via Addition Polymerization

CHAPTER MENU

Chapters 2–8 are devoted to the self-healing methodologies and materials for strength recovery of structural materials developed in the authors' laboratory. Among these works, self-healing epoxy and self-healing epoxy/fiber composites act as the key (but not the only) model materials to be repaired. This is because epoxy has been widely used in industry as surface coatings, structural adhesives, printed circuit boards, insulation materials for electronic devices, and advanced composites matrices, owing to its superior mechanical and electrical properties, good thermal and dimensional stability, and excellent chemical and corrosion resistance, in addition to ease of handling and processing. We believe that it is a reasonable starting point to work with self-healing epoxy, as viewed from both academic and engineering points.

The present chapter is focused on the preparation and characterization of self-healing epoxy with embedded dual encapsulated healant, i.e. two types of microcapsules that, respectively, include epoxy and its hardener (Figure 2.1), and self-healing epoxy/woven glass fabric composites with micropipelines containing epoxy and its hardener, respectively. The healing chemistry is dominated by addition polymerization.

2.1 Design and Selection of Healing System

As mentioned in Chapter 1, extrinsic self-healing represents one of the main strategies of self-healing. It enables autonomic crack rehabilitation of polymers at room temperature and its commercialization is in prospect. In general, there are two ways to implement the concept. The first one fills fragile pipelines (such as glass capillaries, hollow glass fibers, and 3D microvascular networks) with polymerizable monomers and then embeds them in the target materials [1–7]. The monomers can be bled into damage sites upon fracture and then polymerized to heal cracks. The second one uses liquid resins or solvents that are either microencapsulated or phase-separated when being compounded with the matrix [8–16]. The curable resins would be released into the path of a propagating crack and consolidated due to the catalysis of the neighboring predispersed hardeners that present themselves in the form of solid particles or capsules [8–15], while the solvents activate residual matrix reaction, as a simple, and potentially robust alternative to the recovery of virgin properties of a material after crack damage [16].

Extrinsic and Intrinsic Approaches to Self-Healing Polymers and Polymer Composites, First Edition. Ming Qiu Zhang and Min Zhi Rong.

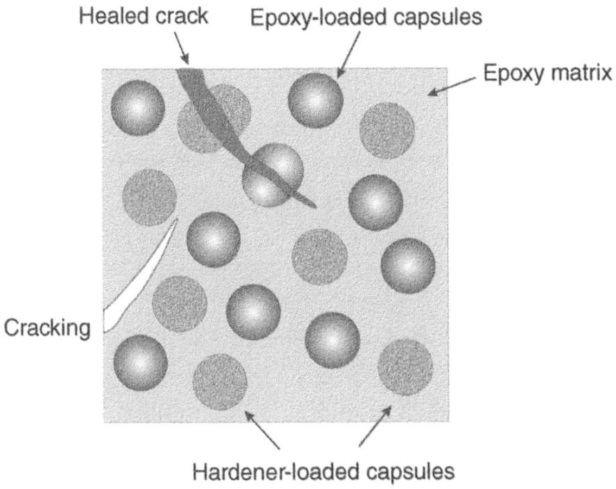

Figure 2.1 Schematic drawing of a self-healing epoxy specimen with embedded dual microcapsules and the principle of crack healing. Both epoxy monomer and its hardener are well protected by the capsules, which favors long term storage of the composites. When a crack (on the upper part, for example) ruptures the capsules, the released epoxy monomer and its hardener (i.e. the fluidic healing agent) would migrate to the damage zone due to capillary effect and then meet with each other. After polymerization, the consolidated healing agent filled in the interstitial volume re-bonds the crack. In case the reaction between the components of the healing agent can take place without the necessity of external stimulation, the healing can proceed in an autonomic way.

Considering the feasibility of mass production and broadness of application, the routes based on microcapsulation [8–16] are very promising. Under the framework of this methodology, three representative self-healing systems, dicyclopentadiene (DCPD)/Grubbs catalyst, polydimethylsiloxane (PDMS)/tin catalyst pairs and organic solvents, proved to be able to automatically repair cracks in different polymers. From these pilot works, the following criteria for design and selection of proper self-healing systems with high healability can be concluded. (i) The healing agent stored in microcapsules should be chemically stable and no degradation and/or self-polymerization take place when the capsules are kept for long time. (ii) The encapsulated healing agent should also be physically stable, possessing low freezing point, low viscosity, high flowability and low vapor pressure, so that it would not evaporate and/or diffuse out of the capsules shell. (iii) Having been released, the healing agent should exhibit high reactivity and rapidly consolidate at ambient temperature. The required dosage proportion of polymerizable component to hardener/catalyst has a wider operation window. (iv) After consolidation, the healing agent should form strong chemical and/or physical interaction with the materials to be repaired, which is necessary for yielding firm adhesion between cracks. (v) Shell thickness of the microcapsules should be thin enough to ensure high core content even in the case of very tiny capsules. (vi) The wall shell of the microcapsules should be compact and thermally stable. (vii) The microcapsules should have balanced mechanical performance. They can not only endure the hard process of composites manufacturing, but also break as soon as cracks appear in the composites. (viii) The system should be low price, easy to prepare, and non-toxic to the environment.

Having been screened by the above criteria, low-viscous highly active epoxy monomer and its hardener should be a potential candidate system. To the authors' knowledge, however, epoxy resins have not been applied in the field of self-healing for a long time in the form of microcapsules except for two-part epoxy adhesives being employed in the aforementioned hollow pipelines strategy [2–6]. It is mainly due to the difficulty in microencapsulation of suitable hardener for epoxy, but not in microencapsulation of epoxy itself. The latter can be easily synthesized by in-situ polymerization or complex coacervation [17–20], while the conventional amine-type curing agents are amphoteric and highly active, and hence hard to be encapsulated in water or solvent by chemical methods. For example, they cannot be encapsulated by poly(urea-formaldehyde) (PUF) under acidic condition. Although physical extrusion art was used to produce some hardener-loaded capsules [21, 22], such as the capsules containing a mixture of diethylenetriamine and nonyl phenol with alginate wall and the capsules containing diethylamine with thermoplastic wall, they were not suitable for fabricating self-healing composites. The shell wall of the former capsules might suffer from bacterial degradation, while that of the latter was too thick (~100 μm) to be sensitive to propagating cracks. Moreover, special equipment had to be involved.

Without proper protection, the embedded hardener in a self-healing composite might be easily deactivated. Jung et al. used polyoxymethylene urea-walled microspheres to store an epoxide monomer to be released into cracks and rebond the cracked faces in a polyester matrix [23]. Solidification of the epoxy resin (i.e. the repair action) was triggered by the naturally occurring functional sites or embedded amine in the composites. White et al. indicated that the method was not feasible as the amine groups did not retain sufficient activity [24]. Zako and Takano proposed an intelligent material system using 40% volume fraction unmodified epoxy particles to repair microcracks and delamination damage in a glass/epoxy composite laminate [25]. By heating to 120 °C, the embedded epoxy particles (~50 μm) would melt, flow to the crack faces, and repair the damage with the help of the excessive amine in the composite. In addition to the poor activity of the amine as mentioned above, manual intervention (i.e. heating) was necessary in this case.

Epoxy is an important universal adhesive and has good affinity to many materials. So long as the hardener of epoxy and epoxy can be microencapsulated, a high performance two-component healing agent would thus be available. In fact, polythiols can also be used for curing epoxy besides amine. Epoxy resin with polythiol as the hardener and strong base as the hardening accelerator is well known for its low-temperature fast curability [26]. Accordingly, microencapsulated epoxy and polythiol might constitute a group of self-healing agents that is able to take effect at or below room temperature. In this context, self-healing can be done free of manual intervention (i.e. heating). It is worth noting, however, that microencapsulation of polythiol is also very difficult by means of the existing techniques such as in-situ polymerization of urea-formaldehyde or interface polymerization due to the very high activity of the hydrosulfide group of polythiol. Liquid polythiol can be cured by a wide variety of compounds including aldehyde, ketone, peroxides, epoxy resins, isocyanates, acrylic and anhydrides through a number of reaction paths such as oxidation, addition and substitution reactions of hydrosulfide terminal groups [27–29]. In most cases, hollow spheres with a very thick wall rather than the desired capsules have to be obtained.

To solve the problem, we planned to try an improved in-situ polymerization of melamine-formaldehyde, which is inert toward mercaptan, in an oil-in-water emulsion. Compared with urea-formaldehyde that has been used to encapsulate healing agent [17], polymerization of melamine-formaldehyde proceeds faster and poly(melamine-formaldehyde) (PMF)-walled microcapsules are superior to PUF-walled microcapsules in terms of mechanical properties, stability, and chemical resistance [30–32]. In this context, there is a possibility that the consumption of polythiol might be substantially reduced. In-situ polymerization with PMF or methanol etherified PMF has been applied to encapsulate liquids such as dye solution [30], dicyclopentadiene [31], lauryl alcohol [33], fragrant oil [34, 35], and methylparathion [36]. To obtain microcapsules with sufficient strength, the encapsulation process was usually conducted at a higher pH value of 3.5–6.5 and higher temperature of 60–80 °C or longer reaction time of three to eight hours. Although these rigorous processing parameters had little effect on the aforementioned inert core materials, they may lead to failure of microencapsulation or excessive consumption of the active core material, e.g. polythiol in the present work. Therefore, it is necessary to explore suitable reaction conditions.

As for the epoxy-loaded microcapsules, the same substance will play the role of their shell wall. This is necessary for homogeneous dispersion of the capsules, which contain epoxy and polythiol, in the composites.

Meanwhile, micropipelines [4, 37] prove to be able to store/provide a larger amount of healing agent and allow for multiple healing function when external supply is connected, which are advantages over microcapsules (Figure 1.11). Accordingly, a variety of micropipeline configurations have been proposed [3, 38–47]. They have worked well in different composites (mostly continuous fiber reinforced thermosetting polymer composites) in terms of mechanical properties restoration.

It is worth noting that mobility of the healing agent inside materials released upon damage is a key issue no matter which type of micropipelines are applied. It determines whether the healing agent can approach the cracked sites and the envisaged healing reaction can be effectively triggered (for a two-component healant system). In fact, the capillary effect, the main driving factor of healing fluid in self-healing materials with microcracks, becomes rather weak for fiber composites with open fracture [6, 43–45, 48, 49]. Moreover, the cracks in fiber composites are rather complicated, which may curve and interconnect. Therefore, improvement of flowability of healing agent becomes a very important issue. Otherwise, overlong healing duration or lower healing efficiency would be observed. In an extreme situation, for example, the process of healing may even last 12 months [3]. To solve the problem, assistant measures based on manual intervention, such as evacuation [45], heating [43, 48, 49], dilution with solvent [6, 45], and external pressurization [50, 51], were used during authentic healing.

Here we describe a novel method by encapsulating the mixture of a foaming agent with the above healing agent in polymeric tubes [52]. So long as the foaming agent takes effect, internal pressure is greatly increased, and the healing agent would burst out of the reservoir soon after breakage. This self-pressurized system can improve flowability of the healing agent free of manual intervention. More interestingly, the healing agent filled micropipelines might thus be less dense, favoring maintaining original properties of the composites to a higher extent.

2.2 Microencapsulation of Mercaptan and Epoxy by In-Situ Polymerization

2.2.1 Microencapsulation of Mercaptan

The typical encapsulation procedures are listed below [53]. Polythiol (pentaerythritol tetrakis (3-mercaptopropionate) [PETMP], Scheme 2.1, boiling point = 275 °C at 1 mmHg, density = 1.28 g ml^{-1} at 25 °C and hydrosulfide group content = 26.55%) was added into an aqueous solution of the emulsifier sodium styrene-maleic anhydride copolymer

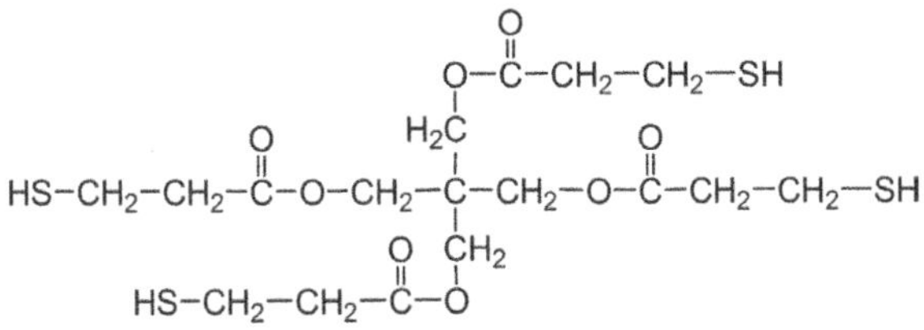

Scheme 2.1 Chemical structure of PETMP.

(SMANa) (150 ml). The mixture was vigorously dispersed by a homogenizer for five minutes at a selected rate and then two drops of 1-octanol were added to eliminate surface bubbles of the emulsion. The pre-condensate of melamine (0.04 mol) and formaldehyde (0.12 mol) was prepared at 70 °C for 30 minutes and the pH value of the solution was kept at about 9–10 by adding triethanolamine (Scheme 2.2a). Subsequently, the pre-condensate solution was added to the above emulsion with 450 rpm continuous mechanical agitation by a two-bladed stirring paddle while the pH value of the solution was kept at about 2.7–3.5 by adding citric acid. Eventually, the reaction mixture was cooled to room temperature. The resultant slurry was neutralized by sodium carbonate solution, and then diluted with deionized water. The deposit of microcapsules was separated through a Buchner funnel and rinsed with deionized water and acetone, and then vacuum dried. Scheme 2.2b shows the condensation reaction scheme of melamine-formaldehyde resins. Table 2.1 lists the influences of the processing parameters on the resultant microcapsules, and the sample IDs as well.

Formaldehyde mostly presents itself as methylene glycol in formalin and has two active functional groups, while melamine is a weakly basic reagent with six functional groups. Therefore, they can react with each other forming melamine-formaldehyde polymer. In general, the reaction consists of two stages as described in the following (Scheme 2.2). (i) The nucleophilic addition stage in basic solution proceeds at about 70 °C, when melamine reversibly and successively reacts with formaldehyde to give nine different methylol melamines ranging from mono- to hexamethylol melamines (a typical reaction with trimethylol melamines is shown in Scheme 2.2a). (ii) The condensation stage proceeds in acidic solution at about 50 °C, during which two mechanisms are involved: formation of ether bridges via the reaction between two methylol groups; and formation of methylene bridges via the reaction between a methylol group and an amino group with elimination of water (cf. steps I and II in Scheme 2.2b) [54–57]. The condensation stage overlaps in time with the in-situ polymerization or the microencapsulation process.

During the in-situ polymerization, the melamine-formaldehyde monomers and pre-condensate reacted with each other in the acidic water phase to form low molecular weight oligomers. The pre-condensate and oligomers possess surface activity like surfactants because their molecular structure contains both hydrophilic and hydrophobic groups [58, 59]. Consequently, the melamine-formaldehyde pre-condensate and oligomers tended to gather at the phase boundary between the dispersed water insoluble core substance PETMP droplets and the aqueous volume phase. The driving force for the enrichment at the surface of the dispersed phase was the surfactant character of the melamine-formaldehyde pre-condensate and oligomers. The higher concentration of the reactive resin molecules (i.e. the melamine-formaldehyde pre-condensate and oligomers) in the boundary layer must favor the condensation reaction. In other words, the resin condensation proceeded much faster at the interface than in the volume phase. As a result, gel-like structures were built up at the beginning owing to liquid–liquid phase separation, which was further

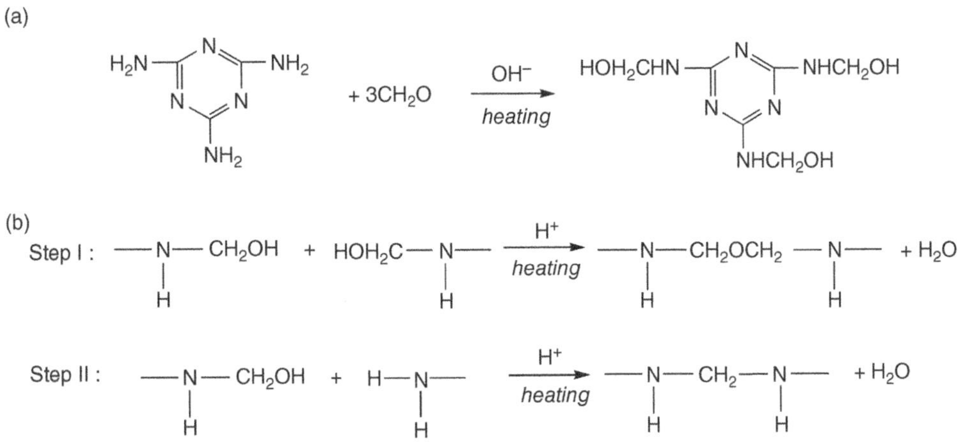

Scheme 2.2 Reaction schemes of melamine-formaldehyde resins. (a) Formation of melamine-formaldehyde pre-condensate. (b) Formation of PMF.

Table 2.1 Descriptions of the PMF-walled microcapsules containing mercaptan (PETMP) prepared using different processing parameters.

Sample ID	Reaction time (min)	Reaction temperature (°C)	System pH	Feeding weight ratio of core/shell monomers	Emulsifier content (%)	Dispersion rate ($\times 10^3$ rpm)	Average diameter (μm)	Core content (%)	Loss factor (%)	Yield (%)
1-1	20	50	3.2	2.3	2	15	–	–	–	41.5
1-2	40	50	3.2	2.3	2	15	11.5	77.6	10.2	80.7
1-3	60	50	3.2	2.3	2	15	11.5	77.3	10.4	80.8
1-4	80	50	3.2	2.3	2	15	11.5	76.4	11.2	81.0
2-1	60	40	3.2	2.3	2	15	23.2	86.2	3.7	77.9
2-2	60	50	3.2	2.3	2	15	11.3	78.5	10.0	79.8
2-3	60	60	3.2	2.3	2	15	9.3	67.1	16.6	86.7
2-4	60	70	3.2	2.3	2	15	16.1	59.0	23.7	90.2
3-1	60	50	2.7	2.3	2	15	12.2	78.8	7.2	82.1
3-2	60	50	2.9	2.3	2	15	12.2	81.7	5.8	80.3
3-3	60	50	3.2	2.3	2	15	11.3	77.6	10.7	80.2
3-4	60	50	3.5	2.3	2	15	11.9	76.9	12.9	79.0
4-1	60	50	3.2	0	2	15	6.0	0	–	36.5
4-2	60	50	3.2	1.2	2	15	9.4	33.1	45.3	88.2
4-3	60	50	3.2	1.7	2	15	8.7	71.8	14.3	83.3
4-4	60	50	3.2	2.3	2	15	11.4	76.9	10.5	81.2
4-5	60	50	3.2	2.9	2	15	13.9	87.4	6.5	79.3
5-1	60	50	3.2	2.3	1	15	36.3	83.5	4.7	79.5
5-2	60	50	3.2	2.3	2	15	11.5	77.6	10.2	80.6
5-3	60	50	3.2	2.3	3	15	5.4	70.8	13.2	85.5
6-1	60	50	3.2	2.3	2	10	36.7	83.8	8.0	76.6
6-2	60	50	3.2	2.3	2	15	11.8	77.6	10.1	80.7
6-3	60	50	3.2	2.3	2	20	5.0	67.7	12.8	91.7

Source: Reprinted from Yuan et al. [53], Copyright 2008, with permission from Elsevier.

hardened to give the capsule walls [59]. Subsequently, integrity and mechanical strength of the shell wall of the capsules was enhanced owing to the increased thickness and degree of cross-linking of PMF. These mean that the condensation reaction kinetics of melamine-formaldehyde must play a key role in the course of microencapsulation.

In fact, the condensation reaction rate and degree are determined by reaction time, reaction temperature, and catalyst H^+ concentration [57]. These factors are investigated in detail hereinafter in an attempt to find out the optimum conditions. The microcapsules prepared at excessively short reaction time (20 minutes, for example), or excessively high pH value (3.5, for sample), or excessively low reaction temperature (40 °C, for sample) cannot maintain their integrity after drying because of the weak shell walls, in spite of holding spherical shape in water. When the reaction time exceeds 40 minutes, both yield and loss factor do not show significant change with the reaction (refer to the samples 1-2–1-4 in Table 2.1).

Comparatively, the influence of pH value on reaction kinetics is quite complex due to interference of the parallel reactions taking place at the condensation stage. When the pH value of solution is reduced, the yield slightly increases (refer to samples 3-1–3-4 in Table 2.1). At pH = 2.9, both loss factor and core content reach their minimum and maximum values of 5.8 and 81.7%, respectively. This implies that higher catalyst concentration may accelerate the condensation reaction, so that less time is left for the reaction between the polythiol and melamine-formaldehyde pre-condensate and oligomers.

It is worth noting that some hollow microcapsules resulting from air bubbles appear in the case of low pH value. The amount of the hollow microcapsules seems to increase with decreasing pH value. This is attributed to the fact that a faster condensation reaction increases the probability of wall formation on the surface of air bubbles. An increased number of hollow microcapsules would consume more wall materials and might reduce the quality of microcapsules containing PETMP. Consequently, an appropriate catalyst concentration (pH = 2.9–3.2) is required to balance the loss factor, core content, as well as the amount of hollow microcapsules.

On the other hand, it was found that a few powder-like PMF nanoparticles adhere to the microcapsules. They should be formed by the melamine-formaldehyde oligomers through successive polymerization, liquid–liquid phase separation and deposition in the aqueous volume phase along with ceaseless decrease of their solubility [35, 60]. Compared with UF microcapsules, the surfaces of the melamine-formaldehyde microcapsules are smoother. In addition, the relative higher yields of melamine-formaldehyde microcapsules indicate that the outgrowth of PMF nanoparticles is not significant. Yield of microencapsulation is defined as the ratio of the mass of the collected microcapsules to the total mass of PETMP core and shell constituents. Table 2.1 shows that the yields of most experiments fall inside the limits of 76–92% except samples 1-1 and 4-1. In fact, the amounts of hollow microcapsules and PMF nanoparticles greatly affect the yield. In the diluted resultant slurry, the microcapsules containing PETMP gradually deposited because their density (1.24 g cm^{-3}) is higher than that of water. However, the hollow microcapsules and some PMF nanoparticles used to be thrown away as they were suspended in the solution.

The emulsifier SMANa adsorbed at the interfaces of the oil droplets is neutral at pH ~3. Its molecules tended to have such directional arrangement with hydrophobic groups oriented into the oil droplet and hydrophilic groups out of the oil droplets, establishing a steric boundary layer around each droplet [58, 61, 62]. This emulsifier layer allowed the melamine-formaldehyde pre-condensate and oligomers to pass through itself and accumulate at the interface because of their surface activity. When the melamine-formaldehyde pre-condensate and oligomers were gradually polymerized and hardened, the newly formed walls began to protect the core materials [58]. No SMANa molecules can be found on the wall of the microcapsules by Fourier transform infrared (FTIR) and nuclear magnetic resonance (NMR) analysis, implying that the external emulsifier has been washed away.

In short, the in-situ polymerization of melamine-formaldehyde can be successfully used to prepare microcapsules containing polythiol. Under certain conditions, satisfactory microcapsules with higher yield and core content but lower loss factor can be obtained. Surprisingly, the loss factor in most cases is about 10% or even lower. This demonstrates that our idea works, i.e. a faster condensation reaction between melamine and formaldehyde would help to reduce polythiol consumption. Besides, the polythiol has four —SH γρουπσ and acts as a curing agent to form a cross-linking network whenever it reacts with melamine-formaldehyde pre-condensate and oligomers. Thus, the inner layer of polythiol core is hard to make contact with the outside melamine-formaldehyde pre-condensate and oligomers because the cross-linking film hinders diffusion of the components from both sides.

If PMF were replaced by PUF for the microencapsulation, no intact microcapsules would be obtained because the slower reaction rate led to a weak shell. Even though the shell strength of PUF-walled microcapsules can be enhanced by increasing reaction time, and/or reaction temperature, and/or catalyst concentration, the core content had to be very low (< 20%) and many hollow spheres with very thick walls were produced. Excessive consumption of the active core material and a different microencapsulation mechanism of urea-formaldehyde resins should take the responsibility.

Since some of the core material PETMP has been involved in the synthesis of the shell wall, it should be known whether the rest of the portion keeps the chemical activity. In addition to the results of micro-Raman spectroscopy demonstrating that PETMP remains unchanged after being encapsulated [53], the reactivity of the core material can be further verified by differential scanning calorimetry (DSC) (Figure 2.2). The heating DSC scans indicate that with a rise in temperature no reaction occurs in either the mixture of epoxy (diglycidyl tetrahydro-*o*-phthalate, DTP) and the catalyst benzyl dimethylamine (BDMA) or the mixture of epoxy (DTP) and PETMP. For the mixture of epoxy (DTP)/PETMP/BDMA, however, an obvious exothermic reaction is detected. The peak temperature is 56.4 °C and the heat of reaction is 466.3 J g^{-1}. When PETMP was replaced by the ground microcapsules containing PETMP, a similar exothermic peak with a heat of reaction of 212.6 J g^{-1} appears at 59.9 °C. The difference might result from: (i) grinding of the microcapsules cannot ensure 100% breakage and the quantity of the released core material failed to reach the stoichiometric ratio of DTP to PETMP; and (ii) both shell wall and intact microcapsules were inert and do not take part in the curing reaction.

When self-healing composites are thermally cured, the embedded microcapsules have to experience heat impact. Therefore, their thermal stability should be known. The pyrolytic behaviors of the related substances show that there are three main phases of weight loss for the shell wall PMF within the temperature range of interest [38]. The first phase with ~2% weight loss from 25 to 100 °C should be due to evaporation of the adsorbed water. The second phase with ~10% weight loss in the range of 100–270 °C is attributed to evaporation of formaldehyde from the deformaldehyde reaction of PMF at elevated temperature [63]. The third one with ~61% weight loss in the range of 270–600 °C originates from decomposition of a portion of melamine and thermal degradation of PMF [64].

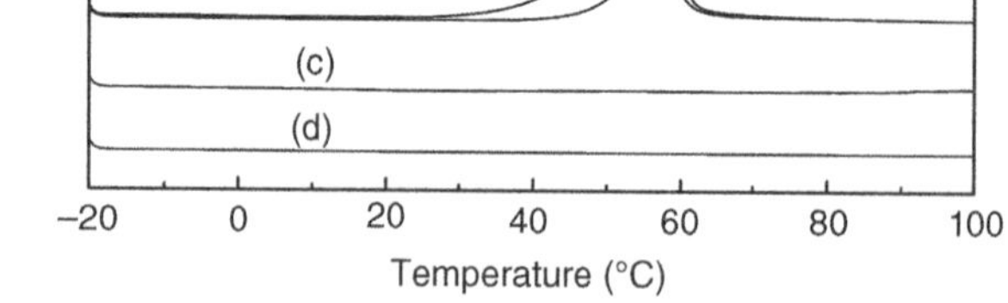

Figure 2.2 DSC heating traces of (a) epoxy (DTP)/PETMP/BDMA = 10.9/8.7/1, (b) epoxy (DTP)/ground PMF-walled microcapsules/BDMA = 10.9/19.2/1, (c) epoxy (DTP)/BDMA = 10.9/1, and (d) epoxy (DTP)/PETMP = 10.9/8.7. The compositions are expressed in terms of weight ratios. *Source:* Reprinted from Yuan et al. [53]. Copyright 2008, with permission from Elsevier.

In contrast to PMF, PETMP is quite stable below 250 °C. It also has three stages of pyrolysis from 250 to 600 °C corresponding to its evaporation and decomposition processes. The detailed mechanisms involved need further investigation. Nevertheless, thermal degradation of the PMF-walled microcapsules containing PETMP is a combination of the contributions made by PMF and PETMP.

In addition to thermal stability, the microcapsules in service should have long-term stability. Figure 2.3 shows the effect of storage time and temperature on weight loss of the microcapsules. Evidently, weight loss of the microcapsules exposed to room temperature for six months is only about 0.14 wt%. When the time is increased to one year, the weight loss is still low (i.e. 0.31 wt%). Visual inspection does not find any leakage from the microcapsules and their appearance also remains unchanged after a long storage time. As the microcapsules were kept in a desiccator and the saturated vapor pressure of polythiol is very low (~6.5×10^{-11} Torr at 25 °C) (www.brunobock.org), it can be deduced that the above weight loss should result from evaporation of the physically adsorbed water.

When the storage temperature is increased, weight loss of the microcapsules becomes more and more significant. Staying in higher temperature for longer time would lead to substantial weight loss. It suggests that the microcapsules could not be timelessly exposed to heat. At elevated temperature, evaporation of adsorbed water, release of formaldehyde, diffusion of the core material out of shell wall, and volatilization of some core material would occur [65, 66]. Compared with the microcapsules using PUF as the wall material, the current PMF-walled microcapsules have obviously higher storage stability and heat resistance because of the difference in their chemical structures [66].

2.2.2 Microencapsulation of Epoxy

The microencapsulation of epoxy resembles the aforementioned process for mercaptan [67]. Typically, the epoxy prepolymer (DTP, 400.0 g, epoxide equivalence weight = 0.65 mol/100 g, density = 1.24 g ml^{-1} at 20 °C, viscosity = 0.36 Pa·s at 25 °C) was added into a 2 wt% aqueous solution of SMANa (1200 ml). The mixture was vigorously stirred for five minutes and then two drops

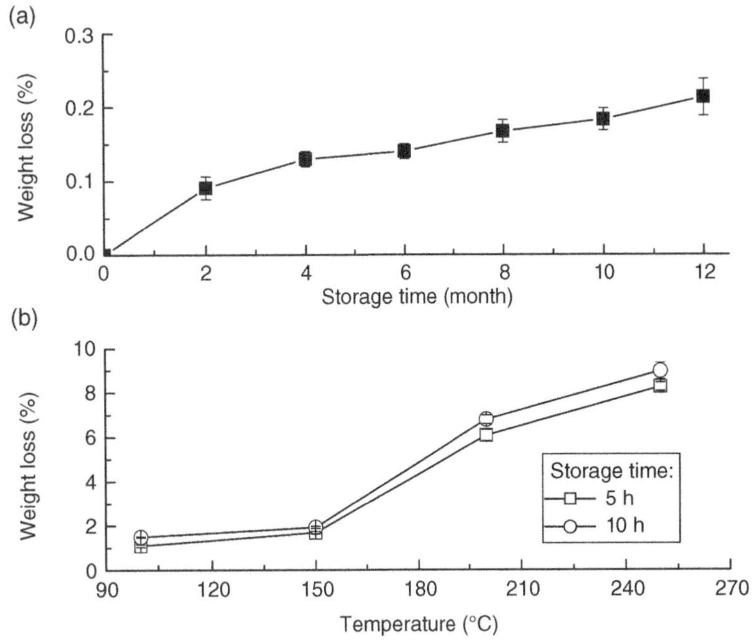

Figure 2.3 Effects of storage time and temperature on stability of PMF-walled microcapsules containing PETMP: (a) weight loss of the microcapsules as a function of storage time at room temperature; and (b) weight loss of the microcapsules as a function of temperature at constant storage times. *Source:* Reprinted from Yuan et al. [53]. Copyright 2008, with permission from Elsevier.

of 1-octanol were added to eliminate surface bubbles of the epoxy emulsion. The prepolymer of melamine (62.5 g) and 37% formaldehyde (135.5 g) was synthesized at 70 °C for 30 minutes and the pH value of the solution was kept at about 9–10 by adding triethanolamine. Subsequently, the prepolymer solution was added to the above epoxy emulsion at 50 °C with continuous agitation for one hour while the pH value of the system was kept at about 3 by adding citric acid. Eventually, the reaction mixture was cooled down to room temperature and the deposit of microcapsules was separated through a Buchner funnel, rinsed with deionized water and vacuum dried. Table 2.2 lists the influences of processing parameters on fabrication of the capsules, which provides the necessary data for optimization.

Pyrolytic analysis of PMF, DTP, and the capsules containing DTP was done to characterize the thermal stability of the capsules. DTP does not decompose until 150 °C. Its peak temperatures of weight loss appear at 248.5 and 384.9 °C, corresponding to evaporation and thermal degradation. As for the microcapsules, their pyrolitic behavior is a combination of those of the PMF shell wall and DTP core. Deformaldehydation starts at around 200 °C, while evaporation and/or decomposition of the core substance take place at ~250 °C. Clearly, the compact shell protects the core well.

When the epoxy-loaded microcapsules were mixed with diglycidyl ether of bisphenol A (EPON 828) and cured to generate a composite, observation of the composite's fracture surface indicated that the spheres are robust enough to survive the composite manufacturing (mechanical stirring, heating, evacuation, etc.). Additionally, a micro-Raman spectrum of the DTP released from the broken capsules embedded in the epoxy composite was recorded for comparison with those of cured epoxy and uncured DTP (Figure 2.4). As seen from the spectrum of cured epoxy (Figure 2.4a), the stretching peak of C=C in the benzene ring and the vibration of epoxide groups appear at 1610 and 915 cm^{-1}, respectively. The spectrum of uncured DTP on cured epoxy (Figure 2.4b) further exhibits the stretching peak of C=O at 1736 cm^{-1} and that of C=C at 1656 cm^{-1}, as well as the vibration of epoxide groups at 906 cm^{-1}. Taking the peak of C=C in benzene at 1610 cm^{-1} as the reference, one might exclude the interference of the uncured epoxide groups of the matrix epoxy. Moreover, when the area of the peak of C=O at 1736 cm^{-1} serves as the internal standard, the ratio of reactive C=C and epoxide to C=O of DTP can be estimated from: $A_{\text{epoxide groups}}/A_{C=C}/A_{C=O} = 1.16/1.33/1.00$. The ratio of liberated DTP calculated from Figure 2.4c is: $A_{\text{epoxide groups}}/A_{C=C}/A_{C=O} = 1.22/1.29/1.00$, which is very close to the value of original DTP. It is thus known that the activity of DTP is unchanged either after being encapsulated or after the capsules are embedded in the epoxy matrix.

DSC heating scans were also conducted in a way similar to Figure 2.2, except that epoxy-loaded capsules rather than mercaptan-loaded capsules were used. The results showed that there is an evident exotherm at about 51.5 °C for the mixture of ground microcapsules containing DTP, mercaptan (PETMP) and catalyst (BDMA). What is more, even when the ground microcapsules containing DTP were heated at 200 °C for 10 hours in advance, the same exothermic reaction was detected. This means that the

Table 2.2 Typical PMF-walled microcapsules containing epoxy (DTP) prepared using different processing parameters.

Sample ID	Reaction time (min)	Reaction temperature (°C)	System pH	Feeding weight ratio of core/shell monomers	Emulsifier content (%)	Dispersion rate ($\times 10^3$rpm)	Average diameter (μm)	Core content (%)	Yield (%)
1-1	40	50	3.2	3	2	18	6.7	83.2	88.3
1-2	80	50	3.2	3	2	18	6.5	82.1	91.5
2-1	40	40	3.2	3	2	18	–	–	–
2-2	40	60	3.2	3	2	18	5.6	81.4	92.0
2-3	40	70	3.2	3	2	18	4.5	76.1	94.8
3-1	40	50	3.5	3	2	18	–	–	–
3-2	40	50	2.9	3	2	18	5.3	80.6	85.7
3-3	40	50	2.5	3	2	18	4.7	78.5	82.4
4-1	40	50	3.2	0	2	18	1.1	0	36.9
4-2	40	50	3.2	2	2	18	6.4	76.6	82.3
4-3	40	50	3.2	4	2	–	–	–	–
5-1	40	50	3.2	3	5	18	2.3	63.5	90.6
5-2	40	50	3.2	3	2	23	2.5	67.6	89.5

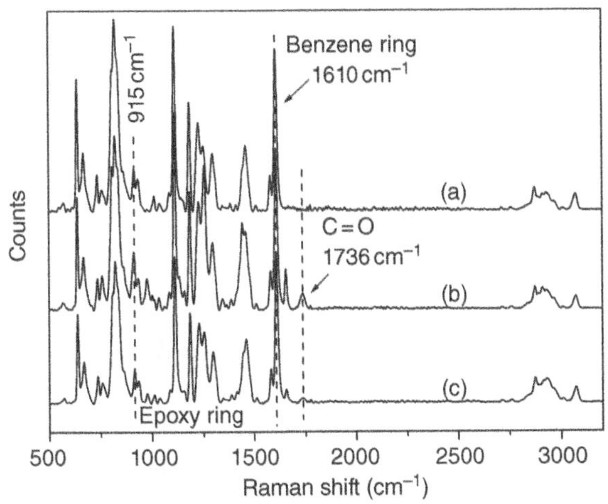

Figure 2.4 Micro-Raman spectra of (a) cured epoxy, (b) a raw DTP drop on cured epoxy resin substrate, and (c) the DTP flowing out from the broken PMF-walled microcapsules embedded in cured epoxy.

encapsulated DTP maintains high reactivity, which does not deteriorate by treatment at high temperature.

Figure 2.5 examines the influence of storage time and temperature on weight loss of the DTP-loaded microcapsules. With increase in time, the weight loss at room temperature of the capsules is only 0.53% after two months, and 0.68% after one year. Microscopic inspection does not find any trace of leakage and morphological change. Since the saturated vapor pressure of DTP is very low, the insignificant weight loss should originate from evaporation of physically adsorbed moisture. On the other hand, Figure 2.5 reveals that the weight loss of the microcapsules is increased with ambient temperature. In the case of treatment at 200 °C for 10 hours, for example, the weight loss is as high as 9.54%, much higher than the value at room temperature. The data suggest that the DTP-loaded capsules should not stay at higher temperature for a long time.

As the solid–liquid transition temperature of DTP is about −51 °C, we can conclude that the operation temperature range of the PMF walled capsules containing DTP is −50 to 200 °C. By varying the species and compositions of the core epoxy, this range can be expanded, favoring improvement of applicability of self-healing composites.

2.3 Filling Polymeric Tubes with Mercaptan and Epoxy

Polypropylene (PP) tubes serve as the micropipelines in the following, which is selected not only because of easy realization of large-scale production, but also owing to the following potential expansibility and advantages (Figure 2.6a). (i) Healing chemicals can be easily filled into larger size tubes first, and then the tube size can be reduced by thermal stretching. This would avoid the difficulty of introducing healing agent into the tubes with fixed small size (glass tubes and fibers, for example), while eventually lead to less deterioration of mechanical properties of the composites compared with the larger size tubes. (ii) The healing agent-loaded PP tubes can be converted into compartmented versions [68] by, for example, ultrasonic welding. This would help to realize

Figure 2.5 Effects of storage time and temperature on stability of PMF-walled microcapsules containing DTP: (a) weight loss of the microcapsules as a function of storage time at room temperature; and (b) weight loss of the microcapsules as a function of temperature at constant storage times.

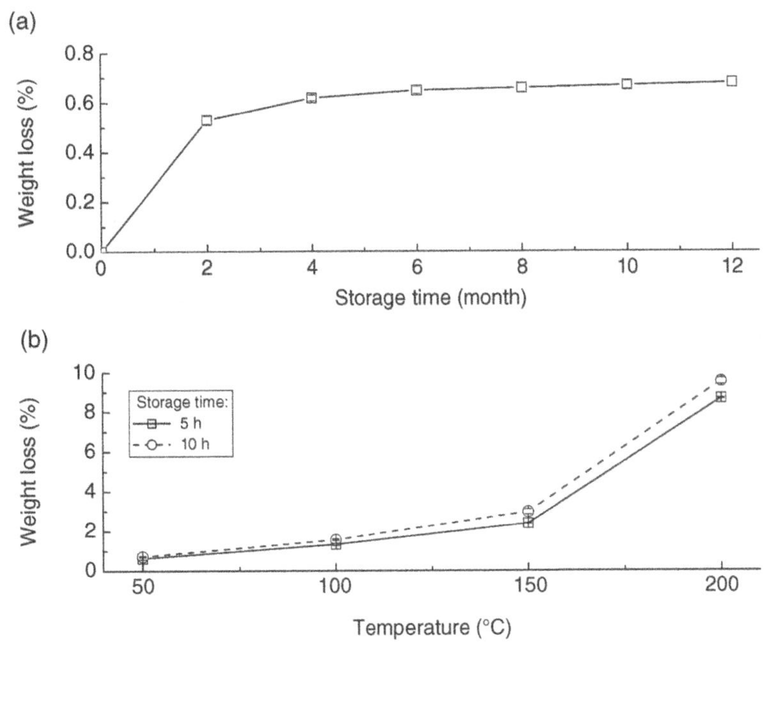

repeated self-healing. (iii) The PP tubes containing healing agent can be purposely deformed due to their flexible nature, so that they can be knitted together with conventional reinforcing fibers. This may favor fabrication of a self-healing structural composite with complicated shapes. (iv) The surface of the PP tubes can be modified by physical or chemical techniques, so that the interfacial interaction in composites could be improved. (v) The chemical inertness of PP ensures that it would not react with the included healing agent.

Since PP is a non-polar semi-crystalline polymer with certain toughness, PP tubes (outer diameter 250 μm, inner diameter 200 μm; Figure 2.6b) were first irradiated by UV light (intensity 95.3 mW cm^{-2}) at 50 °C for four hours for embrittlement and surface polarization, which ensures that the tubes would be preferentially damaged when a crack propagates in the composites. In the meantime, the glass fibers would not be affected because of their much higher strength than the PP tubes. Afterwards, the polymerizable component of the healing agent (i.e. the hand mixing epoxy DTP (100 parts) and foaming agent 2,2′-azobis-(2,4-dimethylvaleronitrile) (ABVN, 3 parts)) and the hardener component (i.e. the hand mixing PETMP (100 parts) and ABVN (3 parts) and the catalyst BDMA (0.4 parts)) were injected into the UV treated PP tubes by a syringe. Then, the tube ends were sealed with an adhesive.

2.4 Characterization of Self-Healing Functionality

2.4.1 Self-Healing Epoxy Materials with Embedded Dual Encapsulated Healant – Healing of Crack Due to Monotonic Fracture

It is worth noting that epoxy would not react with mercaptan in the absence of catalysis (Figure 2.2). Therefore, mercaptan had to be mixed with a proper catalyst inside the

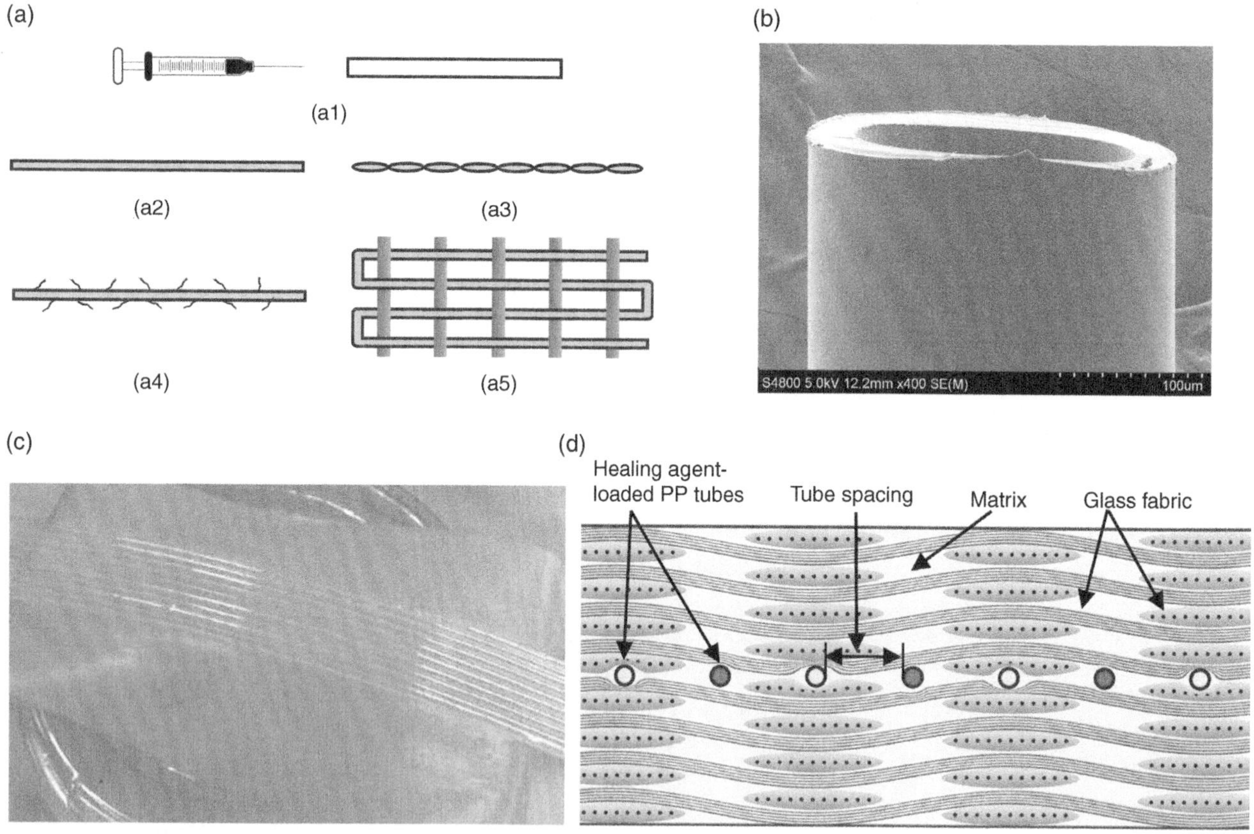

Figure 2.6 (a) Expansibility and advantages of thermoplastic tubing in constructing microvascular self-healing composites: (a1) easy incorporation of healing fluid with thicker tubes; (a2) thinning; (a3) compartmented; (a4) surface modification; and (a5) knitted with fiber reinforcement. (b) Scanning electron microscopy image of a PP tube. (c) Prepreg of healing agent-loaded PP tubes. (d) Arrangement of the healing tubes in the composite laminates. The cross-sections of the tubes containing epoxy and hardener are differently shaded to distinguish them from each other. *Source:* (a, d) Reprinted from Zhu et al. [52]. Copyright 2016, with permission from Elsevier. (b, c) Zhu et al. [52], Figure 1, p. 2/with permission of Elsevier.

microcapsules. Upon fracture of the composite with embedded epoxy oligomer- and mercaptan/catalyst-loaded capsules, the fluids from the dual capsules can be bled into damage sites and then polymerized to repair cracks.

Here a dilemma has to be solved. That is, miroencapsulation with melamine-formaldehyde should be conducted in acidic medium, while the catalyst for mercaptan (i.e. tertiary amine) is basic. If mercaptan was mixed with tertiary amine first, the subsequent encapsulation reaction could not proceed. To produce microcapsules containing both mercaptan and its amine catalyst, an innovative way was by infiltrating the catalyst into the ready-made mercaptan-loaded microcapsules. As the catalyst and mercaptan have good solvability and form a stable strong base–weak acid pair, no reverse bleeding was observed. The catalyst content can be adjusted by changing infiltration conditions and using assistant solvent. A typical process to prepare such mercaptan/catalyst-loaded capsules is as in the following. The PMF-walled microcapsules containing PETMP were uniformly dispersed into BDMA solution at 40 °C for 24 hours, filtrated, rinsed with ethyl ether and dried at room temperature.

Self-healing epoxy composite materials were prepared according to the following procedures [69]. First, the unfilled epoxy specimens were produced as references through mixing 100 parts EPON 828 with 12.5 parts curing agent diethylenetriamine (DETA). Then, the self-healing epoxy composites were prepared by mixing different concentrations and weight ratios of the capsules containing epoxy and its hardener with the aforementioned EPON 828/DETA mixture. Either the unfilled epoxy or the filled version was degassed, poured into a closed silicone rubber mold, and cured for 24 hours at room temperature, followed by 24 hours at 40 °C. The core content of the epoxy-loaded microcapsules (the weight ratio of core to microcapsule) is 96.9 wt%, while the contents of PETMP and BDMA inside the hardener-loaded microcapsules are 90.7 and 3.9 wt%, respectively. The protocol proposed by White et al. was used to evaluate self-healing ability of the materials, with fracture tests carried out on tapered double cantilever beam (TDCB) specimens [8, 9]. Efficiency of healing, η, is defined as the ratio of fracture toughness, K_{IC}, of healed (i.e. K_{IC}^{Healed}) and virgin (i.e. K_{IC}^{Virgin}) materials:

$$\eta = \frac{K_{IC}^{Healed}}{K_{IC}^{Virgin}} \tag{2.1}$$

To highlight the role of the healing agent, three types of control specimens were fabricated and tested. They consisted of: (i) neat epoxy without microencapsulated healing agent; (ii) epoxy with 5 wt% epoxy-loaded microcapsules; and (iii) epoxy with 5 wt% hardener-loaded microcapsules.

The optical microscopic photo in Figure 2.7 shows the dispersion of the epoxy- and hardener-loaded microcapsules in epoxy matrix. The volume mean diameters of the microcapsules are estimated to be 93.9 and 92.8 μm, respectively. Their densities were found to be 1.24 and 1.26 g cm^{-3}, respectively, as indicated by the separated pycnometer tests. The similarity in their surface features, physical properties, and geometry certainly benefits uniform distribution of the microcapsules in the composite. As a result, the contact probability between the released epoxy and curing agent would be increased after damage-induced breakage of the capsules. Furthermore, since the two types of core materials possess good intersolubility, their interdiffusion and homogenization in the subsequent polymerization should be favored.

In fact, the microcapsules can form hydrogen bonding with the matrix due to the polar groups (such as amine and hydroxyl) on their melamine-formaldehyde shells. The strong interaction between the microcapsules and the matrix ensures timely penetration of cracks through the encapsulated healing agent.

Figure 2.8 shows a representative load–displacement curve of a self-healing composite specimen recorded during the TDCB test, demonstrating recovery of 103.3% of the virgin fracture toughness at rather low microencapsulated healing agent content. Compared with the linear (brittle) fracture behavior of the virgin test, failure of the healed self-healing specimen is characterized by a stick–slip process. Additional energy must have been consumed during fracture. In contrast, all three types of control specimens, which experienced the same healing procedure as the self-healing

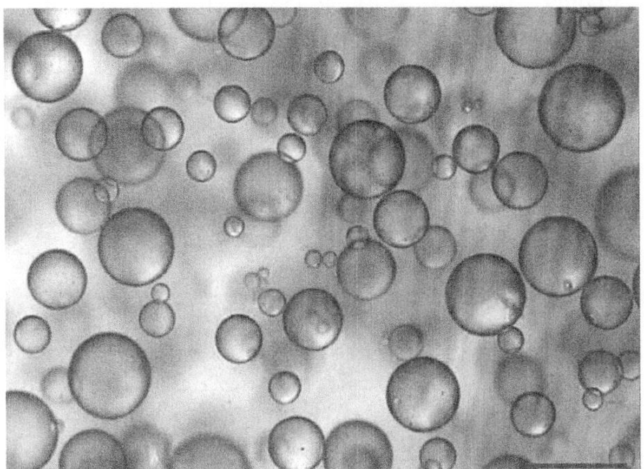

Figure 2.7 Optical microscopy photo taken from a self-healing TDCB specimen, which contains 5 wt% epoxy-loaded capsules and 5 wt% hardener-loaded capsules. To differentiate the two types of microcapsules for visual inspection, the core materials were mixed with 0.1% red and green dyestuff before microencapsulation, respectively.

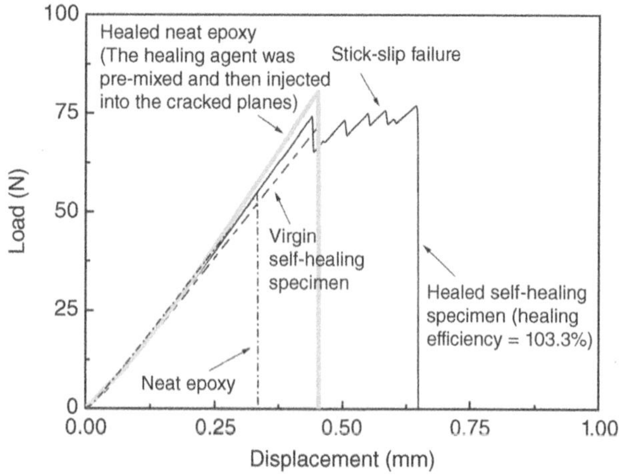

Figure 2.8 Typical load–displacement curves recorded during TDCB tests. The self-healing composite contains 2.5 wt% epoxy-loaded capsules and 2.5 wt% hardener-loaded capsules. The reference specimen was healed by injecting the pre-mixed epoxy/mercaptan (1.26:1) with a mercaptan/amine catalyst ratio of 100:4.3 to the cracked faces. Healing of the fractured specimens was conducted at 20 °C for 24 hours. *Source:* Reprinted from Yuan et al. [69]. Copyright 2008, with permission from American Chemical Society.

specimen after the first failure, showed no healing effect and were unable to carry any load upon reloading.

Since Figure 2.8 reveals the crack healing ability of the healant, the influence of its concentration on healing efficiency should be known. As exhibited in Figure 2.9, at a constant weight ratio of epoxy-/hardener-loaded capsules of 1:1, the healing efficiency drastically increases with capsule concentration from 0 to 5 wt% and then gradually levels off within the capsule content range from 5 to 10 wt%. The dependence of fracture toughness of healed self-healing specimens on capsule content also follows a similar trend of variation. Clearly, the healing effect is correlative to the healing agent quantity offered by the broken capsules on the fracture planes. Further investigation of the influence of weight ratio of epoxy-/hardener-loaded capsules within the capsule concentration range from 1 to 10 wt% indicates that the optimal healing effect is obtained at about 1–1.25 (corresponding to a weight ratio of epoxy/mercaptan of 1.07–1.33). This means that a total amount of the two types of microcapsules of 5 wt% with almost identical fraction of each type of capsule is critical for the current system to achieve the highest healing efficiency.

Incorporation of additives into a polymer would inevitably affect its intrinsic properties. It is seen from Figure 2.9 that the virgin self-healing composites are tougher than the unfilled epoxy. The toughening effect coincides with the observation of Brown et al. [10]. A crack pinning mechanism might be operating as characterized by the tails originating from broken capsules on the fracture plane. As for the static mechanical performance, the continuous reduction in modulus of the specimens with a rise in the capsule content is marginal (e.g. only ~5% reduction in Young's modulus at 10 wt% of the capsules), while the continuous decrease in strength is relatively obvious (e.g. ~28% decrease in tensile strength at 10 wt% of the capsules). In general, it follows the law of other microcapsules or rubbery particle filled polymers [10, 70].

To find out the reason for the healing efficiency higher than 100%, simulative healing experiments were done by injecting a sufficient amount of the model healing reagent, i.e. mixture of epoxy and mercaptan of different proportions with a constant mercaptan/amine catalyst ratio of 100:4.3, into the fracture planes of unfilled epoxy specimen. Both the load–displacement curve of the specimen healed by manual injection (Figure 2.8) and the healing efficiencies of a group of simulative specimens over a wide range of epoxy/mercaptan ratios showed that the recovery of virgin fracture

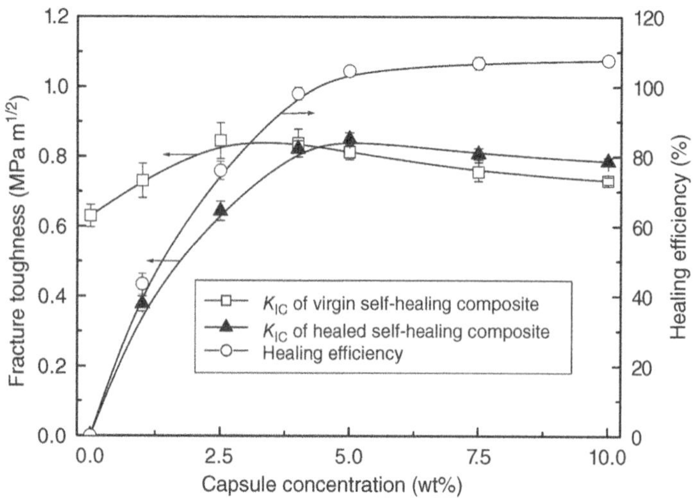

Figure 2.9 Influence of capsule concentration on fracture toughness and healing efficiency. Weight ratio of the two types of capsules in all the self-healing specimens is 1:1. Healing of the fractured specimens was conducted at 20 °C for 24 hours. *Source:* Reprinted from Yuan et al. [69]. Copyright 2008, with permission from American Chemical Society.

toughness exceeds 100%. The independent TDCB fracture measurements of the matrix and the epoxy healing system (produced at epoxy/mercaptan stoichiometric ratio) indicated that their fracture toughness, K_{IC}, are 0.63 and 1.18 MPa m$^{1/2}$, respectively. Therefore, it is known that the bonding material not only heals the cracks but also provides the damaged sites with higher fracture toughness. Under simulative circumstances, uniform spread of the healing agent mixed at optimum epoxy/hardener ratio was guaranteed. The inherent miscibility between the polymerizable component of the healing agent (i.e. epoxy) and the matrix epoxy led to good adhesion between them, so the crack had to deviate from the original fracture plane in reference specimens. The effect resulted in high fracture load since the TDCB specimen thickness is larger away from the groove line. A similar phenomenon has been reported by Keller et al. [15], who found a deviated tear in healed PDMS elastomer that offered healing efficiencies higher than 100%. In practical applications, because the healed portion has higher fracture toughness than the matrix, subsequent cracking would preferably occur in the rest of the composite. Multiple self-healing of the material might thus be possible.

The results of simulative healing experiments further indicate that the maximum healing efficiency was obtained near the stoichiometric weight ratio (i.e. 1.26) of epoxy to mercaptan. It is almost the same as the aforementioned optimum weight ratio of epoxy-/hardener-loaded capsules (1–1.25, equivalent to weight ratio of epoxy/mercaptan of 1.07–1.33) that corresponds to the highest healing efficiency of the authentic specimens. It proves that the two types of capsules must have been well distributed in the authentic specimens. So long as the concentrations of the epoxy- and mercaptan-loaded capsules are the same, their breakage allows delivery of the two-part healing agent with composition close to the stoichiometric ratio to the cracked faces. Nevertheless, healing efficiencies of the authentic specimens are still lower than those of the simulative ones. It means the healing was not as complete as in the control specimens. This is evidenced by the fact that the aforementioned macroscopic crack deviation from the groove line was not perceived in authentic specimens. As seen from energy dispersive spectroscopy (EDS), the contents of the "indicator element" (i.e. sulfur) in different regions on the fracture surface of the healed specimen vary greatly. The heterogeneity might result from the mismatched type of the neighbor capsules, uncertainty of the amount and proportion of broken capsules on the rupture planes, and uneven interdiffusion of the reaction agents. As a result, cohesive failures of cured healing membrane and matrix left their traces on the fracture surfaces of healed specimens. These failures together with opening of a few unbonded surfaces (caused by incomplete coverage of healing membrane) appeared. Accordingly, crack propagated in a zigzag way on a microscopic scale, as reflected by the stick-slip process (Figure 2.8). In principle, such a microscopic zigzag course can be regarded as low-grade crack deviation that also helps to raise fracture load, despite the fact that the crack development looked to be immediate under visual inspection. Therefore, healing efficiencies of authentic specimens are sometimes slightly higher than 100%, but lower than those of reference specimens.

To favor crack healing soon after its formation, fast consolidation of the repair chemicals is required. Epoxy-mercaptan cure is a nucleophilic addition reaction [71]. With the aid of a proper catalyst such as tertiary amine, the reaction involving an intermediate mercaptide-ion mechanism is significantly faster than the epoxy–amine reaction, particularly at low temperatures. However, they hardly react with each other without catalysis (Figure 2.2). In-situ confocal Raman microscopy study of the cracked surface of a self-healing epoxy specimen indicates that epoxide groups quickly react with hydrosulfide groups in the presence of the catalyst BDMA. About 50% epoxide groups are consumed within 30 minutes. Isothermal DSC analysis of the model mixture of the healing agent also shows that the curing is an exothermal reaction, and the degree of cure increases with increasing temperature. At −10 °C, the exothermic peak becomes almost inconspicuous as compared with that at 30 °C. It means the curing reaction proceeds very slowly at low temperature.

Temperature dependence of healing efficiency of the simulative specimens conforms to the reactivity study of the healing agent. Higher healing temperature led to higher healing efficiency. Figure 2.10 collects the results of crack healing in authentic specimens at different temperatures. The healing is very fast at temperatures from 10 to 30 °C as expected. The healing efficiency attains 82–88% after three hours at 20 and 30 °C, and exceeds 100% only after 12 hours. Although the self-repair process is slowed down with decreasing temperature, an 86% recovery at −10 °C is observed after 36 hours. The data are quite satisfactory and demonstrate that our technical route works.

Stability and durability are important for self-healing materials. As shown in the previous subsections, thermogravimetric and DSC tests reveal that no detectable change is found in the components of the healing agent (no matter whether they are encapsulated or not) when temperature is lower than 150 °C. This suggests that the microencapsulated healing agent can be handled at a temperature below 150 °C without any variation in its quality. Moreover, the two types of microcapsules possess long-term durability (Figures 2.3 and 2.5). The microcapsules are embedded in composite matrix in practical usage, which would further improve their storage stability. As a result, healing efficiency of the self-healing

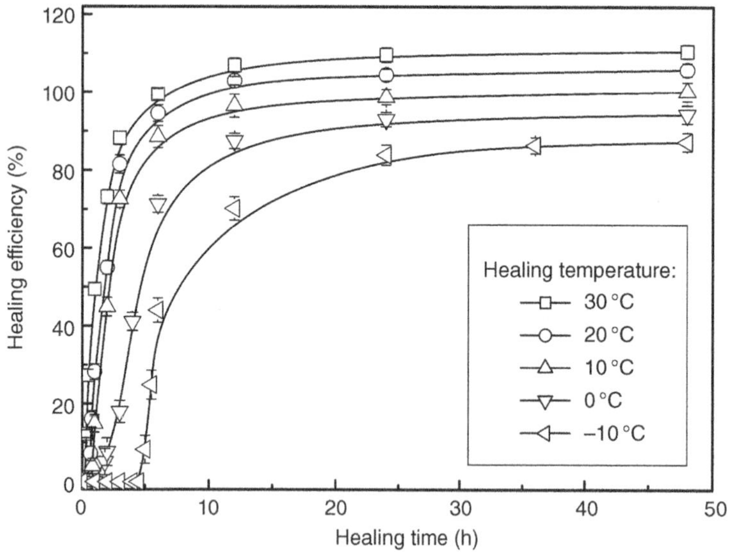

Figure 2.10 Time dependence of healing efficiency at different temperatures. The self-healing composite contains 2.5 wt% epoxy-loaded capsules and 2.5 wt% hardener-loaded capsules. *Source:* Reprinted from Yuan et al. [69]. Copyright 2008, with permission from American Chemical Society.

composite is almost constant within six months. On the other hand, as the freezing points of the components of the healing agent are very low (−51 °C for the epoxy and −68 °C for the hardener), it is expected that the self-healing system might take effect within a relatively wide temperature range.

2.4.2 Factors Related to Performance Improvement

The above discussion demonstrates that because the healing agent possesses high flowability and reactivity and belongs to the same family as the matrix polymer, crack healing is automatically conducted at or below room temperature, offering satisfactory repair effectiveness. Nevertheless, healant features, reactant ratios, and contacts between the components have not yet been optimized.

Therefore, a systematic study on the key issues that are closely related to the chemical reaction of the healant, including species of epoxy, mercaptan and catalyst, component proportions, and capsule size, is necessary [72]. The refined parameters would favor the healing chemistry and help promote future application of the healing system.

Accordingly, three types of epoxy resin (not only the aforementioned DTP and EPON 828 but also diglycidyl phthalate [DP]) were employed as the polymerizable component of the healing agent (Table 2.3), and three types of mercaptan including PETMP were used as the curing agent component of the healing agent (Table 2.4). In the

Table 2.3 Properties of epoxy resins.

Code	Chemical name	Epoxide value (mol/100 g)	Epoxy functionality	Viscosity (25 °C, Pa·s)
EPON 828	Diglycidyl ether of bisphenol A	0.51	2	12.5
DP	Diglycidyl phthalate	0.62	2	0.85
DTP	Diglycidyl tetrahydro-*o*-phthalate	0.63	2	0.53

Source: Reprinted from Yuan et al. [72]. Copyright 2009, with permission from Elsevier.

Table 2.4 Properties of mercaptan.

Code	Chemical name	SH content (w/w %)	SH-functionality	Viscosity (25 °C, Pa·s)
PETMP	Pentaerythritol tetra(3-mercaptopropionate)	26	4	0.45
TMP	Trimethylol-propane tri(3-mercaptopropionate)	24	3	0.12
C38	Tri(bi(2-hydrolpropyl) mercapto) aliphatic ether	11.8	3	1.5

Source: Reprinted from Yuan et al. [72]. Copyright 2009, with permission from Elsevier.

meantime, BDMA, 2,4,6-tris(dimethylaminomethyl)phenol (DMP-30) and triphenylphosphine (TPP) served as the catalysts. The descriptions of the corresponding microcapsules are listed in Tables 2.5 and 2.6.

Healing ability of the current self-healing composite is determined by the reaction between epoxy and mercaptan in the presence of catalyst. The reaction is closely related to activity, adhesivity, viscosity and mixing of the reactants. Therefore, reactivity of the healing system should be investigated at the beginning.

Catalysis of the epoxide–mercaptan reaction with alkali can be considered as a general base catalysis in which mercaptan first reacts with alkali to give the mercaptide ion, and then the ion reacts with epoxide. Alternatively, a nucleophilic catalysis occurs, during which alkali first reacts with epoxide to produce a reactive intermediate, and then the intermediate reacts with mercaptan in a nucleophilic displacement (Scheme 2.3) [26, 71].

As reflected by the kinetic parameters of the cure reaction between epoxy prepolymer and mercaptan and the adhesivity as well (Table 2.7), the alkalinity and content of the catalysts greatly affect the rate of the addition polymerization, curing degree, and bonding strength. The order of alkalinity of the catalysts is DMP-30 > BDMA > TPP [72], which coincides with the orders of reaction rate and adhesivity revealed in Table 2.7. That is, a catalyst with stronger alkalinity would accelerate the reaction and produce a better bonding effect.

It is worth noting that, in the case of a fixed formulation of DTP/PETMP/BDMA, with a rise of BDMA content, the reaction rate reaches a maximum value at 10.2%, while the bonding strength peaks at 3.6%. Clearly, the rate of healing and healing efficiency (here healing efficiency is supposed to be proportional to the bonding strength in the case of the same ingredients but different contents of the healant) cannot be maximized at identical dosage of catalyst. The inconsistency should originate from the different dependence of reaction rate and epoxy network development on catalyst concentration. Accordingly, one has to choose between fast rehabilitation and good repair.

On the other hand, the data in Table 2.7 indicate that a further increase in BDMA content results in lower rate constant and tensile lap-shear strength. For the former, the tertiary amine at higher loading might be directly involved in the consolidation of epoxy (Scheme 2.3b) and hence the system viscosity is greatly raised, which in turn decreases the rate of reaction since the cure becomes controlled by diffusion. With respect to the latter, the reduction in

Table 2.5 Description of the epoxy-loaded microcapsules embedded in self-healing epoxy materials for performance optimization.

ID	Encapsulated epoxy	Average diameter (μm)	Density ($g\,cm^{-3}$)	Core content (wt%)
E_L	DTP	50.7	1.22	87.3
E_S	DTP	10.6	1.19	78.8
E_{DP}	DP	11.3	1.25	80.1
$E_{EPON\ 828}$	EPON 828	11.9	1.28	81.4

Source: Reprinted from Yuan et al. [72]. Copyright 2009, with permission from Elsevier.

Table 2.6 Description of the hardener-loaded microcapsules embedded in self-healing epoxy materials for performance optimization.

ID	Core content (wt%)		Average diameter (μm)	Density ($g\,cm^{-3}$)
	Mercaptan	Catalyst		
H_0	87.3 (PETMP)	0	52.3	1.28
H_B	87.1 (PETMP)	0.7 (BDMA)	52.3	1.27
H_D	86.9 (PETMP)	0.6 (DMP-30)	52.6	1.29
H_L	85.3 (PETMP)	7.2 (BDMA)	52.4	1.25
H_S	63.8 (PETMP)	5.4 (BDMA)	10.2	1.24
H_{TMP}	61.9 (TMP)	5.1 (BDMA)	9.8	1.21
H_{C38}	65.4 (C38)	3.8 (BDMA)	11.1	1.23

Source: Reprinted from Yuan et al. [72]. Copyright 2009, with permission from Elsevier.

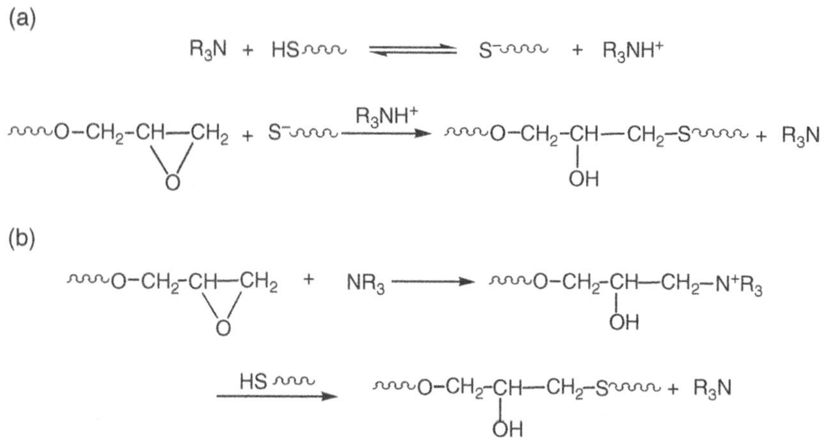

Scheme 2.3 Reactions between epoxide and mercaptan catalyzed by tertiary amine. (a) General base catalysis. (b) Nucleophilic catalysis.

Table 2.7 Curing kinetics (estimated from non-isothermal DSC studies) and adhesivity (characterized in terms of tensile lap-shear strength) of epoxy-mercaptan pairs at the stoichiometric ratios.

Formulation	Catalyst content (wt%)	E_a	k (×10³) at 25 °C	n	ln$A$	Average ΔH (J)	Tensile lap-shear strength (MPa)
DTP/PETMP/BDMA	0	–	–	–	–	–	0
DTP/PETMP/BDMA	1.9	48.29	0.238	0.898	12.23	428.25	2.31
DTP/PETMP/BDMA	3.6	45.57	0.720	0.891	11.14	478.43	4.95
DTP/PETMP/BDMA	7.0	47.19	0.933	0.895	12.06	455.63	2.64
DTP/PETMP/BDMA	10.2	42.55	1.084	0.885	10.34	420.67	1.87
DTP/PETMP/BDMA	13.1	47.17	0.967	0.895	12.09	402.80	1.15
DTP/PETMP/BDMA	18.5	49.94	0.851	0.900	13.08	381.50	0.78
DTP/PETMP/DMP-30	3.6	23.16	1.026	0.807	2.46	461.35	7.46
DTP/PETMP/TPP	3.6	50.45	0.127	0.893	11.38	452.10	4.58

A, pre-exponential factor; E_a, activation energy; ΔH: heat of reaction; k, reaction rate constant; and n, reaction order.
Source: Reprinted from Yuan et al. [72]. Copyright 2009, with permission from Elsevier.

adhesivity might be due to plasticization effect of the excessive catalyst.

To evaluate the role of catalysts in an authentic self-healing process, a series of hardener-loaded capsules carrying different catalysts were prepared (Table 2.6). In terms of infiltration technique, BDMA and DMP-30 were introduced into PETMP-loaded microcapsules. Besides, as TPP cannot be infiltrated into the capsules due to its immiscibility with mercaptan, the catalyst was directly added into the matrix cooperating with PETMP-loaded microcapsules (H_0). Figure 2.11 shows the time dependences of healing efficiency generated by different combinations of the capsules to highlight the influence of the catalysts. With the aid of strong alkaline BDMA or DMP-30, fast repair and high healing efficiency are observed. For weak alkaline TPP, however, slow repair and low healing efficiency are detected. In the case of similar catalyst content, the system with DMP-30 exhibits faster repair rate and higher healing efficiency than that with BDMA as a result of stronger alkalinity of the former. For a given catalyst (e.g. BDMA) with low content (<3.6%), an increase in its content also leads to higher repair rate and healing efficiency. The results agree with the aforementioned influential order of the three catalysts where the epoxide–mercaptan reaction rate is concerned [26, 71]. Evidently, the speed of repair and bonding strength of the healing membrane can be controlled by tuning catalyst species and concentration.

On the basis of the above analysis, simulative healing experiments were conducted to understand the highest healing effect that can be offered by the epoxy-mercaptan pairs (Figure 2.12). The amount of manually injected healing agent (0.03 ml) is enough to fill the cracks (crack volume ≈ 0.0035 ml). In general, in the case of authentic healing, the viscosity of the healing agent would hinder its delivery into cracks and the subsequent mixing, interdiffusion and solidification. Healing efficiency must be lower than that yielded by simulative tests. Here, a mixing efficiency defined as the healing efficiency ratio of authentic

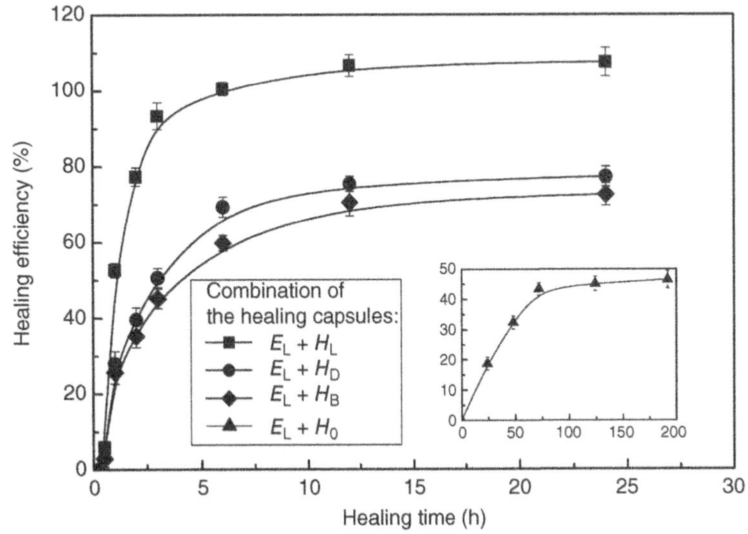

Figure 2.11 Influence of species and content of catalyst on healing efficiency of the self-healing composites containing different combinations of the microencapsulated healing agent. Total weight concentration of the embedded microcapsules is 5 wt% for all specimens, while weight ratios of epoxy- to hardener-loaded capsules follow the corresponding optimal values. When $E_L + H_O$ was employed, 5 wt% TPP was directly added into the matrix cooperating with H_O. Healing of the fractured specimens was conducted at 20 °C. *Source:* Reprinted from Yuan et al. [72]. Copyright 2009, with permission from Elsevier.

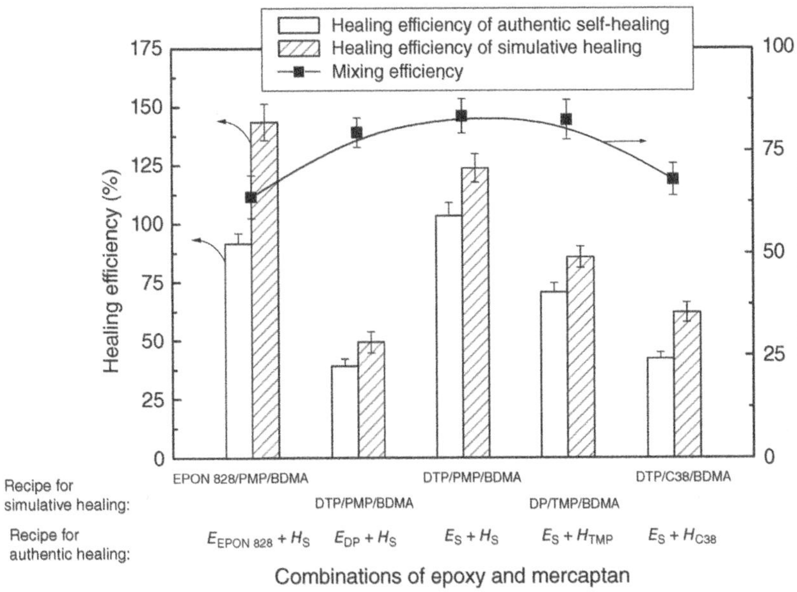

Figure 2.12 Influence of combinations of epoxy and mercaptan on healing efficiency. For authentic self-healing, total weight content of the embedded microcapsules is 5 wt%. For simulative self-healing, the recipes of the healing agent are the same as those formulated by the microcapsules in authentic samples. The ratio of epoxy to mercaptan follows the stoichiometric ratio. The catalyst content in the healing agent for simulative healing is equal to that for authentic self-healing (i.e. 3.6 wt%). Healing of the fractured self-healing was conducted at 20 °C for 24 hours. *Source:* Reprinted from Yuan et al. [72]. Copyright 2009, with permission from Elsevier.

healing to that of simulative healing is introduced. It may be possible to examine the performance of the healant from a comprehensive angle because the impact of its viscosity, bonding performance, and activity is included. As shown in Figure 2.12, the low viscosity epoxy DTP (Table 2.3) achieves the highest mixing efficiency (83.4%), as compared with DP (79.3%) and EPON 828 (63.7%). For mercaptan, the highest mixing efficiency is perceived in the system with PETMP (83.4%), forming a strike contrast to TMP (82.5%) and C38 (67.9%). It implies that hydrosulfide group content plays the leading role.

In short, chemical features of the healing agent are closely related to healing ability and rate of healing. The above discussion suggests that highly flowable epoxy prepolymer, high activity mercaptan and strong alkalinity tertiary amine are prerequisites for improving healing ability of the epoxy-mercaptan-based healant developed in the authors' laboratory. In other words, the healing reaction would be favored when the encapsulated epoxy prepolymer has rather low viscosity, mercaptan carries a sufficiently large amount of hydrosulfide groups, and the catalyst possesses strong alkalinity and/or suitable concentration.

As mentioned above, the healing reaction between epoxy and mercaptan belongs to addition polymerization. Accordingly, a stoichiometric ratio of epoxy/mercaptan and uniform mixing at molecule level are critical for a perfect cure [26, 69]. The size of capsules corresponds to the amount of releasable healing agent (i.e. the core content in Tables 2.5 and 2.6) and spreading area of healing agent delivered into cracked planes, so the embedded epoxy- and hardener-loaded microcapsules with mismatched sizes may lead to localized excess of one ingredient and deviation from the required stoichiometric ratio. Similarly, mismatched fractions of the microcapsules would also bring about the same result. Therefore, a careful study of this aspect should be conducted.

To quantify the aforementioned influence of matching of the capsule sizes and fractions, four combinations of the capsules (i.e. E_L+H_L, E_S+H_S, E_L+H_S, and E_S+H_L, see Tables 2.5 and 2.6) were incorporated into the epoxy composites and tested. The healing efficiency data indicate that the weight ratio of epoxy-/hardener-loaded capsules corresponding to the highest healing effect for each composite is a function of the combination of the capsules, and hardly varies with the concentration of capsules for a given composite (or a given combination of the capsules). These optimal ratios are 1:1, 1:1, 0.68:1, and 1.5:1 for the above four combinations, respectively. According to theoretical estimation [13, 71], the stoichiometric weight ratio for the present healant is 1.26, which can be equivalently converted to the weight ratios of epoxy-/hardener-loaded capsules of 1.18:1, 1.02:1, 0.4:1, and 2.33:1 for the four groups, respectively. Obviously, the value of the composite with E_S+H_S combination well agrees with the prediction, and that of the composite with E_L+H_L takes second place, while those of E_L+H_S and E_S+H_L exhibit severe deviation.

By quantifying the dependence of microcapsule concentration on healing efficiency of the above four groups of capsules at the optimal weight ratios of epoxy- and hardener-loaded capsules, it is found that the combinations of E_L+H_L and E_S+H_S offer higher healing efficiency than those of E_L+H_S and E_S+H_L within the entire concentration range of interest. The different healing effects evidence our analysis of the capsules with mismatched sizes. Even though the concentration of capsules is high enough, the inappropriate combinations of the microcapsules still failed to generate sufficiently high healing efficiency like the proper combinations. The deviation from the required stoichiometric ratio of healants on the fractured surfaces must take responsibility.

The above discussion reveals the importance of stoichiometric ratio of the healing agent as reflected by matching of the diameters of the healing capsules. It is worth noting that when proper combinations of the capsules with similar sizes are selected (e.g. E_L+H_L or E_S+H_S), homogeneous dispersion of the microcapsules should be realized. Otherwise, one could not make the most of the matched capsules.

For preparing the self-healing composites, microcapsules were first dispersed in viscous matrix epoxy prepolymer, then curing agent for the matrix was added, and the matrix started to solidify from gel to stiff material. Spatial distribution of the capsules in composites was fixed basically when the matrix was in the gel state. To homogenize the distribution, the following issues relating to the capsules should be satisfied. (i) Both epoxy- and hardener-loaded microcapsules have similar surface features, physical properties, and geometry. (ii) The microcapsules possess mono-dispersivity and do not adhere to each other. (iii) The microcapsules are robust enough to survive handling during composites manufacturing. Section 2.2 shows that the microcapsules can be made to meet these criteria by improving the encapsulation process. In addition to these conditions, the dispersion rate for mixing the microcapsules and matrix resin is also important.

The experimental results indicate that healing efficiency of the composites is dependent on the stirring speed. The former reaches the maximum level only at optimal dispersion rate. It is because the shearing force at low agitation rate is so weak that the microcapsules could not be well mixed with the matrix epoxy, while that at extremely high dispersion rate would distort or even break the microcapsules. These cases certainly change the actual ratio of the capsules despite being designed according to the stoichiometric ratio. Moreover, either epoxy or its hardener from the unintentionally fractured microcapsules would participate in the curing reaction of the epoxy matrix and change the properties of the composite.

Besides, the optimal dispersion rate for smaller microcapsules is found to be higher than that for larger ones. This should be attributed to the higher specific surface area of the smaller spheres, which leads to higher tendency of agglomeration. As a result, stronger shear must be used to separate them.

The last factor related to the performance of the healing agent involves the crack width. In general, crack healing includes the following stages: surface rearrangement, surface approach, wetting, diffusion, and randomization [73, 74]. Comparatively, the surface approach is considered the most important. No healing occurs if the surfaces are not brought together or the gap between the fractured planes is not filled with healing fluid. White et al. suggested that the healing effect could reach maximum level only when sufficient healing agent was available to entirely fill the crack [13].

To investigate the situation when dual capsules are included, we adjusted crack widths of the TDCB specimens by changing the groove lengths. The modified TDCB specimens incorporate crack-directing grooves that are reduced from 60.25 to 30, 39, 47 and 55 mm long, respectively. As a result, when the specimen was initially fractured, the virgin crack propagated only until it reached the end of the groove. The sample was then unloaded and set aside allowing self-healing to occur. The unfractured portion in front of the crack tip hindered the crack face from complete separation and ensured alignment of the two crack faces. As shown in Figure 2.13, the average crack face separations of the specimens near the loading pin holes are about 4.4, 12.5, 23.6 and 34.8 μm, respectively, which gradually decrease toward the crack tip.

Specimen compliance is a measure to distinctly identify the contribution of self-healing in TDCB tests [13]. If there was no healing effect, the second load–displacement curve would exhibit the same compliance established at the end of the virgin fracture test. When self-healing did occur, the healed load–displacement curve would show similar compliance varying tendency of the virgin fracture test. Therefore, for the TDCB specimens with short groove, both

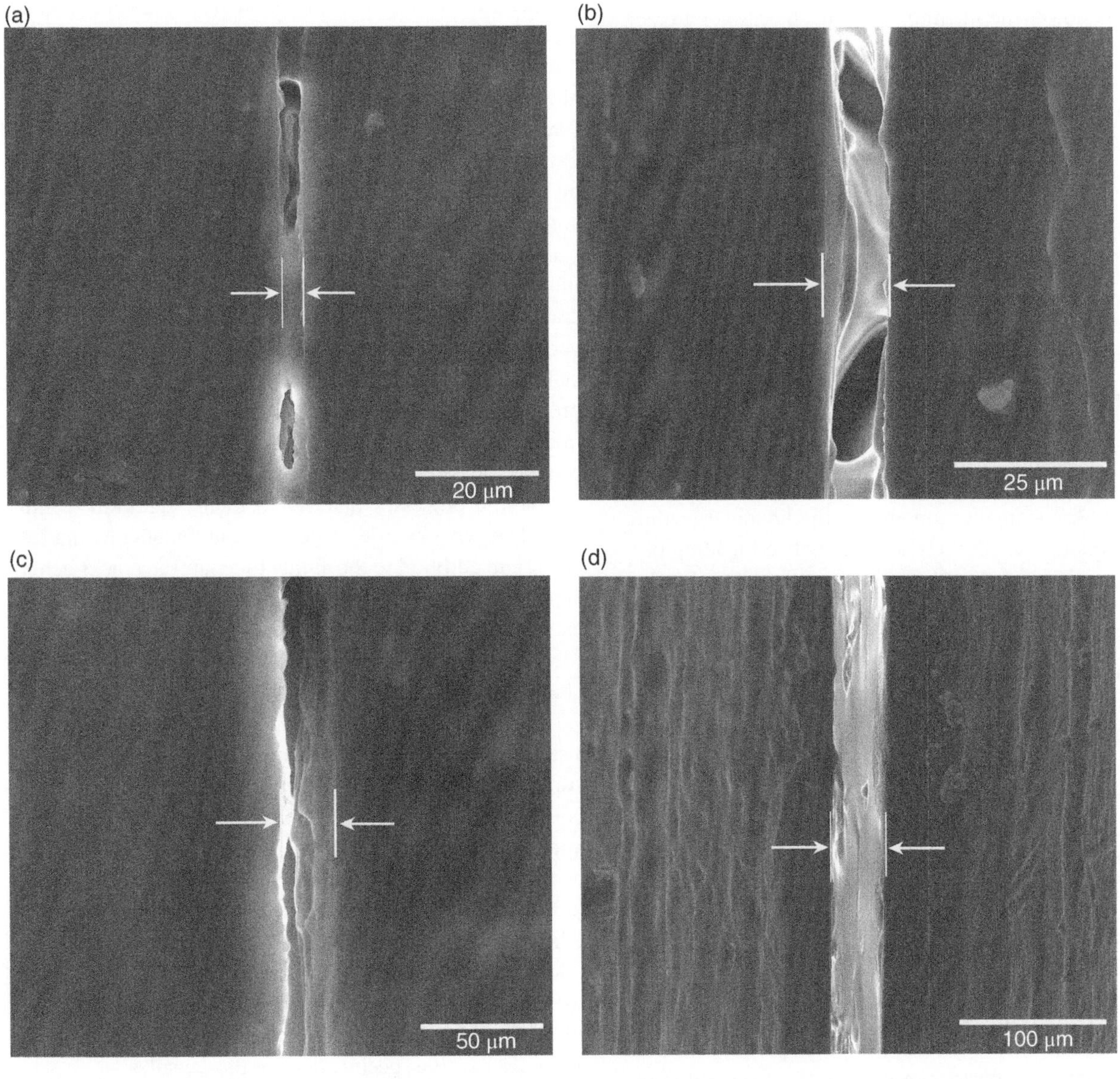

Figure 2.13 Scanning electronic microscopy images of healed cracks near the loading pin holes in the self-healing specimens containing 2.5 wt% epoxy-loaded capsules (E_S) and 2.5 wt% hardener-loaded capsules (H_S). Groove lengths of the TDCB specimens are: (a) 30 mm; (b) 39 mm; (c) 47 mm; and (d) 55 mm. *Source:* Yuan et al. [72], Figure 10, p. 9/with permission of Elsevier.

the virgin and healed fractures are comparable with those of standard TDCB specimens, and the peak loads appear right before crack propagation. That is, our modified TDCB specimens are qualified for evaluating self-healing ability of the composites.

We measured the influence of average crack width on healing effect in the case of different combinations of the epoxy- and hardener-loaded microcapsules and different capsule contents. The greatest healing efficiency is found to be higher than 100%. This is because the healed portions have higher cracking resistance than the matrix (Section 2.4.1). During TDCB tests, crack had to deviate from the original fracture plane. In addition, the data indicate that healing efficiency is very sensitive to the crack width. For the same formulation, the wider the crack, the lower the healing effect. In the case of a crack as wide as 34.8 μm (Figure 2.13d), the quantity of the released healing agent does not meet the need of continuous contact between two fracture planes. Consequently, little bonding effect is perceived. This is true especially when small diameter microcapsules with lower core content are employed. In other words, if the healing fluid only adhered to one surface, little or no healing would occur. This suggests that filling up crack volumes is necessary for crack repair.

On the whole, when the crack widths are greater (i.e. 12.5, 23.6, and 34.8 μm, Figure 2.13b, c, and d), the healing effect is proportional to the size of the capsules at constant capsule content, because the wider cracks need more healing agent to repair and larger microcapsules meet the requirement. For narrower crack width (e.g. 4.4 μm, Figure 2.13a), a smaller amount of the liberated healing agent is enough to fill the crack, so that self-healing performance easily reaches the highest value. The influence of size of capsules is insignificant accordingly. A careful observation reveals that the smaller microcapsules result in higher healing efficiency than the larger ones. This might be attributed to the fact that the tiny healing agent droplets from the small diameter microcapsules facilitate mixing and interdiffusion of the components of the healing agent.

Therefore, microcapsules with larger size and/or higher concentration are needed by larger cracks because a larger quantity of healing agent would be liberated to damaged portions. On the other hand, microcapsules with smaller size are adapted for healing smaller cracks, as the latter only require a small amount of healing agent. In this case, mixing and interdiffusion of the healant's components are critical. The above analysis also implies that the two-part healing agent is better to take effect at the crack initiation stage. When the cracks are developed into larger ones, healing efficiency has to be rather low or even zero.

2.4.3 Self-Healing Epoxy Materials with Embedded Dual Encapsulated Healant – Healing of Fatigue Crack

Fatigue and fatigue failure are critical for polymeric materials used in structural applications. Most polymeric materials suffer from poor fatigue resistance and would fail at stress levels much lower than the critical stress intensity, K_{IC}. Therefore, imparting self-healing capability to polymers and polymer composites should be an effective way to solve the problem [75, 76]. It is hoped that the fatigue cracks can be autonomously eliminated soon after their emergence.

So far, only a few papers of healing of fatigue failure have been reported in the literature for polymeric materials. Daniel and Kim investigated fatigue damage in asphalt by evaluating the changes in the stiffness gain under high temperature during the rest period [77]. Zako and Takano studied crack healing in notched epoxy composite specimens using tensile fatigue tests [25]. The stiffness could almost be fully recovered by application of heat to trigger flow and polymerization of embedded particles of B-staged epoxy resin. Nevertheless, the influence of healing process on neither crack growth rate nor absolute fatigue life was considered in their works [78].

Brown et al. first established a protocol to extend fatigue life of epoxy using DCPD [79]. Viscous flow of DCPD in the crack plane retarded crack growth, and its polymerized version further acted as a wedge at the crack tip for artificial crack closure. On the basis of this pilot research, they prepared self-healing epoxy with embedded DCPD-loaded microcapsules and particulate Grubbs catalyst, which was capable of responding to propagating fatigue cracks by autonomic processes that led to higher endurance limit and life extension, or even complete arrest of cracking [78, 80–82], in addition to the ability to repair the cracks generated by monotonic fracture [8]. The degree of fatigue life extension was found to be dependent on the relative magnitude of mechanical kinetics of crack propagation and chemical kinetics of healing.

In the following, the ability of our healing system consisting of encapsulated epoxy/mercaptan/tertiary amine to suppress and rehabilitate fatigue crack in epoxy materials is evaluated. The PMF-walled capsules containing epoxy and its hardener were made according to the methods described above [53, 67]. The epoxy-loaded microcapsules hold a 1:1 weight ratio mixture of EPON 828 and diglycidyl ether of resorcin (DGER), while the hardener-loaded microcapsules include PETMP and DMP-30. For comparative study, polythiol-loaded microcapsules containing only PETMP without amine catalyst were also synthesized. Table 2.8 lists the specifications of the capsules.

Unfilled epoxy specimens in the form of TDCB configuration with groove length of 55 mm [72] were cast from the

Table 2.8 Description of the microcapsules for making self-healing epoxy materials with extended fatigue life.

			Core substance	
Name	**Average diameter (μm)**	**Density ($g\,cm^{-3}$)**	**Ingredient**	**Content (wt%)**
Epoxy-loaded microcapsules	104.3	1.24	EPON 828	48.7
			DGER	48.7
Hardener-loaded microcapsules	102.1	1.23	PETMP	81.2
			DMP-30	15.7
Polythiol-loaded microcapsules	101.9	1.25	PETMP	97.1

mixture of 100 parts EPON 828 and 12.5 parts DETA, while the self-healing epoxy specimens of the same configuration were prepared by uniformly mixing 10 wt% epoxy-loaded microcapsules and 10 wt% hardener-loaded microcapsules with the aforementioned mixture of EPON 828 and DETA. Control specimens were fabricated with 10 wt% epoxy-loaded microcapsules and 10 wt% polythiol-loaded microcapsules (that exclude tertiary amine catalyst) to inhibit the healability of the healing agent, while maintaining other characteristics of the capsules. The compounds were degassed, poured into a closed silicone rubber mold, and cured for 24 hours at room temperature, followed by 48 hours at 40 °C. Table 2.9 shows the material properties.

Fatigue crack propagation behavior of the TDCB specimens was studied by a fatigue and endurance testing system at room temperature (24 ± 1 °C). The specimens were precracked (~2.5 mm) with a razor blade while ensuring the pre-crack tip was centered in the groove and immediately cyclically loaded. A triangular waveform of frequency 5 Hz was applied with a stress ratio, R, of 0.1 ($R = K_{min}/K_{max}$, where K_{min} and K_{max} denote the minimum and maximum values of the cyclic stress intensity, respectively). Fatigue cracks were grown within constant mode I stress intensity factor range, ΔK_I ($\Delta K_I = K_{max} - K_{min}$). Load line crack opening displacement (COD) was measured by a clip gauge. Mode I fatigue cracks were constrained along the centerline of the specimen because of the use of side grooves molded in the specimen. The optically measured crack tip position and specimen compliance were plotted against the number of cycles. The linear relationship between optically measured crack length and specimen compliance was used to calculate the crack tip position of the specimens during the experiment [78]. Each loading condition was investigated with continuous cyclic loading to specimen failure and/or with a rest period to allow for healing with stationary crack faces. In the latter case, cyclic loading was stopped after a small amount of crack growth and healing was allowed under different steady-state stress intensities for different times. In the case of manual infiltration, about 0.5 μl of pre-mixed epoxy and polythiol (keeping the stoichiometric ratio while excluding catalyst) was injected into the crack plane of unfilled epoxy specimen with a microsyringe after a crack growth increment of ~8 mm, and fatigue loading was not interrupted throughout the process [75].

The healing efficiency, λ, was defined by fatigue life extension [79]:

$$\lambda = \frac{N_{Healed} - N_{Control}}{N_{Control}} \times 100\% \tag{2.2}$$

Table 2.9 Properties of the specimens of cured epoxy and its composites for fatigue testing.

Properties	**Neat epoxy specimens**	**Control specimens**[a]	**Self-healing specimens**[b]
Density ($g\,cm^{-3}$)	1.172	1.182	1.186
K_{IC} ($MPa\,m^{1/2}$)	0.587 ± 0.016	0.692 ± 0.023	0.695 ± 0.038
Young's modulus (GPa)	3.7 ± 0.2	3.2 ± 0.3	3.2 ± 0.1
Paris power law exponent, n[c]	7.12	4.20	–
Paris power law constant, C_0[c]	0.11	4.3×10^{-4}	–

[a] The control specimens contain 10 wt% epoxy-loaded microcapsules and 10 wt% polythiol-loaded microcapsules (excluding tertiary amine catalyst).
[b] The self-healing specimens contain 10 wt% epoxy-loaded microcapsules and 10 wt% hardener-loaded microcapsules.
[c] n and C_0 were obtained according to eq. (3) in ref. [9].

where N_{Healed} and $N_{Control}$ denote the total number of cycles to failure of the self-healing specimen and that of a similar control specimen without healing, respectively. For each test, the result was an average of five specimens.

To evaluate original fracture toughness and development of healing efficiency of epoxy specimens, a natural pre-crack (~2.5 mm) was created on the TDCB specimens by inserting a fresh razor blade and gently tapping into the molded notch starter [9]. Subsequently, the specimen was pin loaded and tested with a universal testing machine under displacement control using a 3 mm min^{-1} displacement rate at room temperature. Specimens were fractured only to the end of the groove. Self-healing specimens were unloaded and left to cure for different times at room temperature. Finally, the healed specimens were tested again following the above procedure.

For successful in-situ self-healing, the healing agent released into the crack plane must have enough time to polymerize [78]. To facilitate the healing process of the cracks soon after their formation, fast consolidation of the repair chemicals is required. In-situ confocal Raman microscopy of the healing agent in the vicinity of crack tip of the self-healing specimen indicates that epoxide groups quickly react with hydrosulfide groups in the presence of catalyst DMP-30. About 70% epoxide groups are consumed within 15 minutes. This means that curing of epoxy healing agent can be completed within a very short time. Accordingly, the development of fracture toughness in self-healing specimens was measured through monotonic fracture tests performed at prescribed times following the initial virgin fracture. It is seen from Figure 2.14 that after an initial dwell period of about six minutes during which no appreciable healing is measured, the fracture toughness increases rapidly, tapering off after about 30 minutes. Moreover, time dependence of curing degree of the healing agents measured by DSC shows a similar exponential relationship. The slight difference between the two groups of data should come from (i) mixing uniformity of the healant ingredients under different circumstances, and (ii) the difference in measurement methods. Overall, it can be considered that the increase in fracture toughness keeps in step with degree of cure of the healing agent. The result well agrees with that reported by Brown et al. [78].

Clearly, the above results manifest that healing proceeds quite fast at room temperature, which should satisfy the prerequisite for in-situ repairing fatigue crack.

Incorporation of healing agent capsules into epoxy would certainly affect the latter's properties. As shown in Table 2.9 and Figure 2.14, the presence of 20 wt% microcapsules (i.e. 10 wt% epoxy-loaded capsules and 10 wt% polythiol-loaded capsules or hardener-loaded capsules) has raised the virgin monotonic fracture toughness of epoxy by up to ~18%. The toughening effect is also reflected by a change in fracture morphology from mirror-like of the unfilled epoxy to hackle markings of the filled capsules version. The result coincides with previous reports on a similar system, where a crack pinning mechanism was believed to be responsible [69, 72, 83]. Such an increase of a material's toughness would inevitably improve its fatigue performance because of the increased resistance to fatigue crack propagation.

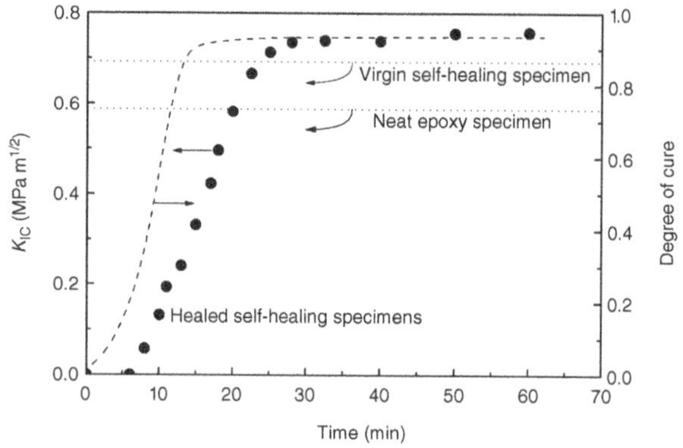

Figure 2.14 Time dependences of fracture toughness (•) developed in healed self-healing specimen and degree of cure (dash line) of the healing agent. The lower horizontal dotted line represents average K_{IC} of neat epoxy specimens and the upper horizontal dotted line represents that of virgin self-healing specimens. All the fracture toughness values of self-healing specimens were obtained from monotonic fracture tests. The data of degree of cure were estimated from isothermal polymerization of the healing agent conducted by DSC at 25 °C. The K_{IC} of virgin self-healing specimen is higher than that of neat epoxy specimen as a result of the toughening effect induced by the embedded microcapsules.

Additionally, under cyclic loading condition, the forces required to squeeze the fluidic healing agent (released from the broken capsules) out of the crack during unloading and to draw the fluid into the crack during loading would shield the crack tip [84]. This hydrodynamic pressure crack tip shielding mechanism can also improve the resistance to fatigue crack propagation. In this work, the effect is first revealed by manually injecting the mixture of epoxy prepolymer and polythiol at the stoichiometric ratio of 1.2:1 by weight (which is the same as the healing agent formulation except that the amine catalyst is excluded) into the crack plane of neat epoxy specimen without interrupting the fatigue experiment (Figure 2.15). The infiltrated liquid mixture can be considered to be chemically stable because epoxy–polythiol would hardly react with each other without proper catalysis. As a result, the subsequent crack propagation keeps steady on the whole like the case of submerged specimens [85]. Compared with the neat epoxy specimen, the infiltration greatly decreases the crack growth rate from 1.8×10^{-3} to 4.0×10^{-4} mm cycle^{-1} and

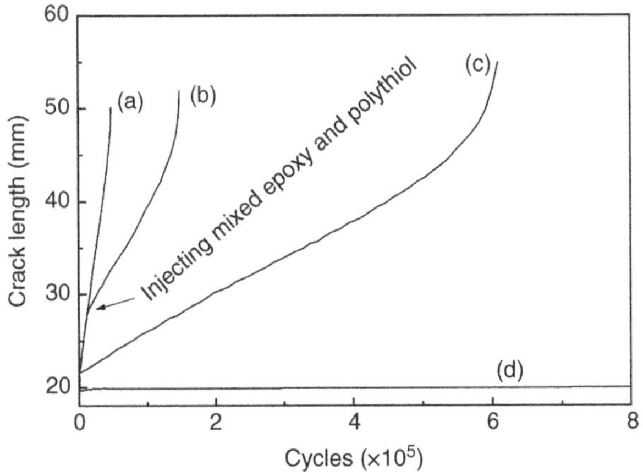

Figure 2.15 Crack length vs. fatigue cycle of (a) neat epoxy specimen, (b) manual infiltration specimen, (c) control specimen, and (d) self-healing specimen. The manual injection specimen was injected by the stoichiometric mixture of epoxy and polythiol excluding the amine catalyst in dynamic infiltration fashion. The testing parameters are $\Delta K_I = 0.504\,\text{MPa}\,\text{m}^{1/2}$, $K_{max} = 0.560\,\text{MPa}\,\text{m}^{1/2}$, $K_{min} = 0.056\,\text{MPa}\,\text{m}^{1/2}$, $R = 0.1$, and $f = 5\,\text{Hz}$.

increases the fatigue life by ~207% (cf. curves a and b in Figure 2.15). The results agree with those observed in the case of infiltration of epoxy prepolymer (or the hardener, i.e. mixture of PMP and DMP-30) alone [76]. Secondly, the dependence of crack length on fatigue cycle of control specimen with 10wt% epoxy-loaded capsules and 10wt% polythiol-loaded capsules (no tertiary amine catalyst is included) was measured (curve c in Figure 2.15). The crack growth rate is further reduced to $1.3 \times 10^{-4}\,\text{mm cycle}^{-1}$, and the fatigue life extension increases by ~1161% compared with neat epoxy specimen. This means that both microcapsule induced-toughening and hydrodynamic pressure crack tip shielding mechanisms have taken effect, with the former making a greater contribution to retardation of fatigue crack than the latter.

When the healing agent flowing out of the broken microcapsules contains catalyst and could rapidly react, polymeric wedge and adhesive bonding mechanisms [69, 72, 76] are involved, which can greatly extend fatigue life of the material (see curve d in Figure 2.15). More details of this aspect will be discussed hereinafter.

Following the traditional crack closure concept [84, 86–92], it is known that if fatigue crack opening load is purposely increased at the crack tip of self-healing specimens by the aforementioned four mechanisms (i.e. microcapsule induced-toughening, hydrodynamic pressure crack tip shielding, polymeric wedge, and adhesive bonding), the effective stress intensity factor range, ΔK_{eff}, would be reduced accordingly:

$$\Delta K_{\text{eff}} = \Delta K_I - \Delta K_{\text{toughening}} - \Delta K_{\text{liquid}} - \Delta K_{\text{bonding}} - \Delta K_{\text{wedge}} \tag{2.3}$$

where $\Delta K_{\text{toughening}}$ denotes the stress intensity due to microcapsule-induced increase of matrix ductility, ΔK_{liquid} the crack-opening and crack-closure stress intensity from viscosity resistance of the liquid, $\Delta K_{\text{bonding}}$ the stress intensity due to the combined (tensile) stresses in adhesives across the crack faces, and ΔK_{wedge} the crack-closure stress intensity due to the wedge from adhesive gelling and hardening.

Considering the competition between polymerization kinetics and crack growth would be a major factor affecting successful healing, three levels of applied range of stress intensity were prescribed for the fatigue tests of self-healing specimens hereinafter: two high cycle fatigue cases and one low cycle fatigue case. The high cycle fatigue refers to the fatigue regime of low ΔK_I, relatively slow crack growth rate and longer fatigue life (even infinite life). Low cycle fatigue refers to the fatigue regime where K_{max} approaches K_{IC} and rapid crack growth causes specimen failure after a very short time.

For low ΔK_I, chemical kinetics of the healing agent dominates, the crack is fully arrested within a short time, and self-healing provides infinite fatigue life extension. In this regime, the specimen's fatigue life exceeds the time for the healing agent to gel and for quasi-static healing efficiency to develop [78]. For example, when $\Delta K_I = 0.504\,\text{MPa}\,\text{m}^{1/2}$ ($K_{max}/K_{IC} = 0.81$), self-healing specimens were precracked and immediately cyclically loaded. A typical plot of crack length vs. fatigue cycle is shown by curve d in Figure 2.15. The initial release of healing agent during pre-cracking first retards the crack growth and leads to a certain amount of crack regression (~2 mm). After a few cycles, the crack does not further progress within the time frame of the test (10^7 cycles) and hence is fully arrested. Accordingly, the healing efficiency is infinite. This behavior was repeatedly observed for all five specimens tested. In contrast, the precrack in the control specimen is more rapidly growing at a constant rate (see curve c in Figure 2.15).

When the self-healing specimen with fully arrested fatigue crack was fractured under static load without precrack, its fracture toughness was found to be $0.963\,\text{MPa}\,\text{m}^{1/2}$, which is much higher than that of virgin self-healing specimen ($0.695\,\text{MPa}\,\text{m}^{1/2}$). SEM observation revealed that the healing agent released during the precrack event has formed a partial polymer wedge at the crack tip. Cohesive failure of the cured healing agent is confirmed by micro-Raman and EDS analysis. Obviously uniform tearing traces appear on the fast fracture plane when the healing agent is in the rubbery state. This implies that stronger adhesion effect is obtained at lower ΔK_I, which favors fatigue crack arrest. Therefore, adhesive bonding and polymeric wedge shielding mechanisms play an important role to indefinitely extend fatigue life of the material in this regime.

With a further rise in ΔK_I, crack growth rate has to be greatly reduced but the crack cannot be fully arrested.

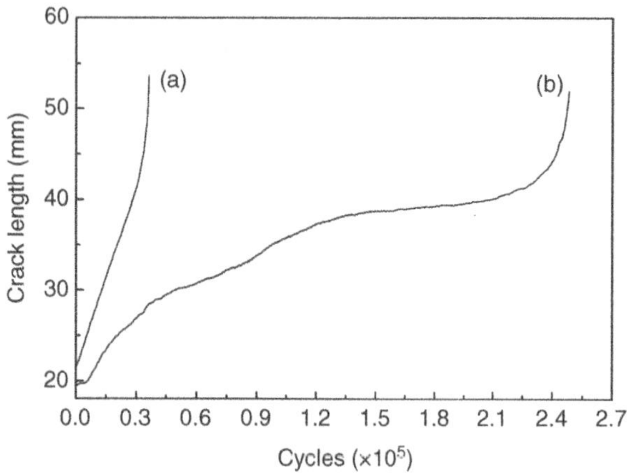

Figure 2.16 Crack length vs. fatigue cycles of (a) control specimen and (b) self-healing specimen. The testing parameters are $\Delta K_I = 0.554\,\text{MPa}\,\text{m}^{1/2}$, $K_{max} = 0.616\,\text{MPa}\,\text{m}^{1/2}$, $K_{min} = 0.062\,\text{MPa}\,\text{m}^{1/2}$, $R = 0.1$, and $f = 5\,\text{Hz}$.

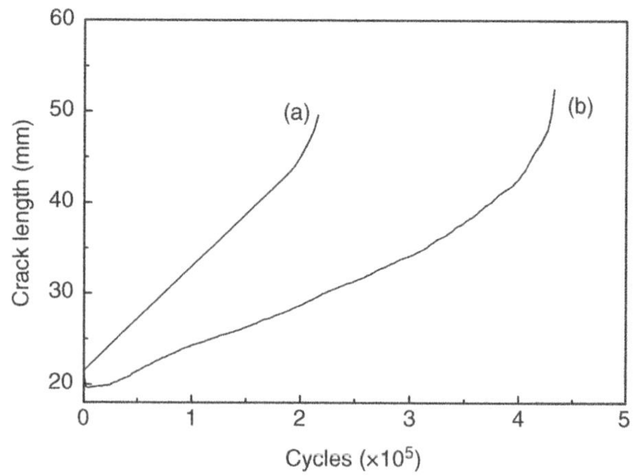

Figure 2.17 Crack length vs. fatigue cycles of (a) control specimen and (b) self-healing specimen. The testing parameters are $\Delta K_I = 0.605\,\text{MPa}\,\text{m}^{1/2}$, $K_{max} = 0.672\,\text{MPa}\,\text{m}^{1/2}$, $K_{min} = 0.067\,\text{MPa}\,\text{m}^{1/2}$, $R = 0.1$, and $f = 5\,\text{Hz}$.

In this regime, the self-healing specimen's fatigue life approaches the time for the healing agent to gel. For example, when $\Delta K_I = 0.554\,\text{MPa}\,\text{m}^{1/2}$ ($K_{max}/K_{IC} = 0.89$), the crack growth is retarded at the beginning (see curve b in Figure 2.16) in a way similar to curve d in Figure 2.15. Following this period of crack arrest, the crack eventually grows past the healed precrack at ~6×10^4 cycles. The subsequent crack growth behavior is a combination of steady growth and retardation, offering a healing efficiency of about 575%. In the course of fatigue crack propagation, as found by SEM, large quantities of epoxy healing agent were squeezed into thin cakes densely covering the fatigue fracture surface, and some were extruded into tiny balls filling in the concaves left by the broken microcapsules. Besides cohesive failure of the cured healing agent membranes, adhesively bonded epoxy from the opposite fracture plane is also discovered in some regions. Both result in retardation of crack growth (Figure 2.16). Evidently, adhesive bonding mechanism governs fatigue life extension in this regime.

Under high ΔK_I (i.e. when K_{max} approaches K_{IC}), the mechanical kinetics of fatigue crack growth is extremely fast and the healing system does not have sufficient time to inhibit the crack propagation. This causes specimen failure within a very short time. In this regime, the specimen fatigue life lags behind the time for the healing agent to gel. For example, when $\Delta K_I = 0.605\,\text{MPa}\,\text{m}^{1/2}$ ($K_{max}/K_{IC} = 0.97$), the fatigue crack in the self-healing specimen shows a constant growth rate with occasional crack retardation (see curve b in Figure 2.17). The healing efficiency is estimated to be about 89%. On the basis of these results, we know that when the mechanical kinetics of fatigue crack growth becomes the controlling factor, the liberated healing agent lacks enough time to produce strong adhesion to inhibit the crack advance.

It is worth noting that, when ΔK_I further increases, as K_{max} overly approaches the fracture toughness of the polymer, K_{IC}, unstable fracture will occur. A summary of life extension values under different loading conditions is given in Figure 2.18. The lower the applied range of stress intensity, the more evident the retardation effect and the higher healing efficiency. This trend is fully in line with the simulative healing results by dynamic manual infiltration [76]. Especially when ΔK_I is not more than $0.504\,\text{MPa}\,\text{m}^{1/2}$, crack can hardly advance for more than 10^7 cycles as a result of gelation and solidification of the healing agent, and hence is considered to be fully arrested. In this context, the value of ΔK_I of $0.504\,\text{MPa}\,\text{m}^{1/2}$ represents a watershed between fatigue crack growth retardation and crack arrest. Compared with the control specimen, the mechanisms of adhesive bonding and polymeric wedge

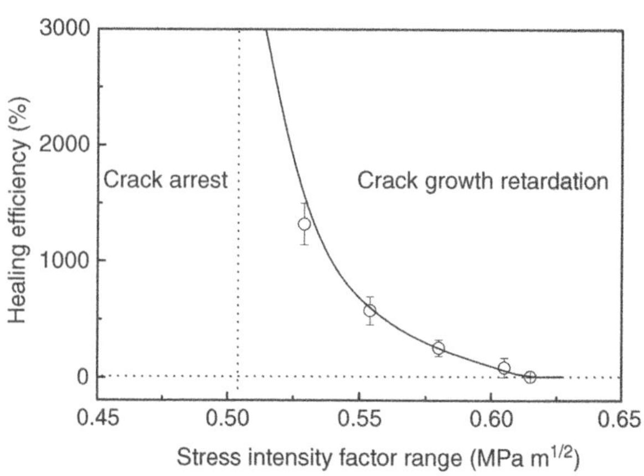

Figure 2.18 Healing efficiency of self-healing specimens as a function of stress intensity factor range, ΔK_I. The data are calculated partly using the results in Figures 2.15–2.17.

involved in crack tip shielding of self-healing specimens are highlighted, regardless of whether ΔK_I is high or low.

From the above expression of ΔK_{eff}, it is known that if the fatigue crack opening load is purposely increased by a wedge with adhesive properties at the crack tip, the effective stress intensity factor range would be reduced. Keeping this in mind, researchers have attempted to artificially introduce crack surface contact. Size and performance of the wedge have been found to be the most crucial factors affecting the crack closure effect. In this work, loading condition is investigated with a rest period to allow (i) healing to be conducted with stationary crack faces and (ii) the healing agent flowing out of the capsules to form a crosslinked epoxy wedge in the crack tip. Meanwhile, the wedge size is adjusted by changing the steady-state stress intensity required for holding the crack open after a small amount of crack growth, while the wedge performance is tuned by changing the steady-state healing time, during which the cyclic loading pauses and the crack is held open.

Figure 2.19 shows that the fatigue life extension due to precrack healing is drastically improved by adding a rest period. In the case of steady-state stress intensity = 0, loading is stopped after a small amount of crack growth and the specimen is allowed to heal unloaded for 10 minutes to ensure partial cure of the healing agent. The crack tip regresses to the approximate position of the TDCB notch as shown by curve b in Figure 2.19. Then, after about 4.1×10^4 cycles, the crack tip slowly progresses through the healed region to its location prior to healing. The crack growth rate of the self-healing specimens is lower than that of control specimens on the whole because the fracture toughness of the healing agent develops rapidly within a short period of time and exceeds that of the matrix as indicated in Figure 2.14. The fatigue healing efficiency for this case is around 33%.

Sharp et al. indicated that the load level or stress intensity applied for holding the crack open plays a decisive role in retarding or arresting fatigue crack [84]. Evidently, it is also valid for the current system. Fatigue crack growth is retarded at different steady-state stress intensities. Higher steady-state stress intensity applied during healing results in more prominent retardation effect (Figure 2.19) and higher healing efficiency (Figure 2.20). The results coincide with those obtained by the simulative healing through static manual infiltration [76]. When the applied stress intensity is not less than 0.336 MPa m$^{1/2}$ (for the steady-state healing time of 10 minutes), crack can hardly advance for more than 10^7 cycles and is considered to be completely arrested. Therefore, the applied stress intensity of 0.336 MPa m$^{1/2}$ acts as a threshold for distinguishing fatigue crack growth retardation from crack arrest (Figure 2.20). The above results can be understood by the fact that a thicker wedge is produced when the specimen is healed at higher applied steady-state stress intensity. Consequently, ΔK_{wedge} increases and ΔK_{eff} decreases. When $\Delta K_{eff} \leq \Delta K_{th}$ (i.e. $\Delta K_{bonding} + \Delta K_{wedge} \geq \Delta K_I - \Delta K_{th}$), fatigue crack is fully arrested.

Consider the specimen with fatigue crack arrested at steady-state stress intensity of 0.336 MPa m$^{1/2}$ as an example. The specimen is eventually fractured at static load without precrack and the fracture toughness reaches 1.089 MPa m$^{1/2}$, much higher than the value of the virgin self-healing specimen (0.695 MPa m$^{1/2}$). SEM observation indicated that the healing agent released during the

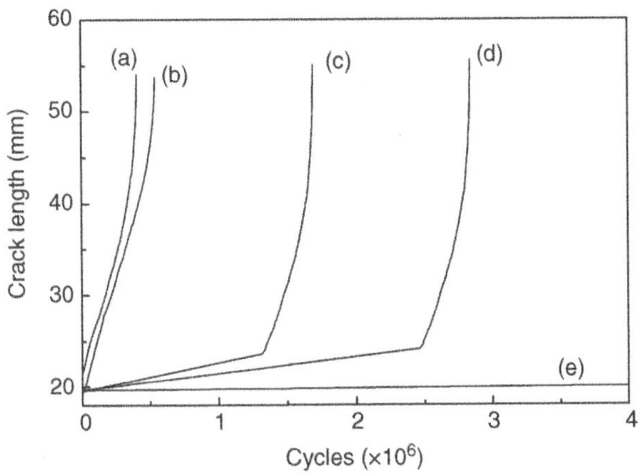

Figure 2.19 Crack length vs. fatigue cycles of self-healing specimens. The applied steady-state stress intensity for holding the crack open and steady-state healing time are (a) 0 MPa m$^{1/2}$, 0 minutes, (b) 0 MPa m$^{1/2}$, 10 minutes, (c) 0.336 MPa m$^{1/2}$, 9.5 minutes, (d) 0.168 MPa m$^{1/2}$, 10 minutes, and (e) 0.336 MPa m$^{1/2}$, 10 minutes. The testing parameters are $\Delta K_I = 0.605$ MPa m$^{1/2}$, $K_{max} = 0.672$ MPa m$^{1/2}$, $K_{min} = 0.067$ MPa m$^{1/2}$, $R = 0.1$, and $f = 5$ Hz.

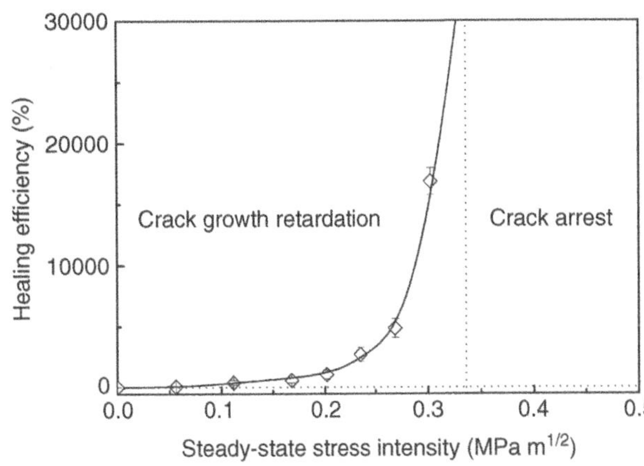

Figure 2.20 Healing efficiency of self-healing specimens with a rest period as a function of the applied steady-state stress intensity. The data are calculated partly using the results in Figure 2.19. The steady-state healing time is set at 10 minutes. The testing parameters are $\Delta K_I = 0.605$ MPa m$^{1/2}$, $K_{max} = 0.672$ MPa m$^{1/2}$, $K_{min} = 0.067$ MPa m$^{1/2}$, $R = 0.1$, and $f = 5$ Hz.

precrack event has formed a thick polymer wedge at the crack tip, leading to cohesive failure of the matrix in the ultimate fracture experiment. Very strong adhesion must have been achieved, which certainly favors crack arrest.

In addition to the above study on the effect of steady-state stress intensity that is related to wedge size, the influence of steady-state healing time that is related to wedge performance is evaluated in the following. For this purpose, the fatigue test paused after a small amount of crack growth and the crack was held open at a constant steady-state stress intensity of 0.336 MPa m$^{1/2}$. After different steady-state healing times from 0 to 1 hour, fatigue loading was reestablished.

The results shown in Figures 2.19 and 2.21 indicate that the healing effect is negligible (healing efficiency = 10–15%) within the steady-state healing time range of 0–8.5 minutes. When the steady-state healing time reaches 9.5 and 9.75 minutes, however, the healing efficiency is greatly increased to 415 and 1263%, respectively. In the case that the steady-state healing time is not less than 10 minutes, crack can hardly advance after more than 10^7 fatigue cycles and is considered to be fully arrested. Similarly, this time can be regarded as the critical threshold to distinguish fatigue crack growth retardation from crack arrest (Figure 2.21). The phenomena have been found in the simulative healing tests based on static manual infiltration [76], but the critical value of steady-state healing time becomes longer for the present in-situ system. This might be because the ingredients ratio, mixing uniformity and amount of the released healing agent in fatigue crack of self-healing specimens are not optimized as in the case of simulative healing [72, 76].

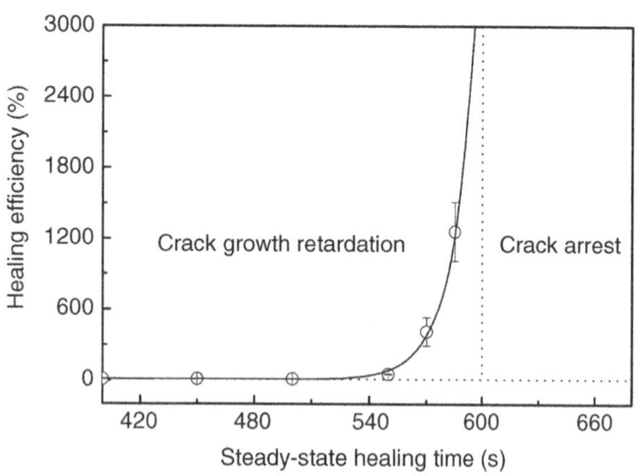

Figure 2.21 Healing efficiency of self-healing specimens with a rest period as a function of steady-state healing time. The data are calculated partly using the results in Figure 2.19. The applied steady-state stress intensity for holding crack open is 0.336 MPa m$^{1/2}$. The testing parameters are ΔK_I = 0.605 MPa m$^{1/2}$, K_{max} = 0.672 MPa m$^{1/2}$, K_{min} = 0.067 MPa m$^{1/2}$, R = 0.1, and f = 5 Hz.

The above results are closely related to the curing process of the healing agent and the development of its fracture toughness. During curing of the healing agent, the system transforms from low molecular liquid mixture into 3D cross-linked macromolecular networks. Molecular dynamics and macroscopic behavior of the material drastically change in the meantime [26]. Previous studies on curing kinetics demonstrated that the viscosity of the healing agent slowly increased in the beginning a few minutes after mixing, and then rapidly increased until it gelled within about one minute [69, 72]. The gelation of the healing agent is essentially a rapid formation and increase process of hardness and adhesive strength as shown in Figure 2.14. Although nearly no adhesion can be detected at a time of less than six minutes, fracture toughness of the healing agent sharply increases as of eight minutes and exceeds that of the epoxy matrix at about 20 minutes. With the rapid development of bonding strength of the cured healing agent, fatigue crack growth would become gradually difficult in the starting position. The deduction is confirmed by the fact that the longer the steady-state healing time corresponds to fewer tearing marks in the gel state appearing at the initial phase of the fatigue crack. Accordingly, evident retardation effect is obtained for the steady-state healing time from 8 to 9.75 minutes and the crack is fully arrested for the steady-state healing time as of 10 minutes.

It is concluded that the effects of microcapsule-induced toughening, hydrodynamic pressure crack tip shielding, polymeric wedge and adhesive bonding of the healing agent are responsible for the extension of fatigue life of the self-healing epoxy composites. Depending on the applied stress intensity range, ΔK_I, and the competition between polymerization kinetics of the healing agent and crack growth rate, the above mechanisms exert different influences on crack retardation.

2.4.4 Self-Healing Epoxy/Glass Fabric Composites with Embedded Dual Encapsulated Healant – Healing of Impact Damage

Although the epoxy-mercaptan healing agent has proved its applicability in self-repair of cured epoxy without manual intervention, whether it works for fiber reinforced composites remains open. In this context, woven glass fabric/epoxy composites containing the epoxy-mercaptan-based self-healing agent were manufactured and characterized to examine their ability of autonomic repair of impact damage hereinafter.

To produce woven glass fabric composites, EPON 828 was used as the matrix resin and DETA acted as the curing agent. Woven glass fabric (C-glass, 13×12 plain weave, 0.2 mm thick, 1000 tows, 200 g m^{-2}) played the role of reinforcement. The unfilled epoxy material was produced by compounding 100 parts EPON 828 with 12.5 parts DETA, while the self-healing epoxy was prepared by uniformly mixing prescribed concentrations of the encapsulated healing agent (Tables 2.10 and 2.11) with the aforementioned mixture of EPON 828 and DETA [93]. For carrying out a comparative study, the microcapsules containing epoxy and its hardener were divided into three groups according to their sizes (Tables 2.10 and 2.11): $E'_S+H'_S$, $E'_M+H'_M$, and $E'_L+H'_L$. Their weight ratios are fixed at 1.1:1 to achieve stoichiometric compositions of the ingredients. Composite laminates were manufactured by hand lay-up and compression molding at 40 °C for 24 hours. The laminates were 2.1 mm thick with the lay-up [45/0/−45/90]$_S$. The woven glass fabric volume fraction of the composites was 41 wt%.

The composites laminates were first impacted and then healed at room temperature. The healing capability was mainly assessed by the change in damage area using scanning acoustic microscopy (SAM) through-scan mode, while the recovery of compressive strength served as an additional validation.

It is seen from Figure 2.22 that when the laminates have been impacted by 1.5 and 3.5 J, the impact damage areas are gradually reduced. Clearly, the impact damage zones are filled by the epoxy healing agent, which are polymerized with time. The rate of damage area reduction for the low impact energy of 1.5 J attains 18% after 30 minutes, exceeds 85% only after six hours, and basically reaches equilibrium at 97% after 12 hours. Under higher impact energy of 3.5 J, however, the equilibrium rate of damage area reduction is about 52%, which is far from full recovery. Almost half of the damage area in the central region could not be repaired (Figure 2.22e–h). In contrast, for the control laminates only containing epoxy-capsules but excluding the hardener-capsules, the damage areas nearly do not change with time under otherwise identical conditions. When the control laminates only contained the hardener-capsules but excluded epoxy-capsules, the same results were observed. In association with the investigation of reactivity of the healing agent at room temperature (Section 2.2) and the recovery of compressive strength after impact as well (which will be shown in the following), the acoustic images are able to reflect the effect of damage healing in the laminates.

The different healing effects under different impact energies are factually related to the damage modes. At low impact energy, the predominant damage characterized by matrix cracking is on the microscopic scale. The released healing agent can easily reach the cracked places for rehabilitation. As for higher impact energy, the damage mode is dominated by delamination and fiber breaking. Repair of this damage has to be less effective on account of the following issues. (i) The healing agent is unable to re-connect the broken fibers. (ii) Delivery of healing agent to the delaminated sites deep inside the fiber-rich regions is difficult due to consumption throughout its journey. (iii) Damage zone volume on the back surface is too large to be filled by the released healing agent. In fact, the cracks are blocked by fiber debris even if lateral pressure is imposed. Consequently, the rate of damage area reduction has to be relatively low.

SEM observation of fracture surface of the composite laminates further shows that the microcapsules can be completely broken regardless of their sizes. The healing agent had flowed off, leaving perfect concaves (Figure 2.23). Clearly, the capsules have good integrity and survive the processing of the composites.

On the other hand, the architecture of the weave fibers provides the composites with large resin-rich regions as natural sites for storage of the microcapsules. Larger microcapsules, such as $E'_L+H'_L$ (~100 μm), are found by the interlacing of undulating warps (Figure 2.23a,b), but few inside fiber bundles because the narrow interstitial space prevents the capsules from entering. In contrast, the smaller capsules $E'_S+H'_S$ (~10 μm) are able to penetrate into the interior of fiber bundles (Figure 2.22d–f) where the larger capsules cannot. As a result, when microcracks are developed in the fiber-rich regions (e.g. fiber/matrix debonding), the healing agent from larger capsules has to be delivered by infiltration (Figure 2.22c), while such an indirect way of transportation is unnecessary for smaller capsules (Figure 2.22e,f).

Since healing of impact damage is quite complicated, the factors influencing the healing process should be analyzed. As demonstrated above, damage mode (dependent on impact energy) of self-healing laminates exerts a significant influence on the repair effect. Moreover, the amount of

Table 2.10 Description of the epoxy-loaded microcapsules for making self-healing epoxy/glass fabric composites[a].

ID	Average diameter (μm)	Density (g cm^{-3})	Core content (wt%)
E'_S	9.8	1.20	78.7
E'_M	50.9	1.25	87.2
E'_L	107.2	1.23	96.9

[a] The core epoxy in the capsules is a 1:1 weight ratio mixture of EPON 828 and DGER. To differentiate between the microcapsules in Table 2.5 and the present ones, the prime symbol (′) is added to the sample IDs.

Table 2.11 Description of the hardener-loaded microcapsules for making self-healing epoxy/glass fabric composites[a].

			Core content (wt%)	
ID	Average diameter (μm)	Density ($g\,cm^{-3}$)	Mercaptan	Catalyst
H'_S	10.2	1.24	73.6 (PETMP)	6.3 (BDMA)
H'_M	51.1	1.27	81.2 (PETMP)	6.9 (BDMA)
H'_L	103.7	1.28	90.1 (PETMP)	7.6 (BDMA)

[a] To differentiate between the microcapsules in Table 2.6 and the present ones, the prime symbol (′) is added to the sample IDs.

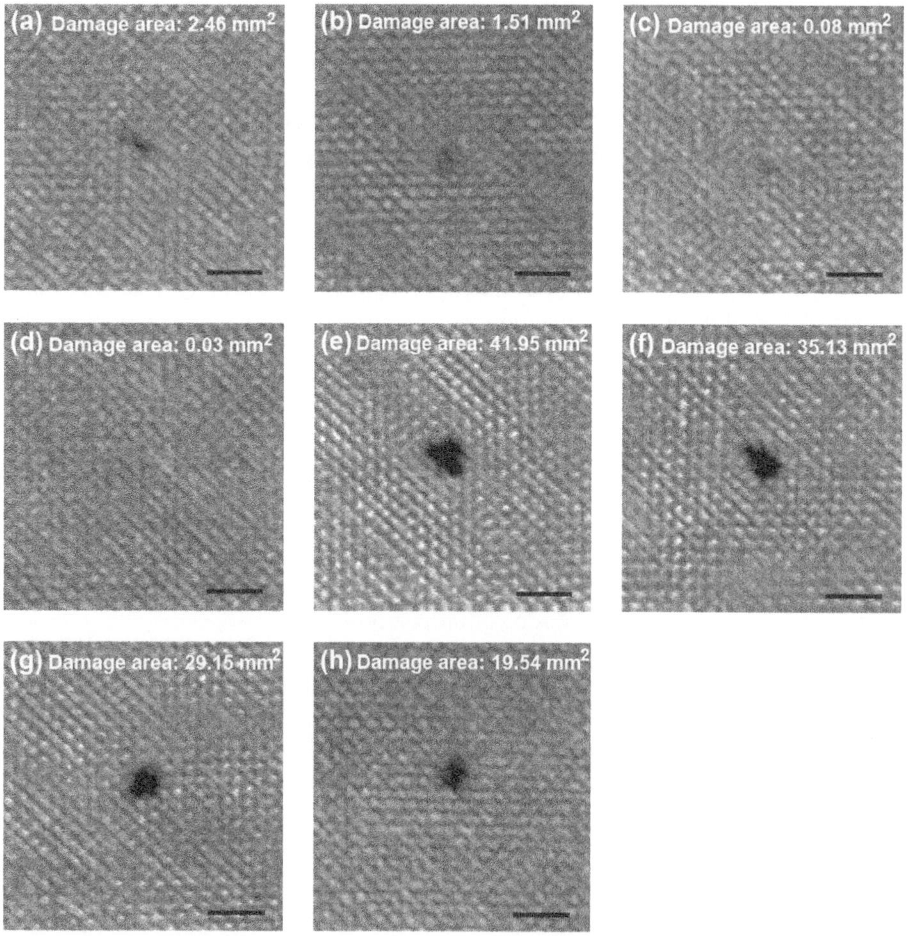

Figure 2.22 Through-scan mode ultrasonic images of the impacted self-healing laminates containing 11.9 wt% dual-capsules E'_L+H'_L, showing variation of damage areas with time at room temperature without manual intervention including lateral pressure. Healing times (1.5 J): (a) 0; (b) 1 hour; (c) 12 hours; and (d) 24 hours. Healing times (3.5 J): (e) 0; (f) 1 hour; (g) 2 hours; and (h) 24 hours. The scale bars represent 10 mm in length.

healing agent (related to capsule size and content) and damage zone volume (adjustable by applying compression to close the cracked gaps) are also closely related to repair efficiency. In order to have a comprehensive view of the effects of these factors on the rate of damage area reduction, experiments based on an orthographic factorial design of four factors (i.e. capsule size, capsule content, impact energy, and lateral pressure) and three levels was conducted (Table 2.12). The results indicate that the order of influence is: impact energy >> capsule size > lateral pressure > capsule content, according to the range of averaged rate of damage area reduction (Table 2.12).

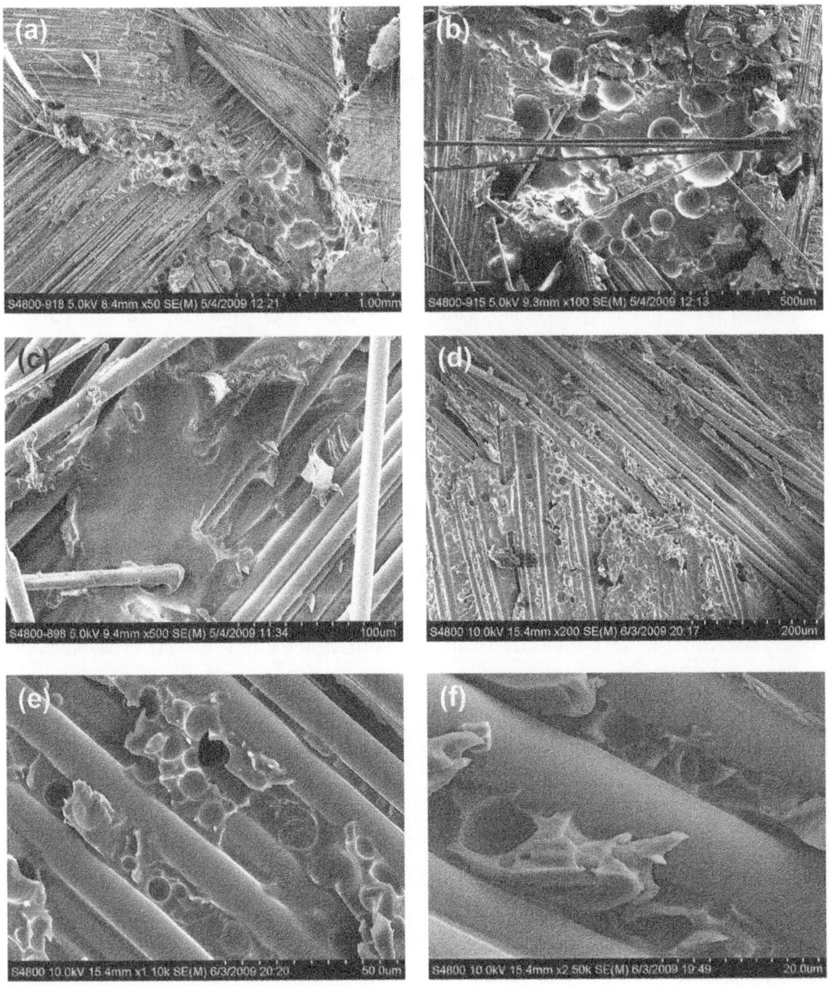

Figure 2.23 SEM micrographs of the fractured surfaces of the self-healing laminates (i.e. the protruded parts due to impact penetration). Healing agent: (a, b, c) E'_L+H'_L; and (d, e, f) E'_S+H'_S. Healing agent content: 11.9 wt%.

Table 2.12 Orthographic factorial design and results of four factors (capsule size, capsule content, impact energy, and lateral pressure) and three levels (1, 2, 3).

Recipe ID	Capsule size	Capsule content (wt%)	Impact energy (J)	Lateral pressure (kPa)	Virgin damage area (mm^2)	Healed damage area (mm^2)	Rate of damage area reduction (%)	η_{CAI} (%)
1#	1 (E'_S+H'_S)	1 (2.9)	1 (1.5)	1 (0)	1.35	0.28	79.26	86.47
2#	1 (E'_S+H'_S)	2 (5.9)	2 (2.5)	2 (60)	30.95	16.58	46.43	–
3#	1 (E'_S+H'_S)	3 (11.9)	3 (3.5)	3 (240)	47.04	26.51	43.64	–
4#	2 (E'_M+H'_M)	1 (2.9)	2 (2.5)	3 (240)	25.13	12.83	48.95	–
5#	2 (E'_M+H'_M)	2 (5.9)	3 (3.5)	1 (0)	43.31	27.24	37.10	51.94
6#	2 (E'_M+H'_M)	3 (11.9)	1 (1.5)	2 (60)	1.85	0.16	91.35	–
7#	3 (E'_L+H'_L)	1 (2.9)	3 (3.5)	2 (60)	40.84	23.15	43.32	–
8#	3 (E'_L+H'_L)	2 (5.9)	1 (1.5)	3 (240)	2.32	0.09	96.12	–
9#	3 (E'_L+H'_L)	3 (11.9)	2 (2.5)	1 (0)	24.53	11.61	52.67	65.12
Results analysis								
m_1 (%)	56.44	57.18	88.91	56.34				
m_2 (%)	59.13	59.88	49.35	60.37				
m_3 (%)	64.04	62.55	41.34	62.9				
R (%)	7.59	5.38	47.56	6.56				

m_1, m_2, m_3, averaged rate of damage area reduction corresponding to levels 1, 2 and 3; R, range of the averaged rate of damage area reduction.

Filling up crack volumes is a prerequisite to crack repair. Therefore, the amount of healing agent and damage zone volume are greatly important. Since the embedded microcapsules can only offer limited amount of healing agent and damage volume is largely dependent on damage mode, impact energy certainly plays the decisive role.

When lateral pressure is applied on the impacted laminates during healing, the apparently separated parts are forced to approach each other, so that the released healing agent can successfully fill the splits and the rate of damage area reduction is increased. This is true for repairing damage generated at higher impact energy. As for the case at low impact energy, the rate of damage area reduction is independent of the applied lateral pressure in the course of healing, because the size of microcracks in the matrix nearly cannot be reduced by the compression. On the basis of the above analysis, we further investigated the relationship among the rate of damage area reduction, capsule size, capsule content, and lateral pressure. As expected, larger capsules or higher content of capsules favors crack healing under higher lateral pressure. Evidently, a larger amount of healing agent ensures supply of sufficient binder to the narrowed cracks. Otherwise, in the case of zero lateral pressure, it roughly makes no difference whether the capsule size and content are changed.

Figure 2.24 gives a clear image of the dependence of rate of damage area reduction on capsule size measured in accordance with orthographic factorial design. With increasing the microcapsule diameter from ~10 to ~100 μm, gradual increase in the rate of damage area reduction is perceived. It is obviously correlated with the amount of healing agent. The larger capsules contain more healing agent (Tables 2.10 and 2.11), and their rupture will deliver more healing agent to the cracked surfaces as compared with the smaller ones [13]. The phenomenon agrees well with the dependence of rate of damage area reduction on capsule content (Figure 2.24). It is worth noticing although smaller capsules are able to settle themselves in fiber-rich regions, they cannot take effect as expected. Previously we believed that the capsules with smaller size would be dispersed inside and outside bundled fibers in composites, and the cracks can be more effectively repaired as compared with larger capsules. It seems not to be the case now. So long as cracks are initiated inside fiber bundles, their coalescence and development would lead to not only matrix cracking and debonding but also fiber rupture. The healing agent is powerless before the latter. As a result, the smaller capsules are not superior to the larger ones.

On the other hand, Table 2.12 also gives a few representative values of healing efficiency calculated on the basis of compression after impact (CAI) tests, η_{CAI}.

$$\eta_{CAI} = \frac{\sigma^{Healed} - \sigma^{Impact}}{\sigma^{Virgin} - \sigma^{Impact}} \times 100\% \tag{2.4}$$

where σ^{Healed} denotes the compressive strength of the impacted specimen after healing at room temperature for 24 hours, σ^{Impact} is the compressive strength of the impacted specimen, and σ^{Virgin} is the compressive strength of the virgin specimen.

The results demonstrate that true mechanical recovery has been achieved. The polymerized product of the infiltrated fluidic healing agent must have sufficient adhesiveness and strength, so that it can rebind the cracks and recover the load-bearing capacity of the laminates to a

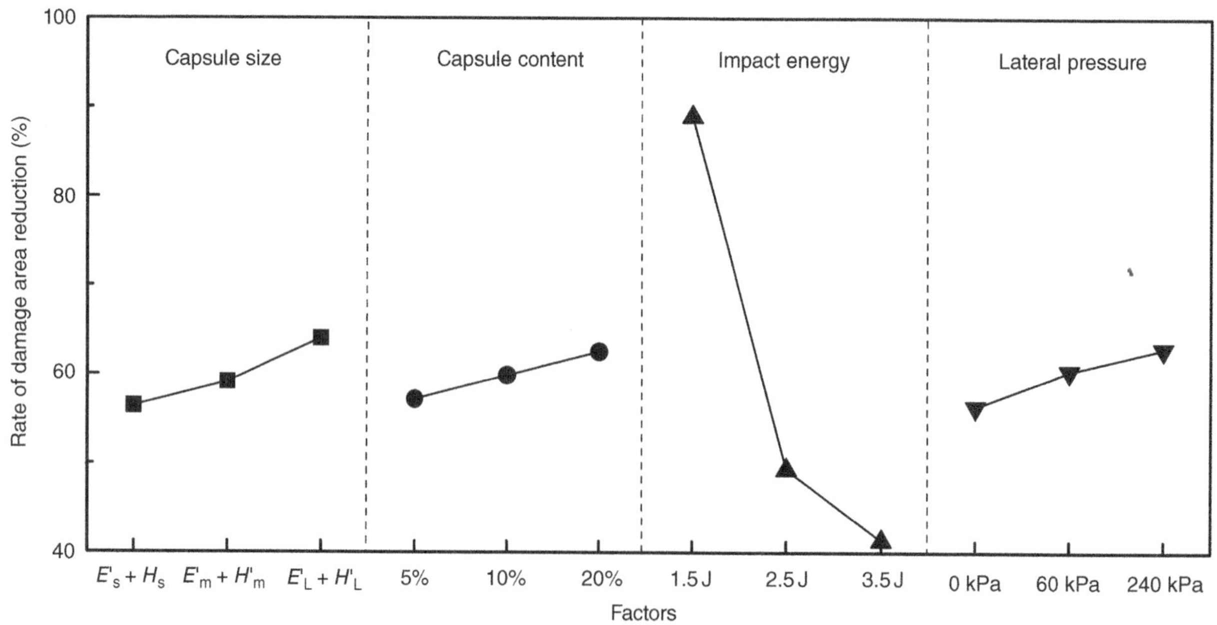

Figure 2.24 Experimental results of the orthographic factorial design.

certain extent. Although the healing efficiency η_{CAI} is generally higher than the values of the rate of damage area reduction for the same laminates (probably because of different characterization methods and definitions as well as other unknown reasons), both η_{CAI} and rate of damage area reduction follow the same regularity. Therefore, it is known that the variation in damage area determined by the ultrasonic non-destructive evaluation technique is closely correlated to mechanical performance recovery. The latter used to act as a measure of healing efficiency of self-healing materials.

2.4.5 Self-Healing Epoxy/Glass Fabric Composites with Self-Pressurized Healing System

Hereinafter the fiber composites with embedded PP tubes containing self-pressurized healing agent are evaluated. To prepare the composite laminates, the woven glass fabric (120 mm × 150 mm) was impregnated in the mixture of EPON 828, tetramethylene pentamine (TEPA) and acrylonitrile. Besides, the prepregs of healing agent loaded PP tubes (Figure 2.6c) were made through impregnation in the mixture of EPON 828 (100 parts) and its hardener (21 parts, consisting of acrylonitrile and TEPA at a molar ratio = 0.7) by hand lay-up. In each prepreg, the tube with epoxy and that with the curing agent were alternately arranged to promote the future mixing and reaction of healing agents after release. Next, eight plies of the glass prepregs with single ply of the PP tube prepreg in the middle (Figure 2.6d) were degassed for 15 minutes, compressed to keep the laminates thickness of 2.5 mm, and cured at 25 °C for 24 hours and 40 °C for 24 hours. The volume fraction of the glass fiber in the composites was estimated to be 41%. To generate higher internal pressure in the PP tubes, the composite laminates were heated to 70 °C for four or eight hours allowing gasification of the blowing agent. Specimens (120 mm × 25 mm × 2.5 mm) were cut from the laminates along with the longitudinal direction of the PP tubes for further mechanical tests. Each specimen contains 6 vol% healing agent (tube spacing = 750 μm).

The control laminates that do not contain any PP tube prepreg were manufactured under the same conditions.

Since foaming agent is incorporated with the healing agent to produce high internal pressure in the PP tubes, it is worth noting whether the reactivity of the healing agent would be changed after addition of the foaming agent. In this context, curing kinetics of the healing agent (epoxy-mercaptan pair) containing the foaming agent should be measured. For this purpose, mixtures of epoxy (DTP)/foaming agent (ABVN, 100 parts/3 parts) and hardener (PETMP)/ABVN (100 parts/3 parts) were pretreated by heating at 70 °C for eight hours to simulate the foaming process. Afterwards, the pretreated DTP/ABVN and PETMP/ABVN were quickly mixed with BDMA (0.4 parts) and then tested by DSC under different heating rates (Figure 2.25). For comparison, non-isothermal curing kinetics of the mixture of DTP (100 parts)/PETMP (100 parts)/BDMA (0.4 parts), which did not contain any foaming agent, was also measured (Figure 2.26). As listed in Table 2.13, the systems with and without the foaming agent exhibit similar curing activation energy and reaction order. Evidently, the foaming agent, foaming process and remainder of foaming

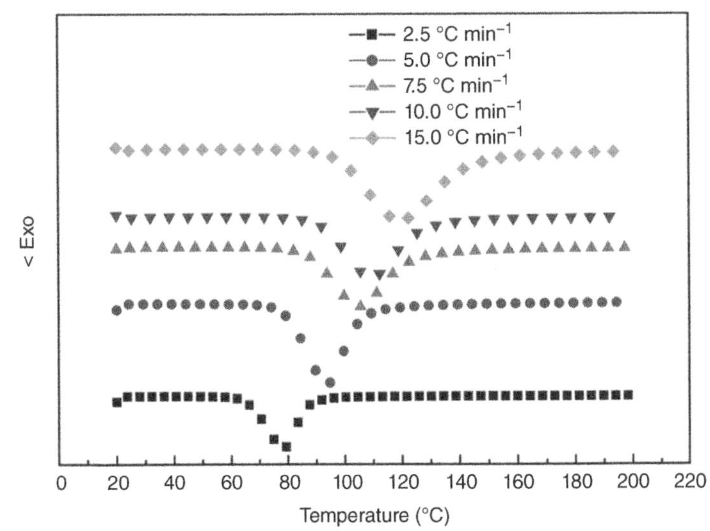

Figure 2.25 Non-isothermal DSC scanning curves of the mixture of 100 parts DTHP (with 3 parts ABVN), 100 parts PETMP (with 3 parts ABVN) and 0.4 parts NBDMA collected at various heating rates (2.5, 5, 7.5, 10, and 15 °C min^{-1}) in N_2. The DTHP with ABVN and PETMP with ABVN were preheated at 70 °C for 8.0 hours before mixing.

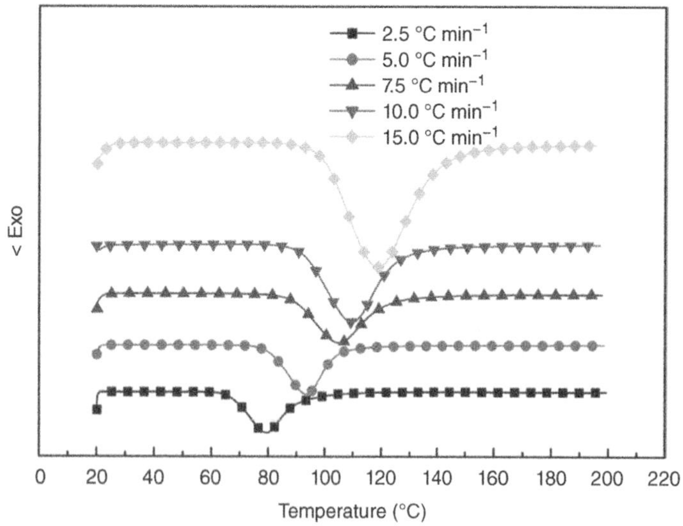

Figure 2.26 Non-isothermal DSC scanning curves of the mixture of 100 parts DTHP, 100 parts PETMP and 0.4 parts NBDMA at various heating rates.

agent almost do not affect curing of the healing agent because of absence of chemical interaction between the foaming agent and healant.

On the basis of the above study, the influence of incorporation of healing agent-loaded PP tubes on mechanical properties of the fiber composites is further examined. As non-polar heterogenous substances in the laminates, the tubes would most likely weaken the interfacial interaction, despite the fact that UV pretreatment might somewhat increase surface polarity. It is interesting to see from Figure 2.27a that the composite containing the PP tubes has similar deformation behavior as the one without the tubes. Moreover, flexural strengths of the composites are only slightly reduced after insertion of the tubes (Figure 2.27b). For example, the highest decrease with a rate of decay of 10.6% is observed in the case of tube spacing of 500 μm. This may be due to the fact that the overall content of the PP tubes is relatively low for the present range of tube spacing of interest, as compared with previous works [6, 43, 45, 48].

According to the data in Figure 2.28, the composite with healing agent-loaded PP tubes spaced 750 μm apart is used for the following characterization unless otherwise specified, because it has the greatest retention of strength.

To assess healability of the system, the composite laminates were damaged by indentation, healed, and evaluated by four-point bend flexural test according to the protocol

Table 2.13 Non-isothermal curing kinetics of the healing agent with and without the foaming agent[a].

Foaming agent	φ (°C min^{-1})	T_i (°C)	T_p (°C)	T_f (°C)	ΔH (J g^{-1})	Characteristic parameters
Excluded	2.5	66.1	79.6	93.6	439.4	$\ln A = 5.08$
	5.0	79.6	93.8	105.5	433.0	$E_a = 46.6\,\text{kJ mol}^{-1}$
	7.5	88.1	104.6	122.9	433.8	$n = 0.88$
	10	91.6	109.1	126.0	438.9	
	15	98.3	118.0	141.8	433.0	
Included	2.5	65.7	78.5	87.5	445.4	$\ln A = 4.57$
	5.0	78.9	93.9	106.4	436.2	$E_a = 45.1\,\text{kJ mol}^{-1}$
	7.5	86.7	104.9	124.3	437.1	$n = 0.89$
	10	91.1	109.0	127.3	431.3	
	15	97.5	119.1	144.8	416.9	

[a] A, pre-exponential factor; E_a, activation energy; ΔH, heat of reaction; n, reaction order; T_f, final temperature of the exothermic peak; T_i, initial temperature of the exothermic peak; T_p, peak exothermic temperature; φ, heating rate.

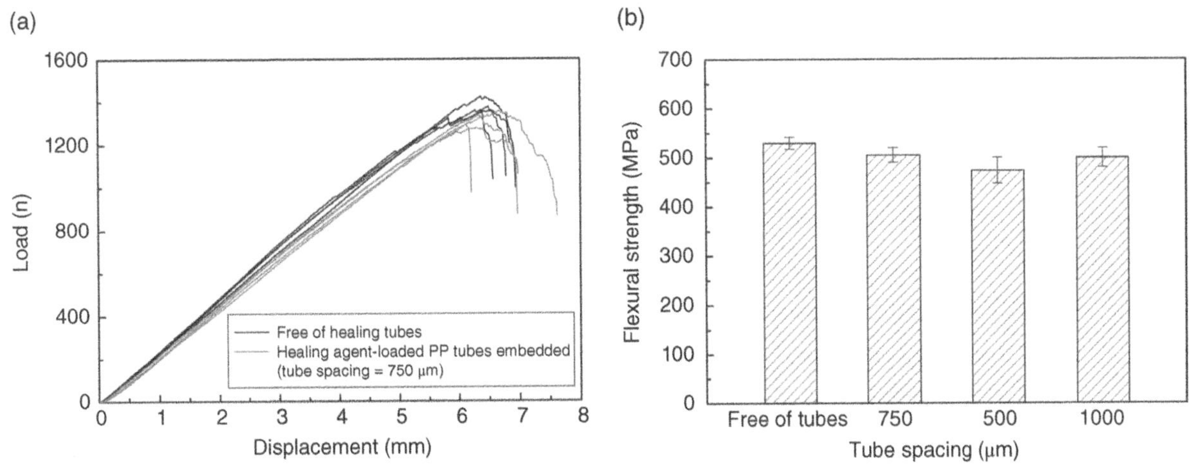

Figure 2.27 (a) Load–displacement curves and (b) flexural strengths of the composite laminates with and without the healing agent-loaded PP tubes measured by four-point bend tests. *Source:* Reprinted from Zhu et al. [52]. Copyright 2016, with permission from Elsevier.

of Bond and co-workers [6, 43, 48] as in the following. The specimen with back face supported by a steel ring (inner diameter 20 mm, outer diameter 35 mm, depth 10 mm) was indented to the displacement of 1.8 mm by a steel hemispherical end (diameter 5 mm) at a crosshead speed of 3 mm min^{-1}. After indentation, the specimen was allowed to heal the internal cracks at room temperature for a period of time without additional lateral pressure. A four-point bend flexural test was conducted at a crosshead rate of 4.0 mm min^{-1} based on ASTM D6272-02. Five specimens were tested for each recipe to get the average value. Healing efficiency was determined by the strength ratio of the healed specimen to the virgin specimen. Typical load–displacement curves of the virgin, damaged and healed laminates containing the PP tubes are shown in Figure 2.28. All the mechanical tests were carried out by a WD-5A universal mechanical testing machine. It is noted that the epoxy healing system not only heals the cracks but also provides the composite with higher displacement, which should be ascribed to the higher ductility of the cured healing material [69].

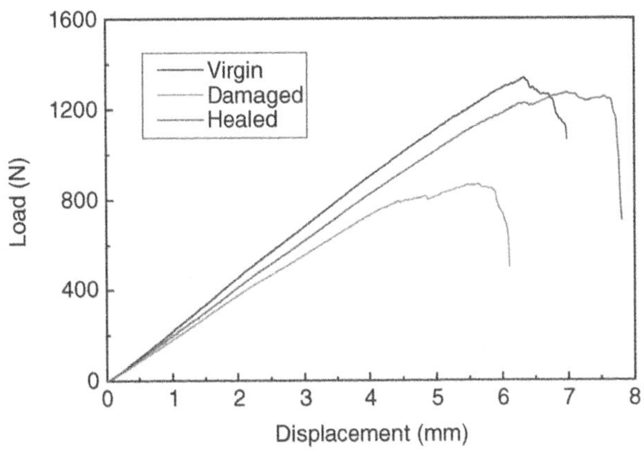

Figure 2.28 Typical load–displacement curves of the composite laminates with the healing agent-loaded PP tubes measured by four-point bend tests (tube spacing: 750 μm; foaming agent content: 3 parts; foaming condition: 70 °C, eight hours; healing condition: 25 °C, seven days).

As mentioned above, the composite laminates have to be damaged by indentation prior to healing. The load–displacement curves in Figure 2.29a indicate that with a rise in the indentation depth, the indentation load gradually increases and then decreases. There is no sudden drop of load on the upward slope. It means that delamination and matrix cracking, rather than glass fiber breakage, predominate the failure process at this stage. Although the fracture of PP tubes is not highlighted in this case, it can be reasonably deduced that they must have ruptured together with the surrounding matrix as a result of UV light pretreatment. In this context, the maximum indentation displacement at 1.8 mm is appropriate. Otherwise, plenty of glass fibers would be broken, which cannot be reconnected by the present healing agent, and the results have to diverge from the theme of this study.

In the meantime, model experiments were conducted, showing the increase of pressure and its effect on outflow of the healing agent due to the activated foaming agent. As shown in Figure 2.30a, PP tubes, which are filled with the mixture of epoxy monomer (DTP) and foaming agent (ABVN) and sealed in advance, are embedded in cured epoxy that has the same formulation as the authentic

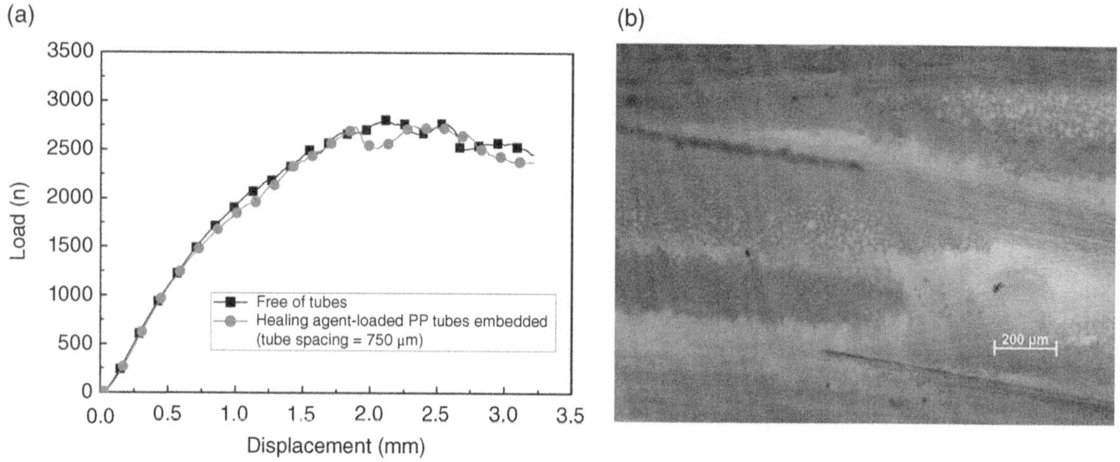

Figure 2.29 (a) Typical load–displacement curves recorded during indentation on the composite laminates with and without the healing agent-loaded PP tubes. (b) Fluorescent microscopy photo of cross-section of the composite laminates with the healing agent-loaded PP tubes. The composite was damaged by indentation. To visualize the microstructure, Rhodamine B (0.2 parts) was introduced in the matrix of the specimen, while another fluorescent penetrant, Ardrox 985 (0.2 parts), was added into the hardener component of the healing agent. The two chemicals emitted orange and purplish blue glows under light illumination of 525 and 365 nm, respectively. Tube spacing: 750 μm; foaming agent content: 3 parts; foaming condition: 70 °C, eight hours. *Source:* (a) Reprinted from Zhu et al. [52]. Copyright 2016, with permission from Elsevier. (b) Zhu et al. [52], Figure 3, p. 4/with permission of Elsevier.

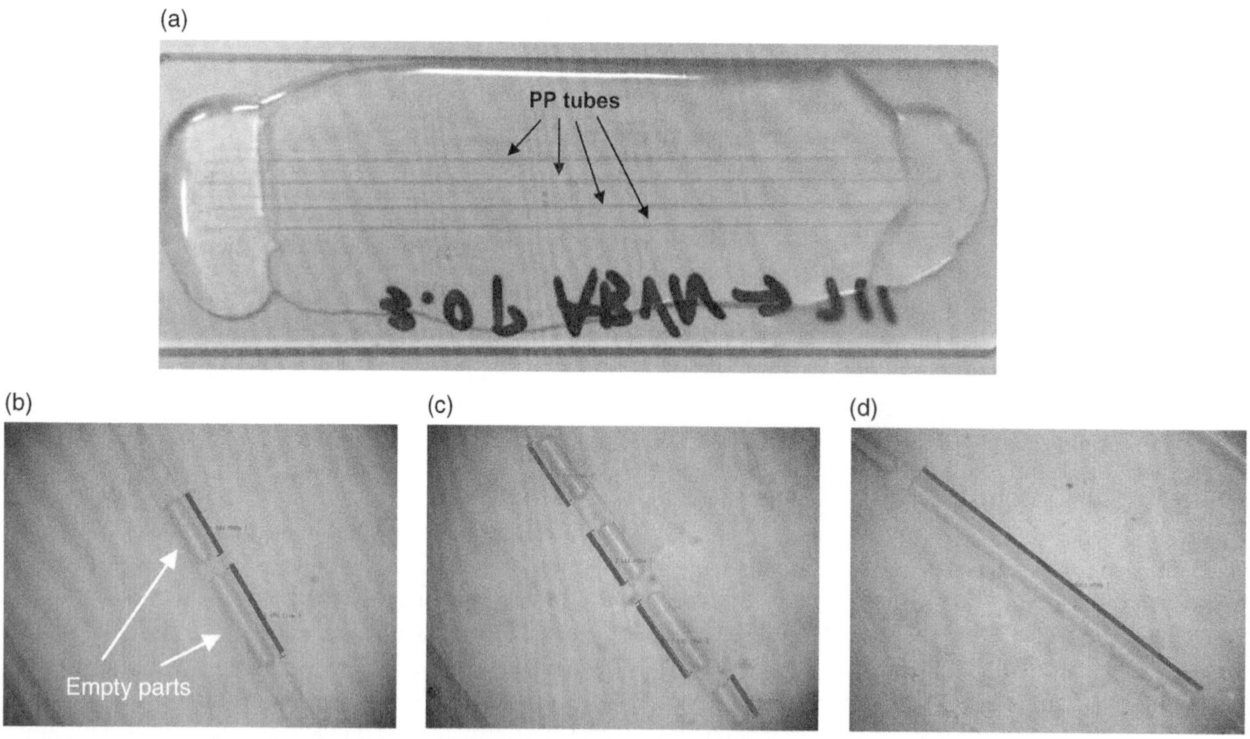

Figure 2.30 (a) Sealed PP tubes containing the mixture of epoxy monomer DTHP (100 parts) and foaming agent (ABVN) are embedded in cured epoxy (the mixture of EPON 828 [100 parts] and its hardener [21 parts, consisting of acrylonitrile and TEPA at a molar ratio = 0.7]) on a glass slide. Average thickness of the matrix epoxy: 1.2 mm. (b–d) The model composites like (a) with different ABVN contents ((b) 1.5 parts, (c) 3 parts, and (d) 4.5 parts) are heated to 70 °C for eight hours allowing gasification of the blowing agent and then sawed in an end region. Most healing agent flows out of the cuts, while some remain in the tubes (the empty parts are labeled). All photos were taken using a digital microscope (Keyence VHX-1000c).

composite matrix used in this work (i.e. the mixture of EPON 828, acrylonitrile and TEPA). Afterwards, the model composite is heated to 70 °C for eight hours allowing gasification of the blowing agent, and then sawed in an end region. As a result, most healing agent flows out of the cuts, while some remains in the tubes (Figure 2.30b,c). Clearly, the higher content of the foaming agent ABVN corresponds to the longer empty part of the tubes, meaning much more healing agent is released. A quantitative evaluation of the experiments based on Figure 2.30 is shown in Table 2.14, which further supports the above analysis. When the model composite is simply stored at room temperature for 10 days without activating the foaming agent at 70 °C, the healing agent can also be released after cutting probably due to gradual decomposition of ABVN. Nevertheless, the outflow is significantly reduced. It is thus known from these results that the foaming agent is able to increase the internal pressure of the PP tubes.

Similarly, the PP tubes filled with DTHP epoxy monomer and the hardener (mixture of PETMP [100 parts] and BDMA [0.4 parts]) are alternately arranged in parallel and wrapped by matrix epoxy on a piece of glass plate. The foaming agent ABVN and fluorescent indicators are included in advance. After treatment at 70 °C for eight hours, the model prepregs are cut and observed by fluorescent microscopy. As exhibited in Figure 2.31a,b, both DTP epoxy monomer and the mercaptan curing agent are liberated and their travelling distances are 500– 750 μm at least. Comparatively, much more mercaptan curing agent flows out (more empty parts are left in the tube) because of its lower viscosity. It is interesting to see from Figure 2.31c that very little healing agent flows out of the cuts in the absence of foaming agent as a result of lack of internal pressure. The results demonstrate that the pressurized healing agents are expelled at longer distances.

In the case of authentic composite, the fluorescent microscopy photo in Figure 2.29b shows that the PP tubes can be indeed broken after the indentation, coinciding with the results of the above model experiments. The healing agent was liberated to the cracks in resin-rich region and bundled fibers/matrix interface. The cured version of the healing agent should help to recover the deteriorated mechanical properties of the composite laminates. The deduction is supported by the data exhibited in Figure 2.32, which clearly demonstrate that the healing agent must have been bled out so that the damaged specimens containing the healing tubes have higher flexural strength after healing.

A careful survey of Figure 2.32a reveals that the healing efficiency of the laminates with PP tubes approaches the maximum at a spacing of 750 μm. The lower value for the tube spacing of 500 μm might result from the harsher damage of the composite under indentation represented by larger amount of broken glass fibers that cannot be reconnected (Figure 2.33). In the meantime, the healing agent released from the tubes spaced 1000 μm might be insufficient for healing, which also leads to poorer healing effect. The competition of these factors yields the highest healing efficiency in the case of tube spacing of 750 μm.

Since the self-pressurized healing system has proved to work in the composite laminates, the effect of foaming agent (ABVN) concentration, which is directly related to the gas volume created inside the tubes and hence the internal pressure, is investigated (Figure 2.32b). Initially, the healing efficiency increases with increasing the amount of ABVN. It means that the higher internal pressure has forced diffusion of the healing agent to wider regions. More and more cracked sites are re-bonded. When the AVBN content exceeds 3 parts, however, healing efficiency starts to decrease instead of further increasing. The internal pressure was so high so that a few healing agents were expelled out of the bulk specimen via the surface damage (Figure 2.34), and the amount of the healing agent remaining inside the composites has to be reduced. Besides, the cracks must have interconnected with each other from the middle to the surface of the composite after indentation damaging. With increasing the dosage of the foaming agent, internal pressure in the PP tubes increases, which expels more healing agent out of the composite. Therefore, the dosage of foaming agent or internal pressure should be optimized in favor of maximum utilization of the healing agent.

Figure 2.35a depicts the dependence of healing efficiency on healing time. It is found that the healing speed is much slower than that of the epoxy composite filled with microencapsulated healing agent (which has the same composition as the present version) [69, 94]. The Raman spectra in Figure 2.35b also show that the epoxide and thiol groups of the released healing agent are almost completely consumed

Table 2.14 Relative outflow[a] of epoxy healing agent from PP tubes.

	Outflow (%)		
Foaming condition	**ABVN (1.5 parts)**	**ABVN (3 parts)**	**ABVN (4.5 parts)**
70 °C, eight hours	36.2 ± 17.3	66.2 ± 22.1	77.7 ± 29.4
25 °C, 10 days	1.5 ± 0.9	13.8 ± 3.0	18.5 ± 2.0

[a] Outflow of the fluidic healing agent is represented by the ratio of the total lengths of the empty parts to the length of the entire tube (Figure 2.30b,c).

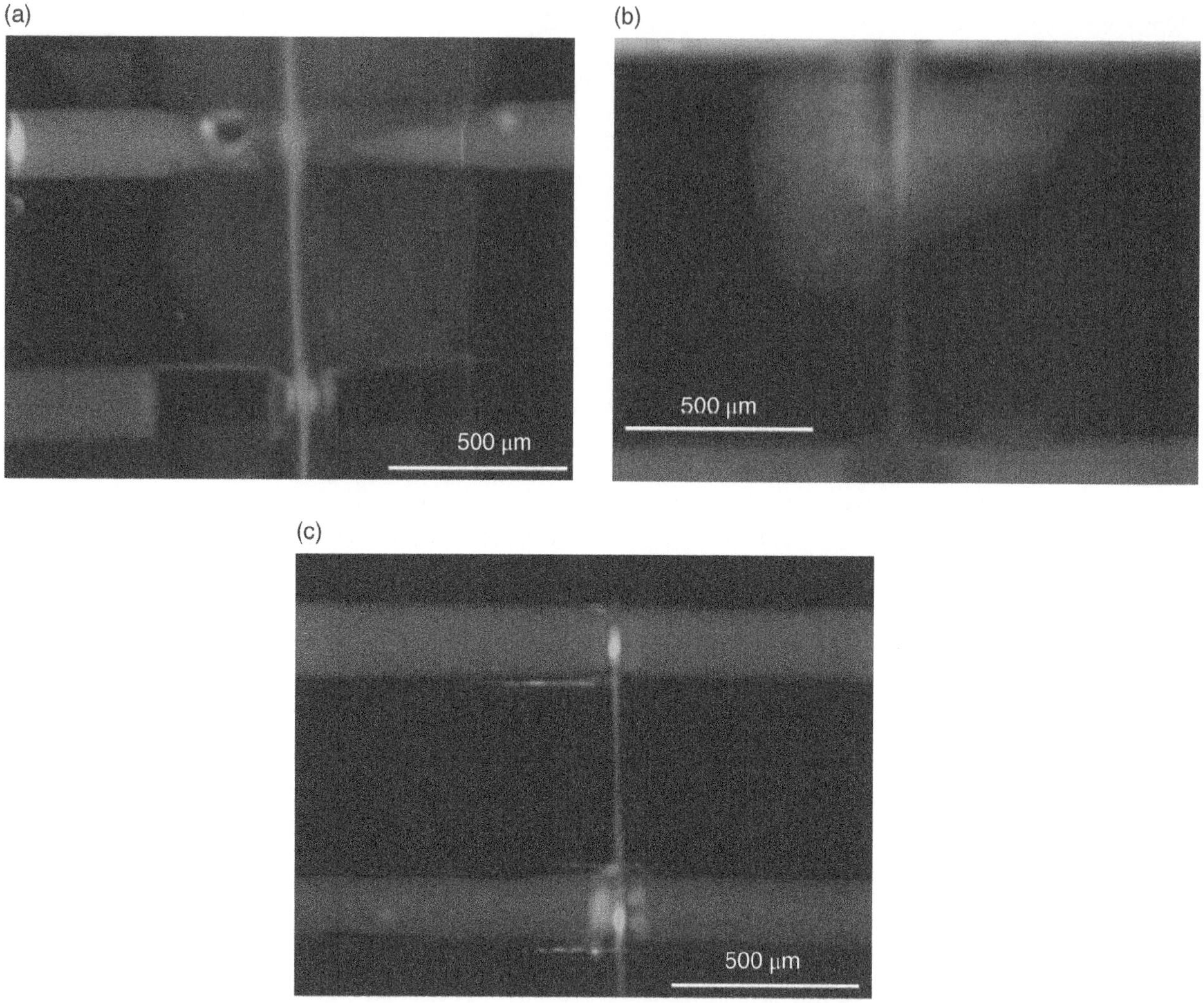

Figure 2.31 Typical fluorescence microscopy images (using an Axio Oberver Z1 fluorescence microscope) of cut model prepregs on a glass slide under light illumination of 525 and 365 nm, respectively. The prepregs consist of PP tubes, which are first filled with DTHP epoxy monomer and the hardener (mixture of PETMP [100 parts] and NBDMA [0.4 parts]), and then alternately arranged in parallel and wrapped by matrix epoxy. The foaming agent ABVN ((a, b) 1.5 parts, (c) 0 part) and fluorescent indicators (Rhodamine B [0.2 parts] for the epoxy monomer and Ardrox 985 [0.2 parts] for the curing agent) are included in advance. After treatment at 70 °C for eight hours, the model prepregs are cut and observed by fluorescent microscopy. Tube space: (a, c) 500 μm and (b) 750 μm.

after seven days, which coincides with the result of the mechanical test in Figure 2.35a. In fact, most of the damaged zones have been repaired within 36 hours (Figure 2.36). Only a very small amount of cracks need a prolonged healing period. According to these observations, we believe that as soon as the released epoxy monomer and hardener meet on the crack plane, curing reaction takes place, which increases viscosity of the healing agent, and hinders movement of the subsequent healing agent. Accordingly, the subsequent healing agent can only travel slowly by penetrating the remaining tiny gaps. As a result, the reaction among these unreacted chemicals has to be considerably postponed. For the microencapsulated healing agent, which possesses shorter inter-sphere distance, polymerization of the released healing agent is localized and there is no similar problem.

Lastly, the effect of room temperature storage time on the healing capability is evaluated. It is related to the durability of the composites. As shown in Figure 2.37, the healing efficiency decreases with the storage time because of gradual leakage of the pressurized gas from the embedded tubes. The depressurization reduces coverage area of the released healing agent and the amount of healed cracks. The analysis is supported by the comparison between Figure 2.37a,b. That is, the decline of healing efficiency becomes milder when the foaming time is decreased from

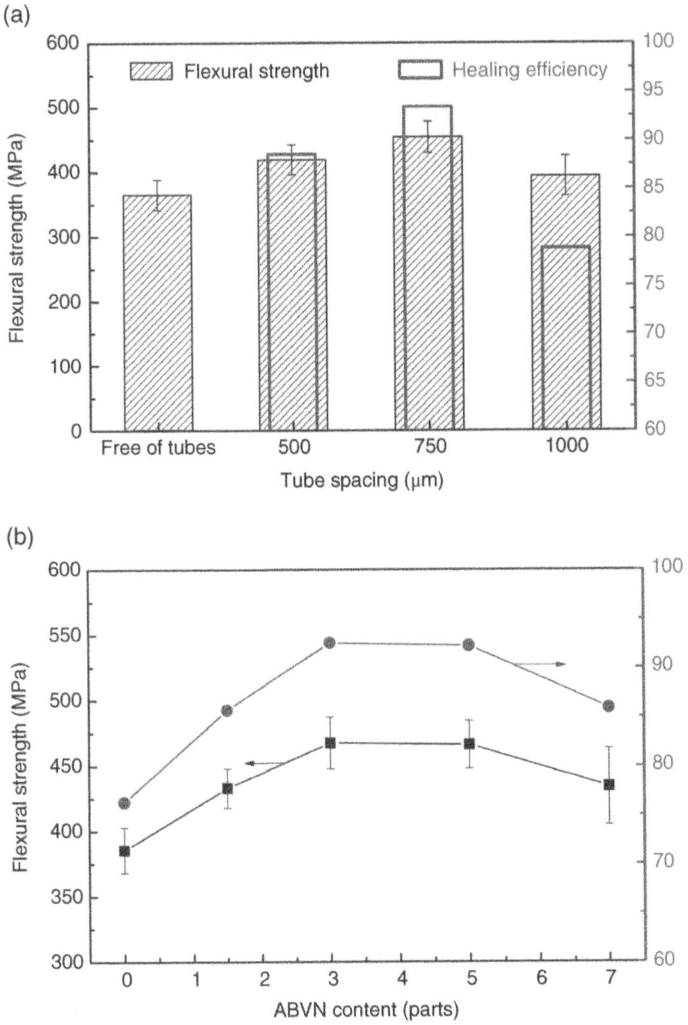

Figure 2.32 Effect of (a) tube spacing and (b) foaming agent content on flexural strength and healing efficiency of the damaged and healed composite laminates with the embedded healing agent-loaded PP tubes. Flexural strength of the damaged laminates without PP tubes is shown in (a) as reference. Tube spacing for (b): 750 μm. Foaming agent content: 3 parts; foaming condition: 70 °C, eight hours; healing condition: 25 °C, seven days. *Source:* Reprinted from Zhu et al. [52]. Copyright 2016, with permission from Elsevier.

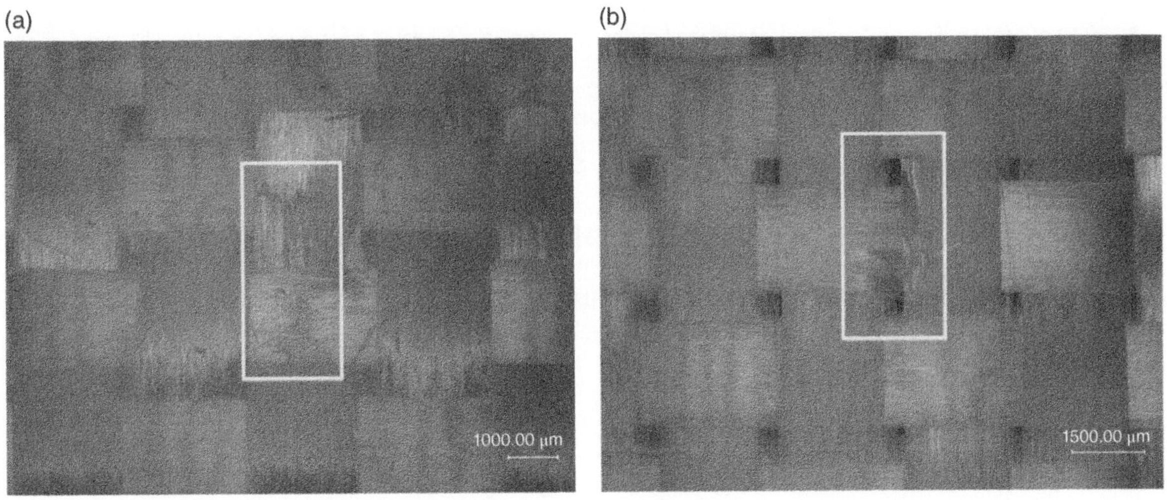

Figure 2.33 Photos of the back surface of the composite laminates with the healing agent-loaded PP tubes after indentation damaging and subsequent calcination for removing the matrix. Tube spacing: (a) 500 μm; and (b) 750 μm. The zones in the rectangular box show the broken glass fibers.

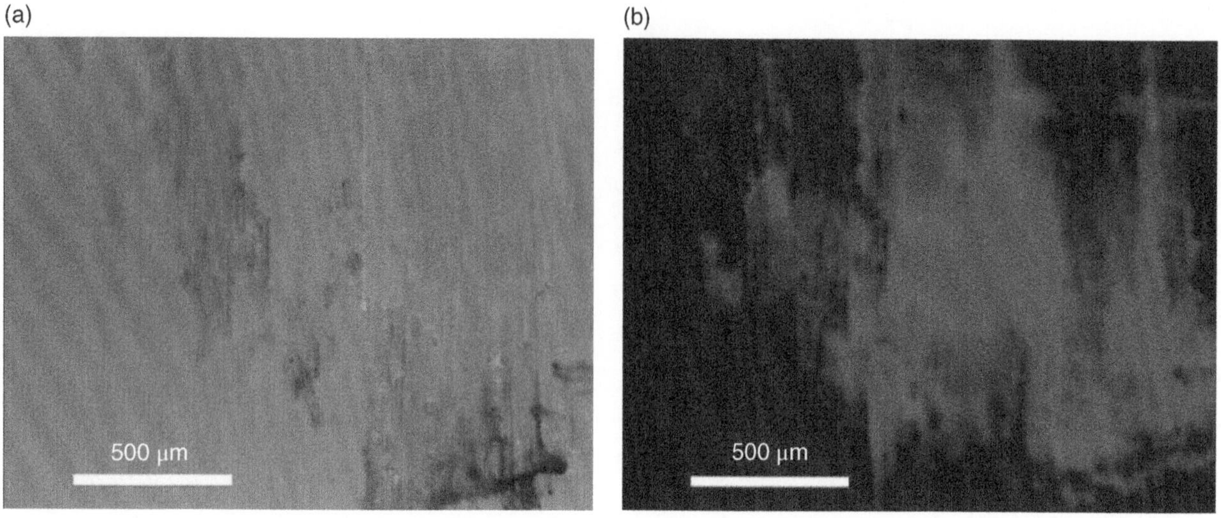

Figure 2.34 Fluorescent microscopy photos of the back face of the composite laminates with the healing agent-loaded PP tubes (tube spacing: 750 μm; foaming agent content: 3 parts; foaming condition: 70 °C, eight hours). The composite was damaged by indentation. The photos were taken under light illumination of (a) 525 nm and (b) 365 nm, respectively. For information on the fluorescent penetrant see Figure 2.29b. *Source:* Zhu et al. [52], Figure 5, p. 4/with permission of Elsevier.

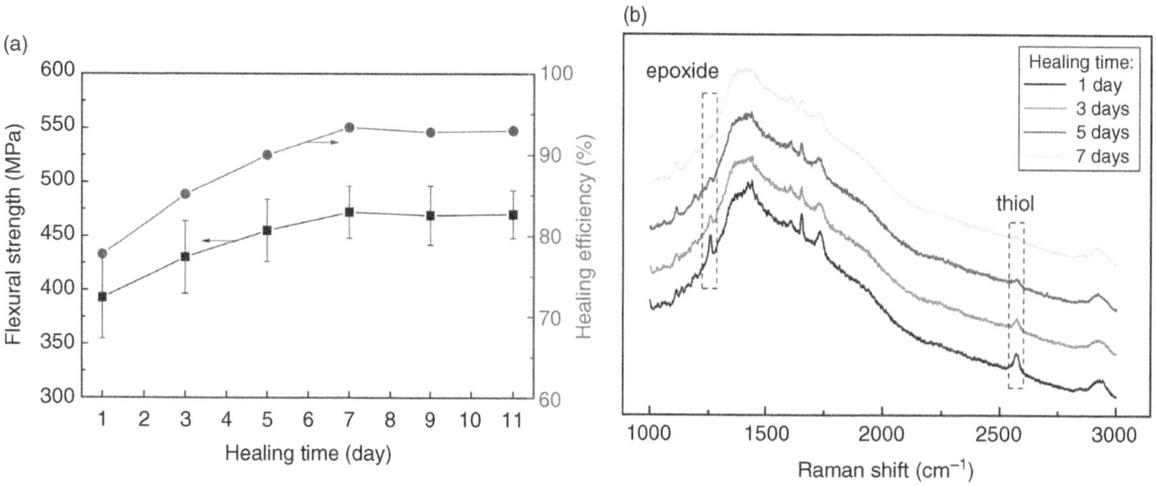

Figure 2.35 (a) Effect of healing time on flexural strength and healing efficiency of the healed composite laminates with the embedded healing agent-loaded PP tubes (tube spacing: 750 μm; foaming agent content: 3 parts; foaming condition: 70 °C, eight hours; healing condition: 25 °C). (b) In-situ confocal Raman microscopy spectra collected from the fractured surface of the above composites. *Source:* Reprinted from Zhu et al. [52]. Copyright 2016, with permission from Elsevier.

eight hours to four hours. Since shorter foaming time corresponds to less decomposition of the foaming agent and lower internal pressure of the tubes, the gas leakage-induced pressure relief is less severe. Accordingly, the healing efficiency plotted in Figure 2.37b takes much longer to approach the equilibrium than that in Figure 2.37a. The results reflect a dilemma. Longer foaming time is beneficial for higher healing efficiency but with poorer durability. The solution lies in the improvement in impermeability of the PP tubes.

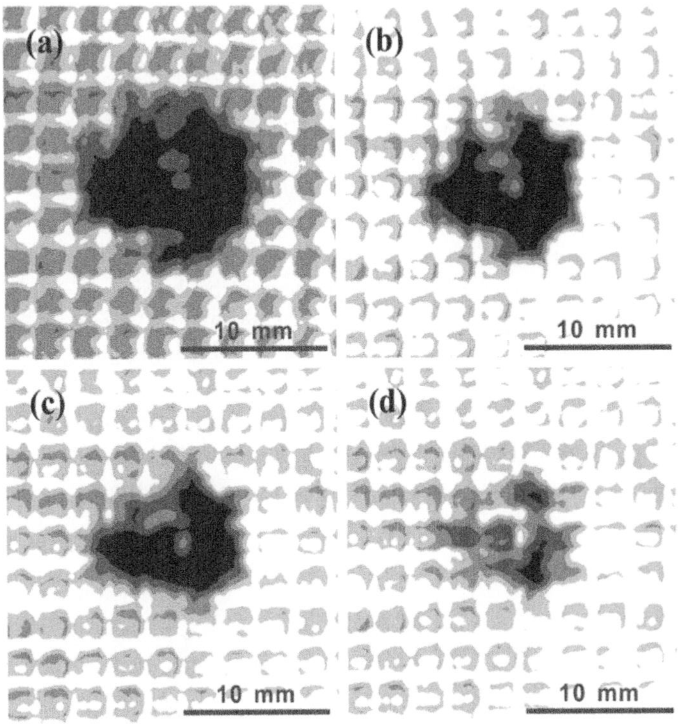

Figure 2.36 Through-scan mode ultrasonic images of the indentation damaged composite laminates with the embedded healing agent-loaded PP tubes (tube spacing: 750 μm; foaming agent content: 3 parts; foaming condition: 70 °C, eight hours; healing temperature: 25 °C). Time of healing: (a) 2 hours; (b) 12 hours; (c) 24 hours; and (d) 36 hours. Damage area: (a) 74.4 mm^2; (b) 53.8 mm^2; (c) 39.1 mm^2; and (d) 5.4 mm^2. *Source:* Zhu et al. [52], Figure 7, p. 5/with permission of Elsevier.

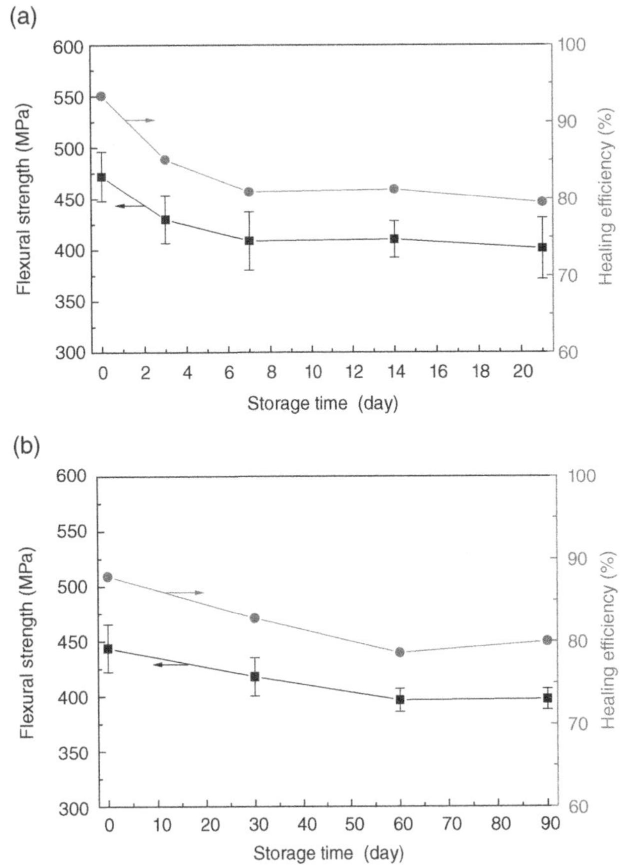

Figure 2.37 Effect of room temperature storage time on flexural strength and healing efficiency of the healed composite laminates with the embedded healing agent-loaded PP tubes (foaming agent content: 3 parts). Healing condition: 25 °C, seven days. Foaming condition: (a) 70 °C, eight hours; and (b) 70 °C, four hours.

2.5 Concluding Remarks

The feasibility of preparing self-healing epoxy composites and self-healing epoxy/glass fabric composites by embedding encapsulated epoxy and its hardener has been demonstrated. The healing system utilizes addition polymerization of epoxy with mercaptan as the curing agent under the catalysis of alkali. On account of its high flowability, high reactivity and intrinsic miscibility with the matrix, healing of cracks generated by monotonic fracture, impact and fatigue can be automatically conducted at or even below room temperature, offering satisfactory repair effectiveness.

The microencapsulated healant is quite stable and durable as a result of the compact wall shell and the stability of the core chemicals as well. This helps to broaden the processing window for fabrication of self-healing composites and prevents deterioration of the healing capability of the composites during storage. In addition, a small amount of the encapsulated healing agent is sufficient to realize the healing action, which greatly reduces the negative influence of the spheres on strength and modulus of the self-healing composites.

On the other hand, by introduction of foaming agent into healing agent-loaded thermoplastic PP tubes, decomposition of the former generates gas inside the sealed tubes and pressurizes the included healing fluid beforehand. Upon damage of the composites, the released healing agent is driven by the high internal pressure and spreads out over larger cracked planes, as compared with the unpressurized version. As a result, higher healing efficiency is detected.

References

1 Dry, C. (1993). Passive tunable fibers and matrices. *International Journal of Modern Physics B* 6: 2763–2771.
2 Dry, C. and Sottos, N.R. (1993). Passive smart self-repair in polymer matrix composites. In: *Proceedings of the SPIE: Smart Structures and Materials*, vol. 1916, 438–444. Albuquerque, NM (1–4 February 1993): SPIE.
3 Dry, C. (1996). Procedures developed for self-repair of polymer matrix composite materials. *Composite Structures* 35: 263–269.
4 Bleay, S.M., Loader, C.B., Hawyes, V.J. et al. (2001). A smart repair system for polymer matrix composites. *Composites Part A: Applied Science and Manufacturing* 32: 1767–1776.
5 Pang, W.C. and Bond, I.P. (2005). A hollow fibre reinforced polymer composite encompassing self-healing and enhanced damage visibility. *Composites Science & Technology* 65: 1791–1799.
6 Pang, W.C. and Bond, I.P. (2005). Bleeding composites'-damage detection and self-repair using a biomimetic approach. *Composites Part A: Applied Science and Manufacturing* 36: 183–188.
7 Toohey, K.S., Sottos, N.R., Lewis, J.A. et al. (2007). Self-healing materials with microvascular networks. *Nature Materials* 6: 581–585.
8 White, S.R., Sottos, N.R., Geubelle, P.H. et al. (2001). Autonomic healing of polymer composites. *Nature* 409: 794–797.
9 Brown, E.N., Sottos, N.R., and White, S.R. (2002). Fracture testing of a self-healing polymer composite. *Experimental Mechanics* 42: 372–379.
10 Brown, E.N., White, S.R., and Sottos, N.R. (2004). Microcapsule induced toughening in a self-healing polymer composite. *Journal of Materials Science* 39: 1703–1710.
11 Rule, J., Brown, E.N., Sottos, N.R. et al. (2005). Wax-protected catalyst microspheres for efficient self-healing materials. *Advanced Materials* 72: 205–208.
12 Mauldin, T.C., Rule, J.D., Sottos, N.R. et al. (2007). Self-healing kinetics and the stereoisomers of dicyclopentadiene. *Journal of the Royal Society Interface* 4: 389–393.
13 Rule, J.D., Sottos, N.R., and White, S.R. (2007). Effect of microcapsule size on the performance of self-healing polymers. *Polymer* 48: 3520–3529.
14 Cho, S.H., Andersson, H.M., White, S.R. et al. (2006). Polydimethylsiloxane-based self-healing materials. *Advanced Materials* 18: 997–1000.
15 Keller, M.K., White, S.R., and Sottos, N.R. (2007). A self-healing poly(dimethyl siloxane) elastomer. *Advanced Functional Materials* 17: 2399–2404.
16 Caruso, M.M., Delafuente, D.A., Ho, V. et al. (2007). Solvent-promoted self-healing materials. *Macromolecules* 40: 8830–8832.
17 Yin, T., Rong, M.Z., Zhang, M.Q., and Yang, G.C. (2007). Self-healing epoxy composites – preparation and effect of the healant consisting of microencapsulated epoxy and latent curing agent. *Composites Science & Technology* 67: 201–212.
18 Yuan, L., Liang, G.Z., Xie, J.Q. et al. (2006). Preparation and characterization of poly(urea-formaldehyde) microcapsules filled with epoxy resins. *Polymer* 47: 5338–5349.
19 Satoshi, M., Ikuzo, U., Makoto, K., and Kunihiko N. (1993). Adhesive containing microcapsules. European Patent 0543675.
20 Ronald, L.H. and Dale, E.W. (1985). Microencapsulated epoxy adhesive system. US Patent 4536524.

21 Schuetze, C.E. and Antonio, S. (1968). Encapsulation technique. US Patent 3396117.

22 Goldsmith, C. (1974). Process for establishing reactive contact between reactive ingredients. US Patent 3791980.

23 Jung, D., Hegeman, A., Sottos, N.R. et al. (1997). Self-healing composites using embedded microspheres. In: *Proceedings of ASME Symposium on Composites and Functionally Graded Materials*, vol. MD-80, 265–275. Dallas, TX: ASME.

24 White, S.R., Sottos, N.R., Geubelle, P.H. et al. (2005). Multifunctional autonomically healing composite material. US Patent 2003060569.

25 Zako, M. and Takano, N. (1999). Intelligent material systems using epoxy particles to repair microcracks and delamination damage in GFRP. *Journal of Intelligent Material Systems and Structures* 10: 836–841.

26 Petrie, E.M. (2006). *Epoxy Adhesive Formulations*. London: McGraw-Hill Co.

27 Xing, Q.Y., Xu, R.Q., and Zhou, Z. (1994). *The Foundations of Organic Chemistry*. Beijing: Higher Education Press (in Chinese).

28 Jencks, W.P. and Gilbert, H.F. (1977). General acid-base catalysis of carbonyl and acyl group reactions. *Pure and Applied Chemistry* 49: 1021–1027.

29 Lowe, G.B. (1997). The cure chemistry of polysulfides. *International Journal of Adhesion and Adhesives* 17: 345–348.

30 Fuchigami, M. (1980). Micro-capsules and method for their production. US Patent 4233178.

31 Yuan, L., Liang, G.Z., Xie, J.Q., and He, S.B. (2007). Synthesis and characterization of microencapsulated dicyclopentadiene with melamine–formaldehyde resins. *Colloid and Polymer Science* 285: 781–791.

32 Sun, G. and Zhang, Z. (2002). Mechanical strength of microcapsules made of different wall materials. *International Journal of Pharmaceutics* 242: 307–311.

33 Su, J.F., Li, R., and Wang, L.X. (2005). Preparation and mechanical properties of thermal energy storage microcapsules. *Colloid and Polymer Science* 284: 224–228.

34 Hong, K. and Park, S. (1999). Melamine resin microcapsules containing fragrant oil: synthesis and characterization. *Materials Chemistry and Physics* 58: 128–131.

35 Lee, H.Y., Lee, S.J., Cheong, I.W., and Kim, J.H. (2002). Microencapsulation of fragrant oil via in situ polymerization: effects of pH and melamine-formaldehyde molar ratio. *Journal of Microencapsulation* 19: 559–569.

36 Dietrich, K., Bonatz, E., Geistlinger, H. et al. (1989). Amino resin microcapsules. II. Preparation and morphology. *Acta Polymerica* 40: 325–331.

37 Saeed, M.-U., Chen, Z.F., and Li, B.B. (2015). Manufacturing strategies for microvascular polymeric composites: a review. *Composites Part A: Applied Science and Manufacturing* 78: 327–240.

38 Motuku, M., Janowski, C.M., and Vaidya, U.K. (1999). Parametric studies on self-repairing approaches for resin infused composites subjected to low velocity impact. *Smart Materials and Structures* 8: 623–638.

39 Kousourakis, A. and Mouritz, A.P. (2010). The effect of self-healing hollow fibres on the mechanical properties of polymer composites. *Smart Materials and Structures* 19: 085021.

40 Norris, C.J., White, J.A.P., McCombe, G. et al. (2012). Autonomous stimulus triggered self-healing in smart structural composites. *Smart Materials and Structures* 21: 094027.

41 Yue, H.-B., Fernández-Blázquez, J.P., Beneito, D.F., and Vilatela, J.J. (2014). Real time monitoring of click chemistry self-healing in polymer composites. *Journal of Materials Chemistry A* 2: 3881–3887.

42 Bond, I.P. (2004). Self-repairing hollow fibre composites. *Reinforced Plastics* 48 (8): 16.

43 Trask, R.S. and Bond, I.P. (2006). Biomimetic self-healing of advanced composite structures using hollow glass fibres. *Smart Materials and Structures* 15: 704–710.

44 Hucker, M., Bond, I., Foreman, A., and Hudd, J. (1999). Optimisation of hollow glass fibres and their composites. *Advanced Composites Letters* 8: 181–189.

45 Pang, J.W.C. and Bond, I.P. (2005). A hollow fiber reinforced polymer composite encompassing self-healing and enhanced damage visibility. *Composites Science and Technology* 65: 1791–1799.

46 Kling, S. and Czigány, T. (2014). Damage detection and self-repair in hollow glass fiber fabric-reinforced epoxy composites via fiber filling. *Composites Science and Technology* 99: 82–88.

47 Kling, S. and Czigány, T. (2013). Analysis of applicability of the hollow carbon fibres for self-repairing composites. *Materials Science Forum* 729: 246–251.

48 Williams, G., Trask, R.S., and Bond, I.P. (2007). A self-healing carbon fibre reinforced polymer for aerospace applications. *Composites Part A: Applied Science and Manufacturing* 38: 1525–1532.

49 Trask, R.S., Williams, G.J., and Bond, I.P. (2007). Bioinspired self-healing of advanced composite structures using hollow glass fibres. *Journal of the Royal Society Interface* 4: 363–371.

50 Williams, H.R., Trask, R.S., and Bond, I.P. (2008). Self-healing sandwich panels: restoration of compressive strength after impact. *Composites Science and Technology* 68: 3171–3177.

51 Hamilton, A.R., Sottos, N.R., and White, S.R. (2012). Pressurized vascular systems for self-healing materials. *Journal of the Royal Society Interface* 9: 1020–1028.

52 Zhu, Y., Ye, X.J., Rong, M.Z., and Zhang, M.Q. (2016). Self-healing glass fiber/epoxy composites with polypropylene tubes containing self-pressurized epoxy and mercaptan healing agents. *Composites Science and Technology* 135: 146–152.

53 Yuan, Y.C., Rong, M.Z., and Zhang, M.Q. (2008). Preparation and characterization of microencapsulated polythiol. *Polymer* 49: 2531–2541.

54 Bauer, D.R. (1986). Melamine/formaldehyde crosslinkers: characterization, network formation and crosslink degradation. *Progress in Organic Coatings* 14: 193–218.

55 Kumar, A. and Katiyer, V. (1990). Modeling and experimental investigation of melamine-formaldehyde polymerization. *Macromolecules* 23: 3729–3736.

56 Jones, F.N., Chu, G.B., and Samaraweera, U. (1994). Recent studies of self-condensation and co-condensation of melamine-formaldehyde resins; cure at low temperatures. *Progress in Organic Coatings* 24: 189–208.

57 Schildknecht, C.E. and Skeist, I. (1977). *Polymerization Processes*. New York: John Wiley & Sons.

58 Seitz, M.E. (1993). Microencapsulation process using melamine-formaldehyde and microcapsules produced thereby. US Patent 5204185.

59 Dietrich, K., Bonatz, E., Nastke, R. et al. (1990). Amino resin microcapsules. IV. Surface tension of the resins and mechanism of capsule formation. *Acta Polymerica* 41: 91–95.

60 Jahromi, S., Litvinov, V., and Geladé, E. (1999). Physical gelation of melamine formaldehyde resin solutions. II. A combined light-scattering and low-resolution relaxation proton NMR study. *Journal of Polymer Science Part B: Polymer Physics* 37: 3307–3318.

61 Braun, D., Sauerwein, R., and Hellmann, G.P. (2001). Polymeric surfactants from styrene-co-maleic-anhydride copolymers. *Macromolecular Symposia* 163: 59–66.

62 Garnier, G., Duskova-Smrckova, M., Vyhnalkova, R. et al. (2000). Association in solution and adsorption at an air-water interface of alternating copolymers of maleic anhydride and styrene. *Langmuir* 16: 3757–3763.

63 Costa, L. and Camino, G. (1988). Thermal behaviour of melamine. *Journal of Thermal Analysis* 34: 423–429.

64 Devallencourt, C., Saiter, J.M., Fafet, A., and Ubrich, E. (1995). Thermogravimetry/Fourier transform infrared coupling investigations to study the thermal stability of melamine formaldehyde resin. *Thermochimica Acta* 259: 143–151.

65 Zhang, X.X., Tao, X.M., Yick, K.L., and Wang, X.C. (2004). Structure and thermal stability of microencapsulated phase-change materials. *Colloid and Polymer Science* 282: 330–336.

66 Yuan, L., Liang, G.Z., Xie, J.Q. et al. (2006). Thermal stability of microencapsulated epoxy resins with poly(urea–formaldehyde). *Polymer Degradation and Stability* 91: 2300–2306.

67 Yuan, Y.C., Rong, M.Z., and Zhang, M.Q. (2008). Preparation and characterization of poly(melamine-formaldehyde) walled microcapsules containing epoxy. *Acta Polymerica Sinica* 5: 472–480.

68 Prajer, M., Wu, X., Garcia, S.J., and van der Zwaag, S. (2015). Direct and indirect observation of multiple local healing events in successively loaded fibre reinforced polymer model composites using healing agent-filled compartmented fibres. *Composites Science and Technology* 106: 127–133.

69 Yuan, Y.C., Rong, M.Z., Zhang, M.Q. et al. (2008). Self-healing polymeric materials using epoxy/mercaptan as the healant. *Macromolecules* 41: 5197–5202.

70 Kinloch, A.J. and Young, R.J. (1983). *Fracture Behaviour of Polymers*. London: Applied Science Publishers.

71 May, C.A. (1988). *Epoxy Resins: Chemistry and Technology*. New York: Marcel Dekker Inc.

72 Yuan, Y.C., Rong, M.Z., Zhang, M.Q., and Yang, G.C. (2009). Study of factors related to performance improvement of self-healing epoxy based on dual encapsulated healant. *Polymer* 50: 5771–5781.

73 Wool, R.P. (2008). Self-healing materials: a review. *Soft Matter* 4: 400–418.

74 Wool, R.P. and O'Connor, K.M. (1981). A theory of crack healing in polymers. *Journal of Applied Physics* 52: 5953–5963.

75 Yuan, Y.C., Rong, M.Z., Zhang, M.Q. et al. (2010). Self-healing of fatigue crack in epoxy materials with epoxy/mercaptan system. *Express Polymer Letters* 5: 47–59.

76 Yuan, Y.C., Rong, M.Z., Zhang, M.Q. et al. (2010). Healing of fatigue crack in epoxy materials with epoxy/mercaptan system via manual infiltration. *Express Polymer Letters* 4: 644–658.

77 Daniel, J.S. and Kim, Y.R. (2001). Laboratory evaluation of fatigue damage and healing of asphalt mixtures. *Journal of Materials in Civil Engineering* 13: 434–440.

78 Brown, E.N., White, S.R., and Sottos, N.R. (2005). Retardation and repair of fatigue cracks in a microcapsule toughened epoxy composite-part II: in situ self-healing. *Composites Science and Technology* 65: 2474–2480.

79 Brown, E.N., White, S.R., and Sottos, N.R. (2005). Retardation and repair of fatigue cracks in a microcapsule toughened epoxy composite-part I: manual infiltration. *Composites Science and Technology* 65: 2466–2473.

80 Jones, A.S., Rule, J.D., Moore, J.S. et al. (2007). Life extension of self-healing polymers with rapidly growing fatigue cracks. *Journal of the Royal Society Interface* 4: 395–403.

81 Maiti, S. and Geubelle, P.H. (2006). Cohesive modeling of fatigue crack retardation in polymers: crack closure effect. *Engineering Fracture Mechanics* 73: 22–41.

82 Maiti, S., Shankar, C., Geubelle, P.H., and Kieffer, J. (2006). Continuum and molecular-level modeling of fatigue crack retardation in self-healing polymers. *Journal of Engineering Materials and Technology, Transactions of the ASME* 128: 595–602.

83 Brown, E.N., White, S.R., and Sottos, N.R. (2006). Fatigue crack propagation in microcapsule toughened epoxy. *Journal of Materials Science* 41: 6266–6273.

84 Sharp, P.K., Clayton, J.Q., and Clark, G. (1997). Retardation and repair of fatigue cracks by adhesive infiltration. *Fatigue & Fracture Engineering Materials & Structures* 20: 605–614.

85 Elber, W. (1970). Fatigue crack closure under cyclic tension. *Engineering Fracture Mechanics* 2: 37–45.

86 James, M.N. and Smith, G.C. (1983). Surface microcrack closure in fatigue: comparison of compliance and crack sectioning data. *International Journal of Fracture* 22: 69–75.

87 Song, P.S., Hwang, S., and Shin, C.S. (1998). Effect of artificial closure materials on crack growth retardation. *Engineering Fracture Mechanics* 60: 47–58.

88 Shin, C.S. and Cai, C.Q. (2008). Surface fatigue crack growth suppression in cylindrical rods by artificial infiltration. *International Journal of Fatigue* 30: 560–567.

89 Shin, C.S., Wang, C.M., and Song, P.S. (1996). Fatigue damage repair-a comparison of some possible methods. *International Journal of Fatigue* 18: 535–546.

90 UR-Rehman, A. and Thomason, P.F. (1993). The effect of artificial fatigue crack closure on fatigue crack growth. *Fatigue & Fracture of Engineering Materials & Structures* 16: 1081–1090.

91 Shin, C.S. and Chen, Z.Z. (2001). Fatigue life extension by electroless nickel infiltration plating. *International Journal of Fatigue* 23: 777–788.

92 Wang, Y.Y., Zho, M.X., and Feng, D.Q. (1993). A study of retarding fatigue crack growth using an artificial wedge. *Fatigue & Fracture of Engineering Materials & Structures* 16: 363–376.

93 Yuan, Y.C., Ye, Y.P., Rong, M.Z. et al. (2010). Self-healing of low-velocity impact damage in glass fabric/epoxy composites using epoxy-mercaptan healing agent. *Smart Materials & Structures* 20: 015024.

94 Yuan, Y.C., Ye, X.J., Rong, M.Z. et al. (2011). Self-healing epoxy composite with heat-resistant healant. *ACS Applied Materials & Interfaces* 3: 4487–4495.

3

Extrinsic Self-Healing Via Cationic Polymerization

CHAPTER MENU

The in-situ self-healing of fatigue cracks discussed in Chapter 2 has shown the importance of chemical kinetics of healing action. In polymeric materials subjected to either fatigue or low-velocity impact, submicrometer cracks (0.009–0.3 μm) would quickly appear and then coalesce forming macroscopic damage. Therefore, a fast healing reaction is critical for practical application so that cracks can be eliminated soon after emergence to avoid possible propagation.

For an epoxy-based healing system able to operate at or below room temperature, acceleration of its rate of crack healing can be achieved by the selection of a proper curing agent. Boron-containing compounds might act as qualified candidates in this regard. Since 1955 when Greenlee reported the applicability of addition products of boron fluoride and compounds carrying an amino nitrogen atom for curing epoxy resin [1], these chemicals have been well developed and commercialized. They are characterized by the following distinctive features [2–8]: (i) low dosage required for solidifying epoxy (1–15%), (ii) high curing activity, (iii) capability of delayed action, and (iv) easy regulation of curing habit and properties of the cured products by the changing composition of boron-containing compounds.

As a typical boron-containing compound, boron trifluoride diethyl etherate ($(C_2H_5)_2O{\cdot}BF_3$) has been commercially used as a hardener for low temperature fast cure epoxy adhesives. The curing reaction of epoxy catalyzed by $(C_2H_5)_2O{\cdot}BF_3$ proceeds so fast at ambient temperature that a tremendous exotherm is produced [9], so it is quite suitable for triggering polymerization of epoxy monomer in a cold environment. Another interesting advantage of $(C_2H_5)_2O{\cdot}BF_3$ lies in the fact that the epoxy-BF_3 cure is dominated by a cationic chain polymerization mechanism (Scheme 3.1). $(C_2H_5)_2O{\cdot}BF_3$ plays the role of catalytic curing agent that does not build itself into the thermoset

Extrinsic and Intrinsic Approaches to Self-Healing Polymers and Polymer Composites, First Edition. Ming Qiu Zhang and Min Zhi Rong.

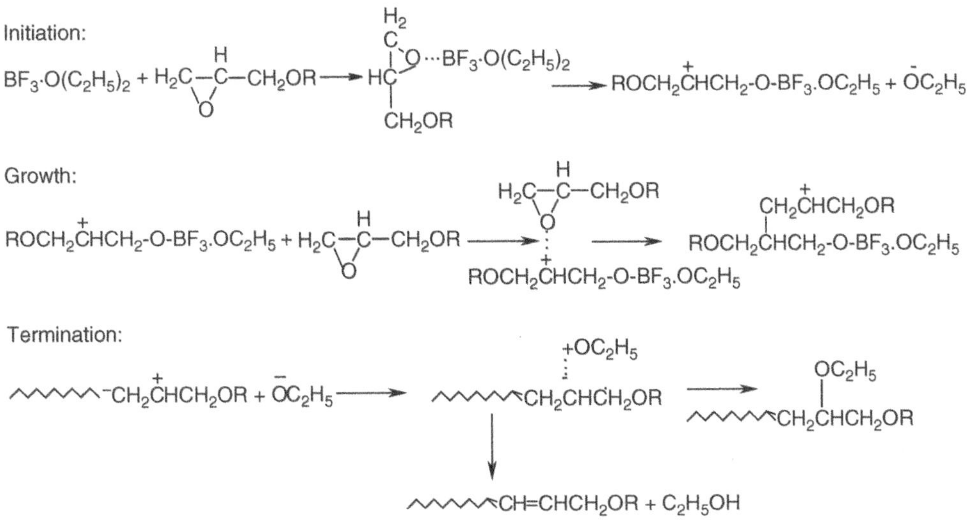

Scheme 3.1 Curing of epoxy in the presence of $(C_2H_5)_2O{\cdot}BF_3$.

structure, which is obviously different from reactive curing agents represented by amines and anhydrides. Owing to the different curing mechanisms, the structures and properties of the cured epoxy are also different accordingly. In the case of reactive curing, the sites where the curing agents are located serve as the reactive loci, and the curing reaction develops radially outward. As a result, a microscopically heterogeneously consolidated product is obtained, in which the cross-linking density near the reactive loci is higher than that away from the reactive loci. Partially reacted or even completely unreacted epoxy monomers might be present adjacent to the reactive loci. In contrast, for catalytic curing, ring-opening of epoxide groups simultaneously takes place throughout the resin. In the vicinity of the reactive sites, curing reaction proceeds very fast leading to higher degree of cross-linking. With the lapse of time, the rate of curing reaction decreases and the degree of cross-linking of the post-cross-linking portions has to be lower. The cured epoxy is eventually comprised of an alternating structure with high and low degrees of cross-linking. Therefore, an increase in concentration of reactive curing agent would increase the number of reactive loci of curing and degree of cross-linking of the resultant, while a higher dosage of catalytic curing agent may result in higher rate of curing reaction, lower amount of the post-cross-linking portions, and higher degree of cross-linking of the cured resin. Owing to these characteristics, the strict stoichiometric ratio of epoxide/hardener required by addition polymerization (the epoxy-mercaptan cure [10–12] mentioned in Chapter 2, for example) is no longer necessary for the epoxy-BF_3 cure. In this context, when $(C_2H_5)_2O{\cdot}BF_3$ is introduced as the partner of epoxy to prepare a two-component self-healing agent, uniform distribution of alternate epoxy- and hardener-loaded capsules (supposing the strategy of microencapsulation of healing agent [13] is followed) is not a must for manufacturing self-healing epoxy composite. Curing of epoxy released from the broken microcapsules would be ignited and spread out quickly so long as the epoxy fluid meets $(C_2H_5)_2O{\cdot}BF_3$. Only a small amount of $(C_2H_5)_2O{\cdot}BF_3$ (1–5 wt% relative to the quantity of epoxy) is sufficient for initiating the reaction. Evidently, the processing and working windows of self-healing materials would thus become much wider.

On the basis of above analysis, fabrication of epoxy-$(C_2H_5)_2O{\cdot}BF_3$ healing agent and evaluation of its effect on crack healing of epoxy materials will be the main concern of this chapter. It is hoped that the healing system would be superior to the epoxy-thiol pair [10–12] as viewed from the above advantages. That is, crack healing in self-healing epoxy with very low content of healant can be conducted at a much faster rate, as compared with the composite containing microencapsulated epoxy and thiol. Not only is early repair of cracks facilitated, but also the influence of the healant on the intrinsic mechanical properties of the matrix is minimized.

To ensure sufficient supply of healing agent during crack rehabilitation, the strategy based on microencapsulation, proposed by White and co-workers [13–26], opened an avenue toward practical application of the technique (see Chapters 1 and 2). The principle lies in the fact that healing agent is encapsulated and embedded in the materials in advance. As soon as cracks destroy the fragile capsules, the healing agent would be released into the crack planes due to capillary effect and then be polymerized, so that the cracks can be re-bound. Taking the advantage of microencapsulation, healing agent is well protected from composite manufacturing and comes into effect only upon cracking due to the applied force. According to this idea, the aforementioned epoxy monomer and $(C_2H_5)_2O{\cdot}BF_3$ fluids are separately encapsulated prior to production of the self-healing epoxy composites.

In addition to the $BF_3{\cdot}O(C_2H_5)_2$-epoxy pair, we plan to develop more fast healing systems. Considering that (i) BF_3 is a Lewis acid, (ii) epoxy can also be cured by other Lewis acids like $TiCl_4$ and $AlCl_3$ via cationic polymerization [27, 28], and (iii) curing speed of epoxy catalyzed by Lewis acids is proportional to their acidity, we decided to formulate another fast healing system consisting of SbF_5 and epoxy [29]. SbF_5 is a high performance fluidizer and has been employed in the pharmaceutical industry. It is the strongest Lewis acid except for $AuF_5{\cdot}F_2$ [30] and can react with almost all known chemical compounds. Therefore, SbF_5 would not only serve as a low temperature hardener of epoxy following the same reaction mechanism as that involved in BF_3-epoxy cure, i.e. cationic polymerization, but also offer much faster curing speed owing to its extremely high activity.

SbF_5 alone has not yet directly been used for curing epoxy probably due to its outstanding reactivity, which makes it difficult to be handled and limits its effective usage. In this context, antimony pentafluoride-ethanol complex ($SbF_5{\cdot}(HOC_2H_5)_n$) is an effective alternative as SbF_5 molecules are stabilized in the complex without losing valuable catalytic activity.

It is worth noting that the small ethanol molecules, which are entrapped in the cured healing membrane, would act as plasticizer and lower mechanical strength of the latter. Moreover, the relatively low boiling point of the ethanol (~78 °C) in the healant capsules limits the curing temperature and working temperature of the target composites for high performance application.

In this context, the trifluoromethanesulfonic acid (TfOH)-epoxy pair is also applied to produce an instant self-healing system [31] in the hope of satisfying the requirements of ultrafast rehabilitation and thermal tolerance for advanced epoxy resin application. TfOH is an ultra-strong Bronsted acid with boiling point of about 162 °C. It is mainly employed in organic synthesis [32–35] such as the Friedel–Crafts reaction [32] and polymerization [34, 35]. Since TfOH is more chemically stable than SbF_5, it is able to cure epoxy as quickly as the ethanol solution of the SbF_5-ethanol complex, without the need of forming a complex with small molecules like ethanol. Consequently, the aforementioned negative effects of ethanol are eliminated. In addition, TfOH has low toxicity (LD50 (oral, mouse): 1605 $mg\,kg^{-1}$; for reference, LD50 (oral, mouse) of aspirin is 1500 $mg\,kg^{-1}$), which is another advantage of the healing system.

With a view to application, imparting self-healing ability to commercially available polymers would be more favorable because there is no need to change their macromolecular structures. However, the relevant research mostly deals with cured epoxy. Comparatively, thermoplastics (especially commodity plastics), which are widely used in diverse applications, are less studied. The reported healing approaches are mainly based on encapsulated solvents [36] or plasticizer [37]. Chain diffusion and entanglement across the interface caused by solvent effect are responsible for crack healing. Although self-healing has been enabled in materials, they used to be fabricated by solution casting rather than thermo-processing already adopted in industry. Evidently, it is a bottleneck that restricts mass production of self-healing thermoplastics.

To tackle the problem, we synthesized multilayer microreactors carrying reactants for room temperature atom transfer radical polymerization (ATRP) and compounded the microreactors with polystyrene (PS) [38]. The microreactors survived melt shear mixing and hot compression molding, and the resultant PS composite was self-healable owing to autonomous ATRP of the released monomer (glycidyl methacrylate, GMA) when the propagating cracks damaged the microreactors [39]. Nevertheless, the healing agent is challenged by the fact that the lower oxidation state copper, Cu(I), of the ATRP catalyst in the microreactors would be easily oxidized to Cu(II). That is, the healing reaction would not proceed due to deactivation of the ATRP catalyst if the fracture surface was exposed to air.

Crivello described a novel cationic redox system with diaryliodonium salt as oxidant and dialkylborane as reducing agent, which can initiate cationic polymerizations of vinyl and heterocyclic monomers at room temperature [40]. It was believed that dialkylborane reduced diaryliodonium salt liberating a strong Lewis acid. Then, the Lewis acid, in combination with a trace of protogen like water, triggers polymerization of the monomer by a cationic process. Compared with ATRP, redox cationic polymerization (RCP) seems to be more suitable for functioning as a healing chemistry for thermoplastics by virtue of the stable redox initiator, as well as its oxygen insensitivity, compatibility with a wider range of monomers, and fast polymerization.

Inspired by these advantages of RCP, we propose the following healing approach for thermoplastics [41], which tolerates the high temperature processing as well as overcomes the oxygen inhibition. Iodonium bis(4-methylphenyl) hexafluorophosphate (IBH) cooperates with $NaBH_4$, constituting a redox cationic initiator for polymerization of GMA. In actual operation, IBH is dissolved in GMA, and then the solution is encapsulated by poly(melamine-formaldehyde) (PMF). Finally, the IBH/GMA-loaded capsules and $NaBH_4$ particles dispersed in PS matrix are compression molded to form self-healing PS (Figure 3.1). Like other extrinsic self-healing materials, polymerization product of the released GMA upon capsule rupture is supposed to reconnect cracks in PS.

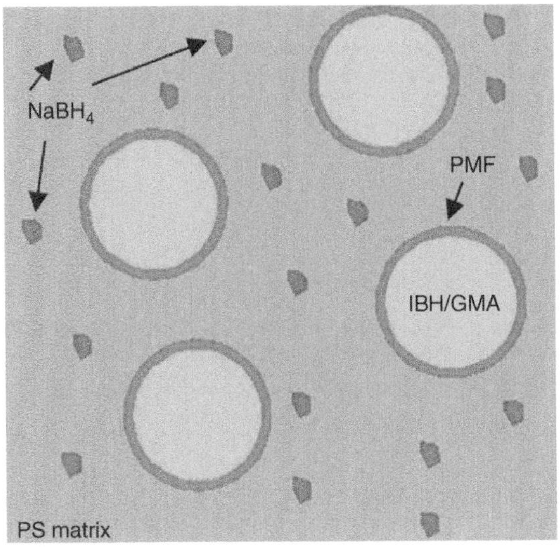

Figure 3.1 Design of the self-healing PS.

3.1 Thermosetting

3.1.1 Microencapsulation of Epoxy by Ultraviolet Irradiation-Induced Interfacial Copolymerization

Interfacial phenomena related to physicochemical processes (such as adsorption/desorption, deposition, and mass transport) and chemical reactions (e.g. polycondensation, in-situ polymerization, and polyaddition) are the major procedures for carrying out microencapsulation [42]. The methods often used in practice include spray drying, phase separation from a continuous phase, solvent extraction, layer-by-layer addition, solvent evaporation, and interfacial polymerization utilizing emulsions and microemulsions [43]. So far, however, only a few works have reported ultraviolet (UV) irradiation-induced radical polymerization for preparing microcapsules. Quek et al. [44] established a cross-linking structure using UV light in the exterior wall made from polyelectrolyte copolymer to enhance the mechanical strength and chemical stability of microcapsules. Similarly, Shen et al. [45] improved the stability of alginate hydrogel microcapsules by developing a photo-cross-linked interpenetrating polymer network with photopolymerized sodium acrylate and *N*-vinylpyrrolidone. In these works, photo-cross-linking was only used to strengthen the wall of the capsules.

Photopolymerization offers some distinct advantages such as rapid reaction, lower energy consumption, and less environmental pollution. Thereby, we propose herein a route to produce epoxy resin-containing microcapsules by means of UV-initiated radical copolymerization between epoxydiacrylates and a polymerizable emulsifier (sodium 3-methacryloyloxy-2-hydroxypropane sulfonate (HPMAS)) [46]. It might be eventually developed into an industrializable approach for producing microencapsulated healing agent.

Epoxy diacrylate is chosen as the wall former because it is a derivative from epoxy (the core substance to be encapsulated) and is able to establish strong interaction with the emulsifier. Besides, it is photosensitive and often used in UV cured coatings [47]. Furthermore, the polymerizable emulsifier plays a key role in this technique. It not only stabilizes the emulsion but also takes part in the formation of the shells. To favor accumulation of epoxy diacrylates around the epoxy colloids, an additional emulsifier, styrene-maleic anhydride (SMA) copolymer, is incorporated to increase the interfacial interaction between epoxy diacrylates and emulsifier.

Hereinafter, two kinds of epoxy resins, diglycidyl ether of bisphenol A (EPON 828) and diglycidyl tetrahydro-*o*-phthalate (DTP are used as the core materials of the microcapsules. Benzoyl peroxide (BPO) acts as the radical initiator. Four photoinitiators are used: 2-isobutoxy-2-phenyl-acetophenone (benzoin butyl ether), 2-benzyl-2-dimethylamino-1-(4-morpholin-4-ylphenyl)butan-1-one (Irgacure 369), 2,2-dimethoxy-1,2-diphenyl-ethanone (Irgacure 651), and 2-methyl-1-(4-methylthiophenyl)-2-morpholinopropan-1-one (Irgacure 907). Pentaerythritol triacrylate (PETA) is employed as the cross-linking agent. Scheme 3.2 shows chemical structures of the aforementioned epoxy resins, epoxy diacrylates (to be synthesized as described below), and polymerizable emulsifier.

Epoxy diacrylates, the photosensitive oligomers, are prepared from the reaction between the epoxy resin (EPON 828 or DTP) and acrylic acid. The typical synthesis procedure is given in the following. The reaction proceeded at 110 °C for seven hours in the presence of 0.05% (w/w) catalyst *N,N*-dimethylbenzylamine (DMBA) until the acid number reached a value of 4 mg KOH g^{-1}. For the convenience of discussion, the resultants were denoted as EPON 828-epoxy diacrylates or DTP-epoxy diacrylates. The acid number was determined by the following procedure. The sample (0.5–1 g) was placed into a conical flask with 40 ml ethanol and 1 ml phenolphthalein color indicator. The mixture was titrated with 0.1 mol l^{-1} KOH solution until a pink color lasted for 30 seconds. The acid number was calculated by:

$$\frac{N \times V \times 56}{G} \tag{3.1}$$

where N denotes the concentration of KOH solution, V is the volume of the consumed KOH solution, and G is the sample weight.

The emulsifier sodium styrene-maleic anhydride (SMANa) copolymer was prepared in accordance with the following steps. Styrene (32.5 g, 0.306 mol), BPO (0.31 g) and 300 ml toluene were charged into a 1000 ml round bottom flask fitted with a nitrogen bubbler, magnetic stir bar, and condenser. After bubbling with nitrogen for 30 minutes, the reaction mixture was heated to 80 °C, and maleic anhydride

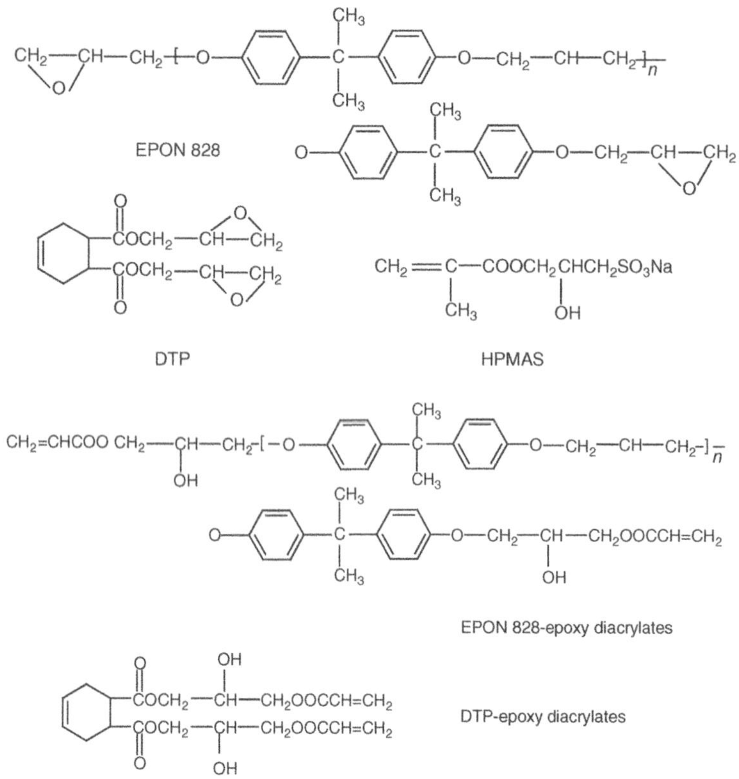

Scheme 3.2 Molecular structures of epoxy resins (EPON 828 and DTP), polymerizable emulsifier HPMAS, and epoxy diacrylates (EPON 828-epoxy diacrylates and DTP-epoxy diacrylates).

(30g, 0.306 mol) and BPO (0.31 g) dissolved in 300 ml toluene were added dropwise over one hour. The reaction proceeded for an additional four hours under stirring. Eventually, the alternating SMA copolymer was filtered, washed with toluene and methanol, and dried at 80 °C under reduced pressure for 48 hours. Details of the above procedures can be found in ref. [43]. The resultant's number average molecular weight (M_n = 11 000) and its distribution (M_w/M_n = 2.5) were determined by gel permeation chromatography (GPC). The produced SMA (5 g) was saponified in 100 g NaOH aqueous solution (10%) for five hours at 80 °C with stirring, and then the extra NaOH was neutralized by acid to give a light yellow solution of SMANa.

To prepare epoxy-loaded microcapsules, 3 g SMANa solution was mixed with 100 g deionized water in a 200 ml beaker; and then, the mixture of 8.5 g epoxy resin (EPON 828 or DTP), 1.5 g epoxy diacrylates, 3 g HPMAS, including 0.5% PETA and 1.0% benzoin butyl ether, was added dropwise under vigorous stirring over one minute to the aqueous phase to form an oil-in-water emulsion. After that, the resultant emulsion was stirred with a homogenizer at 12 000 rpm for an additional five minutes. The emulsion was irradiated by UV light (3 kW high-pressure mercury lamp) for 3–15 minutes at room temperature. By precipitating the irradiated emulsion into a fivefold excess of water, the microcapsules were isolated, and then collected in a centrifuge, and washed with water several times. At last, the powder-like microcapsules were generated after drying in a vacuum freeze dryer.

The core content of the resultant microcapsules was determined by extraction with acetone. First, the microcapsule samples were ground with a mortar and pestle at room temperature. A certain amount of the dried and ground microcapsules, W_0, was sealed in a filter paper bag. Then, the bag with the ground microcapsules was dried for five hours in a vacuum oven at 55 °C, cooled in a vacuum desiccator and precisely weighed, W_1. Afterwards, the sample bag was placed in a Soxhlet apparatus, extracted with acetone for 24 hours, and dried in a vacuum oven at 55 °C. After cooling in a vacuum desiccator, the sample bag was weighed again, W_2. Eventually, the core content of the microcapsules, α, was calculated by:

$$\alpha = \frac{W_1 - W_2}{W_0} \times 100\% \qquad (3.2)$$

To optimize the effect of UV irradiation conditions on the yield of microcapsules, an orthographic factorial design of five factors (i.e. depth of emulsion, irradiation time, UV lamp power, distance between UV lamp and the upper surface of the emulsion, and initiator concentration) and four levels was used (Table 3.1). The microcapsule yield was calculated from:

$$\text{Yield} = W_c / W_{\text{feed}} \times 100\% \qquad (3.3)$$

where W_{feed} represents the theoretical weight of the resultant microcapsules, i.e. the input weight of epoxy and wall

Table 3.1 Creation of an orthographic factorial design of five factors and four levels for microcapsule preparation under UV irradiation [46].

Level	Factor				
	Emulsion depth[a], *A* (cm)	Irradiation time, *t* (min)	Lamp power, *P* (W)	Distance[b], *D* (cm)	Initiator concentration[c], *C* (%)
1	1	15	3000	10	2.0
2	3	10	2500	15	1.5
3	5	5	2000	20	1.0
4	7	3	1500	25	0.5

[a] Distance from the top to the bottom of the emulsion.
[b] Distance between UV lamp and the upper surface of the emulsion.
[c] Initiator: benzoin butyl ether.
Source: Reprinted from Xiao et al. [46]. Copyright 2007, with permission from Elsevier.

formers, and W_C is the weight of the obtained microcapsules after removing the unreacted wall formers, homopolymer and unencapsulated epoxy. It is presumed that the wall materials are sufficient to encapsulate all the epoxy resin. Thus, the weight of the epoxy-containing microcapsules is proportional to their wall weight; the latter was determined by extracting the ground microcapsules with acetone for 48 hours.

For comparison, a control experiment was done as in the following using the same emulsion but a thermal initiator. Here, 100 ml emulsion was charged into a three-neck flask (~250 cm^3) fitted with a nitrogen bubbler, magnetic stir bar and condenser. Under protection of nitrogen and stirring at 500 rpm, 10 ml initiator (0.6% ammonium persulfate solution) was added dropwise to the emulsion. The reaction proceeded for three to seven hours at 70 °C. The product was poured into a large amount of water to remove the emulsifier. The microcapsules were collected in a centrifuge and washed with water several times. Finally, the powder-like microcapsules were obtained after drying in a vacuum freeze dryer.

Emulsification of the epoxy phase containing epoxy diacrylates and HPMAS as wall formers in the continuous aqueous phase offers the possibility of forming capsules by virtue of interfacial polymerization induced by UV irradiation. During emulsification, each wall former, especially HPMAS, tends to partition into the aqueous phase while minute amounts of water diffuse into the oil phase. The epoxy diacrylates and HPMAS, which are not completely soluble in epoxy resin, tend to separate and congregate at the interface between the emulsion colloid and water. This process is ascribed to the desirable solvent-induced phase separation. That is, the hydrophilic wall formers are driven to form colloidal shells.

The mechanism of the solvent-induced phase separation can be further understood by the interaction among the emulsifier, epoxy diacrylates and HPMAS. Meanwhile, HPMAS plays a dual role as emulsifier and polymerizable wall former. As a water-soluble chemical, HPMAS has an affinity for water and acts as an emulsifier at the colloid surface. On the other hand, since two hydroxyl groups will be formed during the reaction of epoxy with acrylic acid, epoxy diacrylates are more likely located at the out layer of the emulsion colloid due to the interaction between hydroxyl and carbonyl groups attached to the emulsifiers, i.e. SMANa and HPMAS. As a result, whenever the emulsion is exposed to UV light, free radical copolymerization of epoxy diacrylates with HPMAS would be initiated to form the shell polymer. Even if the epoxy diacrylates cannot accumulate completely near the aqueous phase, the copolymer of epoxy diacrylates with HPMAS would be more hydrophilic than the epoxy resin. Eventually, microcapsules with epoxy as core and the copolymer as shell are produced.

The capsule formation analyzed above can be proved by the following experimental results. Under UV irradiation, the shell phase containing the wall formers was cured into a solid. As shown in Figure 3.2a, the resulting microcapsules have relatively uniform sizes. No destruction of the capsule walls due to mechanical agitation is perceived. When the capsule has been artificially crushed, a core–shell structure rather than solid particle or porous structure can be identified Figure 3.2c. The wall thickness is estimated at ~300 nm according to Figure 3.2d that shows a microcapsule ~30 μm in diameter. Besides, both the inner and exterior surfaces of the microcapsule are quite smooth (Figure 3.2b,f), but actually the shell is composed of many layers (Figure 3.2e). These morphological features correlate well with the prediction of capsule formation made on the basis of solvent-induced phase separation.

Fourier transform infrared (FTIR) spectra of the uncured epoxy (EPON 828), EPON 828-containing microcapsules, and the shells that were obtained by grinding the

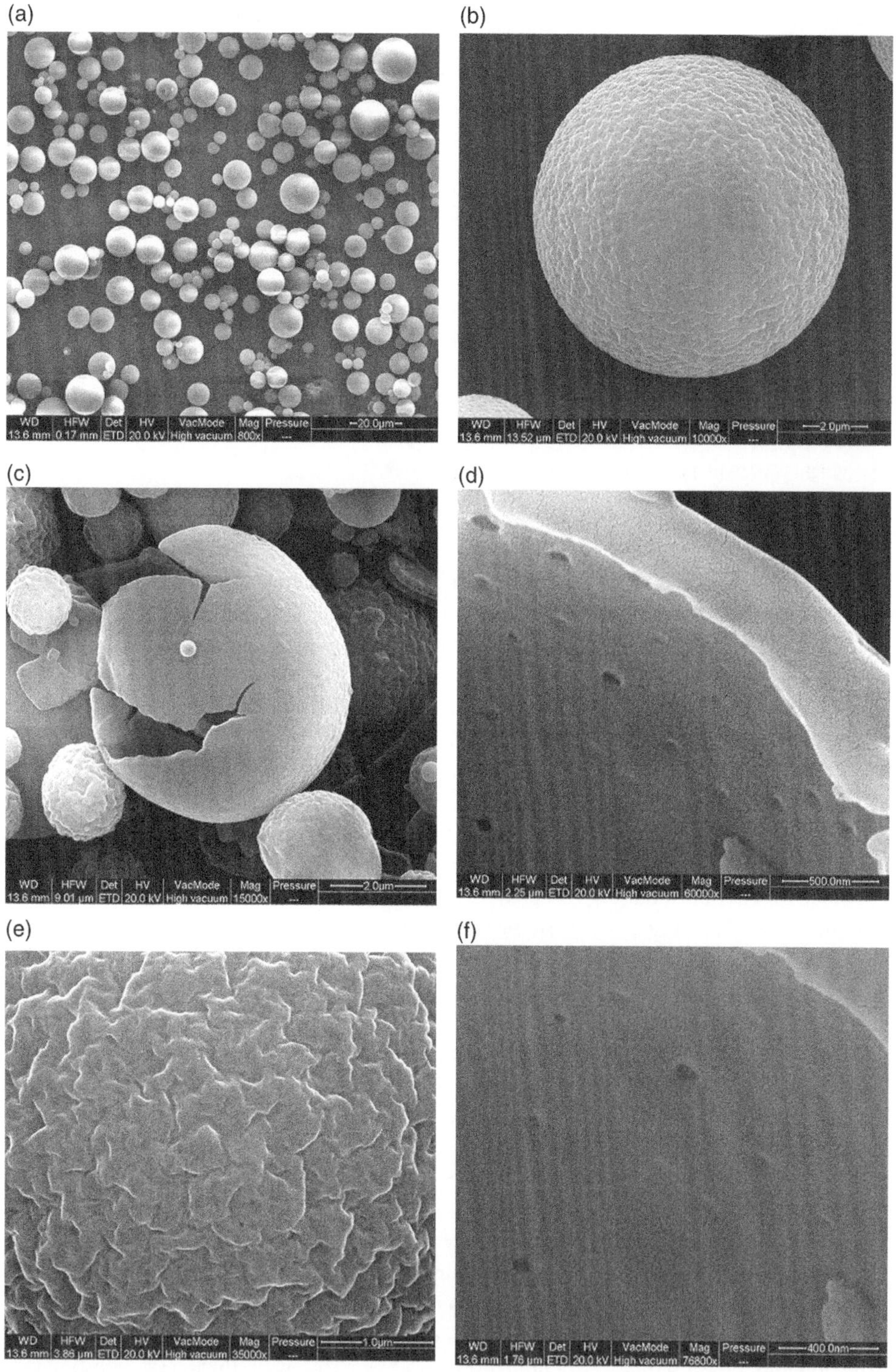

Figure 3.2 Scanning electron microscopy photos of DTP-loaded microcapsules [46]. (a, b) Appearance of the microcapsules; (c) crushed microcapsule; (d) close up view of the capsule shell; (e) exterior surface of a microcapsule; and (f) interior surface of a microcapsule. *Source:* Xiao et al. [46], Figure 3, p. 5/with permission of Elsevier.

microcapsules and then acetone extraction, are shown in Figure 3.3. Evidently, the epoxide groups inside the microcapsules are detectable as characterized by their distinct peaks at 980, 910, and 771 cm^{-1}. These results suggest that the structure of epoxy resin remains unchanged after encapsulation. On the other hand, for both the shells and microcapsules, the characteristic absorption peaks of carbonyl and sulfonic groups were moderately visible at 1739

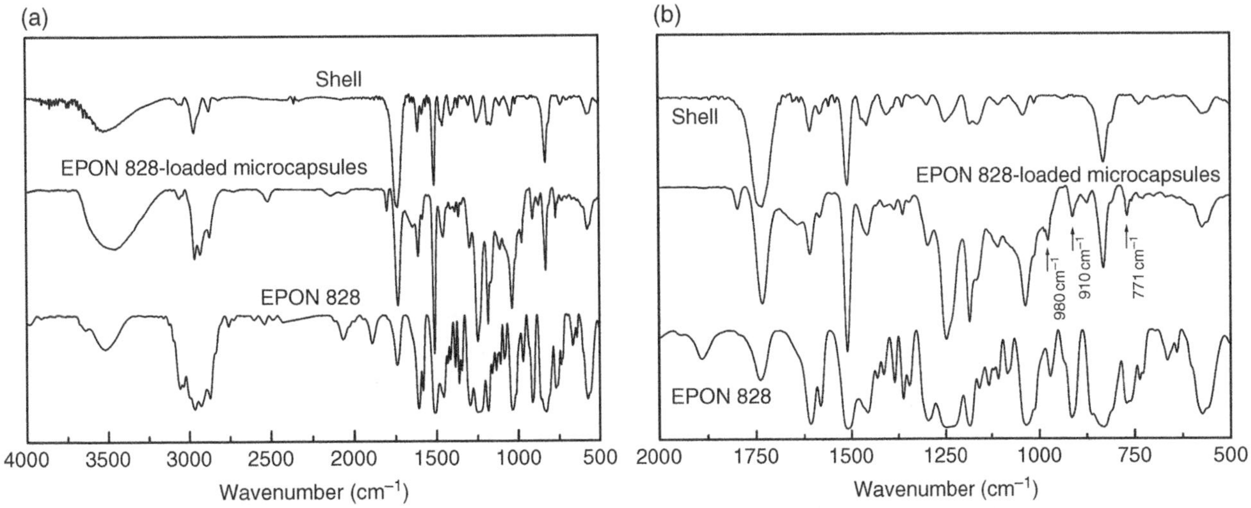

Figure 3.3 Fourier transform infrared spectra of EPON 828, EPON 828-loaded microcapsules and the shells made from EPON 828-epoxy diacrylates/HPMAS. (a) The entire spectra and (b) amplified portion of the spectra from 500 to 2000 cm^{-1}. *Source:* Reprinted from Xiao et al. [46]. Copyright 2007, with permission from Elsevier.

Table 3.2 Pyrolytic characteristics of epoxy, epoxy-containing microcapsules and capsule shells made from EPON 828-epoxy diacrylates/HPMAS.

	First pyrolytic stage			Second pyrolytic stage		
Material	$T1_{onset}$[a] (°C)	$T1_p$[b] (°C)	$T1_{end}$[c] (°C)	$T2_{onset}$[a] (°C)	$T2_p$[b] (°C)	$T2_{end}$[c] (°C)
EPON 828	306	327	342	–	–	–
DTP	242	276	288	–	–	–
EPON 828-containing microcapsules	314	339	358	420	436	450
DTP-containing microcapsules	268	294	328	399	421	449
Shells	–	–	–	362	398	418

[a] $T1_{onset}$ and $T2_{onset}$, onset temperatures of the first and second pyrolytic stages, respectively.
[b] $T1_p$ and $T2_p$, peak temperatures of the first and second pyrolytic stages, respectively.
[c] $T1_{end}$ and $T2_{end}$, end temperatures of the first and second pyrolytic stages, respectively.
Source: Reprinted from Xiao et al. [46]. Copyright 2007, with permission from Elsevier.

and 1175 cm^{-1}, respectively, indicating that the shells are composed of epoxy diacrylates and HPMAS as expected.

The core content of the microcapsules can be measured by subtracting the shell weight from that of the capsules and lies in the range from 54 to 71% for capsules 5–35 μm in diameter. The larger microcapsules contain a larger amount of epoxy. This might result from the fact that the shells of the microcapsules become thinner with the expansion of capsules. Therefore, the microcapsules with larger size should contain larger proportions of the core substance (i.e. epoxy).

The core–shell structure of the microcapsules is further revealed by their pyrolytic behavior. As illustrated in Table 3.2, both neat epoxy and shells show one stage of thermal decomposition, while the latter start to decompose at higher temperature owing to their cross-linked structure. Accordingly, the microcapsules exhibit two stages of weight loss with a rise in temperature. The first stage is attributed to the degradation of core epoxy, while the second corresponds to the shell decomposition. Compared with neat epoxy, the thermal stability of the epoxy encapsulated in the microcapsules is markedly increased owing to the protection of the shells. Meanwhile, decomposition of the shells of the microcapsules (the second pyrolytic stages) also occurs at higher temperature than neat shells, implying that certain reactions might exist between the core and shell at high temperature. It demonstrates that the core and shell materials become more stable when they are combined into the microcapsules.

It is worth noting that EPON 828-containing microcapsules possess higher thermal stability than that of DTP-containing microcapsules. In this context, the former is more appropriate for manufacturing self-healing epoxy composites with high temperature curing agent.

The effect of epoxy diacrylates is illustrated in Figure 3.4. Without epoxy diacrylates, the surface of the microcapsules made from amine acrylate and HPMAS copolymer is rough and wizened (Figure 3.4a). A few broken capsules are observed, indicating that they are somewhat fragile. In fact, the crimpy surface reflects high shrinkage resulting from copolymerization between the low molecular weight amine acrlyate and HPMAS. In contrast to the case shown in Figure 3.4a, the microcapsules based on epoxy diacrylates and HPMAS have much smoother surfaces (Figure 3.4b,c). In addition, EPON 828-epoxy diacrylates perform better than DTP-epoxy diacrylates in constructing the microcapsule surface. This phenomenon may be due to the fact that EPON 828-epoxy diacrylates possess more hydroxyl groups than DTP-epoxy diacrylates and hence acquire higher hydrophilicity in favor of building uniform shell layers.

With respect to the cross-linking agent, PETA, its role in improving thermal stability of the microcapsules is also shown by the pyrolytic behaviors of the latter. The peak thermal decomposition temperatures of the capsules with PETA at the two pyrolytic stages (Table 3.2) are obviously higher than those without PETA. Although epoxy diacrylate might also act as a cross-linking agent owing to the two double bonds at its molecular ends, introduction of PETA that possesses three double bonds would undoubtedly increase the cross-linking density of the shells.

Since the epoxy-containing microcapsules are prepared for producing self-healing epoxy composites, reactivity of the encapsulated epoxy should be explored. Accordingly, a model mixture consisting of stoichiometric encapsulated epoxy (obtained by grinding the microcapsules) with

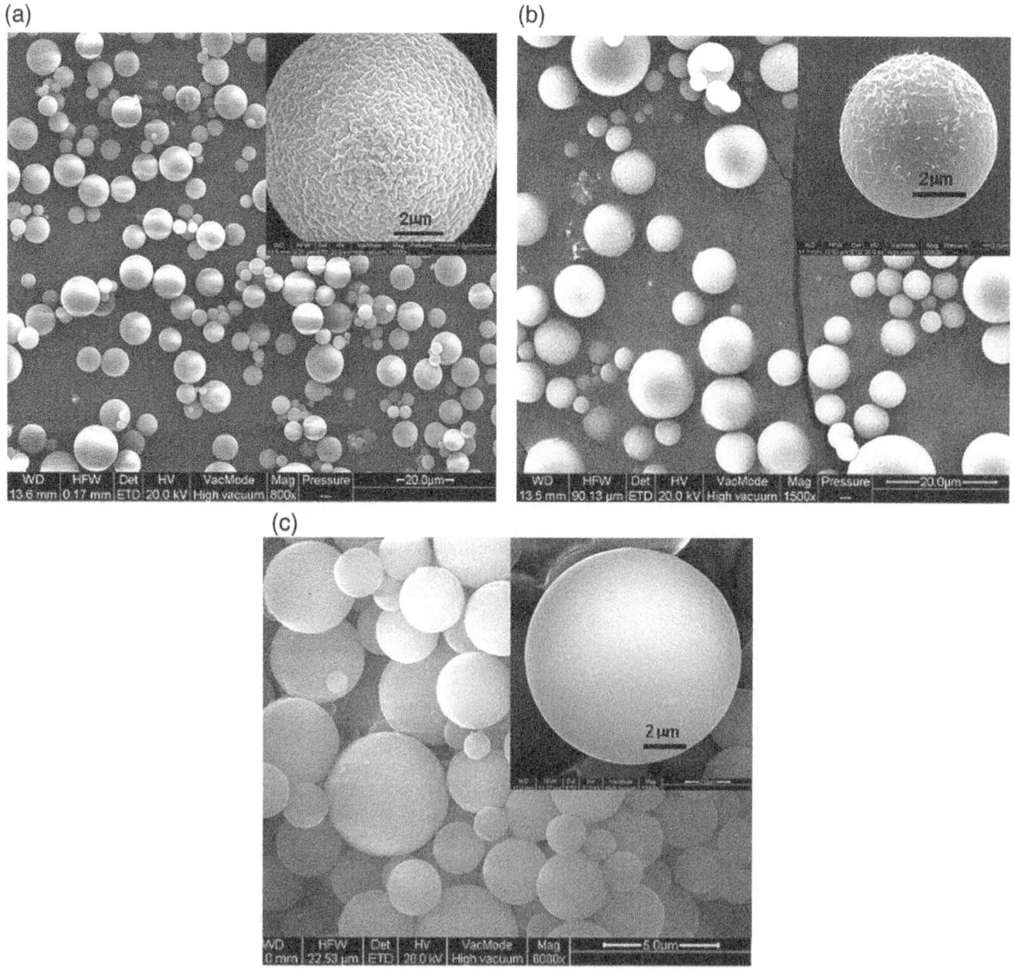

Figure 3.4 Scanning electron microscopy photos of DTP-containing microcapsules. The shells were made from: (a) amine acrylate/HPMAS; (b) DTP-epoxy diacrylates/HPMAS; and (c) EPON 828-epoxy diacrylates/HPMAS. *Source:* Xiao et al. [46], Figure 6, p. 7/with permission of Elsevier.

curing agent 4,4′-diaminodiphenyl sulfone (DDS) was heated and monitored by differential scanning calorimetry (DSC) at different heating rates. It was found that the exothermic peaks of encapsulated epoxy/DDS are similar to the control epoxy/DDS system. By using the well-known Kissinger equation [48], the activation energy of the reaction of the encapsulated epoxy/DDS pair was estimated at $66.9\,kJ\,mol^{-1}$, which is almost the same as the value of the epoxy/DDS system ($66.0\,kJ\,mol^{-1}$), which means that the epoxy resin inside the microcapsules retains its reactivity.

As discussed above, the epoxy-containing microcapsules are produced via emulsification of epoxy resin followed by UV-induced in-situ polymerization of the wall formers. Therefore, all the factors that determine size and size distribution of colloid particles should affect the size of the microcapsules. In the following, the variation in size of the microcapsules is investigated by changing agitation rate, content of SMANa, feeding weight ratio of core–shell monomers, and epoxy diacrylates/HPMAS weight ratio.

Figure 3.5a shows the influence of agitation rate during emulsification on size distribution of the microcapsules. Clearly, increasing the agitation rate reduces the size of the microcapsules and narrows the size distribution. This effect is less significant when the agitation rate exceeds 14 000 rpm. Scanning electron microscopy (SEM) observation further shows the phenomenon that a higher agitation rate leads to smaller microcapsules. Nevertheless, the surface appearance of the microcapsules of different sizes is similar. These results are likely a reflection of the fact that the colloid size decreases along with increasing agitation rate. Figure 3.5b shows that the size and size distribution of the microcapsules decreases with increasing emulsifier dosage. This proves that the size of the microcapsules is indeed correlated with the emulsion colloid size.

In addition, the size of the microcapsules is proportional to the core content. Raising the feeding weight ratio of core–shell monomers would enlarge the capsules and broaden their size distribution (Figure 3.5c). This phenomenon is

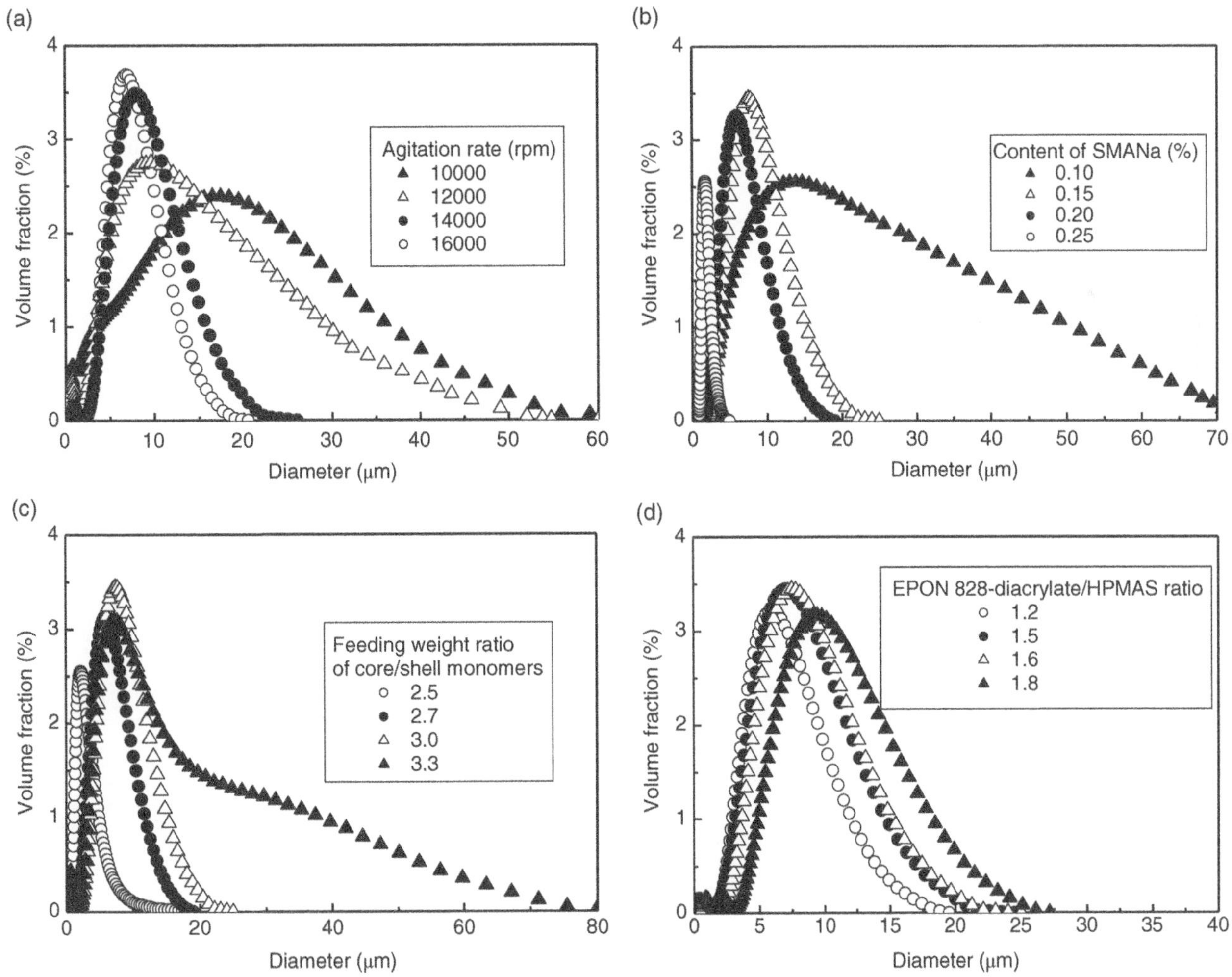

Figure 3.5 Size distribution of DTP-containing microcapsules. Preparation conditions: (a) feeding weight ratio of core/shell monomers = 3.0, EPON 828-epoxy diacrylates/HPMAS ratio = 1.6, SMANa content = 0.15%; (b) feeding weight ratio of core/shell monomers = 3.0, EPON 828-epoxy diacrylates/HPMAS ratio = 1.6, agitation rate = 14 000 rpm; (c) EPON 828-epoxy diacrylates/HPMAS ratio = 1.6, SMANa content = 0.15%, agitation rate = 14 000 rpm; and (d) feeding weight ratio of core/shell monomers = 3.0, SMANa content = 0.15%, agitation rate = 14 000 rpm. *Source:* Reprinted from Xiao et al. [46]. Copyright 2007, with permission from Elsevier.

consistent with the general law of capsule preparation. In the experiments conducted in this work, feeding weight ratios of core–shell monomers are sufficient for constructing microcapsules with core–shell structure. In the case of an extremely low feeding weight ratio of core–shell monomers, however, porous particles might be prepared. When the feeding weight ratio of core–shell monomers is too high, in contrast, a certain amount of the core substance might not be encapsulated. The upper and lower limits of the feeding weight ratio of core–shell monomers should be studied further.

Dependence of the size of the microcapsules on epoxy diacrylates/HPMAS ratio is shown in Figure 3.5d. With a rise in epoxy diacrylates/HPMAS ratio, size and size distribution of the microcapsules both moderately increase. This should be related to the functions of the two chemicals. Epoxy diacrylate is an oligomer, while HPMAS plays dual roles as monomer and emulsifier. Less HPMAS might reduce emulsion stability, leading to larger capsules and wider size distribution of the capsules.

The present work deals with the preparation of epoxy-containing microcapsules using UV irradiation. Undoubtedly, the influence of the conditions of UV irradiation should be carefully studied. For this purpose, orthographic factorial design was applied, and the microcapsule yield was set as the response of the designed experiments. Table 3.3 and Figure 3.6 show the results and analysis of the effects of the five factors at four levels (see Table 3.1).

From Figure 3.6, it is found that emulsion depth has a very evident influence on the capsule yield. The emulsion can absorb and reflect UV light. Accordingly, in the depths of the emulsion the light intensity has to be weakened so

Table 3.3 Results analysis of the orthographic factorial design for microcapsule preparation under UV irradiation [46].

	Factor[a]					
	A	t	P	D	C	Microcapsule yield (%)
Level	1	2	3	2	3	80.2
	3	4	1	2	2	60.4
	2	4	3	3	4	66.3
	4	2	1	3	1	63.1
	1	3	1	4	4	76.5
	3	1	3	4	1	61.8
	2	1	1	1	3	81.5
	4	3	3	1	2	58.4
	1	1	4	3	2	79.9
	3	3	2	3	3	63.3
	2	3	4	2	1	68.9
	4	1	2	2	4	64.8
	1	4	2	1	1	76.4
	3	2	4	1	4	66.5
	2	2	2	4	2	74.8
	4	4	4	4	3	50.9
Results analysis						
[b]K_{1j} (%)	313.0	288.0	281.6	282.8	270.2	–
K_{2j} (%)	291.6	284.7	279.2	274.2	273.4	–
K_{3j} (%)	251.9	267.1	266.8	272.6	275.9	–
K_{4j} (%)	237.1	253.9	266.1	264.1	274.2	–
[c]Y_{1j} (%)	78.3	72.0	70.4	70.7	67.6	–
Y_{2j} (%)	72.9	71.2	69.8	68.6	68.4	–
Y_{3j} (%)	63.0	66.8	66.7	68.1	69.0	–
Y_{4j} (%)	59.3	63.5	66.5	66.0	68.5	–

[a] The symbols denoting the five factors have the same meaning as those in Table 3.1.
[b] K_{ij} denotes the sum of microcapsule yield with level i and factor j.
[c] Y_{ij} denotes the average of K_{ij}. $Y_{ij} = K_{ij}/N_j$, where N_j is the number of level i with the same factor j.
Source: Reprinted from Xiao et al. [46]. Copyright 2007, with permission from Elsevier.

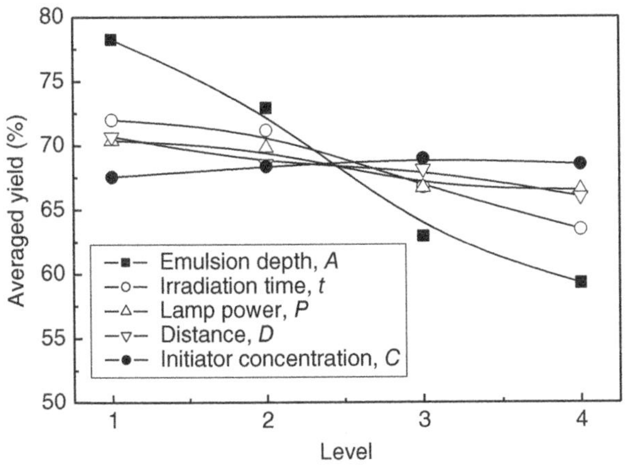

Figure 3.6 Influence of different levels of the five factors (see Table 3.1) on the averaged capsule yield (that was taken from the values of Y_{ij} listed in Table 3.3). *Source:* Reprinted from Xiao et al. [46]. Copyright 2007, with permission from Elsevier.

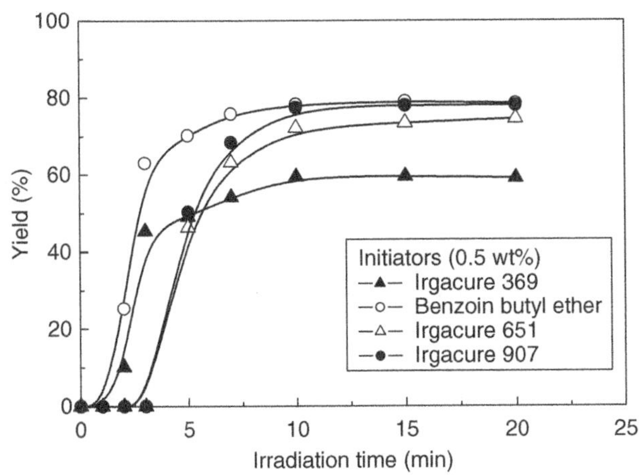

Figure 3.7 Ultraviolet irradiation time dependence of microcapsule yield initiated by different photoinitiators. Irradiation conditions: emulsion depth = 3 cm, irradiate time = 10 minutes, lamp power = 2.5 kW, distance between UV lamp and the emulsion = 10 cm, and initiator concentration = 0.5%. *Source:* Reprinted from Xiao et al. [46]. Copyright 2007, with permission from Elsevier.

that the yield is reduced. Similarly, lower yields are attained by reducing irradiation time, or lowering the lamp power, or increasing the distance between the lamp and the emulsion, decreasing UV light intensity. In general, the differences in the yield at different levels are moderate except for the emulsion depth. In particular, the initiator concentration has nearly no influence on the capsule yield within the range of interest.

The data in Figure 3.6 indicate that the experiments carried out at level 1 give the highest capsule yield. However, it does not mean that optimum conditions can thus be concluded because of the complicated situation in practice. For example, the emulsion depth is the most important factor for the capsule yield and a thinner emulsion depth results in higher yield, but an emulsion with too thin depth (1 cm) does not favor its stirring. As a result, a 3 cm depth of the emulsion in a reverse conical beaker is suggested. For irradiation time and lamp power, since levels 1 and 2 show similar results, level 2 is chosen for energy saving. The distance between the UV lamp and the emulsion is selected at level 1, while the initiator concentration is fixed at level 4 for cost control. In summary, the optimum conditions for the UV irradiation-aided encapsulation are: 3 cm for emulsion depth, 10 minutes for irradiation time, 2.5 kW for lamp power, 10 cm for the distance between the UV lamp and the emulsion, and 0.5 wt% for the photoinitiator.

To reveal formation kinetics of the microcapsules, dependencies of capsule yield on UV irradiation time are shown in Figure 3.7. For all the systems, regardless of the photoinitiator species, the reaction rates are rather high within the first few minutes and then level off. Compared with the systems initiated by benzoin butyl ether, Irgacure 651 and Irgacure 907 that have similar equilibrium capsule yields of about 80%, the system with Irgacure 369 presents a significantly lower yield. The results might be related to the structures of the radicals originating from photodecomposition (Scheme 3.3). One of the radicals from Irgacure 369 has a phenyl group, which should set up a higher steric hindrance to monomer access.

A careful survey of Figure 3.7 indicates that there is no induction period at the early stage of polymerization for the systems with benzoin butyl ether and Irgacure 369, while an induction period of three minutes is required for the systems with Irgacure 651 and Irgacure 907. In the former cases, under the irradiation of UV light, a large amount of free radicals are produced in a short time and start to initiate copolymerization of epoxy diacrylates with HPMAS immediately. As a result, the inhibitor in the system becomes negligible and has no retardation effect. The induction period observed in conventional free radical polymerization is instead replaced by a high speed conversion period. In contrast, for the latter cases, the amount of free radicals is not enough to quickly consume the inhibitor so that the induction period appears. It seems that the ultimate capsule yield has nothing to do with the induction period.

To highlight the advantage of the emulsion encapsulation induced by UV light, a control experiment was carried out using the same emulsion except that a water-soluble initiator, $(NH_4)S_2O_8$, was added instead of photoinitiators. The microcapsules emerge after three hours of reaction, and five hours are needed for completion of the reaction. In comparison with the encapsulation process induced by UV irradiation that is completed within 10 minutes, the thermally initiated encapsulation takes much longer.

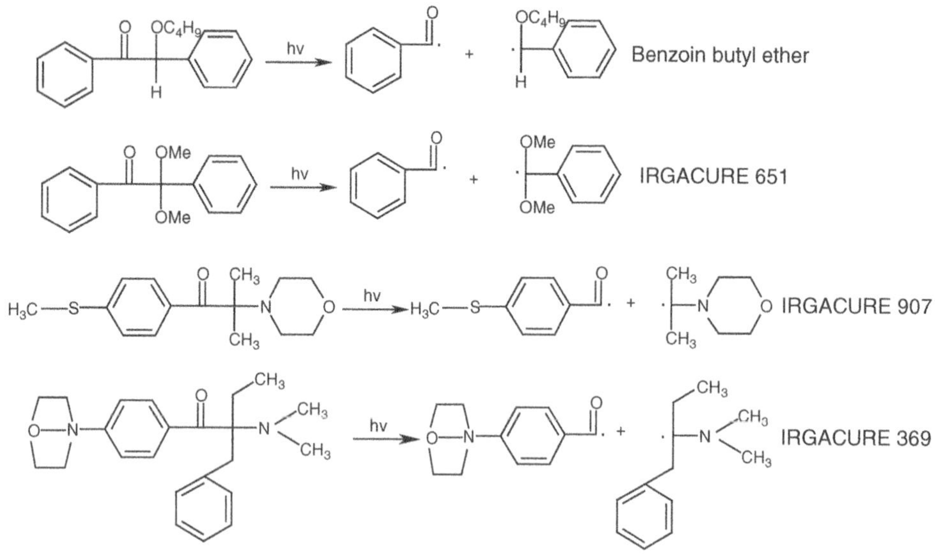

Scheme 3.3 Photodecomposition of the initiators.

Besides, the microcapsules produced via thermal initiation have larger size and wider size distribution. This may also be a result of the prolonged reaction. The emulsion colloids have enough time to collide with each other and to merge before curing of the shell layer. For the UV-induced reaction, the colloids can be solidified within a few minutes, leaving no time to change the colloid size.

3.1.2 Encapsulation of Boron-Containing Curing Agent

The boron-containing curing agent $(C_2H_5)_2O{\cdot}BF_3$, which is planned to work as the mate of epoxy monomer in the healing agent, is highly active and hydroscopic. Its activity would easily be lost if conventional encapsulation approaches are applied. Since it should be properly shielded from the composites manufacturing to avoid any possible deactivation effect, either before or after being introduced into the composites, as stated in the beginning of this chapter, two methods have been developed in our laboratory. The first one avoids the difficulty of including the chemical in the microspheres and simply utilizes porous materials as absorbents. The second one differs from the traditional way of microencapsulation and infiltrates the fluidic $(C_2H_5)_2O{\cdot}BF_3$ into polymeric hollow microcapsules instead. The following subsections will be devoted to the processing details of these two methods.

3.1.2.1 Loading Boron-Containing Curing Agent onto Porous Media

It is well known that a material that is porous may absorb gases and/or liquids into porosities (particularly if they are inter-connecting) through capillary action. The interaction between the absorbed substances and the porous material is governed by either physical or chemical attraction forces [49]. For physical absorption, van der Waals force plays the leading role to align the adsorbed species into layers on the existing surface, and the processes of absorption and desorption are fully reversible with high speed. By using this characteristic, therefore, the highly active $(C_2H_5)_2O{\cdot}BF_3$ can be loaded with porous materials, wrapped by a polymer shell, and released under proper circumstances for conducting crack healing. Evidently, this strategy is much simpler than that based on making microcapsules.

Hereinafter, three types of porous media are employed. They are carbon black, rayon-based activated carbon fiber and sisal fibers, which tend to physically absorb $(C_2H_5)_2O{\cdot}BF_3$ instead of chemisorption.

Carbon black [50] is a form of amorphous carbon produced by incomplete combustion of heavy petroleum products with a small amount from vegetable oil. It can be used as absorbent due to its aggregated structure consisting of numerous tiny primary carbon particles, which offers large surface area.

Activated carbon fibers [51] are made from natural or artificial fibers by high-temperature carbonization and activation, characterized by large specific surface area, well-developed porosity, high absorptive capacity, and fast desorption speed. It has been widely applied in fields such as air and water purification.

Sisal is a lignocellulose fiber [52], which has been used as reinforcement [53], replacing artificial fibers, in polymer composites as a result of ecological considerations. A piece of sisal fiber is not a single filament like carbon or glass fibers but is a bundle of cellular aggregates that consist of more than 100 irregular hexagonal hollow cells (Figure 3.8).

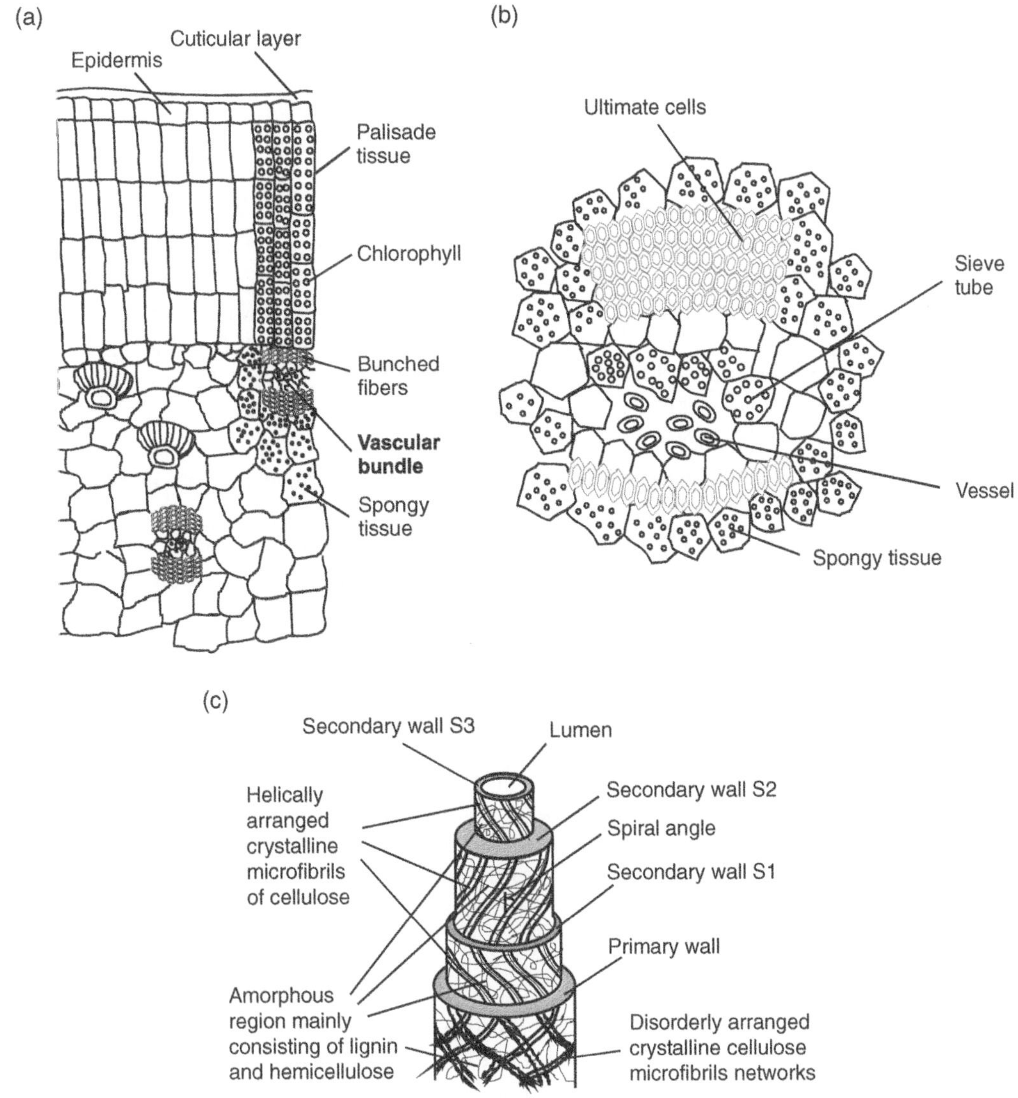

Figure 3.8 Schematic drawing of (a) cross-section of a piece of sisal leaf, (b) inner construction of bunched sisal fibers, and (c) structural constitution of a sisal fiber cell.

Each hollow cell is built up from a number of cellulose microfibrils that are embedded in lignin and hemicellulose and helically wound along the fiber axis. Therefore, there are enormous cavities of different sizes in the fibers that are available for holding hydrophilic fluids.

The typical loading procedures are described in the following. Having been vacuum dried, 1 g of the porous media (carbon black, rayon-based activated carbon fiber, or sisal fibers) were put into a sealed vessel and the system was evacuated again. Then, $(C_2H_5)_2O{\cdot}BF_3$ was introduced into the vessel via an injector until the porous media were completely submerged. After 25 hours, the extra liquid was removed and 10 ml 2.5% PS solution in dichloromethane was added with stirring. The obtained suspension was added into 100 ml anhydrous ethanol by dropwise feeding. Finally, the porous media were collected by filtration, washed with anhydrous ethanol three times, and dried naturally. The loading rate of $(C_2H_5)_2O{\cdot}BF_3$ absorbed by the porous media was determined by the analysis method shown below. The porous media carrying $(C_2H_5)_2O{\cdot}BF_3$ were thoroughly ground to destroy the PS protective coating, and dispersed in water for 24 hours allowing the reaction with water:

$$(C_2H_5)_2O\,BF_3 = BF_3 + (C_2H_5)_2O$$

$$4BF_3 + 6H_2O = 3H_3O^+ + 3BF_4^- + B(OH)_3$$

The pH value of the system was then determined. In the meantime, a control experiment was carried out by dispersing the same weight of porous media excluding

$(C_2H_5)_2O·BF_3$ in water. The pH value of the control system acted as the blank, and the loading rate of $(C_2H_5)_2O·BF_3$ on the porous media was calculated by:

$$4/3\left[H^+\right]VM_{BF_3 \times O(C_2H_5)}/W_{Carrier} \tag{3.4}$$

where $[H^+]$ denotes the acidity deducted from the blank value, V is the volume of water for hydrolysis, $M_{BF_3 \times O(C_2H_5)}$ is the molecular weight of $(C_2H_5)_2O·BF_3$, and $W_{Carrier}$ is the weight of the porous media.

Figure 3.9 illustrates the effect of infiltration time on the weight gain of the porous media during absorption of $(C_2H_5)_2O·BF_3$. All three materials exhibit similar uptake behaviors, characterized by fast absorption at the beginning and gradually leveling off with a further rise in time. Considering the purpose of this work, the most interesting part is not how these materials absorb $(C_2H_5)_2O·BF_3$ but how much $(C_2H_5)_2O·BF_3$ they can eventually absorb. In this context, equilibrium absorption capacities of the porous media are compared. It is clearly seen from Figure 3.9 that sisal has the highest equilibrium absorption. This should be related to the specific microstructure of the plant fibers, which are different from those of carbon black and activated carbon fibers. Sisal consists of many pipeline-like lumens (30 μm in diameter) contributing to water and nutrient uptake [54], unlike carbon black and activated carbon fibers, which are rich in micropores. Accordingly, when sisal is submerged in $(C_2H_5)_2O·BF_3$, not only absorption on a molecular scale but also direct filling with the liquid along the open channels occurs. As a result, the weight gain of sisal due to $(C_2H_5)_2O·BF_3$ uptake reaches the highest equilibrium value at a much faster speed than the other two types of porous media.

As mentioned above, on one hand, the mixture of epoxy and a small amount of $(C_2H_5)_2O·BF_3$ can be cured within a very short time under lower temperature. On the other hand, the rapid physical absorption of $(C_2H_5)_2O·BF_3$ must correspond to rapid desorption. Therefore, application of polymeric coating on the porous media loaded with $(C_2H_5)_2O·BF_3$ in advance is necessary to prevent direct contact between epoxy monomer and $(C_2H_5)_2O·BF_3$ when fabricating self-healing epoxy composites. Here we used PS as the coating material. The coating thickness can be adjusted by changing concentration and applied dosage of the solution of PS in dichloromethane. It was found that a 2.5% PS solution four times the weight of the porous media carrying $(C_2H_5)_2O·BF_3$ gave the optimum shielding effect.

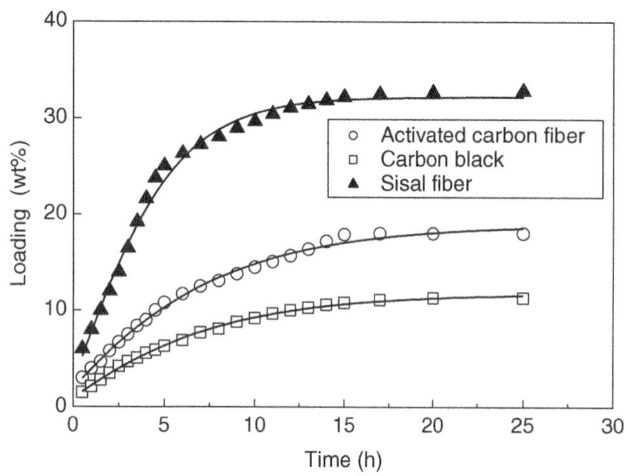

Figure 3.9 Infiltration time dependence of the loading of $(C_2H_5)_2O·BF_3$ on different porous media.

The appearances of the porous media before and after PS coating are shown in Figure 3.10. On the whole, the materials surfaces with the protective layer are less rough. The agglomerated carbon black particles are bound together forming larger clusters, while the grooves on either activated carbon fibers or sisal become less discernible. In fact, the surface polarity of the porous materials determines the coating's homogeneity. At first, the $(C_2H_5)_2O·BF_3$-loaded porous media are wetted by PS dissolving in CH_2Cl_2. As soon as ethanol is added to the system, PS is precipitated covering the surfaces of the porous media. Since both PS and carbon black are low polar substances, the resultant polymer layer is rather even and compact. As for activated carbon fibers and sisal fibers that contain large amounts of polar groups, however, the PS coatings are characterized by evident heterogeneity due to uneven spreading of the PS molecules.

3.1.2.2 Microencapsulation of Boron-Containing Curing Agent Via the Hollow Capsules Approach

Hollow microcapsules are gas-filled spherical particles with advantageous properties, such as low effective density and high specific surface area. They are often used as opacifying plastic pigments for various coatings and gloss-enhancing plastic pigments, lightweight reinforcements, light scattering materials, etc. [55]. Fabrication of tailor-made hollow spherical structures has been carried out by many processes, including spray drying and dripping [56, 57], emulsion/interfacial polymerization [58], and self-assembly [59]. In general, the template technique is involved. That is, microcapsules form with the core (liquid droplet or solid particle) as template. Subsequently, hollow spheres are acquired by decomposing the liquid or solid core either by dissolution, evaporation, or thermolysis. It is worth noting that, however, the shell of the capsules used to be more or less damaged during core decomposition [60–62].

In fact, when a microbubble acts as the template, the integrity of the hollow microcapsules can be ensured. Of course, the shell formation rate should be fast enough to encapsulate microbubbles before their disappearance, as the survival

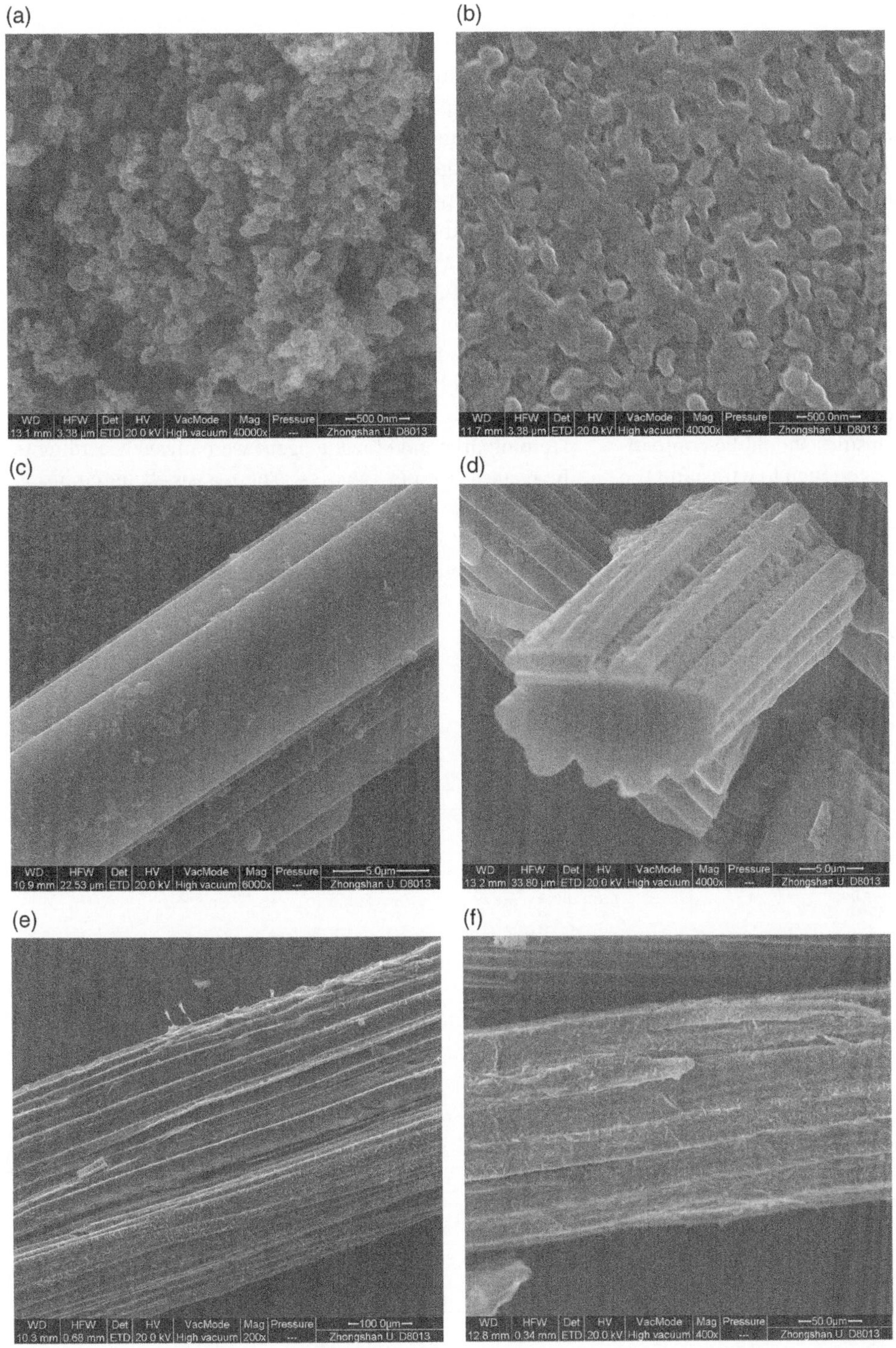

Figure 3.10 Morphologies of (a) as-received carbon black, (b) $(C_2H_5)_2O{\cdot}BF_3$-loaded carbon black with PS coating, (c) as-received rayon-based activated carbon fiber, (d) $(C_2H_5)_2O{\cdot}BF_3$-loaded rayon-based activated carbon fiber with PS coating, (e) as-received sisal, and (f) $(C_2H_5)_2O{\cdot}BF_3$-loaded sisal with PS coating. The loading rates of $(C_2H_5)_2O{\cdot}BF_3$ and weight percentages of PS coating, respectively, are: (b) 11 wt%, 8.3 wt%; (d) 18 wt%, 7.8 wt%; and (f) 32 wt%, 7.3 wt%.

time of microbubbles in liquid is significantly shorter than that of dispersed liquid or solid cores. Harris et al. prepared hollow microcapsules from n-butyl-2-cyanoacrylate (NBCA) using microbubble template [63]. NBCA monomers can be rapidly polymerized on the surface of microbubbles in the presence of water. Daiguji et al. described a technique for generating hollow melamine-formaldehyde microcapsules via direct encapsulation of microbubbles [64].

It is noteworthy that we have developed a novel UV-irradiation method to produce microcapsules containing epoxy resin (Section 3.1.1) [46]. The distinct advantage of this method lies in its rapid reaction rate at room temperature, which just meets the requirement of the microbubble template technique stated above. Thereby, here the hollow microcapsules to be used for holding $(C_2H_5)_2O{\cdot}BF_3$ are first prepared by UV-initiated radical copolymerization in emulsion [65].

To facilitate the subsequent introduction of $(C_2H_5)_2O{\cdot}BF_3$ into the hollow microcapsules, performance of the polymeric wall should be optimized by the proper choice of monomers. It is desired that the shell wall possesses similar polarity as $(C_2H_5)_2O{\cdot}BF_3$ for intimate contact between them. In addition, the microcapsules should be compatible with epoxy matrix and robust enough to withstand the manufacturing of epoxy composites in future applications. According to these criteria, the following monomers are chosen for building up the microcapsules: epoxy diacrylate (EPON 828-epoxy diacrylate), methyl methacrylate (MMA) and monobutyl itaconate (MBI), while PETA and 1,6-hexanediol diacrylate (HDDA) serve as the cross-linking agents. The monomers are characterized by photosensitivity and compatibility with epoxy, and contain many polar groups (such as hydroxyl, carbonyl and carboxylic groups). In addition, the following chemicals also take part in the synthesis: HPMAS (polymerizable emulsifier), SMANa (polymeric surfactant), methyl acrylic acid (MAA)/$NaHCO_3$ (blowing agent), and benzoin butyl ether (photoinitiator).

The hollow microcapsules were prepared via UV-initiated radical copolymerization on a microbubble template (Figure 3.11). First, the monomers, cross-linking agent, photoinitiator and blowing agent were mixed according to the recipes given in Table 3.4. Then, a gas–water emulsion was created by vigorously stirring surfactant aqueous solution (100 ml) containing dissolved $NaHCO_3$ (0.2 g), while dropping the above monomer mixtures within ~1 minute.

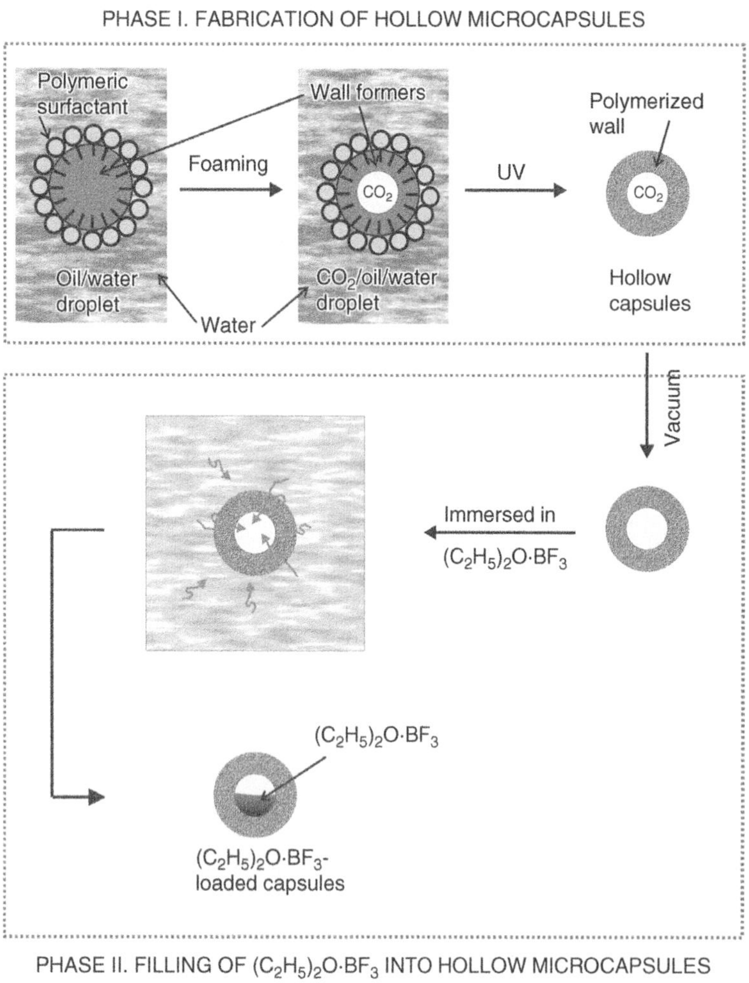

Figure 3.11 Schematic drawing of the proposed route for preparing hollow microcapsules and filling $(C_2H_5)_2O{\cdot}BF_3$ into hollow microcapsules. *Source:* Reprinted from Xiao et al. [65] Copyright 2009, with permission from Elsevier.

Table 3.4 Recipes for preparing hollow microcapsules in emulsion via UV-irradiation method[a].

Recipe ID	EPON 828-epoxy diacrylates (g)	MMA (g)	MBI (g)	HPMAS (g)	HDDA (g)
1#	3	0	0	3	0
2#	0	4	0	0	2
3#	0	2	0	3	1
4#	0	0	2	3	1

[a] SMANa, 3 wt%; PETA, 0.5 wt%; MAA, 2 wt%; photoinitiator, 0.5 wt% based on 100 ml water; agitation rate, 1.2×10^4 rpm (unless otherwise specified).
Source: Reprinted from Xiao et al. [65]. Copyright 2009, with permission from Elsevier.

Afterwards, the resultant emulsion was homogenized at 1.2×10^4 rpm with a homogenizer for an additional five minutes. Inside a UV oven, the emulsion was irradiated by UV light for 1–10 minutes under magnetic stirring at room temperature. The other synthetic parameters, such as concentrations of surfactant and initiator, and UV lamp power, followed the optimum conditions worked out previously (Section 3.1.1).

By pouring the irradiated emulsion into 500 ml of water, hollow microcapsules floated and accumulated on the solution surface. They were collected by filtration and washed with water and acetone alternately several times to remove residual emulsifier, monomers, and oligomers. Finally, powder-like hollow microcapsules were obtained after drying in a vacuum freeze dryer. The yield of the hollow microcapsules was roughly estimated by the ratio of the weight of the resultant hollow microcapsules to the feeding weight of the wall formers.

To highlight the role of UV-initiated polymerization, a comparative experiment was carried out using a thermal initiator instead. The emulsion (100 ml) prepared according to recipe 3# (Table 3.4, excluding the photoinitiator) was charged into a three-neck flask (about 250 cm^3) fitted with a nitrogen bubbler, magnetic stir bar, and condenser. Under protection of nitrogen and stirring at 500 rpm, 10 ml of thermal initiator (0.6% ammonium persulfate water solution) was dropped onto the emulsion. The reaction proceeded for three hours at 70 °C. The product was poured into a large amount of water to remove the emulsifier. Then, the floating particles were collected by filtration, and washed with water and acetone alternately several times, and finally dried in a vacuum freeze dryer.

To introduce $(C_2H_5)_2O{\cdot}BF_3$ into the resultant hollow microcapsules, the gas (i.e. CO_2) inside the hollow microcapsules was evacuated for one hour under vacuum. In addition, a $Ca(OH)_2$ solution was connected to the exporting pipeline to monitor whether CO_2 had been completely removed. After that, $(C_2H_5)_2O{\cdot}BF_3$ was vented into the system to submerge the microcapsules for 1–50 hours. Eventually, the $(C_2H_5)_2O{\cdot}BF_3$-loaded microcapsules were obtained by filtration and purged with ethyl ether. The loading rate of $(C_2H_5)_2O{\cdot}BF_3$ in the hollow microcapsules was determined by the chemical analysis described in the previous subsection for the $(C_2H_5)_2O{\cdot}BF_3$-loaded porous media.

As shown in Figure 3.11, the key step prior to UV irradiation is the formation of microbubbles, which are surrounded with a monomer layer and stabilized by surfactant. To achieve this objective, a special blowing system consisting of MAA and $NaHCO_3$ was applied. It can give out CO_2 as a result of the reaction between the two components. In practice, MAA was mixed with monomers in advance, while $NaHCO_3$ was dissolved in water. As the mixture of the monomers was dropped into the surfactant water under stirring, an emulsion of monomers (oil/water) formed accordingly. In the meantime, MAA in the monomer mixture tended to be dissolved in water due to its water solubility, and minute amounts of water diffused into the oil phase. As soon as MAA contacted $NaHCO_3$, a large amount of microbubbles was generated just inside the droplets of the monomer mixture or at the oil/water interface. Consequently, the original monomer emulsion (oil/water) turned into a CO_2/oil/water emulsion. Although a few oil droplets might remain unchanged during the foaming process, they were likely absorbed by the oil layer of the CO_2/oil/water droplets, because the latter possessed large specific surface area and were miscible with the monomers.

Having been irradiated by UV light, polymerization of the monomers took place and hollow microcapsules were produced. Figure 3.12a indicates that the capsules have relatively uniform size (2–10 μm), which is similar to that of the microbubble templates. This suggests that the shell was successfully in-situ consolidated before the microbubbles lost their stability. On the other hand, destruction of the capsules by mechanical agitation was not found for these collected floating microcapsules, meaning that the wall was strong. Figure 3.12b shows the artificially crushed microcapsules. A hollow structure (filled with CO_2) rather

Figure 3.12 Scanning electron microscopy micrographs of (a) hollow microcapsules made by UV-initiated polymerization, (b) crushed microcapsules made by UV-initiated polymerization, and (c) solid microspheres made by thermally initiated polymerization. Recipe: (a, (b) 3#; and (c) 3# except that the photoinitiator was replaced by a thermal initiator. *Source:* Xiao et al. [65], Figure 3, p. 4/with permission of Elsevier.

than a solid particle or porous structure can be identified. The wall thickness is estimated to be 200–300 nm for a sphere 8 μm in diameter.

Influence of shear on size of the hollow microcapsules was studied. As the stirring speed increased from 1.0×10^4 to 1.5×10^4 rpm during formation of the CO_2/oil/water emulsion, the capsules became obviously smaller, and the size distribution also narrowed. There is a linear reduction in mean diameter with stirring speed on log–log axes, while the specific surface area varies inversely. The results demonstrate that the hollow microcapsules can retain the original size of the microbubbles so long as the microbubbles keep stable during the hardening process.

To clarify the relationship between the sizes of the microbubbles and hollow microcapsules, the microbubbles generated at different stirring speeds were monitored. Clearly, the size of the microbubbles decreased with a rise in stirring speed, which coincides with the aforementioned effect of agitation rate on size distribution of hollow microcapsules. Moreover, since CO_2 would gradually diffuse into water, all the microbubbles have a tendency to decrease in size with time. Comparatively, the smaller microbubbles in the aqueous solution are more stable than the larger ones. On the whole, so long as the shell formation via UV-initiated polymerization is completed within five minutes, a solidified hollow structure of the microbubbles can be obtained. Otherwise, the resultant hollow microcapsules might be much smaller than the initial microbubbles.

Figure 3.13 gives the dependencies of yield of the hollow microcapsules on UV irradiation time. For all the systems, the yields increase quickly within the first several minutes and then level off. Each system attains its highest yield at

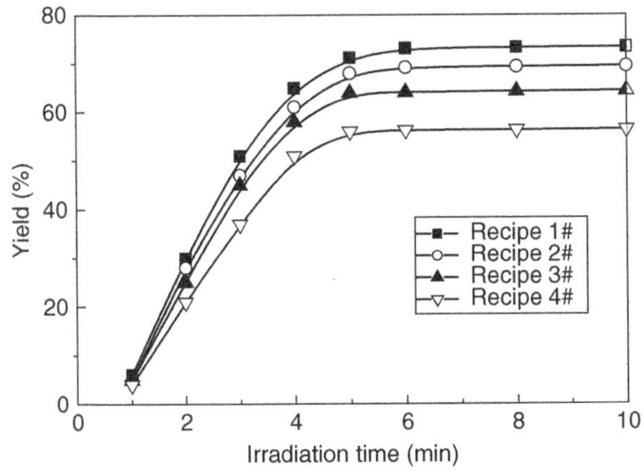

Figure 3.13 Yields of hollow microcapsules as a function of UV irradiation time. *Source:* Reprinted from Xiao et al. [65]. Copyright 2009, with permission from Elsevier.

about five minutes, which just falls in the stable stage of the microbubbles. The absence of an induction period in the early stage of polymerization resulting from the specific photoinitiator [47] must have made a critical contribution to this result.

On the other hand, the equilibrium yields of the hollow microcapsules range between 56% and 73% (Figure 3.13), indicating that a certain amount of monomer droplets has not been turned into microbubbles covered with the oil layer during blowing. In addition, these monomer droplets may prefer to stay in the bulk solution rather than migrate into the oil layer of microbubbles. After UV irradiation, they were converted into solid spheres and sank in the solution. There is another possibility that some monomers were dissolved in water and polymerized in the solution. Also, the broken capsules would be precipitated in the solution.

Figure 3.13 also shows that the equilibrium yields of the hollow microcapsules are different for different recipes. As the system without polymerizable emulsifier HPMAS (recipe 2# in Table 3.4) still presents a higher yield, dosage of the emulsifier might not be the key factor. Naturally, the water solubility of monomers is considered. It is seen from Figure 3.13 that recipe 3# containing MBI offers the lowest equilibrium yield. MBI is a water-soluble monomer owing to its carboxyl acid group and can act as a blowing agent by reacting with $NaHCO_3$. Its reactant MBINa (product of the reaction between MBI and $NaHCO_3$) is also soluble in water. These are certainly detrimental to the production of hollow microcapsules. In contrast, epoxy diacrylate and MMA are basically water-insoluble substances. The difference in the polar component of the solubility parameter between the monomer and water, $|\delta_{p,monomer} - \delta_{p,H2O}|$, is more remarkable in the case of epoxy diacrylate (Table 3.5). As water is a strong polar solvent, similarity in δ_p between water and its solute is closely related to dissolution. It suggests that diffusion of epoxy diacrylate monomers into water might be more difficult as compared with MMA, so that recipe 1# exhibits the highest equilibrium yield (Figure 3.13). In fact, the difference in the equilibrium yields between recipes 2# and 3# provides supporting evidence for the above analysis. Since MMA has poor water solubility, an increase in its concentration in the system would increase the hydrophobicity. Hence, the equilibrium yield of recipe 2# is higher than for recipe 3#.

To understand the effect of blowing agent, a control experiment was conducted, in which recipe 3# in Table 3.4 without MAA/$NaHCO_3$ was used. The equilibrium yield of the hollow microcapsules was only 7%, indicating that the in-situ CO_2 generated by blowing agent plays a key role in producing hollow microcapsules through forming a CO_2/oil/water emulsion. On the contrary, the contribution of monomer coated, entrained air bubbles is limited. Another important factor in the synthesis of hollow microcapsules lies in the surfactant dissolved in water. Stabilization of microbubbles is more difficult than that of monomer droplets. Thereby, a macromolecular surfactant, SMANa, was employed to form a stronger film surrounding the microbubbles and stabilize the microbubbles for a longer time.

Table 3.5 Three-dimensional solubility parameters of MMA and EPON 828-epoxy diacrylates in comparison with those of water[a] (unit: $J^{1/2}/cm^{3/2}$).

Substancet	δ_d[b]	δ_p[c]	δ_h[d]	$\|\delta_{d,monomer} - \delta_{d,H2O}\|$	$\|\delta_{p,monomer} - \delta_{p,H2O}\|$	$\|\delta_{h,monomer} - \delta_{h,H2O}\|$
MMA	18.0	5.2	8.6	5.0	26.1	25.6
E51-AA	19.8	2.9	12.3	6.8	28.4	21.9
H_2O	13.0	31.3	34.2	–	–	–

[a] The solubility parameters of water are cited from ref. [66], while those of the monomers are estimated by the group contribution method [66].
[b] Dispersive or non-polar component of solubility parameter.
[c] Polar component of solubility parameter.
[d] Hydrogen bonding component of solubility parameter.
Source: Reprinted from Xiao et al. [65]. Copyright 2009, with permission from Elsevier.

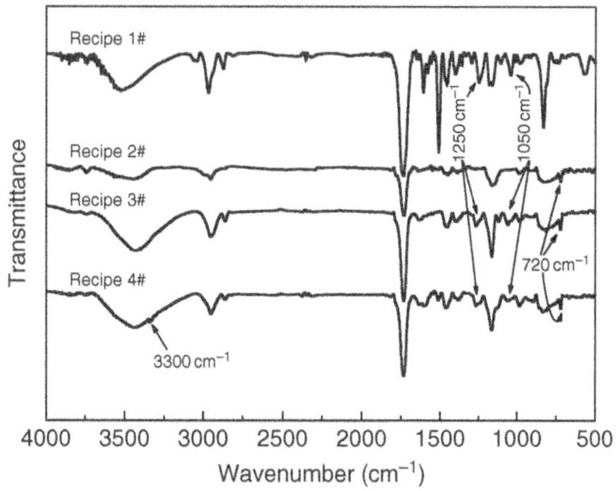

Figure 3.14 Fourier transform infrared spectra of hollow microcapsules. *Source:* Reprinted from Xiao et al. [65]. Copyright 2009, with permission from Elsevier.

Low-molecular weight surfactants, such as Tween-40, Span-40, sodium dodecyl sulfonate, and sodium dodecyl benzene sulfonate, were also tried, but failed to produce hollow microcapsules.

The size of the hollow microcapsules and wall thickness are affected by the feeding concentration of wall formers. In the case of a given group of components (see recipe 3# in Table 3.4), the influence of feeding concentration of wall formers on average radius, wall thickness and aspect ratio of resultant hollow microcapsules was investigated. It is found that the aspect ratio decreases with increasing feeding concentration of wall formers despite the fact that both average radius and wall thickness change in opposite ways. In general, a thicker wall can be attributed to the fact that a larger amount of monomer was absorbed on the oil layer of microbubbles. Meanwhile, a thicker oil layer would help to stabilize CO_2/oil/water droplets, leading to a larger capsule. The descending aspect ratio demonstrates that the increase in wall thickness of the capsules is faster than the increase in their size. Nevertheless, the aspect ratios are not high (6.4–7.5), meaning that the hollow microcapsules possess quite a large cavity with a relatively thin wall.

Overall, the approach based on UV-irradiation polymerization proves to be effective in producing hollow microcapsules. When the photoinitiator was replaced by a thermal initiator (i.e. $(NH_4)S_2O_8$), a much longer time (~2 hours) was needed for the appearance of microspheres with most of them being solid (Figure 3.12c). This is because during the longer thermal polymerization the microbubbles tended to either disappear or only hold a small amount of CO_2. Thus, the yield of hollow microcapsules after reaction for three hours was as low as 20%, and the mean diameter is significantly larger than that of the capsules made by UV irradiation.

As mentioned earlier, the hollow microcapsules are made to hold $(C_2H_5)_2O{\cdot}BF_3$ for application as a sealed self-healing agent. It is thus necessary to tailor the wall features for optimum loading. Figure 3.14 gives FTIR spectra of the walls of the capsules. Distinguished absorption peaks of the sulfonic group at 1250 and 1050 cm^{-1} are observed for the microcapsules prepared according to recipes 1#, 3#, and 4#, demonstrating that HPMAS has been incorporated in the shell. Multi peaks in relation to the phenyl ring of bisphenol A at 1600, 1585, 1500, and 1450 cm^{-1} in association with the peak of carbonyl group at 1745 cm^{-1} confirm that the capsules from recipe 1# are made by epoxy diacrylate. For the capsules based on recipes 2#, 3#, and 4#, in which HDDA molecules are included, the characteristic peak at 720 cm^{-1} corresponding to the -$(CH_2)_n$- group (where $n \geq 4$) is perceivable. In addition, the carbonyl peaks at 1745 and 1760 cm^{-1} also appear on the spectra of these samples. Especially, a small carboxyl peak at 3300 cm^{-1} is distinguished for the microcapsules from recipe 4#, implying that MBI has been attached to the shell. Characteristic absorption peaks of hydroxyl group at 3500 cm^{-1}, and methyl and methylene at 2960 and 2870 cm^{-1} are visible for the four types of hollow microcapsules. From the above analysis, it is concluded that the hollow microcapsules designed by the recipes in Table 3.4 have been produced as expected.

Besides chemical structures, the geometry and morphology of the hollow microcapsules would also affect infiltration of $(C_2H_5)_2O{\cdot}BF_3$. SEM observation indicates that the surface of the microcapsule resulting from recipe 2# is smoother and more compact than those of the microcapsules containing HPMAS. The former is only composed of the hydrophobic monomers MMA and HDDA, and apt to contract after polymerization. On the other hand, HPMAS must have been assembled at the surface of microbubbles with their hydrophilic groups (such as $-SO_3Na$ and hydroxyl) toward the aqueous phase. As a result, the surface of the hollow microcapsule with HPMAS became coarse due to extension of the HPMAS molecules when the capsules were built up. With respect to the microcapsule derived from recipe 4#, MBI also took the role of polymerizable emulsifier and hence the surface roughness of the capsules was further increased.

Since the hollow microcapsules are made from materials that are compatible with $(C_2H_5)_2O{\cdot}BF_3$, the ones with rougher surface representing loose accumulation of molecules should favor infiltration of $(C_2H_5)_2O{\cdot}BF_3$, as intuitively expected. The loading curves of the hollow capsules shown in Figure 3.15 demonstrate this is the case. With a rise in infiltration time, the loading rate is gradually increased until the equilibrium at 14–24% is reached after 40 hours. The capsules made from recipe 4# are indeed

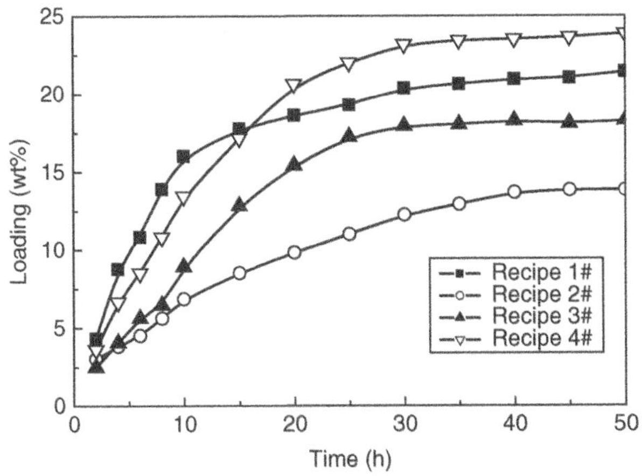

Figure 3.15 Loading rate of $(C_2H_5)_2O{\cdot}BF_3$ in hollow microcapsules as a function of immersion time.
Source: Reprinted from Xiao et al. [65]. Copyright 2009, with permission from Elsevier.

filled with much more $(C_2H_5)_2O{\cdot}BF_3$ than the others. The smooth capsules from recipe 2# have the lowest equilibrium loading rate. The data show that the proposed route for making hollow microcapsule works, and the loading amount of $(C_2H_5)_2O{\cdot}BF_3$ in the capsules can be adjusted by changing the shell composition.

To find out the reason for the different equilibrium loading rates of the different microcapsules, the porous characteristics of the walls of the hollow microcapsules were examined by nitrogen adsorption isotherms at liquid nitrogen temperature. Both mesopores and micropores are recognized on the wall surface, which should have an important contribution to the infiltration of $(C_2H_5)_2O{\cdot}BF_3$. The detected pores of all the walls of the capsules range from several to hundreds of angstroms. They are large enough for the travel of $(C_2H_5)_2O{\cdot}BF_3$, because the latter's size is about 6.86 Å. It seems that the content of the pores plays a critical role. The experimental data show that the amounts of mesopores and micropores in the range of interest decrease in the order: recipe 1# > 4# > 3# > 2#. By surveying Figure 3.15, it is evident that the hollow microcapsules from recipes 1# and 4# with more pores facilitate infiltration of $(C_2H_5)_2O{\cdot}BF_3$, while those from recipes 3# and 2# having fewer pores are bound to exert negative effect on the loading. However, careful observation of Figure 3.15 reveals that the microcapsules from recipe 4# have higher equilibrium loading rate than those from recipe 1#, even though the former possesses less pores than the latter. Other things may influence the loading performance, so the surface chemistry should be analyzed.

It is known that boron trifluoride diethyl etherate is a polar molecule. Therefore, various polar groups, such as –COOH, $-SO_3Na$, –OH, and –COOR, were introduced onto the shell of the capsules. Although the wall formers have met the requirement for molecular miscibility between the capsules and $(C_2H_5)_2O{\cdot}BF_3$, surface characteristics of the capsules have not yet been quantified. For this purpose, Table 3.6 lists surface free energies of the hollow microcapsules. The dispersive component of surface free energy, γ^d, and polar component of surface free energy, γ^p, of the microcapsules rank in different orders: recipe 1# > 4# > 3# > 2# for γ^d and recipe 4# > 1# > 3# > 2# for γ^p. The former sequence resembles that of the pore content, while the latter sequence is similar to that of the loading ability of the hollow microcapsules (Figure 3.15). This implies that higher polarity of the wall of the capsules leads to higher equilibrium loading of $(C_2H_5)_2O{\cdot}BF_3$. The phenomenon should be related to the interaction between the polar groups and $(C_2H_5)_2O{\cdot}BF_3$.

On the other hand, there might exist a reverse infiltration of $(C_2H_5)_2O{\cdot}BF_3$ from the $(C_2H_5)_2O{\cdot}BF_3$-loaded microcapsules, which determines service life of the microcapsules. Accordingly, weight loss of the microcapsules stored at room temperature was recorded as a function of time. It is found that the weight loss is very low (e.g. less than 0.1 wt% for two months). That is, the reverse infiltration proceeds very slowly and $(C_2H_5)_2O{\cdot}BF_3$ in the microcapsules can be used for long time. Such a low reverse infiltration rate relative to the infiltration rate can be understood from the different environments during filling of hollow microcapsules and storage of $(C_2H_5)_2O{\cdot}BF_3$-loaded microcapsules. Prior to infiltration of $(C_2H_5)_2O{\cdot}BF_3$, the gas (i.e. CO_2) inside the hollow microcapsules was evacuated under vacuum, which led to a high infiltration rate within a short time. However, the reverse infiltration has to be carried out under atmospheric pressure, so that very slow weight loss was perceived.

Since $(C_2H_5)_2O{\cdot}BF_3$ is a highly active agent, it should be known whether the filled version in the microcapsules

Table 3.6 Surface free energies of the hollow microcapsules.

	Surface free energy[a] ($mJ\,m^{-2}$)	
Recipe ID	γ^d	γ^p
1#	34.1	9.0
2#	27.3	5.2
3#	31.9	7.7
4#	33.9	10.9

[a] γ^d, dispersive component of surface free energy; and γ^p, polar component of surface free energy.
Source: Reprinted from Xiao et al. [65]. Copyright 2009, with permission from Elsevier.

keeps its chemical activity. Accordingly, in-situ FTIR spectroscopy was used to study the reaction of epoxy resin (EPON 828) coated on the surface of a KBr tablet containing ground $(C_2H_5)_2O{\cdot}BF_3$-loaded hollow microcapsules (Figure 3.16). It is seen that the characteristic peak of epoxide groups at 910 cm^{-1} diminishes with time. When the peak area of the phenyl ring at 835 cm^{-1} serves as the internal standard, the peak area ratios of epoxide group to phenyl ring can be used to estimate the degree of reaction (see the inset of Figure 3.16). Evidently, about 80% of the epoxide groups are consumed within 20 minutes at room temperature. The results suggest that the encapsulated $(C_2H_5)_2O{\cdot}BF_3$ should be able to act as the hardener of uncured epoxy for crack healing in epoxy composites.

3.1.3 Characterization of Self-Healing Functionality

The preparation of a two-component self-healing agent through methods developed in the authors' laboratory is discussed above. The results have paved the way for the subsequent manufacture of self-healing composites. Considering that the hardener $(C_2H_5)_2O{\cdot}BF_3$ can be held by two types of carriers (porous media and hollow microcapsules), two groups of self-healing epoxy materials are fabricated accordingly. One is filled with epoxy-loaded microcapsules and $(C_2H_5)_2O{\cdot}BF_3$-loaded porous media, while the other with epoxy-loaded microcapsules and $(C_2H_5)_2O{\cdot}BF_3$-loaded microcapsules. Their self-healing functionality and the mechanisms involved are studied in the following. By comparing with the system of epoxy-mercaptan shown in Chapter 2, one might determine the features of different healing chemistries and their suitability for practical usage.

3.1.3.1 Self-Healing Epoxy Materials with Embedded Epoxy-Loaded Microcapsules and $(C_2H_5)_2O{\cdot}BF_3$-Loaded Sisal

As Figure 3.9 reveals that sisal has the highest equilibrium loading of $(C_2H_5)_2O{\cdot}BF_3$, in the following work the vegetable fibers are used for constructing self-healing epoxy as a representative of porous media, together with epoxy-loaded microcapsules.

The unfilled epoxy specimens were produced by mixing 100 parts of epoxy (EPON 828) with 12 parts of curing agent, triethylene tetramine (TETA). The self-healing epoxy composites were prepared by uniformly mixing epoxy-loaded capsules (30 μm in diameter, having a core content [i.e. the weight ratio of core to microcapsule] of 70 wt%, see Section 3.1.1 for more information on the synthesis details) and $(C_2H_5)_2O{\cdot}BF_3$-loaded sisal (2.5–3 mm in length, carrying 32 wt% $(C_2H_5)_2O{\cdot}BF_3$ and 7.3 wt% PS coating, see Section 3.1.2.1 for more information on the preparation) together with the aforementioned mixture of epoxy and TETA [67]. To obtain cured versions, both the unfilled epoxy and filled compounds were degassed and poured into closed silicone rubber molds and cured for 72 hours at room temperature, followed by 48 hours at 50 °C. All the specimens were stored at room temperature for one month prior to the assessment of healing efficiency, unless otherwise specified. It is worth noting that $(C_2H_5)_2O{\cdot}BF_3$ has been recognized as an extremely active catalyst for epoxy polymerization. In the case where the $(C_2H_5)_2O{\cdot}BF_3$-loaded

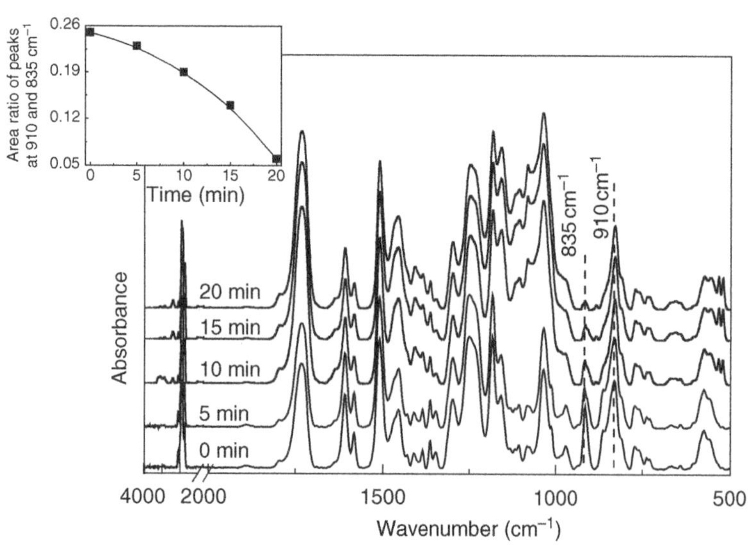

Figure 3.16 In-situ FTIR spectra tracking the reaction between epoxy and ground $(C_2H_5)_2O{\cdot}BF_3$-loaded microcapsules (recipe 3# in Table 3.4; loading rate: 18%). The inset shows the peak area ratio of epoxide group to phenyl ring as a function of time.
Source: Reprinted from Xiao et al. [65]. Copyright 2009, with permission from Elsevier.

sisal fibers were not coated and directly mixed with epoxy, the viscosity of the system would immediately increase and there would not be enough time even for molding.

The self-healing ability of the epoxy composites containing epoxy-loaded capsules and $(C_2H_5)_2O{\cdot}BF_3$-loaded sisal was evaluated by Izod impact tests, according to the approach proposed by Hayes et al. [68] Healing efficiency, η, was described by the percentage recovery in impact strength:

$$\eta = \frac{\sigma_{\text{healed}}}{\sigma_0} \tag{3.5}$$

where σ_{healed} and σ_0 denote impact strengths of the healed and virgin specimens, respectively.

Figure 3.17 illustrates the dependence of healing efficiency on content of the healing agent. Evidently, healing of cracks is available when both epoxy-loaded capsules and $(C_2H_5)_2O{\cdot}BF_3$-loaded sisal are added to the matrix epoxy. The system filled with only a single component of the healing agent cannot offer any healing effect. It is worth noting that although healing efficiency increases with a rise in healant content in the regime of low healant content, the former is nearly independent of healant content above a certain threshold. This coincides with the rule of curing of epoxy catalyzed by BF_3. That is, the $(C_2H_5)_2O{\cdot}BF_3$/epoxy weight ratio is not bound to be kept at a constant value, but can vary within a relatively wide range of 1–5% [4]. Once dosage of $(C_2H_5)_2O{\cdot}BF_3$ reaches the prescribed minimum, cationic chain polymerization of released epoxy begins. The stoichiometric composition for every centimeter of repair region required by the healing agent based on epoxy-amine or epoxy-thiol pairs is unnecessary.

The data in Figure 3.17a indicate that for 0.25 wt% $(C_2H_5)_2O{\cdot}BF_3$-loaded sisal, the highest healing efficiency is only around 50% even at 15 wt% epoxy-loaded capsules. As the loading of $(C_2H_5)_2O{\cdot}BF_3$ in sisal is 32 wt%, the highest $(C_2H_5)_2O{\cdot}BF_3$/epoxy weight ratio should be 0.76 wt% supposing all the absorbed $(C_2H_5)_2O{\cdot}BF_3$ is desorbed out of sisal fibers. The value is still lower than the aforementioned floor level. Curing of the released healant has to be incomplete, leading to low bonding strength.

Similarly low healing efficiency is found in Figure 3.17b, where the theoretical $(C_2H_5)_2O{\cdot}BF_3$/epoxy weight ratio can be as high as 27 wt% for the composite with 2.5 wt% epoxy-loaded capsules and 1.5 wt% $(C_2H_5)_2O{\cdot}BF_3$-loaded sisal. The result might originate from (i) the insufficient amount of epoxy delivered to the cracked faces, which led to smaller bonding area, and (ii) the excessively high $(C_2H_5)_2O{\cdot}BF_3$/epoxy ratio, which brought about a huge exotherm and hence degradation of the reaction mixture [9]. The first explanation receives support from SEM observation of the fractured surface of a healed specimen, which shows the significantly discontinuous healing membranes.

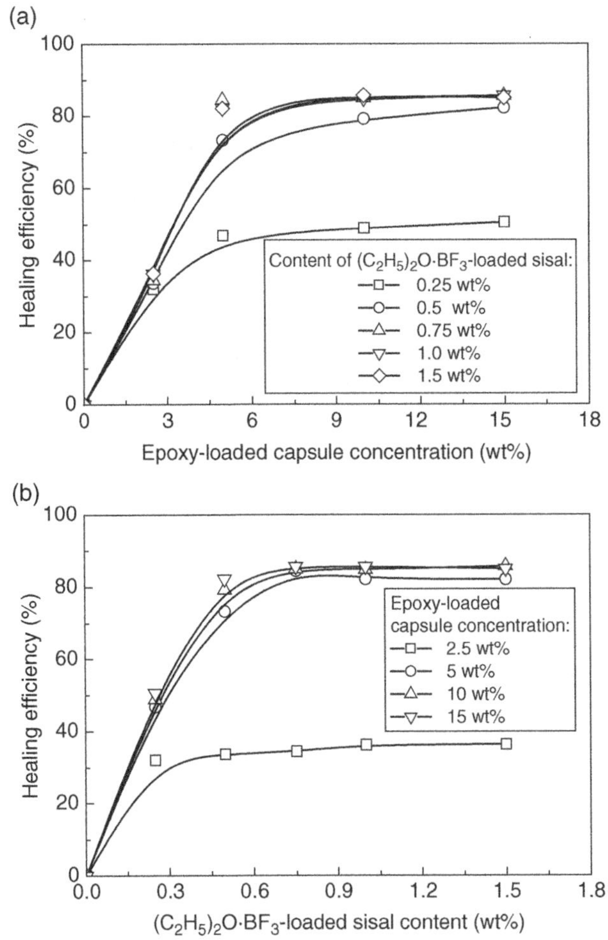

Figure 3.17 (a, b) Dependences of healing efficiency of the self-healing epoxy composites on content of the healing agent. Healing of the fractured specimens was conducted at 20 °C for two hours. *Source:* Reprinted from Xiao et al. [67]. Copyright 2009, with permission from John Wiley and Sons.

In contrast, when the contents of epoxy-loaded capsules and $(C_2H_5)_2O{\cdot}BF_3$-loaded sisal are higher than 5 wt% and 0.75 wt%, respectively, the repair effect is nearly insensitive to the healant concentration, offering a ~85% recovery of impact strength. Such a low critical fraction of healant must have met the requirements of proper $(C_2H_5)_2O{\cdot}BF_3$/epoxy weight ratio and full coverage of the damaged area. It should also be beneficial to retaining the original mechanical properties of the matrix epoxy.

According to the method of Rule et al. [25], the average volume of the crack in the fractured specimens and the amount of epoxy required to deliver on the fracture surface can be evaluated and compared. With a crack face separation of 1.1 μm (from microscopic examination), it is known that an epoxy delivery per unit crack area of 0.13 mg cm^{-2} is needed to completely fill the crack (density of epoxy = 1.16 g cm^{-3}). When the content of epoxy-loaded capsules is 2.5 wt%, the total mass of epoxy available for delivery per unit crack area is estimated to be about

0.09 mg cm^{-2}, which is less than the minimum required dose (0.13 mg cm^{-2}). In the case of 5 wt% epoxy-loaded capsules, the amount of available epoxy becomes 0.18 mg cm^{-2}, meaning the crack volume can be completely filled. The assessment agrees with the above analysis.

On the other hand, Caruso et al. have found that self-healing can be promoted by solvent [69]. To understand whether such a solvent effect is also involved in the present system, three control experiments were conducted. For the first one, epoxy composites containing 5 wt% epoxy-loaded capsules and 0.75 wt% $(C_2H_5)_2O$-loaded sisal were prepared. The embedded sisal had absorbed $(C_2H_5)_2O$ (instead of $(C_2H_5)_2O{\cdot}BF_3$) and was coated with PS following the same procedures for making $(C_2H_5)_2O{\cdot}BF_3$-loaded sisal. The subsequent evaluation tests indicated that no healing was available in this case. The second control experiment was done by manually injecting $(C_2H_5)_2O$ onto the fractured surfaces of epoxy composites containing 5 wt% epoxy-loaded capsules. Having experienced the healing and re-impact steps like the authentic specimens, the broken materials failed to be re-connected, offering zero healing efficiency. The third control experiment resembled the second one, except that the specimens were made of unfilled epoxy. Again, the fractured specimens could not be healed. On the basis of these control experiments and the data of Figure 3.17 (i.e. the specimens with only a single component of the healing agent do not have healing ability), it can be concluded that solvent effect is absent from the materials developed in this work.

Figure 3.18 examines flexural and impact performance of the composites. Because sisal carries $(C_2H_5)_2O{\cdot}BF_3$ fluid, it loses load carrying ability to a great extent and is turned into soft matter like epoxy-loaded capsules. As a result, the addition of both $(C_2H_5)_2O{\cdot}BF_3$-loaded sisal and epoxy-loaded capsules lowers the static mechanical properties of epoxy. This agrees with the common rule of microcapsules or

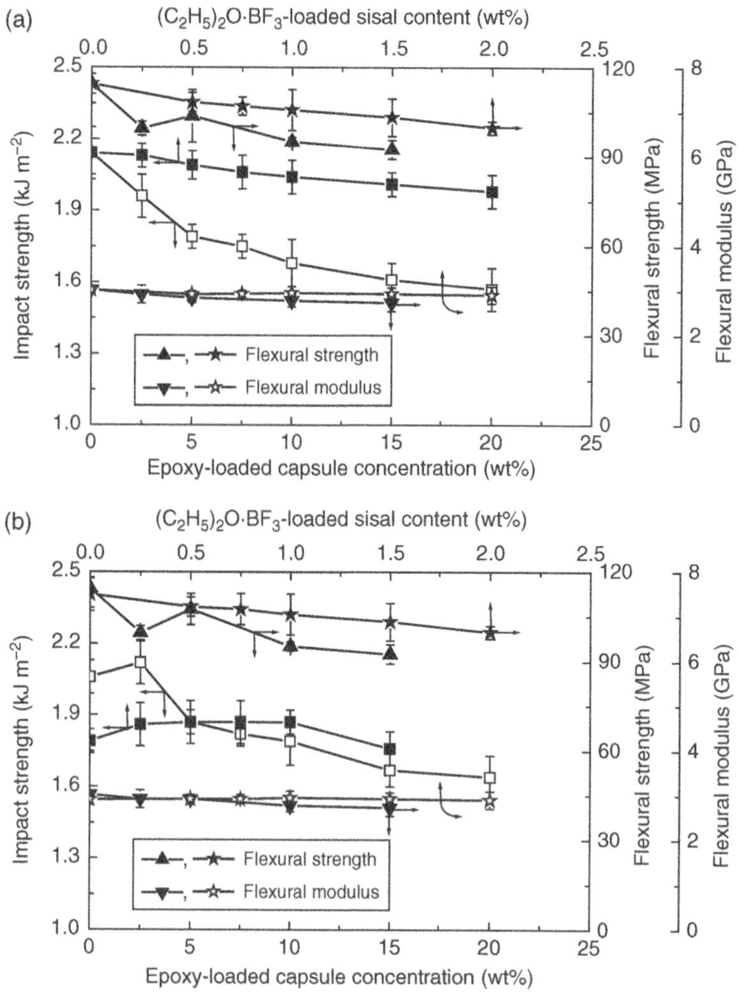

Figure 3.18 Influence of content of the healing agent on mechanical properties of epoxy composites. (a) The composites labeled with (★, ■, ☆) only contain $(C_2H_5)_2O{\cdot}BF_3$-loaded sisal, and those labeled with (▲, □, ▼) only contain epoxy-loaded capsules. (b) The composites labeled with (★, ■, ☆) contain different proportions of $(C_2H_5)_2O{\cdot}BF_3$-loaded sisal and 5 wt% epoxy-loaded capsules, and those labeled with (▲, □, ▼) contain different proportions of epoxy-loaded capsules and 0.75 wt% $(C_2H_5)_2O{\cdot}BF_3$-loaded sisal. The performance data of unfilled epoxy are given for reference: flexural strength (114.5 ± 3.4 MPa), flexural modulus (3.0 ± 0.1 GPa), and impact strength (2.14 ± 0.11 kJ m^{-2}).

rubbery particle filled polymers [15, 70]. With respect to impact strength, the capsules and $(C_2H_5)_2O{\cdot}BF_3$-loaded sisal fibers do not impede crack propagation. Similar embrittling effect can be found in ref. [71]. On the whole, the decay in mechanical properties is not severe, and can be limited to a small degree so long as the amounts of the fillers are not high. In the case of 5 wt% epoxy-loaded capsules and 0.75 wt% $(C_2H_5)_2O{\cdot}BF_3$-loaded sisal, for example, the reductions in flexural strength, flexural modulus and impact strength of epoxy matrix are only 6.5, 3, and 14%, respectively.

To check the healing mechanism involved, energy dispersive spectroscopy (EDS) analysis of the specimens was carried out. Fluorine atoms are found on the surface of as-manufactured specimens (undamaged), and the amount of fluorine increases with increasing storage time. This suggests that $(C_2H_5)_2O{\cdot}BF_3$ carried by sisal fibers must have spread out due to concentration diffusion during storage. This is understandable in view of the volatility of $(C_2H_5)_2O{\cdot}BF_3$ (saturated vapor pressure at 20 °C = 0.56 kPa). The PS coating helps to retard the diffusion and to reserve the activity of $(C_2H_5)_2O{\cdot}BF_3$ in the course of composites manufacturing, but it would not stop the diffusion over a long time span. Eventually, $(C_2H_5)_2O{\cdot}BF_3$ is dispersed in epoxy matrix on a molecular scale and acts as a latent hardener being able to trigger polymerization of any uncured epoxy. EDS analysis of the fractured surface further proves the above deduction (Figure 3.19). The concentration of fluorine detected on the split sisal fiber (region A) is only slightly higher than that in the matrix (region B) at some distance. Owing to the special structure of sisal fiber, when a crack passes across it, $(C_2H_5)_2O{\cdot}BF_3$ could not burst out like capsulated fluid and extend through the entire crack plane. If the diffusion occurred only after fiber damage, it would take a long time to reach the sites far away from the fiber, which contradicts the fact that healing takes effect within a few minutes (see the following discussion).

Since $(C_2H_5)_2O{\cdot}BF_3$ has already been distributed in the matrix, the released epoxy should be easily cured. This is verified by in-situ confocal Raman microscopy observation, which shows that the consumption of uncured epoxy healing agent is completed within about 30–40 minutes. Clearly, the result demonstrates that when the specimen is broken after impact, uncured epoxy would flow over the fractured surface, and is quickly cured with a lapse of time. A polymerized membrane was thus formed as the binder. Time dependence of healing efficiency is given in Figure 3.20. The data show that the crack healing proceeds at a faster rate as compared with the healing agent consisting of epoxy and thiol [10]. It takes only 30 minutes at 20 °C for the present recipe to attain a healing efficiency of ~76%, but the "healed" specimen using epoxy/thiol as healant cannot carry any load under the same healing temperature (i.e. healing efficiency = 0). Moreover, the time for yielding maximum healing efficiency at 20 °C for the former is about two hours, while that for the latter is longer than 24 hours. Another attractive feature revealed by Figure 3.20 lies in the fact that crack healing is available over a wider temperature range from −10 to 30 °C. Even at −10 °C, a healing efficiency of 78% can be achieved after five hours. This forms a sharp contrast to the epoxy/thiol system, which offers nearly no healing within the identical time range at −10 °C (Figure 3.10). With a rise in temperature,

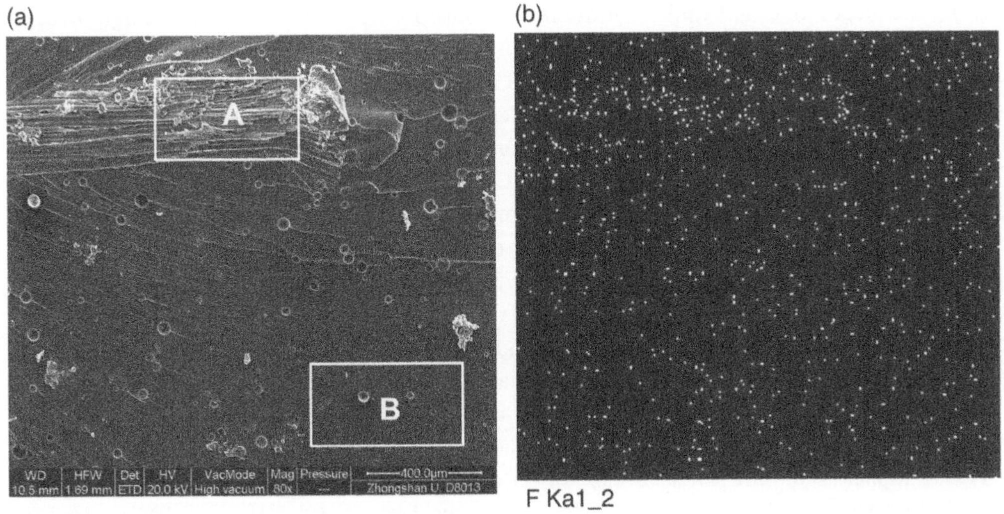

Figure 3.19 (a) Scanning electron microscopy image in conjunction with (b) EDS analysis (with fluorine as the indicator element) of the fracture surface of a self-healing epoxy composite specimen, which contains 5 wt% epoxy-loaded capsules and 0.75 wt% $(C_2H_5)_2O{\cdot}BF_3$-loaded sisal. Detected fluorine contents in selected regions: (A) 2.03 wt%; and (B) 1.27 wt%. *Source:* Xiao et al. [67], Figure 4, p. 5/with permission of John Wiley & Sons, Inc.

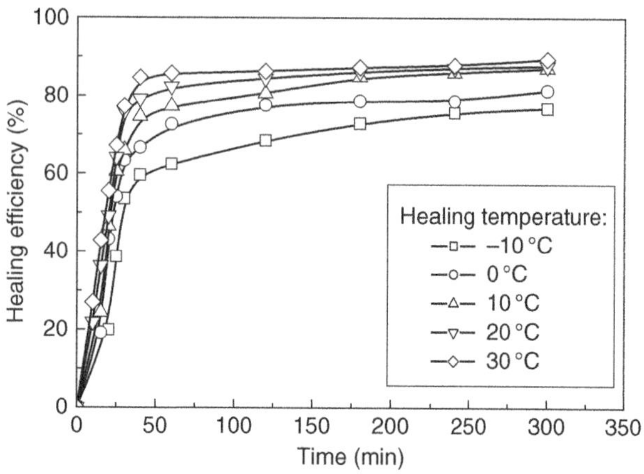

Figure 3.20 Time dependence of healing efficiency of the self-healing epoxy composites containing 5 wt% epoxy-loaded capsules and 0.75 wt% $(C_2H_5)_2O{\cdot}BF_3$-loaded sisal. Healing of the fractured specimens was conducted at different temperatures.

the speed of recovery as well as healing efficiency of the present self-healing composites increases (Figure 3.20). This should be attributed to the accelerated healing reaction and reduced viscosity of healing agent at elevated temperature. The latter effect favors increasing flowability of the healing agent. In conclusion, the results given in Figure 3.20 demonstrate the application potential of the epoxy/BF_3 pair for self-healing especially in the cold as expected at the beginning of this chapter.

The rapid healing process must result from rapid polymerization of the released epoxy. A non-isothermal curing kinetics study demonstrates that the reaction between epoxy and $(C_2H_5)_2O{\cdot}BF_3$ exhibits two exotherms, one at subambient temperature and the other at ambient temperature, depending on heating rate. They are related to nucleation of the microgel and growth of the microgel during consolidation. Quantitative treatment of the results estimates activation energies for the two stages to be 34.5 and 36.0 kJ mol^{-1}, respectively. The values are obviously smaller than those for epoxy-thiol (48.5 kJ mol^{-1} [10]) and epoxy-amine (68.9 kJ mol^{-1} [72]) pairs. This means that cure of epoxy catalyzed by $(C_2H_5)_2O{\cdot}BF_3$ needs low energy.

To have an idea of the maximum healing effect of the system, four types of reference repair were done as in the following. (i) Reference 1: a mixture of epoxy and $(C_2H_5)_2O{\cdot}BF_3$ was applied to the fractured surfaces of unfilled epoxy specimen. (ii) Reference 2: epoxy was first applied to the fractured surfaces of unfilled epoxy specimen, and then $(C_2H_5)_2O{\cdot}BF_3$ was applied onto the epoxy layer. (iii) Reference 3: $(C_2H_5)_2O{\cdot}BF_3$ was applied to the fractured surfaces of epoxy specimen filled with 5 wt% epoxy-loaded capsules. (iv) Reference 4: epoxy was applied to the fractured surfaces of epoxy specimen filled with 1.5 wt% $(C_2H_5)_2O{\cdot}BF_3$-loaded sisal. The weight ratio of epoxy to $(C_2H_5)_2O{\cdot}BF_3$ was kept at 100:5, and healing was conducted at 20 °C for two hours. The experimental results show that the highest recovery of impact strength of 98% can be obtained for Reference 4, while the lowest healing efficiency of 87% is detected in the case of Reference 3. The differences should be understood from the curing nature of the healing agent. The reaction takes place very quickly, resulting in gelation within dozens of seconds at room temperature, so partial polymerization must be completed during the premixing step for Reference 1 before the adhesive was applied. This would certainly lower the wetting and bonding strength to the matrix. In addition, too fast a reaction might lead to incomplete epoxy curing. Reference 2 reduced the preparation time of the adhesive, but homogeneity was still difficult to be controlled. In the case of Reference 3, the amount of uncured epoxy delivered from the broken capsules might be insufficient to cover the crack plane. For Reference 4, $(C_2H_5)_2O{\cdot}BF_3$ had been well distributed in the matrix due to diffusion, so that the curing proceeded in a much milder way as compared with References 1 and 2, leading to the highest healing efficiency. Returning to the data of authentic specimens (Figure 3.17), the lower maximum healing efficiency (~85%) might result from the viscosity-induced incomplete coverage of the adhesive membrane on the cracked surfaces. Application of an epoxy with higher flowability might be a solution. This deduction can be partly supported by the following phenomenon. That is, when healing temperature was increased to 30 °C, the maximum healing efficiency of the composite can be raised to ~89%.

Durability is important for self-healing materials. Having inspected repair effect of the specimens stored for different times, we found that the healing ability is quite stable within 12 months. This means that in the composites no bleeding from the epoxy-loaded capsules occurs and the diffused $(C_2H_5)_2O{\cdot}BF_3$ keeps its activity. In addition, the influence of moisture was also verified as it might deactivate $(C_2H_5)_2O{\cdot}BF_3$ during the storage of composites. It is found that the self-healing ability of the composites is almost independent of environmental humidity. The composites containing 5 wt% epoxy-loaded capsules and 0.75 wt% $(C_2H_5)_2O{\cdot}BF_3$-loaded sisal exhibited rather low equilibrium moisture absorbability (~0.78%) at room temperature even under extreme conditions (i.e. 100% relative humidity), as a result of the favorable water resistance of the matrix resin. It is worth noting that although the composites might absorb a small amount of moisture during storage in ambient air, the possible reaction between the absorbed water and $(C_2H_5)_2O{\cdot}BF_3$ cannot lead to decomposition and accompanying deactivation of the latter as in the case of an aqueous solution. Instead, new complexes of $H_2O \rightarrow BF_3$ might be generated due to (i) insufficient amount of water and (ii) a dynamic barrier for diffusion of the absorbed water molecules in the matrix resulting from the hydrogen bonds between the water molecules and hydroxyl groups of the epoxy matrix [73]. Moreover, the robust interaction between $(C_2H_5)_2O{\cdot}BF_3$ and the epoxy matrix through either coordination bonds (Lewis acid–base interaction) or hydrogen bonds also hinders $(C_2H_5)_2O{\cdot}BF_3$ from contacting and reacting with the absorbed water molecules [73]. In this context, BF_3 always presents itself in the form of complex molecules, and the high activity of the catalyst in the composites can be well maintained after long-term storage.

3.1.3.2 Self-Healing Epoxy Materials with Embedded Dual Encapsulated Healant

As both epoxy- and $(C_2H_5)_2O{\cdot}BF_3$-loaded microcapsules are available as discussed above, self-healing epoxy materials can be made by embedding the two types of microcapsules into the matrix [74], following the concept of dual encapsulated healant.

The epoxy-loaded microcapsules used hereinafter are the same as those used in Section 3.1.3.1, except that their average sizes are adjusted from 5 to 45 μm, and the core content (i.e. weight ratio of core to microcapsule) varies from 54 to 79 wt% (see Section 3.1.1 for more information on the synthesis). The $(C_2H_5)_2O{\cdot}BF_3$-loaded microcapsules are made by infiltration of the chemical into hollow capsules (see Section 3.1.2.2 for more information on the preparation). Their core content and average size are controlled to 20 wt% and 10 μm, respectively.

The self-healing epoxy composites containing the dual capsules were prepared through the same procedures described in Section 3.1.3.1. The protocol suggested by Hayes et al. was also applied to assess healing ability of the composites in terms of impact test [68]. Healing efficiency is defined as the ratio of impact strength of healed and virgin materials. Izod notched tests were conducted at 20 °C. After testing, the specimens that had been broken into two pieces were kept in alignment and intimate contact for healing at 20 °C for different times. Then, the healed specimens were tested again following the above procedure to check the effect of healing.

As mentioned earlier, weight ratios of $(C_2H_5)_2O{\cdot}BF_3$/epoxy ranging from 1 to 5 wt% are sufficient for curing epoxy. However, healing of cracks in composites depends on the curing reaction of the crack-released epoxy with the bled $(C_2H_5)_2O{\cdot}BF_3$. That means, exact proportions of the two types of capsules taking part in the reaction can hardly be predicted for this specific case. The suitable recipe has to be determined from actual tests. For this reason, a series of epoxy composites filled with various contents of epoxy- and $(C_2H_5)_2O{\cdot}BF_3$-loaded capsules at different ratios were prepared and evaluated.

As shown in Figure 3.21 (which resembles Figure 3.17), healing of cracks is available only when both epoxy- and $(C_2H_5)_2O{\cdot}BF_3$-loaded capsules are added to the composites. The system filled with only a single component of the healing agent cannot offer any healing functionality. When the dosage of $(C_2H_5)_2O{\cdot}BF_3$-loaded capsules is fixed, healing efficiency of the epoxy composites increases with the content of epoxy-loaded capsules up to 5 wt%, and then tends to level off (Figure 3.21a). The maximum healing efficiency is about 88%. At lower content of epoxy-loaded capsules, the released epoxy fluid would be insufficient to cover the broken surface, and hence poor healing effect is detected. This deduction can be evidenced from another angle by the data in Figure 3.21b, in which similarly high healing efficiencies are perceived for all of the epoxy composites

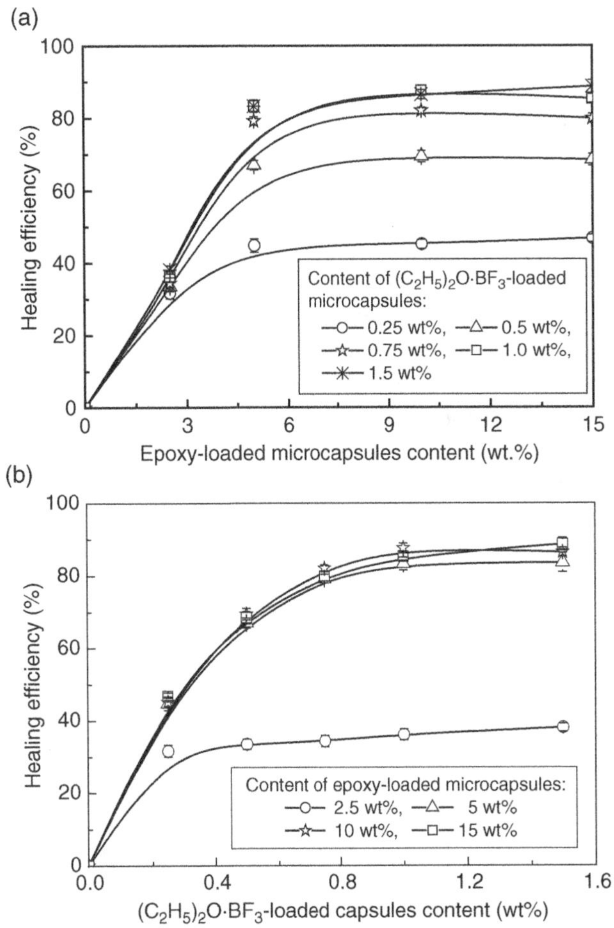

Figure 3.21 (a, b) Dependence of healing efficiency of the self-healing epoxy composites on content of the healing agent. Epoxy-loaded microcapsules: core content = 70 wt %, average diameter = 30 μm. $(C_2H_5)_2O{\cdot}BF_3$-loaded microcapsules: core content = 20 wt%, average diameter = 10 μm. Healing of the fractured specimens was conducted at 20 °C for two hours. *Source:* Reprinted from Xiao et al. [74]. Copyright 2009, with permission from Elsevier.

containing epoxy-loaded capsules over 5 wt% and $(C_2H_5)_2O{\cdot}BF_3$-loaded capsules over 1 wt%. In contrast, the composite with 2.5 wt% epoxy-loaded capsules only shows healing efficiency of ~30% regardless of the content of $(C_2H_5)_2O{\cdot}BF_3$-loaded capsules.

Figure 3.21 also illustrates the importance of content of $(C_2H_5)_2O{\cdot}BF_3$-loaded capsules. Healing efficiency of the composites increases with a rise in addition of $(C_2H_5)_2O{\cdot}BF_3$-loaded capsules in the low content regime, but it becomes nearly independent of the content of $(C_2H_5)_2O{\cdot}BF_3$-loaded capsules at 1 wt% and above.

Supposing that the dual capsules are homogeneously distributed throughout the composites, and all the capsules at the crack surfaces can be fractured, the corresponding $(C_2H_5)_2O{\cdot}BF_3$/epoxy weight ratio of the released healant can be estimated (Table 3.7). Evidently, for the low content epoxy-loaded capsules (i.e. 2.5 wt%), $(C_2H_5)_2O{\cdot}BF_3$-loaded capsules with contents from 0.25 to 1.5 wt% are sufficient, because the estimated $(C_2H_5)_2O{\cdot}BF_3$/epoxy weight ratios range between 2.9 wt% and 17.1 wt%, agreeing with the standard proportions of 1–5 wt%. Therefore, the aforementioned low healing efficiency of the composites filled with 2.5 wt% epoxy-loaded capsules has to result from the insufficient supply of the released epoxy. On the other hand, there are too few 0.25 wt% $(C_2H_5)_2O{\cdot}BF_3$-loaded capsules to induce a complete cure of epoxy healing agent for the composites containing 10–20 wt% epoxy-loaded capsules, since the estimated $C_2H_5)_2O{\cdot}BF_3$/epoxy weight ratios are only 0.4–0.7 wt%, lower than the lower limit of the recipe (i.e. 1 wt%). The analysis explains the poor healing ability of the composites with low content $(C_2H_5)_2O{\cdot}BF_3$-loaded capsules (Figure 3.21).

It is interesting to note that for the composites with 5 wt% epoxy-loaded capsules and 0.25 wt% $(C_2H_5)_2O{\cdot}BF_3$-loaded capsules, the calculated $(C_2H_5)_2O{\cdot}BF_3$/epoxy weight ratio is 1.4%, which exceeds the lower limit of the recipe. Theoretically, the corresponding healing efficiency should also be high. Evidently, it is not the case. The deviation implies that the assumption of the calculation might not always be consistent with the actual case. Heterogeneous distribution of the capsules and their fracture must be responsible for the deviation.

On the whole, the results in Figure 3.21 and Table 3.7 suggest that the $(C_2H_5)_2O{\cdot}BF_3$/epoxy weight ratios can vary over a relatively wide range. The stoichiometric composition at every centimeter of repair region required by the healing agent based on epoxy-amine or epoxy-thiol pairs is unnecessary. Actually, so long as the contents of epoxy- and $(C_2H_5)_2O{\cdot}BF_3$-loaded loaded capsules are, respectively, higher than 5 and 1 wt%, the repair effect is insensitive to the healant concentration, offering a ~85% recovery of impact strength. Such low critical fractions of healant must be beneficial to retaining the original mechanical properties of the matrix epoxy.

To check the speed of crack healing, healing efficiencies of the composites were recorded as a function of healing time (Figure 3.22). It is seen that the healing proceeded quite fast at 20 °C, meaning that the curing of epoxy healing agent under catalysis of $(C_2H_5)_2O{\cdot}BF_3$ must have been completed within a short time as expected. The healing efficiency attains 28% after 5 minutes, exceeds 76% only after 20 minutes, and reaches equilibrium at ~80% after 30 minutes. It should be emphasized that the healing effect is acquired at very low content of healing agent (5 wt% epoxy-loaded microcapsules and 1.0 wt% $(C_2H_5)_2O{\cdot}BF_3$-loaded microcapsules) as compared with our previous work (typically 5 wt% epoxy-loaded microcapsules and 5 wt% thiol-loaded microcapsules [10]). Moreover, the speed of healing is much faster than that of the thiol catalyzed one, which required 24 hours to approach the

Table 3.7 Estimation of $(C_2H_5)_2O{\cdot}BF_3$/epoxy weight ratio of released healant according to the contents and core contents of epoxy- and $(C_2H_5)_2O{\cdot}BF_3$-loaded capsules[a].

$(C_2H_5)_2O{\cdot}BF_3$-loaded capsules		Epoxy-loaded capsules		Calculated $(C_2H_5)_2O{\cdot}BF_3$/epoxy weight ratio (wt%)	
Content (wt%)	Core content (wt%)	Content (wt%)	Core content (wt%)	Corresponding to the content range of $(C_2H_5)_2O{\cdot}BF_3$-loaded capsules	Corresponding to 1 wt% $(C_2H_5)_2O{\cdot}BF_3$-loaded capsules
0.25–1.5	20	2.5	70	2.9–17.1	11.4
0.25–1.5	20	5	70	1.4–8.6	5.7
0.25–1.5	20	10	70	0.7–4.3	2.9
0.25–1.5	20	15	70	0.5–2.9	1.9
0.25–1.5	20	20	70	0.4–2.1	1.4

[a] Here the contents of the microcapsules mean their percentage loading in the composites, while the core contents represent the weight ratios of core substance to microcapsule.

Source: Reprinted from Xiao et al. [74]. Copyright 2009, with permission from Elsevier.

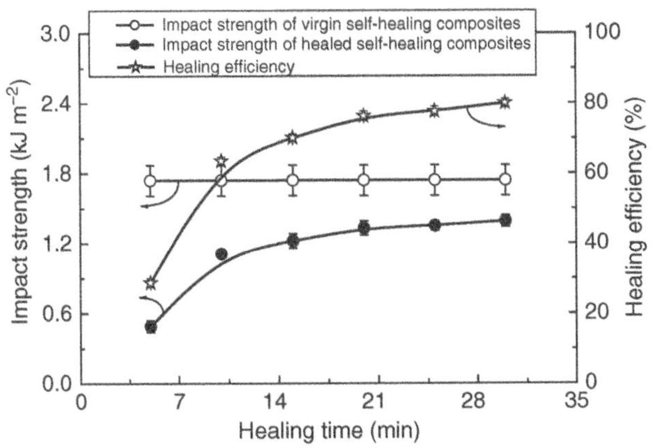

Figure 3.22 Dependences of impact strength of healed self-healing epoxy composites and healing efficiency on healing time at 20 °C. Contents of epoxy- and $(C_2H_5)_2O{\cdot}BF_3$-loaded capsules are 5 and 1 wt%, respectively. Epoxy-loaded microcapsules: core content = 70 wt%, average diameter = 30 μm. $(C_2H_5)_2O{\cdot}BF_3$-loaded microcapsules: core content = 20 wt%, average diameter = 10 μm. *Source:* Reprinted from Xiao et al. [74]. Copyright 2009, with permission from Elsevier.

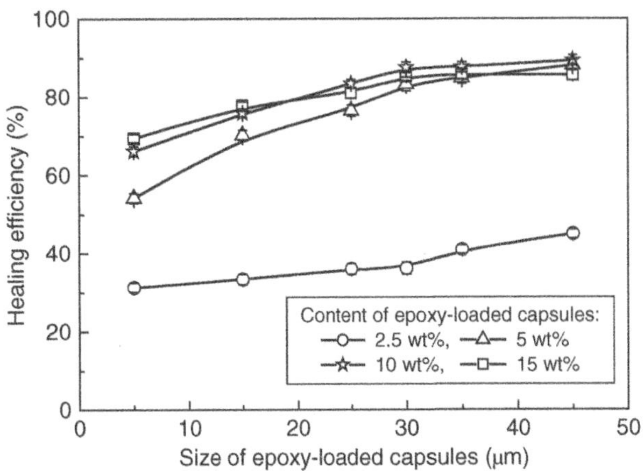

Figure 3.23 Dependence of healing efficiency of the self-healing epoxy composites on the average size of epoxy-loaded capsules. Content of $(C_2H_5)_2O{\cdot}BF_3$-loaded capsules (core content: 20 wt%; average diameter: 10 μm) is 1 wt%. Core contents of epoxy-loaded capsules with average diameters of 5, 15, 25, 30, 35, and 45 μm are 54, 60, 68, 70, 75, and 79 wt%, respectively. Healing of the fractured specimens was conducted at 20 °C for two hours. *Source:* Reprinted from Xiao et al. [74]. Copyright 2009, with permission from Elsevier.

maximum healing efficiency [10]. The result coincides with the data in Figure 3.20, further demonstrating that the healing agent is suitable for quick repair of cracks in epoxy.

Considering that a complete cure might hardly be attained at 20 °C and the present equilibrium healing efficiency might be further increased after longer rehabilitation period, we tried a longer healing time, such as 24 hours. However, the equilibrium healing efficiency turned out to be nearly unchanged. This can be understood from the viscosity change of the healing system. According to the general law of epoxy curing, the viscosity of the reaction system increases with the development of cross-linkages, while contacts between the components are reduced due to diffusion hindrance. Usually, step curing or a slow curing process is applied to maximize the degree of reaction. However, in this work, curing of the healing agent proceeds extremely fast at the test temperature and no measure can be taken to slow the reaction down. As a result, the equilibrium healing efficiency of ~80% is a trade-off between full recovery and rapid repair.

Besides contents and proportions of the dual capsules, the size of the capsules, which is closely related to the core content [46], should also be an important factor in determining healing efficiency of the composites. As shown in Figure 3.23, for composites containing 5, 10, and 15 wt% epoxy-loaded capsules, an increase in size of epoxy-loaded capsules tends to improve healing ability of the composites up to 30 μm, while further increase in the size of the capsules almost does not influence the healing efficiency. This should be ascribed to the fact that larger capsules possess higher core content so that the amount of the released epoxy is increased accordingly. When the capsules are large enough to deliver a sufficient quantity of epoxy fluid for covering the damaged surfaces, further increase in the size of the capsules makes no difference. Considering larger capsules would bring out more negative effect on the mechanical properties of epoxy composites, the capsules with diameters of 30 μm can be considered as the optimum choice. As for the composites containing 2.5 wt% epoxy-loaded microcapsules, an increase in the size of the capsules leads to continuous improvement of the healing efficiency. Evidently, the shortage of released epoxy on the fractured plane should account for this.

Fractured surfaces of the specimens before and after healing are illustrated in Figure 3.24. It is seen that after failure of the composites, both epoxy- and $(C_2H_5)_2O{\cdot}BF_3$-loaded microcapsules are fractured but not pulled out (Figure 3.24a), which implies strong bonding between the microcapsules and epoxy matrix. Having been healed, the surface became rough, and some smaller cavities on the virgin surface were covered by the polymerized healing agent (Figure 3.24b). Under high magnification, membrane from the cross-linked healing agent can be clearly identified, confirming polymerization of the healing agent (Figure 3.24c). Furthermore, if the content of epoxy-loaded capsules is high enough (such as 15 wt%), only the cured film can be observed on the fracture surface and all the cavities from ruptured capsules are shielded. EDS mapping analysis of the fractured surface indicates that fluorine is clearly visible. This agrees with the fact that the cured film is initiated

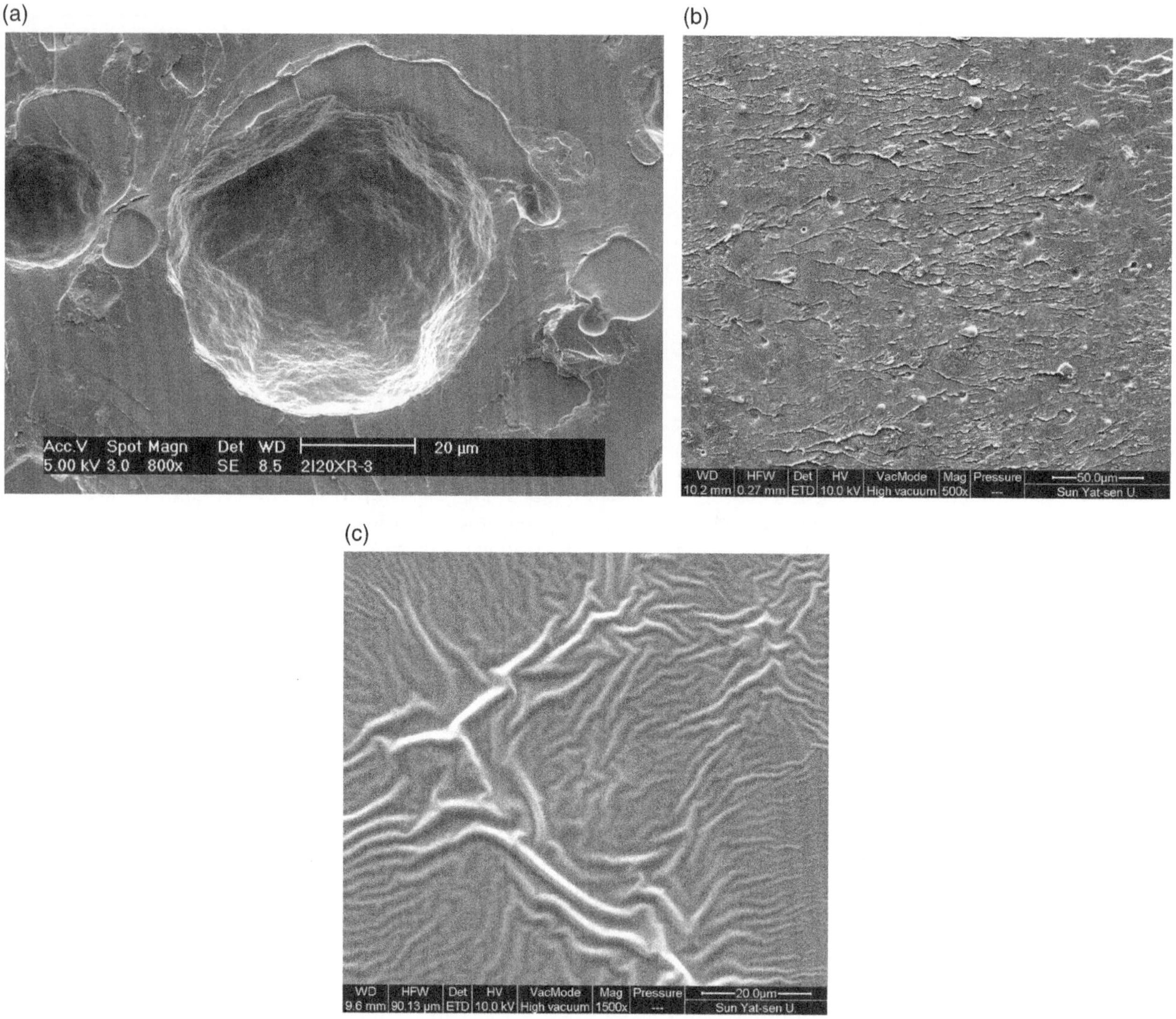

Figure 3.24 Scanning electron microscopy micrographs of fractured surfaces of (a) virgin self-healing epoxy composite, and (b, c) healed self-healing epoxy composite (photo (c) is a partial magnification of (b)). Contents of epoxy-loaded microcapsules (30 μm) and $(C_2H_5)_2O{\cdot}BF_3$-loaded hollow microcapsules (10 μm) are 5 and 1.0 wt%, respectively. *Source:* Xiao et al. [74], Figure 6, p. 5/with permission of Elsevier.

by $(C_2H_5)_2O{\cdot}BF_3$. The above results prove that healing of cracks in the epoxy composites should be attributed to the adhesion of polymerized epoxy healing agent coming from the broken microcapsules.

To have a quantitative idea of the amount of healing agent delivered, the average thickness of bled epoxy, $\bar{d}$, is calculated according to the derivation of Rule et al. [25]:

$$\bar{d} = d_c \cdot c \cdot \phi \tag{3.6}$$

where d_c, c, and φ stand for the average diameter, core content and weight content of epoxy-loaded capsules, respectively. Here it is assumed that the epoxy-loaded capsules were homogeneously distributed in the composites, and all the capsules at the crack surfaces can be fractured. In addition, the contribution of $(C_2H_5)_2O{\cdot}BF_3$-loaded microcapsules is neglected because of its low content. Table 3.8 lists the calculated values of $\bar{d}$ and the measured healing efficiency in the case of different sizes and contents of epoxy-loaded capsules. It is seen that when $\bar{d}$ approaches 1 μm, the healing efficiency is also close to the highest value (i.e. ~80%). If we set $\bar{d} \approx 1\,\mu m$ as the critical value, the requirement is met at least when the content of the epoxy-loaded capsules is ≥5 wt%. This agrees with the results given in Figures 3.21 and 3.23. Additionally, Table 3.8 reveals that although the $\bar{d}$ value of the composites containing 2.5 wt% larger size epoxy-loaded capsules (d_c = 45 μm) is similar to that of the composites containing 5 wt% smaller size

Table 3.8 Estimated average thickness of the bled healing agent delivered to crack planes and the measured healing efficiency (η)[a].

		ϕ (2.5 wt%)		ϕ (5 wt%)		ϕ (10 wt%)		ϕ (15 wt%)	
d_c (μm)	c (%)	$\bar{d}$ (nm)	η (%)	$\bar{d}$ (nm)	η (%)	$\bar{d}$ (nm)	η (%)	$\bar{d}$ (nm)	η (%)
5	54	67	31.3	135	54.2	270	66.1	405	69.4
15	60	225	33.3	450	70.2	900	75.9	1350	77.7
25	68	425	36.0	850	76.5	1700	83.4	2550	81.0
30	70	525	36.2	1050	83.3	2100	87.6	3150	85.4
35	75	656	41.0	1313	84.9	2625	87.7	3938	85.8
45	79	889	45.0	1778	88.2	3555	89.4	5332	85.6

[a] The content of $(C_2H_5)_2O{\cdot}BF_3$-loaded microcapsules in the authentic self-healing specimens from which healing efficiency was measured: 1 wt%.

Source: Reprinted from Xiao et al. [74]. Copyright 2009, with permission from Elsevier.

epoxy-loaded capsules (d_c = 25 μm), the measured healing efficiencies of the two composites are greatly different. This might result from a geometric effect of the capsules. That is, the 2.5 wt% loading of larger capsules might lead to a localized bleeding of epoxy [75], with not enough or no catalyst being present at this location. The more capsules that exist, the higher the probability of both constituents meeting and initiation of the reaction.

Stability and durability are important for self-healing materials. During storage at room temperature for five months, weight losses of epoxy- and $(C_2H_5)_2O{\cdot}BF_3$-loaded microcapsules are only around 0.5 and 0.25 wt%, respectively, probably due to evaporation of adsorbed moisture. Since the microcapsules are embedded in the composite matrix in practical usage, this would further improve their storage stability. As a result, the healing efficiency of the self-healing composite is almost constant within five months. In fact, epoxy-loaded capsules are thermally stable in nature, showing no detectable weight loss at a temperature lower than 200 °C [46]. Although the boiling point of $(C_2H_5)_2O{\cdot}BF_3$ is 126–129 °C and its saturated vapor pressure at 20 °C is 0.56 kPa, meaning that it has certain volatility, our experimental results demonstrate that the catalyst remains active for quite a long time. This should be attributed to the protection of the catalyst offered by both the capsules and the epoxy matrix.

It is worth noting that the curing temperatures used in the present work are relatively low, but it does not mean that the healing agent is not applicable for moderate or high temperature curable epoxy systems. According to the above thermal stability data, it can be reasonably deduced that the healing agent developed in this work might also be valid in the case of higher curing temperatures. Especially when $(C_2H_5)_2O{\cdot}BF_3$ is not encapsulated but pre-diffused into the matrix as a latent hardener (see Section 3.1.3.1), its high activity can be maintained in a more reliable way due to the interaction between the catalyst and matrix epoxy through either coordination bonds (Lewis acid–base interaction) or hydrogen bonds.

Incorporation of fillers into polymers would change intrinsic mechanical properties of the matrices. It is hoped that the embedded healing agent capsules would not bring about a significantly negative influence. The results of mechanical tests demonstrate that addition of either $(C_2H_5)_2O{\cdot}BF_3$- or epoxy-loaded capsules has nearly no influence on tensile and flexural properties of epoxy composites within the filler content range of interest, except a slight decrease in flexural strength and modulus of the composites with epoxy-loaded capsules. This demonstrates that the interfacial interaction between the capsules and the matrix resin is quite strong and the microcapsules are able to carry a certain load transferred by the interface under static testing conditions. On the other hand, the impact strength of the composites gradually decreases with a rise in the content of microcapsules, while the decrement is still insignificant. After all, the capsules are not able to hinder crack propagation in the course of rapid deformation.

To reveal joint action of the two types of microcapsules, mechanical properties of the composites containing both capsules were measured. Figure 3.25 shows tensile and flexural properties of the composites with a constant content of epoxy-loaded capsules of 5 wt%, which is chosen because the above healing tests indicate it is the critical value for having high healing efficiency. Similar to the results of addition of individual types of capsules, incorporation of both types of capsules still does not lead to remarkable variation in the static mechanical properties (Figure 3.25). Nevertheless, addition of the dual capsules can lead to an evident decrease in impact strength of the composites. The trend of embrittlement is more obvious at lower content of epoxy-loaded capsules. For the composite containing 5 wt% epoxy-loaded capsules

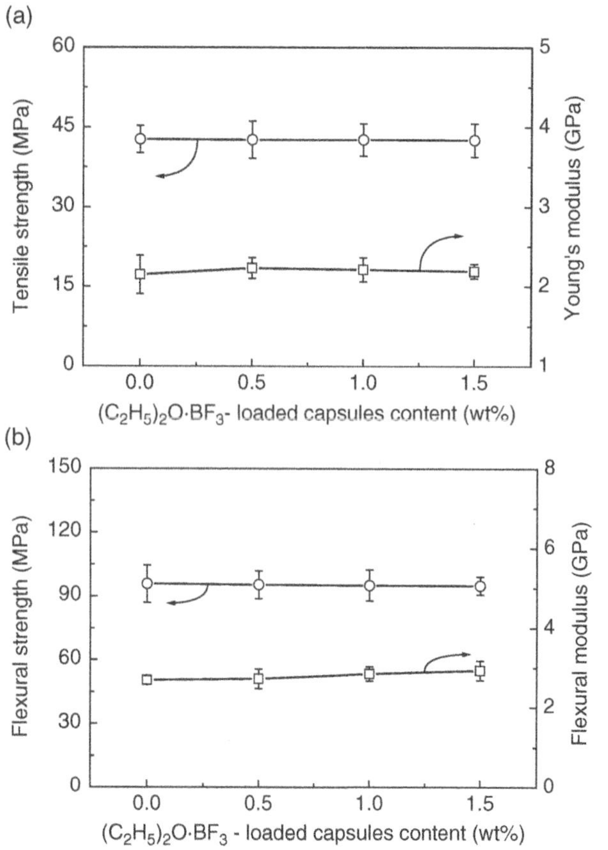

Figure 3.25 Influence of content of $(C_2H_5)_2O{\cdot}BF_3$-loaded capsules on (a) tensile and (b) flexural properties of the self-healing epoxy composites containing 5 wt% epoxy-loaded capsules. Epoxy-loaded microcapsules: core content = 70 wt%, average diameter = 30 μm. $(C_2H_5)_2O{\cdot}BF_3$-loaded microcapsules: core content = 20 wt%, average diameter = 10 μm.
Source: Reprinted from Xiao et al. [74]. Copyright 2009, with permission from Elsevier.

and 1 wt% $(C_2H_5)_2O{\cdot}BF_3$-loaded capsules, its impact strength is about 81% of that of unfilled epoxy.

We have further examined impact performance of the composites as a function of the size of the epoxy-loaded capsules to understand its effect at constant filling concentrations. The capsules cannot arrest the development of cracks, so the larger capsules would definitely cause higher stress concentration [76–78] and greater extent of reduction in impact strength. On the whole, the decay in impact strength is not severe, and can be limited to a marginal range so long as the amounts of the embedded capsules are not high.

3.1.4 Preparation of Silica Walled Microcapsules Containing $SbF_5{\cdot}HOC_2H_5/HOC_2H_5$

First, monodisperse poly(styrene-methacrylate) (PSMA) particles were prepared by dispersion polymerization. Deionized water (28 g), ethanol (140 g) and polyvinylpyrrolidone (3 g) were stirred at a rate of 200 rpm in a 500 ml three-necked flask for 60 minutes at 75 °C under argon purge. Afterwards, a solution of azobisisobutyronitrile (2.3 g), styrene (74 g) and methyl acrylate (6 g) was added into the flask. After 12 hours, the resultant PSMA particles were centrifuged, washed three times with water, and dried at 40 °C.

Then, the precursor tetraethyl orthosilicate (TEOS) was coated onto the polymer microspheres by a sol–gel process as in the following. PSMA particles (5 g) were dispersed into ethanol (70 g) under ultrasonication. Deionized water (10 g), ammonia (4 ml) and TEOS (10 g) were incorporated into the system. The mixture was kept at ambient temperature for 24 hours and then filtered to obtain silica coated particles. After removing the PSMA template through high temperature calcination at 600 °C, hollow silica microcapsules (~6 μm) were obtained (Figure 3.26).

In the meantime, both absolute ethanol and an ice bag were cooled to −80 °C in advance. In the glove box filled with pure nitrogen gas, the ethanol (20 g) was poured into a polytetrafluoroethylene bottle in the ice bag, and then SbF_5 (5 g) was slowly added under stirring. The solution temperature was kept below 5 °C by controlling the dropping. Afterwards, the solution was continuously stirred until it was warmed to room temperature offering $SbF_5{\cdot}HOC_2H_5/HOC_2H_5$, in which the fraction of SbF_5 was 20 wt%.

With the aid of vacuum, $SbF_5{\cdot}HOC_2H_5/HOC_2H_5$ was infiltrated into the hollow silica capsules to produce hardener-loaded capsules. Typically, hollow silica capsules (0.3 g) were submerged by the above $SbF_5{\cdot}HOC_2H_5/HOC_2H_5$ in a 25 ml one-necked flask. Having been sealed and cooled to −20 °C, the system was evacuated to a vacuum degree of 0.001–0.002 mbar. After 50 hours, the infiltration reached equilibrium, and the obtained $SbF_5{\cdot}HOC_2H_5/HOC_2H_5$-loaded capsules (core content of 39 wt%) were filtered, and purged with ethanol. The reason why $SbF_5{\cdot}HOC_2H_5/HOC_2H_5$ was not encapsulated through in-situ polymerization of melamine-formaldehyde in an oil-in-water emulsion like the epoxy counterpart lies in the fact that the former is quite active and would be easily deactivated in water and common organic solvents except ethanol.

3.1.5 Self-Healing Epoxy Materials with Embedded Epoxy-Loaded Microcapsules and $SbF_5{\cdot}HOC_2H_5/HOC_2H_5$-Loaded Silica Capsules

To cooperate with $SbF_5{\cdot}HOC_2H_5/HOC_2H_5$-loaded silica capsules in the target self-healing epoxy composites, epoxy monomer was encapsulated by PMF as mentioned in ref. [10]. The average diameter and core content of the resultant epoxy-loaded capsules (determined by extraction method) are 130 μm and 91 wt%, respectively. Unfilled epoxy specimens were cast from the mixture of 5 parts

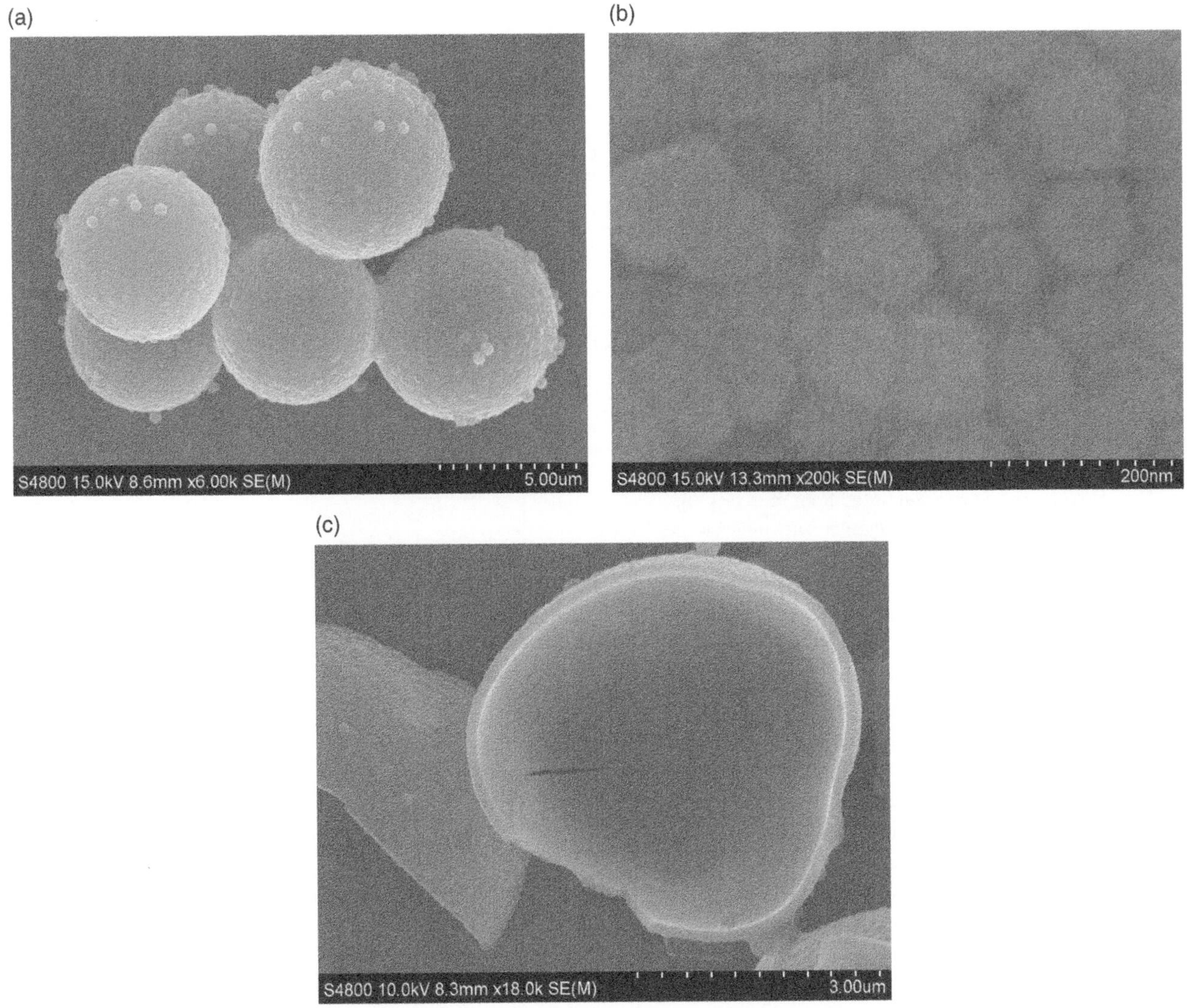

Figure 3.26 (a) Micrograph of hollow silica walled microcapsules, (b) high magnification view of the capsule surface, and (c) squashed capsules (showing thickness of the shell wall is about 200 nm).

boron trifluoride-2,4-dimethylaniline complex (BF_3·DMA), 80 parts methyl hexahydrophthalic anhydride (MHHPA) and 100 parts EPON 828, while the self-healing epoxy specimens were prepared by uniformly blending 10 wt% epoxy-loaded microcapsules and 0.6 wt% $SbF_5 \cdot HOC_2H_5$/HOC_2H_5-loaded microcapsules with the aforementioned mixture of BF_3·DMA, MHHPA, and EPON 828. The compounds were degassed, poured into a preheated closed silicone rubber mold, and cured at 50 °C for 60 hours and 70 °C for 12 hours. The control specimen has the same composition as the self-healing one except that the hardener capsules are replaced by ethanol capsules.

Since the curing behavior of epoxy in the presence of $SbF_5 \cdot HOC_2H_5$ has not been revealed, optical pyrometry [79] and infrared microscopy are used. Conventional DSC is no longer applicable as considerable polymerization proceeds before a homogeneous mixture of the components is obtained. The data of thermal imaging (Figure 3.27) show that soon after the ethanol solution of the antimony pentafluoride-ethanol complex ($SbF_5 \cdot HOC_2H_5$/HOC_2H_5) is added to the epoxy, rapid solidification occurs as characterized by the drastic increase in temperature from ambient temperature to the maximum (i.e. ~200 °C) within several seconds. When the dosage of $SbF_5 \cdot HOC_2H_5$/HOC_2H_5 increases from 1 to 10 μl, the time to reach the maximum temperature is greatly reduced and the maximum temperature is obviously increased. When the hardener dosage is further raised from 10 to 15 μl, however, both the peak position and height are basically the same. The behavior follows the general rule of curing of epoxy in the presence of Lewis acid. That is, once the dosage of Lewis acid reaches the prescribed minimum, cationic chain polymerization of epoxy

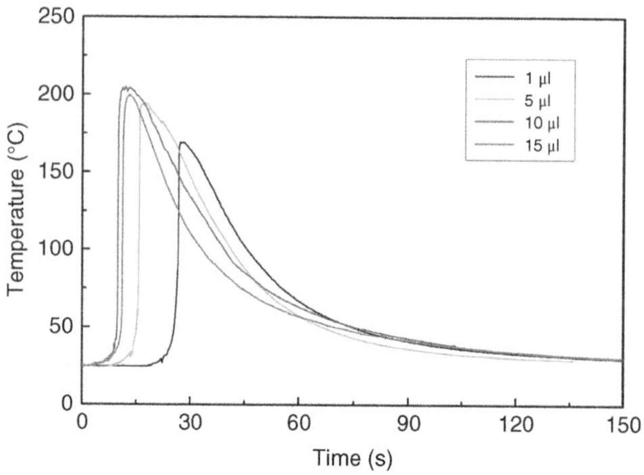

Figure 3.27 Time dependence of temperature of epoxy cured by ethanol solution of antimony pentafluoride-ethanol complex ($SbF_5 \cdot HOC_2H_5/HOC_2H_5$). *Source:* Reprinted from Ye et al. [29]. Copyright 2014, with permission from American Chemical Society.

begins. The curing reaction is nearly independent of Lewis acid content when the latter exceeds a certain value. Scheme 3.4 illustrates the reaction mechanism involved.

A more detailed kinetics study is given by the real-time micro-infrared spectra (Figure 3.28), which show that the amount of epoxide groups decreases with time. About 94% epoxide groups are consumed after 10 seconds when the chemicals are contacted. The spectra were continuously collected from the same site of the sample, so the possible influence of sample thickness, heterogeneity, etc. can be neglected when calculating the degree of curing based on the peak areas.

Now $SbF_5 \cdot HOC_2H_5/HOC_2H_5$ has been proved to be able to initiate fast curing of epoxy, its capability for fast repairing cracks in polymer in cooperation with unreacted epoxy monomer is worth verifying by means of the dual microcapsules approach. Although $SbF_5 \cdot HOC_2H_5$ becomes less

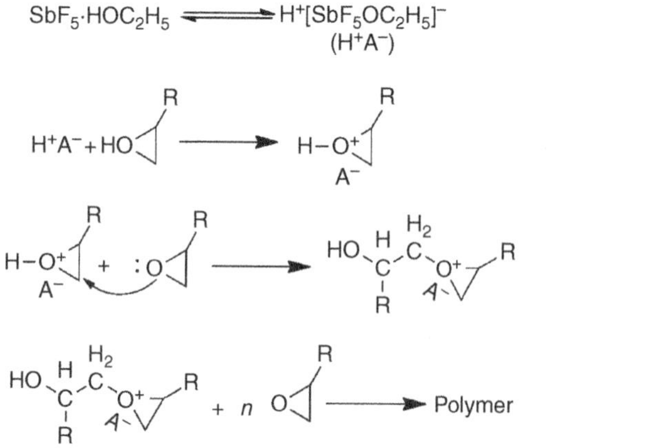

Scheme 3.4 Mechanism of the reaction between $SbF_5 \cdot HOC_2H_5$ and epoxy.

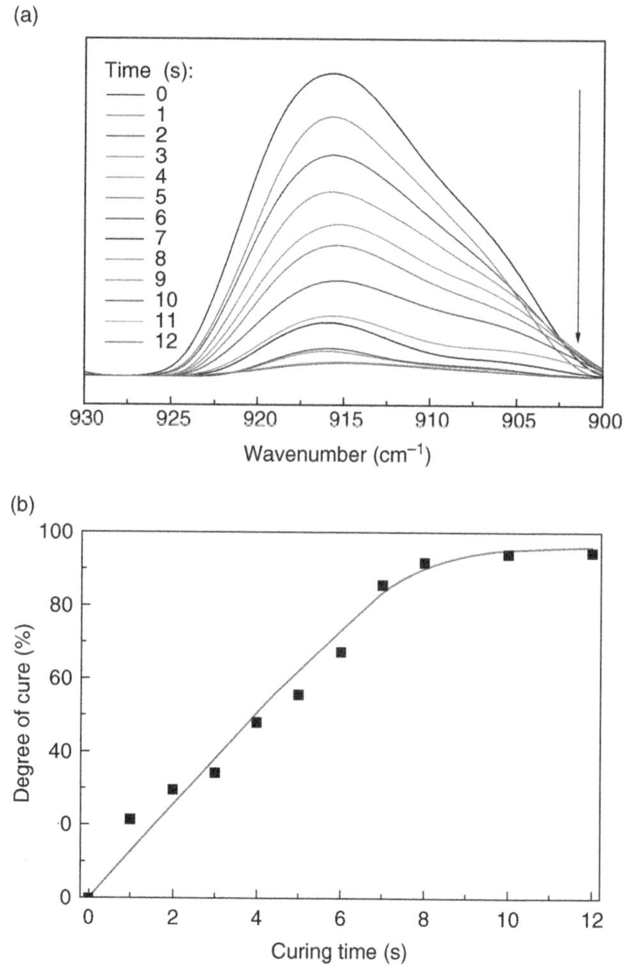

Figure 3.28 (a) Infrared absorption of epoxide groups as a function of time when epoxy meets $SbF_5 \cdot HOC_2H_5/HOC_2H_5$ (20 wt%). (b) Evolution of curing degree of epoxy calculated with the peak areas in (a). *Source:* Reprinted from Ye et al. [29]. Copyright 2014, with permission from American Chemical Society.

corrosive than SbF_5, it is still highly active and could not be encapsulated by a conventional technique like emulsion polymerization. Accordingly, silica walled microcapsules containing $SbF_5 \cdot HOC_2H_5/HOC_2H_5$ were prepared and embedded in epoxy matrix together with epoxy capsules to build up a self-healing epoxy composite.

On the basis of the protocol proposed by Hayes et al. [68], notched impact tests were carried out to evaluate the healing effect of cured epoxy with embedded capsules, which contain epoxy monomer and $SbF_5 \cdot HOC_2H_5/HOC_2H_5$, respectively. The efficiency of healing is defined as the ratio of impact strengths of healed and virgin materials. As expected, the healing agent can be liberated upon fracture of the specimen, spreading over the fractured surface, and rejoining the broken specimen at room temperature (Figure 3.29). In contrast, the impacted control specimen,

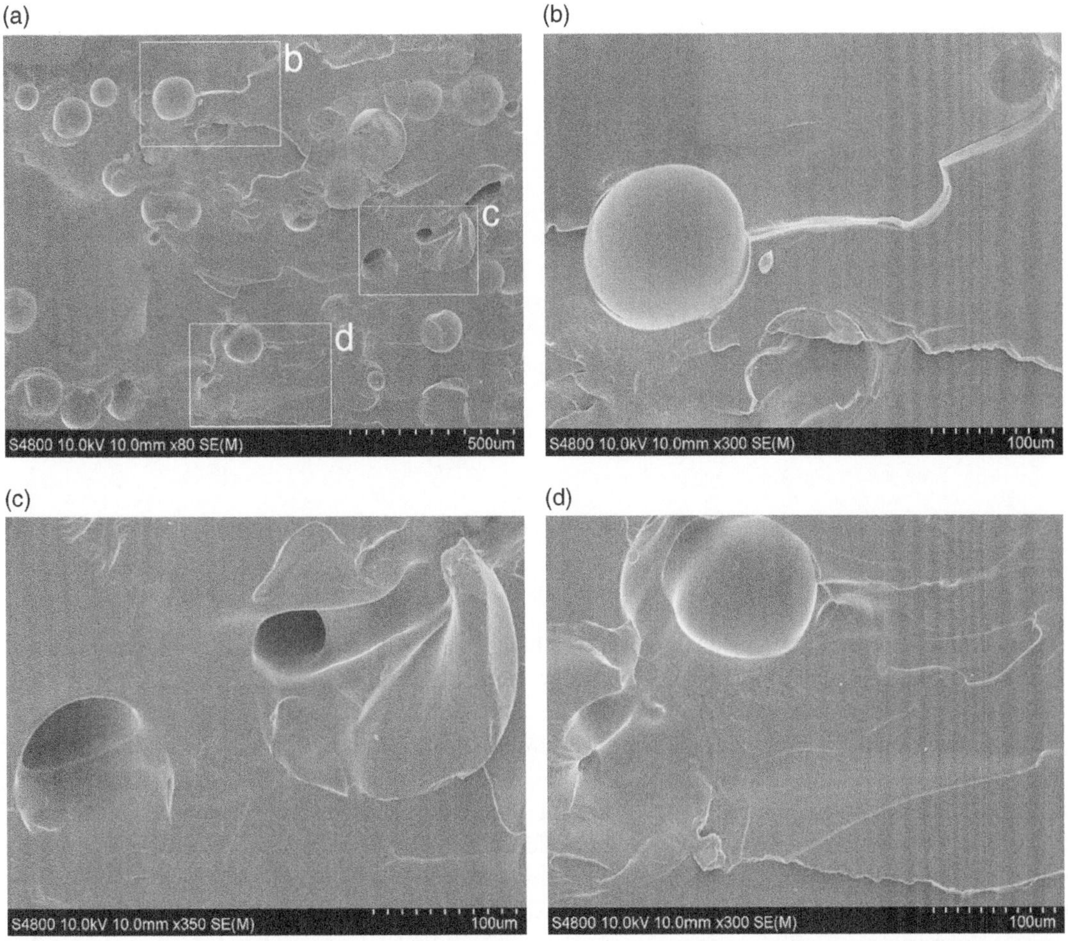

Figure 3.29 (a) Scanning electron microscopy image of the impact fracture surface of healed self-healing specimen. The specimen had been fractured by the first impact test, healed at 25 °C for 100 seconds, and then fractured again by the second impact test. (b–d) Magnified views of different parts of (a). *Source:* Ye et al. [29], Figure 6, p. 7/with permission of American Chemical Society.

which has the same composition as the self-healing specimen except $SbF_5 \cdot HOC_2H_5/HOC_2H_5$ microcapsules are replaced by ethanol microcapsules, could not be re-bonded. More importantly, the dependence of healing efficiency on healing time (Figure 3.30) shows that about 8.3% impact strength is recovered after 20 seconds and the maximum healing efficiency is acquired after 100 seconds of healing time. The results demonstrate that the healing agent not only allows for autonomic healing of cured epoxy without manual intervention, but also obviously accelerates the healing process owing to the translation of curing kinetics.

To understand whether the SbF_5 dispersed in the matrix would affect long-term stability of the healed epoxy, we carried out the following experiments. First, the self-healing specimens were impacted to failure, and then the broken halves were reconnected for healing at room temperature. After 10 months, the healed specimens were tested by impact again. The average healing efficiency was $71.2 \pm 1.5\%$, which equals the steady-state healing efficiency of freshly made specimens (Figure 3.30). Clearly, SbF_5 does not exert detectable influence on the stability of healed epoxy. The result is understandable because the hardener $SbF_5 \cdot HOC_2H_5$ is milder than SbF_5 and even inert to air. Moreover, the content of SbF_5 dispersed in the self-healing composite is low due to the low concentration of the hardener capsules (i.e. 0.6 wt%).

Nevertheless, it is worth noting that curing of the SbF_5-epoxy system is approximately completed within 10 seconds as revealed by infrared spectroscopy (Figure 3.28), which is roughly equivalent to the time for collecting the impact broken specimen and reassembling for crack healing (8–10 seconds). This means that a certain degree of polymerization must have taken place prior to the occurrence of healing reaction across the broken surfaces (i.e. contact of the broken surfaces), despite the fact that the fluidic healing chemicals from the ruptured capsules need time to flow out and meet each other for the subsequent reaction. In fact, healing of cracks is multi-convoluted time

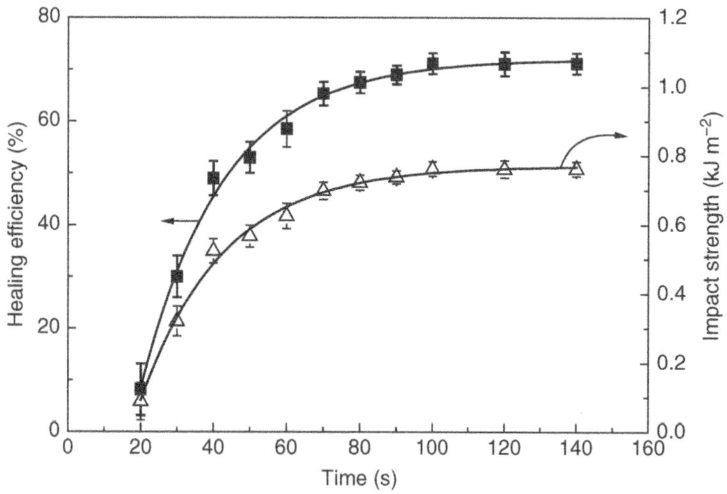

Figure 3.30 Healing efficiency versus healing time of self-healing specimens determined by impact test. Impact strength of the virgin self-healing specimen = 1.070±0.025 kJ m^{-2}.

dependent [80]. With regard to the present fast curing system, the data of impact tests (Figure 3.30) only give a rough description of the healing kinetics. For purposes of more precisely identifying the (initial) healing time or healing speed, the time taken for recombination of broken specimens before healing should be shortened to the greatest extent.

By screening the available destructive testing methods (e.g. tensile and tearing), we found that fatigue measurement seems to be the most appropriate to fulfill the task as the cracked planes are rapidly opened and closed under cyclic stress without the necessity of "time consuming" manual reassembly. The frequent opening and closing also promote transportation and mixing of the healing fluids out of the broken capsules. Meanwhile, the time-dependent self-healability of fatigue crack offered by the healing agent can be determined.

To minimize the influence of curing reaction of the healing agent released from the broken capsules after pre-cracking but prior to fatigue test, the time for pre-cracking and attaching the specimen to the fixing jig of the fatigue tester was managed to be as short as possible (5–7 seconds). Moreover, removal of the razor blade for making a pre-crack was synchronized to the start of the fatigue test. It is found that when the blade is removed after making the pre-crack, a large portion of the pre-cracked faces is allowed to touch each other. In contrast, the two faces are mostly separated so long as the blade stays in place. Therefore, for the pre-cracked self-healing specimen containing microencapsulated healing agent without removal of the blade, polymerization of the released healing agent would be greatly hindered in the interface region. In the case where the fatigue test starts immediately after removal of the blade, the healing time can be roughly reckoned from the start of the fatigue test, supposing that the healing reaction across the crack faces mainly occurs after the beginning of the fatigue test.

Figure 3.31a plots crack length against fatigue cycle of unfilled epoxy, control, and self-healing specimens. As the effects of (i) microcapsule-induced toughening, (ii) hydrodynamic pressure crack tip shielding, (iii) polymeric wedge made of cured healing agent, and (iv) adhesive bonding of the healing agent are responsible for the extension of fatigue life of self-healing polymer with encapsulated healant [12, 21, 22, 81], the control specimen was prepared according to the same formulation as the self-healing one except that the hardener capsules are replaced by ethanol capsules to highlight factors (iii) and (iv) of the self-healing specimen. Compared with both unfilled epoxy and control specimens, the self-healing specimen exhibits remarkably reduced crack growth rate (Figure 3.31a). This demonstrates that the latter two mechanisms of the healing agent mentioned above play the leading role. The room temperature healing efficiency defined by fatigue life extension [22] is estimated to be 167% and 304% relative to the control specimen and the unfilled epoxy specimen, respectively. Furthermore, the healability remains unchanged within eight months (Figure 3.32).

More detailed information about the fatigue retardation behavior of the healing agent can be found in the load–displacement curves (Figure 3.32b,c). The viscous flow of the released healing agent on the crack plane results in the appearance of hysteresis at the beginning of the test. With a rise in time, the hysteresis loop area increases and then decreases to constant. This implies that the healing agent liberated by pre-cracking starts to be polymerized,

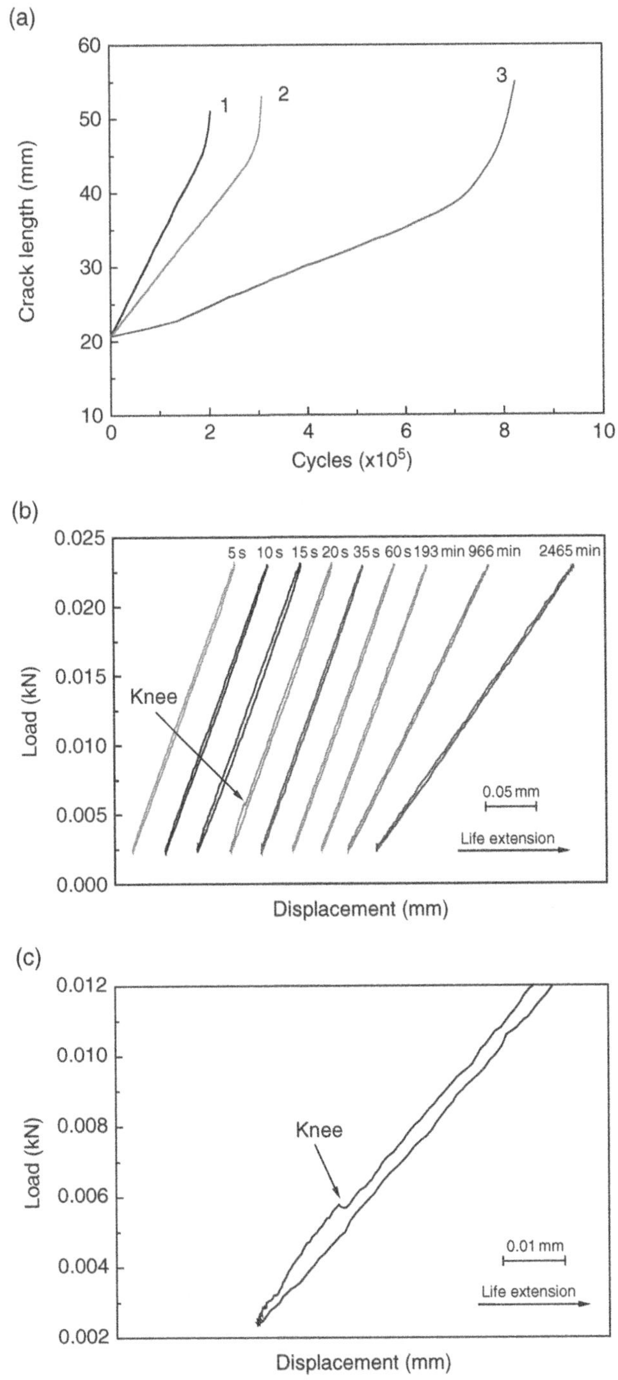

Figure 3.31 (a) Crack length versus fatigue cycle of (1) neat epoxy, (2) control and (3) self-healing specimens. (b) Load–displacement curves of self-healing specimen measured during selected cycles in (a). The times counted from starting of the fatigue test are shown on top of the corresponding curves. (c) Enlarged view of the curve of 20 seconds in (b) showing the knee phenomenon. Testing conditions: K_{max} = 0.258 MPa m$^{1/2}$, K_{min} = 0.026 MPa m$^{1/2}$, and ΔK_I = 0.232 MPa m$^{1/2}$.
Source: Reprinted from Ye et al. [29]. Copyright 2014, with permission from American Chemical Society.

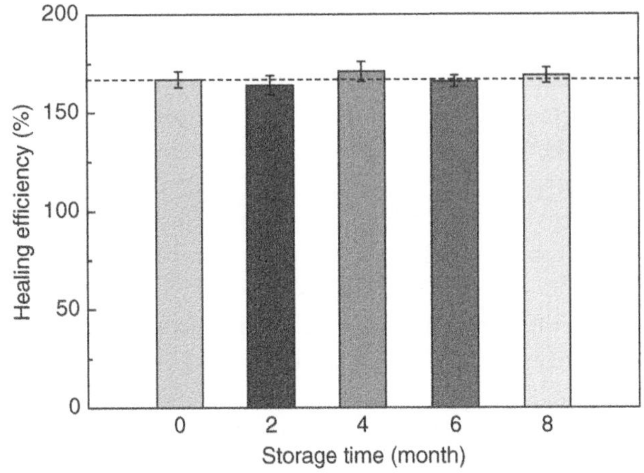

Figure 3.32 Effect of storage time in ambient air on healing efficiency of the self-healing specimens determined by fatigue test. Testing conditions: K_{max} = 0.258 MPa m$^{1/2}$, K_{min} = 0.026 MPa m$^{1/2}$, and ΔK_I = 0.232 MPa m$^{1/2}$.

enhancing the hydrodynamic pressure effect ascribed to the increased viscosity. Afterwards, this effect is reduced due to (i) further polymerization (hardening) of the healing agent released during pre-cracking and (ii) less amount of fluidic healing agent refilled by the gradually broken capsules on the crack growth path as compared with that offered in one go by the pre-cracked portion (~2.5 mm long). In this context, it is known that the first mechanically detectable gel (healing) time is 15 seconds (see the loop of 15 seconds in Figure 3.31b showing the largest area).

In addition, a knee point is observed on the loading part of the loop of 20 seconds (Figure 3.31b,c), representing crack closure effect contributed by the polymeric wedge and adhesion at the crack tip [82]. Such a phenomenon is absent in the case of the control specimen, in which the released epoxy monomer cannot be cross-linked due to the lack of hardener. Since the results in Figure 3.31a reveal that (i) the mechanisms of polymeric wedge and adhesive bonding, rather than microcapsule-induced toughening and hydrodynamic pressure, predominate crack retardation and (ii) the crack grows steadily prior to failure, the time when the knee point is perceived (i.e. 20 seconds, see Figure 3.31c) can be regarded as the time when the healing effect reaches the maximum under cyclic loading. That is, the solidified healing agent provides the largest closure effect at the knee point. Subsequently, the crack grows past the wedge from the cured healing agent due to higher $\Delta K_I = K_{max} - K_{min}$ (where K_{min} and K_{max} denote the minimum and maximum values of the cyclic stress intensity, respectively). Although the above data are acquired within the pre-crack region, which might generate experimental errors of fatigue performance measurement due to slight deflection of the pre-crack, the rapid healability has been fully demonstrated.

The analysis is supported by the in-situ temperature measurement of the crack tip of the self-healing specimen during the fatigue test (Figure 3.33). With increasing test time, the temperature at the crack tip region increases and then keeps at a slightly lower level. This coincides with the aforementioned rapid exotherm of epoxy cured by $SbF_5{\cdot}HOC_2H_5/HOC_2H_5$ (Figures 3.27 and 3.28). That is, because the released healing agent in the pre-crack tip is polymerized, the reaction temperature peaks within a very short time. When the pre-crack does not obviously move forward, no additional healing agent can be refilled so that the on-site temperature decreases within the time range of interest. Here, the highest reaction temperature is found at ~20 seconds, in agreement with the time of the knee point observation on the load–displacement curve (Figure 3.28b).

3.1.6 Preparation of Silica Walled Microcapsules Containing TfOH

TfOH is quite active and would be readily deactivated in water or common organic solvents, so it could not be encapsulated through conventional techniques such as in-situ polymerization in an oil-in-water emulsion [10]. To encapsulate TfOH, infiltration of the chemical into silica walled hollow microcapsules was carried out as TfOH is inert with silica. The silica capsules were made following the method described in Section 3.1.4.

TfOH was infiltrated into the hollow silica capsules under vacuum to produce hardener-loaded capsules. Typically, hollow silica capsules (0.3 g) were submerged by TfOH in a 25 ml one-necked flask. The system was sealed

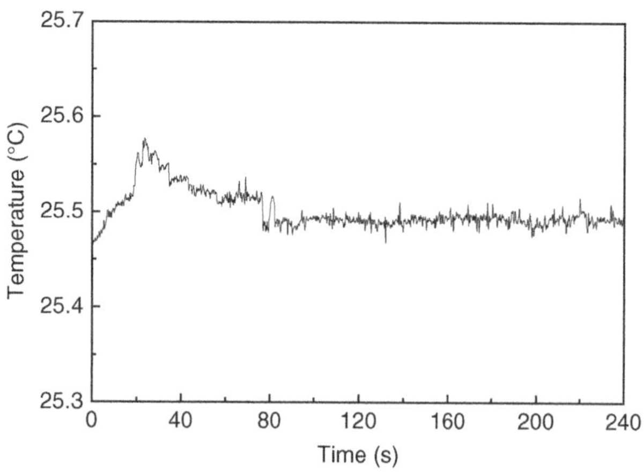

Figure 3.33 Temperature of crack tip of self-healing specimen as a function of time recorded by infrared camera during fatigue test. The time is counted from the start of the fatigue test. *Source:* Reprinted from Ye et al. [29]. Copyright 2014, with permission from American Chemical Society.

and cooled to −30 °C and evacuated to a vacuum degree of 0.001–0.002 mbar. After five days, TfOH-loaded capsules were produced and purged with cool acetonitrile. Accordingly, the core content of TfOH-loaded microcapsules was found to be 42 wt% by comparing the weight of the capsules before and after the infiltration.

To examine the reactivity of the encapsulated TfOH, TfOH-loaded microcapsules (0.05 g) were ground together with epoxy monomer (1 g). The temperature of the mixture increased from room temperature to 180 °C within four seconds, as revealed by optical pyrometry, meaning that the TfOH stored in the capsules maintains its high reactivity of curing epoxy.

3.1.7 Self-Healing Epoxy Materials with Embedded Epoxy-Loaded Microcapsules and TfOH-Loaded Silica Capsules

Similar to the case described in the previous section, PMF walled microcapsules containing epoxy monomer (EPON 828) were also prepared for the present model composites. The average diameter and core content of the epoxy-loaded capsules are 130 μm and 91 wt%, respectively.

Unfilled epoxy specimens were cast from the mixture of 54 parts MHHPA, 46 parts high performance cycloaliphatic epoxy resin CY 179 and 0.5 parts AO-4 catalyst (which is based on an activated oxo-centered-trinuclear Cr^{3+} complex), while the self-healing epoxy specimens were prepared by uniformly blending 10 wt% epoxy-loaded microcapsules and 0.6 wt% TfOH-loaded microcapsules with the aforementioned mixture of MHHPA, CY 179 and AO-4. The compounds were degassed, poured into a preheated closed silicone rubber mold, and cured at 80 °C for five hours, 100 °C for five hours, 150 °C for five hours, and 180 °C for five hours. The control specimen has the same composition as the self-healing one except that the hardener capsules are replaced by hollow silica capsules.

The curing mechanism of epoxy catalyzed by strong Bronsted acid belongs to cationic chain polymerization (Scheme 3.5) [28]. An exact stoichiometric proportion of the components required by addition polymerization (e.g. amine-epoxy cure) is no longer necessary. The stronger acid corresponds to the faster curing speed. Since TfOH is a super-strong Bronsted acid, fast curing should be achieved at room temperature with only a small amount of TfOH. This inference is supported by the result of optical pyrometry characterization (Figure 3.34a), which uses an infrared camera to record the real-time temperature change of epoxy resin during curing. The plots show a drastic increase in temperature from ambient temperature to the maximum (i.e. ~200 °C) within less than five seconds due to the exothermic reaction of curing after the addition of

Scheme 3.5 The reaction mechanism involved in TfOH-epoxy cure.

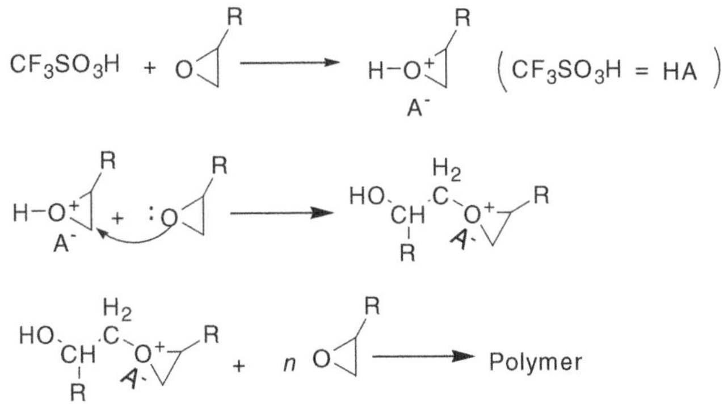

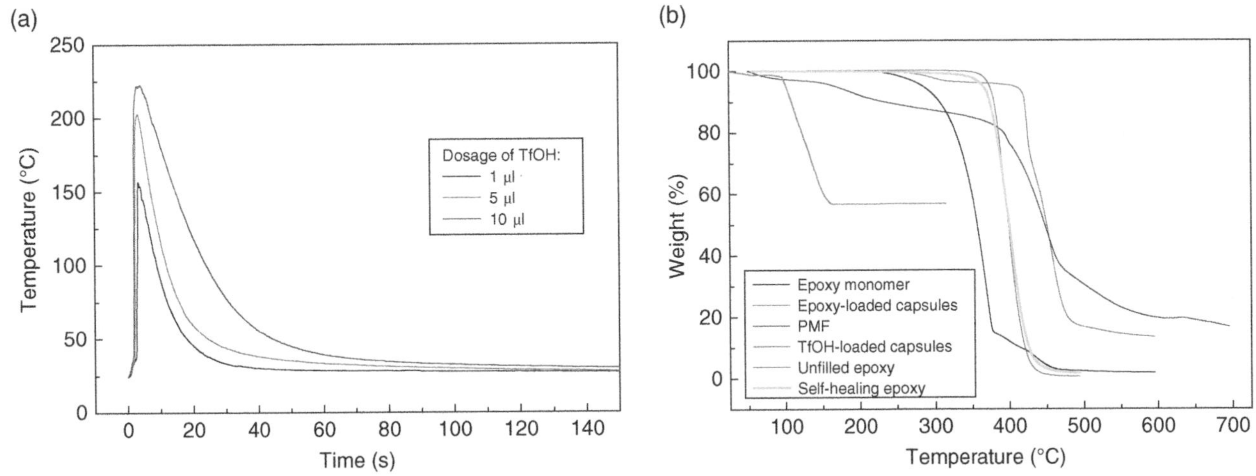

Figure 3.34 (a) Time dependence of temperature of epoxy cured by different dosages of TfOH (epoxy monomer: 1 g). (b) Weight loss as a function of temperature of microencapsulated healing agent, self-healing epoxy composite and the related substances. *Source:* Reprinted from Ye et al. [31]. Copyright 2014, with permission from Elsevier.

TfOH into the epoxy monomer. When the dosage of TfOH increases from 1 to 10 μl, the time to reach the maximum temperature is basically the same even though the maximum temperature is obviously increased. The reason is that once the dosage of TfOH reaches the prescribed minimum, cationic chain polymerization of epoxy begins and develops rapidly. The curing speed becomes independent of the hardener content above the threshold.

The above experiment has demonstrated that TfOH-epoxy cure proceeds very quickly. Since the aim of the current work is the development of self-healing high glass transition temperature (T_g) epoxy with ultrafast and heat-resistant healing agent, the thermal properties of the epoxy and its composites should be characterized.

Figure 3.35 shows that the T_g, represented by the peak temperature of tan δ, of the epoxy matrix studied in this work is as high as 244 °C. Besides, the thermogravimetric analysis in Figure 3.34b indicates that TfOH-loaded capsules start to obviously lose their weight at 99.6 °C. This may be due to volatilization of TfOH below its boiling point through the nanopores on the shell wall. As for the epoxy monomer, epoxy-loaded microcapsules and PMF (the wall material of the epoxy capsules), a detailed discussion about the thermal degradation performance is available in ref. [83]. The most attractive result of Figure 3.34b is that the self-healing epoxy containing both epoxy- and TfOH-loaded capsules does not exhibit a detectable change in weight below 300 °C, like the unfilled epoxy. This means that the matrix epoxy has blocked the escape channels of the volatile TfOH and the composite can withstand high temperature exposure. The epoxy matrix was cured at elevated temperatures step-by-step, being first gelled at 80 °C and then completely cured at 180 °C. Though the final curing temperature is higher than the boiling point of TfOH, TfOH can still be enclosed in the capsules under the double protection of the silica shell and epoxy matrix.

In the following, the self-healing ability of the epoxy composite filled with epoxy and TfOH capsules is verified

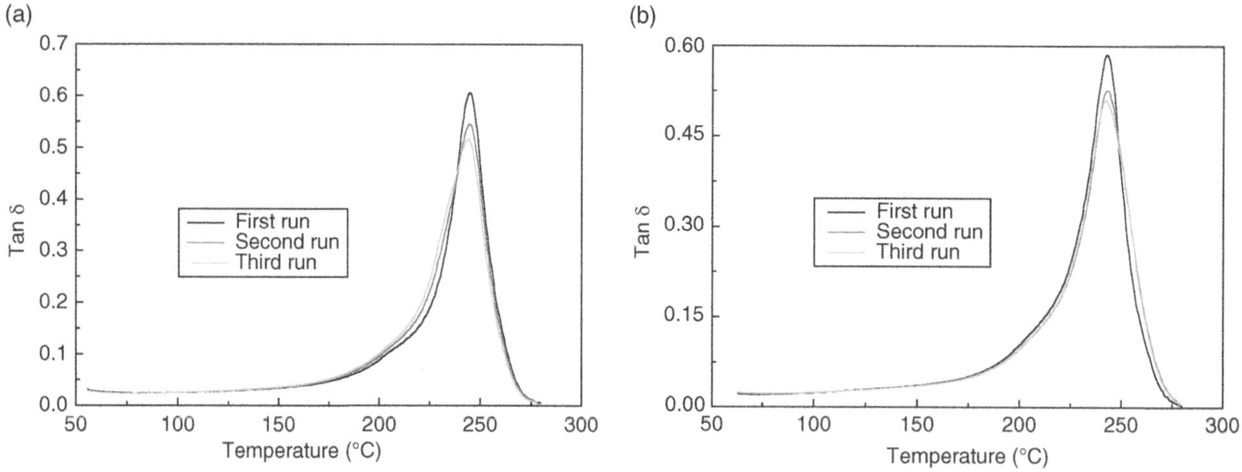

Figure 3.35 Temperature dependence of tan δ of (a) unfilled epoxy and (b) self-healing epoxy composite measured by repeated dynamic mechanical analysis scans.

to reveal whether the dual-capsule healing chemistry is robust enough to satisfy the rigorous demands for high performance polymeric materials.

First, an impact test was conducted [68]. Figure 3.36 indicates that the self-healing system consisting of 10 wt% epoxy loaded-microcapsules and 0.6 wt% TfOH loaded-microcapsules exerts favorable repairing effect in a short time. The measurable healing efficiency, about 9.7%, is observed at 20 seconds, and the maximum healing efficiency (~79%) is acquired at 100 seconds. SEM observation of the fracture surface of the self-healing specimen shows that the healing agent stored in the capsules has been released, spreading over the fractured surface and rejoining the broken specimen at room temperature. Crack propagates in a cohesive manner through the cured healing agent. On the contrary, no re-bonding effect can be observed in the control specimens subjected to the same impact and healing treatments, which have the same composition as the self-healing specimens except that the TfOH-loaded silica microcapsules are replaced by hollow silica microcapsules. These results show evidence that the epoxy composite with the dual-microcapsules system based on TfOH-epoxy chemistry not only can self-heal without manual intervention but also has coupled with ultrafast healability. The reactivity of the encapsulated healing agent remains after composite fabrication. What is more, the maximum healing efficiency offered by the present healing agent is higher than the value offered by the healant containing SbF_5 (~71%) [29]. Clearly, the absence of ethanol in the cured healing membrane should account for this improvement.

To know whether the appearance of TfOH in the matrix would lead to degradation of the healed epoxy, the following examination was made. First, the self-healing specimens were impacted to failure, and then the broken halves were reconnected for healing at room temperature. After 10 months of open air storage, the healed specimens were tested by impact again. The healing

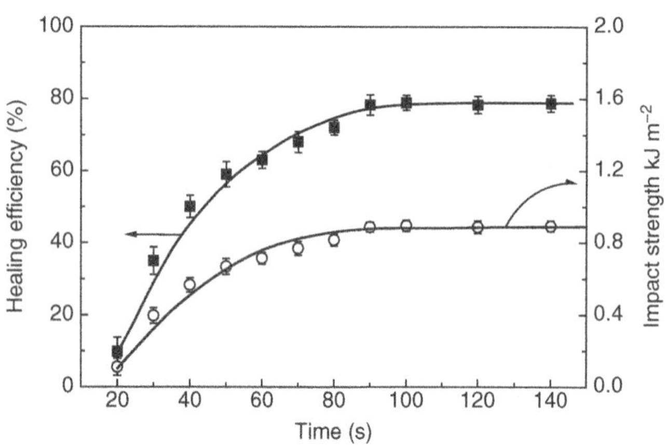

Figure 3.36 Healing efficiency versus room temperature healing time of self-healing specimens determined by impact test. Impact strength of the virgin self-healing specimen is $1.13 \pm 0.03\,kJ\,m^{-2}$. *Source:* Reprinted from Ye et al. [31]. Copyright 2014, with permission from Elsevier.

efficiency was 79.4 ± 1.6%, which equals the value measured from the freshly healed specimens (Figure 3.36). This suggests that TfOH does not exert detectable negative influence on the healed epoxy. In fact, even if TfOH might cause deterioration of the healed epoxy, its effect would be insignificant because of the low concentration of the hardener capsules (i.e. 0.6 wt%).

Although the healing agent with TfOH has been shown to work, a comparison between Figures 3.36 and 3.34a reveals that the healing speed is apparently slower than the curing speed of the same material system. This is because 8–10 seconds has to be taken for collecting and reassembling of the broken specimen before crack healing. A certain amount of released healing agent is polymerized prior to intimate contact of the two halves of the broken specimen, which would also decrease the ultimate healing efficiency. Clearly, the impact test is not an ideal method to evaluate the fast healing agent.

For purposes of reducing the time for recombination of broken specimen, the fatigue test is employed. Under cyclic stress, the cracked specimen is rapidly recombined without the necessity of manual reassembling. The mixing and delivery of the healing agent released from the broken capsules are promoted by the frequent opening and closing of the crack. In this context, the measured healing speed and healing efficiency would be more reasonable.

Figure 3.37a depicts crack length as a function of fatigue cycle of related materials. It is evident that the self-healing specimen possesses much longer fatigue life than the unfilled and control specimens. The room temperature

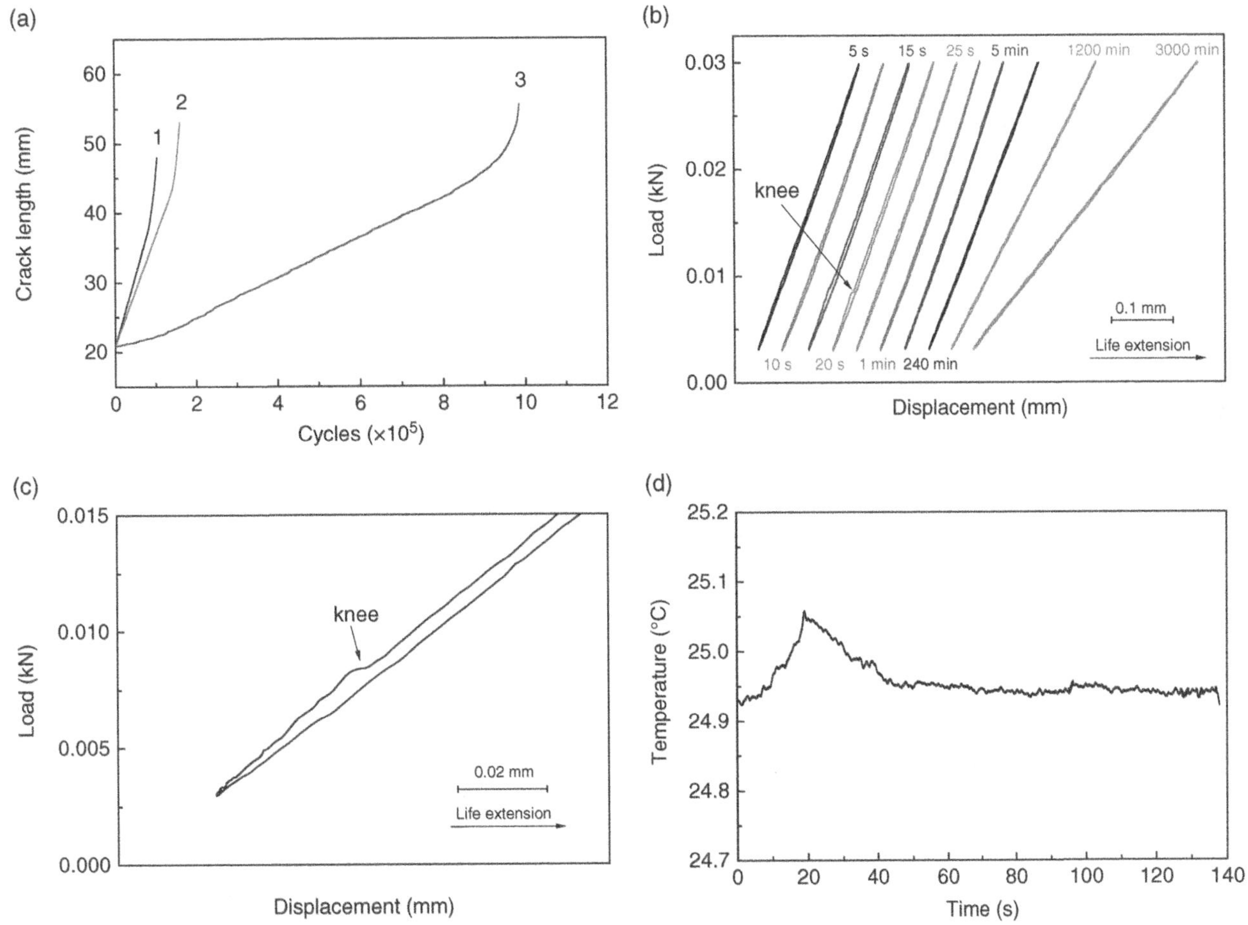

Figure 3.37 (a) Crack length versus fatigue cycle of (1) neat epoxy, (2) control, and (3) self-healing specimens (testing conditions: K_{max} = 0.258 MPa m$^{1/2}$, K_{min} = 0.026 MPa m$^{1/2}$, and ΔK_I = 0.232 MPa m$^{1/2}$). (b) Load–displacement curves of self-healing specimen measured during selected cycles in (a). The times counted from the start of the fatigue test are shown on top of the corresponding curves. (c) Enlarged view of the curve of 20 seconds in (b) showing the knee phenomenon. (d) Crack tip temperature of self-healing specimen as a function of time recorded by infrared camera during the fatigue test. The time is counted from the start of the fatigue test. *Source:* Reprinted from Ye et al. [31]. Copyright 2014, with permission from Elsevier.

fatigue life extension is estimated to be 509% relative to the control specimen and 849% relative to the unfilled epoxy specimen.

In general, incorporation of healing capsules into epoxy would increase its toughness and hence fatigue performance due to the reduced sensitivity of fatigue crack rate to ΔK_I [21, 22]. Moreover, the uncured healing agent fluid or gel would shield the crack tip via the hydrodynamic pressure effect, which also contributes to the extension of fatigue life. When looking at the plots in Figure 3.37a, however, it is known that these two effects are not mainly responsible for the improved fatigue resistance of the self-healing specimen, since the control specimen has the same composition as the self-healing one except that TfOH is excluded. Therefore, the extended fatigue life of the self-healing specimen should be attributed to the adhesive bonding presented by the healing agent and polymeric wedge formed by the cured healing agent. Accordingly, fatigue crack grows much slower in the self-healing specimen.

Figure 3.37b shows load–displacement curves of the self-healing specimen, providing more detailed information about the fatigue retardation behavior of the healing agent. The hysteresis loop originating from the healing agent at the crack tip becomes more and more obvious with time, meaning continuously increased viscosity of the fluid and greater reduction in crack growth rate [84]. At 15 seconds, the loop area becomes the largest, implying that the healing agent released from the broken capsules mostly during pre-cracking starts to gel. Naturally, this time (i.e. 15 seconds) represents the first mechanically detectable gel (healing) time.

With further increasing time, the healing agent is converted to solid due to polymerization, so that a knee point appears on the curve at 20 seconds (Figure 3.37b,c) because of the crack closure effect contributed by the polymeric wedge and its adhesiveness [82]. This phenomenon is absent in the case of the control specimen as the released epoxy monomer cannot be cured to form a polymeric wedge due to the lack of the hardener TfOH. Afterwards, a smaller amount of fluidic healing agent is refilled by the gradually broken capsules on the crack growth path as compared with that offered in one go by the pre-cracked portion (~2.5 mm long). The hysteresis loop area decreases to constant (Figure 3.37b). Accordingly, the load–displacement relationship does not change during the test until failure. Since the crack grows steadily prior to failure, it is believed that the predominant mechanisms of crack retardation remain the same. The appearance of the knee point (Figure 3.37b,c) represents the maximum healing effect provided by the healing system. In other words, the healing effect reaches the maximum after 20 seconds of the fatigue test.

The in-situ temperature measurement of the crack tip of the self-healing specimen during the fatigue test supports the above analysis (Figure 3.37d). The healing agent released during pre-cracking is polymerized and a large amount of heat is released by the ring opening polymerization of epoxy (healing chemistry), so the temperature at the crack tip region has to rise rapidly and then keeps at a slightly lower level. Figure 3.37d shows that the highest temperature is detected at ~20 seconds, coinciding with the presence of the knee phenomenon on the load–displacement curve (Figure 3.37b) and the aforementioned rapid exotherm of epoxy cured by TfOH (Figure 3.34a) as well.

On the whole, the results of impact and fatigue tests are mutually corroborated. It can thus be concluded that the TfOH-epoxy system is allowed for ultrafast self-healing within seconds. Further investigations of the stability of the healing system and the heat resistance of the healing agent were also conducted. The results indicate that the healing ability of the self-healing specimen stored at room temperature in air is quite stable within six months (Figure 3.38). In addition, the self-healing epoxy specimens exposed to either 150 or 180 °C for five hours offer the same healing efficiency or fatigue life extension as the as-manufactured versions (Figure 3.39). In this context, it is known that the healing system is not only stable for long term storage at room temperature but also thermally stable in a high temperature environment.

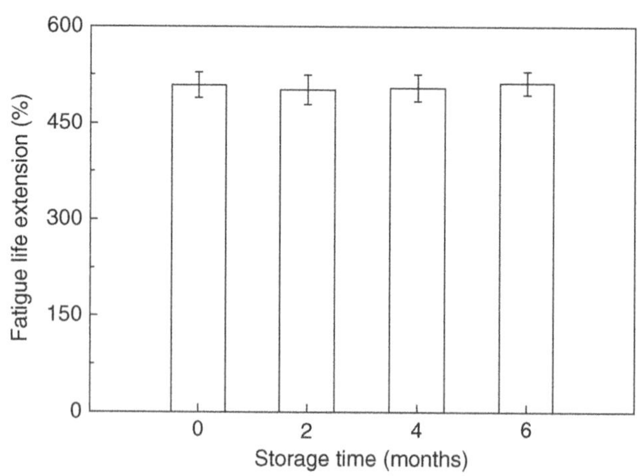

Figure 3.38 Dependence of fatigue life extension on time of storage in ambient air of the self-healing epoxy composite containing 10 wt% epoxy-loaded capsules and 0.6 wt% TfOH-loaded capsules measured by fatigue tests. Testing conditions: K_{max} = 0.336 MPa m$^{1/2}$, K_{min} = 0.034 MPa m$^{1/2}$, and ΔK_I = 0.302 MPa m$^{1/2}$. *Source:* Reprinted from Ye et al. [31]. Copyright 2014, with permission from Elsevier.

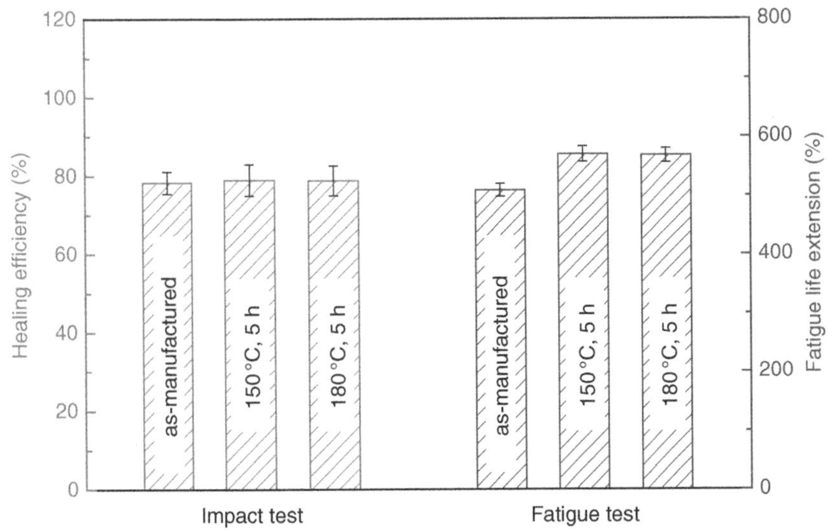

Figure 3.39 Influence of thermal exposure of self-healing epoxy composite specimens on their room temperature healing efficiency determined by impact and fatigue life extension measured by fatigue tests. For the conditions of the fatigue test refer to the caption of Figure 3.37. *Source:* Reprinted from Ye et al. [31]. Copyright 2014, with permission from Elsevier.

3.2 Thermoplastics

3.2.1 Preparation of IBH/GMA-Loaded Microcapsules

The IBH/GMA-loaded microcapsules were prepared using in-situ polymerization of melamine and formaldehyde. IBH was first dissolved in GMA, and then the solution was encapsulated as in the following.

The prepolymer of melamine (25 g) and 37% formaldehyde (50 g) was synthesized at 70 °C for 20 minutes and the pH value of the solution was kept at about 8–9 by adding triethanolamine. Afterwards, IBH/GMA solution (80 g) was incorporated into the surfactant aqueous solution containing sodium styrene-maleate copolymer (320 ml) and emulsified for five minutes with a pH value of the emulsion of about 3.8 through adding acetic acid solution (10 wt%). Subsequently, the above prepolymer solution was added to the emulsion at 35 °C under agitation, and the system temperature was gradually raised to 65 °C. Having been agitated for an additional three hours, the reaction mixture was cooled to room temperature. The deposit was washed with water and dried to obtain IBH/GMA-loaded PMF microcapsules (average diameter 75 μm, core content 66 wt%, and shell thickness 1.2 μm).

3.2.2 Self-Healing PS Composites Filled with IBH/GMA-Loaded Microcapsules and $NaBH_4$ Particles

Before examining the effectiveness of the healing system, we should ascertain whether room temperature RCP of GMA is allowed to proceed under the catalysis of IBH and $NaBH_4$. Accordingly, rheological measurement of the mixture of IBH/$NaBH_4$/GMA was conducted. From Figure 3.40a it is seen that with increasing time the storage shear modulus, G', loss shear modulus, G'', and complex viscosity, η^*, initially decrease and then increase. Originally, the high viscosity results from the uneven distribution of $NaBH_4$ particles in the system, which gradually decreases together with G' and G'' because of the shear-induced homogenization. Afterwards, these parameters increase sharply with time, which must be attributed to the polymerization of the mixture. The deduction is supported by the fact that (i) loss tangent, tan δ, peaks up at the turning point, and (ii) the tested suspension between the parallel-plate clamps was solidified after the experiment. It is thus known that GMA can be polymerized by the IBH-$NaBH_4$ initiator at room temperature within ~3 hours.

To understand the chemical structure of the polymerized product of the mixture of IBH/$NaBH_4$/GMA, its FTIR spectrum is shown in Figure 3.40b with those of GMA and IBH/GMA solution as references. By comparing the curves of GMA and IBH/GMA, it can be seen that the characteristic absorptions of GMA at 1720 cm^{-1} (C=O), 1640 cm^{-1} (C=C), and 908 cm^{-1} (epoxy group) remain unchanged in IBH/GMA, suggesting that GMA keeps stable with IBH. As for the polymerized product of IBH/$NaBH_4$/GMA, however, its peak profiles are different. For purposes of quantitative estimation of the changes, we define the area ratio of ν(C=C)/ν(C=O) as α, and that of ν(epoxy)/ν(C=O) as β. Accordingly, α_0 and β_0 represent the corresponding peak area ratios of the spectrum of IBH/GMA, while α_1 and β_1 denote those of the spectrum of IBH/$NaBH_4$/GMA. The calculation results in Figure 3.40b indicate that C=C has been greatly consumed, and the epoxy group also has a small diminution after the reaction. This means that C=C of GMA is much more active than the epoxy group during RCP under the current conditions, yielding the random copolymer (Scheme 3.6).

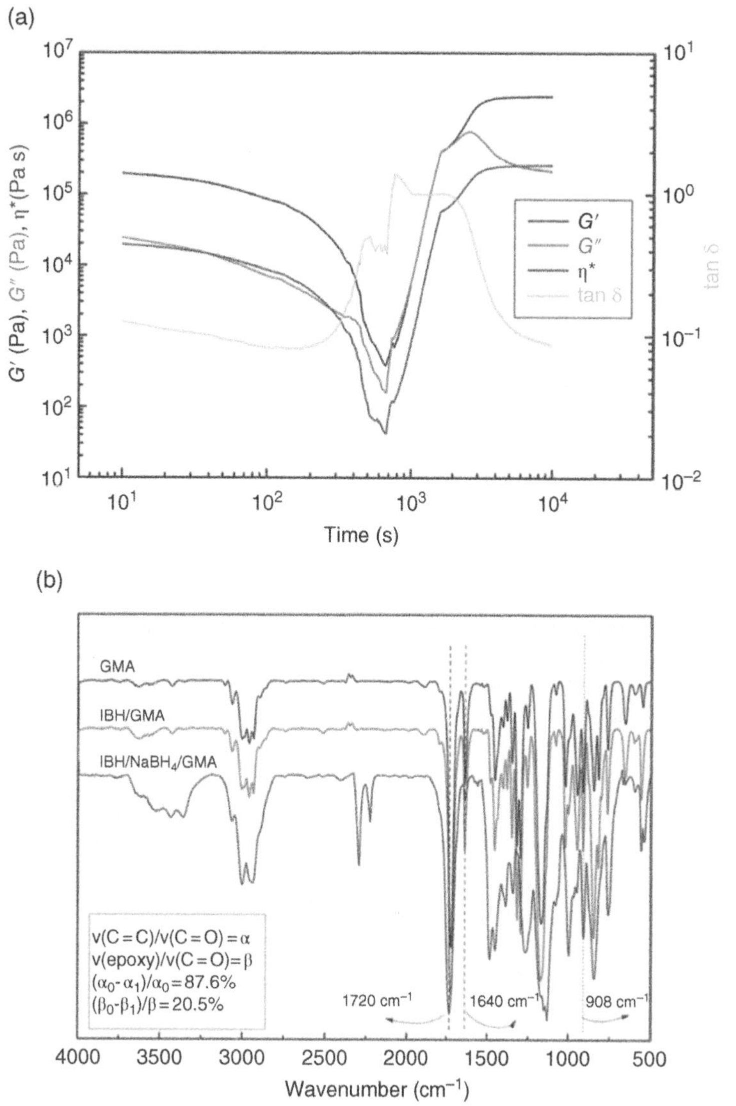

Figure 3.40 (a) Time-dependent rheological behaviors of IBH/$NaBH_4$/GMA mixture (IBH:$NaBH_4$:GMA = 3:3:100 by mol). (b) Fourier transform infrared spectra of GMA, IBH/GMA solution (IBH:GMA = 3:100 by mol) and the polymerized product of IBH/$NaBH_4$/GMA mixture (IBH:$NaBH_4$:GMA = 3:3:100 by mol) that had reacted at 25 °C for three hours. *Source:* Reprinted from Zhu et al. [41]. Copyright 2015, with permission from the Royal Society of Chemistry.

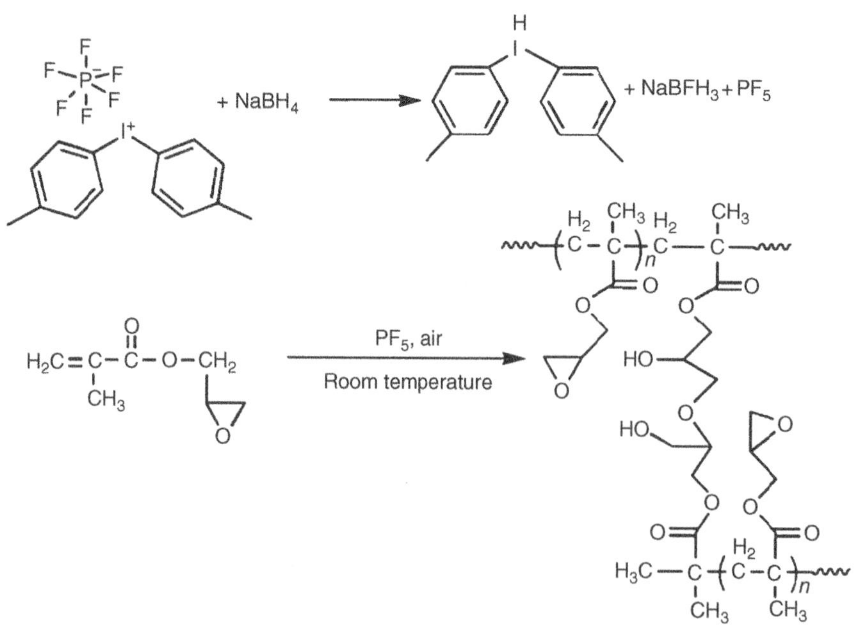

Scheme 3.6 Redox cationic polymerization of GMA.

Although IBH/$NaBH_4$/GMA has shown its capability of polymerization, the optimal proportions of the redox cationic initiators in the self-healing system are still unknown. Therefore, two groups of verification tests were done, during which manual infiltration of the chemicals to the fracture surface of PS specimens for healing plays the key step. In this study, healing ability was assessed by impact test [68], and healing efficiency was characterized by the percentage recovery of impact strength. Figure 3.41a shows the influence of $NaBH_4$ content embedded in PS matrix on the healing efficiency at a constant concentration of IBH in GMA. It is seen that the healing efficiency increases with increasing $NaBH_4$ content, and almost levels off at above 3 wt%. Figure 3.41b further reveals the effect of IBH concentration in GMA under a fixed $NaBH_4$ content in PS. The maximum healing efficiency of 100% is obtained for 2 wt% IBH. Accordingly, it can be concluded that the self-healing PS should be made by incorporating 3 wt% $NaBH_4$ particles, while the encapsulated IBH/GMA should contain 2 wt% IBH. It should be mentioned that the measurement error of the healing efficiency at $NaBH_4 = 1$ wt% in Figure 3.41a is much greater than the others. This results from the difficulty of even distribution of the low concentration $NaBH_4$ particles in PS, which affects the uniformity of polymerization degree of the healing agent at the interface.

In fact, Figure 3.41a shows a healing efficiency of around 25% in the absence of $NaBH_4$. A similar result is also shown in Figure 3.41b when the IBH concentration is zero. This demonstrates that the solvent effect of GMA can only arouse a minor healing effect [39, 85], while the polymerization of GMA mainly accounts for the rehabilitation of damaged PS.

On the basis of the above investigation, PMF walled microcapsules containing IBH/GMA were prepared by in-situ polymerization. Figure 3.42a shows the FTIR spectrum of the microcapsules as well as those of the core and shell substances. The peaks assigned to PMF at 3390 cm^{-1} (N—H, O—H), IBH at 1400 and 1475 cm^{-1}, and GMA (refer to the discussion of Figure 3.40b for its detailed peak positions) appear on the spectrum of IBH/GMA loaded microcapsules. Moreover, elements F, P and I are found in the remaining inclusion of broken capsule by EDS analysis, confirming the presence of IBH inside the microcapsules. These results demonstrate that the microcapsules have been made as designed.

We also measured the thermal degradation behavior of the microcapsules. There is no obvious weight loss below 200 °C (Figure 3.42b). This ensures the thermal stability of the healing capsules during thermo-processing with PS. Meanwhile, when the IBH/GMA-loaded microcapsules were ground with a small amount of $NaBH_4$, the released fluid was quickly solidified. Evidently, the activity of IBH in GMA is maintained after encapsulation.

Figure 3.43 illustrates the dependences of impact strength and healing efficiency of self-healing PS composite on the content of IBH/GMA-loaded microcapsules. Here the content of $NaBH_4$ particles is fixed at 3 wt% as suggested by Figure 3.41a. With increasing capsule content, impact strength of the composite slightly decreases. Basically, the reduction of impact strength is not significant. In the case of 20 wt% IBH/GMA-loaded microcapsules and 3 wt% $NaBH_4$ particles, for example, the impact strength of the PS composite is 20% lower than that of neat PS. A similar embrittlement effect was reported in previous works

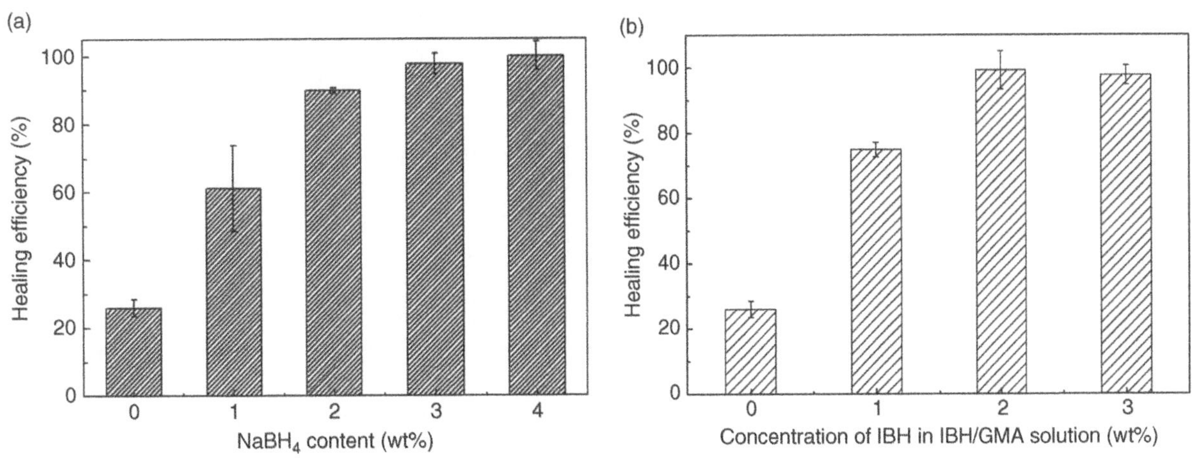

Figure 3.41 (a) Dependence of healing efficiency of PS on content of pre-embedded $NaBH_4$ particles. IBH/GMA solution (concentration of IBH = 3 wt%) was injected onto the fractured surfaces of the specimens for healing. (b) Dependence of healing efficiency of PS on concentration of IBH in IBH/GMA. 3 wt% $NaBH_4$ particles were pre-embedded in PS. IBH/GMA solution was injected onto the fractured surfaces of the specimens for healing. Healing of the fractured specimens was conducted at 25 °C in air for 24 hours. *Source:* Reprinted from Zhu et al. [41]. Copyright 2015, with permission from the Royal Society of Chemistry.

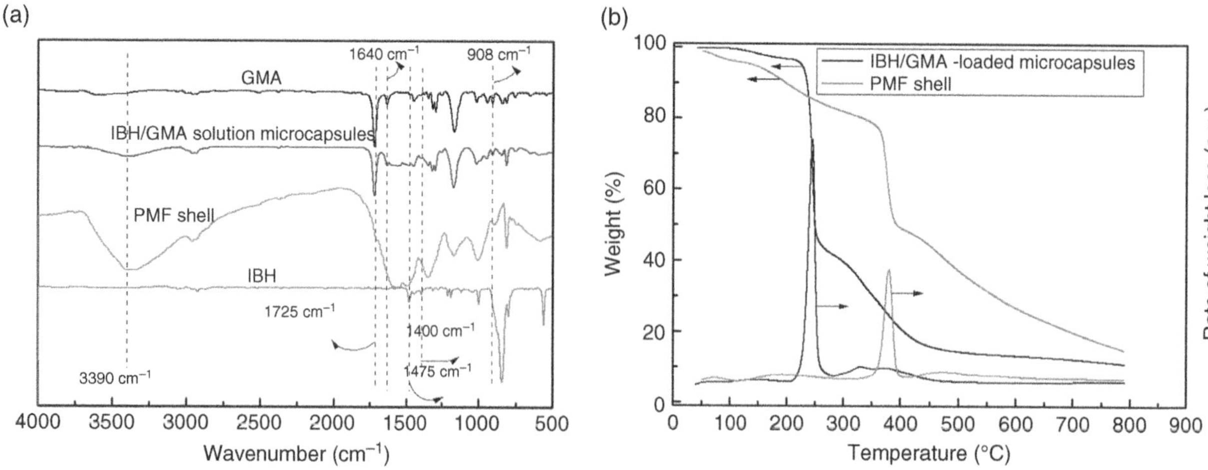

Figure 3.42 (a) Fourier transform infrared spectra of GMA, IBH, PMF, and IBH/GMA-loaded microcapsules. Concentration of IBH in IBH/GMA: 2 wt%. (b) Thermogravimetric curves of IBH/GMA-loaded microcapsules and PMF shell. *Source:* Reprinted from Zhu et al. [41]. Copyright 2015, with permission from the Royal Society of Chemistry.

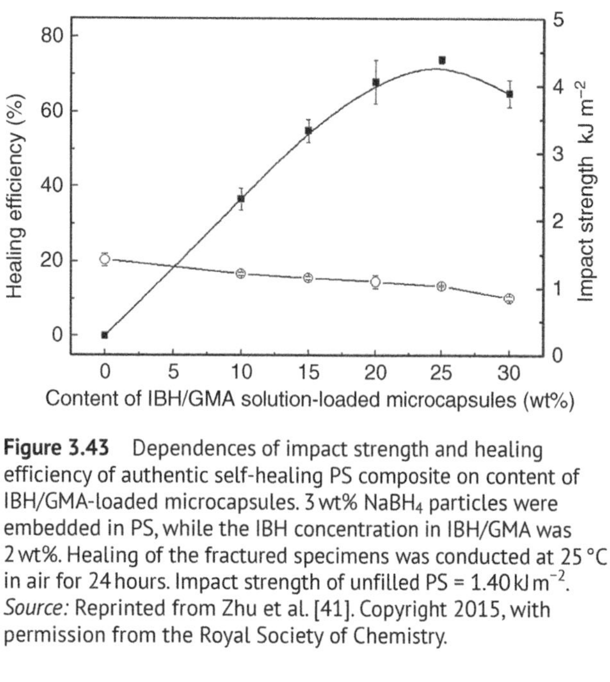

Figure 3.43 Dependences of impact strength and healing efficiency of authentic self-healing PS composite on content of IBH/GMA-loaded microcapsules. 3 wt% $NaBH_4$ particles were embedded in PS, while the IBH concentration in IBH/GMA was 2 wt%. Healing of the fractured specimens was conducted at 25 °C in air for 24 hours. Impact strength of unfilled PS = 1.40 kJ m^{-2}. *Source:* Reprinted from Zhu et al. [41]. Copyright 2015, with permission from the Royal Society of Chemistry.

concerning fluid-loaded microcapsules/polymer composites [39, 85], but the percentage decline of impact strength of the solution cast PS composite was much more remarkable, i.e. ~60% compared with unfilled PS [85]. When looking at the dependences of the flexural properties of the composites on healing capsules content, we find that at filler loadings of 20 wt% IBH/GMA-loaded microcapsules and 3 wt% $NaBH_4$, the rates of reduction are ~18 and ~20% for flexural modulus and flexural strength, respectively. This means that the present self-healing composites still have moderate mechanical properties for possible application, and the appearance of the healing agent nearly does not change the deformation habit of PS. It is worth noting that the strength of the base polymer should not be greatly decreased due to incorporation of the healing capsules. Otherwise, the technique is unlikely to be used in engineering practice. Therefore, we should alleviate the negative influence of the addition of capsules on the mechanical properties of the matrix to the greatest extent.

As for the self-healing efficiency, it increases linearly with increasing content of the microcapsules until the highest value of ~78% is achieved at 25 wt% microcapsules. The data support that the thermo-molded self-healing PS composite can indeed take effect at room temperature. The redox cationic initiators and the healing system are sufficiently stable under hot compression and air exposure. Besides, the effect of healing time on healing efficiency was also measured. It is found that the healing proceeds rapidly at the beginning and gradually slows down with time. The healing efficiency approaches ~40% within 45 minutes and reaches equilibrium at about 180 minutes. According to these results, the healing time for our experiments set at 24 hours can ensure complete polymerization of the healing fluid.

The SEM image in Figure 3.44a shows that there are quite a few concaves on the fracture surface of the PS composite, which are the halves left by the broken microcapsules. The intact edges of the concaves imply that the microcapsules were not damaged when the composite was made. In contrast, the refractured surface of a healed PS composite seems to be covered by a thin membrane (Figure 3.44b), which represents the polymerized healing agent. It is this film that reconnects the cracked PS and restores the strength of the latter. EDS of the membrane confirms the existence of IBH and $NaBH_4$. That is, the RCP product of GMA successfully provides PS with self-healability.

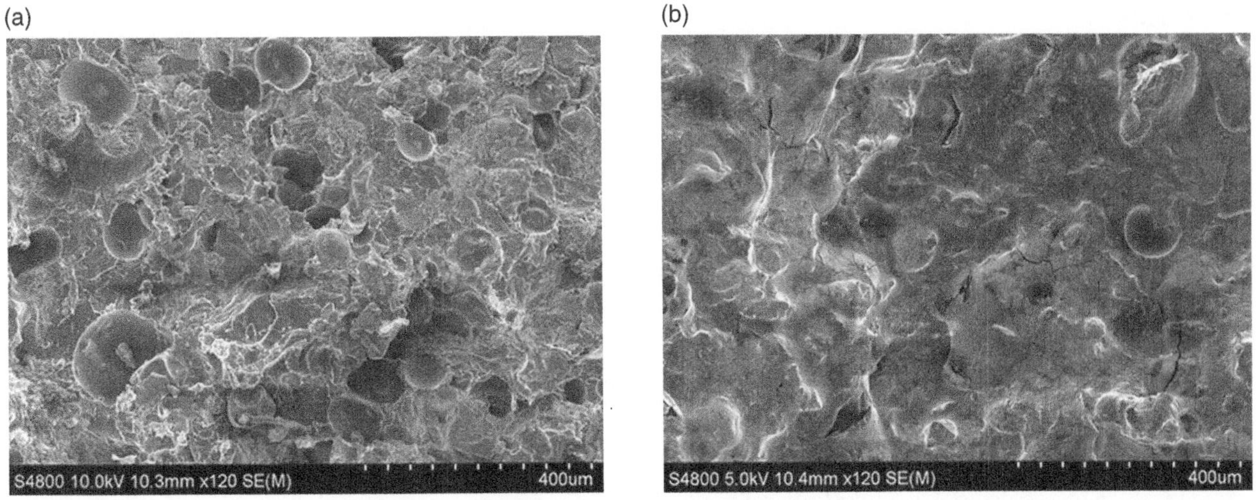

Figure 3.44 (a) Scanning electron microscopy micrograph of the fracture surface of a virgin self-healing PS composite. The healing agent released from the broken microcapsules was immediately removed by ethanol. (b) Scanning electron microscopy micrograph of the refracture surface of a healed self-healing PS composite. IBH/GMA-loaded microcapsules: 10 wt%; and $NaBH_4$ particles: 3 wt%.

RCP proves to be an effective healing chemistry for air-insensitive room temperature self-healing of commercially available thermoplastics. Oxidant IBH together with its boron-based reducing agent $NaBH_4$ creates an ideal redox cationic initiator. By using the thermally stable IBH/$NaBH_4$ initiator system, commodity plastics PS filled with IBH/GMA-loaded microcapsules and $NaBH_4$ particles can be fabricated by conventional compression molding. Additionally, the damaged PS composite can be healed in air at room temperature without manual intervention. This study expands the family of healing chemistries for preparing self-healing thermoplastics processable at elevated temperature.

Further effort should be made to use the residual (unreacted) epoxide groups of the polymerization product of IBH/$NaBH_4$/GMA, so that the bonding strength of the healing membrane is increased and fewer healing capsules are required. When the repaired part is stronger than the matrix, repeated healing would be available at the sites with intact healing capsules. Moreover, the composite with lower concentration capsules would be less weakened than before.

3.3 Concluding Remarks

Although the healing systems discussed still consist of two components or more, the replacement of healing chemistry based on addition polymerization by cationic polymerization brings about a significant improvement. Perfectly homogeneous dispersion of the pre-embedded encapsulated polymerizable chemicals and catalysts is no longer necessary. This is especially true for the epoxy-based composites. Only a few curing agents can trigger the envisaged healing reaction, and the healant composition does not need to follow a stoichiometric ratio. Autonomic recovery of mechanical strength of self-healing epoxy composites can be completed even under cold circumstances at a rate much faster than that observed in the case of epoxy/mercaptan combination.

On the other hand, the technical strategy of using porous media as carriers of healing agent opens up a means to encapsulate highly active chemicals in a facile way. It acts as an alternative approach instead of making microcapsules. The latter is difficult to conduct in some cases. Moreover, the proposed absorption–diffusion route shows a novel manner to introduce the hardener component of the self-healing agent into the polymeric matrix.

The work has revealed the importance of developing a new method to characterize the effect of the fast repair of cracks. The present technique that compares mechanical performance of virgin and healed specimens and gives a measure of healing efficiency does not match the speed of curing. This is because collecting and rejoining the broken specimens takes some time, during which the released healing agent might have been partly polymerized. Therefore, a new method that is able to determine recovery of mechanical strength of damaged materials as soon as the healing reaction starts to take effect needs to be developed.

References

1 Greenlee, S.O. (1955). Compositions. US Patent 2717885.
2 Haworth, D.T. and Pollnow, G.F. (1962). Boron curing agents for epoxy resins. *Industrial & Engineering Chemistry Product Research and Development* 1: 185–187.
3 Lin, S.C. and Pearce, E.M. (1979). Epoxy resins. I. The stability of the epoxy-trimethoxyboroxine system. *Journal of Applied Polymer Science* 23: 3355–3374.
4 Xu, X.M. and Li, J.Z. (1989). Research progress in boron-containing compounds curing agents for epoxy. *Thermosetting Resin* 3: 32–37. (in Chinese).
5 Lopata, E.S. and Riccitiello, S.R. (1975). The influence of extent of polymerization on the thermal behavior of an epoxy-based polyether resin. *Journal of Applied Polymer Science* 19: 1127–1133.
6 Chen, C.S., Bulkin, B.J., and Pearce, E.M. (1982). New epoxy resins. I. The stability of epoxy-trialkoxyboroxines triaryloxyboroxine system. *Journal of Applied Polymer Science* 27: 1177–1190.
7 Harris, J.J. and Temin, S.C. (1966). Proposed mechanism for the curing of epoxy resins with amine-Lewis acid complexes or salts. *Journal of Applied Polymer Science* 10: 523–534.
8 Happe, J.A., Morgan, R.J., and Walkup, C.M. (1985). ^{1}H, ^{19}F and ^{11}B nuclear magnetic resonance characterization of BF_3: amine catalysts used in the cure of C fibre-epoxy prepregs. *Polymer* 26: 827–836.
9 Wagner, H.B. (1957). Controlled boron trifluoride catalysis of epoxy polymerizations. *Journal of Polymer Science* 26: 329–332.
10 Yuan, Y.C., Rong, M.Z., Zhang, M.Q. et al. (2008). Self-healing polymeric materials using epoxy/mercaptan as the healant. *Macromolecules* 41: 5197–5202.
11 Yuan, Y.C., Rong, M.Z., Zhang, M.Q., and Yang, G.C. (2009). Study of factors related to performance improvement of self-healing epoxy based on dual encapsulated healant. *Polymer* 50: 5771–5781.
12 Yuan, Y.C., Rong, M.Z., Zhang, M.Q. et al. (2010). Healing of fatigue crack in epoxy materials with epoxy/mercaptan system via manual infiltration. *Express Polymer Letters* 4: 644–658.
13 White, S.R., Sottos, N.R., Geubelle, P.H. et al. (2001). Autonomic healing of polymer composites. *Nature* 409: 794–797.
14 Brown, E.N., Sottos, N.R., and White, S.R. (2002). Fracture testing of a self-healing polymer composite. *Experimental Mechanics* 42: 4372–4379.
15 Brown, E.N., White, S.R., and Sottos, N.R. (2004). Microcapsule induced toughening in a self-healing polymer composite. *Journal of Materials Science* 39: 1703–1710.
16 Rule, J.D., Sottos, N.R., White, S.R., and Moore, J.S. (2005). The chemistry of self-healing polymers. *Education in Chemistry* 42 (5): 130–132.
17 Rule, J., Brown, E.N., Sottos, N.R. et al. (2005). Wax-protected catalyst microspheres for efficient self-healing materials. *Advanced Materials* 72: 205–208.
18 Jones, A.S., Rule, J.D., Moore, J.S. et al. (2006). Catalyst morphology and dissolution kinetics of self-healing polymers. *Chemistry of Materials* 18: 1312–1317.
19 Kessler, M.R. and White, S.R. (2001). Self-activated healing of delamination damage in woven composites. *Composites Part A: Applied Science and Manufacturing* 32: 683–699.
20 Kessler, M.K., Sottos, N.R., and White, S.R. (2003). Self-healing structural composite material. *Composites Part A: Applied Science and Manufacturing* 34: 743–753.
21 Brown, E.N., White, S.R., and Sottos, N.R. (2005). Retardation and repair of fatigue cracks in a microcapsule toughened epoxy composite – part I: manual infiltration. *Composites Science and Technology* 65: 2466–2473.
22 Brown, E.N., White, S.R., and Sottos, N.R. (2005). Retardation and repair of fatigue cracks in a microcapsule toughened epoxy composite – part II: in situ self-healing. *Composites Science and Technology* 65: 2474–2480.
23 Jones, A.S., Rule, J.D., Moore, J.S. et al. (2007). Life extension of self-healing polymers with rapidly growing fatigue cracks. *Journal of the Royal Society Interface* 4: 395–403.
24 Wilson, G.O., Moore, J.S., White, S.R. et al. (2008). Autonomic healing of epoxy vinyl esters via ring opening metathesis polymerization. *Advanced Functional Materials* 18: 44–52.
25 Rule, J.D., Sottos, N.R., and White, S.R. (2007). Effect of microcapsule size on the performance of self-healing polymers. *Polymer* 48: 3520–3529.
26 Keller, M.W., White, S.R., and Sottos, N.R. (2008). Torsion fatigue response of self-healing poly(dimethylsiloxane) elastomers. *Polymer* 49: 3136–3145.
27 Pencezek, S., Kubisa, P., and Matyjaszewski, K. (1985). *Cationic Ring-Opening Polymerization Part II: Synthetic Applications*. Berlin: Springer.
28 Matyjaszewski, K. (1996). *Cationic Polymerization: Mechanisms, Synthesis, and Applications*. New York: Marcel Dekker.
29 Ye, X.J., Zhang, J.L., Zhu, Y. et al. (2014). Ultrafast self-healing of polymer toward strength restoration. *ACS Applied Materials & Interfaces* 6: 3661–3670.
30 Olah, G.A., Prakash, G.K.S., Wang, Q., and Li, X.Y. (1995). *Fluorosulfuric acid*. In: *Encyclopedia of Reagents for Organic Synthesis* (ed. L. Paquette), 2715–2716. New York: Wiley.

31 Ye, X.J., Song, Y.X., Zhu, Y. et al. (2014). Self-healing epoxy with ultrafast and heat-resistant healing system processable at elevated temperature. *Composites Science and Technology* 104: 40–46.
32 Howells, R.D. and McCown, J.D. (1977). Trifluoromethanesulfonic acid and derivatives. *Chemical Reviews* 77: 69–92.
33 Paquette, L.A., Crich, D., Fuchs, P.L., and Molander, G.A. (2006). *Encyclopedia of Reagents for Organic Synthesis*. New York: Wiley.
34 Walker, F.H., Dickenson, J.B., Hegedus, C.R., and Pepe, F.R. (2002). Cationic polymerization of emulsified epoxy resins. *Progress in Organic Coatings* 45: 291–303.
35 Wang, Y.X. and Ishida, H. (1999). Cationic ring-opening polymerization of benzoxazines. *Polymer* 40: 4563–4570.
36 Caruso, M.M. (2010). Solvent-based self-healing polymeric materials. PhD thesis. University of Illinois at Urbana-Champaign.
37 Jackson, A.C., Bartelt, J.A., and Braun, P.V. (2011). Transparent self-healing polymers based on encapsulated plasticizers in a thermoplastic matrix. *Advanced Functional Materials* 21: 4705–4711.
38 Zhu, D.Y., Rong, M.Z., and Zhang, M.Q. (2013). Self-healing polyvinyl chloride (PVC) based on microencapsulated nucleophilic thiol-click chemistry. *Polymer* 54: 4227–4236.
39 Zhu, D.Y., Wetzel, B., Noll, A. et al. (2013). Thermo-molded self-healing thermoplastics containing multilayer microreactors. *Journal of Materials Chemistry A* 1: 7191–7198.
40 Crivello, J.V. (2009). Redox initiated cationic polymerization. *Journal of Polymer Science Part A: Polymer Chemistry* 47: 1825–1835.
41 Zhu, D.Y., Guo, J.W., Cao, G.S. et al. (2015). Thermo-moldable self-healing commodity plastics with heat resisting and oxygen-insensitive healant capable of room temperature redox cationic polymerization. *Journal of Materials Chemistry A* 3: 1858–1862.
42 Deasy, P.B. (1984). *Microencapsulation and Related Drug Processes*, 119–219. New York: Marcel Dekker.
43 Chang, T.M.S. (1995). Artificial cells with emphasis on bioencapsulation in biotechnology. *Biotechnology Annual Review* 1: 267–295.
44 Quek, C.H., Li, J., Sun, T. et al. (2004). Photo-crosslinkable microcapsules formed by polyelectrolyte copolymer and modified collagen for rat hepatocyte encapsulation. *Biomaterials* 25: 3531–3540.
45 Shen, F., Li, A.A., Cornelius, R.M. et al. (2005). Biological properties of photocrosslinked alginate microcapsules. *Journal of Biomedical Materials Research Part B: Applied Biomaterials* 75B: 425–434.
46 Xiao, D.S., Rong, M.Z., and Zhang, M.Q. (2007). A novel method for preparing epoxy-containing microcapsules via UV irradiation-induced interfacial copolymerization in emulsions. *Polymer* 48: 4765–4776.
47 Chattopadhyay, D.K., Panda, S.S., and Raju, K.V.S.N. (2005). Thermal and mechanical properties of epoxy acrylate/methacrylates UV cured coatings. *Progress in Organic Coatings* 54: 10–19.
48 Kissinger, H.E. (1957). Reaction kinetics in differential thermal analysis. *Analytical Chemistry* 29: 1702–1706.
49 Rouquerol, J., Rouquerol, F., and Sing, K.S.W. (1999). *Adsorption by Powders and Porous Solids: Principles, Methodology and Applications*. London: Academic Press.
50 Donnet, J.B., Bansal, R.C., and Wang, M.J. (1993). *Carbon Black: Science and Technology*, 2e. New York: Marcel Dekker.
51 Inagaki, M. (2000). *New Carbons – Control of Structure and Functions*. Oxford: Elsevier.
52 Authorial Collective of the Dongfanghong State Farm in Guangdong (1976). *Sisal Fibers*. Guangzhou: People's Press of Guangdong (in Chinese).
53 Lu, X., Zhang, M.Q., Rong, M.Z. et al. (2001). All-plant fibre composites: self reinforced composites based on sisal. *Advanced Composites Letters* 10: 73–79.
54 Mukherjee, P.S. and Satyanarayana, K.G. (1984). Structure and properties of some vegetable fibres. Part 1. Sisal fibre. *Journal of Materials Science* 19: 3925–3934.
55 Ma, G.H., Chen, A.Y., Su, Z.G., and Omi, S. (2003). Preparation of uniform hollow polystyrene particles with large voids by a glass-membrane emulsification technique and a subsequent suspension polymerization. *Journal of Applied Polymer Science* 87: 244–251.
56 Netting, D. (1974). Hollow spheres consisting of alkali metal silicate and a polysalt. US Patent 3794503.
57 Sands, B.W. (1983). Manufacturing process for hollow microspheres. US Patent 4421562.
58 Omi, S., Katami, K., Yamamoto, A., and Iso, M. (1994). Synthesis of polymeric microspheres employing SPG emulsification technique. *Journal of Applied Polymer Science* 51: 1–11.
59 Fujimoto, K., Toyoda, T., and Fukui, Y. (2007). Preparation of bionanocapsules by the layer-by-layer deposition of polypeptides onto a liposome. *Macromolecules* 40: 5122–5128.
60 Velev, O.D., Furusawa, K., and Nagayama, K. (1996). Assembly of latex particles by using emulsion droplets as templates. 1. Microstructured hollow spheres. *Langmuir* 12: 2374–2384.
61 Yang, M., Ma, J., Niu, Z.W. et al. (2005). Synthesis of spheres with complex structures using hollow latex cages as templates. *Advanced Functional Materials* 15: 1523–1528.

62 Hu, Y.X., Ge, J.P., Sun, Y. et al. (2007). A self-templated approach to TiO_2 microcapsules. *Nano Letters* 7: 1832–1836.
63 Harris, J.R., Depoix, F., and Urich, K. (1995). The structure of gas-filled n-butyl-2-cyanoacrylate (BCA) polymer particles. *Micron* 26: 103–111.
64 Daiguji, H., Makuta, T., Kinoshita, H. et al. (2007). Fabrication of hollow melamine–formaldehyde microcapsules from microbubble templates. *Journal of Physical Chemistry B* 111: 8879–8884.
65 Xiao, D.S., Yuan, Y.C., Rong, M.Z., and Zhang, M.Q. (2009). Hollow polymeric microcapsules: preparation, characterization and application in holding boron trifluoride diethyl etherate. *Polymer* 50: 560–568.
66 van Krevelen, D.W. (1990). *Properties of Polymers*, 3e. Amsterdam: Elsevier.
67 Xiao, D.S., Yuan, Y.C., Rong, M.Z., and Zhang, M.Q. (2009). A facile strategy for preparing self-healing polymer composites by incorporation of cationic catalyst-loaded vegetable fibers. *Advanced Functional Materials* 19: 2289–2296.
68 Hayes, S.A., Zhang, W., Branthwaite, M., and Jones, F.R. (2007). Self-healing of damage in fibre-reinforced polymer-matrix composites. *Journal of the Royal Society Interface* 4: 381–387.
69 Caruso, M.M., Delafuente, D.A., Ho, V. et al. (2007). Solvent-promoted self-healing materials. *Macromolecules* 40: 8830–8832.
70 Kinloch, A.J. and Young, R.J. (1983). *Fracture Behavior of Polymers*. London: Applied Science Publishers.
71 Yin, T., Rong, M.Z., Zhang, M.Q., and Yang, G.C. (2007). Self-healing epoxy composites – preparation and effect of the healant consisting of microencapsulated epoxy and latent curing agent. *Composites Science and Technology* 67: 201–212.
72 Zhang, M.Q., Rong, M.Z., Yu, S.L. et al. (2002). Effect of particle surface treatment on the tribological performance of epoxy based nanocomposites. *Wear* 253: 1086–1093.
73 Miessler, G.L. and Tarr, D.A. (2004). *Inorganic Chemistry*. Beijing: Pearson Education Asia Ltd and Higher Education Press.
74 Xiao, D.S., Yuan, Y.C., Rong, M.Z., and Zhang, M.Q. (2009). Self-healing epoxy based on cationic chain polymerization. *Polymer* 50: 2967–2975.
75 Yin, T., Rong, M.Z., Zhang, M.Q., and Zhao, J.Q. (2009). Durability of self-healing woven glass fabric/epoxy composites. *Smart Materials and Structures* 18: 074001.
76 Landon, G., Lewis, G., and Boden, G.F. (1977). The influence of particle size on the tensile strength of particulate-filled polymers. *Journal of Materials Science* 12: 1605–1613.
77 Langer, F.F. (1971). Fracture energy and strength behavior of a sodium borosilicate glass-Al_2O_3 composite system. *Journal of the American Ceramic Society* 54: 614–620.
78 Okuno, K. and Woodhams, R.T. (1974). Mechanical properties and characterization of phenolic resin syntactic foams. *Journal of Cellular Plastics* 10: 237–244.
79 Crivello, J.V. (2009). Redox initiated cationic polymerization: reduction of diaryliodonium salts by 9-BBN. *Journal of Polymer Science Part A: Polymer Chemistry* 47: 5639–5651.
80 Wool, R.P. and O'Connor, K.M. (1981). A theory of crack healing in polymers. *Journal of Applied Physics* 52: 5953–5963.
81 Yuan, Y.C., Rong, M.Z., Zhang, M.Q. et al. (2011). Self-healing of fatigue crack in epoxy materials with epoxy/mercaptan system. *Express Polymer Letters* 5: 47–59.
82 Wolf, E. (1970). Fatigue crack closure under cyclic tension. *Engineering Fracture Mechanics* 2: 37–45.
83 Yuan, Y.C., Ye, X.J., Rong, M.Z. et al. (2011). Self-healing epoxy composite with heat-resistant healant. *ACS Applied Materials & Interfaces* 3: 4487–4495.
84 Tzou, J.-L., Hsueh, C.H., Evans, A.G., and Ritchie, R.O. (1985). Fatigue crack induced by viscous fluids. *Acta Metallurgica* 33: 117–127.
85 Yao, L., Yuan, Y.C., Rong, M.Z., and Zhang, M.Q. (2011). Self-healing linear polymers based on RAFT polymerization. *Polymer* 52: 3137–3145.

4

Extrinsic Self-Healing via Anionic Polymerization

CHAPTER MENU

The work in Chapters 2 and 3 has shown the feasibility of developing self-healing systems capable of crack repair at or below room temperature without the necessity of manual intervention. In addition to their promising academic and commercialization values, one should notice that the healed model materials are epoxy and epoxy composites cured by a room temperature curing agent.

It is known that many properties of epoxy resins are determined by the processing variables such as reactant ratio and cure condition. Comparatively, the temperature at which the resins are cured is one of the most important factors [1]. In a systematic study of cure temperature on the microstructure and performance relationship of epoxy (diglycidyl ether of bisphenol A (EPON 828)) cured by 4,4-diaminodiphenyl sulfone (DDS), for example, Min et al. [2] found that the glass transition temperature (which depends on the extent of cure) shows a linear increase with increasing cure temperature and levels off toward the cure temperature of 200 °C. Water and solvent uptake tended to increase with increasing cure temperature and, hence, with cross-linking level. Fracture toughness, tensile strength, strain to break, and tensile fracture energy also increased significantly with increasing final cure temperature regardless of cure cycle (one-step or two-step cure). In contrast, the small strain properties such as modulus, yield stress, and yield strain were not influenced greatly by cure temperature and cure cycle.

From these results, it is easily understood that the epoxy materials for advanced engineering applications that require high performance should be cured by higher temperature curing agent, besides molecular modification or introduction of the secondary components. Similarly, when imparting self-healing functionality to these epoxy materials, the corresponding epoxy-based healing agent should also be cured at higher temperature after triggering to provide the repaired damage sites with comparable properties.

Based on the above consideration, we proposed making a new two-component healant consisting of epoxy-loaded microcapsules and latent curing agent. The recipe is different from those reported previously. Epoxy is microencapsulated as polymerizable healing resin in the hope of guaranteeing miscibility between the healing agent and the epoxy-based composites. Additionally, the complex of $CuBr_2$ and 2-methylimidazole ($CuBr_2(2\text{-MeIm})_4$, Scheme 4.1) is synthesized as the latent hardener of the epoxy healing agent. The complex possesses long-term stability and would be dissociated into $CuBr_2$ and 2-methylimidazole again at around 130–170 °C [3–5]. Taking advantage of this habit, anionic polymerization of the released epoxy healing agent

Extrinsic and Intrinsic Approaches to Self-Healing Polymers and Polymer Composites, First Edition. Ming Qiu Zhang and Min Zhi Rong.

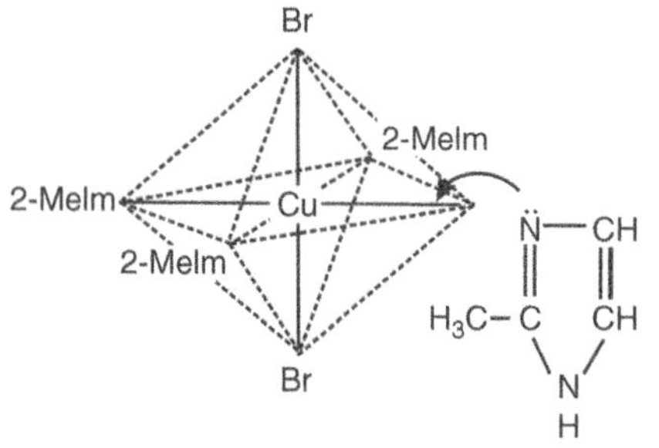

Scheme 4.1 Structural scheme of the latent curing agent $CuBr_2(2\text{-MeIm})_4$.

catalyzed by 2-methylimidazole (i.e. cracks healing) can be triggered at the dissociation temperature of $CuBr_2(2\text{-MeIm})_4$, which is higher than the curing temperature for making the composites. Another advantage of $CuBr_2(2\text{-MeIm})_4$ lies in its dissolubility in uncured epoxy. As a result, the latent curing agent can be homogenously pre-dispersed (dissolved) in the matrix of the composites on a molecular scale. It is believed that this might increase the probability of contact between the epoxy resin from the ruptured microcapsules and the dissociated imidazole. That is, the released epoxy healing agent can be activated wherever it is. Consequently, higher adhesion strength and better repair effect can be expected. Figure 4.1 shows the self-healing epoxy material made in accordance with the aforementioned concept.

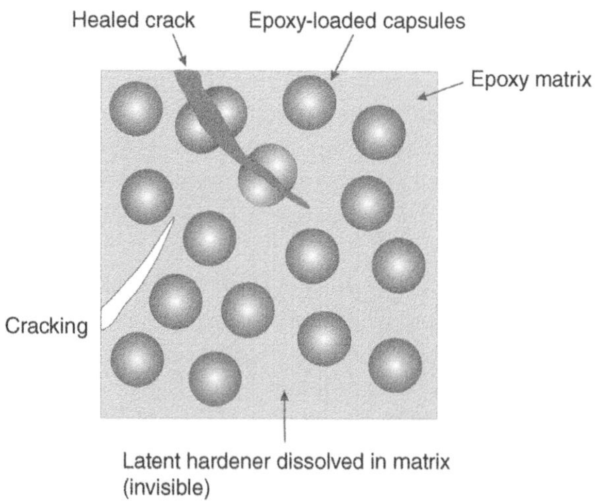

Figure 4.1 Schematic drawing of the principle of self-healing epoxy material using embedded epoxy-loaded microcapsules and pre-dissolved latent hardener. The crack induced breakage of the epoxy monomer-loaded microcapsules (see the upper part, for example) leads to liberation of the stored fluid, which flows to the crack as a result of capillary effect. Under certain heat stimulation, curing of the epoxy monomer is initiated by the pre-dissolved latent hardener and the cracked faces are re-connected.

Previously, we tried to fabricate self-healing epoxy with both encapsulated epoxy and imidazole particles [6]. The latter has smaller size (≤10 μm) and higher softening point (95–115 °C). The included imidazole granules can be flowable at a temperature higher than its softening point and then catalyze consolidation of the counterpart epoxy. Although full recovery of fracture toughness was observed after healing at 120–140 °C, the encapsulated imidazole particulates with irregularly angular appearances were the sources of defects or stress concentration in the matrix of the composites after all. Therefore, utilization of the above $CuBr_2(2\text{-MeIm})_4$ as latent hardener should be interesting progress, which excludes the shortcoming of the solid imidazole particles, while still bringing the higher temperature curing characteristics of imidazole into play.

In fact, the healing method studied here depends on manual intervention (i.e. heating). As mentioned above, it is more suitable for healing of advanced composites that usually possess higher glass transition temperature. Besides, its application is related to the development of intelligent structural health monitoring [7] or damage self-sensing techniques [8]. By accurate positioning of the damage, site-specific heating-induced crack healing can be conducted. On the other hand, since the healing is triggered by heating, the resultant self-healing composites can be used in an environment subjected to circulated cooling–heating, so that the internal micro-damage would be self-repaired during service.

4.1 Preparation of Epoxy-Loaded Microcapsules and Latent Hardener

4.1.1 Microencapsulation of Epoxy by In-Situ Condensation

EPON 828 acts as the inclusion of epoxy-loaded microcapsules. A two-step in-situ condensation approach with urea-formaldehyde as the wall material was applied [9]. The typical synthesis procedures are described in the following. First, 20.0 g urea was mixed with 50.4 ml formaldehyde (37 wt%) and the solution was adjusted to pH = 8.0 with 10% NaOH. After reaction for one hour at 70 °C, transparent water-soluble methylol urea prepolymer was yielded. On the other hand, 40.0 g epoxy was added into 800 ml sodium polyacrylate (PAANa) solution (1.5 wt%, pH = 8.0) together with 4.0 g resorcinol, 4.0 g NaCl and 0.8 g poly(vinyl alcohol) (PVA). Under mechanical stirring at a speed of 16 000 rpm, an oil-in-water emulsion was formed. Secondly, the epoxy emulsion was compounded with methylol urea prepolymer to dissolve the latter in the water phase of the former. Afterwards, 10% HCl solution was continuously added into the system by drip feed, and the system was heated when pH = 4.0. Eventually, the system was heated to 70 °C while the pH value reached 2.8–3.0. Having been kept at this temperature for one hour, the system was further adjusted to pH = 1.5–2.0. One hour later, the system was neutralized to pH = 7 by NaOH, cooled down, filtrated and dried, giving rise to urea-formaldehyde resin encapsulated epoxy healing agent (Figure 4.2).

The technique of microencapsulation has been developed rapidly since its emergence in the 1950s [10–12]. Although many studeis have been conducted using urea-formaldehyde resin as wall material, the use of microcapsules containing epoxy monomer has not been widely reported [13]. In our laboratory, urea and formaldehyde were pre-polymerized and then formed microcapsules via in-situ condensation as mentioned above. It was found that the pH value of the reaction system exerted a critical influence on the products.

When water-soluble methylol urea prepolymer was synthesized in alkaline circumstances, excessive basicity and reaction time would lead to precipitation of white solids as the methylol urea prepolymer began to be polymerized. It would hinder the subsequent in-situ polymerization in the presence of epoxy emulsion and generate a large amount of unwanted cured prepolymers that cannot be used as the wall material in the ultimate product.

Having been mixed with epoxy emulsion, elimination reactions among methylol urea molecules took place due to catalysis of alkali or acid. As a result, linear or branched prepolymers with low relative molecular mass were connected through methylene, methylene ether, as well as cyclic bridges between the urea units, evolving into water-insoluble polymer networks with time, depositing onto oil-soluble epoxy droplets and leading to microencapsulated epoxy. It is noteworthy that if the condensation polymerization was carried out in alkali medium, methylol urea molecules would not react with each other forming a methylene linkage but a dimethylene ether linkage. It would lower functionality of the system, and hence the cross-linking density of the product, which is detrimental to the wall strength of microcapsules. In acidic environment, the product of polycondensation of methylol urea was mainly bonded by methylene linkages, which facilitated chain growth and formation of a highly cross-linked structure. Considering that high acidity at the initial stage would result in too high a reaction rate to be controlled, a gradual decrease in pH value was thus followed to prepare the microcapsules containing epoxy.

The Fourier transform infrared (FTIR) spectra of uncured epoxy, microcapsules containing epoxy, and urea-formaldehyde resin that was synthesized under the same conditions as those employed for making the microcapsules are shown in Figure 4.3. Clearly, strong absorptions appear at 3300–3500 cm^{-1} in the spectrum of the microcapsules, which represent the stretching modes of —OH and —NH of urea-formaldehyde resin. Besides, the other characteristic peaks of urea-formaldehyde resin, like amine bands (at 1600–1630 cm^{-1} and 1530–1600 cm^{-1}), and those of epoxy including the bands of the terminal epoxide group at 914 cm^{-1} and —CH_2— at 2873 and 2929 cm^{-1} can also be seen in the spectrum of the microcapsules. The results demonstrate that the urea-formaldehyde microcapsules contain epoxy as expected.

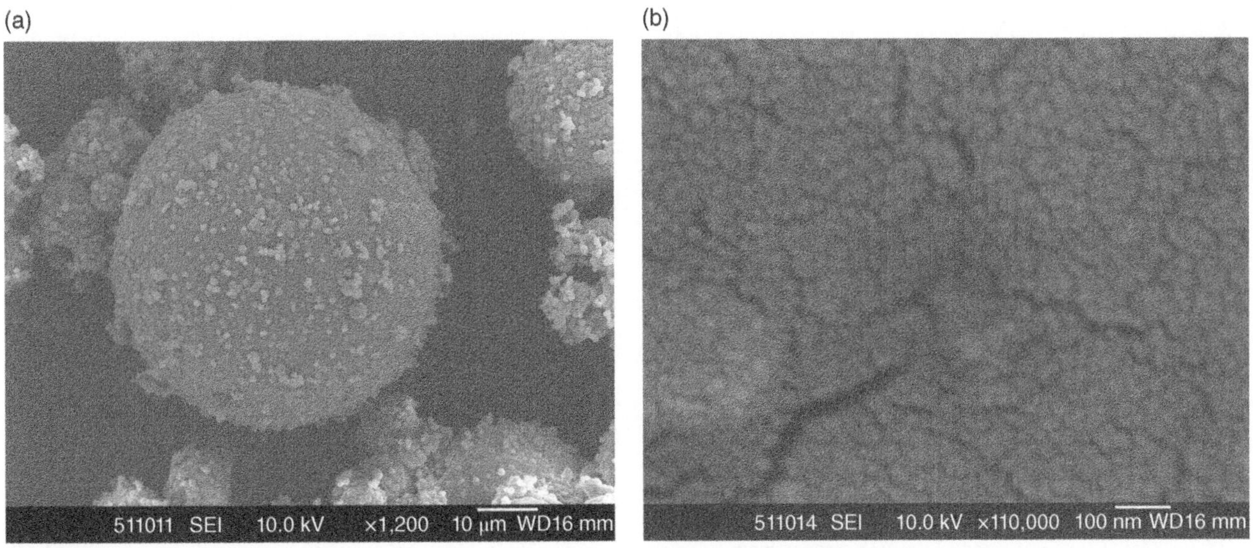

Figure 4.2 Scanning electronic microscopy (SEM) micrograph of urea-formaldehyde encapsulated epoxy. (a) General view. (b) Close-up view of the capsule's surface. *Source:* (a) Yin et al. [9], Figure 2, p. 3/with permission of Elsevier.

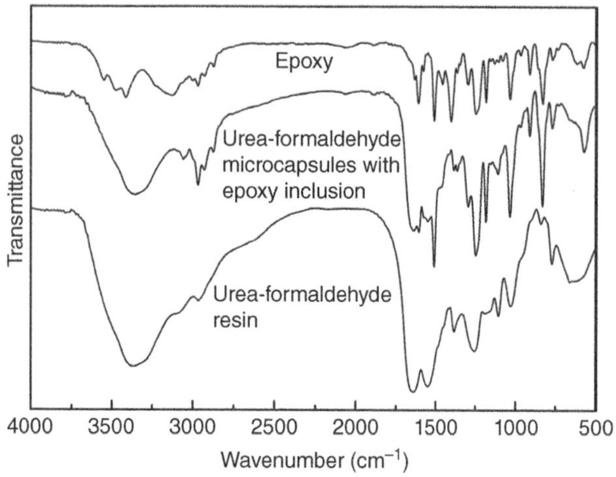

Figure 4.3 FTIR spectra of bisphenol-A epoxy, urea-formaldehyde encapsulated epoxy, and urea-formaldehyde resin. *Source:* Reprinted from Yin et al. [9]. Copyright 2007, with permission from Elsevier.

The thermal degradation behaviors of the materials give supporting evidence for the above analysis. As illustrated in Figure 4.4, the onset temperature of urea-formaldehyde resin lies at about 242 °C and the maximum rate of pyrolysis appears at 272 °C, while epoxy starts to be degraded at 359 °C and exhibits a maximum decomposition rate at 383 °C. The curves of the microcapsules consist of two stages of weight loss with a rise in temperature. Accordingly, the temperatures of the maximum rates of pyrolysis are about 266–269 °C and 404–415 °C, respectively. They should result from the thermal degradation of the shell

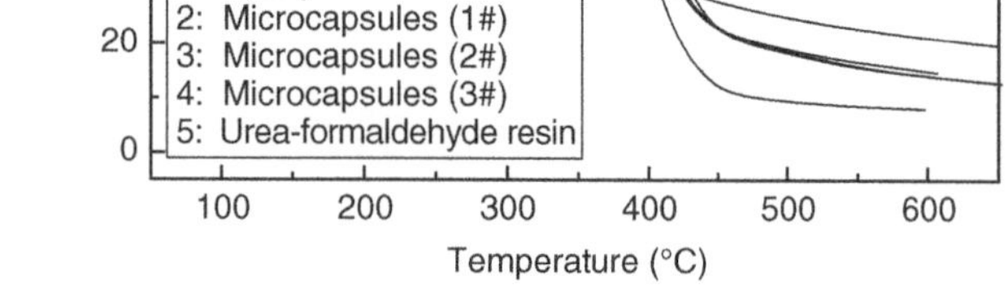

Figure 4.4 Pyrolytic behaviors of bisphenol-A epoxy, urea-formaldehyde encapsulated epoxy, and urea-formaldehyde resin. Microcapsules (1#), microcapsules (2#) and microcapsules (3#) were produced in different batches by the same method. Their average diameters and the contents of the encapsulated epoxy are: 36 μm, 64%; 37 μm, 64%; and 31 μm, 61%, respectively. *Source:* Reprinted from Yin et al. [9]. Copyright 2007, with permission from Elsevier.

(urea-formaldehyde resin) and the core (epoxy) of the microcapsules. Compared with the pristine epoxy (curve 1 in Figure 4.4), thermal stability of the epoxy in the microcapsules (curves 2, 3, and 4 in Figure 4.4) is evidently increased owing to the protection of the shield of the urea-formaldehyde wall.

Since the epoxy emulsion was produced by high-speed stirring, the ultimate size of the microcapsules should be a function of the stirring speed used for preparing the epoxy emulsion. The experimental results indicate that with increasing speed, the microcapsules become smaller, and the size distribution is narrower. That is, both size and size distribution are inversely proportional to the stirring speed. This obeys the common understanding of agitation. In addition, the data showed that the stirring time is another influencing factor. In general, five minutes of stirring is sufficient for the present system to obtain microcapsulated epoxy with stable size.

By weighing the core and shell of the microcapsules (before and after extraction), it was revealed that microcapsules of different sizes contain different epoxy contents. The larger microcapsules possess a larger amount of epoxy. This might result from the fact that the thickness of the shells of the microcapsules is similar, regardless of their sizes. Therefore, the microcapsules with larger size should have a larger portion of the core (i.e. epoxy). On the basis of this relationship, one is able to adjust the concentration of the healing agent.

4.1.2 Preparation of Imidazole Latent Hardener

The latent curing agent $CuBr_2(2\text{-MeIm})_4$ was typically prepared as follows: 11.2 g $CuBr_2$ was dissolved in 50 ml methanol and then the solution was gradually added to a methanol solution (25 ml) of 2-methylimidazole (16.4 g). Having been stirred for a period of time, the above mixture was diluted by 150 ml acetone, resulting in precipitation. The sludge (i.e. the complex of $CuBr_2$ and 2-methylimidazole) was filtrated, washed, and dried. The rate of production is around 97.1%.

In accordance with ref. [4], the possible structure of $CuBr_2(2\text{-MeIm})_4$ is shown in Scheme 4.1. To verify the synthesized $CuBr_2(2\text{-MeIm})_4$, the content of the element N in it was measured. The value of 20.43% is close to the theoretical one of 20.31%, suggesting that the target substance has been prepared. Also, the FTIR spectrum of $CuBr_2(2\text{-MeIm})_4$ is compared with 2-MeIm in Figure 4.5. It is seen that all the characteristic bands of imidazole can be found in the former, including N—H stretching at 3400 cm^{-1}, N—H wagging vibration at 756 cm^{-1}, C—N stretching at 1110 cm^{-1}, C—N stretching at 1600 cm^{-1}, C=C stretching at 1680 cm^{-1}, and

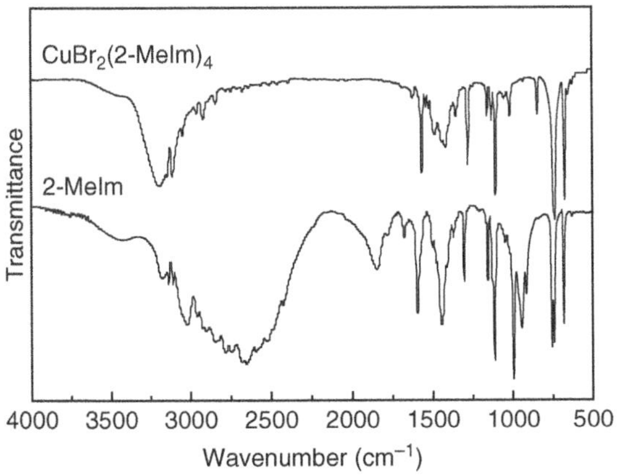

Figure 4.5 FTIR spectra of 2-MeIm and $CuBr_2(2\text{-MeIm})_4$. *Source:* Reprinted from Yin et al. [9]. Copyright 2007, with permission from Elsevier.

=CH rocking vibration at 1440 cm^{-1}. Therefore, there is evidence that the imidazole in $CuBr_2(2\text{-MeIm})_4$ keeps its original structure as desired. On the other hand, the most significant difference in the two spectra lies in that the stretching absorptions of C—H in the imidazole ring and methyl at 2500–3200 cm^{-1} disappear in the $CuBr_2(2\text{-MeIm})_4$ spectrum. This demonstrates that the coordination might have obstructed certain vibration modes.

The mechanism responsible for the reaction between epoxy and a complex of metal salt and imidazole [14] is given in Scheme 4.2, which takes $CuCl_2(Im)_4$ (the complex of $CuCl_2$ and imidazole) for example and is applicable for the system studied in this work. Having been heated, the complex is dissociated into $CuCl_2$ and imidazole. First, the active hydrogen of the secondary amine in imidazole reacts with the epoxide group yielding affixture. Then the reaction of the affixture with another epoxide group is performed and an ionic complex is obtained. When the anionic portion further reacts with epoxide group, epoxy can be cured by ring-opening polymerization via chain reaction. As the anionic polymerization of imidazole is restricted by the affixture, the rate of epoxy curing is lower in comparison with the case where tertiary amine family is used as the hardener.

Figure 4.6 shows the curing processes of the $CuBr_2(2\text{-MeIm})_4$/epoxy system in terms of conversion versus temperature estimated from the non-isothermal differential scanning calorimetry (DSC) scans. It is seen that curing of epoxy occurred at about 130 °C. The exothermic peaks appear at 141–176 °C, while the corresponding conversions are still lower than 50%. This means that $CuBr_2(2\text{-MeIm})_4$ is a mild curing agent. By using Kissinger [15] and Crane [16] equations, the characteristic parameters of the curing kinetics including the activation energy, E_a, and the order of reaction, n, of $CuBr_2(2\text{-MeIm})_4$ (1 wt%)/epoxy were calculated to be 83.7 kJ mol^{-1} and 0.92, respectively. As n is non-integral, the curing of this system should be a complicated process. On the other hand, the temperature dependence of the rate constant of the curing reaction was obtained from the Arrhenius equation. Clearly, the curing reaction proceeds very slowly at lower temperatures (e.g. 120 °C). This is supported by the isothermal DSC scans of the system. At a constant temperature of 120 °C, there is no detectable exothermic peak within 90 minutes, implying nearly no curing takes place. In accordance with this finding, the curing temperature for preparing self-healing composites should be lower than the healing temperature to avoid any change in the healing agent and the latent hardener when the composites are cured. In other words, the

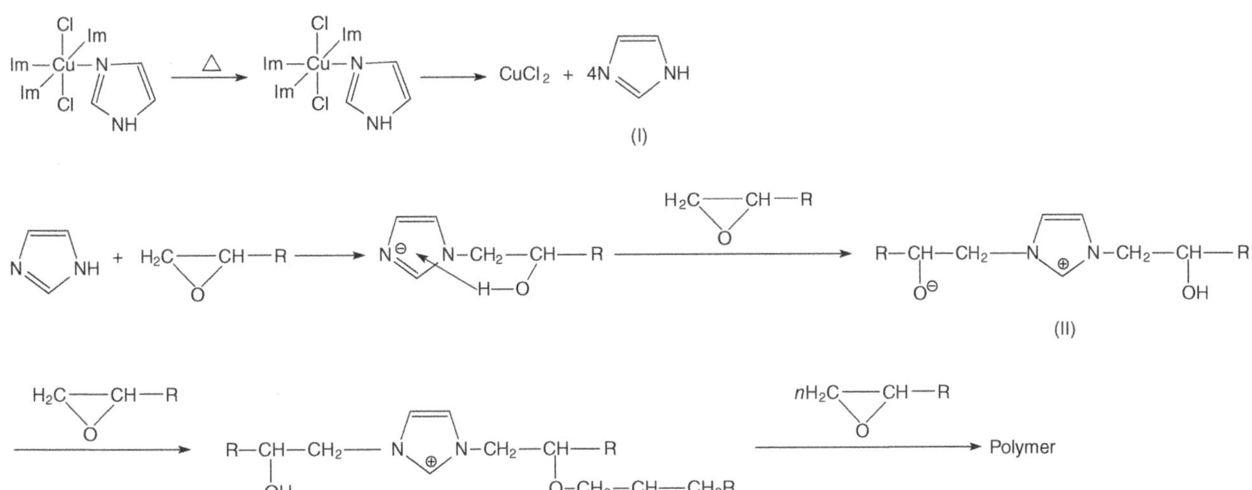

Scheme 4.2 Mechanism of the curing reaction between epoxy resin and the complex $CuCl_2(Im)_4$.

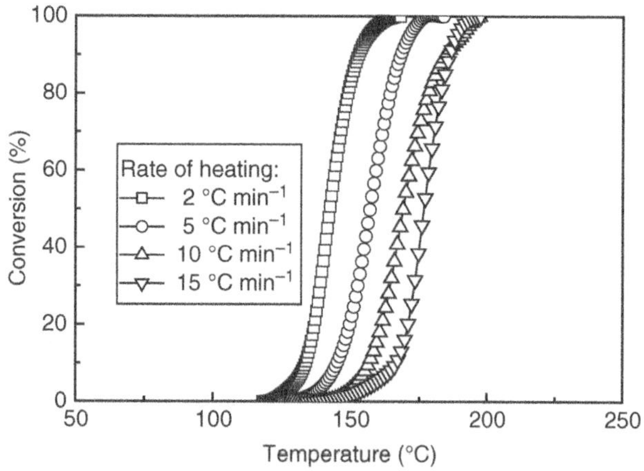

Figure 4.6 Temperature dependence of conversion of curing reaction of epoxy activated by $CuBr_2$(2-MeIm)$_4$ (1 wt%) at different heating rates. *Source:* Reprinted from Yin et al. [9]. Copyright 2007, with permission from Elsevier.

developed healing system proves to be durable for long-term application under moderate temperature.

4.2 Self-Healing Epoxy Materials with Embedded Epoxy-Loaded Microcapsules and Latent Hardener

The self-healing epoxy material was prepared according to the following procedures. A certain amount of $CuBr_2$(2-MeIm)$_4$ latent hardener was mixed with EPON 828 at 60 °C with stirring. Then the mixture was cooled down to 40 °C and compounded with a given proportion of the aforementioned urea-formaldehyde resin encapsulated epoxy monomer as the healing agent. Under ultrasonic agitation, the compound was mixed with 15.2 phr tetraethylenepentamine (TEPA) curing agent and degassed. Eventually, the mixture was cast and cured for two hours at 60 °C, followed by two hours at 80 °C, and two hours at 100 °C.

Since the microencapsulated epoxy and the latent curing agent have been filled into the matrix of the composites, it is worth first understanding their influence on the basic mechanical performance of the matrix. Hence, the tensile properties of epoxy containing the self-repair system were measured as a function of the concentration of the microcapsules at a fixed content of $CuBr_2$(2-MeIm)$_4$.

The plots in Figure 4.7a show that tensile strength of the compounds keeps almost unchanged with a rise in content of epoxy-loaded microcapsules. This is different from the results of Brown et al. [17], who reported a continuous reduction in the strength of epoxy with embedded microcapsules. Generally, the addition of either rigid particles or rubbery particles into polymers would lead to substantial

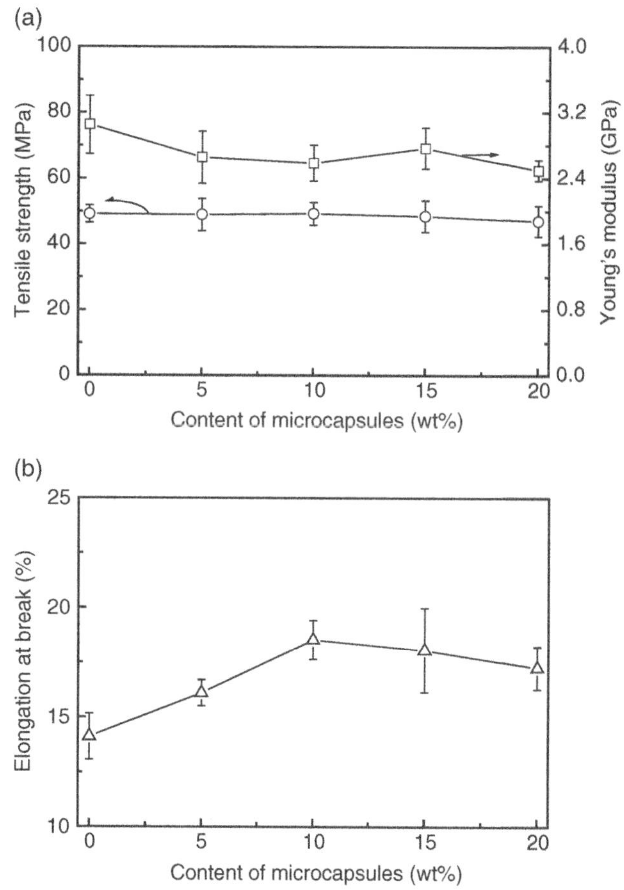

Figure 4.7 Influence of the content of epoxy-loaded microcapsules on the tensile properties of epoxy. Content of $CuBr_2$(2-MeIm)$_4$: 2 wt%; average diameter of the epoxy-loaded microcapsules: 37–42 μm; and content of the epoxy monomer inside the microcapsules: 63–68%. *Source:* Reprinted from Yin et al. [9]. Copyright 2007, with permission from Elsevier.

decrease in tensile strength of the matrices [18, 19]. Nevertheless, counter examples can be found in inorganic particulate composites when the bonding between the fillers and the matrix is strong enough [20], or the size of the particles is in the nanometer range [21]. The strength data in Figure 4.7a suggest that the shell material of the epoxy-loaded microcapsules, urea-formaldehyde resin, is compatible with epoxy and a strong interfacial interaction was established during curing. Besides, the epoxy-loaded microcapsules behave unlike soft rubber, but can carry a certain load transferred by the interface. These account for the dependence of tensile strength on the content of epoxy-loaded microcapsules illustrated in Figure 4.7a.

The above analysis receives support from the variation trend of Young's modulus (Figure 4.7a). As the stiffness of the epoxy-loaded microcapsules ranks among rigid particles and soft rubber, the decrease in modulus of the epoxy specimens is not remarkable, even at a filler concentration as high as 20 wt%. On the other hand, the increase in failure strain (Figure 4.7b) should be attributed to the fact that the epoxy-loaded microcapsules have induced interfacial viscoelastic deformation and matrix yielding. The decrease in elongation at break for the highly loaded specimens might be due to the uneven distribution of the epoxy-loaded microcapsules, which led to stress concentration in some parts of the specimen.

Similarly, fracture toughness of the epoxy with the healing agent was also evaluated by means of single-edge notched bending (SENB) tests. Figure 4.8a indicates that the incorporation of the latent hardener $CuBr_2(2\text{-MeIm})_4$ tends to toughen cured epoxy up to 4 wt%. It is believed that the latent curing agent dissolved in epoxy plays the role of plasticizer. With increasing content of $CuBr_2(2\text{-MeIm})_4$, the $CuBr_2(2\text{-MeIm})_4$ that cannot be well dissolved is present in the form of tiny particles (Figure 4.9) and reduces the toughening effect. This explains the reduced toughening

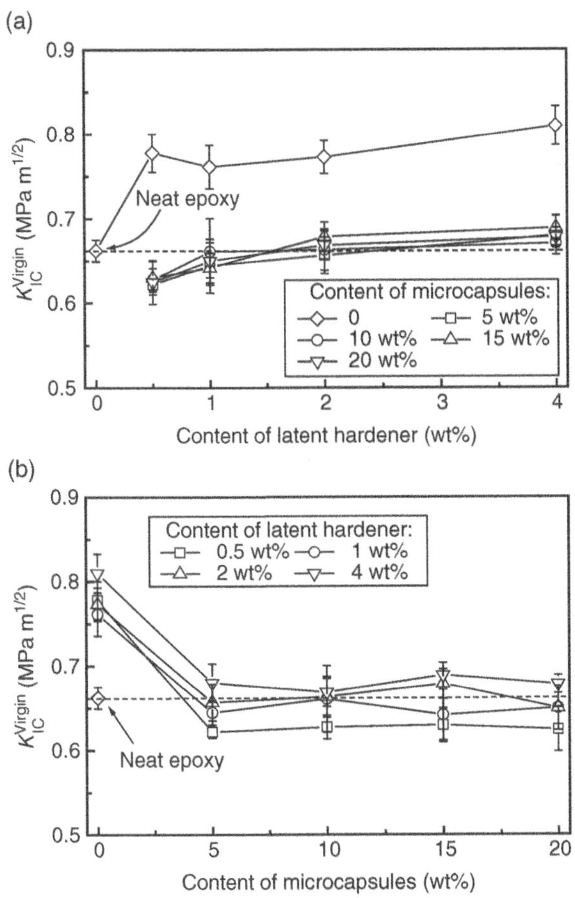

Figure 4.8 Influence of (a) content of latent hardener and (b) content of epoxy-loaded microcapsules on fracture toughness of epoxy. Average diameter of the epoxy-loaded microcapsules: 37–42 μm; and content of the epoxy monomer inside the microcapsules: 63–68%. *Source:* Reprinted from Yin et al. [9]. Copyright 2007, with permission from Elsevier.

efficiency from 0.5 to 4 wt% of $CuBr_2(2\text{-MeIm})_4$, as compared with the drastic increase in the fracture toughness of the original specimen, K_{IC}^{Virgin}, when the latent hardener content increases from 0 to 0.5 wt% (Figure 4.8a).

In the case that both epoxy-loaded microcapsules and the latent curing agent were added, the fracture toughness of the system is slightly lower than that of neat epoxy for certain proportions of composition. Obviously, the epoxy-loaded microcapsules can neither hinder the crack propagation nor result in energy consumption. Owing to the positive effect of the latent hardener that counteracted the negative effect of the toughening, however, fracture toughness of the blends is similar to that of the neat epoxy on the whole (Figure 4.8).

From the above results, it is concluded that the above two-component self-healing system would not significantly change the mechanical properties of epoxy. Instead, it introduces toughening in some cases.

Crack healing of the polymer composites depends on the polymerization of the crack released healing agent (i.e. epoxy monomer in the present work) activated by the latent hardener. Since a precise proportion of the ingredients cannot be pre-estimated for this specific case, a series of materials with different epoxy-loaded microcapsule/latent hardener ratios have to be prepared for comparing the fracture toughness of the specimens before and after healing (Figure 4.10a,b).

Figure 4.10a shows the dependence of fracture toughness of the healed specimen, K_{IC}^{Healed}, on the content of the latent hardener. As the epoxy matrix containing different amounts of the epoxy monomer-loaded microcapsules and latent hardener possesses similar K_{IC}^{Virgin} (Figure 4.8), the dependence of healing efficiency on the content of the latent hardener is similar to the corresponding curves of K_{IC}^{Healed} (Figure 4.10a). Here the healing efficiency, η, is defined as [22]:

$$\eta = \frac{K_{IC}^{Healed}}{K_{IC}^{Virgin}} \tag{4.1}$$

When the epoxy-loaded microcapsules content is 5 wt%, the values of K_{IC}^{Healed} are relatively low over the entire loading range of the latent curing agent. This results from insufficient dosage of the epoxy-loaded microcapsules, leading to an insufficient quantity of released epoxy monomer to cover the broken surface. Therefore, some cracked portions have to be left unhealed and the apparent fracture toughness is not as high as expected. In the case of 10 wt% of epoxy-loaded microcapsules and 2 wt% of the latent hardener, the highest K_{IC}^{Healed} and η are observed, meaning the optimum proportion has been reached. Further increase in the content of the latent hardener

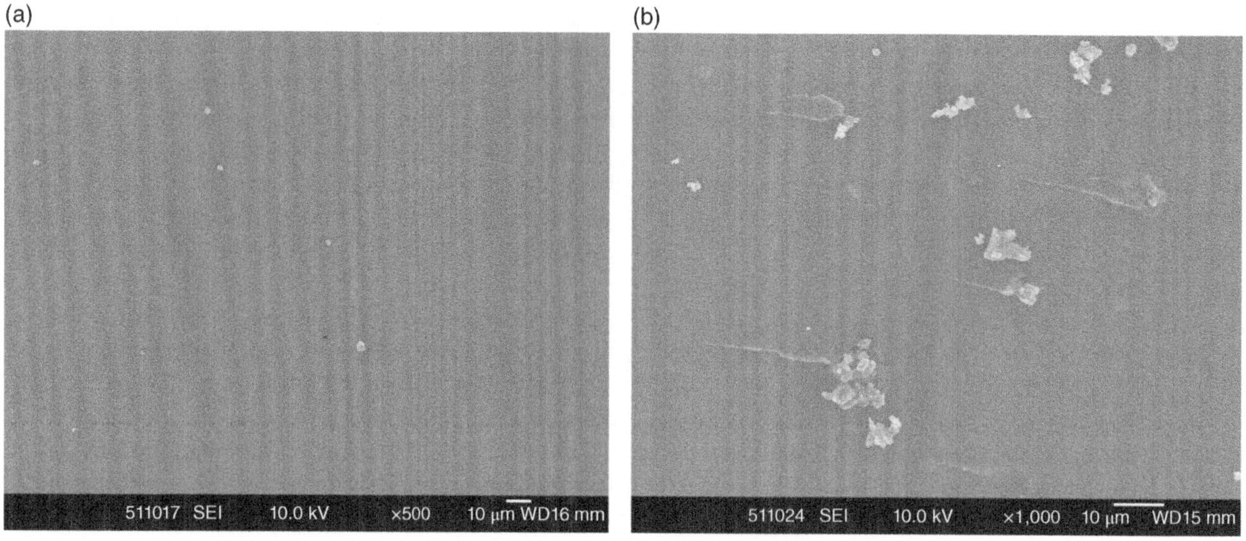

Figure 4.9 SEM micrographs of cured $CuBr_2$(2-MeIm)$_4$-epoxy system. Content of $CuBr_2$(2-MeIm)$_4$: (a) 0.5 wt%; and (b) 2 wt%. *Source:* Yin et al. [9], Figure 16, p. 9/with permission of Elsevier.

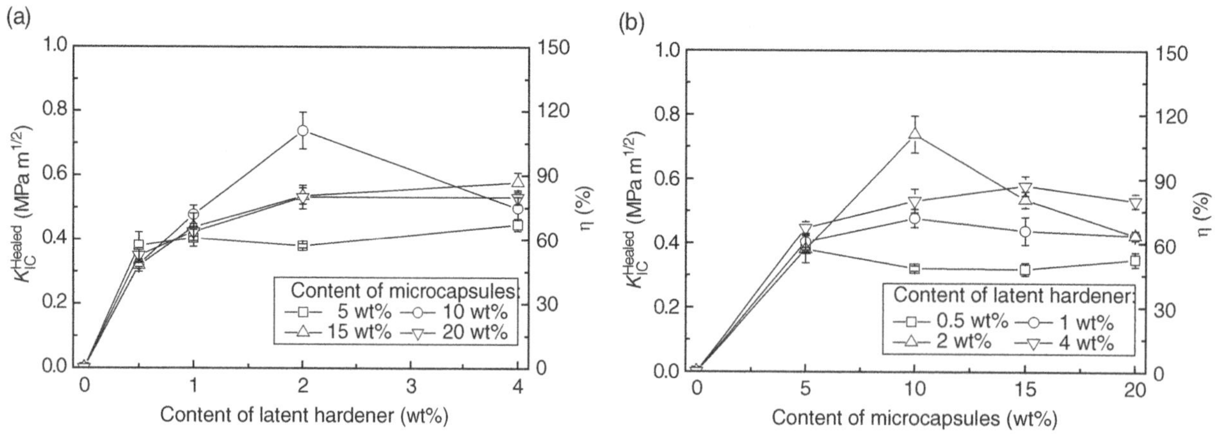

Figure 4.10 Influence of (a) content of latent hardener and (b) content of epoxy-loaded microcapsules on self-healing ability of epoxy. Average diameter of the epoxy-loaded microcapsules: 37–42 μm; and content of the epoxy monomer inside the microcapsules: 63–68%. Healing was conducted at 130 °C for one hour. *Source:* Reprinted from Yin et al. [9]. Copyright 2007, with permission from Elsevier.

could not ensure the best degree of cross-linking, and the effect of repair has to be lower accordingly. For the system containing 15 and 20 wt% epoxy-loaded microcapsules, the areas of the cracked planes that can be healed are reduced accordingly. The amount of latent hardener that might contact the released epoxy becomes insufficient, which also leads to lower repair efficiencies than the maximum. It is interesting to note that the aforementioned highest healing efficiency is 111%, which implies that the fracture toughness of the healed sample is higher than that of the virgin one. To understand the cause, the fracture toughness of epoxy cured only by $CuBr_2$(2-MeIm)$_4$ (5 wt%) was measured. The value 0.81 MPa m$^{1/2}$ is 1.23 times higher than that of bulk epoxy, which was cured by TEPA. Therefore, the bonding material not only healed the cracks but also provided the damaged sites with higher fracture toughness. The phenomenon resembles the outcome of dual capsules of the epoxy–mercaptan system described in Chapter 2 [23].

Figure 4.10b shows the influence of the content of epoxy-loaded microcapsules at certain dosage of the latent hardener. With increasing content of epoxy-loaded microcapsules, K_{IC}^{Healed} first increases and then decreases. Besides, the content of the epoxy-loaded microcapsules corresponding to the maximum K_{IC}^{Healed} increases with the latent hardener content. This demonstrates that the highest healing

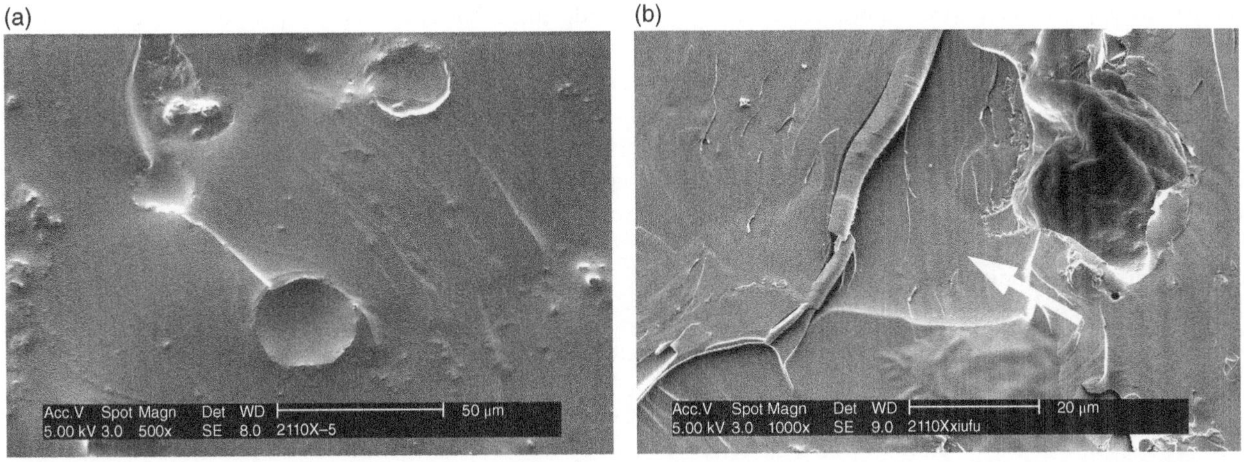

Figure 4.11 SEM micrographs of the fractured surfaces of the self-healing epoxy specimens: (a) virgin specimen; and (b) healed specimen (the arrow indicates the direction of crack propagation. Contents of microcapsules and $CuBr_2(2\text{-MeIm})_4$:10 and 2 wt%, respectively. *Source:* Yin et al. [9], Figure 19, p. 10/with permission of Elsevier.

efficiency can be obtained only at the optimum microcapsule/latent hardener ratio, as found in Figure 4.10a.

The SEM micrograph of the fractured surface illustrates that after failure of the first SENB test the epoxy-loaded microcapsules were damaged and the healing agent had flowed off, leaving the ring-like concaves (Figure 4.11a). Having been healed and undergone the second SENB test, the specimen is characterized by more complicated fracture patterns (Figure 4.11b). The thin layers on the fracture surface represent the result of crack propagation through the cured healing agent.

4.3 Self-Healing Epoxy/Woven Glass Fabric Composites with Embedded Epoxy-Loaded Microcapsules and Latent Hardener – Healing of Interlaminar Failure

Woven glass fiber reinforced polymer composites have been widely used as structural materials in many fields. They are susceptible to damage in the form of delamination generated due to manufacturing defects or low-velocity impact during service. Since the above work has proved that the two-component healant consisting of urea-formaldehyde walled microcapsules containing epoxy and 2-methylimidazole/metal complex latent hardener is applicable to crack rehabilitation in unfilled epoxy, both the self-healing ability of epoxy-based composite laminates using this healing system and the influencing factors should be studied. The systematic knowledge would facilitate optimization of formulation and processing for future practical application of the composites.

In the following investigation, epoxy EPON 828 acts as the matrix of the composite laminates as before. Meanwhile, one of the most common woven materials, 0/90 woven glass roving (C-glass, 13 × 12 plain weave, 0.2 mm thick, 1000 tows, $200\,g\,m^{-2}$), is selected to reinforce the composites. The specimens for the investigation are 12-ply laminates with 27 vol% glass fiber, in which the epoxy-loaded microcapsules (with average diameter of 37–42 µm and core content of 63–68%) were embedded and the latent hardener ($CuBr_2(2\text{-MeIm})_4$) was dissolved in advance [24].

Two types of composite panels were manufactured by hand lay-up and compression molding, namely, laminates for interlaminar fracture toughness tests and laminates for tensile tests. The former only had self-healing agent in several middle planes of the matrix, while the latter contained the latent hardener and microencapsulated epoxy throughout the matrix. In addition, laminates without the self-healing agent were also prepared for comparison.

To prepare self-healing laminates for interlaminar fracture toughness tests, first EPON 828 was mixed with $CuBr_2(2\text{-MeIm})_4$ catalyst in certain proportions. A portion of the mixture was set aside to be used for impregnation of the exterior eight fabric layers. The other portion of the mixture was slowly blended with the epoxy-loaded microcapsules by stirring and ultrasonic dispersion for impregnating the central four fabric layers. Then, the addition product of TEPA and acrylonitrile (molar ratio = 1:2) was incorporated into all of the aforementioned resin mixtures as the curing agent of the composite matrix. Afterwards, the resin was degassed in a vacuum chamber for approximately five minutes to thoroughly remove the entrapped air bubbles.

Twelve 140 mm × 160 mm rectangles were cut from the woven glass fabric. During laminate lay-up the first four woven glass fabric piles were impregnated with the epoxy mixture excluding the microencapsulated epoxy. Then, the central four layers were impregnated with the filled resin mixture. The final four piles were impregnated again with the epoxy mixture excluding the epoxy-loaded microcapsules and stacked on top to produce 12-ply laminates (Figure 4.12a). Meanwhile, a 13 μm thick Teflon film was placed at the midplane of the laminates to create the initial delamination crack. The laminates were cured at room temperature for seven hours and then compressed in a hot press at 60 °C for two hours, 80 °C for two hours, and 100 °C for two hours. The fiber volume fraction was estimated to be 27% for the laminates. The lower fiber content ensured higher concentration of the epoxy-loaded microcapsules in the composites (for achieving high healing efficiency) and avoided rupture of the microcapsules during the composite manufacturing. When the composite laminates were completely cured, they were machined into rectangle specimens 20 mm wide, 150 mm long, and 4.5 mm thick. The Teflon insert extended approximately 60 mm along the midplane of the specimens. A pair of loading blocks was bonded to the end of each specimen with an epoxy adhesive for the subsequent double cantilever beam (DCB) test (Figure 4.12b).

Preparation of laminates for tensile tests followed the same procedures described above except that (i) the epoxy-loaded microcapsules and the latent hardener $CuBr_2(2\text{-MeIm})_4$ were dispersed throughout the matrix of the composites and (ii) there was no Teflon film at the midplane.

Since the self-healing composites would be used as structural materials, it is worth studying their basic mechanical performance. Typical tensile stress–strain curves of the composite laminates are shown in Figure 4.13. Evidently, incorporation of the latent hardener into the matrix resin almost does not change the tensile deformation habit (i.e. brittle failure) of the composites (cf. curves 1 and 2 in Figure 4.13). The curves are roughly linear prior to the failure point at first glance. Nevertheless, the laminates with the pre-embedded epoxy-loaded microcapsules show slightly different behaviors (see curves 3, 4 and 5 in Figure 4.13). Higher loading of the microcapsules leads to earlier appearance of the knee point at which the stress–strain curve starts to deviate from linearity. This means that the stable crack

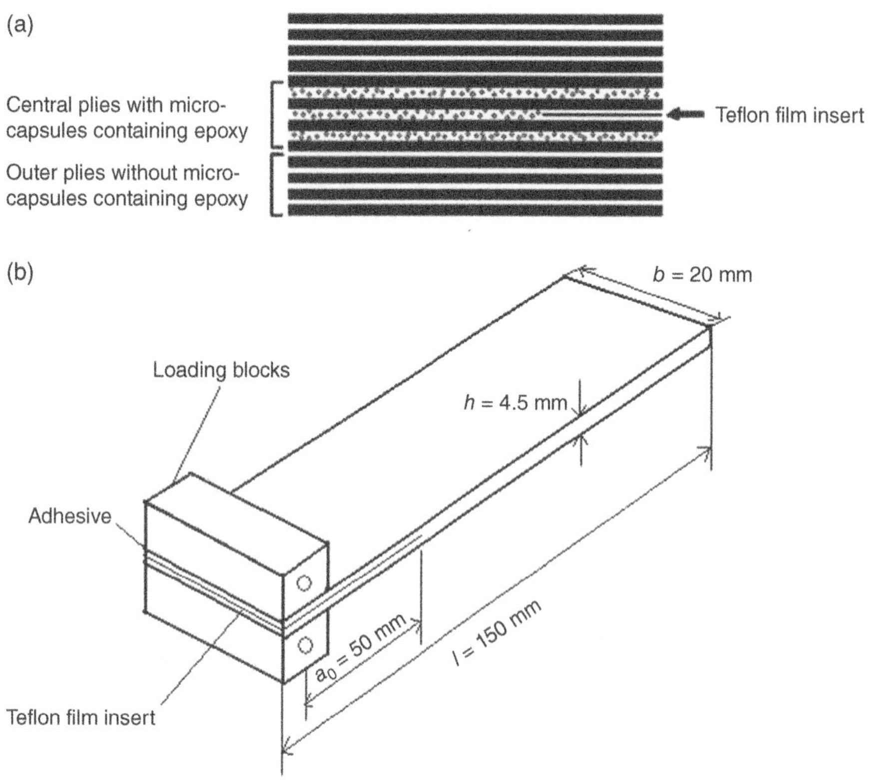

Figure 4.12 Schematic drawing of (a) the cross-section of the laminates and (b) the double cantilever beam specimen.

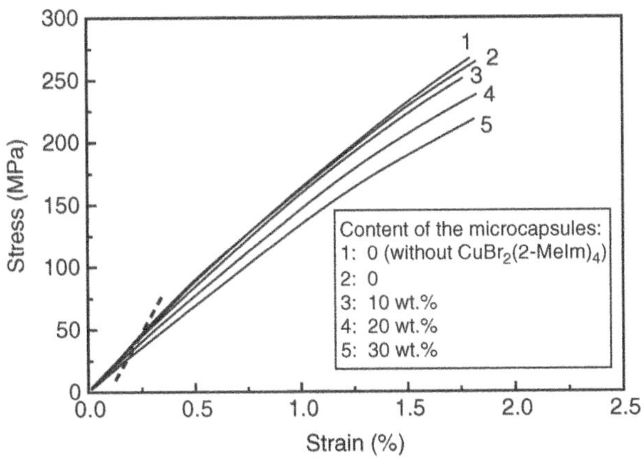

Figure 4.13 Typical tensile stress–strain curves of the laminates with different contents of the epoxy-loaded microcapsules in the central plies. The intersections between the curves and the dashed line indicate the knee points at which the curves deviate from the straight line. Content of $CuBr_2(2\text{-MeIm})_4$: 2 wt%. *Source:* Reprinted from Yin et al. [24]. Copyright 2008, with permission from IOP Publishing.

Table 4.1 Porosity of self-healing woven glass fabric/epoxy composites.

Microcapsule content (wt%)	0^b	10^c	20^c	30^c
Porosity^a (%)	3.07	3.47	3.73	5.76

^{a} Porosity $= \frac{\rho_0 - \rho_1}{\rho_0} \times 100\%$, where ρ_0 and ρ_1 denote the theoretical and experimental density of the laminates, respectively. ρ_0 was estimated from $\rho_0 = \frac{W_{EP} + W_{TEPA\text{-}acrylonitrile} + W_{CuBr_2(2\text{-}MeIm)_4} + W_{MC} + W_{GF}}{\frac{W_{EP} + W_{TEPA\text{-}acrylonitrile} + W_{CuBr_2(2\text{-}MeIm)_4}}{\rho_{matrix}} + \frac{W_{MC}}{\rho_{MC}} + \frac{W_{GF}}{\rho_{GF}}}$, where W_{EP}, $W_{TEPA\text{-}acrylonitrile}$, $W_{CuBr_2(2\text{-}MeIm)_4}$, W_{MC} and W_{GF} denote weights of epoxy, the addition product of TEPA and acrylonitrile, $CuBr_2(2\text{-MeIm})_4$, epoxy-loaded microcapsules and the woven glass fabric, respectively. ρ_{matrix} (1.192 g cm^{-3}) represents the density of the epoxy mixed with $CuBr_2(2\text{-MeIm})_4$ and cured by the addition product of TEPA and acrylonitrile. ρ_{MC} (1.101 g cm^{-3}) and ρ_{GF} (2.378 g cm^{-3}) stand for the densities of the epoxy-loaded microcapsules and the woven glass fabric, respectively.

^{b} Content of $CuBr_2(2\text{-MeIm})_4$: 0.

^{c} Content of $CuBr_2(2\text{-MeIm})_4$: 2 wt%.

Source: Reprinted from Yin et al. [24]. Copyright 2008, with permission from IOP Publishing.

propagation in the laminates is closely related to the matrix and fiber/matrix interfacial characteristics in addition to the fact that tensile behavior of the composites is mainly dependent on the fabric reinforcement [25]. The knee points on the stress–strain curves represent the first cracking, and the nonlinear stable propagation between the knee points and the failure points reflects the process of multiple cracking that offers pseudo-plasticity of materials [26]. Such pseudo-plasticity often reduces the notch sensitivity of the materials and improves reliability. When the cracking starts at the interface between the fibers and the matrix, the microcapsules could behave like stress concentration sites at that interface and lead to the knee occurring at lower load levels with increasing microcapsule content.

The dependence of Young's modulus and tensile strength on the content of the microcapsules further reveals the effect of tiny particles. Due to the lower stiffness of the microcapsules, the rigidity of the matrix has to be decreased and hence the composites exhibit lower modulus with a rise in the fraction of microcapsules. On the other hand, the microcapsules might act as hollow voids in the matrix, which in turn become the sites of stress concentration. The poor load-bearing capacity of the microcapsules would eventually result in lower strength of the composites. Consequently, the tensile strength of the composites with higher microcapsule loading has to be remarkably reduced. Nevertheless, 95% of the strength of the composites excluding the microcapsules can be retained so long as the content of the microcapsules is lower than 20 wt%. This might be related to the change of the matrix resin's viscosity. It was found that the addition of the epoxy-loaded microcapsules at a content lower than 20 wt% only slightly influenced the viscosity of the laminating resin. However, further addition of the microcapsules drastically increased the resin's viscosity. It must severely deteriorate the impregnation and the interfacial coupling as indicated by the greatly increased porosity of the laminates at 30 wt% of microcapsules (Table 4.1). Therefore, the laminate's tensile performance became worse.

To evaluate the self-healing capability of the composites, interlaminar fracture toughness of the composite laminates was measured in terms of DCB tests [27–32]. One edge of the specimen was coated with a thin layer of white opaque fluid to aid visualization of the crack tip, where it was marked every millimeter to make a reference for crack length measurement. A magnifier was used to observe propagation of the crack tip. The specimen was initially loaded by displacement control at a rate of 3 mm min^{-1} until reaching 10 mm crack length. Subsequently, the specimen was immediately unloaded at a rate of 30 mm min^{-1} until the load decreased by half. The load–unload process was repeated five times. Then, the specimen was completely unloaded, allowing the crack faces to come back into close contact. Healing was conducted by putting the tested specimen in an oven preset at 130 °C for one hour. Afterwards, the healed specimen was taken out of the oven and rested at room temperature for 24 hours. Finally, the healed sample was tested again according to the above steps. Figure 4.14 shows the typical load–crack opening displacement curves of the specimens. For the virgin laminates, the behavior is approximately linear up to the onset of crack growth near the peak load of about 27 N. Upon further displacement, the crack propagates along the midplane in a stable mode and the load falls slowly. Through seven load–unload cycles, i.e. 70 mm crack extension, the crack opening displacement attains 105 mm. In contrast, the healed specimen closely follows the original loading curve until the crack begins to advance again near a load of 20 N. Then, the load drops fast along with the crack growth, which is quite unstable compared with the virgin crack propagation. When the crack length reaches 70 mm, the crack-opening displacement is only 65 mm. That is to say, the delamination resisting ability of the healed midplane is reduced.

The analysis receives support from SEM micrographs of the fractured surfaces of the composites (Figure 4.15). Figure 4.15a shows that after failure due to the first DCB test, fiber breakage and debonding are obvious besides rupture of the epoxy-loaded microcapsules. These damage modes must have consumed substantial energy, resulting in higher interlaminar performance. The magnified view of the resin-rich region (Figure 4.15b) further indicates that the healing agent had flowed out of the broken microcapsules as expected, leaving ring-like concaves. After the second DCB test, the repaired specimen exhibits complicated fracture patterns. Figure 4.15c illustrates that the main areas that have been repaired should be the resin-rich region. The cured healing agent has formed thin layers on the fractured surface (Figure 4.15d). It is worth noting that the quantity of released healing agent might be insufficient to build up a compact coverage on the fiber-rich region. In addition, the broken fibers cannot be reconnected by the present self-healing agent. Consequently, the contribution of fiber bridging to hinder crack development has to be lowered.

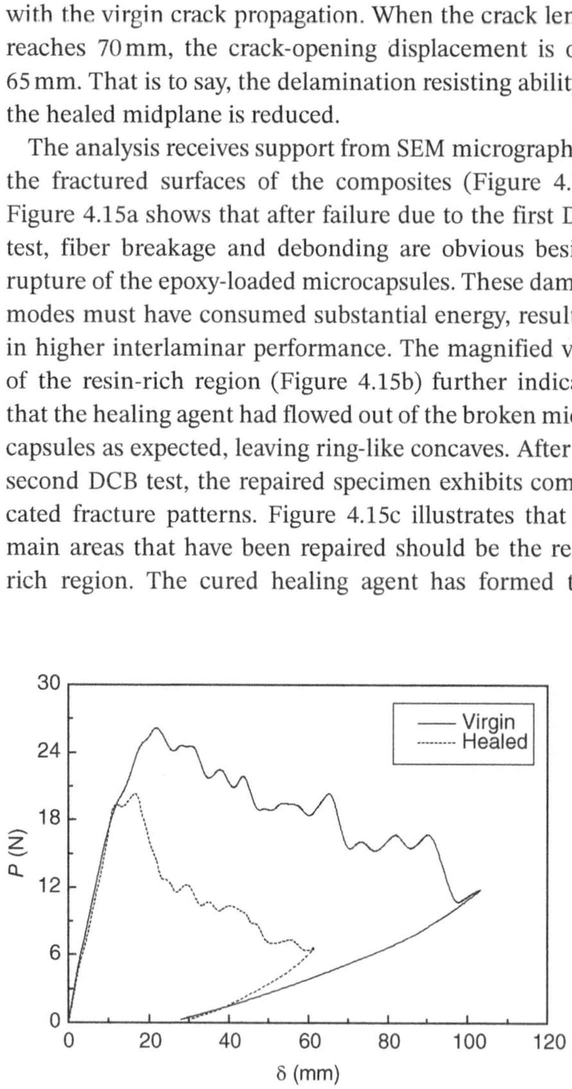

Figure 4.14 Typical load (*P*)–crack opening displacement (*δ*) curves of the self-healing laminates before and after healing. (Note: unloading curves are not shown.) Contents of the epoxy-loaded microcapsules and $CuBr_2(2\text{-MeIm})_4$ in the central plies: matrix epoxy/microcapsules/$CuBr_2(2\text{-MeIm})_4$ = 100/20/2. *Source:* Reprinted from Yin et al. [24]. Copyright 2008, with permission from IOP Publishing.

In Figure 4.16 the typical delamination resistance curves (*R*-curves) of the composite laminates are presented. The mode I critical strain energy release rate, G_{IC}, was calculated according to the modified beam theory (MBT) method. For the virgin laminates with and without the self-healing agent, the tendencies of their *R*-curves and G_{IC} values are similar. G_{IC} increases with crack propagation until reaching a steady state after crack extension of about 10 mm. The plateau values are about twice the initiation ones. The results imply that the pre-embedded microcapsules and pre-dissolved catalyst nearly have no effect on interlaminar fracture toughness of the composites because fiber bridging is the main influencing factor during the first DCB tests. For the healed specimen, G_{IC} shows a decline after the maximum until the values are roughly the same as the initiation ones. Clearly, lack of in-plane fiber bridging in front of the crack should account for this phenomenon. It confirms the above estimation based on morphology observation (Figure 4.15). That is, the open areas of the weave were rich in the epoxy-loaded microcapsules probably due the fact that the microcapsules were filtered to the specific sites during compression molding, so that the healing took place mostly in the open weave areas. The repaired portions failed to provide sufficiently strong closure force as a result.

Table 4.2 systematically compares the effects of the microcapsule concentration, catalyst concentration and laminates thickness on G_{IC} and healing efficiency. The healing efficiency, η, average healing efficiency, η_{avg}, and maximum healing efficiency, η_{max}, are calculated from:

$$\eta = \frac{K_{IC}^{Healed}}{K_{IC}^{Virgin}} = \sqrt{\frac{G_{IC}^{Healed}}{G_{IC}^{Virgin}}} \tag{4.2}$$

$$\eta_{avg} = \sqrt{\frac{G_{IC}^{Healed,avg}}{G_{IC}^{Virgin,avg}}} \tag{4.3}$$

$$\eta_{max} = \sqrt{\frac{G_{IC}^{Healed,max}}{G_{IC}^{Virgin,avg}}} \tag{4.4}$$

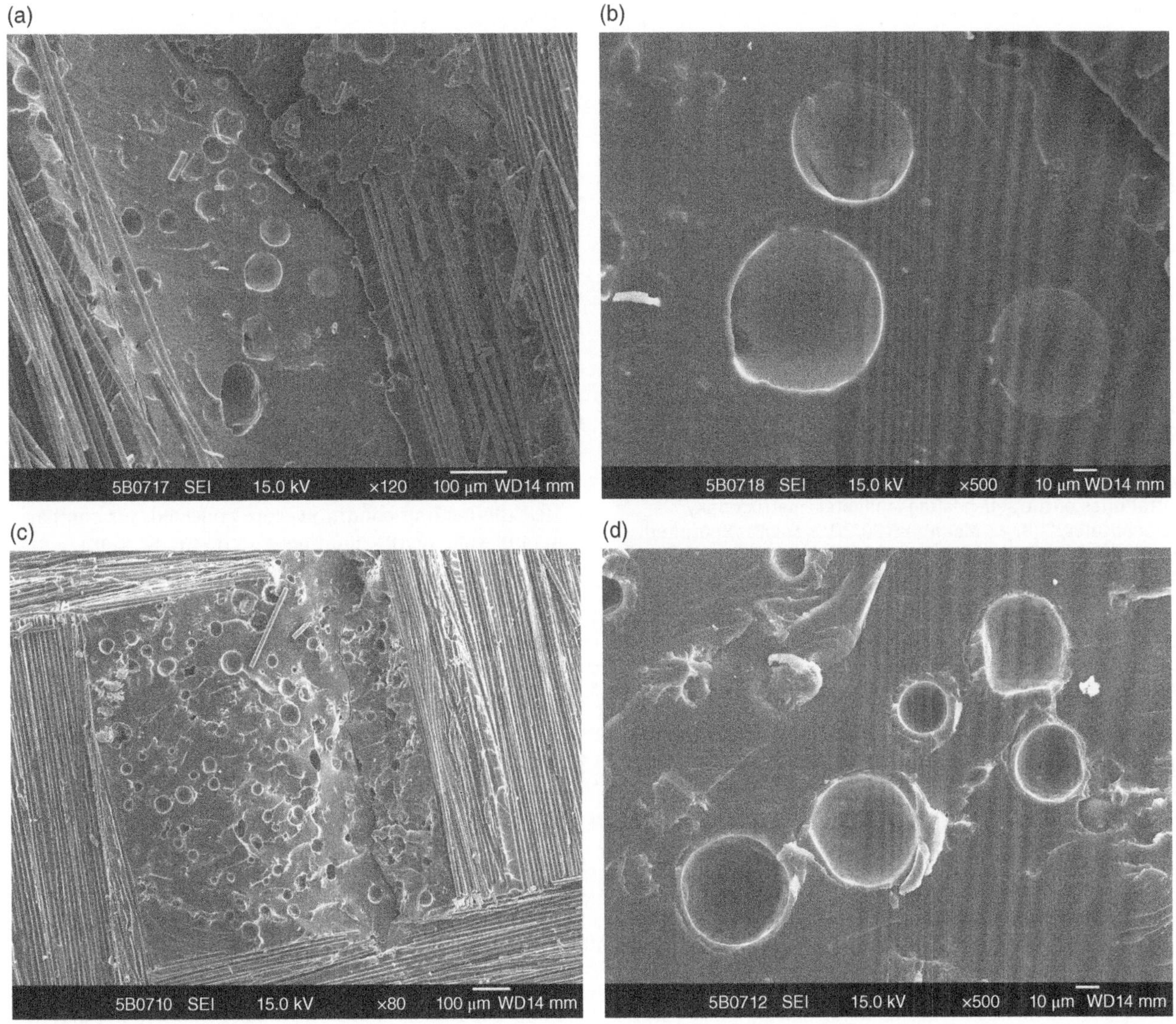

Figure 4.15 SEM micrographs of (a, b) the fractured surfaces of the self-healing laminates obtained after the first DCB test, and (c, d) those obtained after the second DCB test. Contents of the epoxy-loaded microcapsules and $CuBr_2(2\text{-MeIm})_4$: 30 and 2 wt%, respectively.

Equations (4.3) and (4.4) are based on the average and maximum of critical strain energy release rate G_{IC}, respectively. In addition, engineering healing efficiency, η^{eng}, is employed to check crack healing efficiency from an engineering point of view. It is calculated by normalizing critical stress intensity factor of the healed specimen against that of the virgin specimen without healing agent, $K_{IC}^{Virgin,0}$:

$$\eta^{eng} = \frac{K_{IC}^{Healed}}{K_{IC}^{Virgin,0}} = \sqrt{\frac{G_{IC}^{Healed}}{G_{IC}^{Virgin,0}}} \tag{4.5}$$

Accordingly, the average engineering healing efficiency, η_{avg}^{eng}, and maximum engineering healing efficiency, η_{max}^{eng}, are obtained from:

$$\eta^{eng} = \frac{K_{IC}^{Healed}}{K_{IC}^{Virgin,0}} = \sqrt{\frac{G_{IC}^{Healed}}{G_{IC}^{Virgin,0}}} \tag{4.6}$$

$$\eta_{max}^{eng} = \sqrt{\frac{G_{IC}^{Healed,max}}{G_{IC}^{Virgin,avg,0}}} \tag{4.7}$$

where $G_{IC}^{Virgin,0}$ denotes the critical strain energy release rate of the virgin specimen without healing agent.

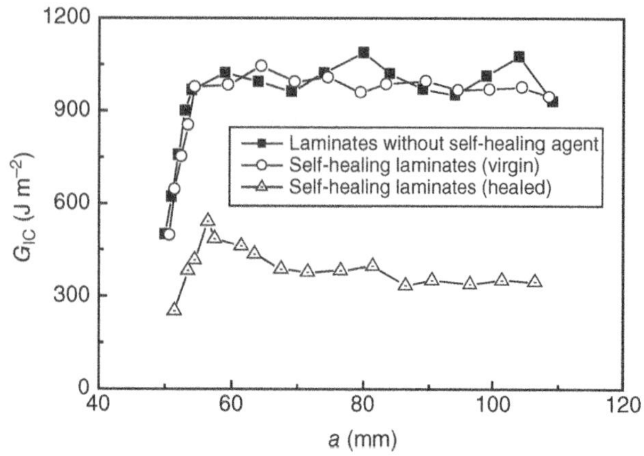

Figure 4.16 Interlaminar fracture toughness of the laminates, G_{IC}, as a function of crack length, *a*. Contents of the epoxy-loaded microcapsules and $CuBr_2(2\text{-MeIm})_4$ in the central plies of the self-healing laminates: matrix epoxy/ microcapsules/$CuBr_2(2\text{-MeIm})_4$ = 100/20/2. *Source:* Reprinted from Yin et al. [24]. Copyright 2008, with permission from IOP Publishing.

It is seen that critical strain energy release rate in the virgin specimen, G_{IC}^{Virgin}, decreases with increasing content of the epoxy-loaded microcapsules, while the influence of the catalyst content is negligible. This might be due to the fact that the appearance of the epoxy-loaded microcapsules raised the matrix viscosity and hence caused deterioration of the impregnation. The worse interfacial adhesion would facilitate crack propagation. With respect to the average and maximum critical strain energy release rate in the healed specimens, $G_{IC}^{Healed,avg}$ and $G_{IC}^{Healed,max}$, however, the catalyst content has more significant influence than the microcapsules content. For the laminates with 20 wt% microencapsulated epoxy, for example, both $G_{IC}^{Healed,avg}$ and $G_{IC}^{Healed,max}$ increase with increasing the catalyst concentration from 0.5 to 2 wt%. Accordingly, the maximum healing efficiency of the composites is raised by nearly twofold. It is interesting to see from Table 4.2 that a further increase in the dosage of the catalyst at constant microcapsule content could not continuously increase the healing efficiency. Similarly, the dependence of the self-healing efficiency on the microcapsule content also varies in this way. This demonstrates that the microcapsules-to-catalyst ratio is a key factor in determining the self-healing efficiency of the laminates. If the concentrations of the microcapsule and catalyst were mismatched, a satisfactory repair result could not be acquired. The data in Table 4.2 reveal that the highest healing efficiency is seen at 30 wt% microcapsules and 2 wt% catalyst. This implies that higher loading of the microcapsules is generally required for delivering sufficient polymerizable healing agent to the cracked portions to obtain higher healing efficiency. However, it is worth noting that a rise in the concentration of microcapsules would lead to significant reduction in tensile properties of the laminates (Figure 4.13). Therefore, a balance between strength and toughness restoration should be considered when manufacturing self-healing composites for practical applications.

To confirm that the healing at 130 °C only comes from the microcapsules plus the latent hardener, and not from unreacted epoxy monomer within the host laminates, FTIR spectra of the fractured surfaces of the self-healing laminates before healing and after heat treatment that simulated the healing condition were collected, for comparison with that of the laminates without the self-healing agent. As shown in Figure 4.17, the band of the terminal epoxide group appears at 914 cm^{-1} of the spectrum of the fracture surface of the self-healing composites (curve 2), while it is absent on the spectrum of the laminates without the self-healing agent (cure 1). This proves that there is no unreacted epoxy in the host laminates and the broken microcapsules are able to provide uncured epoxy on the facture surface. Having been heated at 130 °C for one hour, this epoxide peak disappears (curve 3), indicating that the epoxy from the broken microcapsules has been consolidated by the latent hardener. The stronger absorptions of the C—O—C stretching mode (at 1100–1140 cm^{-1}) and the weaker absorptions of the —OH stretching mode (at 3300–3500 cm^{-1}) on curve 2 than those on curve 3 support the above analysis. Therefore, it can be concluded that the pre-embedded epoxy-loaded microcapsules and the latent hardener are the only resources responsible for the crack healing.

In fact, the laminates thickness is also critical for self-healing efficiency. As shown in Table 4.2, the thinner specimens (4 mm thick) offer lower healing efficiency. This should be related to the laminate processing. That is, the microcapsules might have burst during the compression molding and the effective amount of microcapsules in the composites had to be reduced. In the case of thicker specimens (4.8 mm thick), the healing efficiency is also not high. Removal of air bubbles deep inside the composites by degasification was somewhat difficult. The healed areas might be reduced and hence the healing efficiency was lowered. Only when the laminates have optimum thickness (4.5 mm, for instance) can higher healing efficiency be achieved.

Table 4.2 Dependence of healing efficiency of self-healing woven glass fabric/epoxy composites on the influencing factors.

Variables				G_{IC}^{Healed} (J m^{-2})		η^{a} (%)		$\eta^{eng b}$ (%)	
Microcapsules (wt %)	Catalyst (wt %)	Thickness (mm)	$G_{IC}^{Virgin,avg}$ (J m^{-2})	avg.	max.	avg.	max.	avg.	max.
0	0	4.3	1059 (±137)	0	0	0	0	0	0
20	0.5	4.5	905 (±165)	102 (±56)	138 (±119)	34	39	31	36
	1		931 (±110)	185 (±81)	327 (±85)	45	59	42	56
	2		950 (±112)	206 (±38)	478 (±54)	47	71	44	67
	3		927 (±142)	198 (±55)	431 (±62)	46	68	43	64
10	2	4.5	967 (±187)	135 (±65)	236 (±213)	37	49	36	47
30			838 (±151)	205 (±83)	517 (±123)	49	79	44	70
35			799 (±138)	211 (±114)	446 (±05)	51	75	45	65
10	3	4.5	942 (±100)	116 (±93)	238 (±199)	35	50	33	47
30			856 (±170)	197 (±78)	489 (±97)	48	76	43	68
35			787 (±154)	203 (±109)	442 (±141)	51	78	44	65
20	2	4.0	989 (±162)	84 (±57)	154 (±145)	29	39	28	38
		4.8	935 (±136)	154 (±124)	225 (±198)	41	49	38	46

[a] Healing efficiency estimated from Eq. (4.2).
[b] Engineering healing efficiency estimated from Eq. (4.5).
Source: Reprinted from Yin et al. [24]. Copyright 2008, with permission from IOP Publishing.

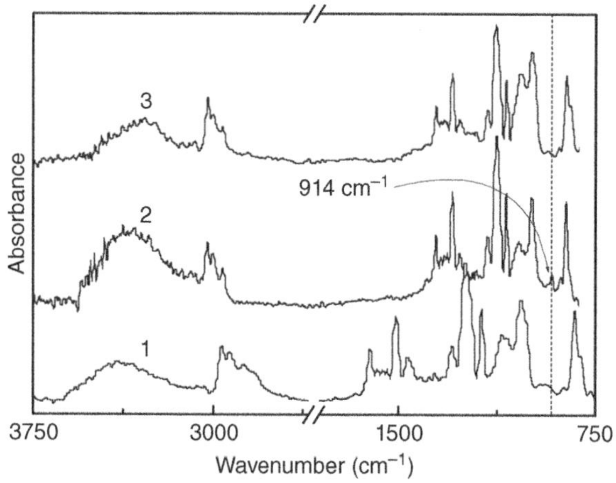

Figure 4.17 FTIR spectra of the fractured surface of (1) the laminates without the self-healing agent, (2) the laminates with the self-healing agent (contents of the epoxy-loaded microcapsules and $CuBr_2(2\text{-MeIm})_4$ are 20 and 2 wt%, respectively), and (3) specimen (2) heated at 130 °C for one hour simulating the healing procedure. *Source:* Reprinted from Yin et al. [24]. Copyright 2008, with permission from IOP Publishing.

4.4 Durability of Healing Ability

In view of practical applications, the self-healing function of composite materials should last for a long time [33]. Although there is not a long history of extensive study on self-healing polymers and polymer composites, some pilot studies have been done on the latter. Kessler et al. [31], for example, examined the influence of different amine-type hardeners for consolidating epoxy composites on the reactivity of the pre-embedded healant (consisting of Grubbs catalyst particles and microencapsulated dicyclopentadiene [DCPD]). They found that in contrast with primary amine curing agent, the tertiary amine system had little chemical interaction with Grubbs catalyst and can thus mitigated degradation of the catalyst activity. In a paper dealing with the same healant by Rule et al. [34], it was shown that wax could protect Grubbs catalyst from the detrimental interaction with diethylenetriamine. As a result, healing efficiency was evidently increased. In addition, the research of Jones et al. [35, 36] further demonstrated that the use of wax-protected and recrystallized Grubbs catalyst led to a fourfold increase in the rate of polymerization of bulk DCPD and extended the fatigue life of a polymer specimen by over 30 times longer than a control non-healing specimen.

On the other hand, Pang et al. [37] assessed the rate of degradation of the repair resin effectiveness over storage time in the healing agent consisting of epoxy and hardener, respectively, filled into hollow glass fibers. The specimens were stored for different periods (0, 3, 6, and 9 weeks) before being damaged. The results showed that the efficiency of repair markedly deteriorated over a nine-week period because the rise of viscosity precluded penetration of the repair resin in damage crack paths. The possible reason was attributed to the fact that the repair resin mixed with acetone and fluorescent dye might experience discernible deterioration with time, or it was unsuited to storage in an uncured state for a long period.

Returning to the self-healing system based on epoxy-loaded microcapsules and latent hardener $CuBr_2(2\text{-MeIm})_4$, although it proved to be able to recover interlaminar fracture toughness of the composites to a certain extent (refer to the previous subsection), it was revealed that healing efficiency of aged composite specimens is obviously lower than that of newly made ones. This is unfavorable for practical usage. In this context, clarification of the key factor that leads to reduction in healing ability of the laminates is necessary. This would help to increase service life of the composites in terms of the crack healing capability.

Table 4.3 lists interlaminar fracture toughness and healing efficiency of the laminates containing 30 wt% epoxy-loaded

Table 4.3 Effect of storage time on healing efficiency of self-healing woven glass fabric/epoxy composites[a].

Storage time (months)	$G_{IC}^{Virgin,avg}$ (J m^{-2})	G_{IC}^{Healed} (J m^{-2})		η (%)		η^{eng} (%)	
		avg.	max.	avg.	max.	avg.	max.
<0.25	838 (±151)	205 (±83)	517 (±123)	49	79	44	69
0.5	864 (±161)	183 (±76)	458 (±154)	46	73	42	66
2	861 (±185)	167 (±65)	401 (±125)	44	68	40	62
6	795 (±134)	155 (±58)	388 (±79)	44	70	38	61
12	815 (±177)	165 (±68)	378 (±51)	45	68	39	60
18	867 (±130)	159 (±54)	392 (±163)	43	67	39	61

[a] Contents of epoxy-loaded microcapsules and $CuBr_2(2\text{-MeIm})_4$ are 30 and 2 wt%, respectively. The definitions of the healing efficiencies, η and η^{eng}, are the same as those listed in Table 4.2.
Source: Reprinted from Yin et al. [33]. Copyright 2009, with permission from IOP Publishing.

microcapsules and 2 wt% $CuBr_2(2\text{-MeIm})_4$ as a function of storage time. The composites are selected because they have the highest healing efficiency as compared with those with other proportions of the healant (Table 4.2). Evidently, G_{IC}^{Healed} of the laminates decreases with an increase in storage time, while the values of G_{IC}^{Virgin} are nearly unchanged. As a result, healing efficiency has to decrease with increasing storage time. A careful survey of the data in Table 4.3 indicates that the reduction in both the interlaminar fracture toughness of healed specimens and the corresponding healing efficiency is remarkable for the first two months, which is similar to the findings of Pang et al. [37]. For the specimens stored for longer, however, their self-healing efficiency no longer changes. This implies that the factor resulting in reduction in G_{IC}^{Healed} and healing efficiency only takes effect within a relatively short period after production of the laminates.

As discussed previously, the healing chemistry of the present system lies in the curing reaction between epoxy and $CuBr_2(2\text{-MeIm})_4$. In this context, the aforementioned deterioration in healing ability must be related to performance of the reaction.

With respect to the latent hardener $CuBr_2(2\text{-MeIm})_4$, its long-term stability has been generally approved [3–5], so it should not be affected by the surrounding environment. Our previous investigation also demonstrated that its reactivity was not influenced by composite fabrication at a temperature below 120 °C (Section 4.1). Since the highest curing temperature for the present laminates is 100 °C, the possibility that deactivation of the latent hardener during composite manufacturing and storage at room temperature can be excluded.

Naturally, the uncured epoxy inside the microcapsules has to be considered. Owing to the protection of the poly(urea-formaldehyde) (PUF) shell, the loaded resin would not be attacked by active species in the composites. Furthermore, the uncured epoxy possesses high thermal stability as characterized by its onset pyrolytic temperature of 359 °C (Figure 4.4). Therefore, reactivity of the encapsulated epoxy should not deteriorate during manufacture of the laminate and storage at room temperature.

Table 4.4 lists linear shrinkage rates of cured epoxy containing the healing agent. It is seen that incorporation of 2 wt% of the latent hardener has no effect on linear shrinkage rate of the cured epoxy. The low dosage of $CuBr_2(2\text{-MeIm})_4$ might account for the result. When the epoxy-loaded microcapsules are added, however, the linear shrinkage rate of the cured epoxy slightly increases, regardless of the presence of the latent hardener. This suggests that the microcapsules embedded in the composite matrix must suffer compression during curing.

To check stability of the microcapsules from another viewpoint, the weight change of the spheres was examined. As shown in Figure 4.18, when the epoxy-loaded capsules experienced the thermal treatment identical to the curing procedures (i.e. 60 °C for two hours, 80 °C for two hours, and 100 °C for two hours), a 2.62% weight loss is observed. In the meantime, the core content of the microcapsules increased from 64.2 to 66.0 wt%, as revealed by solvent extraction and weighing. Accordingly, the reduction in weight of the wall of the capsules is estimated to be 5%. Since the value is close to the measured one of the wall of the capsules alone originating from the same heating steps, it is known that the above weight loss of 2.62% of the microcapsules is wholly contributed by the urea-formaldehyde wall. On the other hand, Figure 4.18 also indicates that weight of the capsules only slightly decreases at room temperature (e.g. a loss of 0.3 wt% for three

Table 4.4 Linear shrinkage rates of TEPA cured epoxy resin[a].

Materials	Linear shrinkage rate ($cm\,cm^{-1}$)
Unfilled epoxy	0.015
Epoxy with 2 wt% $CuBr_2(2\text{-MeIm})_4$	0.015
Epoxy with 10 wt% epoxy-loaded microcapsules	0.016
Epoxy with 10 wt% epoxy-loaded microcapsules and 2 wt% $CuBr_2(2\text{-MeIm})_4$	0.016

[a] The resin of interest was added into the mold, which was made for examining cure shrinkage according to Chinese chemical engineering standard HG/T 2625–2694, and then cured by TEPA at 60 °C for two hours, 80 °C for two hours, and 100 °C for two hours. Linear shrinkage rate of cured resin was calculated by:

Linear shrinkage rate $\left(cm\,cm^{-1}\right) = \frac{L - L'}{L}$, where L and L' denote the mold length and the cured specimen length, respectively.

Source: Reprinted from Yin et al. [33]. Copyright 2009, with permission from IOP Publishing.

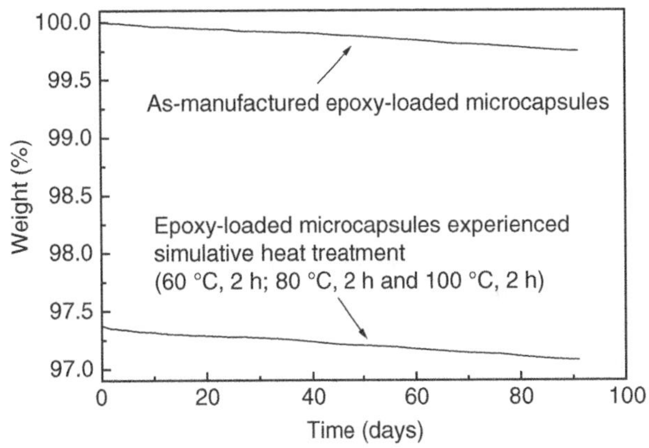

Figure 4.18 Time dependence of weight of epoxy-loaded microcapsules measured at room temperature. *Source:* Reprinted from Yin et al. [33]. Copyright 2009, with permission from IOP Publishing.

months), despite whether they were heat treated or not in advance. As visual inspection did not find any trace of uncured epoxy on the surface of the capsules, the negligible weight loss under the condition free of compression should be the result of desorption of the physically adsorbed moisture. In summary, the thermogravimetric analysis reveals that partial degradation occurs in the shell of the epoxy-loaded capsules when the composites are cured. This must somewhat weaken shielding of the capsules.

In fact, the wall of the epoxy-loaded microcapsules is constructed by aggregation of PUF particles with an average diameter of about 20 nm (Figure 4.2). This coincides with the colloid theory of urea-formaldehyde resin [38]. Such an aggregated structure is far from compact and has the potential to contain microcracks. When the microcapsules are compressed during curing, the loaded uncured epoxy resin might be squeezed out through the microcracks.

On the basis of the above results and discussion, it can be reasonably deduced that diffusion of uncured epoxy inside the microcapsules might take place when the composite laminates are produced because of volumetric contraction of the matrix. The diffusion should proceed very slowly after initiation and last several weeks at room temperature. In the meantime, the diffused epoxy would be gradually cured by the remnant amine curing agent in the matrix and then block the diffusing sites on the surface of the microcapsules. As a result of reduction in the amount of uncured epoxy in the microcapsules, the repair agent delivered to the cracking planes is insufficient, leading to lower healing efficiency than the case of recently made laminates. The decline is eventually terminated when permeability of the wall of the microcapsules for uncured epoxy is greatly reduced by the cured version of the diffused epoxy. Consequently, decrease in healing efficiency of the laminates is no longer observed in the case of longer storage time.

To verify the above hypothetical mechanism, dynamic mechanical analysis (DMA) tests were conducted on the following model materials: (i) epoxy with 10 wt% epoxy-loaded microcapsules, (ii) epoxy with 2 wt% $CuBr_2$(2-MeIm)$_4$, (iii) epoxy with 10 wt% epoxy-loaded microcapsules and 2 wt% $CuBr_2$(2-MeIm)$_4$, and (iv) unfilled epoxy. In addition, neat epoxy cured by $CuBr_2$(2-MeIm)$_4$ was also produced as the control. The woven glass fabric was excluded because it has almost nothing to do with the decay of healing efficiency.

DMA has been known as an effective means to study thermomechanical properties of thermosets [39–41]. If epoxy monomer indeed diffused from the embedded microcapsules in the composites, its curing behavior catalyzed by the surrounding curing agent would be easily detected by DMA. Here no fiber reinforcement was added to the model composites to simplify the influencing factors. Considering that in the composites of interest two types of curing agent are added, TEPA for curing the matrix and $CuBr_2$(2-MeIm)$_4$ for curing the epoxy monomer released from the microcapsules, basic DMA characteristics of epoxy cured by the curing agents should be known in advance. As shown in Figure 4.19, the storage modulus of both systems exhibits a drastic decrease within the glass transition region. The peak temperature of the loss factor (i.e. glass transition temperature) of the $CuBr_2$(2-MeIm)$_4$-cured version is much higher than that of the TEPA-cured one, so the storage modulus of the former is much higher than that of the latter from 130 to 180 °C, the working temperature range of $CuBr_2$(2-MeIm)$_4$.

On the basis of these findings and knowledge of the healant, time dependences of storage modulus of the model composites were measured at 140 °C to monitor variation in the stiffness of the specimens caused by curing of possibly diffused epoxy. Prior to this, the specimens were kept at 30 °C for 10 minutes to equalize the starting conditions. The reason for choosing 140 °C as the temperature at which DMA spectra were collected is that both TEPA and $CuBr_2$(2-MeIm)$_4$ can take effect at this temperature.

Figure 4.20a shows the results of unfilled epoxy. In addition to the time dependence of the storage modulus, the specimen's temperature is also given as a function of time. When the specimens are heated from 30 to 140 °C, an evident decrease in the modulus is perceived because the TEPA-cured material falls into a rubbery state (Figure 4.19). With the lapse of time, storage moduli of the specimens slightly increase during the isothermal process at 140 °C. This phenomenon should be indicative of the post cure of epoxy. Namely, a small quantity of incompletely cured epoxy reacts with the remnant TEPA. Consequently, the cure degrees of the specimens increase with time, as reflected by a rise in their moduli. On the other hand,

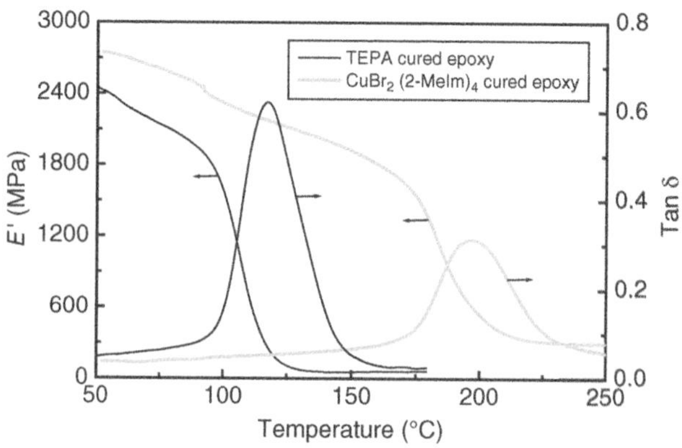

Figure 4.19 Temperature dependence of storage modulus, *E'*, and loss factor, tan δ, of epoxy cured by TEPA and $CuBr_2$(2-MeIm)$_4$, respectively. *Source:* Reprinted from Yin et al. [33]. Copyright 2009, with permission from IOP Publishing.

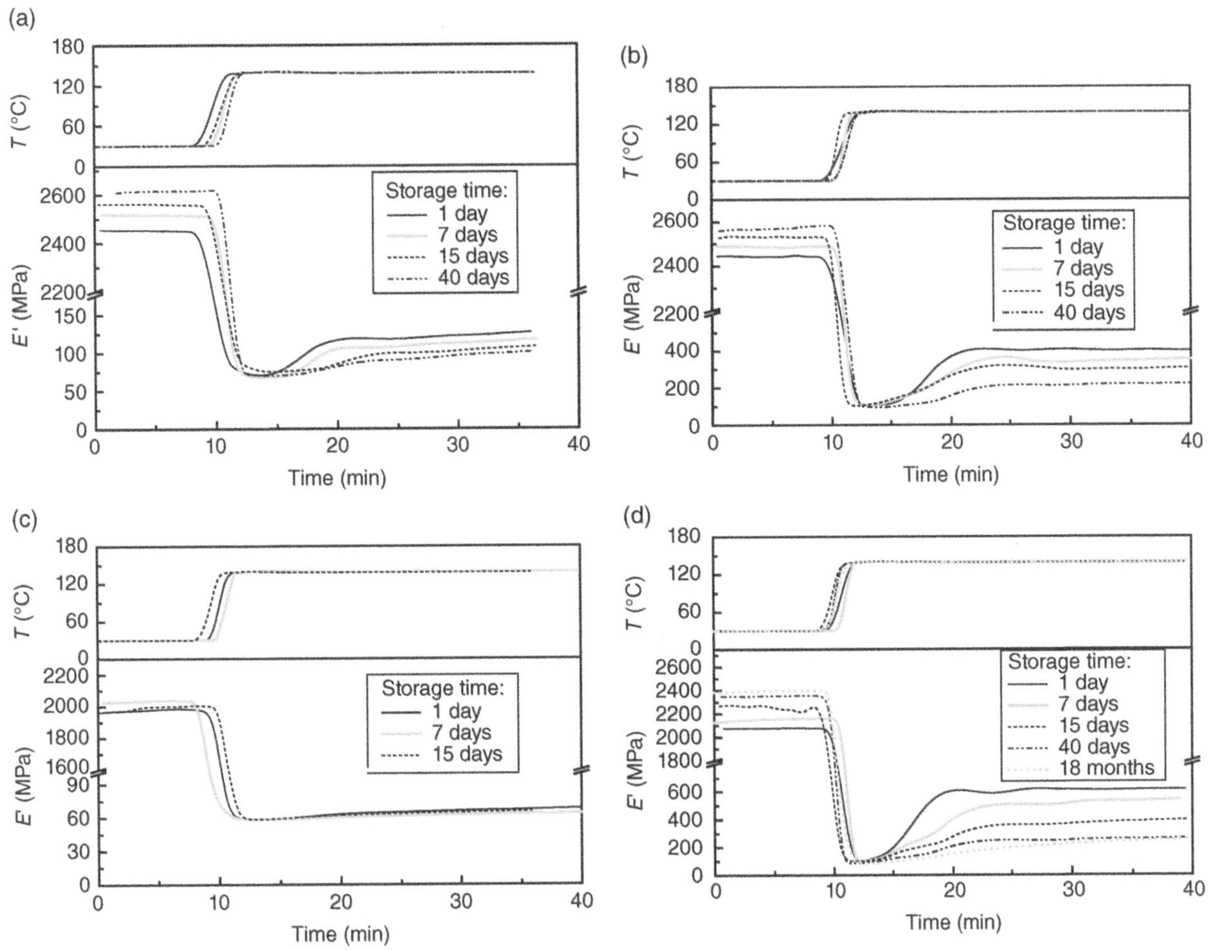

Figure 4.20 Time dependence of storage modulus, *E'*, of (a) unfilled epoxy, (b) epoxy filled with 2 wt% $CuBr_2$(2-MeIm)$_4$, (c) epoxy filled with 10 wt% epoxy-loaded microcapsules, and (d) epoxy filled with 10 wt% epoxy-loaded microcapsules and 2 wt% $CuBr_2$(2-MeIm)$_4$ at 30 and 140 °C. Test temperature, *T*, is shown as a function of time for reference. Prior to the DMA measurements, the specimens had been stored at room temperature for different periods of time.

because the post cure reaction with TEPA can also proceed at room temperature, cure degrees of the previously made specimens have to be higher than the newly made ones prior to the DMA measurements. This explains the fact that the room temperature modulus of the former is higher than that of the latter, while the modulus increment of the former at 140 °C is smaller than that of the latter.

For the epoxy filled with $CuBr_2$(2-MeIm)$_4$ (Figure 4.20b), its storage modulus at 30 °C is about 2.5 GPa, almost the same as that of the unfilled epoxy (Figure 4.20a), but the modulus increment at 140 °C is much more obvious. It is clear that the latent hardener must have participated in the reaction with incompletely cured epoxy at 140 °C. Since $CuBr_2$(2-MeIm)$_4$-cured epoxy is much stiffer than TEPA-cured one (Figure 4.19), the system exhibits higher modulus at 140 °C after aging at this temperature. Besides, as the cure degree of the specimens slightly increases with time due to the reaction between incompletely cured epoxy and remnant TEPA at room temperature, the modulus increment at 140 °C also has to rank in the sequence of storage time.

When epoxy-loaded microcapsules are incorporated into the model composites, the room temperature modulus of the composites is lower than that of the unfilled epoxy (cf. Figure 4.20a,c). This is because of the soft nature of the capsules. Similar results were reported by Brown et al. [17], who found a continuous reduction in Young's modulus of epoxy with embedded DCPD-loaded microcapsules. Interestingly, the evident increase in storage modulus at 140 °C observed in Figure 4.20a,b is replaced by a marginal increase in the present system (Figure 4.20c). The reason is that the composites contain not only incompletely cured matrix epoxy but also uncured epoxy diffused from the microcapsules. As the remnant TEPA in the composites is certainly insufficient for conducting a thorough curing, the diffused resin penetrating the matrix might act as a plasticizer and tend to lower the stiffness of the composites. This counteracts the aforementioned stiffening effect originating from post cure, so that the increase in modulus is far from significant.

The above analysis is supported by the results of the composites filled with both epoxy-loaded microcapsules and $CuBr_2$(2-MeIm)$_4$ (Figure 4.20d). Compared with the case shown in Figure 4.20c, a new component $CuBr_2$(2-MeIm)$_4$ is added. If there was diffused epoxy in the composites, the well dispersed latent hardener must be able to trigger its consolidation at 140 °C. The data in Figure 4.20d demonstrate this is the case. The modulus increments at 140 °C are much greater than those for the system containing $CuBr_2$(2-MeIm)$_4$ without microcapsules (Figure 4.20b) and the system containing microcapsules without $CuBr_2$(2-MeIm)$_4$ (Figure 4.20c) as well. This means that additional material with high modulus is present, which has to be the $CuBr_2$(2-MeIm)$_4$-cured version of diffused epoxy.

It is worth noting that because the diffused epoxy might react with the remnant TEPA during the storage at room temperature and gradually block the diffusing points, the diffusion would eventually be stopped. This is proved by the fact that the modulus increment at 140 °C of the composites stored at room temperature for 40 days is close to that of the composites stored at room temperature for 18 months. In summary, the extent of diffused epoxy resin in both composites is almost identical.

The above systematic analysis has confirmed that volumetric contraction in the course of composite manufacturing may have resulted in gradual diffusion of uncured epoxy resin from the microcapsules. The diffused resin was then cured by the remnant amine curing agent in the matrix. Consequently, the polymerizable component of the healing agent was insufficient to rebind cracks, leading to lower healing efficiency. As the cured version of diffused resin in turn kept on blocking the diffusing points on the surface of the microcapsules, diffusion of epoxy had to be eventually terminated. Therefore, healing efficiency no longer changed for the laminates stored at room temperature for a longer length of time.

On the basis of these findings, it is understood that enhancement of compactness of the microcapsules containing epoxy would be critical for improving stability of the self-healing composites. Innovative synthetic approaches, new materials and specific characterization techniques are required to fulfill this task.

4.5 Self-Healing Epoxy/Woven Glass Fabric Composites with Embedded Epoxy-Loaded Microcapsules and Latent Hardener – Healing of Impact Damage

Woven fabric reinforced thermosetting composites are widely used as structural components in many fields. Low velocity impact damage is often generated during their use [42, 43]. The damage (such as matrix microcracking and interfacial debonding) used to be on a microscopic level and invisible. When the microcracks propagate, mechanical performance of the materials declines rapidly, particularly compression strength [44]. For example, damage resulting from impact can cause a loss of 50% of compressive strength. Therefore, impact damage is one of the primary factors for the design of structural composites [45].

In recent years, research on self-healing of impact damage in fiber reinforced polymers has been conducted. Motuku et al. [46] studied the effect of low velocity impact on self-healing materials. Their experiments showed that insertion of glass capillary pipettes as storage tubes did not alter the impact response of the composites, and the healing agents could be released in the case of cracking. Bleay et al. [47] investigated the compression performance of glass fabric self-healing composites after low velocity impact. When the impact-treated specimens were exposed to a combination of high temperature and vacuum, an improvement of ~10% in compression strength was observed. Jones et al. [48, 49] reported a thermoplastic self-healing system, which was predissolved in glass fiber reinforced epoxy composites. When the composites were subjected to repeated falling weight impact, the efficiency of healing was determined by analyzing the growth of delamination. On the other hand, Bond et al. [37, 50–53] studied flexural strength of specimens with resin filled hollow glass fibers before and after repairing the damage from indentation tests by a four-point bending test.

The Illinois group of White et al. pioneered the development of self-healing polymeric materials [22]. They prepared and evaluated self-healing neat polymers and fiber reinforced polymers containing microencapsulated healing agents. In terms of ring-opening metathesis polymerization of DCPD triggered by Grubbs catalyst, healing of delamination damage in glass fiber/epoxy and carbon fiber/epoxy composites was successfully carried out [31, 32], and the damage generated by impact was also repaired [54].

The discussion in the last section has shown the ability of the self-healing agent (encapsulated epoxy monomer + latent hardener $CuBr_2$(2-MeIm)$_4$) in recovering interlaminar fracture toughness of epoxy laminates. Therefore, it is worth further exploring its capability of healing impact damage in laminates. Compression after impact (CAI) [55], a measure of damage tolerance of composite laminates, is employed as the main characterization means hereinafter. In addition, the factors that influence the healing efficiency are also analyzed.

Here the woven glass fabric composites are made of EPON 828 epoxy as the matrix, 2-ethyl-4-methylimidazole (2E4MIm) as the curing agent, and woven glass fabric (C-glass, 13 × 12 plain weave, 0.2 mm thick, 1000 tows, 200 g m^{-2}, see Section 4.3) as the reinforcement [56]. EPON 828 was first mixed with $CuBr_2$(2-MeIm)$_4$ in a proportion of 100:2 (by weight). Then, the microcapsules containing epoxy were slowly incorporated by stirring and ultrasonic

Table 4.5 Specifications of the epoxy-loaded microcapsules.

Average diameter (μm)	140	65	40
Core content (wt%)	81.2	73.9	63.9

dispersion. Afterwards, 2E4MIm was added at a curing agent-to-epoxide ratio of 2:100 (by weight). Eventually, the compounds were degassed in a vacuum chamber for five minutes to remove air bubbles. The microcapsules 40 μm in diameter (Table 4.5) were added to the epoxy resin in two proportions (10 and 20 wt%). The composites with 10 wt% microcapsules 65 and 140 μm in diameter were also prepared and tested in the same way.

Composite laminates were manufactured by hand lay-up and compression molding. The laminates were 2.1 mm thick with the lay-up $[45/0/-45/90]_S$, which were laid up at 60 °C, and then successively molded at 80 °C for two hours, 120 °C for two hours, and 140 °C for two hours. The woven glass fabric volume fraction of the composites was 30 vol%, as calculated by the data of thermogravimetric analysis and pycnometry. The above curing procedures were designed (i) to ensure complete consolidation of the epoxy matrix due to the reaction between epoxy and 2E4MIm, and (ii) to inhibit the reaction between matrix epoxy and the latent hardener $CuBr_2(2\text{-MeIm})_4$. Our previous work demonstrated that no curing would take place in an epoxy/$CuBr_2(2\text{-MeIm})_4$ system at a temperature below 120 °C (Section 4.1). That is, although both 2E4MIm and $CuBr_2(2\text{-MeIm})_4$ were mixed with the epoxy matrix, $CuBr_2(2\text{-MeIm})_4$ would not react with epoxy when the composites were successively molded at 80 °C for two hours and 120 °C for two hours. In the meantime, only the reaction between epoxy and 2E4MIm occurred. During the post curing stage at 140 °C for two hours, the unreacted epoxy would be preferentially cured by 2E4MIm because 2E4MIm has higher reactivity than 2-methylimidazole (2-MeIm). In this way, the matrix epoxy can be completely cured, while the latent hardener would not be deactivated when the composite laminates were cured.

SEM observation of the fracture surface of the composite laminates shows that the epoxy-loaded microcapsules can be damaged as expected, and the healing agent had flowed off, leaving the perfectly ring-like concaves (Figure 4.21). This suggests that the microcapsules have good integrity and survive processing.

The impact tests were performed using a drop-weight tower at room temperature. Four levels of impact energy were selected, i.e. 1.5, 2.0, 2.5 and 3.5 J. The impact energy levels were chosen because they would result in damage modes of the laminates varying from matrix cracking to fiber facture, which are sufficiently representative. Referring to the work by Sánchez-Sáez et al. [55], an improved CAI set-up was used. The device was designed for thin specimens, which did not require tabs or any modification of the specimen geometry. It can assure laminate failure occurring in the mode of delamination and propagation of the delamination takes place predominantly in the direction perpendicular to the loading direction. To check the self-healing capability of the composites, the specimens were first impacted and then healed at 140 °C for 30 minutes under (i) no added pressure, (ii) 60 kPa, or (iii) 240 kPa. Figure 4.22 shows the flowchart of the tests.

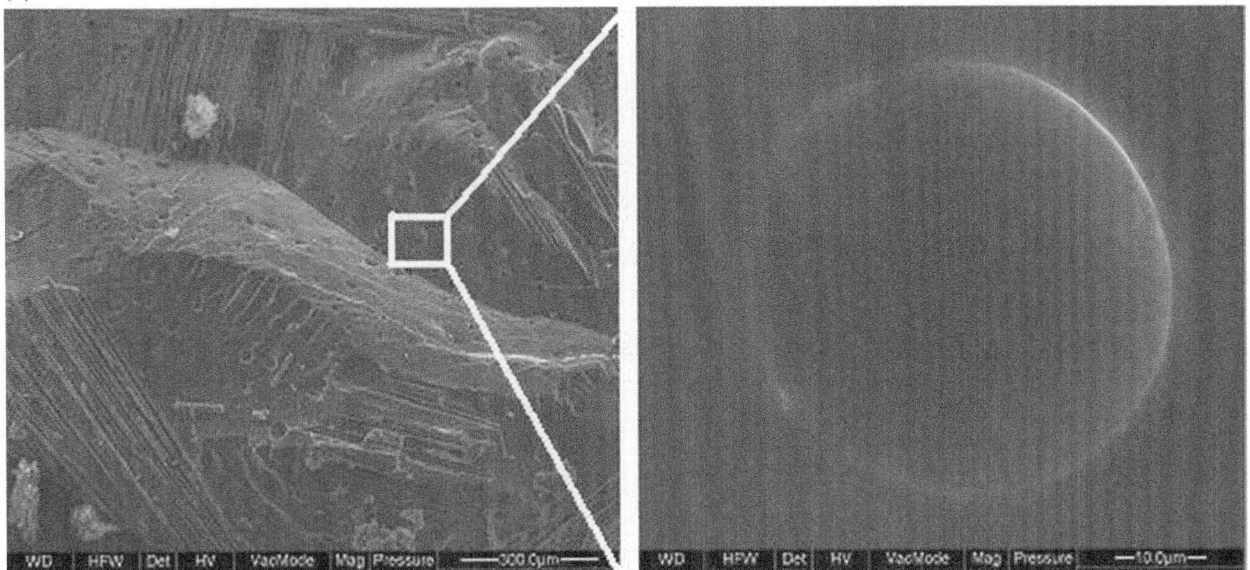

Figure 4.21 SEM micrographs of the fractured surfaces of the laminates containing healing agent. (b) Magnified region from (a). Content and size of the microencapsulated epoxy: 10 wt% and 40 μm, respectively. Content of $CuBr_2(2\text{-MeIm})_4$: 2 wt%.
Source: Yin et al. [56], Figure 1, p. 3/with permission of Elsevier.

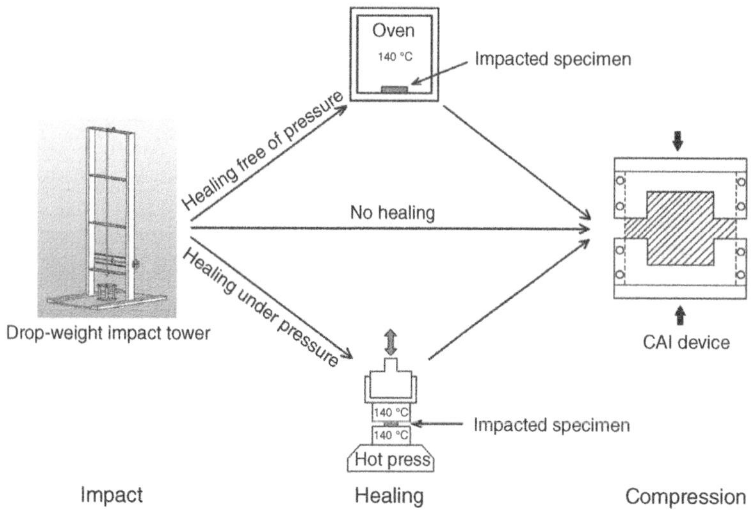

Figure 4.22 Flowchart of the damage healing procedures and CAI tests. *Source:* Reprinted from Yin et al. [56]. Copyright 2009, with permission from Elsevier.

The low velocity impact of fiber-reinforced plastics has been the subject of many experimental and analytical investigations, for example [43–45, 55, 57–66]. Impact energy and laminate lay-up are the crucial factors in determining the damage modes of composites, which include matrix cracking, delamination, and fiber fracture.

Figure 4.23 shows the morphologies of the impact damage zones taken from the back surface of the composite laminates subjected to different impact energies. At low impact energy level of 1.5 J, slight matrix cracking is the main damage mode (Figure 4.23a,b). With a rise in impact energy, the visible damage size extends, and the damage mode becomes more complicated. This is consistent with the results of Hiral et al. [61]. In Figure 4.23c,d, substantial matrix cracking appears in the immediate vicinity of the impact contact area, and initial delamination is also perceivable. At 2.5 J impact, there is obvious deformation on the unimpacted part due to growth of delamination and slight fiber rupture (Figure 4.23e,f). At a higher impact energy level of 3.5 J, the dominant damage mechanisms are extensive delamination and fiber fracture (Figure 4.23g,h and the insets).

Compared with the composites containing the self-healing agent, the composites free of self-healing agent were severely damaged under the same impact condition. This suggests that the microencapsulated epoxy and the latent hardener might have helped in absorbing some impact energy.

Effect of impact energy on residual compressive strength of the composite laminates with and without self-healing agent is shown in Figure 4.24a. Although the results are somewhat scattered, they are characteristic of CAI tests of fiber reinforced composites [43, 58, 64]. The 5–10% reduction in the compressive strength at 1.5 J impact coincides with the results of Khondker et al. [65], who reported that the decrease of CAI strength at 1.75 J impact energy was about 0–10% in glass textile composites. Incorporation of the healing agent obviously enhanced the compressive strength of the composites. This is supported by the fact that the curve for composites with healing agent is above that of the conventional composites within the impact energy region of interest. This might probably result from the reduced stiffness of the matrix epoxy. Brown et al. [17] reported that Young's modulus of epoxy was decreased with a rise in content of the embedded microencapsulated DCPD. The decrease of matrix modulus could reduce the residual stress in fiber reinforced epoxy composites and help to enhance the compressive strength of the composites [67]. On the other hand, the residual compressive strength of the impacted composites decreases with increasing impact energy, which matches the increase in the damage area [44, 61].

To highlight the influence of impact energy on the residual compressive strength of the composites in an objective way, the strength data of the impacted specimens are normalized with respect to those of the virgin ones. As shown in Figure 4.24b, introduction of healing agent into the composites slightly retarded the rate of decline of the residual compressive strength with impact energy. The result coincides with the variation in both damage area and damage mode of the composites revealed by Figure 4.23. With a rise in impact energy, the influence of impact damage on the performance of the composites was more severe, so that the aforementioned effect of change in matrix property became insignificant once the impact energy reached 3.5 J (Figure 4.24b).

Compressive strength of the healed impacted laminates is plotted as a function of impact energy in Figure 4.25.

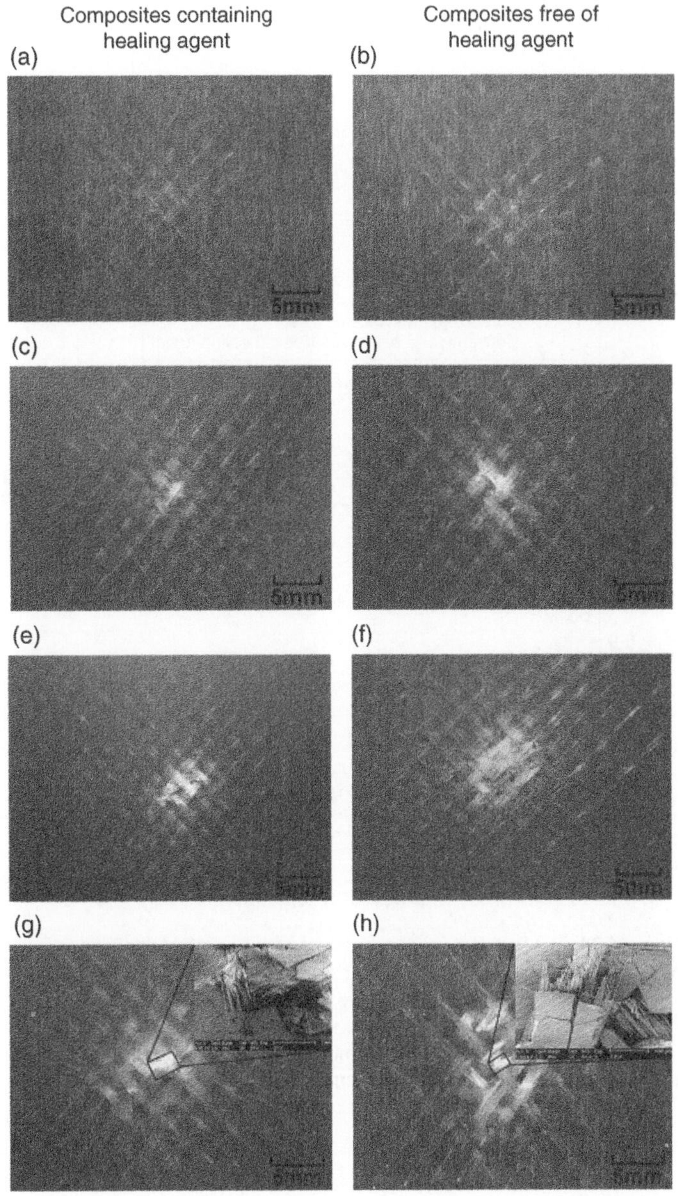

Figure 4.23 Photographs showing the impact damage zones on the woven glass fabric/epoxy composite laminates with and without the healing agent. Content and size of the microencapsulated epoxy: 10 wt% and 40 μm, respectively. Content of $CuBr_2(2\text{-MeIm})_4$: 2 wt%. Impact energy: (a, b) 1.5 J; (c, d) 2.0 J; (e, f) 2.5 J; and (g, h) 3.5 J. The insets in (g) and (h) show the broken fibers. *Source:* Yin et al. [56], Figure 3, p. 4/with permission of Elsevier.

Besides, the healing efficiency of the specimens is also calculated from Eq. (4.8) and given in Figure 4.26 to quantify the effect of crack repair.

$$\eta_{\text{CAI}} = \frac{\sigma^{\text{Healed}} - \sigma^{\text{Impact}}}{\sigma^{\text{Virgin}} - \sigma^{\text{Impact}}} \times 100\% \tag{4.8}$$

where η_{CAI} is the healing efficiency characterized by the CAI test, σ^{Healed} is the compressive strength of the impacted specimen after healing procedures, σ^{Impact} is the compressive strength of the impacted specimen, and σ^{Virgin} is the compressive strength of the virgin specimen.

At 1.5 J impact energy, σ^{Healed} values of the laminates are very close to the σ^{Virgin} values. This means that the cracks in the matrix of the laminates were almost completely healed. Accordingly, the damage zone is no longer visible in the T-scan ultrasonic images (Figure 4.27a–c), which indicate zero damage area in the repaired laminates. In fact, the healing of cracked matrix can be further shown by the simulative indentation experiment. As shown in Figure 4.28, having been healed in an oven, the split in the resin-rich region of the edge-cracked laminates is filled with consolidated healing agent that binds the cracked planes together. The porous structure of the seam might result from the pressure-free healing procedure.

When impact energy is increased, both σ^{Healed} and η_{CAI} are reduced. The main reason might be due to the change of the impact damage mode. At low impact energy level,

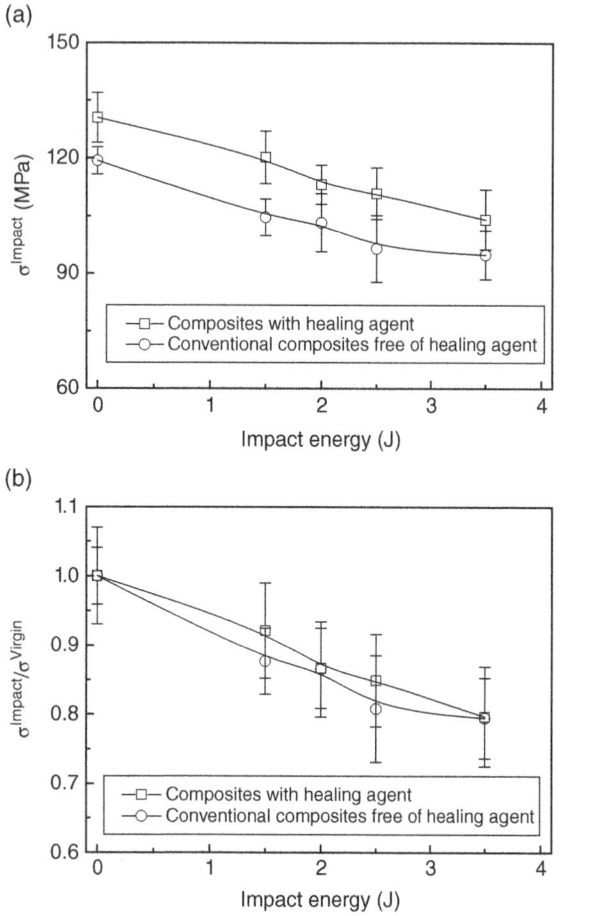

Figure 4.24 (a) Residual compressive strength and (b) normalized residual compressive strength of the composite laminates as a function of impact energy. σ^{Impact} and σ^{Virgin} denote compressive strengths of the impacted specimen and virgin specimen, respectively. Content and size of the microencapsulated epoxy: 10 wt% and 40 μm, respectively. Content of $CuBr_2(2\text{-MeIm})_4$: 2 wt%.

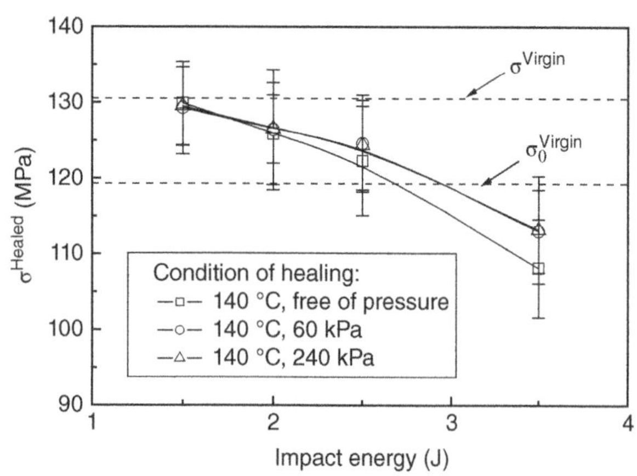

Figure 4.25 Influence of impact strength on residual compressive strength of the repaired impacted composite laminates, σ^{Healed}. The two dashed lines represent the compressive strength of the unimpacted laminates, σ^{Virgin}, and that of the virgin specimen without healing agent, σ_0^{Virgin}. Content and size of the microencapsulated epoxy: 10 wt% and 40 μm, respectively. Content of $CuBr_2(2\text{-MeIm})_4$: 2 wt%. *Source:* Reprinted from Yin et al. [56]. Copyright 2009, with permission from Elsevier.

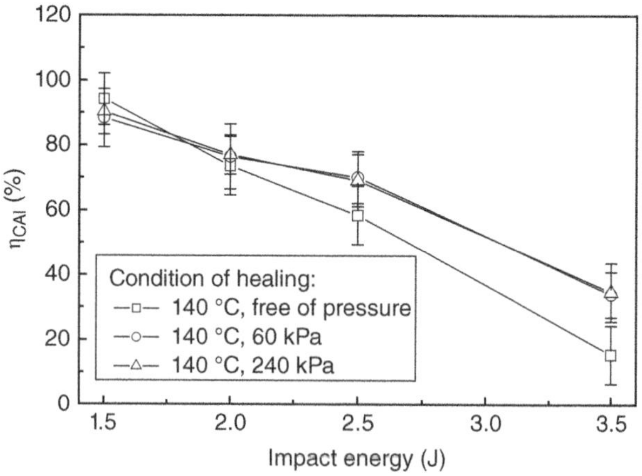

Figure 4.26 Influence of impact energy on the healing efficiency of CAI specimens, η_{CAI}, under different healing conditions. Content and size of the microencapsulated epoxy: 10 wt% and 40 μm, respectively. Content of $CuBr_2(2\text{-MeIm})_4$: 2 wt%. *Source:* Reprinted from Yin et al. [56]. Copyright 2009, with permission from Elsevier.

the predominant damage characterized by matrix cracking is on the microscopic scale. The released healing agent can easily reach the cracked places for rehabilitation. However, the damage mode is dominated by delamination and fiber breaking as the impact energy level is raised. Repair of these cracks becomes less effective (Figure 4.27d–i). First, the healing agent is unable to re-connect the broken fibers. Secondly, delivery of enough healing agent to the delaminated sites deep inside the fiber-rich regions might be difficult. As a result, the healing efficiency has to be low.

The above analysis received support from the dependence of σ^{Healed} on pressure (Figure 4.29), which was applied to the impacted specimens during healing treatment (Figure 4.22). The aid of lateral pressure can close the cracked gaps in the course of repair, so that the amount of the healing agent required for filling the splits would be less compared with the healing free of lateral pressure. As the impact damage is directly related to the impact energy (Figure 4.27), the effect of lateral pressure on healing efficiency is more prominent in the laminates subjected to higher impact energy. Therefore, when the lateral pressure is raised from 0 to 60 kPa, the increment in σ^{Healed} of the laminates that experienced a 3.5 J impact is greater than that of the laminates that experienced a 2.5 J impact (Figure 4.29).

On the other hand, Figure 4.29 indicates that σ^{Healed} of all the composite laminates is unchanged with a rise in pressure from 60 to 240 kPa. This shows that the external pressure is irrelevant to the mechanism of healing but closes the two crack surfaces. It is interesting to note that σ^{Healed} of the specimens subjected to 1.5 J impact has nothing to do with the pressure. Evidently, absence of severe damage like delamination and fiber fracture should explain this phenomenon. It can thus be concluded that applying

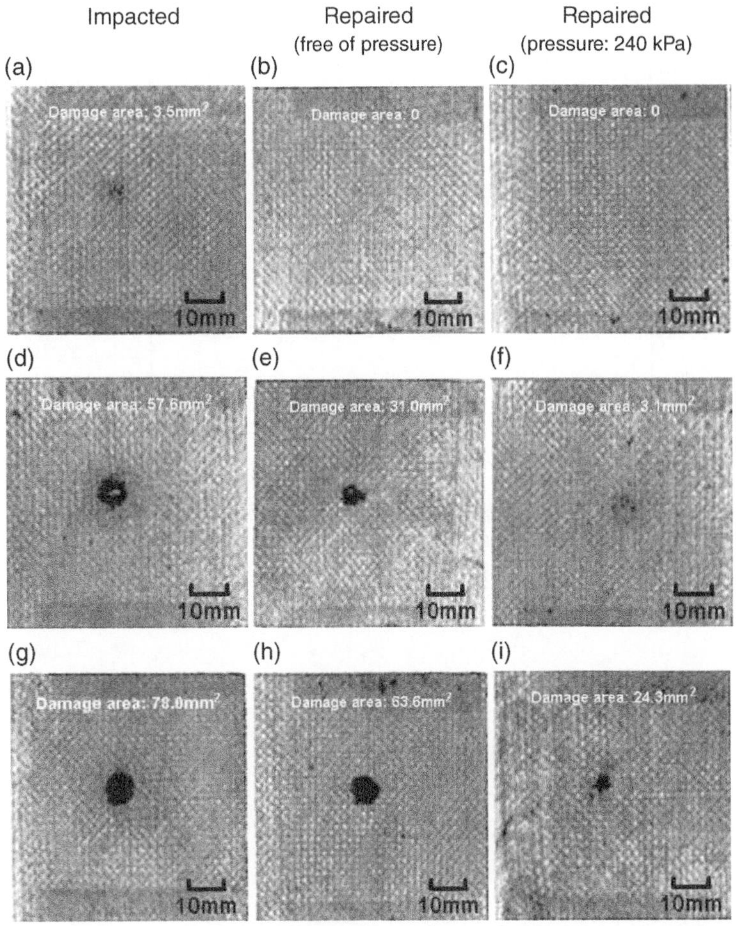

Figure 4.27 T-scan ultrasonic images of the impacted repairable composite laminates before and after being healed. Content and size of the microencapsulated epoxy: 10 wt% and 40 μm, respectively. Content of $CuBr_2(2\text{-MeIm})_4$: 2 wt%. Impact energy: (a, b, c) 1.5 J; (d, e, f) 2.5 J; and (g, h, i) 3.5 J. *Source:* Yin et al. [56], Figure 8, p. 6/with permission of Elsevier.

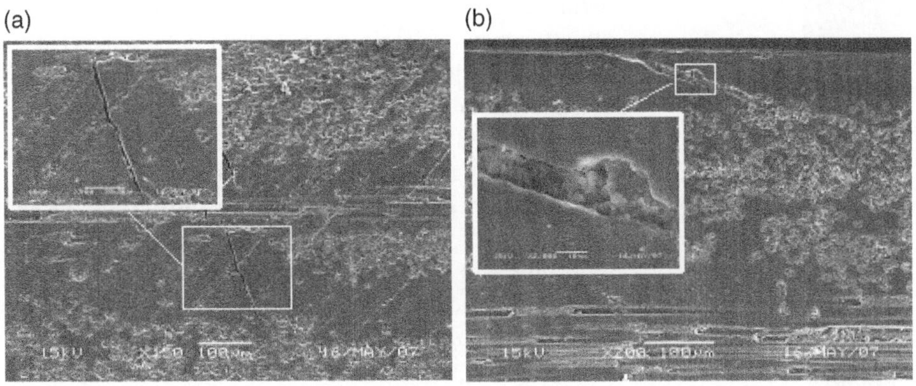

Figure 4.28 SEM side views of (a) indented composite laminates and (b) indented composite laminates healed in an oven at 140 °C for 30 minutes. Here the indentation experiments were done to simulate preliminary damage formation due to low velocity impact. This is because it is very hard to section the impacted specimens without introducing artifacts. Content and size of the microencapsulated epoxy: 10 wt% and 40 μm, respectively. Content of $CuBr_2(2\text{-MeIm})_4$: 2 wt%. *Source:* Yin et al. [56], Figure 9, p. 6/with permission of Elsevier.

pressure on the specimen during healing is necessary for composites with severe damage to improve the efficiency of repairing.

In fact, failure modes of the CAI specimens are greatly affected by the healing. As illustrated in Figure 4.30a, failure of the impacted composite laminates is primarily caused by the crack propagating from the existing damage on the back face [61]. The fracture plane is approximately perpendicular to the compression direction (Figure 4.30b). When the impacted specimen is healed and then subjected to compression, however, shear buckling takes place as in conventional laminates [68]. The fracture plane makes an angle of about 45° with the loading direction and the broken parts become non-coplanar

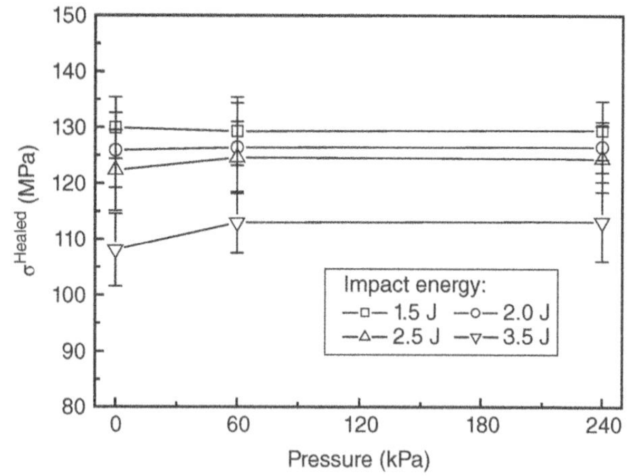

Figure 4.29 Influence of lateral pressure on σ^{Healed} of the composite laminates. Content and size of the microencapsulated epoxy: 10 wt% and 40 μm, respectively. Content of $CuBr_2$(2-MeIm)$_4$: 2 wt%. *Source:* Reprinted from Yin et al. [56]. Copyright 2009, with permission from Elsevier.

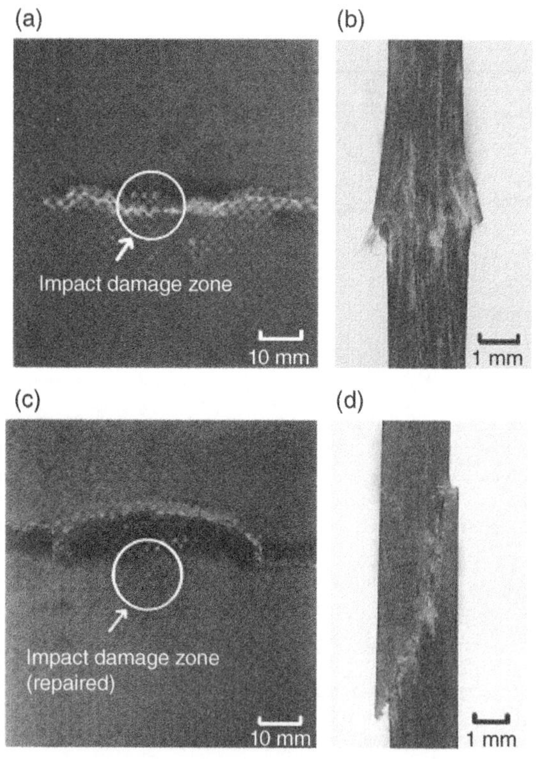

Figure 4.30 Photographs of (a, b) composite laminates that were impacted at 2.0 J and then compressed to failure, and (c, d) composite laminates that were impacted at 2.0 J, repaired in a hot press under 60 kPa at 140 °C for 30 minutes and then compressed to failure. (a, c) Front views; and (b, d) side views. Content and size of the microencapsulated epoxy: 10 wt% and 40 μm, respectively. Content of $CuBr_2$(2-MeIm)$_4$: 2 wt%. *Source:* Yin et al. [56], Figure 11, p. 7/with permission of Elsevier.

(Figure 4.30c,d), implying that the compression failure starts with the kink zone formation [69] that propagates into the fracture. This demonstrates that the impact damage must have been recovered to a great extent prior to the compression test. In addition, Figure 4.30c indicates that the compression failure site is not located in the original impact damage zone (i.e. the central region) of the healed specimen, which is different from the case of Figure 4.30a. The mechanism is unclear. As the average facture toughness of the bulk epoxy cured by 2E4MIm (i.e. the cured matrix of the laminates) is 0.46 MPa m$^{1/2}$, which is only 57% of the bulk epoxy cured by $CuBr_2$(2-MeIm)$_4$ (i.e. the cured healing agent), it might be inferred that the higher fracture toughness of the healed portion than the matrix leads to higher resistance to compressive failure of the former. Consequently, the ultimate compression failure has to occur in a place other than the healed portion. Further study is needed to prove this.

The effect of content of the healing agent is shown in Table 4.6. When the composite laminates do not contain any healing agent, compressive strengths of the impacted specimens are unchanged after treatment with either heat or heat plus pressure, so that $\eta_{CAI} \approx 0$. The phenomenon is also found for the composite laminates with microencapsulated epoxy alone. This proves that the host matrix resin of the composites has no healing capability and the recovery in mechanical properties is due to the reaction between the epoxy released from the broken microcapsules and the latent hardener. On the other hand, the data in Table 4.6 show that the increase in healing efficiency is marginal for the laminates subjected to 1.5 J impact when the content of the microencapsulated epoxy is raised from 10 to 20 wt% at a fixed loading of $CuBr_2$(2-MeIm)$_4$. This implies that 10 wt% microencapsulated epoxy might be enough for repairing the small-scale damage generated by the low impact energy. In contrast, in the case of impact energy of 2.5 J, the increase in healing efficiency is obvious with a rise in the content of epoxy-loaded microcapsules from 10 to 20 wt% whether external pressure is applied or not during the healing process. Evidently, 10 wt% microencapsulated epoxy is insufficient to heal the greater damage under these circumstances. The increased amount of released epoxy ensures much more crack volume can be filled up by the binder.

Besides healing agent content, the influence of the size of the microcapsules containing epoxy on the healing efficiency was also studied. With increasing the microcapsule size from 40 to 140 μm at a given healing agent content, the healing efficiency does not change much for the laminates impacted at the same energy level (Figure 4.31). This might

Table 4.6 Healing efficiencies of the composite laminates.

Variables		Impacted at 1.5 J		Impacted at 2.5 J	
Microcapsules[a] (wt%)	Catalyst (wt%)	η_{CAI}[b] (%)	η_{CAI}[c] (%)	η_{CAI}[b] (%)	η_{CAI}[c] (%)
0	0	~0	~0	~0	~0
10	0	~0	~0	~0	~0
10	2	94±8	94±9	58±9	70±8
20	2	100±7	99±8	71±6	90±8

[a] Average diameter: 40 μm.
[b] Conditions of crack repair: pressure free, 140 °C, 30 minutes.
[c] Conditions of crack repair: 60 kPa, 140 °C, 30 minutes.
Source: Reprinted from Yin et al. [56]. Copyright 2009, with permission from Elsevier.

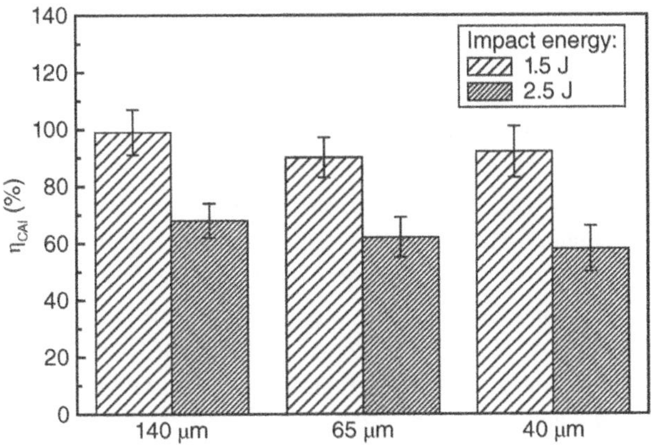

Figure 4.31 Influence of size of the microcapsules containing epoxy on healing efficiency of CAI specimens. Contents of the microencapsulated epoxy and $CuBr_2(2\text{-MeIm})_4$: 10 and 2 wt%, respectively. Conditions of crack repair: pressure free, 140 °C, 30 minutes.

be the result of the effects of two factors. The larger capsules contain more epoxy (Table 4.5), and their rupture will deliver more healing agent to the cracked surfaces as compared with the smaller ones. On the other hand, the smaller capsules might be able to penetrate into the fiber-rich regions where the larger capsules cannot and heal cracks there. Nevertheless, much more work should be done before coming to any conclusion.

4.6 Concluding Remarks

To prepare self-healing epoxy materials for advanced engineering applications, high temperature healing agent has to be developed. Accordingly, anionic polymerization is introduced as the healing reaction by formulating a two-component healing agent consisting of epoxy-loaded microcapsules and imidazole latent hardener. The latent hardener possesses long-term stability and is hardly affected by the surrounding environment. Moreover, it can be well pre-dissolved in the matrix of the uncured composites, leading to homogenous distribution of the reagent on a molecular scale. The epoxy released from the ruptured microcapsules can thus meet the latent hardener everywhere.

The healing system proves to be workable in both neat epoxy and fiber reinforced epoxy composites. The one-capsule strategy remarkably reduces the content of healing microcapsules required by the dual-capsule approach mentioned in Chapters 2 and 3. The possible deterioration of the original properties of the composites due to incorporation of the capsules is thus limited.

Since elevated temperature is essential for many advanced engineering applications, healing of internal microcracks in terms of the present healing system might be automatically completed during service without manual intervention in the case where the environmental temperature is high enough to initiate polymerization of the healing agent.

References

1 Sinclair, J.W. (1992). Effects of cure temperature on epoxy resin properties. *The Journal of Adhesion* 38: 219–234.
2 Min, B.-G., Hodgkin, J.H., and Stachurski, Z.H. (1993). The dependence of fracture properties on cure temperature in a DGEBA/DDS epoxy system. *Journal of Applied Polymer Science* 48: 1303–1312.
3 Dowbenko, R., Anderson, C.C., and Chang, W.H. (1971). Imidazole complexes as hardeners for epoxy adhesives.

Industrial & Engineering Chemistry Product Research and Development 10: 344–351.

4 Ibonai, M. and Kuramochi, T. (1975). Curing of epoxy resin by use of imidazole/metal complexs. *Purasuchikkusu* 26 (7): 69–73. (in Japanese).

5 Bi, C.H., Gan, C.L., and Zhao, S.Q. (1997). Study on synthesis, curing reaction and properties of imidazole salt. *Thermosetting Resin* 12 (1): 12–15. (in Chinese).

6 Rong, M.Z., Zhang, M.Q., and Zhang, W. (2007). A novel self-healing epoxy system with microencapsulated epoxy and imidazole curing agent. *Advanced Composites Letters* 16: 167–172.

7 Su, Z. and Ye, L. (2004). An intelligent signal processing and pattern recognition technique for defect identification using an active sensor network. *Smart Materials and Structures* 13: 957–969.

8 Fernando, G.F., Degamber, B., Wang, L.W. et al. (2004). Self-sensing fibre reinforced composites. *Advanced Composites Letters* 13: 123–129.

9 Yin, T., Rong, M.Z., Zhang, M.Q., and Yang, G.C. (2007). Self-healing epoxy composites – preparation and effect of the healant consisting of microencapsulated epoxy and latent curing agent. *Composites Science & Technology* 67: 201–212.

10 Gardner, G.L. (1966). Manufacturing encapsulated products. *Chemical Engineering Progress* 62 (4): 87–91.

11 Fanger, G.O. (1974). What good are microcapsules? *Chemtech* 7: 397–405.

12 Song, J., Chen, L., and Li, X.J. (2001). *Microencapsulation and Its Application*. Beijing: Chemical Engineering Press (in Chinese).

13 Yuan, L., Liang, G.Z., Xie, J.Q. et al. (2006). Preparation and characterization of poly(urea-formaldehyde) microcapsules filled with epoxy resins. *Polymer* 47: 5338–5349.

14 Eilbeck, W.J., Holmes, F., Taylor, C.E., and Underhill, A.E. (1968). Cobalt (II), nickel (II), and copper (II) complexes of 2-methylimidazole. *Journal of the Chemical Society* 1: 128–132.

15 Kissinger, H.E. (1957). Reaction kinetics in differential thermal analysis. *Analytical Chemistry* 29: 1702–1706.

16 Crane, L.W., Dynes, P.J., and Kaelble, D.H. (1973). Analysis of curing kinetics in polymer composites. *Journal of Polymer Science – Polymer Letters Edition* 11: 533–540.

17 Brown, E.N., White, S.R., and Sottos, N.R. (2004). Microcapsule induced toughening in a self-healing polymer composite. *Journal of Materials Science* 39: 1703–1710.

18 Ahmed, A. and Jones, F.R. (1990). A review of particulate reinforcement theories for polymer composites. *Journal of Materials Science* 25: 4933–4942.

19 Kinloch, A.J. and Young, R.J. (1983). *Fracture Behaviour of Polymers*. London: Applied Science Publishers.

20 Jancar, J., Dianselmo, A., and Dibenedetto, A.T. (1992). The yield strength of particulate reinforced thermoplastic composites. *Polymer Engineering and Science* 32: 1394–1399.

21 Rong, M.Z., Zhang, M.Q., Zheng, Y.X. et al. (2000). Irradiation graft polymerization on nano-inorganic particles: an effective means to design polymer based nanocomposites. *Journal of Materials Science Letters* 19: 1159–1161.

22 White, S.R., Sottos, N.R., Geubelle, P.H. et al. (2001). Autonomic healing of polymer composites. *Nature* 409: 794–797.

23 Yuan, Y.C., Rong, M.Z., Zhang, M.Q. et al. (2008). Self-healing polymeric materials using epoxy/mercaptan as the healant. *Macromolecules* 41: 5197–5202.

24 Yin, T., Zhou, L., Rong, M.Z., and Zhang, M.Q. (2008). Self-healing woven glass fabric/epoxy composites with the healant consisting of micro-encapsulated epoxy and latent curing agent. *Smart Materials and Structures* 17: 015019.

25 Hull, D. (1981). *An Introduction to Composite Materials*. New York: Cambridge University Press.

26 Leung, C.K.Y. and Li, V.C. (1990). Applications of a two-way debonding theory to short fibre composites. *Composites* 21: 305–317.

27 Bazhenov, S.L. (1991). Strong bending in the DCB interlaminar test of thin, E-glass woven-fabric-reinforced laminates. *Composites* 22: 275–280.

28 Wang, Y. and Zhao, D. (1995). Characterization of interlaminar facture behaviour of woven fabric reinforced polymeric composites. *Composites* 26: 115–124.

29 Kotaki, M. and Hamada, H. (1997). Effect of interfacial properties and weave structure on mode I interlaminar facture toughness behaviour of glass satin woven fabric composites. *Composites Part A: Applied Science and Manufacturing* 28: 257–266.

30 Kuboki, T., Jar, P.Y.B., and Cheng, J.J.R. (2004). Interlaminar fracture toughness and the associated fracture behavior for glass fiber reinforced polymers (GFRP). *Journal of Materials Science* 39: 1419–1423.

31 Kessler, M.R., Sottos, N.R., and White, S.R. (2003). Self-healing structural composite materials. *Composites Part A: Applied Science and Manufacturing* 34: 743–753.

32 Kessler, M.R. and White, S.R. (2001). Self-activated healing of delamination damage in woven composites. *Composites Part A: Applied Science and Manufacturing* 32: 683–699.

33 Yin, T., Rong, M.Z., Zhang, M.Q., and Zhao, J.Q. (2009). Durability of self-healing woven glass fabric/epoxy composites. *Smart Materials and Structures* 18: 074001.

34 Rule, J., Brown, E.N., Sottos, N.R. et al. (2005). Wax-protected catalyst microspheres for efficient self-healing materials. *Advanced Materials* 72: 205–208.

35 Jones, A.S., Rule, J.D., Moore, J.S. et al. (2007). Life extension of self-healing polymers with rapidly growing fatigue cracks. *Journal of the Royal Society Interface* 4: 395–403.

36 Jones, A.S., Rule, J.D., Moore, J.S. et al. (2006). Catalyst morphology and dissolution kinetics of self-healing polymers. *Chemistry of Materials* 18: 1312–1317.

37 Pang, W.C. and Bond, I.P. (2005). A hollow fibre reinforced polymer composite encompassing self-healing and enhanced damage visibility. *Composites Science & Technology* 65: 1791–1799.

38 Pratt, T.J., Johns, W.E., Rammon, R.M., and Plagemann, W.L. (1985). A novel concept on the structure of cured urea-formaldehyde resin. *Journal of Adhesion* 17: 275–295.

39 White, S.R., Mather, P.T., and Smith, M.J. (2002). Characterization of the cure-state of DGEBA-DDS epoxy using ultrasonic, dynamic mechanical, and thermal probes. *Polymer Engineering and Science* 42: 51–67.

40 Steeman, P.A.M., Dias, A.A., Wienke, D., and Zwartkruis, T. (2004). Polymerization and network formation of UV-curable systems monitored by hyphenated real-time dynamic mechanical analysis and near-infrared spectroscopy. *Macromolecules* 37: 7001–7007.

41 Wu, L.X., Hoa, S.V., Tan, M., and That, T. (2006). Effects of composition of hardener on the curing and aging for an epoxy resin system. *Journal of Applied Polymer Science* 99: 580–588.

42 Horton, R.E. and McCarty, J.E. (1987). Damage tolerance of composites. In: *Engineered Materials Handbook*, vol. 1 (ed. T.J. Reinhart), 259–267. Metals Park, OH: ASM International.

43 Soutis, C. and Curtis, P.T. (1996). Prediction of the post-impact compressive strength of CFRP laminated composites. *Composites Science & Technology* 56: 677–684.

44 Pritchard, J.C. and Hogg, P.J. (1990). The role of impact damage in post-impact compression testing. *Composites Part A: Applied Science and Manufacturing* 21: 503–511.

45 Baker, A.A., Jones, R., and Callinan, R.J. (1985). Damage tolerance of graphites/epoxy composites. *Composites Structures* 4: 15–44.

46 Motuku, M., Janowski, C.M., and Vaidya, U.K. (1999). Parametric studies on self-repairing approaches for resin infused composites subjected to low velocity impact. *Smart Materials and Structures* 8: 623–638.

47 Bleay, S.M., Loader, C.B., Hawyes, V.J. et al. (2001). A smart repair system for polymer matrix composites. *Composites Part A: Applied Science and Manufacturing* 32: 1767–1776.

48 Hayes, S.A., Jones, F.R., Marshiya, K., and Zhang, W. (2007). A self-healing thermosetting composite material. *Composites Part A: Applied Science and Manufacturing* 38: 1116–1120.

49 Jones, F.R., Hayes, S.A., and Zhang, W. (2007). Self-healing of damage in fibre-reinforced polymer-matrix composites. *Journal of the Royal Society Interface* 4: 381–387.

50 Pang, W.C. and Bond, I.P. (2005). Bleeding composites' damage detection and self-repair using a biomimetic approach. *Composites Part A: Applied Science and Manufacturing* 36: 183–188.

51 Trask, R.S. and Bond, I.P. (2006). Biomimetic self-healing of advanced composite structures using hollow glass fibres. *Smart Materials and Structures* 15: 704–710.

52 Trask, R.S., Williams, G.J., and Bond, I.P. (2007). Bioinspired self-healing of advanced composite structures using hollow glass fibres. *Journal of the Royal Society Interface* 4: 363–371.

53 Williams, G., Trask, R.S., and Bond, I.P. (2007). A self-healing carbon fibre reinforced polymer for aerospace applications. *Composites Part A: Applied Science and Manufacturing* 38: 1525–1532.

54 Patel, A.J., Sottos, N.R., Wetzel, E.D., and White, S.R. (2010). Automatic healing of low-velocity impact damage in fiber-reinforced composites. *Composites Part A: Applied Science and Manufacturing* 41: 360–368.

55 Sánchez-Sáez, S., Barbero, E., Zaera, R., Navarro, C. (2005) Compression after impact of thin composite laminates. *Composites Science & Technology*, 65, 1911–1919.

56 Yin, T., Rong, M.Z., Wu, J.S. et al. (2009). Healing of impact damage in woven glass fabric reinforced epoxy composites. *Composites Part A: Applied Science and Manufacturing* 39: 1479–1487.

57 Bogdanovich, A.E. and Friedrich, K. (1994). Initial and progressive failure analysis of laminated composite structures under dynamic loading. *Composites Structures* 27: 439–456.

58 Davies, G.A.O., Hitchings, D., and Zhou, G. (1996). Impact damage and residual strengths of woven fabric glass/polyester laminates. *Composites Part A: Applied Science and Manufacturing* 27: 1147–1156.

59 Shen, W.Q. (1997). Dynamic response of rectangular plates under drop mass impact. *International Journal of Impact Engineering* 19: 207–229.

60 Naik, N.K. and Sekher, Y.C. (1998). Damage in laminated composites due to low velocity impact. *Journal of Reinforced Plastics and Composites* 17: 1232–1263.

61 Hiral, Y., Hamada, H., and Kim, J.K. (1998). Impact response of woven glass-fabric composites – I. Effect of fibre surface treatment. *Composites Science & Technology* 58: 91–104.

62 Naik, N.K., Sekher, Y.C., and Meduri, S. (2000). Damage in woven-fabric composites subjected to low-velocity impact. *Composites Science & Technology* 60: 731–744.

63 Kim, J.K. and Sham, M.L. (2000). Impact and delamination failure of woven-fabric composites. *Composites Science & Technology* 60: 745–761.

64 Cartié, D. and Irving, P. (2002). Effect of resin and fibre properties on impact and compression after impact performance of CFRP. *Composites Part A: Applied Science and Manufacturing* 33: 483–493.

65 Khondker, O.A., Herszberg, I., and Hamada, H. (2004). Measurements and prediction of the compression-after-impact strength of glass knitted textile composites. *Composites Part A: Applied Science and Manufacturing* 35: 145–157.

66 Baucom, J.N., Zikry, M.A., and Rajendran, A.M. (2006). Low-velocity impact damage accumulation in woven S2-glass composite systems. *Composites Science & Technology* 66: 1229–1238.

67 Wang, H.M. and Liang, W.R. (1992). Effect of properties of epoxy resins on compression failure characteristics of composites. *Thermoset Resin* 1: 31–35. (in Chinese).

68 Chou, T.W. (1993). *Structure and Properties of Composites*. Weinheim: VCH.

69 Carlsson, L.A. and Pipes, R.B. (1997). *Experimental Characterization of Advanced Composite Materials*. Lancaster: Technomic Publishing.

5

Extrinsic Self-Healing Via Miscellaneous Reactions

CHAPTER MENU

In Chapters 2–4, the two-component healing agent was introduced. No matter whether the catalysts or curing agents are encapsulated, or loaded on porous media, or dissolved in the matrix, there is a common ground for these systems. That is, encapsulated epoxy monomer serves as the polymerizable component. In spite of the advantage of epoxy represented by its superior adhesive power that is critical for rebinding cracked faces in damaged materials, relatively poor flowability of the epoxy monomer is a disadvantage in the case of self-healing. On account of its higher molecular weight, epoxy monomer has higher viscosity at ambient temperature. When it is released from ruptured capsules, spreading of fluid over the cracked planes through capillary effect is limited. As a result, not only speed of healing but also healing efficiency related to the coverage of the healing membrane might not be sufficiently satisfactory. The challenge would become more obvious under low temperature since the viscosity of epoxy fluid is highly temperature dependent. In addition, when we look at the issue more carefully, it is known that the higher molecular weight of epoxy monomer would lead to poor wetting on the fracture surface. The wetting stage is an essential step for most of the mechanical property recovery or healing [1–3], which helps to build up an interface in advance. Insufficient wetting certainly disfavors intimate contact between healant molecules and the cracked face. The number of bonds established per unit area at the liquid–solid interface would thus be reduced.

In consideration of the above problems, we believed that utilization of small molecule monomers as a substitute for epoxy is a possible solution, besides addition of specific thinners into epoxy monomer. These candidate monomers should have lower molecular weight than epoxy, and hence lower viscosity. In the case where they replace epoxy monomer in the microcapsules, the healant molecules can quickly travel throughout the cracked planes with good wetting after crack-induced breakage of the containers. They can also easily diffuse deeply into the subsurface so that much more matrix volume is involved in the subsequent crack repair. The resultant joints would be robust accordingly. More importantly, these small molecules possess very low freezing point. This means that their viscosities are nearly temperature independent over a wide range

Extrinsic and Intrinsic Approaches to Self-Healing Polymers and Polymer Composites, First Edition. Ming Qiu Zhang and Min Zhi Rong.

of temperature in a cold environment, which is desired for low temperature self-healing.

It is worth noting that the reactivity of the proposed small molecule monomers is another important factor in addition to their flowability and wettability. It directly determines the formation of new bonds at the cracked faces. In this context, chemicals having carbon–carbon unsaturated double bonds might meet the criteria stated above. As hetero atoms can be added across the double bonds, these compounds are very exciting and reactive. Unsaturated carbon compounds are commercially much more valuable than saturated ones for this reason.

On the basis of these analyses, small molecule monomers such as glycidyl methacrylate (GMA) and styrene are chosen to play the role of healing agent hereinafter, within the microcapsule strategy framework [4]. The specific reaction mechanisms and the effects on crack healing will be revealed. It is hoped that the spectrum of healant species would be broadened accordingly.

5.1 Extrinsic Self-Healing Via Nucleophilic Addition and Ring-Opening Reactions

In this subsection, GMA is employed as the healing agent for epoxy materials [5]. GMA is non-toxic and a low viscosity liquid chemical at ambient temperature (boiling point = 189 °C), possessing reactive C=C bonds and epoxide groups (Scheme 5.1). It intrinsically combines the characteristics of both solvent and bifunctional monomer and is miscible with epoxy resin. In this context, when GMA acts as healant for healing cracks in epoxy material, two healing actions, namely, solvent effect and chemical reactions (i.e. nucleophilic addition and ring-opening reactions between GMA and the residual amine in the matrix), would be integrated. Both would help to reconnect the separated faces during crack healing.

Recently, Caruso et al. demonstrated an approach that employed encapsulated organic solvents as healing agent for cured epoxy [6]. The solvents locally swell the matrix, allowing accessibility of residual amines and further crosslinking with residual epoxy functionality. In addition to the unique working principle that is different from the commonly used polymerization, such a one-capsule design obviously simplifies the fabrication of self-healing materials [7]. In their subsequent work [8], a mixture of epoxy monomer and solvent replaced pure solvent as core substance of the microspheres. Owing to the residual amine in the partially cured epoxy matrix, epoxy monomer released from the epoxy-solvent microcapsules can be cured and adhered to the original matrix interface. Consequently, healing efficiency is significantly higher than that offered by pure solvent capsules, as reflected by complete recovery of fracture toughness.

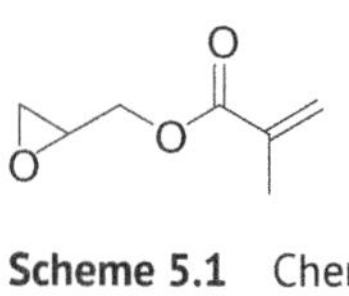

Scheme 5.1 Chemical structure of GMA.

When comparing our approach with that of Caruso et al. using mixture of epoxy and solvent [8], it is known that GMA might simplify the latter's recipe due to its specific structural features, while their one-capsule strategy is still applicable. Accordingly, the conventional two-part healing system of polymerizable monomer and its catalysts or curing agents, which used to be encapsulated in the form of dual capsules [9–11], is also simplified.

5.1.1 Microencapsulation of GMA by In-Situ Polymerization

In-situ emulsion polymerization was applied to encapsulate GMA with poly(melamine-formaldehyde) (PMF) as the wall substance. The typical synthesis procedure is described as follows: 0.06 mol melamine and 0.2 mol formaldehyde were added to a three-neck flask containing 20 ml distilled water. Then, the pH of the solution was adjusted to 8.5–9.0 by incorporation of 2 wt% NaOH solution under mechanical stirring. The system was heated to 65–70 °C. After 30 minutes, prepolymer of melamine and formaldehyde was yielded.

Then, 100 ml sodium dodecyl benzenesulfonate (SDBS, used as the emulsifier) and poly(vinyl alcohol) (PVA) aqueous solution and 25.2 ml GMA were added to the above prepolymer solution, and the mixture was agitated for 30 minutes to get a stable emulsifiable solution. PVA was used to protect the colloid. Drops of 1-octanol were added to the solution to remove bubbles resulting from mechanical stirring. Afterwards, the pH of the emulsifiable solution was slowly adjusted to 3.8–4.1 with 5 wt% acetic acid solution. The reaction was conducted at 65 °C for three hours. Eventually, PMF walled microcapsules containing GMA were produced. The resultants were washed with water to remove the rest of the GMA and the PMF particles attached to the surface of the capsules, and dried at room temperature for 24 hours. The yield of the microcapsules was defined by the ratio of the net mass of GMA core of the collected capsules to the feeding mass of the core material. Table 5.1 lists the influence of processing parameters on the resultant microcapsules.

Since encapsulated GMA is planned to serve as a single-component healant for epoxy materials, preparation of

Table 5.1 Descriptions of PMF-walled microcapsules containing GMA prepared using different processing parameters.

Sample ID	Dispersion rate (rpm)	Feeding weight ratio of core/shell monomers	Emulsifier content (wt%)		Reaction temperature (°C)	Average diameter (μm)	Core content (%) (by extraction)	Yield (wt%)
			SDBS	PVA				
1-1	200	2	1	0.3	65	291.8	95.6	28.9
1-2	300	2	1	0.3	65	193.8	95.2	37.6
1-3	400	2	1	0.3	65	107.6	81.7	22.0
1-4	500	2	1	0.3	65	93.9	79.5	24.6
2-1	300	1	1	0.3	65	182.7	95.5	9.3
2-2	300	2	1	0.3	65	193.8	95.2	37.6
2-3	300	3	1	0.3	65	193.2	92.4	39.0
3-1	300	2	0.5	0.3	65	183.3	95.0	9.7
3-2	300	2	1	0.3	65	193.8	95.2	37.6
3-3	300	2	2	0.3	65	168.7	94.9	31.4
4-1	300	2	1	0.1	65	183.8	92.9	18.0
4-2	300	2	1	0.2	65	210.3	95.6	31. 7
4-3	300	2	1	0.3	65	193.8	95.2	37.6
5-1	300	2	1	0.3	40	–	–	–
5-2	300	2	1	0.3	50	164.9	89.0	18.9
5-3	300	2	1	0.3	65	193.8	95.2	37.6

Source: Reprinted from Meng et al. [5]. Copyright 2010, with permission from the Royal Society of Chemistry.

GMA-loaded microcapsules should be first discussed. The data in Table 5.1 indicate that both core content and average diameter of the microcapsules are not sensitive to the processing parameters except the rate of dispersion. Such a wider processing window would be advantageous for future industrial-scale production. In general, the droplets of GMA in the emulsible solution become smaller with increasing dispersion rate (cf. samples 1-1–1-4), so that the resultant microcapsules have smaller size and narrower size distribution, and the relative quantity of the core material also has to be reduced. Considering that the diameter of the capsules produced in most cases is about 190 μm (Table 5.1), the representative spheres (sample 1-2) are employed for the fabrication of the self-healing epoxy.

The yield of microcapsules is critical for productivity. It is seen that when the feeding weight ratio of core/shell monomers is lower (1, for example), the yield is as low as 9.3 wt% (sample 2-1). That is because GMA can slightly dissolve in water and the solubility is increased with a rise in temperature. In the case of lower feeding ratio of core/shell monomers, the dissolved GMA makes up a larger portion of the total amount of available GMA and there is insufficient GMA in the emulsible solution to be encapsulated. The problem can be solved by simply raising the feeding ratio of core/shell monomers (cf. samples 2-1–2-3). A similar situation can be found for sample 3-1, where the dosage of the emulsification agent SDBS is insufficient. As stable droplets of GMA in the system cannot be achieved, it is hard to form the microcapsules. Eventually, only a few microcapsules are gained. In this context, higher SDBS content favors the production of capsules (cf. samples 3-1–3-3). It is worth noting that the microcapsules would adhere to each other in the course of polycondensation owing to the generation of sticky oligomers from the reaction of formaldehyde, melamine and the precondensate. The PVA in the emulsifiable solution protects the colloids and prevents the newly formed capsules from sticking together. In the case of higher concentration of PVA in the emulsifiable solution, the microcapsules become more stable, and higher yield of the capsules is achieved (cf. samples 4-1–4-3).

In the process of microencapsulation, the pre-condensate of formaldehyde and melamine forms low molecular oligomers in the acidic water phase. Meanwhile, the oligomers gather at the boundary between the GMA droplets and the aqueous volume phase. Therefore, reaction temperature has a significant effect on the polymerization [12]. Here, it is found that the encapsulation fails to provide microcapsules at 40 °C (sample 5-1). Optical microscopic observation indicates that only PMF particles and microcapsule fragments are available (Figure 5.1a). This should be due to the slow reaction rate at the relatively lower temperature. The shell wall cannot form quickly enough to sustain the collision between the capsules during the manufacturing associated with drastic stirring. At a higher temperature of 50 °C, GMA-loaded microcapsules can be successfully made (sample 5-2), and the resultants have a good appearance (Figure 5.1b). When the reaction temperature is further increased to 65 °C (sample 5-3), the yield of microcapsules is greater than that produced at 50 °C, but some of the capsules are not perfect spheres (Figure 5.1c). A possible reason for the partial collapse of the microcapsules might arise from the collision among the capsules when the stiffness of the wall is low at the beginning of capsule formation owing to lower cross-link density.

In addition to the above analysis of capsule manufacture, the structure and properties of the capsules should be studied before application. Figure 5.2 shows the Fourier transform infrared (FTIR) spectrum of the GMA-loaded microcapsules in comparison with those of pristine GMA and the acetone-extracted PMF wall. The spectrum of GMA indicates C=O stretching at 1720 cm^{-1} and epoxide stretching at 908 cm^{-1}. As for the extracted PMF, its spectrum exhibits a broad stretching vibration of N—H and O—H at about 3332 cm^{-1}, stretching of C—N at 1348–1592 cm^{-1}, and a stretching peak of the triazine ring at 814 cm^{-1}. From the spectrum of GMA-loaded capsules, it is seen that all the aforementioned characteristic peaks

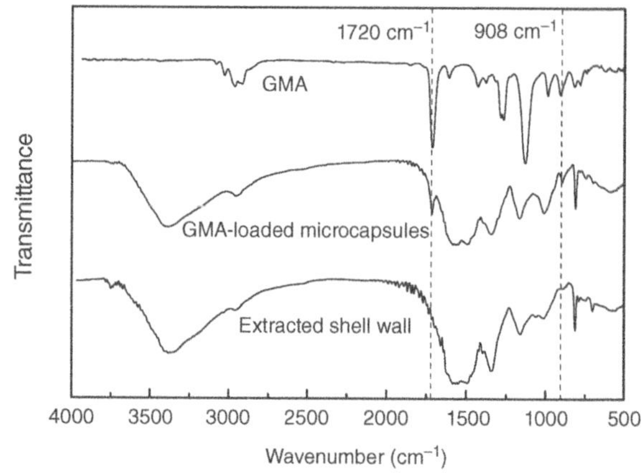

Figure 5.2 FTIR spectrum of PMF-walled microcapsules containing GMA in comparison with those of GMA and acetone-extracted shell wall.

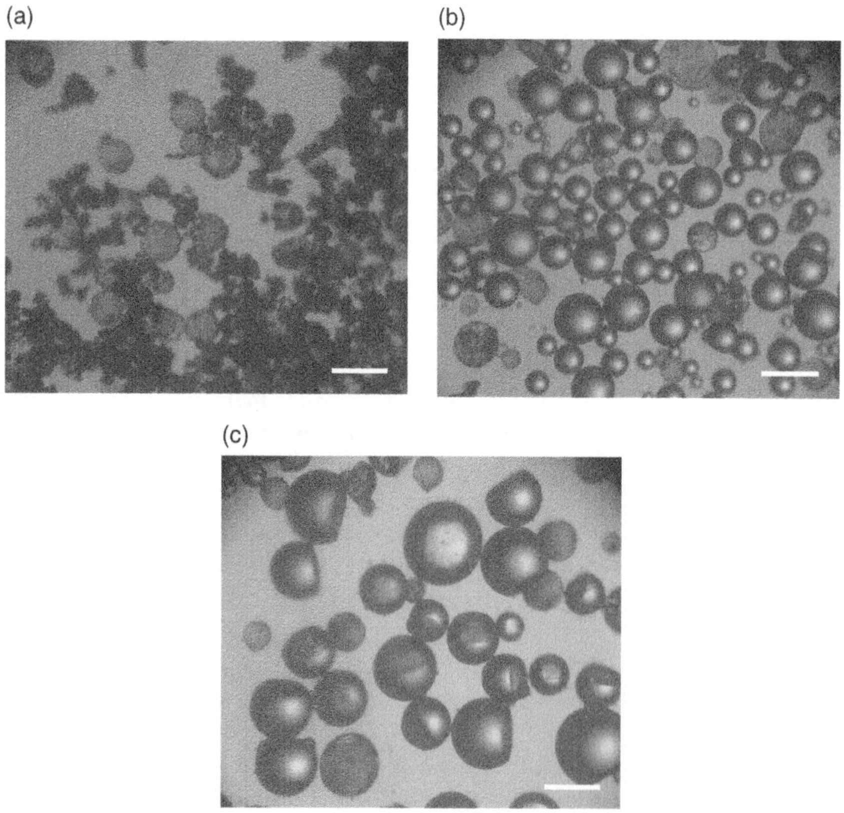

Figure 5.1 Optical microscopic images of GMA-loaded microcapsules prepared at different temperatures: (a) 40 °C, (b) 50 °C, and (c) 60 °C. Other reaction conditions refer to those of samples 5-1–5-3 in Table 5.1. The scale bars represent 200 μm in length. *Source:* Meng et al. [5], Figure 1, p. 4/with permission of Royal Society of Chemistry.

of GMA and PMF appear. This means that GMA has been encapsulated by PMF as expected.

To confirm whether GMA keeps its chemical activity after encapsulation, differential scanning calorimetry (DSC) measurements were conducted (Figure 5.3). The heating DSC scan indicates that no detectable reaction occurs in GMA within the temperature range of interest. As for the reference mixture of GMA and the amine curing agent diethylenetriamine (DETA), a clearly exothermic reaction is observed. The exothermic peak temperature and heat of reaction are 80.8 °C and 697.2 J g^{-1}, respectively. When GMA was replaced by the ground GMA-loaded microcapsules, a similar exothermic peak appeared at 64.9 °C with heat of reaction of 145.1 J g^{-1}. This demonstrates that the core material inside the capsules still has reactivity. The reason for the difference in the two reactions lies in the fact that grinding of the microcapsules cannot ensure entire breakage and the released core material fails to reach the stoichiometric ratio of GMA and DETA. The lower exothermic peak temperature of the ground GMA-loaded microcapsules/DETA system at 64.9 °C results from the fact that the ternary amine and methylol groups in the wall of the capsules promote the reaction between GMA and DETA. Besides, on the curve of the mixture of ground GMA-loaded microcapsules and DETA, it is found that there is another exothermic peak at about 117.3 °C with heat of reaction of 365.6 J g^{-1}. This should reflect the reaction between the free formaldehyde decomposed from the shell material as of ~100 °C and the residual DETA (refer to the discussion hereinafter) [13].

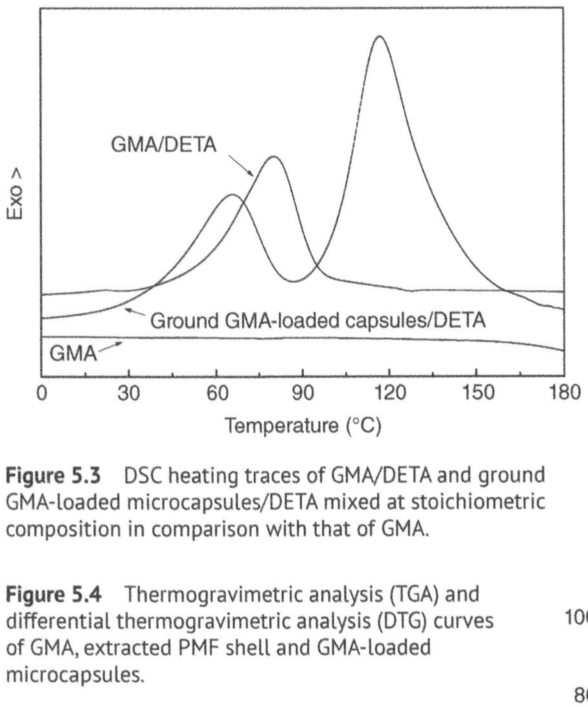

Figure 5.3 DSC heating traces of GMA/DETA and ground GMA-loaded microcapsules/DETA mixed at stoichiometric composition in comparison with that of GMA.

Pyrolytic performance of the capsules is given in Figure 5.4. GMA volatilizes with increasing temperature with full volatilization at about 180 °C (curve a in Figure 5.4). The weight loss of PMF shell and GMA-loaded microcapsules below 100 °C is mainly attributed to evaporation of adsorbed water (curves b and c in Figure 5.4). The weight loss of PMF shell above 100 °C consists of three main stages (curve c in Figure 5.4): (i) deformaldehyde reaction (100–270 °C) producing free formaldehyde [13], (ii) decomposition of melamine (270–450 °C), and (iii) progressive deamination forming a cyameluric structure and extensive degradation of PMF (>450 °C. [14]). The GMA-loaded microcapsules are thermally stable below 150 °C (curve b in Figure 5.4). When the temperature is raised to 150–200 °C, GMA starts to evaporate, making the main contribution to the weight loss from 220 to 310 °C. When the temperature is higher than 310 °C, decomposition of partial melamine and extensive degradation of PMF play the main roles, which resembles the aforementioned pyrolysis of the PMF shell. The results show that GMA can endure higher temperature with the protection of the PMF shell wall.

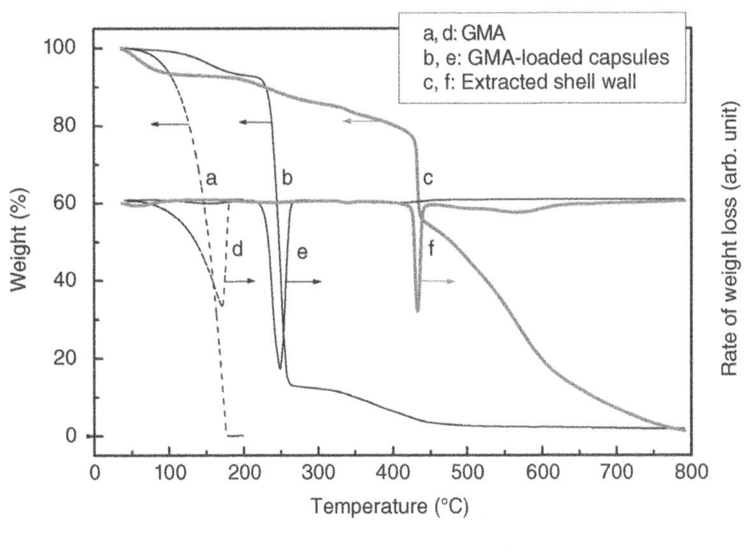

Figure 5.4 Thermogravimetric analysis (TGA) and differential thermogravimetric analysis (DTG) curves of GMA, extracted PMF shell and GMA-loaded microcapsules.

For long-term utilization of the self-healing materials with GMA-loaded microcapsules, the capsules should be sufficiently stable. The effect of storage time on weight loss of the microcapsules was monitored accordingly. When the microcapsules have been exposed to room temperature for six months, there is about 3.5% weight loss. A further increase of the storage time to over one year leads to a significantly reduced rate of weight loss. Microscopic inspection does not find an increased tendency for agglomeration of the capsules and their appearance also remains unchanged after storage for a long time. Therefore, evaporation of the absorbed water might be mainly responsible for the aforementioned weight loss, otherwise the capsules would stick together if GMA (boiling point = 189 °C) leaked out. It is believed that the compact PMF wall has provided significant protection for the core material.

5.1.2 Self-Healing Epoxy Materials with Embedded Single-Component Healant

The unfilled epoxy material was fabricated by mixing 100 parts diglycidyl ether of bisphenol A (EPON 828) with 12 parts (stoichiometric) or 20 parts (excessive) curing agent DETA. The self-healing epoxy materials were prepared by uniformly mixing a certain amount of GMA-loaded microcapsules (sample 1-2 in Table 5.1) with the aforementioned mixtures of EPON 828 epoxy and DETA. Either the unfilled epoxy or the filled version was degassed, poured into a closed silicone rubber mold, and cured for 24 hours at room temperature, followed by 24 hours at 40 °C.

Tapered double cantilever beam (TDCB) samples with a long molded groove of 47 mm were used to evaluate self-healing ability of the materials according to the protocol proposed by White et al. [4] The healing efficiency, η, is defined as the ratio of fracture toughness of healed materials, K_{IC}^{Healed}, to that of virgin ones, K_{IC}^{Virgin}:

$$\eta = \frac{K_{IC}^{Healed}}{K_{IC}^{Virgin}} \tag{5.1}$$

A pre-crack was made by first inserting a sharp razor blade into the groove of a TDCB specimen and gently tapping into the molded notch starter. Then, the specimen was pin-loaded and tested to failure under displacement control using a 5 μm s^{-1} displacement rate. Specimens were tested to failure, while compliance and peak load were determined to estimate the virgin fracture toughness. Load was then removed, allowing the crack faces to come back into contact and to be self-healed at 25 °C for a certain period of time. Finally, the healed specimens were tested again following the above procedure. Each batch included 10 specimens to yield an averaged value. For the reference experiment, 30 μl of GMA was manually injected to the crack plane of unfilled specimen before crack closing. Accordingly, the healed specimen is referred to as manually healed in the following text.

On the basis of TDCB tests, the averaged fracture toughness, K_{IC}, of the cured versions of the unfilled epoxy and the self-healing epoxy specimens with 10 wt% GMA-loaded microcapsules were determined to be 0.58 MPa m$^{1/2}$ and 0.68 MPa m$^{1/2}$ (for the systems cured at a DETA to epoxy ratio of 12:100 [w/w]), and 0.68 MPa m$^{1/2}$ and 0.95 MPa m$^{1/2}$ (for the systems cured at a DETA to epoxy ratio of 20:100 [w/w]), respectively. The values are comparable with those of self-healing epoxy containing mercaptan-loaded capsules (i.e. 0.63–0.73 MPa m$^{1/2}$) [9].

Since GMA-loaded microcapsules were compounded with epoxy as the single-component healing agent, their ability to repair cracks should be studied. As revealed by Figure 5.5, the encapsulated GMA can effectively rebond the broken specimen by nearly full recovery of fracture toughness, which is comparable with the healing systems based on dicyclopentadiene/Grubbs catalyst [4], organic solvents [6, 8], and epoxy/$(C_2H_5)_2O{\cdot}BF_3$ [11, 15]. Although the healing effect of the authentic specimen is poorer than that of the neat epoxy control specimen, probably due to uneven spreading of the healant on the fracture surface of the former, the result is still satisfactory. Evidently, the system is worth studying further.

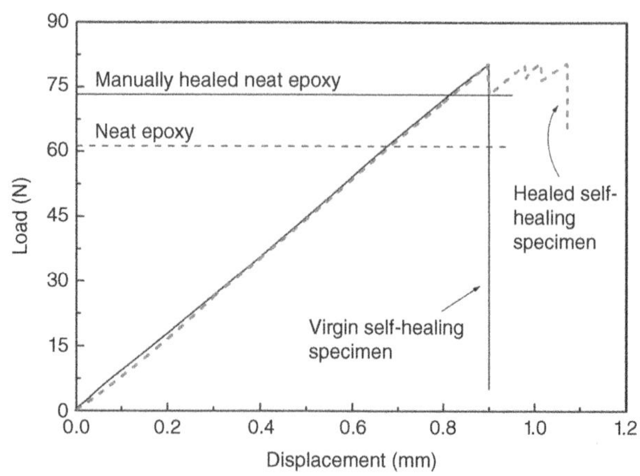

Figure 5.5 Typical load–displacement curves recorded during TDCB tests. The self-healing epoxy specimen contains 10 wt% GMA-loaded microcapsules. The fracture loads of both unfilled epoxy and manually healed unfilled epoxy specimens are given in the form of horizontal lines as references. The critical load of virgin self-healing specimen is higher than that of neat epoxy control specimen as a result of the toughening effect induced by the embedded spheres. Healing of the fractured specimens was conducted at 25 °C for 72 hours. All the specimens were cured at a DETA to epoxy ratio of 20:100 (w/w). *Source:* Reprinted from Meng et al. [5]. Copyright 2010, with permission from the Royal Society of Chemistry.

In consideration of the fact that residual amine in cured epoxy matrix might play an important role in the healing reaction, two typical proportions of hardener/epoxy were employed in this study for curing the epoxy specimens, so as to clarify and discriminate the mechanisms involved. That is, a stoichiometric DETA to epoxy ratio of 12:100 (w/w) and an excess ratio of amine of 20:100 (w/w) were used. For the DETA to epoxy ratio of 20:100 (w/w), there is an excess of 8 g of DETA per 100 g of epoxy resin. Since DETA contains two primary amines (each of which carries two active hydrogen atoms) and one secondary amine (which carries one active hydrogen atom), the activities of the five active hydrogen atoms differ from each other. As a result, the residual amines in the cured epoxy would be primary amine, or secondary amine, or even DETA that does not take part in the curing reaction.

Figure 5.6 compares FTIR spectra of the epoxy cured at different DETA/epoxy ratios with that of the uncured version. It is seen that the peak at 908 cm^{-1} representing skeleton absorption of epoxide group [16] disappears after curing. Epoxide groups of the matrix epoxy resin must have been completely consumed in the course of consolidation, no matter whether the DETA to epoxy ratio is 12:100 or 20:100 (w/w). The possible contribution of epoxide groups offered by the cured matrix itself to subsequent crack healing can thus be excluded.

Figure 5.7 illustrates healing efficiency of the self-healing epoxy materials as a function of content of GMA-loaded microcapsules. The first impression is that the specimens cured at a DETA to epoxy ratio of 20:100 (w/w) possess higher healing efficiency than those at the stoichiometric ratio over the entire range of interest of the content of

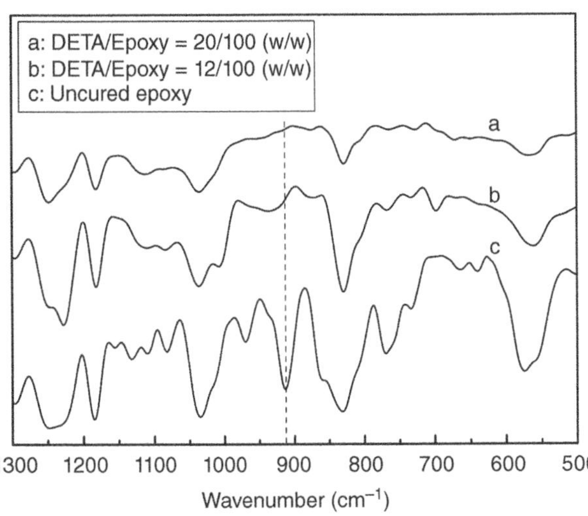

Figure 5.6 FTIR spectra of epoxy resins cured at different DETA/epoxy ratios in comparison with that of the uncured version.

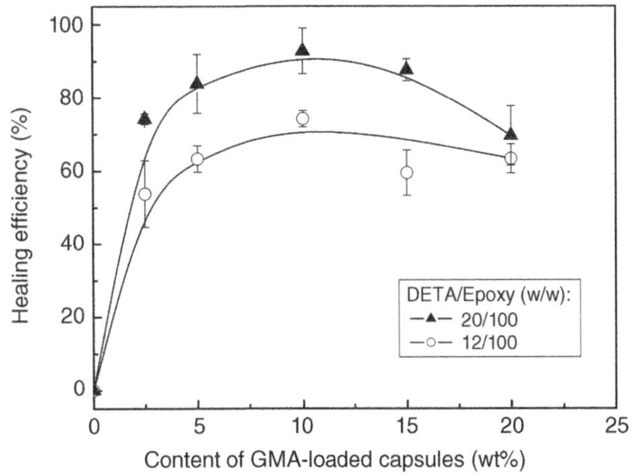

Figure 5.7 Effect of content of GMA-loaded microcapsules on healing efficiency. Healing of the epoxy based self-healing specimens was conducted at 25 °C for 72 hours.

capsules. This means that the presence of residual amine in the matrix must be related to the improvement in the healing effect. To understand the role of released GMA, it is better to start with the simpler system cured at a DETA to epoxy ratio of 12:100 (w/w), in which no unreacted epoxide groups need to be considered and the concentration of residual amine can be neglected compared with the case of DETA to epoxy ratio of 20:100 (w/w).

As mentioned at the beginning of this subsection, GMA is miscible with epoxy by its very nature. Therefore, the above healing behavior is naturally correlated to its solvent effect. In general, when surfaces of broken thermoplastic polymer are brought together at a temperature higher than the glass transition temperature, T_g, interdiffusion and entanglements of macromolecular chains would be activated and lead to gradual strength development at the interface [3]. A similar effect can be realized when a small amount of solvent is added instead of heating. With respect to thermosetting polymers such as cured epoxy, however, the cross-linked molecular networks would only be swollen but not disentangled by the diffused solvent molecules. Consequently, establishment of new entanglements of cross-linked chains across cracked interface is hard to realize except for the broken pendant chains. Other mechanisms should account for the GMA-aided healing shown in Figures 5.5 and 5.7.

The deduction is supported by the following comparison with experiment. If the interface was formed by random walk chains diffusing by reptation, the scaling law $K_{IC}^{Healed} \sim t^{1/4}$ of healed specimens (where K_{IC}^{Healed} represents the stress intensity factor of healed material and t is the healing time) should be obeyed [17, 18]. However, the results show that this is not the case because the exponent for the increase in refracture toughness is around 0.9, which is far from 1/4, despite the fact that K_{IC}^{Healed} linearly changes with t within double logarithmic axes.

On the basis of the above process of elimination, it is thus believed that a secondary force, i.e. hydrogen bonding, may account for healing cracks in the epoxy material cured at the stoichiometric DETA to epoxy ratio (Scheme 5.2). After damage of the epoxy-based composite, GMA is released from the capsules and delivered to the cracked portions, partially diffusing into the matrix. Then, hydrogen bonds are established between GMA and the functional groups (i.e. hydroxyl and ternary amine) of the matrix (at the interface or inside the matrix), and between the functional groups on the fractured surfaces as well. Volumetric expansion resulting from GMA-induced swelling of the matrix facilitates intimate contact of the separated parts, and hence formation of hydrogen bonding between hydroxyl and ternary amine from the matrix itself. Since the fluidic GMA is highly mobile, it can weld the cracks via building up bridge-type hydrogen bonding that is hardly formed by direct contact between the cracked planes due to mismatched locations of the functional groups. In addition, GMA can also form hydrogen bonds with hydroxyl in the subsurface layer. Consequently, not only the epoxy networks on the fracture surface but also those beneath the surface are stapled (pinned) by GMA. A similar conclusion involving hydrogen bond induced healing offered by a solvent in an epoxy matrix was reported by Caruso et al. [6]

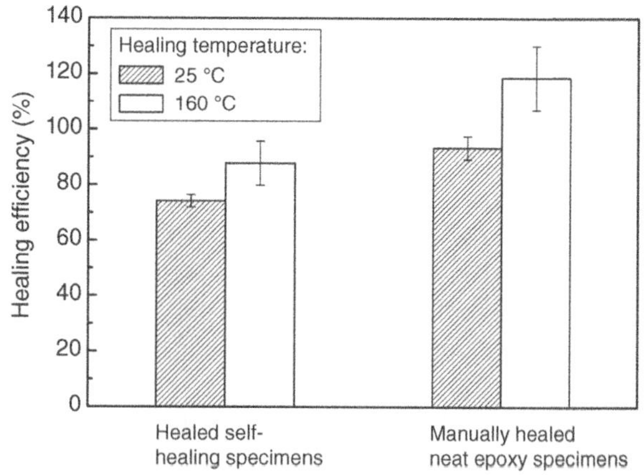

Figure 5.8 Effect of healing temperature on healing efficiency at a constant healing time of 72 hours. The self-healing epoxy specimens contain 10 wt% GMA-loaded microcapsules. All the specimens were cured at a DETA to epoxy ratio of 12:100 (w/w). T_g of the cured epoxy: 130 °C.

The data in Figure 5.8 partly provide evidence for the above analysis from another angle. When the specimens are healed at 160 °C (>T_g of the matrix, 130 °C, and lower than the boiling point of GMA, 189 °C), the healing efficiencies are obviously raised compared with those healed at room temperature. Mechanical interlocking of nodular features on fracture surfaces in the rubbery state [19] must have further promoted intimate contact among crack interfaces and GMA. This in turn helps to build up more hydrogen bonds when the specimens are cooled down. The higher healing temperature also enhances the diffusion of GMA into the matrix. In addition, since the weld is not dominated by covalent bonding between GMA and epoxy matrix, refracture of the healed specimen would not leave remarkable traces of material rupture on the fractured surface. As shown in Figure 5.9, the fractured surface of the virgin self-healing material (Figure 5.9a) resembles that of the healed specimen (Figure 5.9b). Unlike the case with polymerizable resin as healant where cohesively ruptured consolidated membranes appear [9], here the fracture surface of healed

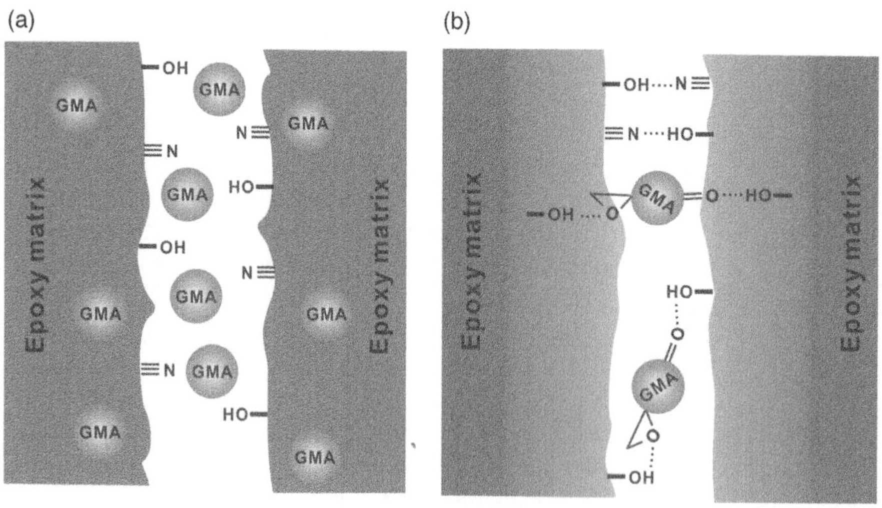

Scheme 5.2 Schematic drawing of crack healing in epoxy composite containing GMA-loaded microcapsules. The matrix has been fully cured at the stoichiometric weight ratio of epoxy to DETA, so that only hydroxyl and ternary amine are exposed on the fractured surfaces. (a) Distribution of the released GMA in the damage region. (b) Hydrogen bonds formed between the cracked faces. *Source:* Reprinted from Meng et al. [5]. Copyright 2010, with permission from the Royal Society of Chemistry.

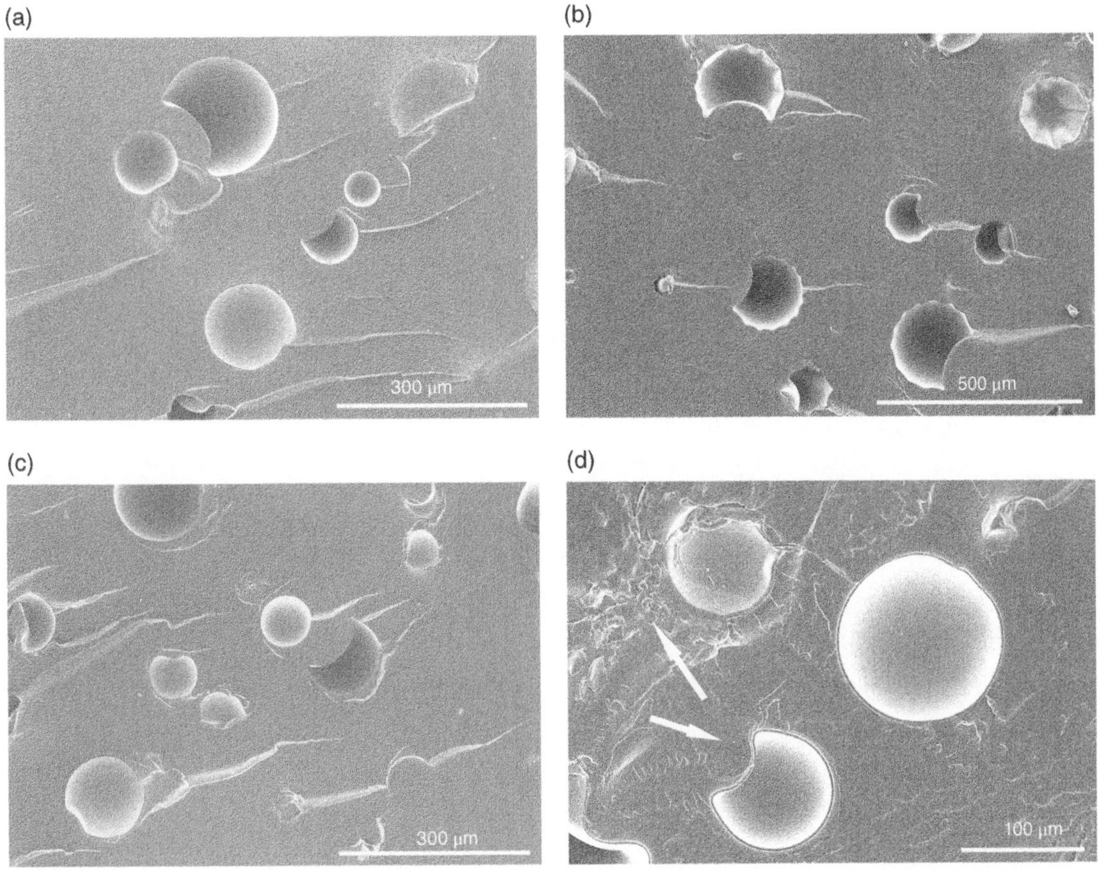

Figure 5.9 SEM micrographs of the fractured surface of (a) virgin self-healing epoxy specimen (cured at a DETA to epoxy ratio of 12:100 [w/w]), (b) healed self-healing epoxy specimen (cured at a DETA to epoxy ratio of 12:100 [w/w]), (c) virgin self-healing epoxy specimen (cured at a DETA to epoxy ratio of 20:100 [w/w]), and (d) healed self-healing epoxy specimen (cured at a DETA to epoxy ratio of 20:100 [w/w]). Content of GMA-loaded microcapsules in all the specimens is 10 wt%. The surface shown in (b) is relatively neat because non-covalent bonding governs the reconnection of the damaged specimens. In contrast, filmy splinters are found to be stuck on the surface shown in (d), as indicated by arrows, owing to stronger covalent adhesion generated during healing. On the other hand, the partially deflated capsules originating from collapses during capsule fabrication (see Section 5.1.1 for further information) should help establish additional interlocking at the interface and ensure timely penetration of cracks through the capsules. *Source:* Meng et al. [5], Figure 9, p. 7/with permission of Royal Society of Chemistry.

specimen is rather smooth (Figure 5.9b), indicating that the adhesion between the cracked planes is not strong enough.

We conducted extraction experiments as in the following to support the model in Scheme 5.2. Manually healed reference unfilled epoxy specimens were tested to failure, and then extracted by ethanol for 24 hours. Both FTIR and Raman inspections of the refractured surface were done and the characteristic peak of C=O was not found. This agrees with the fact that due to the limitation of secondary force, GMA is not firmly linked with the matrix during healing. When the extractant ethanol meets and diffuses into the fractured surface, the hydrogen bonds between GMA and the matrix resin can be easily damaged, freeing GMA from the material.

Returning to the plots of Figure 5.7 indicates that the residual amine resulting from excess hardener DETA can provide the materials with higher healing efficiency (refer to the data of the specimens cured at a DETA to epoxy ratio of 20:100 [w/w]). Chemical bonds established between GMA and the matrix should be responsible. This can be demonstrated by the appearance of the C=O peak on FTIR and Raman spectra of the ethanol extracted fracture surface of a manually healed reference specimen cured with excess DETA. Unlike the case at stoichiometric DETA to epoxy ratio, here GMA is covalently connected with matrix epoxy networks at the cracked planes, so that it is hard to remove by the extractant ethanol. Accordingly, the refractured surface of healed specimen (Figure 5.9d) becomes rougher as a result of cohesive failure of the resultant of the healing reaction.

In this context, the healing mechanisms involved in the specimens containing residual amine are summarized as in the following (Scheme 5.3), where not only hydroxyl groups and ternary amine but also primary and secondary amines are possibly present in the matrix. When the liberated GMA reaches the cracked portions, partial diffusion of its molecules would extract the unreacted DETA from the matrix to some extent, transporting it to different places on

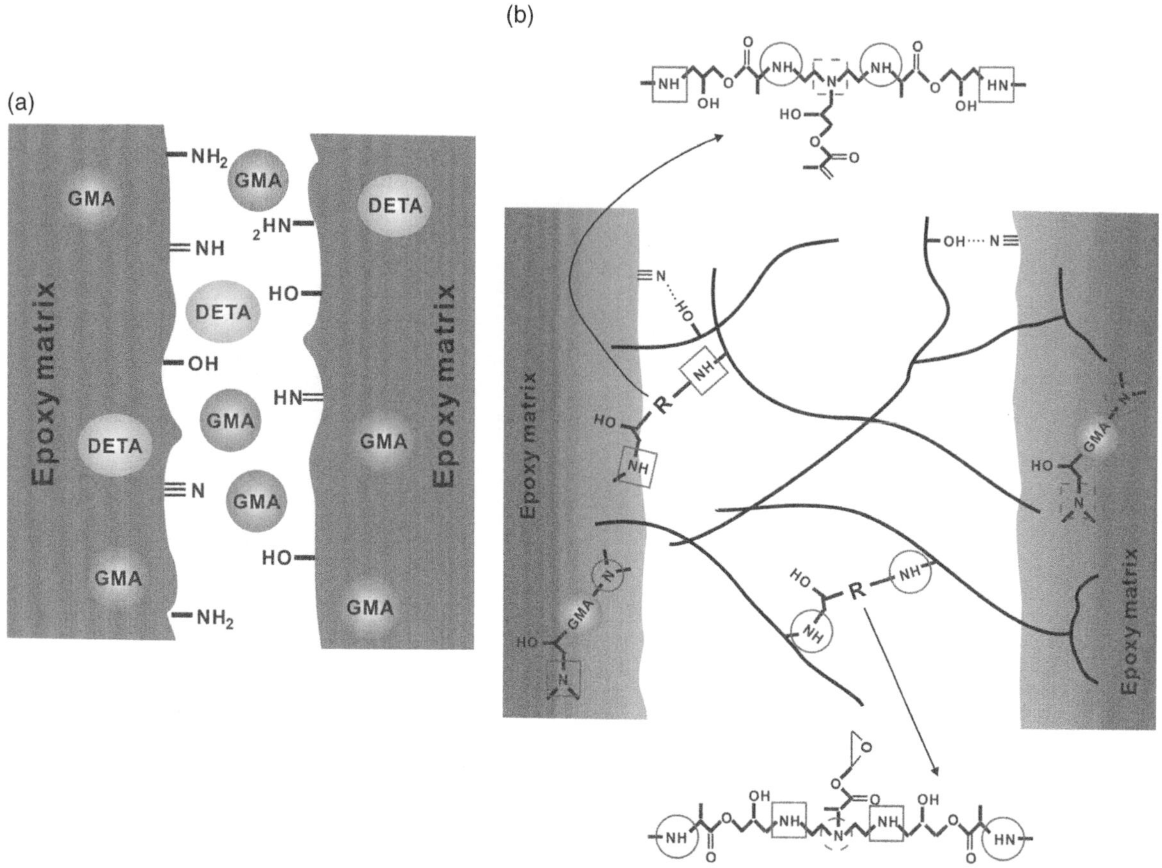

Scheme 5.3 Schematic drawing of crack healing in epoxy composite containing GMA-loaded microcapsules. The matrix has been cured with excessive DETA, so that hydroxyl groups, primary amine, secondary amine, and ternary amine are exposed on the fractured surfaces. (a) Distribution of the released GMA and unreacted DETA in the damage region. (b) Covalent bonds formed between the cracked faces. The curved solid lines represent the interconnected molecules (molecular networks) generated by the reaction between GMA and amines. The secondary amine (denoted by a solid line square) and ternary amine (denoted by a dashed line square) on the newly formed molecules result from a ring-opening reaction between the remaining amine of DETA and epoxide groups of GMA. Moreover, nucleophilic addition reaction between the remaining amine and the C=C double bond of GMA also produces secondary amine (denoted by a solid line circle) and ternary amine (denoted by a dashed line circle) on the newly formed molecules. *Source:* Reprinted from Meng et al. [5]. Copyright 2010, with permission from the Royal Society of Chemistry.

the fracture surfaces. In addition to the hydrogen bonds shown in Scheme 5.2, covalent bonds are built up by the reactions among GMA, DETA, primary amine and secondary amine in the matrix. Due to multifunctionality of both GMA and DETA, there are many possible reactions. The molecules drawn in Scheme 5.3 are only typical resultants of the reactions.

To have a deeper understanding of the influence of the reaction between GMA and amine on healing performance, a model mixture of GMA and DETA was prepared at stoichiometric ratio. In-situ FTIR measurement was carried out to check variations in chemical structure of the mixture. By using the peak area of C=O as the reference, reduction in the amount of epoxide groups versus time can be characterized (Figure 5.10a). It is found that with a rise in time the amount of epoxide gradually decreases, meaning the occurrence of reaction between GMA and DETA. In addition, when the time dependence of healing efficiency of the specimens cured with excess DETA is added to the figure, we can see the two curves are almost inversely symmetric. Clearly, the ring-opening reaction of epoxide groups of GMA with the residual amine shown in Scheme 5.3 has made a greater contribution to the recovery of interfacial adhesion. Similarly, the reduction in intensity of the C=C peak at 1635 cm^{-1} of GMA/DETA with increasing time (Figure 5.10b) should be attributed to the nucleophilic addition between the C=C double bond of GMA and primary amine (Scheme 5.3). A simultaneous increase in healing efficiency with time also confirms the effect of this reaction. Furthermore, the resultant of the abovementioned ring-opening and nucleophilic addition [20] reactions occurring at 25 °C is confirmed by proton nuclear magnetic resonance (^{1}H-NMR) spectroscopy.

An interesting fact revealed by the insets of Figure 5.10 is that although consumption of epoxide groups and C=C starts soon after the mixing of GMA and DETA, the healing efficiency of the specimens cured with excess amine is

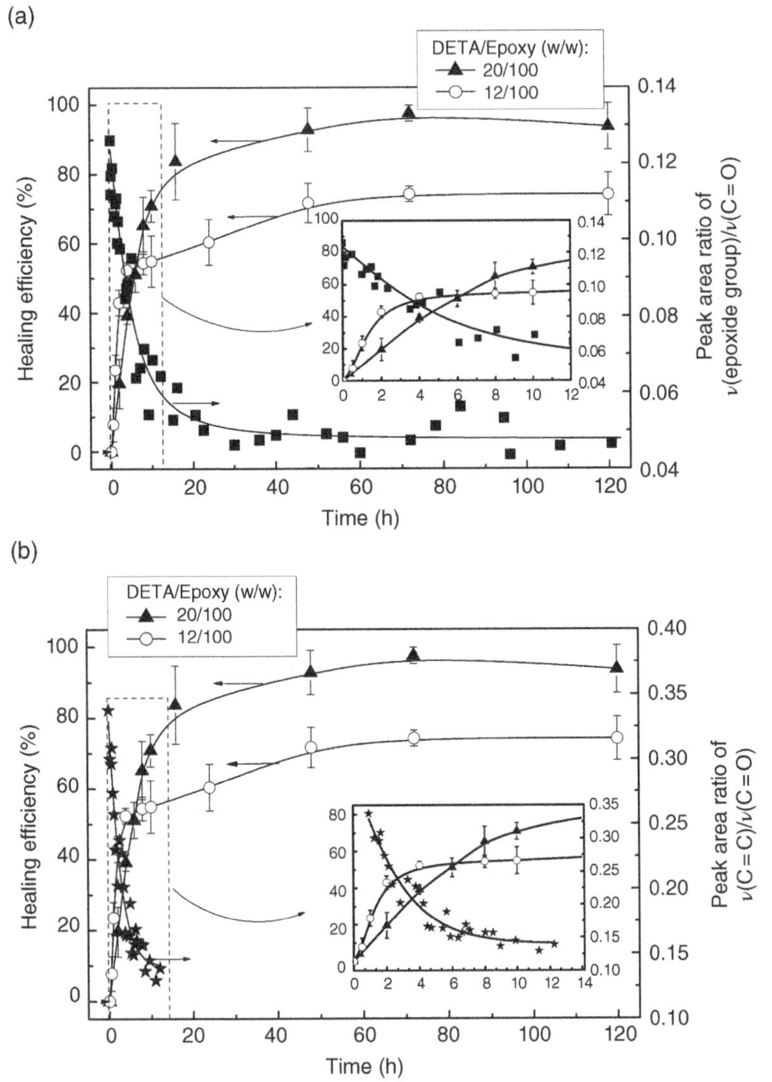

Figure 5.10 Time dependences of healing efficiency in comparison with that of in-situ FTIR peak area of (a) epoxide group and (b) C=C in the stoichiometric mixture of GMA and DETA. The self-healing epoxy specimens (cured at different DETA/epoxy ratios) contain 10 wt% GMA-loaded microcapsules. Healing and in-situ FTIR inspection were conducted at 25 °C. Since carbonyl only originates from GMA, and the reaction between GMA and DETA cannot produce any additional carbonyl, the peak area of the carbonyl group serves as the reference for showing the variations in the peak areas of epoxide and C=C with time. *Source:* Reprinted from Meng et al. [5]. Copyright 2010, with permission from the Royal Society of Chemistry.

lower than that of those cured at the stoichiometric DETA to epoxy ratio in the initial few minutes. This means that at the early stage of crack repair, the solvent effect of GMA plays the leading role, while the reaction of GMA is not bound to successfully connect the cracked planes with each other. On the other hand, Figure 5.10 shows that full consumption of epoxide groups needs a longer time as compared with C=C. In the late stage of healing, the reaction of epoxide groups of GMA should be the main cause of reconnection of cracks.

In fact, the cross-linking density of matrix epoxy is closely related to the healing effect. By carrying out dynamic mechanical analysis (DMA), the molecular weight between cross-links of the cured epoxy, M_c, can be calculated according to the equation of state for rubber elasticity [21]:

$$E = 3\frac{\rho}{M_c}RT\left(1-\frac{2}{\phi}\right) \tag{5.2}$$

where E' stands for storage modulus at rubbery plateau zone, ρ is density, R is the gas constant, T is absolute temperature, and ϕ is functionality of epoxy, respectively. In this work, E' values at $T = T_g + 30\,°C$ were used for the calculation. By plotting M_c against DETA to epoxy ratio (Figure 5.11), we can see both M_c and healing efficiency increase with increasing DETA concentration above the

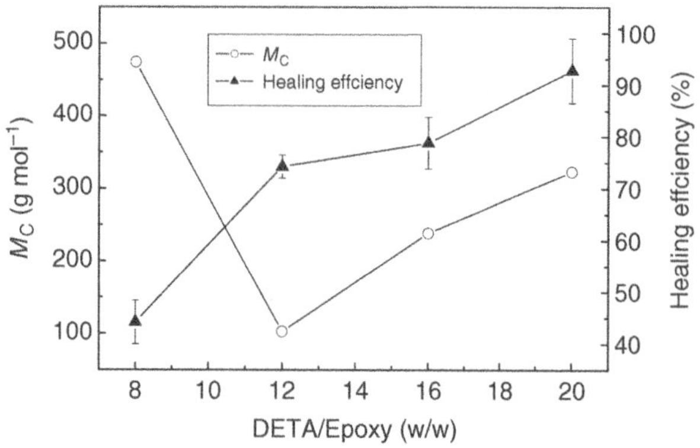

Figure 5.11 Effect of DETA to epoxy ratio on molecular weight between cross-links of the self-healing epoxy specimens, M_c, and healing efficiency. The self-healing epoxy specimens contain 10 wt% GMA-loaded microcapsules. Healing of the specimens was conducted at 25 °C for 72 hours.

stoichiometric value of 12:100 (w/w). The result agrees with ref. [6]. Since M_c is inversely proportional to cross-linking density, a higher M_c corresponds to lower cross-linking density and higher mobility of the molecules. Since GMA is easier to swell the specimens with higher M_c, the establishment of either hydrogen bonds or chemical bonds at the cracked faces is certainly facilitated as mentioned above. As for the specimen cured with insufficient DETA (i.e. DETA/epoxy = 8/100 [w/w]), because of shortage of amine in the matrix, although the molecules are much more mobile than those at and above the stoichiometric point, the GMA/amine interaction or interfacial adhesion is not strong enough on the whole. This further verifies the mechanisms proposed in Schemes 5.2 and 5.3.

5.2 Extrinsic Self-Healing Via Living Polymerization

So far, the approach based on microencapsulation of fluidic healing agent [4] has been successfully used to heal cracks in thermosets or elastomers [4, 9, 15, 22, 23]. However, there are few published attempts to repair damaged thermoplastic polymers in this way.

On the other hand, to ensure sufficient strength for long-term structural applications, establishing covalent bonding between the separated parts during self-healing of cracks in polymers is desirable. This is because the adhesion generated by chemical bonds is much stronger than that produced by secondary forces (such as van der Waals, hydrogen, and electrostatic forces, see the previous subsection). In this context, it is better to base the repair strategy on chemical reaction between healing agent and crack surfaces. For thermoplastic polymers, however, no therapy allowing autonomic self-healing is available. The healing agent such as epoxy monomer that is good at repairing cured epoxy usually cannot react with thermoplastics at room temperature. Even in the case of thermosetting polymers, polymerization of healing agent could not bring about any covalent bond between the healant and completely cured matrix. The existing healing methods do not enable the matrix to take part in the healing reaction. New concepts should be proposed accordingly.

Living polymerization is a process in which chain transfer and termination are removed [24]. It proves to be very useful in obtaining various types of tailor-made polymers with finely controlled molecular weight and molecular weight distribution. The resultant polymer carries living ends, so chain growth is always allowed so long as monomer is available. It is therefore a popular method for synthesizing block copolymers since the polymer can be prepared in stages, each of which contains a different monomer. With knowledge of this interesting characteristic, we suggest mixing the aforementioned microencapsulated GMA monomer (serving as healing agent) with a living poly(methyl methacrylate) (PMMA) as matrix [25]. Owing to the infinitely long lives of the molecules' ends of the matrix, as long as the monomer is released from the spheres as a result of crack initiation or propagation, the polymerization process of the healing agent will be started at ambient temperature wherever the monomer meets the matrix (Scheme 5.4). Then, the newly formed macromolecules, which are covalently attached to the interface, would fill the interstitial space of cracks and fuse with the matrix. In this way, the inert matrix that is passively healed in the case of thermosetting polymers plays an active role, simply as part of the healing agent (initiator). Since no catalyst is required for resuming chain growth in this system, naturally there is no question about catalyst deactivation. In addition, repeated (or multiple) healing might be feasible in the case of sufficient monomer supply because the matrix and the repaired portions still contain living terminals.

Meanwhile, to extend the above healing chemistry to commercial thermoplastics, we put forward a novel concept of multilayered microcapsules as shown in Figure 5.12. First, fluidic monomer acts as the core substance, which is compartmentalized by a thin layer of cross-linked polymer. Then, the microcapsule is coated by a living polymer carrying the necessary catalyst. Finally, an outer shell wraps the double walled capsule for maintaining activity of the living polymer during storage and composite fabrication. When crack is initiated and propagates in a thermoplastic polymer, which is filled with these multilayered microcapsules in advance, monomer liberated by the ruptured microcapsules would run through the middle layer and eventually reach the crack plane. Accordingly, living polymerization of the monomer restarts, leading to reconnection of the cracks. Evidently, the healing can be completed at room temperature without any

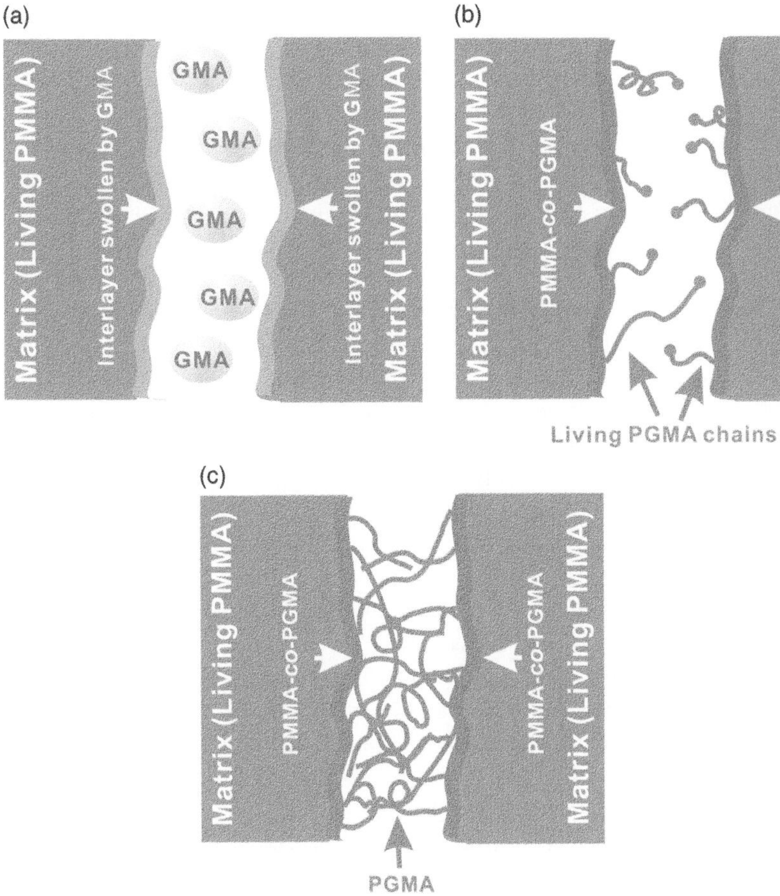

Scheme 5.4 Schematic drawing of crack repair by living copolymerization of GMA from the cracked surfaces of a living PMMA matrix. (a) Diffusion of released GMA monomers into the living PMMA matrix, leading to a swollen interlayer at the cracked surface. (b) Growth of living poly(glycidyl methacrylate) (PGMA) chains from the copolymer of PGMA and PMMA (i.e. PMMA-*co*-PGMA). (c) Crack filled with PMMA-*co*-PGMA.

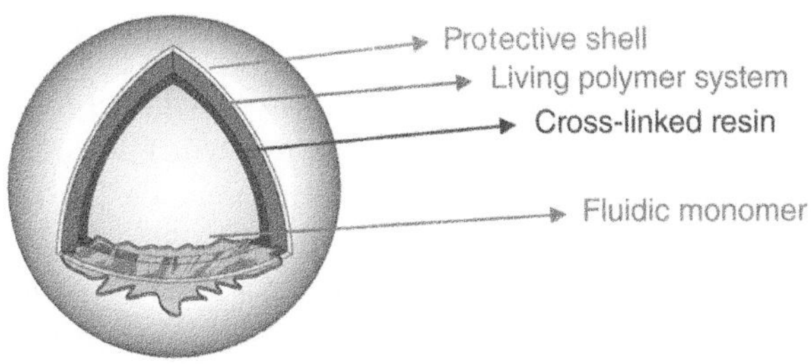

Figure 5.12 Profile of a multilayered microcapsule.

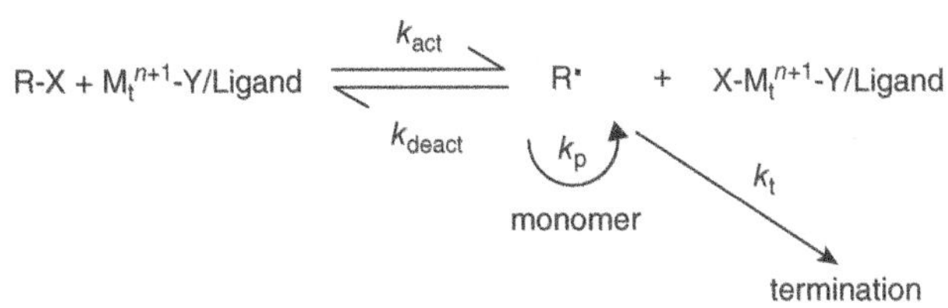

Scheme 5.5 Mechanism for ATRP.

manual intervention, and the polymer to be repaired does not have to be a living one [26]. The matching problems of a dual microcapsule system [10] are also prevented.

5.2.1 Preparation of Living PMMA and Its Composites with GMA-Loaded Microcapsules

According to the abovementioned idea, atom transfer radical polymerization (ATRP), one of the most powerful control/living radical polymerization techniques [27–29], is selected as the healing reaction. It has been known that polymers prepared by ATRP are highly chain end-functionalized and can participate in various post-polymerization modifications [30]. Scheme 5.5 shows the typical mechanism involved in ATRP. The control of the polymerization afforded by ATRP is a result of the formation of radicals that can grow but are reversibly deactivated to form dormant species. Reactivation of the dormant species allows the polymer chains to grow again, only to be deactivated later. Such a process results in a polymer chain that slowly, but steadily, grows and has a well-defined end group. In general, the initiator is a simple, commercially available alkyl halide. The catalyst is a transition metal that is complexed by one or more ligands. The catalyst does not need to be used in a one-to-one ratio with the initiator but can be used in much smaller amounts. The deactivator can be formed in situ, or for better control, a small amount (relative to the catalyst) can be added. Additionally, the catalyst is tolerant of water and trace amounts of oxygen.

So far, a variety of methacrylate monomers possessing carbon to carbon double bonds have been successfully homopolymerized in ATRP systems at ambient temperature, including methyl methacrylate (MMA) [31], benzyl methacrylate (BMA) [32], and GMA [33]. Moreover, copolymerization of GMA with methyl acrylate can also proceed by ATRP at ambient temperature [34]. Considering the need of autonomic self-healing without manual intervention, therefore, we choose PMMA made by ATRP as the matrix polymer, in which microencapsulated GMA is dispersed as one-component healant in a way similar to that applied in making self-healing epoxy composites with GMA-loaded microcapsules (Section 5.1.2). Compared with MMA, GMA is easier to be encapsulated via in-situ interfacial polymerization because of its high boiling point (189 °C), low volatility, and low solubility in water. The similar polarity of PMMA and GMA facilitates wetting of GMA monomer on the surface of PMMA matrix and hence the subsequent copolymerization.

To prepare self-healing PMMA composites with GMA-loaded microcapsules, both living PMMA and the GMA capsules have to be synthesized in advance.

The bulk ATRP of MMA was conducted as in the following to prepare living PMMA by using ethyl 2-bromoisobutyrate (EBiB) as initiator, and cuprous bromide (CuBr)/ *N*,*N*,*N*′,*N*′,*N*″-pentamethyldiethylenetriamine (PMDETA)/ tetrabutylammonium bromide (Bu4NBr) as catalyst. The role of Bu_4NBr is to increase solubility of the catalyst complex, so that homogeneous reaction can be always ensured, because Cu(II)/PMDETA is poorly soluble in MMA monomer. Typically, Bu_4NBr (1.031 g, 3.2 mmol) and CuBr (0.057 g, 0.4 mmol) were added to an argon-purged tube (8 × 2.5 cm) equipped with a B-19 standard joint and a stir bar. Argon purging was continued for 10 minutes, following which the tube was closed with a rubber septum that was sealed by parafilm. MMA (24.0 g, 0.24 mol) previously purged with argon was then injected into the tube with an argon-purged gas-tight syringe, followed by PMDETA (0.069 g, 0.4 mmol). The admixture was magnetically stirred to make a clear homogeneous solution. EBiB (0.078 g, 0.4 mmol) was then added. The tube was kept in an oil bath placed over a magnetic stirrer and maintained at 25 °C. After 12 hours, a conversion of 70% was reached. The resultant polymer was a transparent solid exhibiting a light green color. For kinetic studies, separate runs were performed for different reaction times. After the desired time period, the contents of the tube were diluted with tetrahydrofuran (THF) and precipitated into a large amount of methanol depending on the molecular weight of the resultant polymer. The precipitation was filtrated and dried in a vacuum oven at 45 °C for 48 hours. The monomer conversion was gravimetrically determined. The synthesis route mentioned above is rather convenient as no post-treatment including purification is needed.

GMA was microencapsulated by a two-step approach as mentioned in Section 5.1.1, using PMF as the wall substance.

Since both living PMMA and GMA-loaded microcapsules are available, self-healing PMMA composites can be fabricated. In a typical run, EBiB (0.078 g, 0.4 mmol) was added into a 100 ml argon-purged three-neck round-bottom flask containing a mixture of MMA (24.0 g, 0.24 mol), Bu_4NBr (1.031 g, 3.2 mmol), PMDETA (0.069 g, 0.4 mmol) and CuBr (0.057 g, 0.4 mmol). Argon purging was continued for 10 minutes. Then, the flask was tightly sealed with a rubber septum and immersed in an oil bath at a preset temperature of 25 °C. After five hours, a given proportion of GMA-loaded microcapsules (diameter: 283 μm, core content: 94.6%) were incorporated into the flask during polymerization of the matrix under argon atmosphere and stirring. The mixture was then degassed and poured into a silicone rubber mold in a glove bag with argon, and cured for 48 hours at 25 °C.

5.2.2 Self-Healing Performance of Living PMMA Composites Filled with GMA-Loaded Microcapsules

Since the healing chemistry of the present strategy depends on "living" characteristics of the matrix polymer, features of the polymerization and the resultant should be known first. In the course of living polymerization of PMMA, the dosage of Bu_4NBr is an important factor as it helps to increase the solubility of the catalyst complex. It was experimentally found that the use of a Bu_4NBr to PMDETA ratio lower than 8:1 yielded a homogeneous solution in the beginning, while the system became heterogeneous with the progress of the reaction because of precipitation of the Cu(II) complex. In addition, with a further rise in Bu_4NBr concentration, the initial rate of polymerization decreased and the remains of insoluble Bu_4NBr appeared in the ultimate reaction system. Therefore, the mole ratio of CuBr:PMDETA:Bu_4NBr = 1:1:8 was selected, and it proved to work well for the ambient temperature ATRP of MMA.

According to the mechanism of ATRP, the initiator group should be incorporated into the α-end of the resultant polymer chain, while the ω-end remains at the terminal halide. It can be verified by a ^{1}H-NMR spectrum (Figure 5.13). The resonances at 0.7–1.2 ppm (designated as "a" in Figure 5.13) and 1.25 ppm (designated as "b" in Figure 5.13) are due to $-CH_3$ protons of PMMA. The peaks at 1.35–2.10 ppm (designated as "c" in Figure 5.13) are attributed to $-CH_2-$ protons of PMMA, corresponding to the different tacticity of the macromolecules. The methyl protons (designated as "e" in Figure 5.13) of the terminal methyl ester unit are perceived at around 3.79 ppm, which are away from other methyl protons (designated as "d" in Figure 5.13) of the repeated methyl ester at around 3.60 ppm, because they are affected by the ω-bromine atom in the PMMA chain end through electrophilic effect. Clearly, the ^{1}H-NMR spectrum demonstrates that the resultant PMMA is indeed made following the ATRP mechanism.

The kinetic plot for monomer disappearance during polymerization of MMA is illustrated by Figure 5.14a. It reveals the linear time dependence of $\ln([M]_0/[M])$, indicating that the polymerization is a first-order reaction with respect to monomer, while the concentration of the growing radicals remains constant. Side reactions such as chain termination and transfer are negligible. By using gel permeation chromatography (GPC), Figure 5.14b further shows the variations of number average molecular weight, M_n, and polydispersity, M_w/M_n, with conversion for the ATRP system presented in Figure 5.14a. Clearly, M_n increases linearly with conversion, and the polydispersity

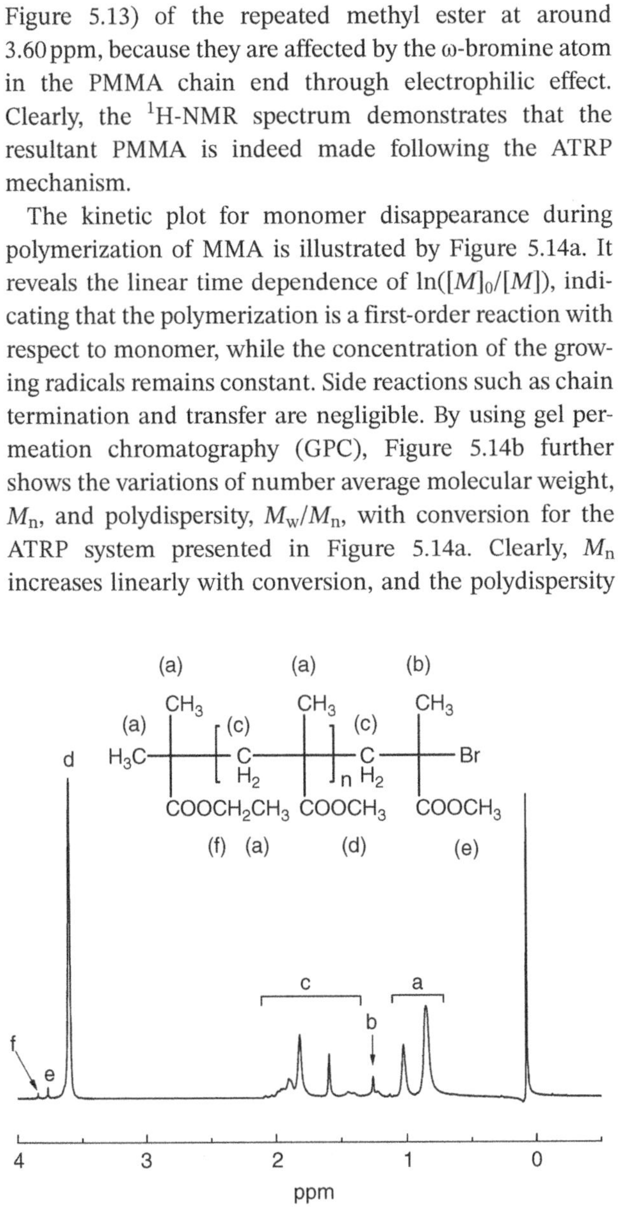

Figure 5.13 ^{1}H-NMR spectrum of resultant PMMA-Br initiated with 2-EBiB/CuBr/PMDETA in bulk at 25 °C ($\bar{M}_n = 9.5 \times 10^3$, $\bar{M}_w/\bar{M}_n = 1.08$). The single signal at 3.79 ppm for the methoxy group next to the halogen chain end is consistent with that reported in ref. [35].

is rather low in all cases (1.08–1.15). This demonstrates that ATRP of MMA is a controlled/living process.

As mentioned earlier, it is expected that the fluidic healing agent (i.e. monomer) issued from the broken capsules is able to polymerize with the living matrix on its fracture surface to mend cracks. Therefore, the initiating activity of the resultant matrix should be verified prior to authentic healing experiments. For this purpose, two verification tests were carried out.

First, fresh MMA monomers were polymerized using PMMA-Br as the macroinitiator under bulk polymerization conditions at ambient temperature as established above. Specifically, the experiment was carried out as follows. Living PMMA ($M_n = 2.3 \times 10^4$, $M_w/M_n = 1.15$, 27.7 g, 1.2 mmol), Bu_4NBr (3.093 g, 9.6 mmol), MMA (5.0 g, 0.06 mmol) and acetone (300 ml) were incorporated into a flask. Under mechanical agitation, both PMMA and Bu_4NBr were completely dissolved and then argon was introduced for 30 minutes. Afterwards, CuBr (0.171 g, 1.2 mmol) was quickly added to the system with stirring for 10 minutes, and finally PMDETA (0.207 g, 1.2 mmol) was injected into the flask. Having been reacted for 12 hours, the polymerized fluidic product was taken out, dissolved in THF, precipitated in excessive methanol, and dried. GPC measurement of the chain-extended product indicates a higher molecule weight PMMA has been yielded ($M_n = 2.8 \times 10^4$, $M_w/M_n = 1.10$, see Figure 5.15a). This proves the living nature of the PMMA as well as the existence of terminal bromine at PMMA chains.

Secondly, a visual inspection was conducted. Bulk PMMA was first prepared by ATRP in a tube. Then, several acrylate monomers were poured onto it in random sequence. The typical procedure is described in the following. Ethyl methacrylate (EMA) was first added onto the aforementioned bulk PMMA, allowing polymerization of the monomer at 25 °C. Similarly, MMA was subsequently added onto the surface of solidified poly(ethyl methacrylate) (PEMA). The procedure was repeated several times by using EMA, MMA and GMA monomers. Eventually, a rod with a multilayer sandwich structure was obtained (Figure 5.15b). This experiment simulates the situation of when released healing monomer meets the matrix and indicates that the macromolecules in either the base PMMA or the resultant polymers (such as PEMA, PMMA, and PGMA shown in Figure 5.15b) remain active. The obscure interfaces between layers are indicative of diffusion of monomers and formation of copolymers. Clearly, the PMMA derived from ATRP can initiate polymerization of proper monomers whenever the monomers come into contact with the living polymer. This satisfies the requirement of the proposed self-healing via living polymerization.

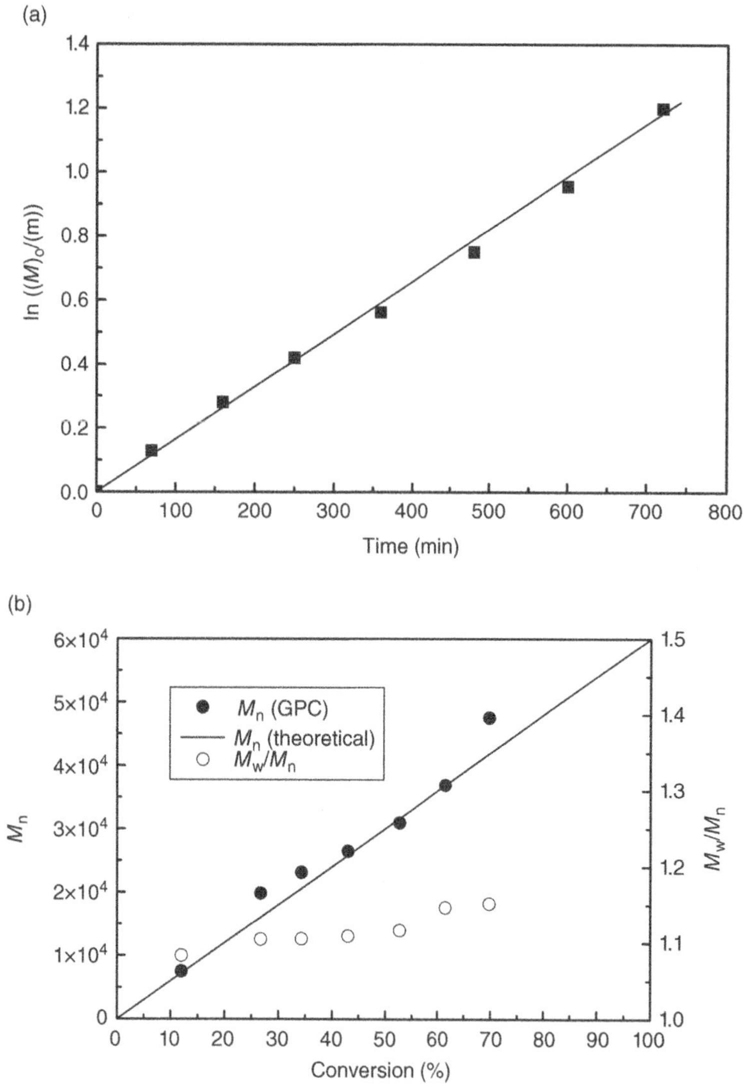

Figure 5.14 Characteristics of living polymerization of MMA. (a) ln($[M]_0/[M]$) versus reaction time for ATRP of MMA in bulk at 25 °C, where $[M]_0$ and $[M]$ denote the overall MMA monomer concentrations at times 0 and *t*, respectively. $[M]_0$ = 0.24 mol, $[EBiB]_0 = [CuBr]_0 = [PMDETA]_0$ = 0.40 mmol, and $[Bu_4NBr]$ = 3.2 mmol. (b) Number average molecular weight, M_n, and polydispersity, M_w/M_n, versus percent conversion for ATRP of MMA using the same recipe as given in (a). $M_n(\text{theoretical}) = \frac{[M]_0}{[I]_0} \times W \times \text{Conversion}$, where $[I]_0$ is the initial concentration of the intiator and *W* is the molecular weight of the monomer. *Source:* Reprinted from Wang et al. [25]. Copyright 2010, with permission from American Chemical Society.

As the main concern of our work lies in the self-healing capability of the proposed system, self-healing PMMA composites were fabricated by embedding GMA-loaded capsules. To assess healing ability of the materials, the method proposed by Jones et al. [36] was employed as in the following. Healing efficiency is defined as the ratio of impact strength of healed and virgin materials. Izod notched impact tests were conducted at 25 °C. After testing, the specimens that had been broken into two pieces were placed into a desiccator with argon and kept in alignment and intimate contact for healing at 25 °C for 24 hours. Then, the healed specimens were tested again following the above procedure to check the effect of healing. It is seen from Figure 5.16a that our technical route works as planned. Initially, the healing efficiency increases rapidly with increasing healing time, attains ~89% after 12 hours, and reaches the equilibrium at about 100% after 21 hours. So, a conservative healing period of 24 hours was used for the rest of the tests to ensure full healing under room temperature conditions (Figure 5.16b).

To obtain more information about the feasibility of achieving in-situ self-healing, four groups of self-healing

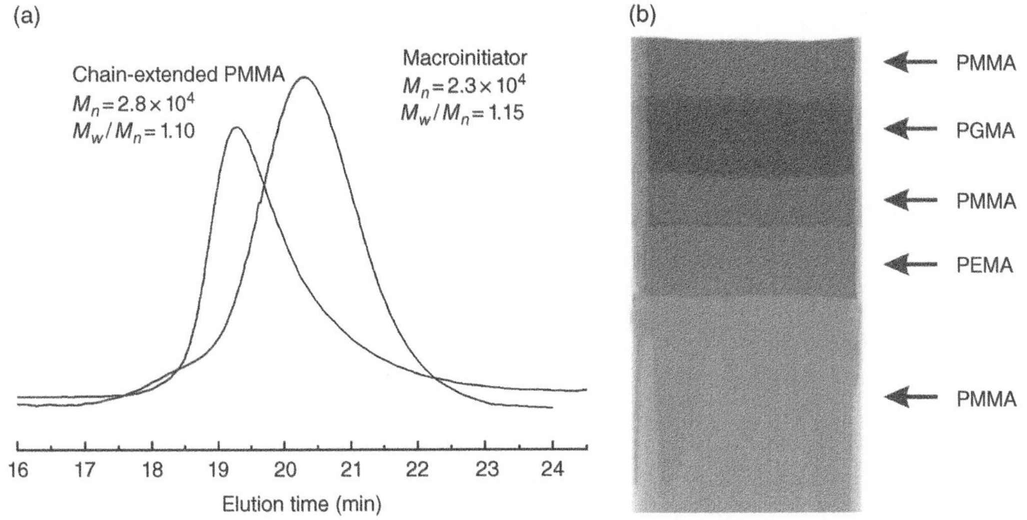

Figure 5.15 (a) GPC curves of PMMA before and after chain extension reaction via the ATRP process. Reaction conditions: $[MMA]_0$:[macroinitiator]$_0$:$[CuBr]_0$:$[PMDETA]_0$:$[Bu_4NBr]_0$ = 50:1:1:1:8 (molar ratio), 25 °C, 12 h. (b) Multilayer sandwich structure formed by successively adding acrylate monomers onto living PMMA. Carbon black and yellow dye were, respectively, incorporated into the layers of PGMA and PEMA for coloring. *Source:* (a) Reprinted from Wang et al. [25]. Copyright 2010, with permission from American Chemical Society. (b) Wang et al. [25], Figure 1, p. 2/with permission of American Chemical Society.

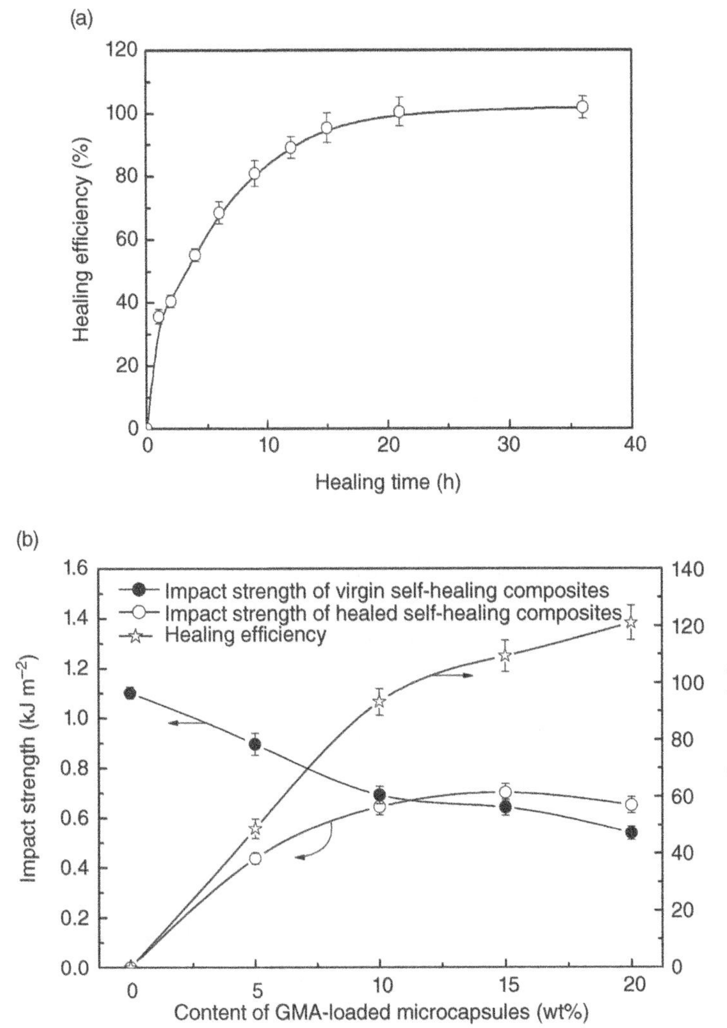

Figure 5.16 (a) Healing efficiency of PMMA composites containing 15 wt% GMA-loaded microcapsules as a function of healing time at 25 °C. (b) Influence of content of GMA-loaded microcapsules on impact strengths of virgin and healed composites and healing efficiency. Healing of the fractured specimens was conducted at 25 °C for 24 hours. The average diameter of the microcapsules is 283 μm. *Source:* (a) Reprinted from Wang et al. [25]. Copyright 2010, with permission from American Chemical Society.

PMMA specimens were manufactured by varying the concentration of GMA-loaded microcapsules from 5 to 20 wt%. Relatively large microcapsules with an average diameter of 283 μm are employed in order to maximize the delivery of healing agent for best healing performance. As exhibited in Figure 5.16b, the healing efficiency drastically increases with capsule concentration from 0 to 10 wt% and then slowly increases within the capsule content range from 10 to 20 wt%. This suggests that the self-healing performance of the present system is determined by the amount of healing agent provided by the broken microcapsules on the facture planes. Besides, it is seen that the impact strength of virgin self-healing specimens gradually decreases with a rise in the content of the microcapsules. The reasons might be as in the following. (i) The soft capsules cannot hinder crack propagation. (ii) An increase of content of the microcapsules reduces effective cross-section area of the load-bearing matrix. (iii) Stress concentration becomes more evident with a rise in volume fraction of the capsules. Nevertheless, the healing efficiency is found to be higher than 100% when the capsule content exceeds 15 wt%. Perhaps this can be attributed to the fact that a large quantity of healing agent has been issued from the broken capsules and undergone polymerization on the crack planes including the area taken up by the capsules. As a result, the cross-sectional area of the load-bearing matrix is increased, which possibly leads to higher impact strengths of the healed specimens as compared with those of the virgin specimens.

Scanning electron microscopy (SEM) observation of fracture surfaces of the specimens indicates that the microcapsules are well adhered to the matrix (Figure 5.17a). After failure of the specimen, the spheres must be rapidly cleaved but not pulled out, giving off GMA, which was then converted into solidified membrane on the fracture surface (Figure 5.17b). Figure 5.18 further gives a live record of the surface polymerization of the released GMA on the living PMMA matrix in terms of in-situ Raman spectroscopy. By using the C=O band at 1725 cm^{-1} that is unaffected by the polymerization as internal standard, the time dependence of the peak intensity of C=C at 1638 cm^{-1} representing the methacrylate group is plotted in Figure 5.18b. Evidently, the amount of C=C decreases with time, which coincides with the trend exhibited in Figure 5.16 and can serve as a measure of the healing polymerization kinetics. It reveals that GMA flowed out of the broken microcapsules and can undergo polymerization on the fracture surface of PMMA, accounting for the healing effect.

Considering that solvent might also help to heal cracks [6, 37, 38] in thermoplastics through swollen driven chain entanglement across the cracked planes and GMA is a good solvent for PMMA, the contribution of solvent effect of GMA to the healing capability should be understood. Accordingly, a mixture of GMA monomer and its polymerization inhibitor mequinol (which prevents GMA from participating in polymerization) was manually injected onto (i) fracture surfaces of living PMMA specimens, and (ii) fracture surfaces of PMMA specimens that had been deactivated by hydrogen peroxide in advance (X-ray photoelectron spectroscopy [XPS] measurements prove the deactivation because univalent copper on the facture surface has been oxidized into Cu(II) in the form of $CuBr_2$). As a reference, GMA alone was also injected onto fracture surfaces of living PMMA specimens. All the control PMMA

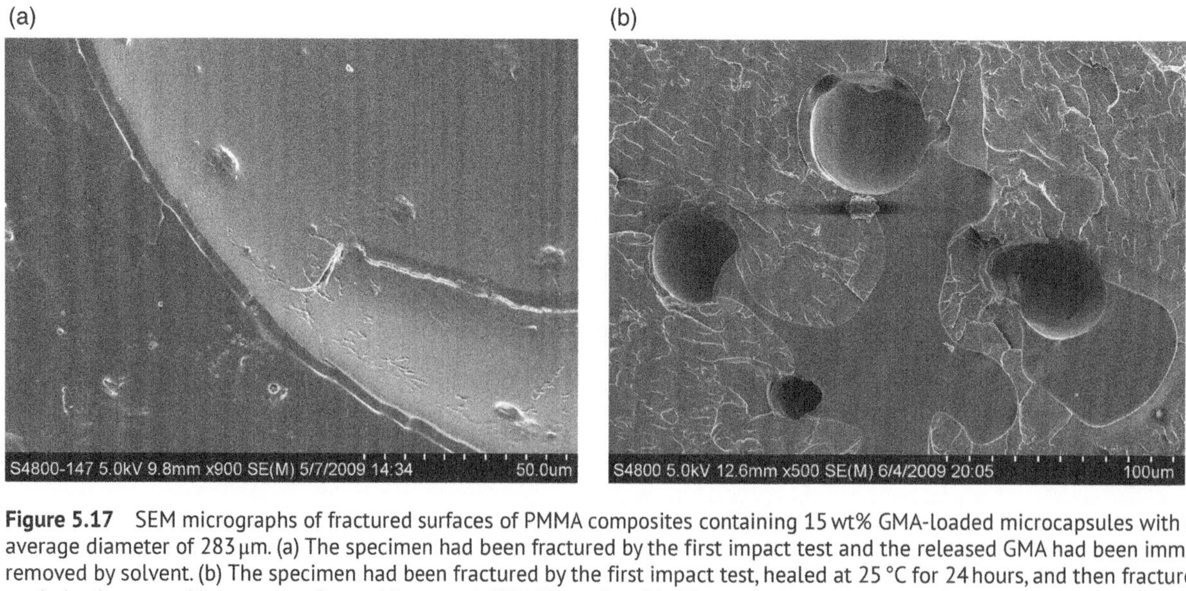

Figure 5.17 SEM micrographs of fractured surfaces of PMMA composites containing 15 wt% GMA-loaded microcapsules with an average diameter of 283 μm. (a) The specimen had been fractured by the first impact test and the released GMA had been immediately removed by solvent. (b) The specimen had been fractured by the first impact test, healed at 25 °C for 24 hours, and then fractured again by the second impact test. *Source:* Wang et al. [25], Figure 3, p. 3/with permission of American Chemical Society.

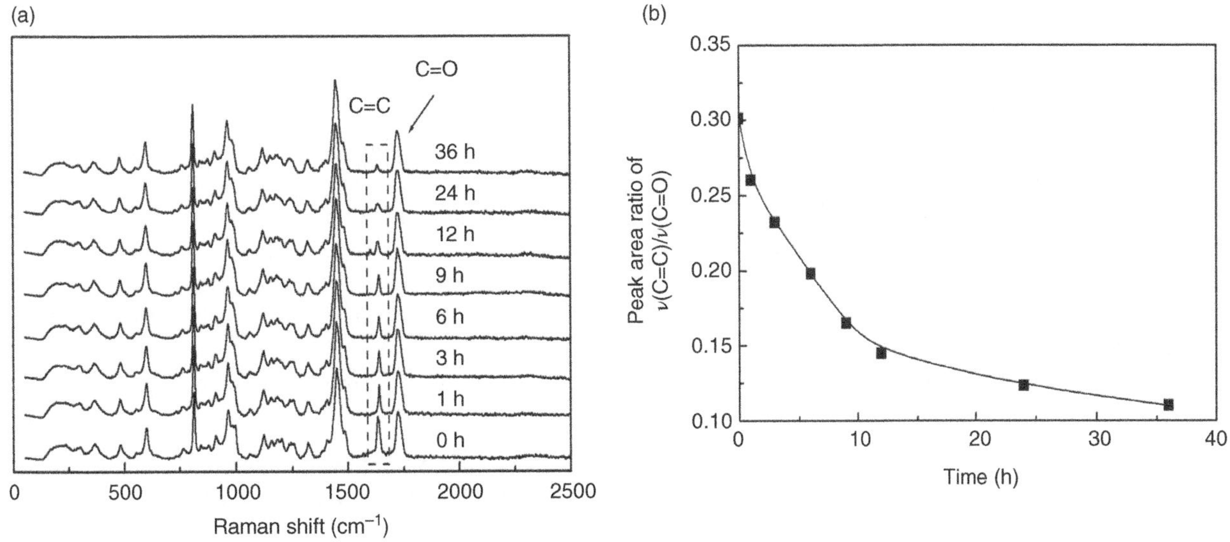

Figure 5.18 In-situ confocal Raman microscopy observation [25]. (a) The typical Raman spectra collected from the center of the fracture surface of a self-healing specimen with 15 wt% GMA-loaded microcapsules (average diameter: 283 μm). (b) Time dependence of the characteristic Raman peak area ratio, showing the consumption of GMA during polymerization. *Source:* Reprinted from Wang et al. [25]. Copyright 2010, with permission from American Chemical Society.

specimens used in this part of the experiment do not contain any GMA-loaded capsules. After rehabilitation at 25 °C for 24 hours like the authentic samples, healing efficiencies (Table 5.2) offered by the solvent effect (i.e. the first two cases) were found to contribute little to the overall value (refer to the last case). This means that chemical reaction rather than solvent effect is the main contributor to crack repair. In fact, crack healing of thermoplastics simply using solvent might be not an appropriate option. Regardless of the low healing efficiency shown above, the solvent absorbed deep inside thermoplastics would hardly be desorbed, leading to obvious plasticization and property deterioration. Here, GMA plays the role of solvent at the beginning and then is very soon polymerized. Therefore, only the positive side of the solvent effect (i.e. wetting) is included.

Table 5.2 Control tests showing the solvent effect of GMA[a].

Healing agent	Fracture surface	Healing efficiency (%)
GMA/mequinol (500 ppm)	Active	38.2 ± 2.0
GMA/mequinol (500 ppm)	Inactive[b]	35.1 ± 2.2
GMA	Active	98.9 ± 4.0

[a] GMA or GMA/mequinol mixture (5 μl) was manually injected into the fracture surfaces of control PMMA specimens that did not contain any capsules. Healing of the fractured control samples was conducted at 25 °C for 24 hours. Since the area of crack damage on the specimen is 10.1 mm × 12.3 mm = 124.23 mm^2, the thickness of injected GMA or GMA/mequinol mixture (5 μl) is about 40.25 μm. According to the derivation of Rule et al. [39], the average thickness of bled healing agent in a capsules embedded composite, $\bar{d}$, can be calculated from: $\bar{d} = d_c \cdot c \cdot \varphi$, where d_c, c and ϕ stand for the average diameter, core content and weight content of capsules, respectively. When $\phi = 15$ wt%, $c = 95\%$, $d_c = 283$ μm, $\bar{d}$ is estimated to be 40.33 μm. This means that the injection of 5 μl of GMA or GMA/mequinol mixture corresponds to a capsule loading of 15 wt%. In fact, the authentic self-healing specimens with 15 wt% GMA-loaded microcapsules offer full recovery of impact strength (see Figure 5.16).
[b] The fractured surface was treated with 15% hydrogen peroxide (w/v) for two hours for deactivation [40, 41].

In summary, room temperature living polymerization proves to be an effective way to impart autonomic self-healing capability to thermoplastic polymers. Such a healing system helps to build up chemical bonding from the cracked planes and does not need the addition of catalyst in a polymeric matrix and manual intervention as well. Our preliminary tests demonstrate that the healing function of the present system remains unchanged after storage for one year as characterized by full recovery of impact strength. An additional potential advantage lies in the fact that the living matrix itself might allow self-healing of radiation (like ionizing, UV and electromagnetic radiation) induced degradation because of recombination between the new free radicals and the macromolecular radicals on the matrix chain ends [42]. Consequently, multi-scale self-healing at both micrometer and molecular levels that is critical for space applications could possibly be developed.

It is worth noting that in addition to GMA, other vinyl monomers might also be applicable for crack repair due to the living polymerization mechanism. In this context, styrene was encapsulated by PMF and embedded in living

PMMA matrix. To reduce the high volatility of styrene and maintain the amount of core substance in capsulated healant for long time usage, microcapsules containing a mixture of styrene and epoxy diacrylate were also prepared. The active double bonds of epoxy diacrylate would also take part in the living polymerization during crack healing. Figure 5.19 shows the results of mechanical tests of the self-healing specimens. Compared with the composites with GMA-loaded capsules, the PMMA composites filled with either styrene-loaded microcapsules or styrene/epoxy diacrylate-loaded microcapsules exhibit poorer effect of impact strength restoration in the case of identical content of capsules of 15 wt%. This means that the polymerized GMA at the cracks offers the highest inherent strength and adhesion. The lowest healing efficiency of the composites with styrene-loaded microcapsules should result from rapid volatilization of styrene after breakage of the specimens (including the embedded microcapsules), which greatly decreases the effective dosage of styrene. As for the composites filled with styrene/epoxy diacrylate-loaded microcapsules, the healing efficiency is higher than those with styrene-loaded microcapsules, not only because of the reduced volatility of styrene as a result of the incorporation of epoxy diacrylate, but also because epoxy diacrylate reacts with the living matrix and growing polystyrene (PS) like a cross-linking agent.

The above data demonstrate that although the small molecules styrene and epoxy diacrylate can react with living PMMA at room temperature, other key factors related to the healing ability (e.g. volatility and strength of the formed polymer) should be considered and optimized when screening the possible candidates. Moreover, solubility difference between healing agent and the target material to be repaired as well as rate of polymerization of the healing agent, which control healing kinetics but are not discussed here, are also important parameters.

5.2.3 Preparation of GMA-Loaded Multilayered Microcapsules and their PS-Based Composites

The multilayered microcapsules are manufactured according to the processes shown in Figure 5.20. GMA is also chosen as the core material and PMF plays the role of the innermost shell. As for the intermediate shell, PMMA generated from ATRP constitutes the majority, because of the fast reaction rate of ATRP and commercial accessibility of the raw materials. Specifically, the second layer of the

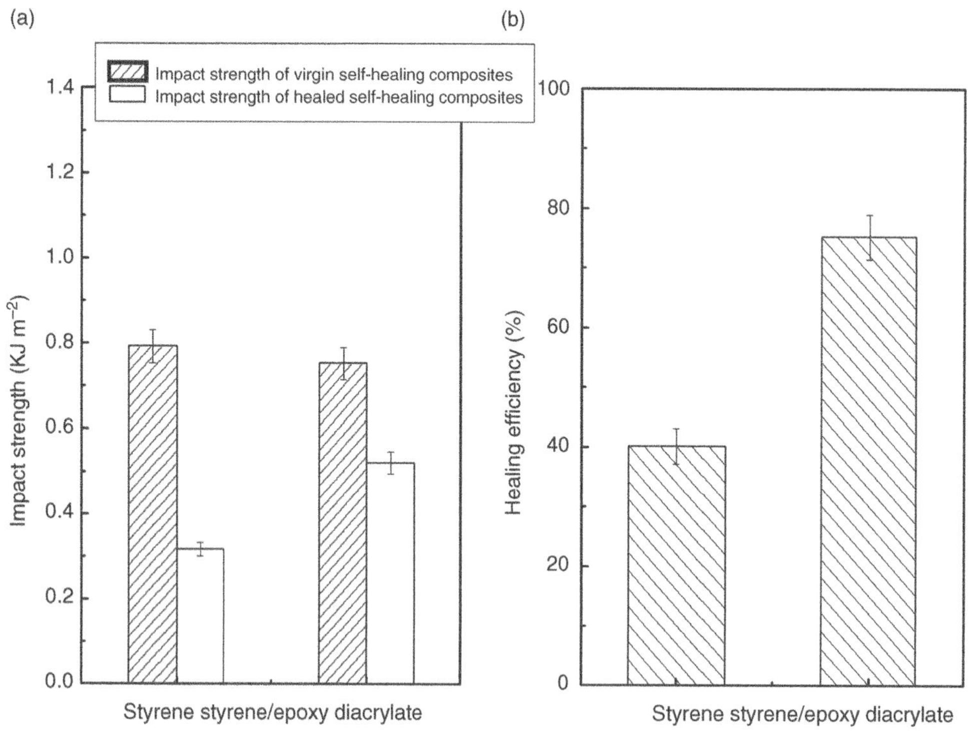

Figure 5.19 Healing effect of PMMA composites with styrene- and styrene/epoxydiacrylate-loaded microcapsules. (a) Impact strength. (b) Healing efficiency. Content of styrene- and styrene/epoxy diacrylate-loaded microcapsules in the composites: 15 wt%. Healing of the fractured specimens was conducted at 25 °C for 24 hours. The average diameter and core content of styrene-loaded microcapsules are 39 μm and 59.9%, respectively, and those of styrene/epoxy diacrylate-loaded microcapsules are 110 μm and 84.5%, respectively (the content of styrene is 67.6%).

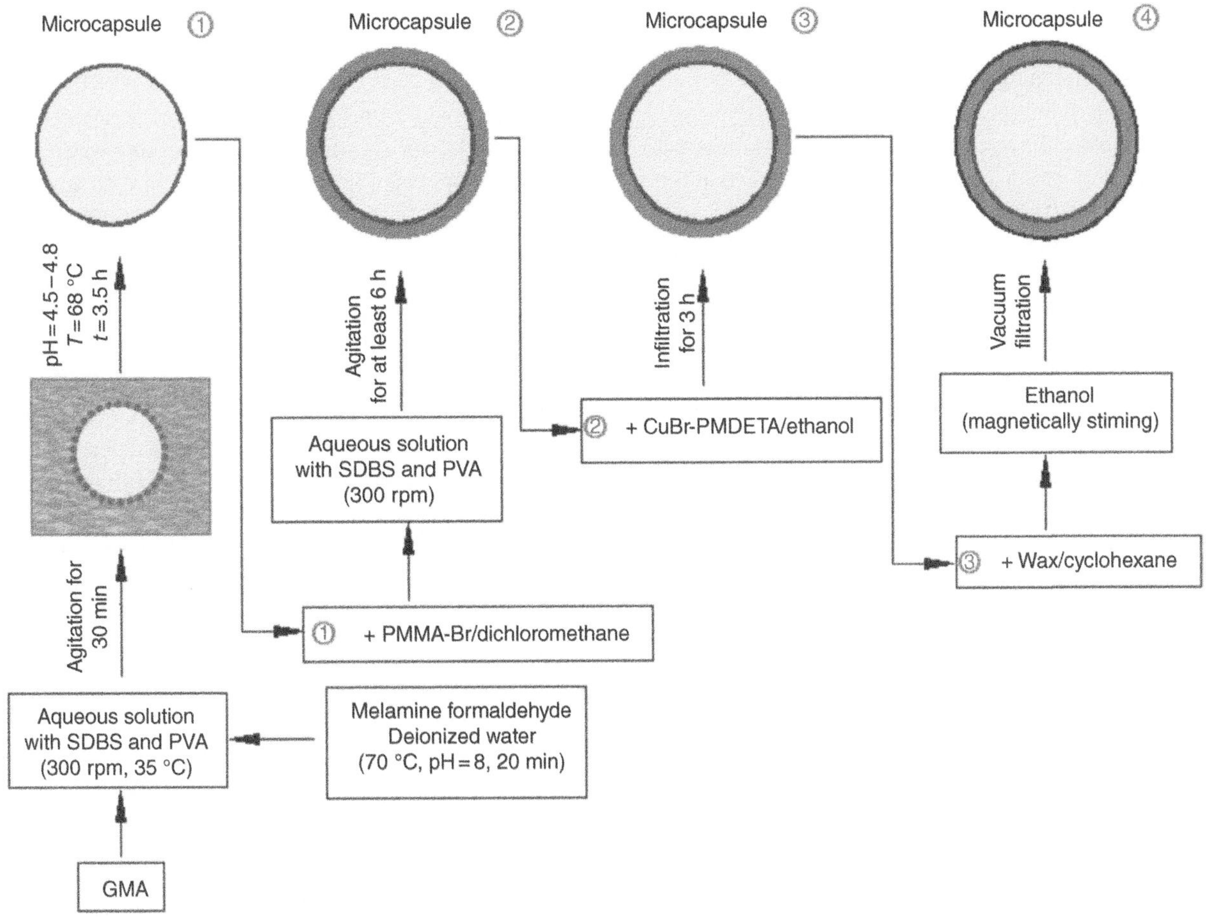

Figure 5.20 Schematic drawing of the route for preparing the multilayered microcapsule.

microcapsule, made of ATRP derived living PMMA together with the catalyst CuBr/PMDETA, is formed by the emulsion solvent evaporation method. Considering that CuBr/PMDETA would be easily dissolved in an aqueous system during the encapsulation and tend to be deactivated, the catalyst in the living polymer mixture is first removed after ATRP reaction. When the GMA-loaded PMF microcapsules have been coated by the living PMMA, the catalyst CuBr/PMDETA is incorporated once again via infiltration. Then, the outermost shell is built up by paraffin wax through dissolution–precipitation. The layer is believed to be capable of preventing Cu(I) in the catalyst from being oxidized to Cu(II).

For preparation of microcapsules ① (Figure 5.20), melamine (25 g, 0.198 mol), formaldehyde (37 wt%, 45 g, 0.554 mol) and 50 ml deionized water were mixed in a 250 ml three-neck round-bottom flask equipped with a magneton. Two drops of triethanolamine were added to the mixture. Then, the system was stirred at 70 °C for 15 minutes until a clearly transparent MF prepolymer solution was obtained. Afterwards, it was cooled down to room temperature. In another three-necked round-bottom flask (1000 ml) equipped with a mechanical stirring paddle, 320 ml surfactant aqueous solution containing 0.80 g SDBS and 0.40 g PVA were added. The flask was set into an oil bath of 35 °C and stirred at 300 rpm before the MF prepolymer solution and 80 g GMA were added into the solution. Then, 16 ml 5.0 wt% acetic acid was dropped into the flask to tune the pH value to about 4.8, meanwhile the system was maintained at 35 °C for at least 30 minutes to form a stable oil-in-water emulsion. The reaction temperature was then raised to 68 °C for over 40 minutes and maintained for 200 minutes under continuous agitation. Another 10% of the original MF prepolymer solution was added after the reaction proceeded at 68 °C for 30 minutes, and 8 ml 5.0 wt% acetic acid was added to the system every 60 minutes (twice in total). Once the isothermal reaction was completed, the flask was removed from the oil bath and cooled down to ambient temperature. The slurry of microcapsules was poured into a large amount of water to remove the emulsifier. The deposits were filtered through 300 mesh sieve and rinsed with deionized water several times. Having been dried at room temperature in a draught cupboard, the GMA-loaded PMF microcapsules (microcapsule ①, see Figure 5.20) were obtained as white fluidic powder.

For reparation of microcapsules ② (Figure 5.20), the second layer of the multilayered microcapsules was fabricated by the method of emulsion solvent evaporation. For the sake of visual identification, the second layer was created with red colorant except where otherwise specified. First, the living PMMA synthesized by ATRP was dissolved in methylene chloride with a content of 10% ($g\,ml^{-1}$), and a little Sudan IV was added to form a red solution. Then, 1 g GMA-loaded PMF microcapsules were immersed into certain amount of the solution in a 10 ml beaker sealed by preservative film for 5–60 minutes. Meanwhile, 100 ml aqueous solution containing 0.3 g SDBS and 0.15 g PVA was charged into a 250 ml three-neck round-bottom flask and vigorously agitated by an impeller blade. Afterwards, the slurry containing GMA-loaded PMF microcapsules and living PMMA solution was gradually fed into the above aqueous solution under continuous stirring. Having been agitated for at least six hours in a draught cupboard to make sure that methylene chloride was completely volatilized, the floating particles accumulated on the surface of the aqueous solution were carefully removed, while the deposition was washed three times by deionized water and collected by vacuum filtration. After being dried in atmosphere at room temperature, the resultant double-walled microcapsules (microcapsule ②, see Figure 5.20) were available as red powder.

To incorporate a CuBr/PMDETA catalyst system into the second layer to form microcapsules ③, a solution of CuBr/PMDETA in ethanol was made first as in the following. In a N_2 filled glove box, CuBr (1.425 g, 10 mmol) was added into ethanol, and PMDETA (1.725 g, 10 mmol) was dropped into the mixture under magnetic stirring to form a transparent solution with a light green color. Then, the prepared double-walled microcapsules ② were submerged into the catalyst solution for one to five hours. Intermittent shakes were given to the system to ensure the capsules were equally bathed in the solution. Eventually, the double-walled microcapsules including both living PMMA and CuBr/PMDETA within the second layer (microcapsule ③, see Figure 5.20) were obtained after filtration and dried in N_2 atmosphere.

For preparation of microcapsules ④, in a N_2-filled glove box, 1 g microcapsules ③ were immersed in 2 ml 6% ($g\,ml^{-1}$) solution of paraffin wax in cyclohexane for 10 minutes. The resultant suspension was added to 100 ml ethanol under continuous stirring, and then immediately vacuum filtrated. The ultimate multilayered microcapsules (microcapsule ④, see Figure 5.20) were produced after being naturally dried.

Before the microcapsules are compounded with PS to produce self-healing PS composites, modifications were made as follows. First, a silica coating was added to the PMMA-Br layer by fluoride-catalyzed silica condensation chemistry [43]. Fabrication of the silica coating was based on the procedure by Jackson [44]. A solution with cetyltrimethylammonium bromide (CTAB) (0.35 g), monobasic sodium phosphoric acid (0.7 g) and tetrabutylammonium fluoride (TBAF) solution (42 ml) was prepared. Then the mixture of the double-walled microcapsules (5 g) and Millipore water (140 ml) was poured into the above solution under slow agitation. The suspension was magnetically stirred at a low rate for five minutes before adding tetraethyl orthosilicate (TEOS) (3.5 g). The reaction continued at a low stirring rate for 24 hours. The resultant three-layered microcapsules were finally filtered and washed with deionized water followed by ethanol. After that, the catalyst CuCl/PMDETA was incorporated into the second shell following the same procedure mentioned above, and chloroparaffin instead of paraffin wax was then deposited onto the silica shell.

The self-healing PS composites were fabricated through compression molding. Prior to this, powdered PS was premixed with the multilayered microcapsules by dry blending and melt blending. For dry blending, PS particles were mixed with the multilayered microcapsules. For melt blending, PS particles were firstly melted at 170 °C for 10 minutes in the mixer, and then the multilayered microcapsules were added and mixed with the PS melt for 5 minutes under a rotation speed of 30 rpm. When the PS/multilayered microcapsule blends were available, they were compression molded at 150 °C using different molds to produce specimens for different measurements. Unfilled PS specimens were manufactured under the same conditions for comparison.

5.2.4 Self-Healing Performance of PS Composites Filled with GMA-Loaded Multilayered Microcapsules

The self-healing ability of PS composites containing these multilayered microcapsules was evaluated by Izod notched impact tests [36]. The healing efficiency, H_E, was described by the percentage recovery in impact strength: $H_E = \sigma_{Healed}/\sigma_o$, where σ_{Healed} and σ_o denote impact strengths of healed and virgin specimens for a given concentration of healing multilayered microcapsules, respectively. To eliminate the influence of multilayered microcapsules on impact strength of the polymer, apparent healing efficiency, H_E', was also calculated, which employs the impact strength of unfilled PS as the virgin property.

The composites specimens for the impact test were fabricated by compression molding at 150 °C after dry blending of both PS and the multilayered microcapsules. The 3D micro-computed tomography (μCT) image in Figure 5.21 shows that the multilayered microcapsules are well dispersed in the PS matrix, maintaining their spherical shape. Evidently, the multilayered microcapsules weathered the processing procedures and should be able to take effect after breakage.

Figure 5.21 3D distribution of the GMA-loaded multilayered microcapsules in thermally molded PS composites (content: 15 wt%, diameter: 101 μm, specimen size: 3 × 3 × 3 mm^3).

Figure 5.22 illustrates the time dependence of healing efficiency. It takes 12 hours at 25 °C for the present recipe to attain a healing efficiency of ~86%. Moreover, the curves begin to approximately level off after 24 hours approaching the equilibrium value of about 95% at 48 hours. In this context, a conservative healing period of 48 hours is used for the subsequent tests to ensure sufficient healing under room temperature conditions. The calculated healing efficiencies are similar to the apparent ones due to insignificant variation in impact strength of PS after the addition of the multilayered microcapsules.

The influence of the microreactor's content on healing efficiency is systematically investigated (Figure 5.23). It is found that healing efficiency increases with a rise in microreactor's content in the range studied owing to the increased amount of healing agent. At a microreactor's content of 20 wt%, the average healing efficiency amounted to 95%. Moreover, the composites also exhibit attractive durability within a couple of months. The results suggest that the designed multilayered microcapsules indeed work in self-healing of the thermo-molded PS composites. Considering that the solvent effect of the released fluidic GMA might possibly contribute to crack healing of the thermoplastic material, control experiments were conducted on the specimens filled with "inactive" multilayered microcapsules, which have nearly the same composition and structure as the conventional multilayered microcapsules but exclude CuCl/PMDETA in the catalyst system. As a result, the healant GMA offered by the "inactive" multilayered microcapsules could not be polymerized when the PS composites are ruptured and only serves as solvent. The data in Figure 5.23 indicate that the composites

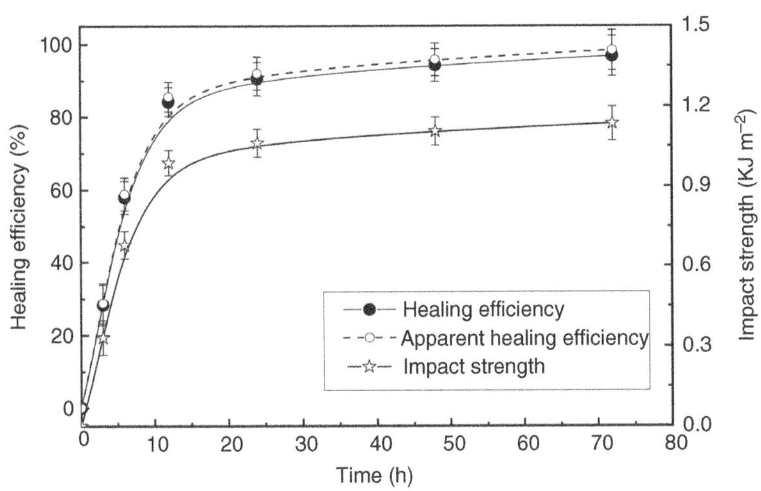

Figure 5.22 Time dependence of healing efficiency of the thermally molded self-healing PS composites containing 20 wt% GMA-loaded multilayered microcapsules (diameter: 101 μm, healing temperature: 25 °C). Impact strengths of the healed specimens are also plotted as reference. The average impact strengths of unfilled PS and the virgin specimens with 20 wt% multilayered microcapsules are 1.154 and 1.173 kJ m^{-2}, respectively. *Source:* Reprinted from Zhu et al. [26]. Copyright 2013, with permission from the Royal Society of Chemistry.

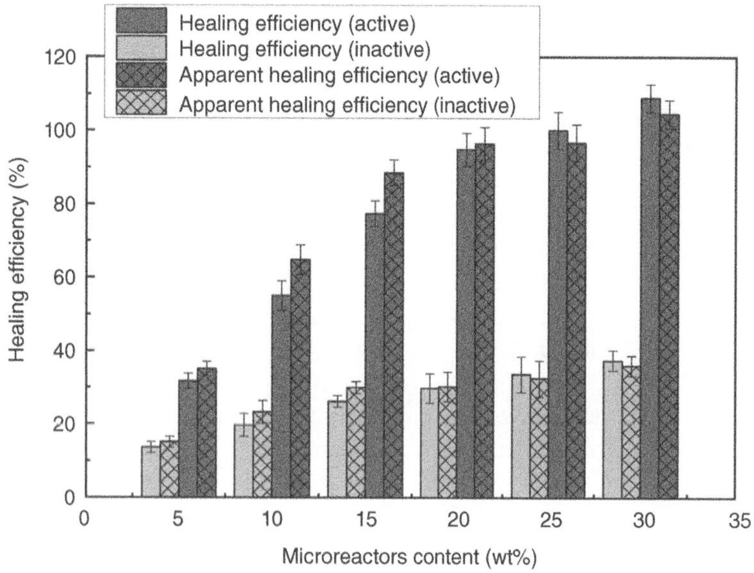

Figure 5.23 Healing efficiencies of the thermally molded self-healing PS composites containing different contents of GMA-loaded multilayered microcapsules (diameter: 101 μm, healing temperature and time: 25 °C, 48 hours). The "inactive" multilayered microcapsules were incorporated to highlight the effect of solvent healing (see Section 5.1.2 for details of the specimen preparation). *Source:* Reprinted from Zhu et al. [26]. Copyright 2013, with permission from the Royal Society of Chemistry.

based on the conventional multilayered microcapsules are able to recover impact strength to a much higher extent than those based on "inactive" multilayered microcapsules. In the case of higher content of multilayered microcapsules, the solvent effect of GMA is only responsible for a small part of the healing effect. In the meantime, we carried out another two groups of control healing experiments to check the solvent effect. A certain amount of active and inactive healing agent (estimated by the method of Rule et al. [20]) was manually injected onto impact fracture surfaces of unfilled PS specimens. The so-called active healing agent is a mixture of GMA, PMMA-Br and CuCl/PMDETA, while the inactive healing agent is a mixture of GMA and PMMA-Br. In principle, the active healing agent should be polymerizable whilst the inactive one cannot. After rehabilitation at 25 °C for 48 hours like the authentic specimens, healing efficiency offered by the inactive healing agent is only about one-third of that of the active version. The results coincide with those shown in Figure 5.23, and confirm that polymerization of GMA, rather than its solvent effect, plays the leading role in crack repair of PS. This is understandable as GMA is not a good solvent for PS so that only low degree swelling and chain entanglement arise.

Encouraged by the fact that the multilayered microcapsules can withstand the compression molding of PS and provide PS with self-healing performance, we further compounded PS with the multilayered microcapsules through melt blending (using a Haake torque rheometer) instead of dry blending before compression molding. Figure 5.24 compares the healing ability of PS composites containing multilayered microcapsules of different diameters at a fixed content of 20 wt%, which were fabricated by different methods. It is seen that for powder blending, the healing efficiency increases with increasing size of multilayered microcapsules. As for the composites manufactured by melt blending, the healing efficiency increases with a rise in size of multilayered microcapsules up to 52 μm and reaches a maximum of about 70%. When the size of multilayered microcapsules exceeds 52 μm, the healing efficiency significantly decreases and even no healing effect is detected in the case of larger multilayered microcapsules (e.g. 153 and 215 μm). Since the above investigation of solvent effect reveals that even if the released GMA fails to be polymerized the composites can still be healed at lower level, the zero healing efficiency observed here should result from the fact that the larger multilayered microcapsules are mostly ruptured during melt blending by the shear force in the torque rheometer. Because the healant has been released to the matrix in advance, no healant is available for the subsequent crack repairing. The deduction is supported by DMA of PS composites that were subjected to successive melt blending and compression molding. That is, the glass transition temperatures, T_g, of the specimens containing larger microreactors are slightly lower than those of unfilled PS and the specimens containing smaller microreactors due to the plastification by the released healant.

Figure 5.25 depicts the impact and flexural performances of the composites. The impact strength of PS composites

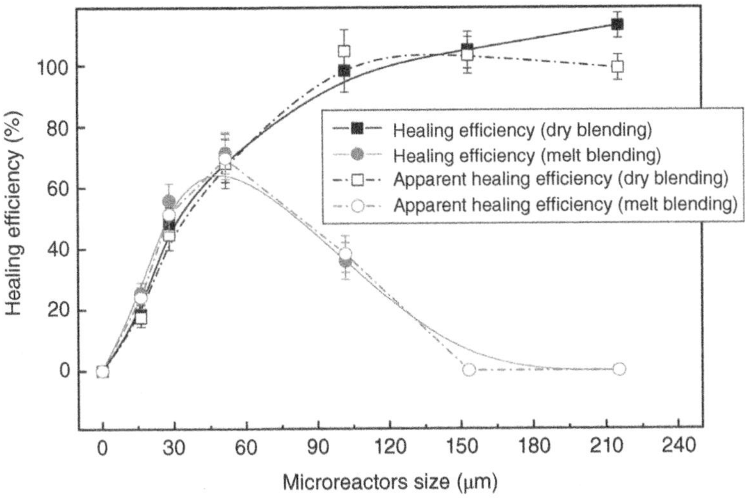

Figure 5.24 Healing efficiencies of the thermally molded self-healing PS composites containing 20 wt% GMA-loaded multilayered microcapsules as a function of premixing technique and size of multilayered microcapsules (healing temperature and time: 25 °C, 48 hours). *Source:* Reprinted from Zhu et al. [26]. Copyright 2013, with permission from the Royal Society of Chemistry.

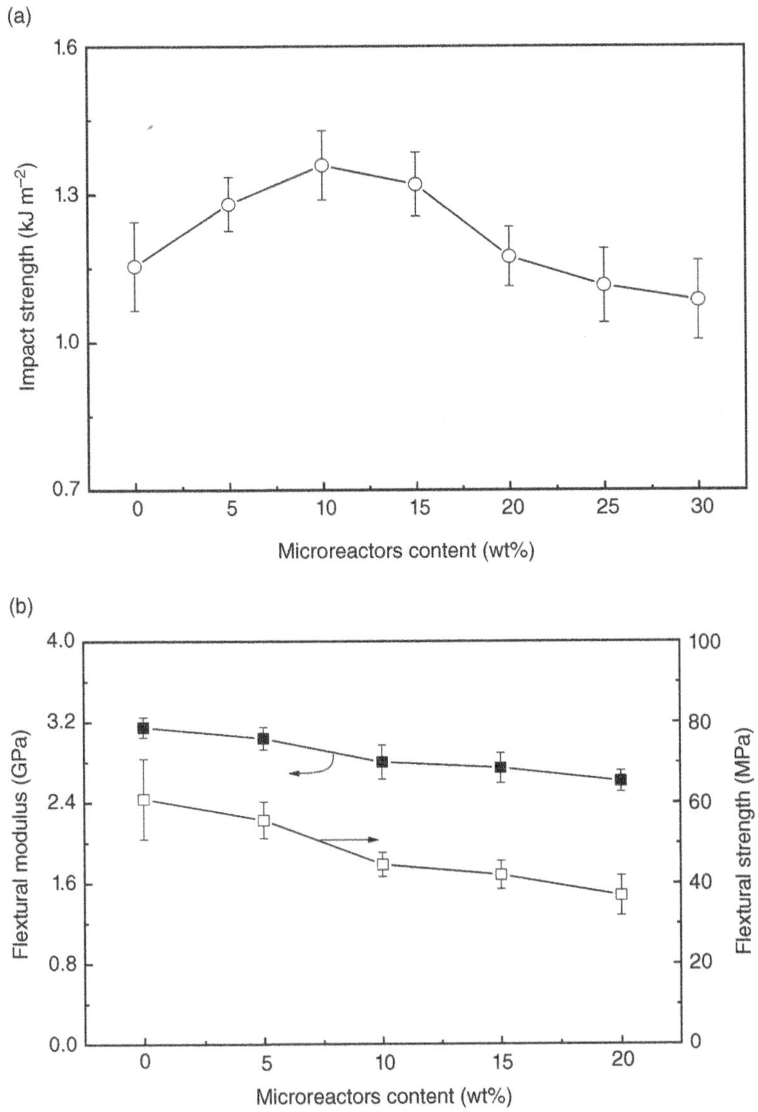

Figure 5.25 Effect of the content of the GMA-loaded multilayered microcapsules on (a) impact strength and (b) flexural properties of self-healing PS composites (diameter: 101 μm). All the specimens were fabricated by compression molding at 150 °C after dry blending.

increases first and then decreases with a rise in content of multilayered microcapsules (Figure 5.25a). A similar toughening effect has been reported in rigid polymer with liquid-loaded microcapsules [9, 45]. As for the static mechanical performance, the reduction in modulus of the materials with a rise in content of the multilayered microcapsules is marginal, while the decrease in strength is relatively obvious (Figure 5.25b). In general, this agrees with the law of microcapsules or rubbery particle filled polymers [45, 46].

In-situ confocal Raman microscopy was used to study the re-cracked surface of a healed PS composites specimen containing multilayered microcapsules (Figure 5.26a). By using the area of the peak at 1730 cm^{-1} (related to the stretching vibration of the carbonyl group) as the reference, variation in the area of the peak at 1640 cm^{-1} (assigned to the stretching vibration of the double bond) quantifies consumption of C=C. As the area ratio of ν(C=C)/ν(C=O) of the released GMA monomer spectrum approximately equals 0.9 (see the trace denoted by 0 hours in Figure 5.26a), the remarkably reduced area ratio of ν(C=C)/ν(C=O) of the fractured surface spectrum (see the trace denoted by 48 hours in Figure 5.26a) is indicative of consumption of the double bonds. This means that the healant GMA has been polymerized, covering the crack plane. Moreover, the confocal Raman intensity depth

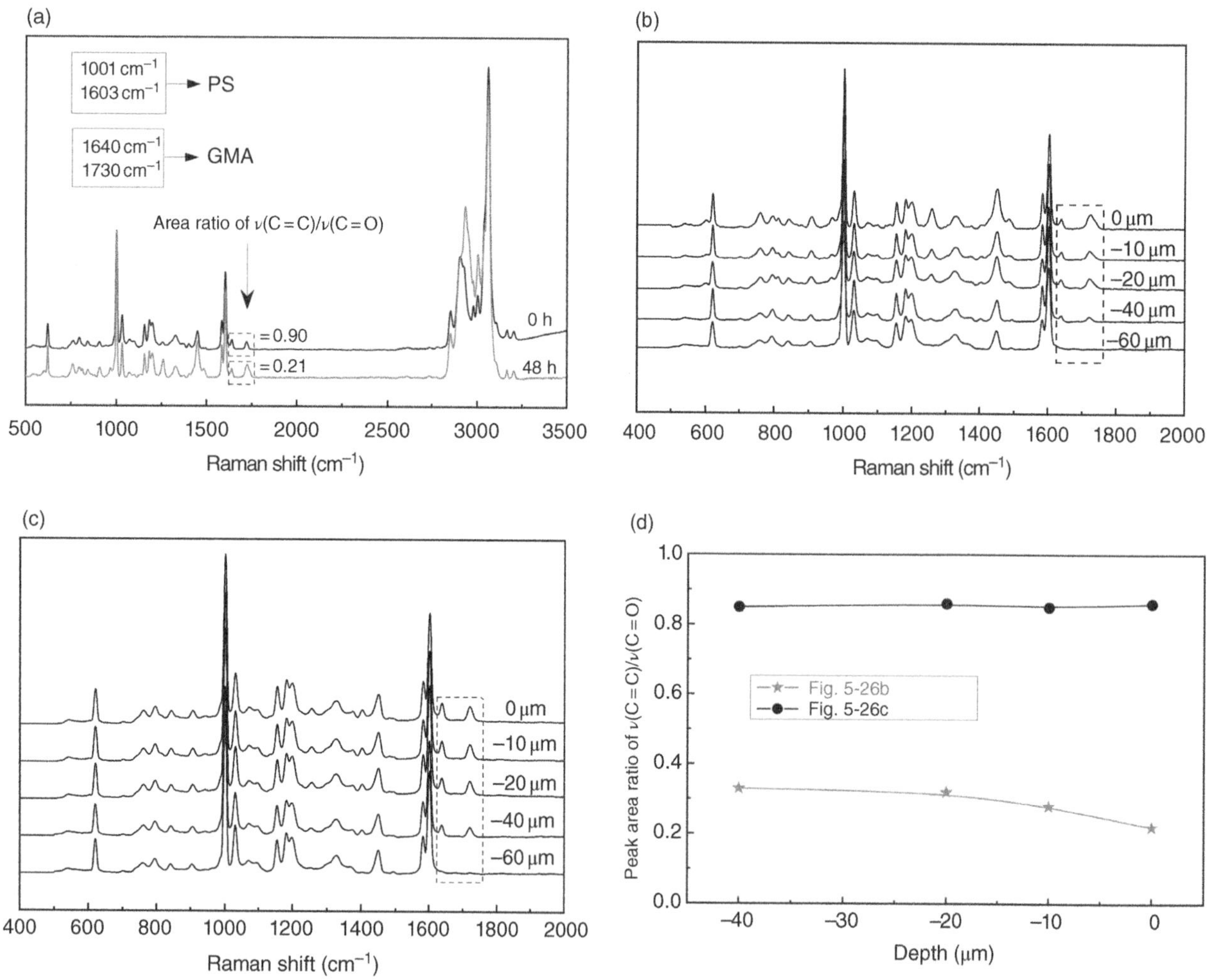

Figure 5.26 (a) Raman spectrum collected from the re-fractured surface of a healed PS specimen with 20 wt% GMA-loaded multilayered microcapsules (healing temperature and time: 25 °C, 48 hours) in comparison with that from the newly fractured surface of the composite (i.e. healing time: 0 hours). (b) Raman spectra collected from the re-fractured surface of a healed PS specimen with 20 wt% GMA-loaded multilayered microcapsules at different depths (healing temperature and time: 25 °C, 48 hours). (c) Raman spectra from the fractured surface of a control PS specimen with 20 wt% "inactive" multilayered microcapsules recorded at different depths. (d) Area ratio of ν(C=C)/ν(C=O) versus scanning depth from the fractured surface of the PS specimens. *Source:* Reprinted from Zhu et al. [26]. Copyright 2013, with permission from the Royal Society of Chemistry.

profiles demonstrate that the healant GMA, or more exactly PGMA, has permeated into the matrix for a certain distance (Figure 5.26b,d). For the control PS composites containing the same content of "inactive" multilayered microcapsules, however, the area ratio of ν(C=C)/ν(C=O) of the fractured surface spectrum keeps at ~0.9 at different depths (Figure 5.26c,d). This implies that only the GMA released from active multilayered microcapsules is able to form macromolecules going under the surface of the PS matrix about 40 μm deep after diffusion. Such a mechanical interlocking must help to improve re-bonding of cracks. On the basis of the above discussion, the healing mechanism can be summarized as in the following. With the crack initiation and propagation in the PS composites, the embedded multilayered microcapsules are damaged, releasing GMA monomer into the crack plane through capillary action. Meanwhile, polymerization of the GMA is triggered by the second shell (i.e. the catalyst system) of the multilayered microcapsules (Figure 5.12). Prior to the formation of PGMA macromolecules, GMA has enough time to infiltrate the matrix, slightly swell the matrix and induces marginal physical entanglement of matrix polymer across the crack interface to contribute to some crack repair. With the lapse of time, PGMA chains gradually grow. They not only fill the crack volume but also are firmly rooted in the subsurface of cracks by entangling both PGMA and PS molecules. Consequently, the cracks are re-bonded, and the mechanical strength of the material is restored. This explains why the increment of healing efficiency of the PS composites containing conventional multilayered microcapsules with a rise in content of multilayered microcapsules is far more remarkable than that of the composites containing inactive multilayered microcapsules (Figure 5.23). After all, solubility of PS in GMA is not that high and the effect of solvent healing must be limited accordingly, less dependent on the solvent dosage. In contrast, the polymerization product of GMA fastens the cracked portions together like a staple. Its effect should be strengthened with increasing amount of the healant.

5.3 Extrinsic Self-Healing Via Free Radical Polymerization

In the above subsections, the feasibility of using GMA as healing agent has been shown. Owing to its bifunctional feature, nucleophilic addition, ring-opening reaction, and ATRP are involved in the crack healing processes.

Besides, styrene has also shown its applicability in crack repair of living PMMA. Styrene is a reactive monomer with a vinyl group and has been employed to make a variety of styrene plastics [47, 48]. In particular, it has much lower viscosity than epoxy monomer, and is used to mix with epoxy as diluent. Owing to these characteristics and its compatibility with epoxy, styrene should also serve as a healing agent for manufacturing self-healing epoxy composite based on microencapsulation strategy. Jung et al. [49] prepared poly(oxymethylene urea) (PMU) walled microcapsules containing styrene and PS for making self-healing polyester composite. PS in the microcapsules helped to increase the viscosity of the core material. The corresponding repair mechanism used naturally occurring functional sites in the polyester matrix network to initiate the repair action. In our work, another healing chemistry (i.e. free radical polymerization) will be applied. By embedding the microcapsules containing styrene and those with its radical initiator benzoyl peroxide (BPO) in epoxy matrix, a new self-healing epoxy material will be produced. Following the extrinsic self-healing mechanism, the product of room temperature polymerization of styrene initiated by BPO, i.e. PS, would fill up the cracked portions of epoxy matrix and recover the mechanical performance of the latter.

5.3.1 Microencapsulation of Styrene and BPO

In accordance with the above idea, styrene- and BPO-loaded microcapsules should be prepared as the first step for making the self-healing composites. Here, PMF acts as the wall substance of the two types of microcapsules. Since both styrene- and BPO-loaded microcapsules have the same wall shell, homogeneous distribution of the spheres in epoxy matrix can be ensured.

To synthesize styrene-loaded microcapsules, 4.4 g cobaltous naphthenate (an accelerant for the subsequent polymerization of released styrene) was dissolved in 44.48 g styrene together with 7.2 g epoxy acrylate (which reduced volatility of styreneduring encapsulation). Afterwards, 14.6 ml sodium styrene-maleic anhydride (SMANa) copolymer emulsifier and 2.3 g gum arabic were incorporated into the mixture under agitation. The prepolymer of melamine (17.5 g) and formaldehyde (28.5 g) formed at 80 °C and pH = 8–9 was poured into the system with stirring. By dropwise addition of 5% citric acid, the pH value of the system became about 5–6. The reaction proceeded at 50 °C for three hours. Eventually, the resultant spheres were thoroughly washed with deionized water three times, and dried.

The BPO-loaded capsules were prepared in a way similar to that for making styrene-loaded capsules. BPO is solid at room temperature, so the core substance of the BPO-loaded capsules has to be a toluene solution of BPO. To be precise, 4.42 g BPO was dissolved in 39.8 g toluene together with 400 ml deionized water, 14.5 ml SMANa emulsifier and 2 g gum arabic. Having been rigorously stirred at room temperature for eight minutes, the prepolymer of melamine and formaldehyde was added. Similarly, 5% citric acid was dropwise incorporated into the system, leading to pH value of 5–6. After four to five hours at 50 °C, the produced capsules were collected, washed, and dried.

Since styrene is highly volatile and BPO is thermally sensitive, the encapsulation reactions cannot be carried out at higher temperature. Therefore, the mild two-step approach using melamine-formaldehyde resin as the wall material was applied. To obtain the optimal products, some important processing parameters should be investigated.

It is known that each melamine molecule contains three amino groups. In principle, 1 mol melamine can react with 6 mol formaldehyde yielding hexamethylol melamine. The subsequent condensation of hexamethylol groups results in cross-linked polymer consisting of either methylene or dimethylene ether linkages. It is understood that the cross-link density and contents of methylene or dimethylene ether linkages depend on the molar ratio of melamine to formaldehyde. Figure 5.27 shows the pyrolytic temperatures at 1% weight loss of the polymerization products of melamine and formaldehyde as a function of their molar ratios, which can serve as a measure of cross-link density and composition of the cross-linked polymer. It is seen that when the formaldehyde to melamine molar ratio is in the range of 1–3, the pyrolytic temperature of the resultant resin increases as a result of raised cross-link density. In principle, a further increase in the molar ratio (4–6) may lead to higher cross-link density, but the gradual decrease of pyrolytic temperatures is indicative of increased dimethylene ether linkages that can be easily cut off at higher temperature. Consequently, the proper formaldehyde to melamine molar ratio should be 2.5–3.

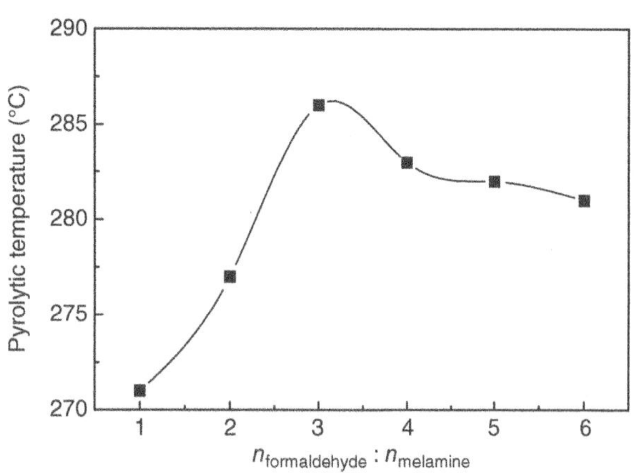

Figure 5.27 Thermal decomposition temperatures at 1% weight loss of melamine-formaldehyde resin versus molar proportions of the monomers.

The pH value of the reaction system is an important factor for quality control of the ultimate product. It was found that it was better to synthesize the prepolymer of melamine and formaldehyde at a pH value of 8–9. The obtained dimethylol melamine and trimethylol melamine favored further condensation at lower pH value. For the subsequent condensation process, higher catalyst concentration (i.e. lower pH value) may accelerate the reaction, so that there is not enough time for the polymer to encapsulate styrene or BPO solution, leading to formation of melamine resin depositions. Contrarily, higher pH would delay the condensation reaction, and hence induce incomplete encapsulation. Accordingly, the encapsulation should also proceed at a suitable pH value, i.e. 5–6.

Our previous work suggested that the size of microencapsulated healing agent is closely related to the healing performance [10]. In this regard, species and dosage of emulsifier should be carefully chosen. First, we used SDBS) and found its emulsification is not satisfied as characterized by the heterogeneous coverage of the wall materials on the capsules. We also tried Tween 80, which resulted in extremely low reaction rate and thin wall shell of the capsules, because it is electroneutral and has poor affinity for the positively charged prepolymer of melamine and formaldehyde.

Comparatively, SMANa copolymer that consists of hydrophobic backbone, hydrophilic carboxylic side groups and hydrophobic side aromatic rings can bring about much better emulsification. When SMANa copolymer and the core substance are mixed in water, oriented arrangement of the emulsifier molecules surrounding the hydrophobic core substance would take place. That is, an adsorption layer is built up on each core droplet with inward aromatic rings and outward carboxylic groups. Consequently, a negative electric field appears at the outer surface of the core droplets, which prevents the droplets from coalescing and attracts the wall formers with positive charges. Polycondensation of the prepolymer of melamine and formaldehyde happens in this region. During emulsification, adequate emulsifier dosage should be used to separate the hydrophobic core substance into tiny droplets. Besides, gum arabic is a nonionic surfactant and is able to reduce oil–water interfacial tension. Oil droplets can be stabilized by the gum arabic adsorbed on their surfaces. A certain amount of gum arabic in the system helps improve the stability of the initial emulsion. We have measured the influence of emulsifier concentration on size and size distribution of the resultant microcapsules. With an increase in emulsifier concentration, the size of the capsules is reduced and the size distribution becomes narrower, and then the changes are not significant. This can be explained by the role of emulsifier as discussed above. That is, when all the core droplets are covered by the emulsifier layer with sufficient thickness, the excessive emulsifier cannot have any effect.

The morphology of the microcapsules is shown in Figure 5.28. Many small particles adhere to the surface of the capsules. This reflects that the formation of the microcapsules is actually the accumulation of tiny particles of melamine-formaldehyde resin onto the emulsified core droplets. Figure 5.28c clearly shows the wave-like surface originating from compact stacking of the particles. According to the observation of the broken capsules (Figure 5.28e,f), the wall thickness is about 0.1 μm. Overall, the microcapsules with melamine-formaldehyde resin as the wall material have a relatively rough exterior surface, which would help enhance interfacial interaction with the matrix in the subsequent manufacturing of self-healing materials.

Figure 5.29 shows FTIR spectra of the microcapsules, and core and wall substances. The wall substance was obtained from the ground microcapsules that had been

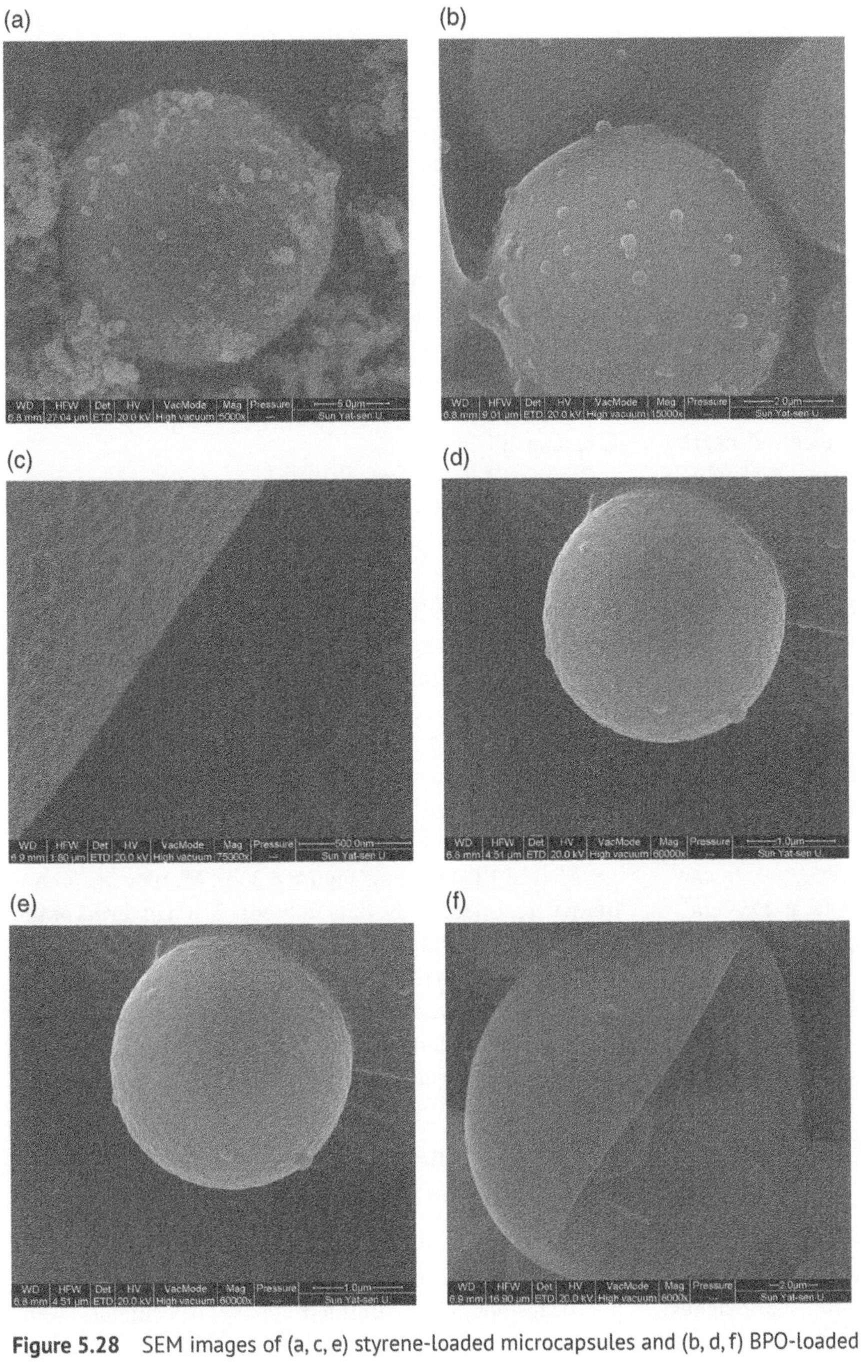

Figure 5.28 SEM images of (a, c, e) styrene-loaded microcapsules and (b, d, f) BPO-loaded microcapsules.

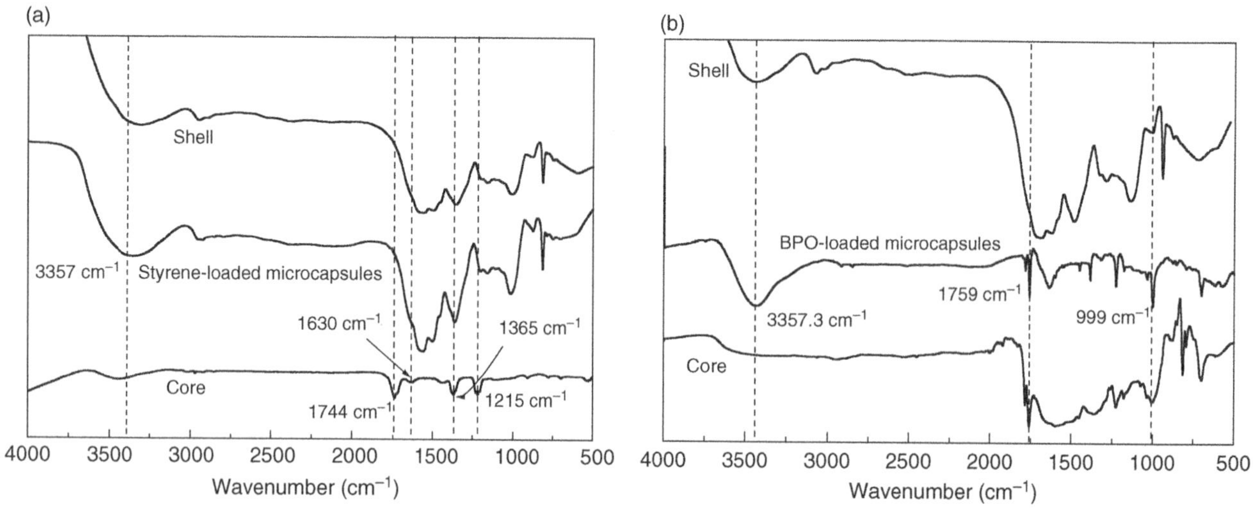

Figure 5.29 FTIR spectra of (a) styrene- and (b) BPO-loaded microcapsules in comparison with those of the wall shell and core substances.

thoroughly extracted and dried. Meantime, the core substance was isolated from the extractant (i.e. acetone). On the spectra of both styrene-loaded microcapsules and their wall shell (Figure 5.29a), there is a strong absorption at 3357.3 cm^{-1} due to the stretching mode of N—H. It is worth noting that a weak peak is also perceived nearby (~3357.3 cm^{-1}) on the spectrum of the core material, which can be attributed to the characteristic absorption of vinylbenzene. In addition, the peak at 1630 cm^{-1} represents the stretching of C=C, while those at 1365 cm^{-1} are assigned to the aromatic ring. The absorptions at 1744 and 1215 cm^{-1} result from C=O and C—O of epoxy acrylate (which is mixed with styrene to increase viscosity of the latter to certain extent). The results suggest that the produced styrene-loaded capsules are indeed melamine-formaldehyde resin walled with the inclusion of styrene. On the other hand, we can also find the peak at 3357.3 cm^{-1} on the spectra of BPO-loaded microcapsules and the wall shell, which is absent on the spectrum of the core (Figure 5.29b). In contrast, the characteristic absorption of C=O at 1759 cm^{-1} and that of O—O stretching at 999 cm^{-1} appear on the spectra of the capsules and core. Evidently, the BPO/toluene solution has been encapsulated by melamine-formaldehyde resin as expected.

Thermal stability of the microcapsules is closely related to the manufacturing of self-healing composites. By measuring pyrolytic behaviors of the capsules, we find that both styrene- and BPO-loaded microcapsules exhibit low onset temperatures. This is because of the high volatility of the core substances, which tend to penetrate the wall shells and evaporate at elevated temperature. As a result, gradual weight loss is perceived. When temperature approaches 400 °C, a drastic decrease in weight appears due to the decomposition of the wall material. It is interesting to find that the onset temperatures of the BPO-loaded microcapsules near 400 °C are higher than that of neat melamine-formaldehyde resin. Probably the presence of the core BPO is responsible.

Compared with styrene, BPO is more active as viewed from their responses to heating. Considering that the microcapsules are prepared at 50 °C, it should be confirmed whether the activity of BPO is maintained after encapsulation. Accordingly, BPO-loaded microcapsules were ground at room temperature and then mixed with styrene at 0 °C. By heating the mixture in DSC pans, the exotherm resulting from the thermal polymerization of styrene was recorded (Figure 5.30a). Clearly, the BPO inside the microcapsules is able to initiate the polymerization of styrene as reflected by the obvious exothermic peaks at different heating rates, which resemble the curves of the styrene/as-received BPO system (Figure 5.30b). Moreover, a quantitative treatment of the data in Figure 5.30a in terms of the well-known Kissinger method [50] gives the activation energy of 64.0 kJ mol^{-1}, which is comparable with the value estimated from Figure 5.30b, i.e. 57.7 kJ mol^{-1} [51]. It can thus be concluded that the activity of BPO is almost unaffected by the encapsulation process.

5.3.2 Self-Healing Performance of Epoxy Composites Filled with Dual Capsules

Based on the above study, we prepared self-healing epoxy composites with embedded styrene- and BPO-loaded microcapsules. The unfilled epoxy specimens were

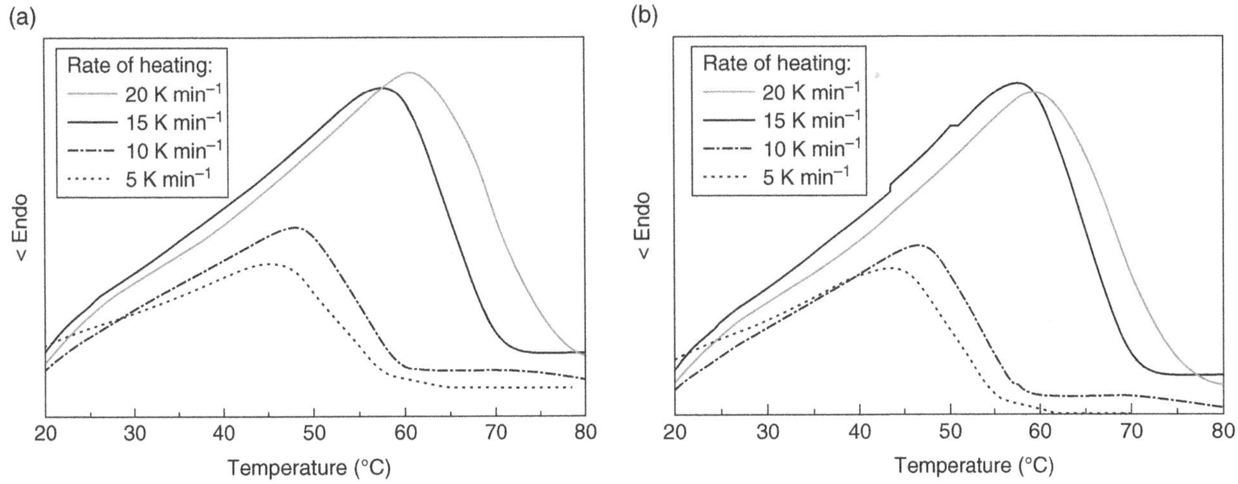

Figure 5.30 DSC heating curves of the polymerization of styrene initiated with (a) BPO released from the ground BPO-loaded microcapsules (core content = 50 wt%) and (b) as-received BPO. The weight ratio of the BPO-loaded microcapsules to styrene is 1:5.

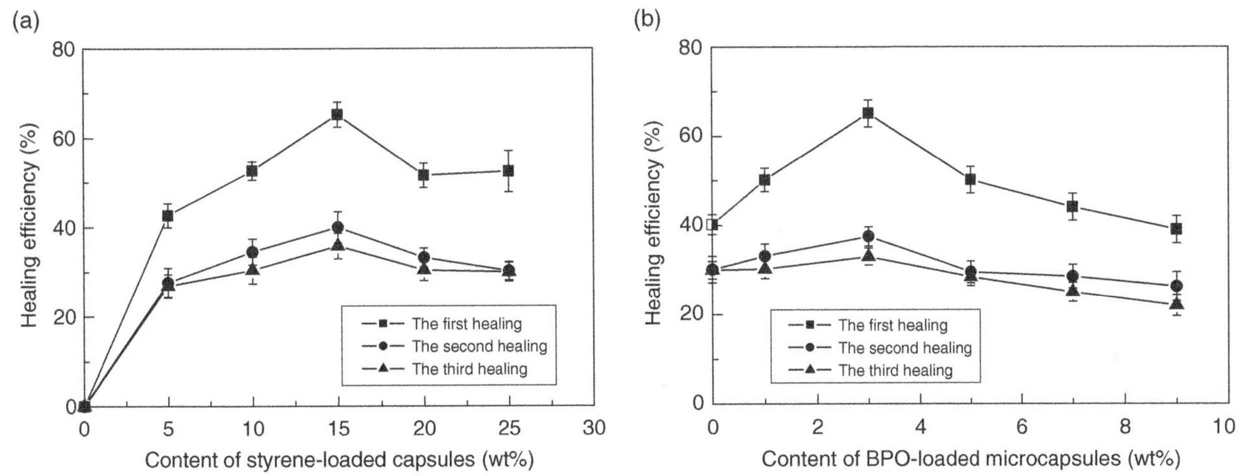

Figure 5.31 Healing efficiency of self-healing epoxy composites containing styrene- and BPO-loaded microcapsules as a function of (a) content of styrene-loaded capsules at a fixed content of BPO-loaded capsules of 3 wt% and (b) content of BPO-loaded capsules at a fixed content of styrene-loaded capsules of 15 wt%. Size and core content of the styrene-loaded capsules: 30 μm and 40 wt%, respectively. Size and core content of the BPO-loaded capsules: 50 μm and 45 wt%, respectively. Healing was conducted at 25 °C for 24 hours.

fabricated by mixing EPON 828 epoxy with its curing agent tetraethylenepentamine (TEPA) at a weight proportion of 100/14. The self-healing epoxy composites were prepared by uniformly mixing a certain amount of styrene- and BPO-loaded microcapsules together with the aforementioned mixture of EPON 828 and TEPA. To obtain cured versions, both the unfilled epoxy and filled compounds were degassed and poured into closed silicone rubber molds and cured for seven days at room temperature.

The healing efficiency of the materials was determined by comparing impact strengths of virgin and healed specimens according to the method proposed by Jones et al. [36] (see Section 5.2.2 for more details). Figure 5.31 shows the plots of healing efficiency of the epoxy composites as a function of the content of the capsules. The results indicate that the healing system works. In addition, the healing effect is controlled by the relative amount of the two types of microcapsules. Basically, a maximum healing efficiency of about 65% is detected at 15 wt% styrene-loaded capsules and 3 wt% BPO-loaded capsules. This can be explained by the fact that when BPO dosage is excessive the resultant polystyrene has short chains and

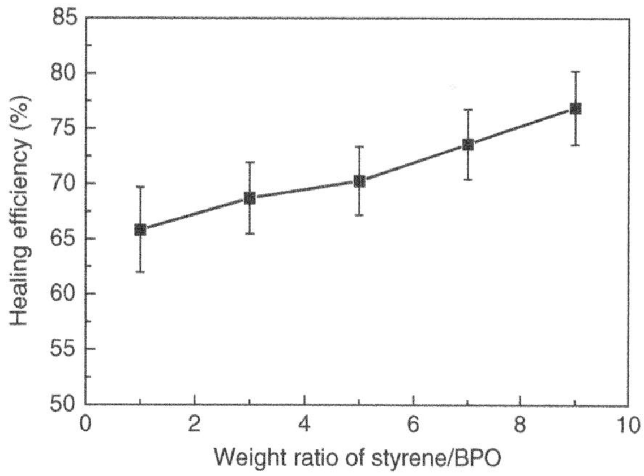

Figure 5.32 Healing efficiency of control unfilled epoxy specimens that were repaired by manually injecting premixed styrene and BPO. Healing was conducted at 25 °C for 24 hours.

low molecular weight. The adhesive strength of the bonding material at the cracked interface has to be low, so that the healing efficiency is not that high. In the case of excessive styrene-loaded microcapsules, the insufficient BPO leads to rather low reaction rate or even oligomer instead of macromolecules. The measured healing efficiency has to be low again.

We also investigated the ability of multi-repair of the materials. It is seen that the self-healing epoxy offers detectable healing effect even after the third failure. Compared with the data of the first healing, the healing efficiency of the second and the third repair is reduced by about 20–30%. This means that some healed parts might be stronger than the matrix. Under the second or third impact, cracks have to go round the healed portions and travel along a new route with unbroken microcapsules. As a result, these capsules take effect, and the cracks are partly healed showing the capability of multi-repair.

Figure 5.32 shows the healing efficiency of control unfilled epoxy specimens, which were healed by manually injecting premixed styrene and BPO onto the impact fracture surfaces. The results of the reference specimen testing are believed to provide an upper limit for the healing efficiency under ideal conditions and a benchmark for comparison with specimens in which the healing agent is embedded [52]. The data in Figure 5.32 indicate that the healing efficiency increases with increasing styrene content within the range of interest and is obviously higher than the values of authentic healing. This is because styrene can be well mixed with BPO under the simulation condition, which allows a complete reaction. Nevertheless, the healing efficiency of the reference specimens has not yet approached 100%. It seems that (i) the high volatility of styrene and (ii) the lower strength of polystyrene than that of matrix epoxy might account for the unsatisfactory healing effect. The problems can be solved by increasing the content of thickening reagent like epoxy acrylate and incorporation of reinforcements like carbon nanotubes into the styrene-loaded microcapsules, which will be investigated in future work.

5.4 Concluding Remarks

This chapter is devoted to expansion of the family of healing agents, and hence the family of materials to be repaired and healing chemistries. Small molecule monomers including GMA and styrene have shown their capability of rehabilitating damage in epoxy and PMMA. The monomers locally swell the matrix, allowing the reaction between the monomers and active groups in the matrix or living chain ends of the matrix. Accordingly, nucleophilic addition, ring-opening reaction, ATRP, and free radical polymerization are involved in the healing processes.

The application of small molecule monomers combines different effects into the healant, such as low viscosity and solvent effect. The different healing mechanisms might take effect under different conditions or in different portions of a composite, so that high healing efficiency is always ensured. This improved adaptive capability is meaningful for practical usage because fabrication and shelf aging would lead to heterogeneous change of the properties of materials. Compared with an epoxy monomer based healing system where viscosity is more temperature dependent, the flowability of the small molecule monomers does not greatly change with temperature and hence they might work more effectively at lower temperature.

Since both GMA and styrene contain C=C bonds, it is also worth checking other similar chemicals to see whether they are applicable for self-healing formulations. There are far more species of alkene compounds carrying one or more than one C=C bond connected with electron withdrawing groups than those of molecules with specific functional groups like epoxide. In this context, there are plenty of potential candidates for healing agents waiting to be developed for different matrices.

References

1 Wool, R.P. and O'Connor, K.M. (1981). A theory of crack healing in polymers. *Journal of Applied Physics* 52: 5953–5963.

2 Wool, R.P. and O'Connor, K.M. (1982). Time dependence of crack healing. *Journal of Polymer Science, Polymer Letters Edition* 20: 7–15.

3 Wool, R.P. (2008). Self-healing materials: a review. *Soft Matter* 4: 400–418.

4 White, S.R., Sottos, N.R., Geubelle, P.H. et al. (2001). Autonomic healing of polymer composites. *Nature* 409: 794–797.

5 Meng, L.M., Yuan, Y.C., Rong, M.Z., and Zhang, M.Q. (2010). A dual mechanism single-component self-healing strategy for polymers. *Journal of Materials Chemistry* 20: 6030–6038.

6 Caruso, M.M., Delafuente, D.A., Ho, V. et al. (2007). Solvent-promoted self-healing materials. *Macromolecules* 40: 8830–8832.

7 Yang, J., Keller, M.W., Moore, J.S. et al. (2008). Microencapsulation of isocyanates for self-healing polymers. *Macromolecules* 41: 9650–9655.

8 Caruso, M.M., Blaiszik, B.J., White, S.R. et al. (2008). Full recovery of fracture toughness using a nontoxic solvent-based self-healing system. *Advanced Functional Materials* 18: 1898–1904.

9 Yuan, Y.C., Rong, M.Z., Zhang, M.Q. et al. (2008). Self-healing polymeric materials using epoxy/mercaptan as the healant. *Macromolecules* 41: 5197–5202.

10 Yuan, Y.C., Rong, M.Z., Zhang, M.Q., and Yang, G.C. (2009). Study of factors related to performance improvement of self-healing epoxy based on dual encapsulated healant. *Polymer* 50: 5771–5781.

11 Xiao, D.S., Yuan, Y.C., Rong, M.Z., and Zhang, M.Q. (2009). Self-healing epoxy based on cationic chain polymerization. *Polymer* 50: 2967–2975.

12 Yuan, Y.C., Rong, M.Z., and Zhang, M.Q. (2008). Preparation and characterization of microencapsulated polythiol. *Polymer* 49: 2531–2541.

13 Costa, L. and Camino, G. (1988). Thermal behaviour of melamine. *Journal of Thermal Analysis and Calorimetry* 34: 423–429.

14 Devallencourt, C., Saiter, J.M., Fafet, A., and Ubrich, E. (1995). Thermogravimetry/Fourier transform infrared coupling investigations to study the thermal stability of melamine formaldehyde resin. *Thermochimica Acta* 259: 143–151.

15 Xiao, D.S., Yuan, Y.C., Rong, M.Z., and Zhang, M.Q. (2009). A facile strategy for preparing self-healing polymer composites by incorporation of cationic catalyst-loaded vegetable fibers. *Advanced Functional Materials* 19: 2289–2295.

16 N'oskov, A.M. (1975). Infrared spectroscopic study of molecular associates in epoxy oligomers hardened by amines. *Journal of Applied Spectroscopy* 22: 183–185.

17 Kim, Y.H. and Wood, W.P. (1983). A theory of healing at a polymer-polymer interface. *Macromolecules* 16: 1115–1120.

18 Jud, K., Kausch, H.H., and Williams, J.G. (1981). Fracture mechanics studies of crack healing and welding of polymers. *Journal of Materials Science* 16: 204–210.

19 Aflal, M., Rahmathullah, M., and Palmese, G.R. (2009). Crack-healing behavior of epoxy–amine thermosets. *Journal of Applied Polymer Science* 113: 2191–2201.

20 Liang, S.Y. (1980). *Addition Reactions*, 37–40. Beijing: People's Education Press (in Chinese).

21 He, M.J. (1990). *Polymer Physics*, 334. Shanghai: Press of Fudan University (in Chinese).

22 Keller, M.K., White, S.R., and Sottos, N.R. (2007). A self-healing poly(dimethyl siloxane) elastomer. *Advanced Functional Materials* 17: 2399–2404.

23 Keller, M.W., White, S.R., and Sottos, N.R. (2008). Torsion fatigue response of self-healing poly(dimethylsiloxane) elastomers. *Polymer* 49: 3136–3145.

24 Szwarc, M. (1956). "Living" polymers. *Nature* 178: 1168–1169.

25 Wang, H.P., Yuan, Y.C., Rong, M.Z., and Zhang, M.Q. (2010). Self-healing of thermoplastics via living polymerization. *Macromolecules* 43: 595–598.

26 Zhu, D.Y., Wetzel, B., Noll, A. et al. (2013). Thermo-molded self-healing thermoplastics containing multilayer multilayered microcapsules. *Journal of Materials Chemistry A* 1: 7191–7198.

27 Matyjaszewski, K. and Xia, J. (2001). Atom transfer radical polymerization. *Chemical Reviews* 101: 2921–2990.

28 Wang, J.S. and Matyjaszewski, K. (1995). Controlled/"living" radical polymerization. Atom transfer radical polymerization in the presence of transition-metal complexes. *Journal of the American Chemical Society* 117: 5614–5615.

29 Kato, M., Kamigaito, M., Sawamoto, M., and Higashimura, T. (1995). Polymerization of methyl methacrylate with the carbon tetrachloride/dichlorotris-(triphenylphosphine)ruthenium(II)/methylaluminum bis(2,6-di-tert-butylphenoxide) initiating system: possibility of living radical polymerization. *Macromolecules* 28: 1721–1723.

30 Matyjaszewski, K. (2001). Macromolecular engineering by controlled/living ionic and radical polymerizations. *Macromolecular Symposia* 174: 51–68.

31 Chatterjee, D.P., Chatterjee, U., and Mandal, B.M. (2004). Atom transfer radical polymerization of methyl methacrylate at ambient temperature using soluble Cu(I) complex catalysts formed with mixed ligands of multidentate amines and halide ions. *Journal of Polymer Science, Part A: Polymer Chemistry* 42: 4132–4142.

32 Munirasu, S. and Dhamodharan, R. (2004). Very rapid copper-mediated atom transfer radical polymerization of benzyl methacrylate at ambient temperature. *Journal of Polymer Science, Part A: Polymer Chemistry* 42: 1053–1057.

33 Krishnan, R. and Srinivasan, K.S.V. (2003). Controlled/“living” radical polymerization of glycidyl methacrylate at ambient temperature. *Macromolecules* 36: 1769–1771.

34 Cañamero, P.F., de la Fuente, J.L., Madruga, E.L., and Fernández-García, M. (2004). Atom transfer radical polymerization of glycidyl methacrylate: a functional monomer. *Macromolecular Chemistry and Physics* 205: 2221–2228.

35 Kamigaito, M., Ando, T., and Sawamoto, M. (2001). Metal-catalyzed living radical polymerization. *Chemical Reviews* 101: 3689–3745.

36 Hayes, S.A., Zhang, W., Branthwaite, M., and Jones, F.R. (2007). Self-healing of damage in fibre-reinforced polymer-matrix composites. *Journal of the Royal Society Interface* 4: 381–387.

37 Lin, C.B., Lee, S., and Liu, K.S. (1990). Methanol-induced crack healing in poly(methyl methacrylate). *Polymer Engineering and Science* 30: 1399–1405.

38 Wang, P.P., Lee, S., and Harmon, J. (1994). Ethanol-induced crack healing in poly(methyl methacrylate). *Journal of Polymer Science, Part B: Polymer Physics* 32: 1217–1227.

39 Rule, J.D., Sottos, N.R., and White, S.R. (2007). Effect of microcapsule size on the performance of self-healing polymers. *Polymer* 48: 3520–3529.

40 Neyens, E. and Baeyens, J. (2003). A review of classic Fenton's peroxidation as an advanced oxidation technique. *Journal of Hazardous Materials* 98: 33–50.

41 Sharma, V.K. and Millero, F.J. (1989). The oxidation of Cu(I) with H_2O_2 in natural waters. *Geochimica et Cosmochimica Acta* 53: 2269–2275.

42 Chipara, M. and Wooley, K. (2005). Molecular self-healing processes in polymers. *Materials Research Society Symposium Proceedings* 851: 127–132.

43 Bégu, S., Durand, R., Lerner, D.A. et al. (2003). Preparation and characterization of siliceous material using liposomes as template. *Chemical Communications* 5: 640–641.

44 Jackson, A.C. (2011). Novel encapsulation technologies for small size-scale self-healing applications. PhD thesis. University of Illinois at Urbana-Champaign.

45 Brown, E.N., White, S.R., and Sottos, N.R. (2004). Microcapsule induced toughening in a self-healing polymer composite. *Journal of Materials Science* 39: 1703–1710.

46 Kinloch, A.J. and Young, R.J. (1983). *Fracture Behavior of Polymers*. London: Applied Science Publishers.

47 Schroeder, W.F., Auad, M.L., Barcia Vico, M.A. et al. (2005). Thermodynamic, morphological, mechanical and fracture properties of poly(methyl methacrylate) (PMMA) modified divinylester (DVE)/styrene (St) thermosets. *Polymer* 46: 2306–2319.

48 Mimura, K., Ito, H., and Fujioka, H. (2001). Toughening of epoxy resin modified with in situ polymerized thermoplastic polymers. *Polymer* 42: 9223–9233.

49 Jung, D., Hegeman, A., Sottos, N.R. et al. (1997). Self-healing composites using embedded microspheres. In: *Proceedings of ASME Symposium on Composites and Functionally Graded Materials*, vol. MD-80, 265–275. Dallas, TX: ASME.

50 Kissinger, H.E. (1957). Reaction kinetics in differential thermal analysis. *Analytical Chemistry* 29: 1702–1075.

51 Jiang, X.B. (2008). Self-healing polymer material with embedded two types of microcapsules. MSc thesis. Zhonghan University, Guangzhou.

52 Kessler, M.R. and White, S.R. (2001). Self-activated healing of delamination damage in woven composites. *Composites Part A: Applied Science and Manufacturing* 32: 683–699.

6

Intrinsic Self-Healing Via the Diels–Alder Reaction

CHAPTER MENU

Extrinsic self-healing with the help of embedded healing agent has been discussed in Chapters 2–5, typically using epoxy as the model material to be repaired. Although the healing efficiency, healing speed and healing conditions can be conveniently tailored and tuned by choosing the proper healing chemistry, the most obvious challenge lies in the strategy itself – additional healing agent has to be incorporated. The complexity of materials manufacturing might thus be increased in the case where the function of self-healing is planned to be integrated. In this context, synthesis of intrinsic self-healing materials seems to be a promising solution, besides further optimization and improvement of the healing agent for extrinsic self-healing.

As reviewed in Chapter 1, the development of intrinsic self-healing polymers, in which the ruptured macromolecules can be repeatedly reconnected by themselves as a result of physical and/or chemical interactions free of healant, somewhat lags behind that of extrinsic versions. This is particularly true where covalently bonded polymers for structural applications are concerned, in which intramolecular interaction should be strong enough to provide load bearing capability. The most important issue – initiation of molecular recombination toward crack healing – still needs manual intervention. Accordingly, a large amount of energy (heat or light) has to be input to break and reform covalent bonds, which is apparently unfavorable for self-healing [1]. Comparatively, non-covalently bonded polymers, which are constructed by ionic, hydrophobic, metal coordination and hydrogen bonding, are assembled with lower kinetic stability [2]. Fully autonomic self-healing would be more easily designed and realized in principle.

Nevertheless, ongoing attempts in this field have shown the interesting potential of reversible reactions [3, 4]. Although external stimuli have to be applied as triggers, crack rehabilitation through reversible reactions among macromolecules proves to be feasible. It is reasonably expected that brand new intrinsic self-healing polymers will emerge when the reversible reactions are coupled with mechanochemical sensitivity [5–7].

As the saying goes, Rome was not built in a day. A substantial study of reversible reaction induced self-healing is a necessary step in the development of intrinsic self-healing materials. The work based on the Diels–Alder (DA) reaction (Schemes 6.1–6.4) reported by Wudl et al. [8–11] has set up an example for the re-formation of carbon–carbon bonds via a relatively simple route. The DA reaction is an organic chemical reaction between a conjugated diene and a substituted alkene to generate a substituted cyclohexene system. The reversibility of DA cycloaddition can be exploited in the preparation of self-healing polymers carrying functionalities, either as the polymer chain-end (i.e. pendant ends) or in the monomeric repeating units [1, 12]. The decomposition

Extrinsic and Intrinsic Approaches to Self-Healing Polymers and Polymer Composites, First Edition. Ming Qiu Zhang and Min Zhi Rong.

reaction of the cyclic system obtained by the DA reaction (i.e. retro-DA reaction) does not involve free radicals, so side reactions that might prevent reformation are hindered [13]. The group of Wudl produced a DA cross-linked polymer consisting of furan- and maleimide-based monomers [8].

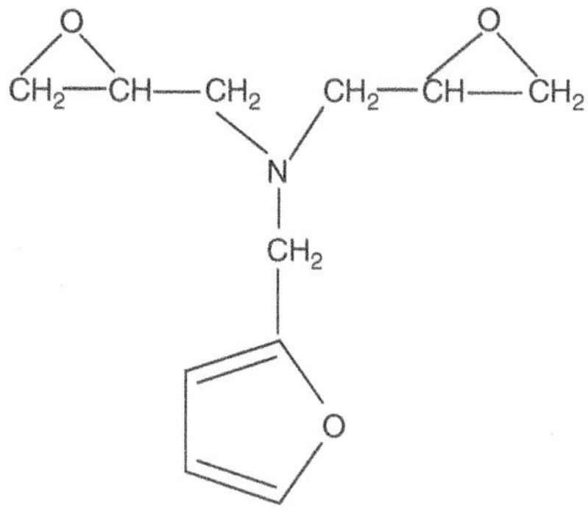

Scheme 6.1 Molecular structure of DGFA.

The appearance of DA bonds on the polymer backbone brings thermal reversibility that enables healing of the fractured surface after being heated to 120 °C and then cooled down to room temperature. Owing to the fact that the DA reaction plays the key role and no healing agent and/or catalyst, monomers and special surface treatment are required, multiple (infinite) healing of cracks would be theoretically possible. On the other hand, since the healing is driven by heat treatment, it is more suitable for structural bulk materials as compared with illumination induced healing based

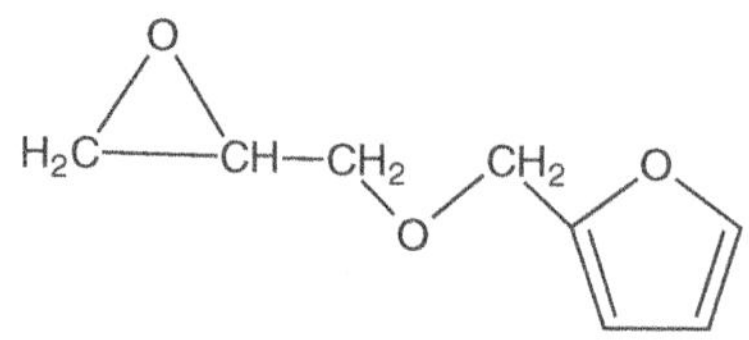

Scheme 6.3 Molecular structure of FGE.

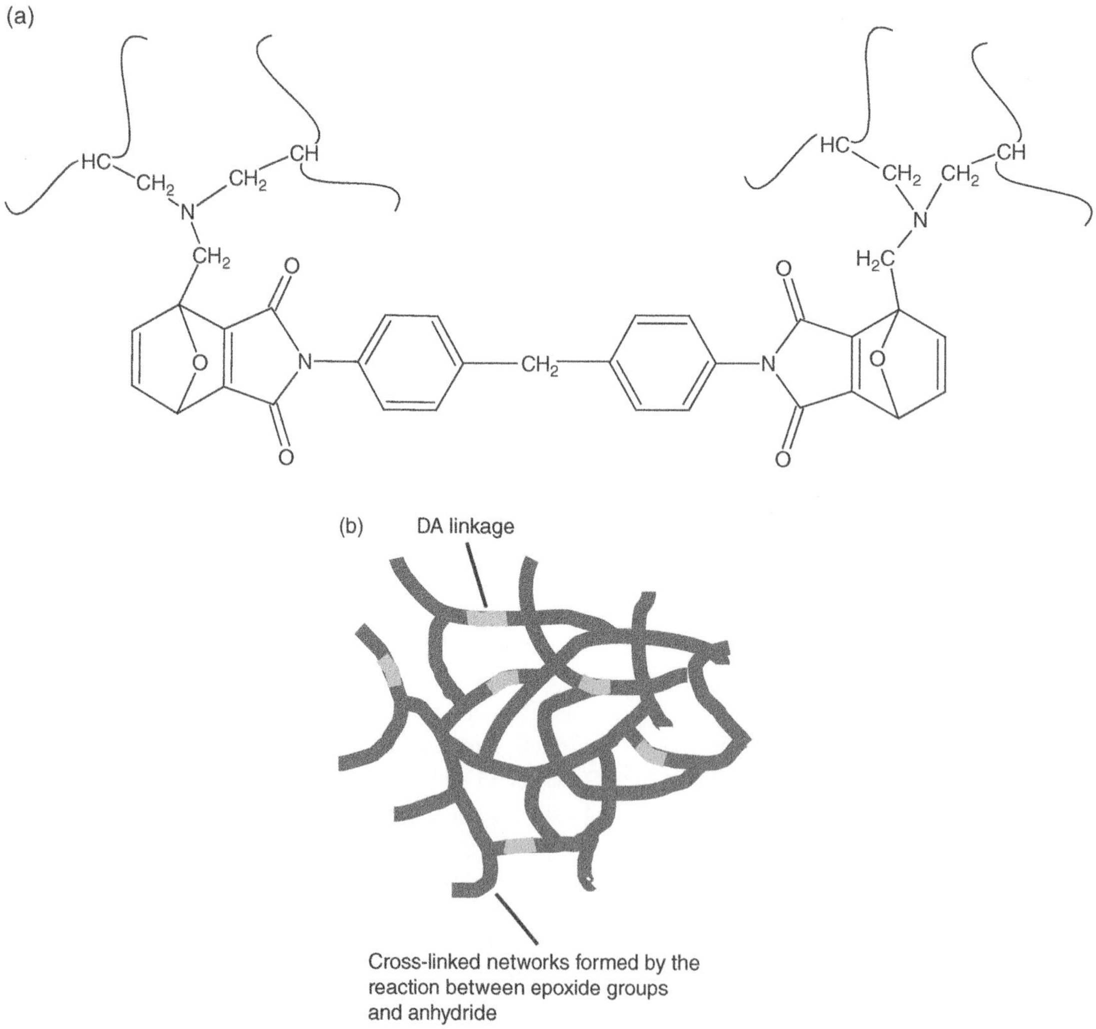

Scheme 6.2 (a) Molecular structure of cured DGFA and (b) schematic drawing highlighting the thermally reversible and irreversible bonds in the polymeric network.

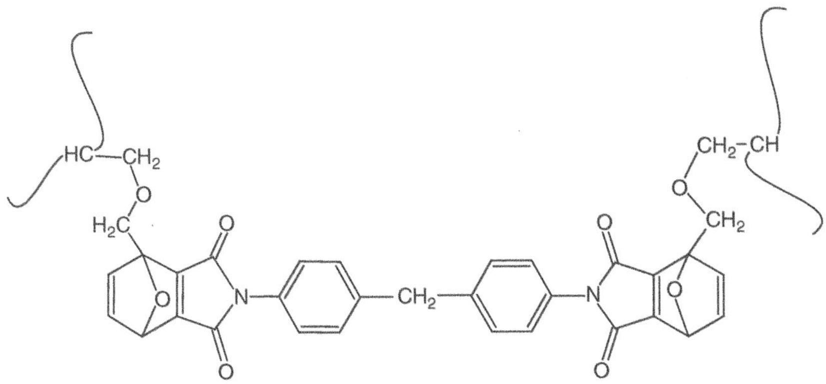

Scheme 6.4 Molecular structure of cured FGE.

on photoreversibility. Due to the limited penetration depth of light, the latter is better employed in the case of surface engineering or filmy materials like a coating.

In this chapter, novel epoxy materials with intrinsic (thermally responsive) self-healing ability developed in our laboratory following the abovementioned concept of DA chemistry are demonstrated. The polymers might not only become new members of the epoxy family but also help to enrich the database of design of the relevant molecules.

Epoxy has been extensively employed in many applications because of its excellent properties [14, 15]. Curing reactions of epoxy with hardeners are generally irreversible. Consequently, conventional cured epoxy can hardly exhibit remendable behavior due to lack of the ability to recombine broken molecules. Since epoxy materials mostly work under severe circumstances where long-term service is required, it would be ideal if they were associated with multiple self-healing capability.

Ober et al. have shown a series of reworkable epoxy compounds that incorporated thermally cleavable groups, such as secondary or tertiary esters, into the networks [16, 17]. These groups were chosen to decompose between 200 °C and 300 °C so that the cured epoxy can be removed without damaging the underlying structure. Also, Small et al. made a thermally removable encapsulant from bismaleimide compound and tris(furan) or tetrakis(furan), which can be easily removed by heating to temperatures higher than 90 °C, preferably in a polar solvent [18]. Although their works were not targeted for the specific production of self-healing materials, the feasibility of synthesis of thermally remendable epoxy was revealed.

6.1 Molecular Design and Synthesis

So far, there are two types of intrinsic self-healing polymers made of maleimide and furan compounds [12]. One is the macromolecular network constructed directly by the copolymerization of multi-furan and multi-maleimide [8]. The polymer is characterized by high density of DA bonds and high healing efficiency. In the case of low-melting point bismaleimide, the synthesis can be conducted without the aid of any solvent. As the entire polymeric network is composed of DA bonds, however, the glass transition temperature and service temperature of the material could not be high. In addition, the components made from this polymer might be easily deformed during crack healing as its healing temperature is higher than the glass transition temperature. The other type of intrinsic self-healing polymer from maleimide and furan is made through DA reactions among the maleimide and furan in the side chains [19, 20]. The skeletons are responsible for mechanical strength, heat resistance, and higher glass transition temperatures. Due to the presence of macromolecular main chains, such a self-healing system has poorer processability, and strong polar solvents have to be used to dissolve the monomers. On the other hand, the low amount of maleimide and furan in the side chains would result in limited healing efficiency.

To obtain intrinsic thermally responsive self-healing polymer with improved healing efficiency, operating temperature and processability, we proposed synthesizing an epoxy monomer, *N,N*-diglycidyl-furfurylamine (DGFA, Scheme 6.1) [21]. Epoxide groups are combined with furan in one molecule. The epoxide groups can react with traditional curing agent of epoxy resin (anhydride, for example) to form an irreversible epoxy network, providing the material with outstanding mechanical properties and thermal resistance as usual. Meanwhile, the furan groups can react with maleimide to introduce thermally reversible DA bonds into the epoxy network, so that the self-healing function is implanted. Eventually, the molecular networks in the cured material are comprised of two types of intermonomer linkage (Scheme 6.2). In this context, the advantages of epoxy and intrinsic self-healing ability are joined. By optimizing the curing conditions, it is hoped that the epoxy material would possess thermal remendability under its glass transition temperature.

It is worth noting that each DGFA molecule contains two epoxide groups and one furan (Scheme 6.1). This means that the fraction of DA bonds in the ultimate material must be lower than that of cross-linked bonds of epoxy (Scheme 6.2). Under the circumstances, the healing efficiency of cured DGFA might not be sufficiently satisfactory. Therefore, another epoxy monomer, furfuryl glycidyl ether (FGE, Scheme 6.3), was synthesized [22]. It has a similar structure to DGFA, while the fraction of furan is higher. Having been cured by anhydride and maleimide (Scheme 6.4), the FGE-based polymer also acquires thermal reversibility and should have higher healing efficiency.

In fact, synthesis of FGE does not represent the optimal solution. Although the concentration of furan is increased (as compared with DGFA), which helps to improve the healability, the cured version of FGE might have lower glass transition temperature and poorer mechanical properties as each FGE molecule carries only one epoxide group. Again, FGE was put in this dilemma by the requirements of thermal remendability and mechanical performance like DGFA.

In consideration of complementarity of structure and properties between DGFA and FGE, blends of DGFA and FGE might be able to offer balanced performance. Blending of polymers is a popular approach used in the plastics industry [23, 24]. Two or more polymers are blended together to create a new material with different properties. It is more cost-effective and time-saving than the design and synthesis of completely new polymers. Polymer blends can be broadly divided into three categories: miscible, partially miscible, and immiscible. Due to the inherent structural similarity between DGFA and FGE (Scheme 6.1 and Scheme 6.3), it is known that the blends of DGFA and FGE should be completely miscible, and would fulfill the envisaged task.

Accordingly, the preparation and characterization of DGFA, FGE and their cured versions will be studied first in the following. Then, the blends of DGFA and FGE will be the main concern.

6.1.1 Synthesis of DGFA

DGFA can be synthesized via a two-step route from epichlorohydrin and furfural amine (Scheme 6.5). In the first step, the ring-opening addition reaction between the oxirane ring of epichlorohydrin and the amine group of 2-furfurylamine occurs to produce a chlorohydrin end group compound (chloroethanolamine). In the second step, DGFA forms through the ring-closing reaction of chlorohydrin groups due to the dehydrochlorination action of NaOH solution.

Epichlorohydrin carries two reactive groups including epoxide and chlorine. As the atomic orbitals of the three-membered epoxide ring cannot be fully overlapped, the eclipsing strain leads to instability of the epoxide group. As a result, it will rapidly react with many chemicals under mild conditions. In addition, the chlorine on epichlorohydrin is highly electronegative, so that the adjacent carbon bears a partial positive charge and can be easily attacked by nucleophilic reagents. Owing to the coexistence of the two

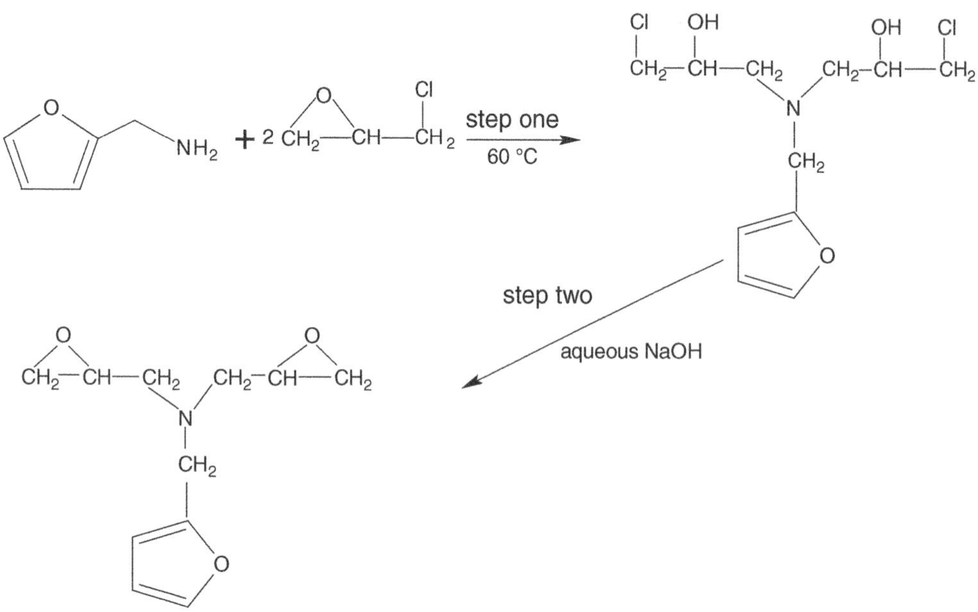

Scheme 6.5 DGFA synthesized by a two-step route.

types of functional groups, which might be mutually influenced, the reactivity of epichlorohydrin is higher than that of conventional epoxy compounds and alkyl chlorides.

The typical synthetic procedures are described as follows: 92.8 g epichlorohydrin (1 mol) was charged into a 250 ml four-neck round-bottom flask equipped with a stirrer, a thermometer, and an inlet of dry nitrogen. When the solution was heated to 40 °C, 48.5 g furfural amine (0.5 mol) was added dropwise while keeping the solution temperature below 60 °C. The reaction mixture was stirred for five hours at 60 °C under N_2 atmosphere and then cooled down to room temperature. After that, 160 ml sodium hydroxide aqueous solution (50% (w/v)) was added dropwise within one hour. The reaction proceeded for an additional five hours at 30 °C. Afterwards, 200 ml ethyl acetate was added, and the organic layer was collected, washed with water several times before being dried with anhydrous sodium sulfate, filtrated, and evaporated to give a 80% liquid. The liquid was further purified by chromatography on a silica gel column using a mixed solvent of ethyl acetate/hexane (1/3). The solvent was removed on a rotary evaporator, and lastly, a light yellow liquid was obtained with a yield of 50%. After purification by chromatography on a silica gel column, the product of DGFA (viscosity at 25 °C = 0.02 Pa·s) was characterized.

The data of Fourier transform infrared (FTIR) spectroscopy, proton nuclear magnetic resonance (^{1}H-NMR) spectroscopy, carbon nuclear magnetic resonance (^{13}C-NMR) spectroscopy and element analysis provide sufficient evidence for the chemical structure of DGFA. It is seen from Figure 6.1 that the oxirane ring of DGFA is present as reflected by the FTIR absorption peaks at 3051 cm^{-1} (C—H in the oxirane ring), 1250 and 851 cm^{-1} (C—O—C), and 918 cm^{-1} (oxirane ring breathing). The furan ring of DGFA is proved by the peaks at 3144 and 3117 cm^{-1} (C—H in the furan ring), 1503 cm^{-1} (C=C in the furan ring), 1148 cm^{-1} (C—O in the furan ring), 1013 cm^{-1} (furan ring breathing), and 752 cm^{-1} (representing monosubstituted furan ring) [14]. Besides, the characteristic peak at 1346 cm^{-1} is attributed to the stretching vibration of C—N.

In the ^{1}H-NMR spectrum of the reaction product (Figure 6.2), the peaks at δ_H = 3.1–2.4 ppm are due to the protons designated as "5" in the oxirane rings. The furan ring is identified by the peaks at δ_H = 6.2–6.3 ppm (=CH—CH=, 2H) and δ_H = 7.3 ppm (—O—CH=C—, 1H) corresponding to protons "2", "3" and "1", respectively. The characteristic peak at δ_H = 3.8 ppm (—NCH$_2$—furan ring, 2H) indicates the existence of group "4" that links glycidyl amine to the furan ring. Figure 6.3 shows the ^{13}C-NMR spectrum of DGFA with the following assignments: 151, 141, 110, 109 (C_{1-4} for furan ring), 66.6, 66.1 (C_5, —N—CH$_2$—oxirane ring), 61.1, 60.9 (C_6, oxirane ring), 60.6 (C_7, —N—CH$_2$—furan), and 44.8, 44.5 (C_8, oxirane ring). Moreover, the result of element analysis (found: C, 62.49; H, 7.15; N, 6.47; $C_{11}H_{15}NO_3$ requires: C, 63.14; H, 7.23; N, 6.69%) also coincides with the chemical structure of DGFA.

To highlight the effect of purification during preparation of DGFA, the FTIR spectrum of raw DGFA product is compared with that of its purified version (cf. Figure 6.1a,b). A broad absorption at 3471 cm^{-1} appears on the former, which is ascribed to the stretching mode of chlorohydrin groups (O—H). This means that the ring-closing reactions of chlorohydrin groups in the second step (Scheme 6.5) were not completely carried out, which explains the necessity of purification by chromatography to obtain high purity DGFA.

Since the structure of DGFA has been confirmed, the reactivity of the attached epoxide groups has to be checked. Accordingly, isothermal curing kinetics of DGFA with

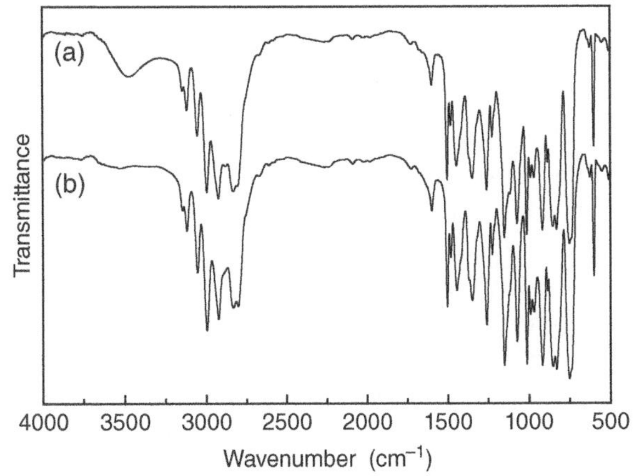

Figure 6.1 FTIR spectra of (a) raw DGFA and (b) its purified version.

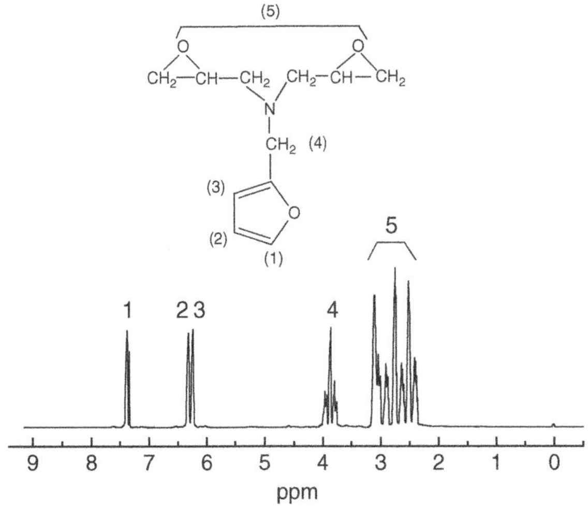

Figure 6.2 ^{1}H-NMR spectrum of DGFA.

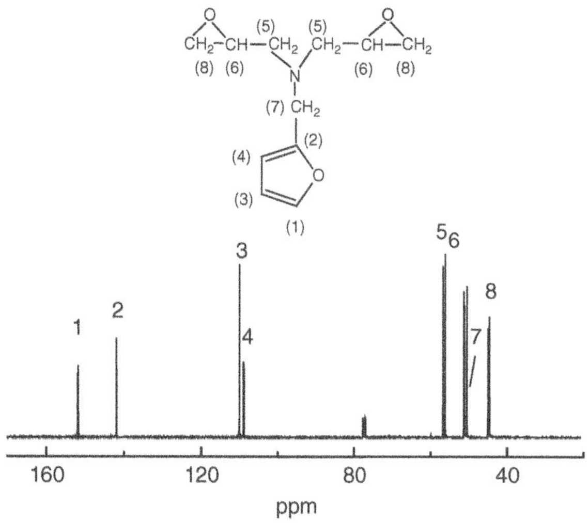

Figure 6.3 ^{13}C-NMR spectrum of DGFA.

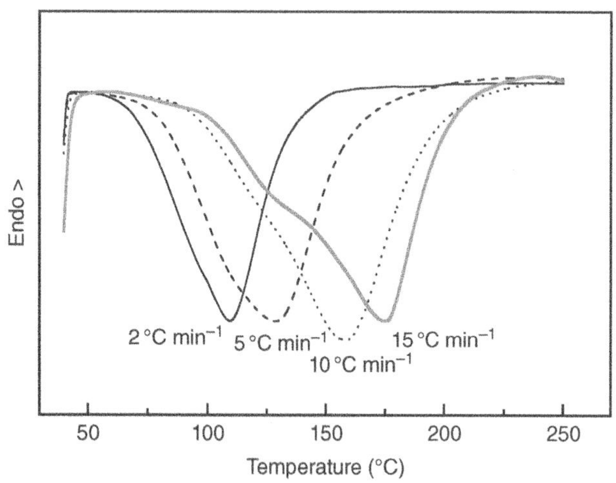

Figure 6.4 DSC heating traces of the mixture of DGFA and MHHPA at stoichiometric ratio.

methylhexahydrophthalic anhydride (MHHPA) at stoichiometric ratio was studied by differential scanning calorimetry (DSC) based on the modified Avrami equation [25]. It is seen that DGFA can perform a curing reaction with MHHPA like conventional epoxy resin (Figure 6.4 and Table 6.1), implying that the synthesized DGFA might be applicable for the field in which conventional epoxy resin is employed. In addition, the activation energy, E_a, 61.8 kJ mol^{-1}, agrees well with that of 71.6 kJ mol^{-1} for similar epoxy–anhydride systems [26]. Moreover, it is found that DGFA can be cured by MHHPA at a temperature as low as ~50 °C during heating DSC scanning, which contrasts sharply with the case of conventional epoxy cured by anhydride even in the presence of accelerant (i.e. 100–150 °C). The catalytic effect of the tertiary amine group on DGFA molecules (Scheme 6.5) is responsible.

Table 6.1 Curing kinetic parameters of stoichiometric DGFA/MHHPA obtained from isothermal DSC tests at different temperatures and calculated according to the modified Avrami equation [25].

Temperature (°C)	n	$k \times 10^{-2}$ (s^{-1})	E_a (kJ mol^{-1})	ln A
100	1.13	10.3	61.8	16
110	1.04	31.3		
120	1.05	48.2		
130	1.06	66.2		

A, pre-exponential factor; E_a, activation energy; k, rate constant; and n, reaction order.

6.1.2 Reversibility of DA Bonds and Crack Remendability of DGFA-Based Polymer

The objective of this work is to develop a thermally remendable epoxy resin by introducing thermally reversible groups into the cured networks. Therefore, reversibility of the DA reaction between DGFA (containing furan ring, electron-rich diene) and maleimide (Scheme 6.2) should be verified in addition to the above-proved reactivity of epoxide groups of DGFA. Here, *N,N′*-(4,4′-diphenylmethane)bismaleimide (DPMBMI, involving maleimide, electron-poor dienophile) acts as the role of curing agent for the reaction with furan groups of DGFA.

According to the literature [27], DA and retro-DA reaction temperatures of the current system were set at 70 and 110 °C (Scheme 6.6), respectively. First, the DA reaction of furan and maleimide groups in an equivalent molar mixture of DGFA and DPMBMI was monitored by both FTIR and ^{1}H-NMR spectroscopy. The mixture solution in acetone (50% w/v) was coated on the surface of a KBr tablet, and then the FTIR spectrum was collected after the tablet had been dried at room temperature. As shown in Figure 6.5a, there is only a negligible peak at 1771 cm^{-1}, which is specific to DA adducts of maleimides [28]. This peak appears obviously after keeping the mixture solution in tetrahydrofuran (THF) at about 66 °C for 24 hours (see details in the following discussion concerning the retro-DA reaction of the DA adduct) (Figure 6.5b). This demonstrates that the DA reaction can take place very slowly at room temperature, while it is accelerated at around 70 °C.

The DA reaction between DGFA and DPMBMI was further evaluated by in-situ ^{1}H-NMR spectroscopy. An equivalent molar mixture of DGFA and DPMBMI was dissolved in dimethyl sulfoxide (DMSO)-d6 (76 mmol l^{-1}) and then sealed in a glass tube under N_2 atmosphere. Then, ^{1}H-NMR spectra were collected at specific time intervals to monitor the DA reaction proceeding in the mixture at 70 °C. The results show that at the beginning of the reaction, no characteristic peaks between 5.1 ppm and 5.3 ppm corresponding

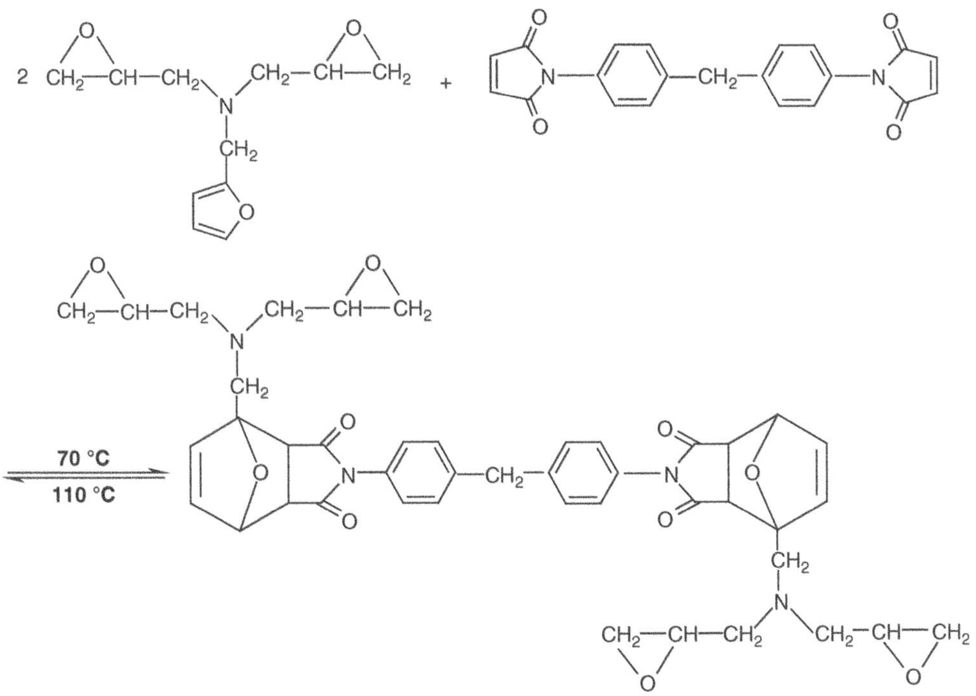

Scheme 6.6 DA and retro-DA reactions between DGFA and DPMBMI.

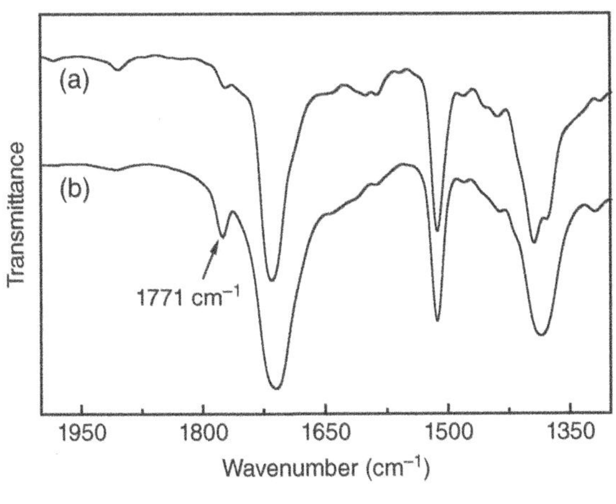

Figure 6.5 FTIR spectra of (a) mixture of DGFA and DPMBMI, and (b) mixture of DGFA and DPMBMI treated at 66 °C for 24 hours.

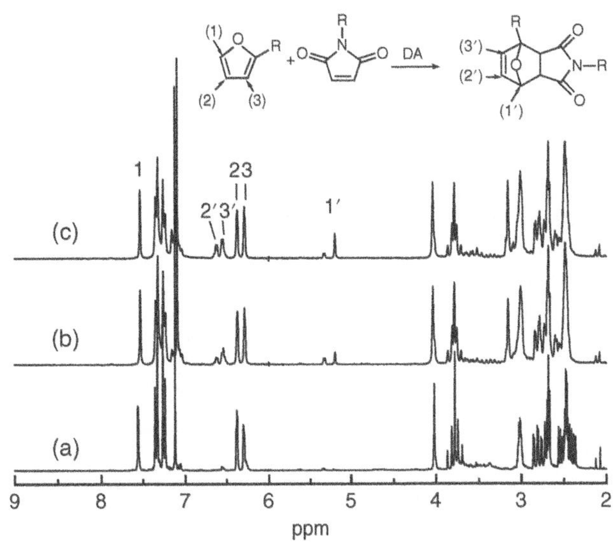

Figure 6.6 ^{1}H-NMR spectra of the mixture of DGFA and DPMBMI kept at 70 °C for (a) zero hours, (b) two hours, and (c) six hours. *Source:* Reprinted from Tian et al. [21]. Copyright 2009, with permission from the Royal Society of Chemistry.

to the DA adduct are observed (Figure 6.6a). After two hours, these peaks appear (Figure 6.6b), and become more evident when the reaction has proceeded for six hours (Figure 6.6c). The feasibility of a DA reaction between DGFA and DPMBMI is thus confirmed.

On the other hand, contents of DA adduct and unreacted DGFA can be characterized by the integrals of H_2' and H_3' peaks, and those of H_2 and H_3 peaks, respectively (Figure 6.6). This is because H_2' and H_3' peaks at 6.6 and 6.5 ppm are due to H_2' and H_3' protons of the DA adduct [29, 30], while the peaks at 6.3 and 6.2 ppm are assigned to H_2 and H_3 protons of the unreacted DGFA. As a result, the time-dependent reaction conversion of DA adduct, x, is known:

$$x = \frac{\text{Integral area of}\left(H_2' + H_3'\right)}{\text{Integral area of}\left(H_2' + H_3' + H_2 + H_3\right)} \tag{6.1}$$

Also, first- and second-order kinetics of the reaction, respectively, can be estimated from:

$$\ln 1/(1-x)=kt \tag{6.2}$$

$$1/(1-x)=kt \tag{6.3}$$

where k stands for the rate constant. Quantification of the DA reaction between DGFA and DPMBMI based on in-situ ^{1}H-NMR spectroscopy indicates that linear fit of the data at 70 °C gives a rate constant with regression coefficient of 0.9882 for the first-order plot and 0.9960 for the second-order plot, suggesting that the second-order kinetics reaction should be more reasonable. Similarly, the reaction at 60 °C also exhibits greater regression coefficient for the second-order plot than that of the first-order one. The activation energy derived from the second-order kinetics is 42.9 kJ mol^{-1}, which falls in the similar ranges reported for the other furan-maleimide DA reactions [19, 31–34].

In order to reveal the temperature range of the DA reaction between DGFA and DPMBMI, a DSC heating experiment was conducted (Figure 6.7). The sample mixture of DGFA and DPMBMI was obtained by dissolving stoichiometric DPMBMI into DGFA at 90 °C for five minutes, and then quickly cooling down to room temperature. Upon being heated in a DSC cell, the mixture performs a DA reaction from 27 to 86 °C with an exothermal peak at 65 °C. Afterwards, an endothermic peak appears at 110 °C representing the retro-DA reaction, followed by two evident exothermal peaks at 160 and 214 °C due to epoxide ring opening reaction and polymerization of maleimide moieties, respectively. This DSC analysis proves that the DA reaction can take place at 70 °C. Therefore, isothermal DSC scanning was subsequently carried out at 70 °C to evaluate the extent of DA reaction. After reaction for 200 minutes, the exothermal enthalpy is found to be 216 J g^{-1}, corresponding to 83.8 kJ mol^{-1} (per DA bond). This value agrees with that reported in the literature [27].

To perceive the retro-DA reaction feature of the adduct of DGFA and DPMBMI, a solution method was chosen to obtain the adduct as entirely as possible. First, 5.71 g (0.016 mol) DPMBMI was dissolved in 50 ml anhydrous THF. The solution was charged into a 100 ml three-neck round-bottom flask equipped with magnetic stirring. Then, 6.68 g (0.032 mol) DGFA was slowly added into the solution. Afterwards, the solution was refluxed at about 66 °C for 24 hours under N_2 atmosphere and cooled down to room temperature. The reaction solution was poured into a large excess of diethyl ether. The precipitate was filtered, washed with methanol, and then dried under vacuum ($<1.3\times10^3$ Pa) at room temperature. The DA adduct was further dissolved in DMSO-d6 in a glass tube and heated at 110 °C for 20 minutes to disconnect the DA adduct. After the retro-DA reaction, the product was immediately cooled down to room temperature and analyzed. From the ^{1}H-NMR spectra in Figure 6.8, it is seen that the peaks at 5.1–5.3 ppm ascribed to the DA adduct (Figure 6.8a) entirely disappear in the spectrum of retro-DA product (Figure 6.8b). Also, the characteristic peaks at 82 and 93 ppm [8], which are clearly found in the ^{13}C-NMR spectrum of DA product (Figure 6.9a), are no longer perceived in the spectrum of retro-DA product (Figure 6.9b). These data demonstrate that the retro-DA reaction can be completed at 110 °C for 20 minutes.

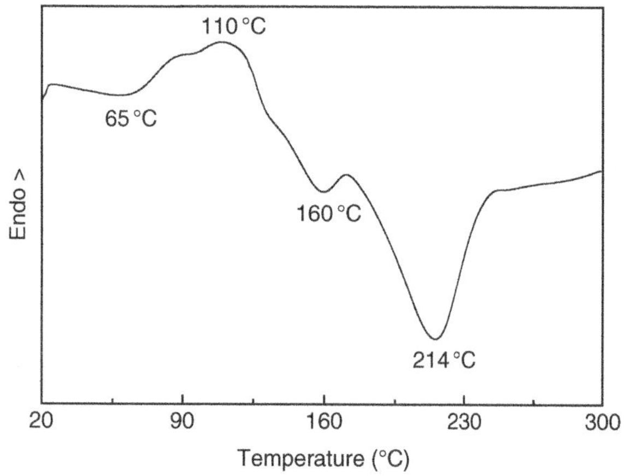

Figure 6.7 DSC heating trace of the mixture of DGFA and DPMBMI. Heating rate: 2 °C min^{-1}.

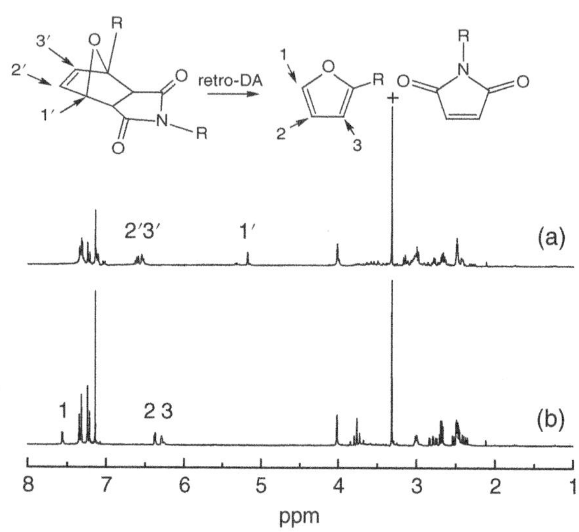

Figure 6.8 ^{1}H-NMR spectra of (a) the adduct of DGFA and DPMBMI and (b) the retro-DA product of the adduct of DGFA and DPMBMI by heating the adduct at 110 °C for 20 minutes. *Source:* Reprinted from Tian et al. [21]. Copyright 2009, with permission from the Royal Society of Chemistry.

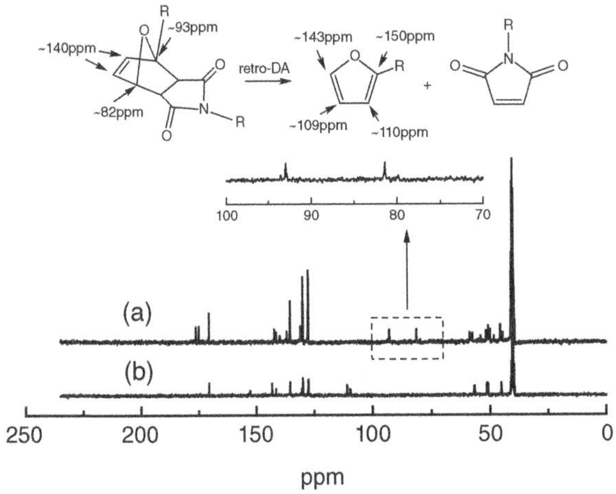

Figure 6.9 ^{13}C-NMR spectra of (a) the adduct of DGFA and DPMBMI and (b) the retro-DA product of the adduct of DGFA and DPMBMI by heating the adduct at 110 °C for 20 minutes. *Source:* Reprinted from Tian et al. [21]. Copyright 2009, with permission from the Royal Society of Chemistry.

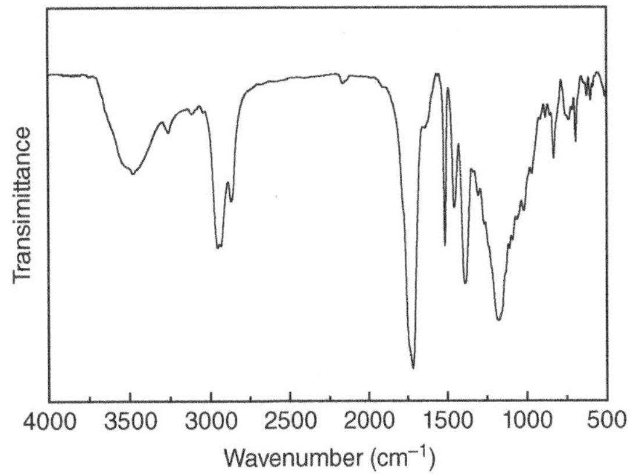

Figure 6.10 FTIR spectrum of DGFA/MHHPA/DPMBMI cross-linked polymer.

The above work has shown not only the reactivity of the epoxide groups of DGFA but also the DA and retro-DA reactions between DGFA and DPMBMI. Therefore, thermal reversibility of the cross-linked polymer of DGFA/MHHPA/DPMBMI should be subsequently studied. To prepare the polymer, 17.13 g (0.048 mol) DPMBMI was dissolved in 20 g (0.096 mol) DGFA liquid under stirring at 90 °C for 10 minutes without the help of solvent. Then, 25.86 g (0.1536 mol) MHHPA was mixed with the above mixture at 80 °C for an additional 10 minutes. The resultant homogeneous liquid was degassed, poured into a closed silicone rubber mold, and cured at 70 °C for 24 hours. This curing temperature was selected based on the optimum temperature for DA bond formation (see above) to favor most of the furan and maleimide groups taking part in the reaction before solidification of the system. Furthermore, a non-stoichiometric ratio of epoxy ring to anhydride of 1:0.8 was used to decrease the curing rate of the system, facilitating sufficient reaction between furan and maleimide groups. In the FTIR spectrum of cured DGFA (Figure 6.10), the peaks attributed to epoxide groups at 918 cm^{-1} (oxirane ring breathing) and 851 cm^{-1} (C—O—C) disappear, indicating that all epoxide groups have reacted with anhydride groups of MHHPA to form epoxy networks. An evident peak of hydroxyl groups emerges at 3500 cm^{-1} as a result of this reaction. Besides, the DA reaction of furan and maleimide groups in the cured sample is also proved by the existence of the peak at 1771 cm^{-1}.

As mentioned earlier, cured DGFA polymer should contain two types of cross-link covalent bonds (Scheme 6.2): (i) thermally stable bonds from the reaction between epoxide and anhydride groups; and (ii) thermally reversible bonds from the DA reaction of furan and maleic groups. Although DA and retro-DA reactions between DGFA and DPMBMI have been proved by the simulation tests (Figures 6.5–6.9), it is still unknown whether these thermally reversible reactions would be inhibited by the stable cross-linked epoxy networks. Therefore, thermal reversibility of the cured DGFA polymer should be studied as it is directly related to thermal remendability of the material.

Figure 6.11 shows the heating DSC curves of the cured DGFA polymer. The first heating curve has an endothermic peak at 126 °C (curve a), while the second one shows no peak except a transition at 128.5 °C (curve b). It is thus reasonable to deduce that the endotherm appearing on the first curve must result from the retro-DA reaction. As there is not enough time for the recovered furan and maleimide moieties to be reconnected during the subsequent cooling and reheating processes, no more retro-DA reaction but glass transition of the cured epoxy is detected by the second

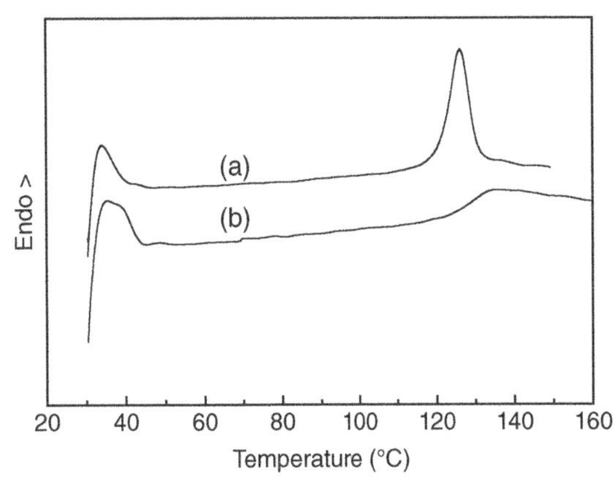

Figure 6.11 DSC heating traces of DGFA/MHHPA/DPMBMI cross-linked polymer (heating rate: 5 °C min^{-1}). The sample was first heated (for recording curve a), and then naturally cooled down to room temperature in the DSC cell, followed by a second heating (to record curve b).

DSC heating scan. In comparison with the retro-DA reaction at 110 °C in solution (Figures 6.8 and 6.9), the higher retro-DA reaction temperature in solid (126 °C) is indicative of the restraint effect of cured epoxy. It can also be concluded that the reconnection of furan and maleic groups in cured DGFA has to proceed at a rate much more slowly than that in solution, due to the same restriction of the epoxy network. Nevertheless, the DSC results demonstrate that the retro-DA reaction is an accessible reaction pathway under heating circumstances that is preferred over bond-breaking degradation reaction in the cured epoxy network.

Reversibility of the cross-linked DGFA polymer was further studied by applying cyclic retro-DA (heat treatment at 140 °C for 20 minutes to attain a retro-DA sample) and DA (heat treatment at 80 °C for 72 hours to attain a DA sample) reactions up to four times. The higher heating temperatures for both retro-DA and DA reactions were chosen to increase the reaction efficiency. Again, the reactions were monitored by DSC (Figure 6.12). The original cured DGFA (DA0) gives an endothermic peak of retro-DA reaction at 126 °C, while the rDA1 only shows a glass transition at 128.5 °C. This implies that the retro-DA reaction has already been completed during the heat treatment at 140 °C for 20 minutes. Subsequently, the rDA1 is allowed to reconnect the cleavage groups by DA reaction, and then an endothermic peak appears again at 131 °C for DA1. It is evident that the DA reaction between the disconnected furan and maleimide moieties has definitely taken place in the thermal treatment process. The recovery efficiency of the second DA reaction in comparison with the original cured DGFA polymer is about 77% as given by the ratio of ΔH_1 (enthalpy of DA1) to ΔH_0 (enthalpy of DA0). It is noted that ΔH_0 is $122.9\,\mathrm{J\,g^{-1}}$, corresponding to $80.2\,\mathrm{kJ\,mol^{-1}}$ (per DA bond), which is quite close to the aforementioned exothermal enthalpy per mole of DA bond formation. This means that all DA bonds can be cleaved during the first retro-DA reaction, and the cross-linked epoxy networks do not block the retro-DA reaction.

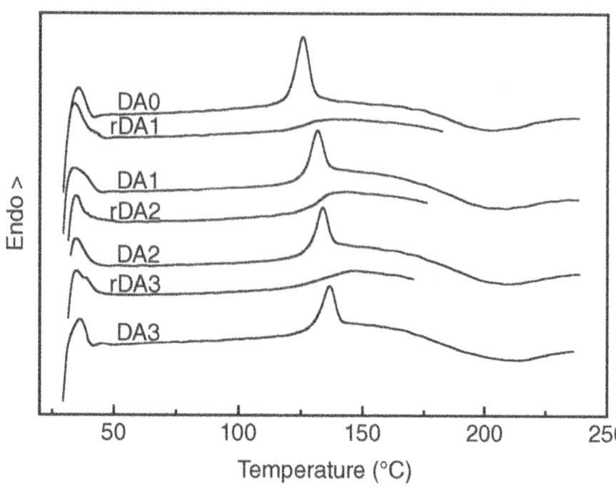

Figure 6.12 DSC heating traces of DGFA/MHHPA/DPMBMI cross-linked polymer (heating rate: $5\,°\mathrm{C\,min^{-1}}$). DA0: as-manufactured sample. rDA1: DA0 treated at 140 °C for 20 minutes, and then quenched to room temperature. DA1: rDA1 treated at 80 °C for 72 hours, and then cooled inside the oven by switching off the electricity supply. rDA2: DA1 treated at 140 °C for 20 minutes, and then quenched to room temperature. DA2: rDA2 treated at 80 °C for 72 hours, and then cooled inside the oven by switching off the electricity supply. rDA3: DA2 treated at 140 °C for 20 minutes, and then quenched to room temperature. DA3: rDA3 treated at 80 °C for 72 hours, and then cooled inside the oven by switching off the electricity supply. *Source:* Reprinted from Tian et al. [21]. Copyright 2009, with permission from the Royal Society of Chemistry.

In this manner, recyclability of retro-DA and DA reactions in the cross-linked DGFA polymer has been verified by the repeated endotherms and glass transitions for the DA and retro-DA samples. The characteristic parameters of the reactions inspected by DSC (Figure 6.12) are summarized in Table 6.2. It is noted that the endothermic peak temperature of the DA sample slightly increases with heat treatment time, implying that the retro-DA reaction becomes more difficult with the heating cycles. This phenomenon may be related to the increase in cross-link density of epoxy resin after heat treatment, as shown by the gradual increase in glass transition temperatures of rDA1 (128.5 °C), rDA2 (132.5 °C), and rDA3 (136 °C). Accordingly, the enthalpy of each endothermic peak, ΔH, of the retro-DA reaction has to gradually decrease from the second heating cycle on, indicating that the increased cross-link density of the epoxy part somewhat hinders recovery of DA bonds. Nevertheless, the decrease of each retro-DA endothermic enthalpy is marginal. The DA and retro-DA reactions in the cured DGFA can still be considered as reversible.

In fact, the cross-link density of epoxy networks would not ceaselessly change after a couple of heating–cooling cycles, so that reversibility of DA and retro-DA reactions in the cured DGFA can eventually be fixed. Moreover, the epoxy network should favor the disconnected furan and maleimide moieties remaining at their original places during cooling. As a result, they can be reconnected very efficiently upon being heated at 80 °C. In other words, the epoxy network would help to retain the integrity of the material when the retro-DA or DA reaction occurs.

Figure 6.12 also shows that there is an exothermic peak at around 225 °C on the heating curves of the DA1–DA3 samples. It should be ascribed to polymerization of the free maleimide moieties resulting from the retro-DA reaction. This reaction is detrimental to thermal reversibility of the cured DGFA polymer because it converts maleimide to irreversible bonds. To avoid the negative effect, therefore, the material should not be heated to elevated temperature, e.g. 180 °C.

Dynamic mechanical analysis (DMA) spectra of cured DGFA polymer are shown in Figure 6.13. For the original

Table 6.2 Quantification of endothermic peaks of cured DGFA polymer measured by DSC heating scan.

Sample ID[a]	Onset temperature (°C)	Peak temperature (°C)	Enthalpy (J g^{-1})
DA0	120.7	126.2	122.9
DA1	126.2	131.6	94.7
DA2	128.4	133.7	92.6
DA3	129.2	136.5	90.0

[a] Descriptions of the samples are listed in the caption of Figure 6.12.

sample, its storage modulus exhibits a two-stage behavior with a rise in temperature. At around 110 °C, a drastic decrease in the modulus starts, and then a plateau appears at 118 °C. When the temperature is further raised, the modulus drops again at 121 °C. Meanwhile, the temperature dependence of tan δ shows a knee point at 119 °C and a peak at 128 °C, respectively. In contrast, after retro-DA reaction at 140 °C for 20 minutes, the sample only shows a single transition on the curve of storage modulus versus temperature. The corresponding tan δ plot also has only one peak and the aforementioned knee point is absent. Clearly, the two-stage variation of storage modulus of the original cured DGFA polymer reflects retro-DA reaction and glass transition, respectively. The retro-DA reaction happens below the glass transition temperature of the material, but there is still overlap between the two processes. Furthermore, the DMA analysis matches previous DSC data, which also show a retro-DA reaction peak at around 120 °C and a glass transition at around 128 °C.

It is interesting to see from the DMA measurements (Figure 6.13) that after retro-DA reaction, the cured DGFA polymer possesses a higher glass transition temperature

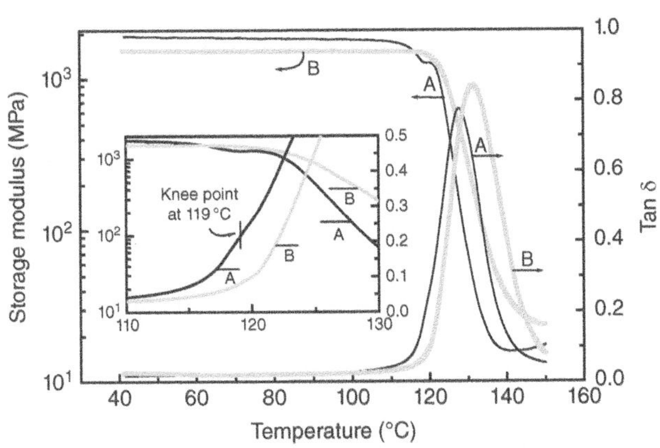

Figure 6.13 DMA spectra of DGFA/MHHPA/DPMBMI cross-linked polymer. (A) As-manufactured sample and (B) as-manufactured sample treated at 140 °C for 20 minutes, and then quenched to room temperature. *Source:* Reprinted from Tian et al. [21]. Copyright 2009, with permission from the Royal Society of Chemistry.

and significantly lower room temperature storage modulus. The former phenomenon should result from the increased cross-link density of epoxy network, which has been discussed when analyzing Figure 6.12. The latter can be explained by the disconnection of DA bonds that may lead to increase of chain ends or low molecular substance, and accordingly is detrimental to the stiffness of materials.

Above the glass transition temperature, all DA bonds have been broken. Thus, the storage modulus in the rubbery region is only related to the cross-link density variations of the epoxy network. It is evident that the sample after retro-DA reaction exhibited a higher storage modulus than that of the as-manufactured sample. This again reveals that the cross-link density of the epoxy network can be further enhanced by heat treatment at higher temperature above the glass transition temperature.

Tensile and flexural properties of DGFA/MHHPA/DPMBMI cross-linked polymer were measured at room temperature, in comparison with those of conventional bisphenol-A epoxy cured by MHHPA (Table 6.3). The cured version of the newly synthesized epoxy has higher modulus but a moderately lower strength as compared with cured bisphenol-A epoxy [35]. On the whole, the data of the cured DGFA are still close to those of commercial epoxy resins. This means that DGFA possesses the potential to replace traditional epoxy resin in future applications.

Thermally remendable processes of artificial damage (made by impact with a falling iron ball) on the cured DGFA are illustrated in Figure 6.14. Since the interfaces of cracks have higher refractive index, they are easily visually monitored. All the samples were thermally treated at different temperatures (100–125 °C) for 20 minutes for disconnecting DA covalent bonds via retro-DA reaction, and then annealed at 80 °C to repair the cracks by DA reaction between the disconnected furan and maleimide moieties.

As shown in Figure 6.14a, no crack repair is observed when the heat treatment temperature (100 °C) is lower than the starting temperature of the retro-DA reaction (110 °C). When heat treatment is carried out at a temperature (119 °C) higher than the onset temperature of the

Table 6.3 Mechanical properties of cured epoxy.

Properties	DGFA/MHHPA/DPMBMI cross-linked polymer	Bisphenol-A epoxy cured by MHHPA [35]
Young's modulus (GPa)	2.5	1.8
Tensile strength (MPa)	53	65
Elongation at break (%)	1.5	—
Flexural modulus (GPa)	4.6	2.6
Flexural strength (MPa)	110	128

retro-DA reaction, most of the cracks can be healed after DA reaction treatment at 80 °C (Figure 6.14b). Some larger cracks remain unhealed probably due to the relatively low retro-DA treatment temperature, and/or material loss during damaging, and/or misalignment of the cracked portions. In the case of higher retro-DA reaction temperature (125 °C), which is close to the glass transition temperature of cured DGFA polymer (128 °C), larger cracks can also be repaired owing to the enhanced mobility of the segments (Figure 6.14c). These results clearly demonstrate that the synthesized epoxy has been provided with thermal remendability as expected.

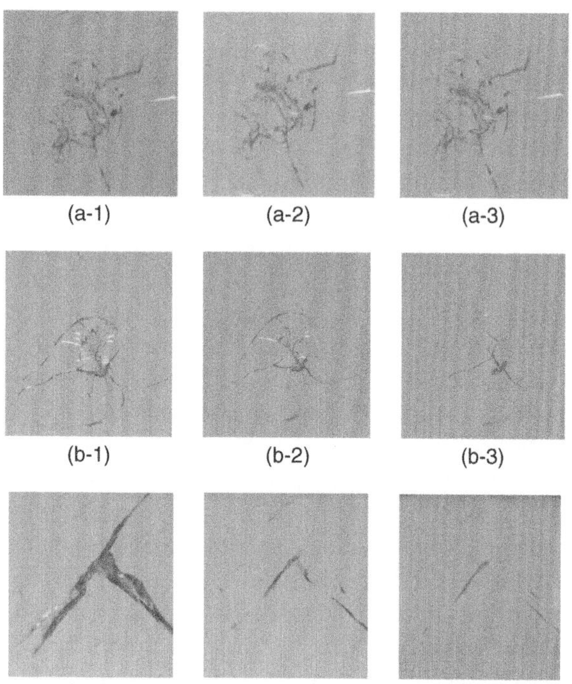

Figure 6.14 Visual inspection of thermal remendability of cured DGFA polymer. The damaged samples were first treated at (a) 100 °C, (b) 119 °C, and (c) 125 °C for 20 minutes. Then, they were move to an oven preset at 80 °C for (1) 0 hours, (2) 12 hours, and (3) 72 hours. *Source:* Tian et al. [21], Figure 10, p. 7/with permission of Royal Society of Chemistry.

6.1.3 Synthesis and Characterization of FGE

In general, glycidyl ether epoxy resin can be synthesized from condensation of epichlorohydrin and phenol (or alcohol) that contains active hydrogen by two approaches as in the following. (i) The one-step method takes advantage of dual catalysis of alkali for etherification and ring-closing. Accordingly, ring-opening addition and ring-closing reaction take place under identical conditions. It is a conventional technique and has been widely used. (ii) The two-step method first produces an intermediate chlorohydrin ether through etherification under phase transfer catalysis, and then completes the ring-closing reaction by incorporating alkali. It reduces hydrolysis of epichlorohydrin and increases the epoxy value of the resultant.

Here, we selected the two-step method (Scheme 6.7). In the first step, ring-opening addition between oxirane ring of epichlorohydrin and hydroxyl group of furfural alcohol occurred with the catalysis of tetrabutylammonium hydrogen sulfate, giving rise to chlorohydrin ether. In the second step, FGE formed through the ring-closing reaction of chlorohydrin ether owing to dehydrochlorination action of NaOH solution.

Typically, 102 g (1.1 mol) epichlorohydrin and 4 g tetrabutylammonium hydrogen sulfate were charged into a 500 ml three-neck round-bottom flask equipped with a stirrer, a thermometer, and an inlet of dry nitrogen. The system temperature was kept at 25 °C, and 98 g (1 mol) furfural alcohol was added dropwise within 30 minutes. The reaction proceeded for four hours and then 160 ml 50% (w/v) sodium hydroxide aqueous solution was added. After two hours the reactant was extracted by diethyl ether and washed with water several times. The organic layer was collected, and the solvent was removed on a rotary evaporator, producing a light yellow liquid with a yield of 90%. The liquid was further purified by chromatography on a silica gel column using mixed solvents of ethyl acetate/hexane (5/1). Eventually, a colorless liquid was obtained after removal of solvents under reduced pressure.

Figure 6.15 shows FTIR spectra of FGE before and after purification. The oxirane ring is present as reflected by the

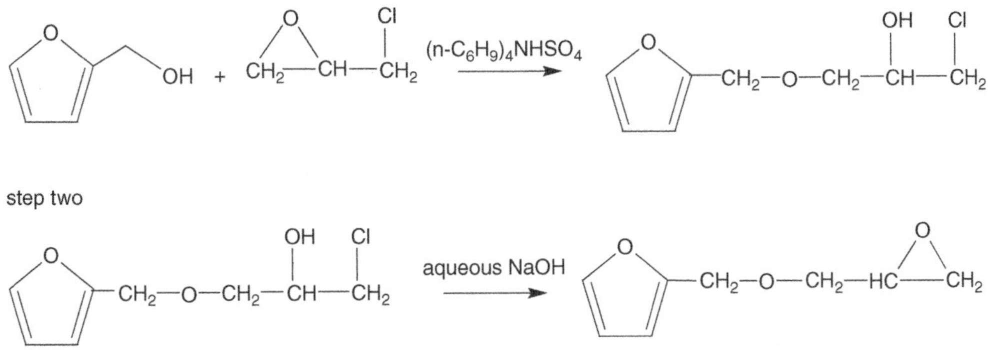

Scheme 6.7 FGE synthesized by a two-step route.

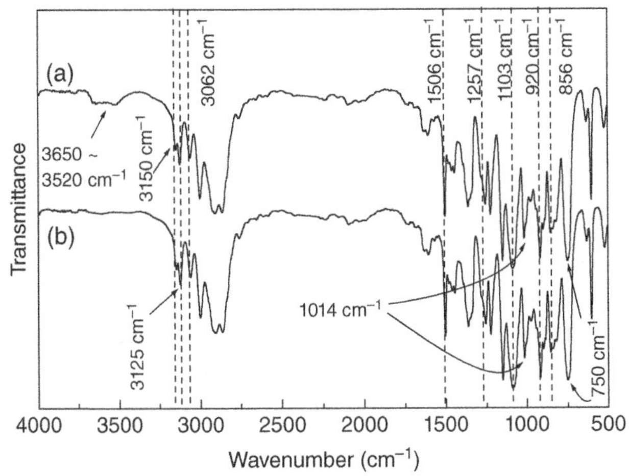

Figure 6.15 FTIR spectra of (a) raw FGE and (b) its purified version. *Source:* Reprinted from Tian et al. [22]. Copyright 2010, with permission from John Wiley and Sons.

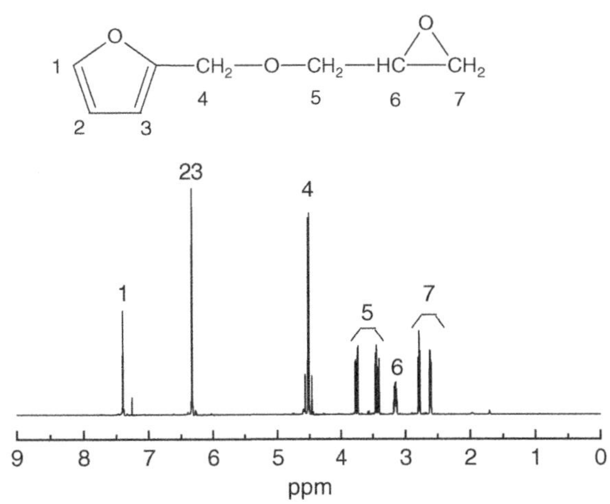

Figure 6.16 ^{1}H-NMR spectrum of FGE. Reprinted from Tian et al. [22]. Copyright 2010, with permission from John Wiley and Sons.

peaks at 3062 cm^{-1} (stretching of C—H in the oxirane ring), 1257 and 856 cm^{-1} (C—O—C), and 920 cm^{-1} (oxirane ring breathing). Moreover, the furan ring is indicated by the peaks at 3150 and 3125 cm^{-1} (stretching of C—H in the furan ring), 1506 cm^{-1} (C=C in the furan ring), 1103 cm^{-1} (C—O in the furan ring), 1014 cm^{-1} (furan ring breathing), and 750 cm^{-1} (representing the monosubstituted furan ring) [14]. On the other hand, the peaks at 1222 and 1085 cm^{-1} are attributed to the stretching vibration of C—O—C in FGE.

For the purified product, the broad absorption at 3520–3650 cm^{-1} on the spectrum of its raw version disappears. The peak is ascribed to O—H of chlorohydrin ether, originating from incomplete ring-closing reaction mentioned above. Clearly, the treatment by chromatography helps to increase the purity of the product.

Since the FTIR study demonstrates that the product contains both oxirane and furan rings, which coincides with the chemical structure of the target resultant, its ^{1}H-NMR spectrum should be analyzed for confirmation. As illustrated in Figure 6.16, the peaks at $\delta = 2.7$–2.6 ppm are due to the protons in the oxirane ring (designated as “7”), while those at $\delta = 3.8$–3.4 ppm reflect the protons “5” of methylene adjacent to the oxirane ring. The peak at $\delta = 3.2$ ppm indicates the existence of methylidyne “6” in the oxirane ring. The furan ring is identified with the peaks at $\delta = 6.2$–6.3 ppm (=CH—CH=, 2H) and $\delta = 7.4$ ppm (—O—CH=C—, 1H) corresponding to the protons “2”, “3”, and “1”, respectively. In addition, the peak at $\delta = 4.5$ ppm (—OCH_2—furan ring, 2H) represents group “4” that links glycidyl ether to the furan ring.

Figure 6.17 shows the ^{13}C-NMR spectrum of FGE with the corresponding assignments: 151, 142, 110, 109 ppm (C_{1-4} for the furan ring), 70 ppm (C_5, —O—CH_2—furan), 65 ppm (C_6, —O—CH_2—oxirane ring), and 50 and 44 ppm ($C_{7,8}$, oxirane ring). Moreover, the result of element analysis (found: C, 62.11; H, 6.40; $C_8H_{10}O_2$ requires: C, 62.33; H, 6.54%) also agrees with the chemical structure of FGE.

It can thus be concluded that the synthetic resultant is indeed FGE as expected.

6.1.4 Reversibility of DA Bonds and Crack Remendability of FGE-Based Polymer

Since the newly synthesized epoxy resin FGE is planned to act as a thermal remendable polymer using the effect of DA and retro-DA reactions, the reversibility of DA bonds in the adduct of FGE (containing furan ring, electron-rich diene) and DPMBMI (involving maleimide, electron-poor dienophile) should be first examined.

Our previous work on DGFA (Section 6.1.1) indicated that the DA reaction involving DPMBMI used to take place within a temperature range 50–70 °C, which would change with the structures of the furan compounds. To understand the temperature range of the DA reaction of the present system, a DSC heating experiment was conducted (Figure 6.18). The sample mixture of FGE and DPMBMI was obtained by dissolving stoichiometric DPMBMI into FGE at 90 °C for five minutes, and then quickly cooling down to room temperature to avoid the occurrence of DA reaction during decreasing temperature. The trace in Figure 6.18 shows a broad exotherm from 27 to 88 °C with the peak at 66 °C caused by the DA reaction. Moreover, an endothermic peak appears at 107 °C because of the retro-DA reaction, followed by an exothermal peak at 137 °C due to the opening reaction of the epoxide ring. When the temperature exceeds 170 °C, evident endothermic processes are observed, which originate from decomposition of low molecular weight glycidyl ether epoxy. This DSC analysis proves that DA reaction can take place at around 70 °C. Accordingly, isothermal DSC scanning was carried out at 70 °C to evaluate the extent of DA reaction. After 180 minutes of reaction, the exothermal enthalpy is found to be 248 J g^{-1}, corresponding to 83.0 kJ mol^{-1} (per DA bond). The value agrees with that reported in the literature [27].

Figure 6.19 shows the FTIR spectrum of the adduct of FGE and DPMBMI generated at 66 °C in comparison with that of FGE/DPMBMI mixture. The adduct was prepared in a similar way to that for making DGFA and DPMBMI (Section 6.1.1). The representative preparation procedures are shown below. First, 7.14 g (0.02 mol) DPMBMI was dissolved in 50 ml THF, which was then charged into a 100 ml three-neck round-bottom flask equipped with magnetic stirring under the protection of N_2. Afterwards, 6.16 g (0.04 mol) FGE was slowly added into the solution. The system was refluxed at about 66 °C for 24 hours and cooled down to room temperature. The reaction solution was poured into a large excess of diethyl ether. The precipitate was filtered, washed with methanol, and then dried under vacuum (<1.3 × 10^3 Pa) at room temperature. From

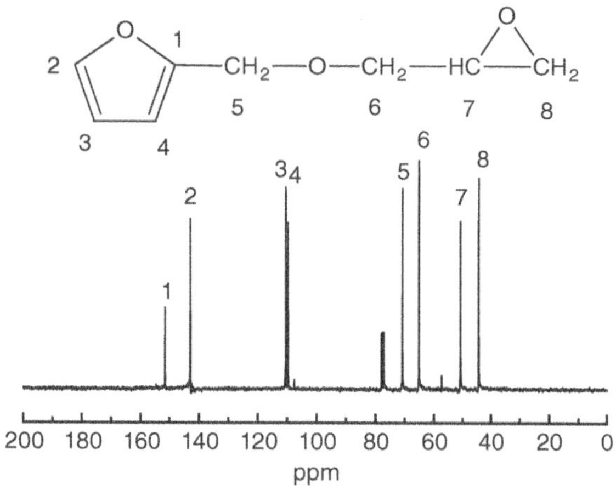

Figure 6.17 13C-NMR spectrum of FGE. *Source:* Reprinted from Tian et al. [22]. Copyright 2010, with permission from John Wiley and Sons.

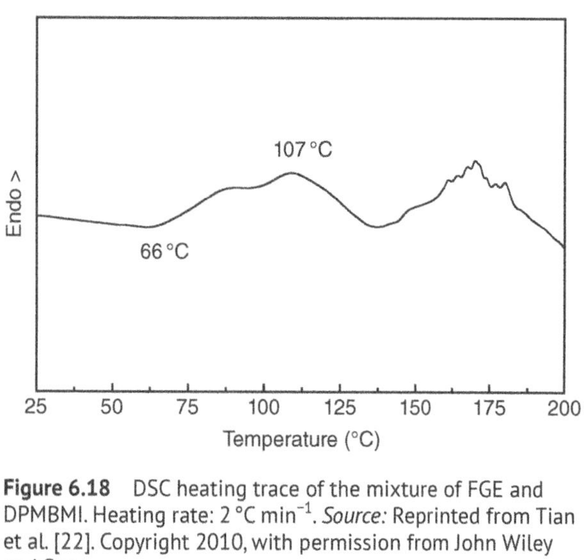

Figure 6.18 DSC heating trace of the mixture of FGE and DPMBMI. Heating rate: 2 °C min^{-1}. *Source:* Reprinted from Tian et al. [22]. Copyright 2010, with permission from John Wiley and Sons.

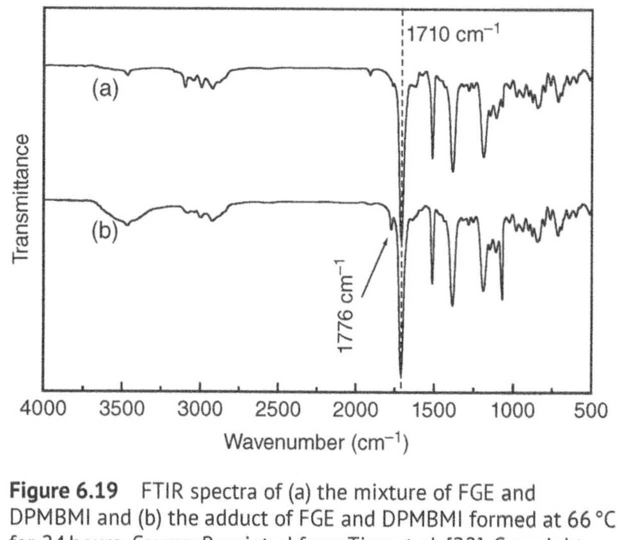

Figure 6.19 FTIR spectra of (a) the mixture of FGE and DPMBMI and (b) the adduct of FGE and DPMBMI formed at 66 °C for 24 hours. *Source:* Reprinted from Tian et al. [22]. Copyright 2010, with permission from John Wiley and Sons.

Figure 6.19 it is seen that a new peak is perceived at $1776\,cm^{-1}$ near the stretching of C=O at $1710\,cm^{-1}$ (curve b in Figure 6.19), which is peculiar to DA adducts of maleimides [28]. This peak is nearly negligible on the spectrum of the FGE/DPMBMI mixture prepared at ambient temperature (curve a in Figure 6.19). Clearly, the DA reaction proceeds very fast at around 70 °C.

Having confirmed the generation of DA bonds in the adduct of FGE and DPMBMI, we further investigate the kinetic process of the reaction using in-situ ^{1}H-NMR spectroscopy. An equivalent molar mixture of FGE and DPMBMI was dissolved in DMSO-d6 ($76\,mmol\,l^{-1}$) and then sealed in a glass tube under N_2 atmosphere. Then, ^{1}H-NMR spectra were collected at specific time intervals to monitor the DA reaction proceeding in the mixture at 60 °C. Figure 6.20 shows that with increasing time the peaks at 6.3 and 6.2 ppm assigned to H_2 and H_3 protons of FGE are weakened, while those at 6.5 and 6.6 ppm corresponding to H_2' and H_3' protons of the DA adduct become intensified. This demonstrates that the reactants are gradually converted to the adduct. Additionally, two new peaks appear at 5.2 and 5.34 ppm after one hour of reaction. They represent the H_1' proton of the DA adduct with different conformations.

On the basis of above analysis and the results of Figure 6.20, time-dependent reaction conversion of DA adduct can be calculated (see Eqs. (6.1)–(6.3) in Section 6.1.1 for details of the calculation). Linear fits of the data give the rate constant with regression coefficient of 0.98316 for the first-order plot and 0.99556 for the second-order one, suggesting that the second-order kinetics reaction should be more reasonable. Similarly, the reaction at 70 °C also exhibits greater regression coefficient for the second-order plot than that for the first-order one. Therefore, the second-order kinetics should be more reasonable for describing the current DA reaction. The activation energy derived from the second-order kinetics is $35.7\,kJ\,mol^{-1}$, which falls in the similar ranges reported for the other furan-maleimide DA reactions [19, 31–34], and is lower than the value of the reaction between DGFA and DPMBMI ($42.9\,kJ\,mol^{-1}$). Compared with DGFA, it is easier for FGE to conduct cycloaddition because of its lower molecular weight and less significant steric hindrance of the furan group.

In addition to the DA reaction, the retro-DA reaction of the adduct of FGE and DPMBMI should also be characterized. For this purpose, ^{1}H-NMR and ^{13}C-NMR were used. Prior to the measurements, the DA adduct was dissolved in DMSO-d6 in a glass tube and heated at 110 °C (which was determined according to Figure 6.18) for 20 minutes to disconnect the DA adduct. After the retro-DA reaction, the product was immediately cooled down to room temperature and analyzed. From the ^{1}H-NMR spectra in Figure 6.21, it is seen that the peaks at 5.19–5.3 ppm ascribed to the H_1' proton of the DA adduct (curve a in Figure 6.21) entirely disappear in the spectrum of the retro-DA product (curve b in Figure 6.21). Instead, the peak of the H_1 proton of the furan ring is perceived at 7.4 ppm. Also, the characteristic peaks at 6.5 and 6.6 ppm assigned to the H_2' and H_3' protons of the DA adduct (curve a in Figure 6.21) have been replaced by the new peaks at 6.3 and 6.4 ppm corresponding to the H_2 and H_3 protons of furan (curve b in

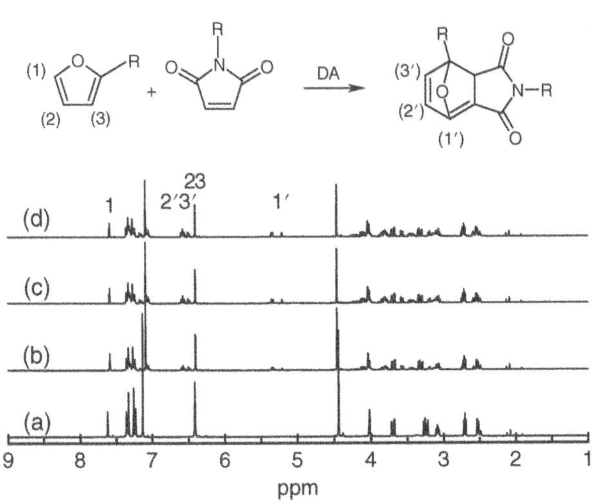

Figure 6.20 ^{1}H-NMR spectra of the mixture of FGE and DPMBMI kept at 60 °C for (a) zero hours, (b) one hour, (c) two hours, and (d) three hours. *Source:* Reprinted from Tian et al. [22]. Copyright 2010, with permission from John Wiley and Sons.

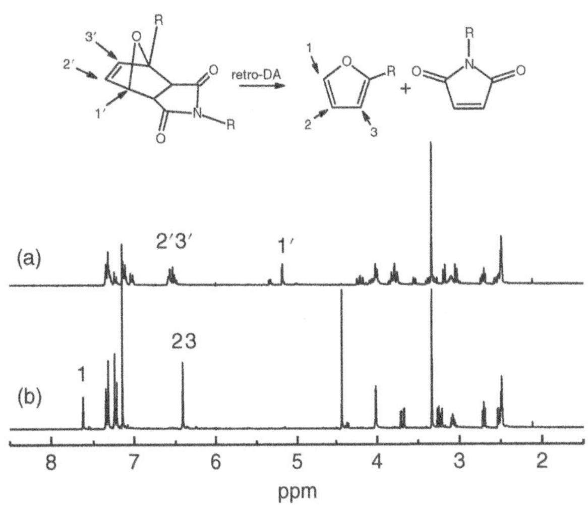

Figure 6.21 1H-NMR spectra of (a) the adduct of FGE and DPMBMI and (b) the retro-DA product of the adduct of FGE and DPMBMI by heating the adduct at 110 °C for 20 minutes. *Source:* Reprinted from Tian et al. [22]. Copyright 2010, with permission from John Wiley and Sons.

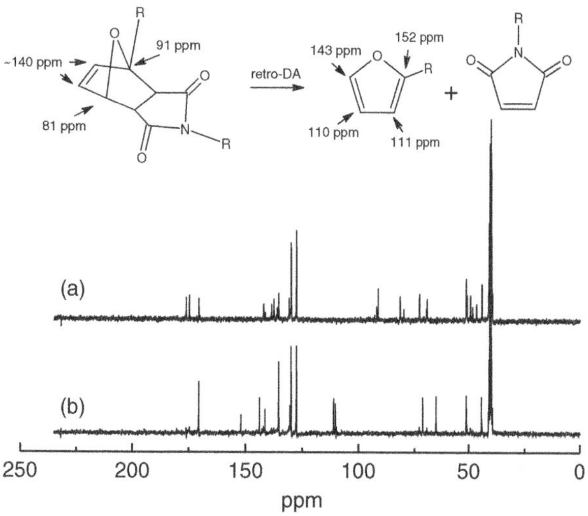

Figure 6.22 13C-NMR spectra of (a) the adduct of FGE and DPMBMI and (b) the retro-DA product of the adduct of FGE and DPMBMI by heating the adduct at 110 °C for 20 minutes. Reprinted from Tian et al. [22]. Copyright 2010, with permission from John Wiley and Sons.

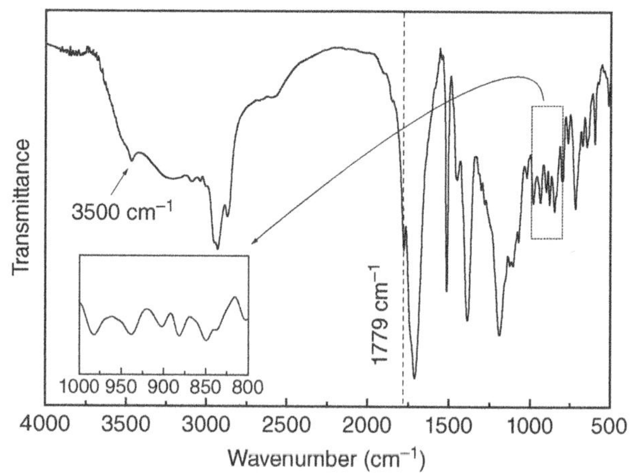

Figure 6.23 FTIR spectrum of cured FGE/MHHPA/DPMBMI polymer. *Source:* Reprinted from Tian et al. [22]. Copyright 2010, with permission from John Wiley and Sons.

Figure 6.21). On the other hand, the peaks of DA product at 82 and 93 ppm of the ^{13}C-NMR spectrum [9] (curve a in Figure 6.22) are no longer perceived in the spectrum of retro-DA product (curve b in Figure 6.22). These data demonstrate that the retro-DA reaction can be completed under the aforementioned conditions (i.e. 110 °C, 20 minutes).

In the above discussion, thermal reversibility of the DA adduct of FGE and DPMBMI is verified. Nevertheless, the epoxide groups of FGE have not yet reacted. As macromolecules constructed of irreversible epoxide bonds are critical for the load bearing ability of cured FGE material and might obstruct the reversible DA reaction, the reversibility of DA bonds in FGE cured by both MHHPA and DPMBMI should be explored.

For this purpose, cured FGE/MHHPA/DPMBMI polymer should be prepared. Typically, 17.13 g (0.048 mol) DPMBMI was dissolved in 14.78 g (0.096 mol) FGE under stirring at 90 °C for five minutes without the help of solvent. Then, 12.9 g (0.077 mol) MHHPA was added to the above mixture and stirred at 80 °C for 10 minutes. The resultant homogeneous compound was degassed, poured into a closed silicone rubber mold, and cured at 70 °C for 24 hours. This curing temperature was selected based on the optimum temperature for DA bond formation [36] to favor most of the furan and maleimide groups taking part in the reaction prior to solidification of the system. In addition, non-stoichiometric ratio of epoxy ring to anhydride of 1:0.8 was also used to delay the curing of the system, allowing completion of the DA reaction as far as possible.

Figure 6.23 shows the FTIR spectrum of FGE cured by both MHHPA and DPMBMI. The peaks attributed to epoxide groups at 920 cm^{-1} (oxirane ring breathing) and 856 cm^{-1} (C—O—C) disappear, indicating that all the epoxide groups of FGE have reacted with the anhydride groups of MHHPA. The evident peak of hydroxyl groups at 3500 cm^{-1} supports the deduction. Besides, the DA reaction between furan and maleimide groups in the cured sample is also proved by the existence of the peak at 1779 cm^{-1}.

Isothermal curing kinetics of FGE with MHHPA at stoichiometric ratio was studied by DSC based on the modified Avrami equation [25]. It was found that FGE can perform curing reaction with MHHPA like conventional epoxy resin. The activation energy was estimated to be 102.5 kJ mol^{-1}, implying that reactivity of the attached epoxide groups is maintained.

To check how the DA bonds behave under thermal stimulus, heating DSC curves of the cured FGE polymer were recorded. The first heating trace shows a glass transition at around 93 °C and an endothermic peak at 113 °C as a result of retro-DA reaction. There is not enough time for the recovered furan and maleimide moieties to be reconnected during the subsequent cooling and reheating processes, so no more retro-DA reaction but glass transition of the cured epoxy is detected by the second DSC heating scan. In comparison with the retro-DA reaction at 107 °C in solution (Figure 6.18), the higher retro-DA reaction temperature in solid (113 °C) suggests that reconnection of furan and maleic groups in cured FGE has to proceed at a much slower rate than that in solution.

By applying cyclic retro-DA (heat treatment at 120 °C for 20 minutes to attain a retro-DA sample) and DA reactions (heat treatment at 80 °C for 72 hours to attain a DA sample), reversibility of the cured FGE polymer was studied by DSC (Figure 6.24, Table 6.4). The higher heating

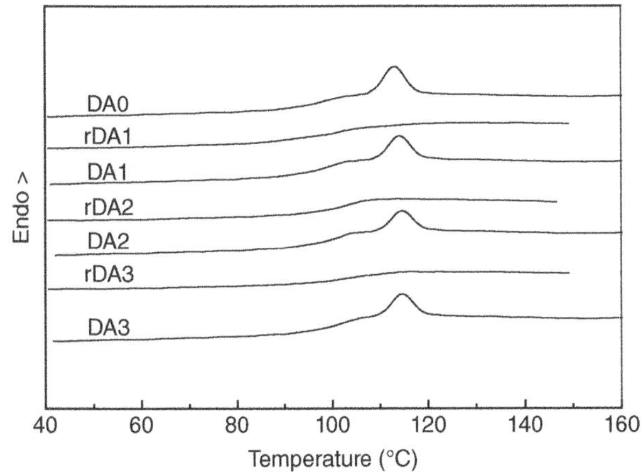

Figure 6.24 DSC heating traces of cured FGE/MHHPA/DPMBMI polymer (heating rate: 5 °C min^{-1}). DA0: as-manufactured sample. rDA1: DA0 treated at 120 °C for 20 minutes, and then quenched to room temperature. DA1: rDA1 treated at 80 °C for 72 hours, and then cooled inside the oven by switching off the electricity supply. rDA2: DA1 treated at 120 °C for 20 minutes, and then quenched to room temperature. DA2: rDA2 treated at 80 °C for 72 hours, and then cooled inside the oven by switching off the electricity supply. rDA3: DA2 treated at 120 °C for 20 minutes, and then quenched to room temperature. DA3: rDA3 treated at 80 °C for 72 hours, and then cooled inside the oven by switching off the electricity supply. *Source:* Reprinted from Tian et al. [22]. Copyright 2010, with permission from John Wiley and Sons.

temperatures for both retro-DA and DA reactions were chosen to increase the reaction efficiency. It is seen that the original cured FGE (DA0) gives an endothermic peak of retro-DA reaction at 113 °C, while the rDA1 only shows a glass transition at 93.5 °C. This implies that the retro-DA reaction has already been completed during the heat treatment at 120 °C for 20 minutes. Subsequently, the rDA1 sample is allowed to reconnect the cleavage groups by DA reaction, and then an endothermic peak appears again at 114.1 °C for DA1. It is evident that the DA reaction between the disconnected furan and maleimide moieties has definitely taken place in the thermal treatment process. The recovery efficiency of the second DA reaction in comparison with the original cured FGE polymer is about 99.3% by the ratio of ΔH_1 (enthalpy of DA1) to ΔH_o (enthalpy of DA0), which is obviously higher than the value of cured DGFA (77%, see Table 6.2). It is noted that the ΔH_o is 176.4 J g^{-1}, corresponding to 83.0 kJ mol^{-1} (per DA bond), which equals the aforementioned exothermal enthalpy per mole of DA bond formation. This means that all DA bonds can be cleaved during the first retro-DA reaction, and the macromolecules formed due to the reaction between epoxy and anhydride do not block the retro-DA reaction.

The data in Table 6.4 indicate that the endothermic peak temperature of the DA sample slightly increases with heat treatment time at the beginning and then remains unchanged. This phenomenon may be related to structure stabilization after the said heat treatment. Accordingly, the enthalpy of endothermic peak, ΔH, of the retro-DA reaction also slightly decreases and becomes nearly constant from the second heating cycle on. On the whole, the decrease of each retro-DA endothermic enthalpy is marginal. The DA and retro-DA reactions in the cured FGE can still be considered as fully reversible.

To quantify thermal remendability of the cured polymer in terms of healing efficiency that has been applied for assessment of mechanical property recovery in self-healing materials, the method suggested by Nemat-Nasser et al. [37], i.e. the double cleavage drilled compression (DCDC) test, was used. The design allows for controlled incremental crack growth so that the cracked specimen remains in one piece after the test, ensuring realignment of the fracture surfaces prior to healing. The DCDC specimen was a column of a rectangular cross section (length, $l = 50$ mm; width, $w = 15$ mm; thickness, $t = 8.9$ mm) with a circular hole (radius, $R = 1.98$ mm) drilled through its center that is subjected to axial compression. Prior to the test, a blade was wedged into the central hole of the DCDC specimen to create pre-cracks 0.5 mm deep at the upper and lower crowns. Then, a restraining clamp was placed around the specimen and axial compression was applied. The lateral confinement offered by the clamp prevented the cracks from growing beyond a length, l ($l/R = 0.8$), from the crowns of the hole. Afterwards, the clamp was removed, and the

Table 6.4 Quantification of endothermic peaks of cured FGE polymer measured by DSC heating scan.

Sample ID[a]	Onset temperature (°C)	Peak temperature (°C)	Enthalpy (J g^{-1})
DA0	103	113	176.4
DA1	104.1	114.1	175.2
DA2	104	114.6	174.8
DA3	104.1	114.6	174.0

[a] Descriptions of the samples are listed in the caption of Figure 6.24.
Source: Reprinted from Tian et al. [22]. Copyright 2010, with permission from John Wiley and Sons.

specimen was reloaded in compression with a universal tester at a crosshead rate of 0.2 mm min^{-1} to record the force, σ_{crit}, and crack length, l/R, of the subsequent fracture event. The test was paused at increments of 60 μm in crosshead displacement and, after an equilibration period of 10 seconds, the peak load over that period was recorded and a photo was taken for determining crack length. After the test, the specimen was healed in a pre-heated oven at 110 °C for 20 minutes (for disconnecting DA bonds) and then at 80 °C for 72 hours (for reconnecting the disconnected furan and maleimide moieties). Finally, the healed specimen was tested again following the above procedure. Healing efficiency, η, was calculated from the ratio of critical stress required to propagate the crack to a given length in the healed material, σ_{Crit}^{Healed}, to the stress required to propagate the crack to the same length in the virgin material, σ_{Crit}^{Virgin}.

$$\eta = \frac{K_{IC}^{Healed}}{K_{IC}^{Virgin}} \approx \frac{\sigma_{Crit}^{Healed}}{\sigma_{Crit}^{Virgin}} \tag{6.4}$$

It is worth noting that the healing efficiency was originally defined as the ratio of the fracture toughness of the healed specimen, K_{IC}^{Healed}, to the fracture toughness of the virgin specimen, K_{IC}^{Virgin} [38], which only depended on material properties. The above simplified calculation based on the ratio of critical stresses, proposed by Nemat-Nasser et al. [37], is related to both material properties and geometry effects, and can be meaningful when the cracks in both the virgin and healed cases had nearly identical length. Accordingly, in order to obtain the healing efficiency in the present work, for a measured critical stress of the virgin specimen at a certain crack length, the corresponding critical stress of the healed specimen at the same crack length was estimated by interpolation from the fitted crack length dependence of the measured critical stress of the healed specimen.

Table 6.5 lists healing efficiencies measured at different crack lengths, and also the average healing efficiency. Coinciding with high thermal reversibility revealed in the repeated heating and cooling, nearly complete rehabilitation is observed for the cured FGE, as characterized by the average healing efficiency of 96%. The crack healability is evidently higher than that of cured DGFA (healing efficiency = 66% [36], also measured by DCDC method, see Section 6.2.2 for more details).

6.2 Blends of DGFA and FGE

In the previous subsections, the synthesis and characterization of DGFA and FGE have been discussed. In their cured versions, the thermally reversible DA bonds offer crack repair effect based on successive retro-DA and DA reactions inducing chain reconnection, while the irreversible epoxy bonds are responsible for the load bearing capacity. In particular, the specifically designed molecular structures enable rehabilitation below the glass transition temperature of the materials, so that possible distortion during crack healing can be avoided. As a result, the new epoxy monomers have preliminarily demonstrated their potential for practical application.

As mentioned at the beginning of this chapter, however, both DGFA and FGE have their shortcomings by their very nature. The former possesses fewer furan groups and hence the healing efficiency of the cured polymer is limited. In contrast, the latter has unsatisfactory mechanical performance and resistance to thermal deformation due to lower concentration of epoxide groups. Therefore, we proposed to prepare blends of the two monomers as a remedy. On the one hand, the concentration of DA bonds can be higher than that in DGFA due to the incorporation of FGE that contains more furan groups, so that the fraction of intermonomer linkages that can be disconnected and reconnected upon specific heat treatment is increased accordingly. Improved crack healing efficiency would thus be detected. On the other hand, the concentration of crosslinked bonds of epoxy can also be higher than that in FGE

Table 6.5 Healing efficiency of cured FGE.

	l/R = 1.11		l/R = 1.72		l/R = 2.18		
Crack length	σ_{crit} **(MPa)**	η **(%)**	σ_{crit} **(MPa)**	η **(%)**	σ_{crit} **(MPa)**	η **(%)**	η_{avg} **(%)**
Virgin[a]	8.51	—	9.94	—	10.68	—	—
Healed[b]	8.16	95.9	9.55	96.0	10.26	96.1	96.0

[a] As-manufactured specimen.
[b] Cracked virgin specimen was treated at 110 °C for 20 minutes, and then at 80 °C for 72 hours. The treatment temperature (110 °C) is close to the peak retro-DA temperature (Figure 6.24) and higher than the glass transition temperature (93.5 °C, measured by DSC, see Figure 6.24). The loss in modulus during glass transition of cured FGE is so severe that no complete DMA spectrum can be recorded, so here we have to use the galss transition temperature determined by DSC.
Source: Reprinted from Tian et al. [22]. Copyright 2010, with permission from John Wiley and Sons.

as a result of addition of DGFA. Consequently, the cross-linked epoxy networks would provide the blends with acceptable mechanical strength. The thermal remendability and mechanical properties of the new epoxy originating from the blends can be optimized [39].

6.2.1 Reversibility of DA Bonds

DGFA and FGE are inherently miscible, so the blends of DGFA and FGE can be easily prepared through conventional mixing without the help of any additional compatibilizers. As for preparation of the cured blends, DPMBMI and MHHPA that have been employed for DGFA or FGE alone still serve as the hardeners since both DGFA and FGE carry the same functional groups (i.e. furan and epoxide). DPMBMI and MHHPA are, respectively, responsible for establishing thermally reversible and irreversible bonds in the cured systems.

Typically, DPMBMI was dissolved in DGFA, or FGE, or DGFA/FGE under stirring at 90 °C for 10 minutes. Then, MHHPA was added to the above mixture, and stirred at 80 °C for 10 minutes. The resultant homogeneous compound was degassed, poured into a closed silicone rubber mold, and cured at 70 °C for 24 hours. Table 6.6 lists the formulations of the blends, in which the molar ratio of epoxide and anhydride groups is fixed at 1:0.8 and that of furan and maleimide groups is 1:1. These conditions were selected to provide the cured DGFA/FGE blends with as many DA bonds as possible.

Figure 6.25 shows FTIR spectra of the cured blends. It is seen that the peaks attributed to epoxide groups at 918 cm^{-1} (oxirane ring breathing) and 851 cm^{-1} (C—O—C) are absent, meaning that all epoxide groups must have reacted with anhydride groups of MHHPA to form epoxy networks. The evident peak of hydroxyl groups at around 3500 cm^{-1} also demonstrates the occurrence of the reaction. Meanwhile, the peak at about 1776 cm^{-1} supports the existence of DA adduct from the reaction between furan and maleimide groups. It is known that in the cured blends the two types of desired macromolecular networks have been built up, regardless of the blending ratio of DGFA and FGE resins.

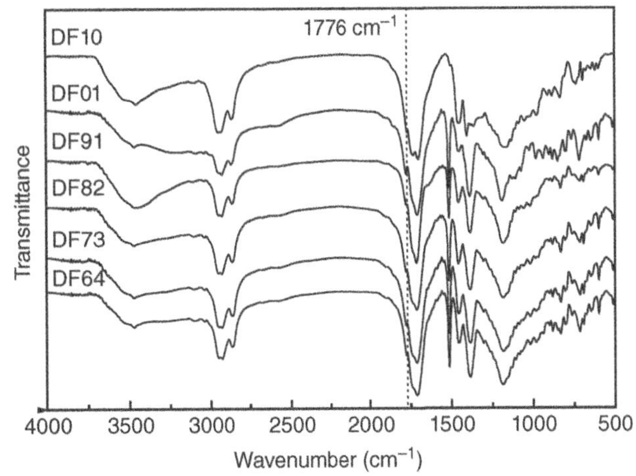

Figure 6.25 FTIR spectra of cured DGFA/FGE blends. The sample IDs and compositions are shown in Table 6.6. *Source:* Reprinted from Tian et al. [39]. Copyright 2010, with permission from Elsevier.

Thermal reversibility of the cured blends is viewed by DSC in terms of repeated heating and cooling treatments. That is, the specimens were treated at 140 °C for 20 minutes to attain retro-DA samples and then at 80 °C for 72 hours to attain DA samples up to three times. The detailed steps are described in the caption of Figure 6.26, which are the same as those used for verifying individual DGFA (Figure 6.12) and FGE (Figure 6.24). It is seen that an endothermic peak appears at around 120 °C on the curves of as-manufactured samples, which results from retro-DA reaction. As for rDA1, only a glass transition at around 120 °C is observed. This means that the treatment at 140 °C for 20 minutes has led to complete disconnection of DA bonds via retro-DA reaction. Subsequently, the rDA1 sample is allowed to reconnect the cleavage groups through DA reaction by rehabilitation at 80 °C for 72 hours, so that an endothermic peak appears again on the DSC curve of DA1. The

Table 6.6 Molar compositions of cured DGFA/FGE blends.

Sample ID	DGFA	FGE	DPMBMI	MHHPA
DF10	1	0	0.5	1.6
DF01	0	1	0.5	0.8
DF91	0.9	0.1	0.5	1.52
DF82	0.8	0.2	0.5	1.44
DF73	0.7	0.3	0.5	1.36
DF64	0.6	0.4	0.5	1.28

Source: Reprinted from Tian et al. [39]. Copyright 2010, with permission from Elsevier.

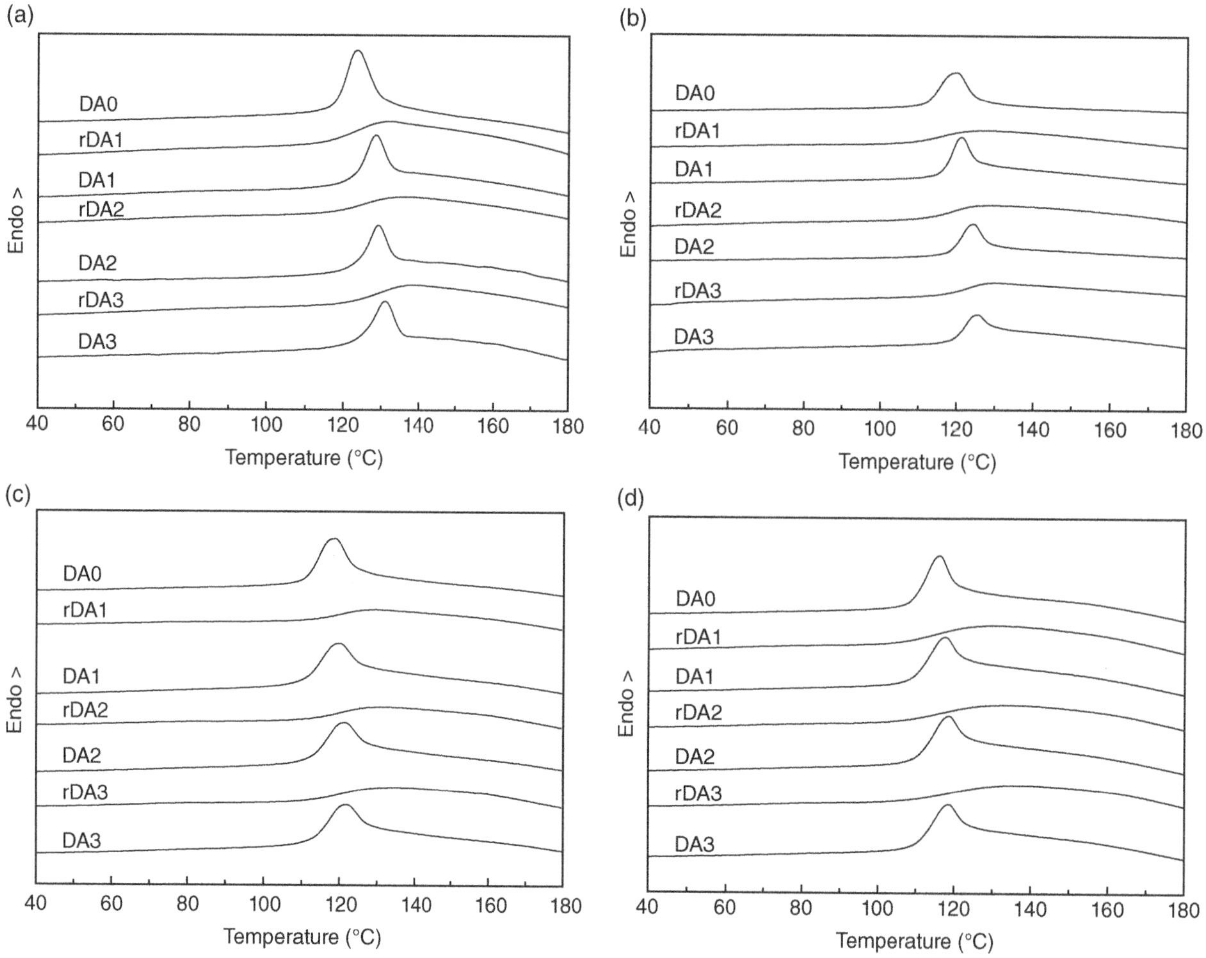

Figure 6.26 DSC heating traces of the cured blends of DGFA and FGE (heating rate: 5 °C min^{-1}): (a) DF91; (b) DF82; (c) DF73; and (d) DF64. DA0: as-manufactured sample. rDA1: DA0 treated at 140 °C for 20 min, and then quenched to room temperature. DA1: rDA1 treated at 80 °C for 72 h, and then cooled inside the oven by switching off the electricity supply. rDA2: DA1 treated at 140 °C for 20 minutes, and then quenched to room temperature. DA2: rDA2 treated at 80 °C for 72 hours, and then cooled inside the oven by switching off the electricity supply. rDA3: DA2 treated at 140 °C for 20 minutes, and then quenched to room temperature. DA3: rDA3 treated at 80 °C for 72 hours, and then cooled inside the oven by switching off the electricity supply. *Source:* Reprinted from Tian et al. [39]. Copyright 2010, with permission from Elsevier.

repeatable emergence of the endotherm indicates thermal reversibility of the DA bonds in the blends.

A careful survey of the data listed in Table 6.7 reveals that with a rise in the content of FGE in the blends, the enthalpy of retro-DA reaction increases. Clearly, an increased contribution from FGE should account for the variations because FGE contains more furan rings. Moreover, Table 6.7 also indicates that the peak temperature of retro-DA reaction decreases with the fraction of FGE. This should be attributed to the relatively lower number of epoxide groups in FGE than DGFA, which reduces the restraint effect of epoxy networks.

It is interesting to note that for cured DGFA (i.e. DF10) the enthalpy of retro-DA reaction significantly decreases after the first cycle of heating and cooling, and this trend becomes less obvious in the subsequent tests. The phenomenon should be related to the increase in cross-link density of the system as a result of heat treatment, which hinders recovery of DA bonds. However, when FGE is incorporated, the decrement becomes smaller. Consequently, recovery efficiency of the DA reaction is raised with increasing concentration of FGE, which would certainly benefit the improvement of thermal remendability of the blends.

In addition to the DSC study under quiescent condition, DMA was also employed to examine thermal reversibility of the materials. The results are more closely related to mobility of the macromolecules. As illustrated in Figure 6.27, the storage moduli of as-manufactured samples drastically decrease with increasing temperature at around 110 °C and then level off. When temperature is further raised, the moduli drop again. In contrast, after retro-DA reaction at 140 °C

Table 6.7 Thermal reversibility of DA reaction in the cured blends of DGFA and FGE.

Blends	DF10			DF01			DF91			DF82			DF73			DF64		
Treatment	T_p[a] (°C)	ΔH[b] (J g^{-1})	R[c] (%)	T_p (°C)	ΔH (J g^{-1})	R (%)	T_p (°C)	ΔH (J g^{-1})	R (%)	T_p (°C)	ΔH (J g^{-1})	R (%)	T_p (°C)	ΔH (J g^{-1})	R (%)	T_p (°C)	ΔH (J g^{-1})	R (%)
DA0	126.2	122.9	—	113.0	176.4	—	124.0	126.6	—	120.0	132.1	—	118.5	136.5	—	116.0	140.4	—
DA1	131.6	94.7	77.1	114.1	175.2	99.3	128.0	100.5	79.4	121.5	113.5	85.9	119.5	118.5	86.8	117.5	122.1	87.0
DA2	133.7	92.6	75.3	114.6	174.8	99.1	129.0	98.7	78.0	124.5	108.4	82.1	121.5	111.2	81.5	118.0	120.0	85.5
DA3	136.5	90.0	73.3	114.6	174.0	98.6	131.0	96.0	75.8	125.5	105.9	80.2	122.0	110.1	80.7	119.0	116.2	82.8

[a] Peak temperature for the retro-DA reaction.
[b] Enthalpy of the endothermic peak due to the retro-DA reaction.
[c] Recovery efficiency of the DA reaction, calculated from the enthalpy ratio of the retro-DA of DAi ($i = 1$, 2 and 3) to the retro-DA of DA0.
Source: Reprinted from Tian et al. [39]. Copyright 2010, with permission from Elsevier.

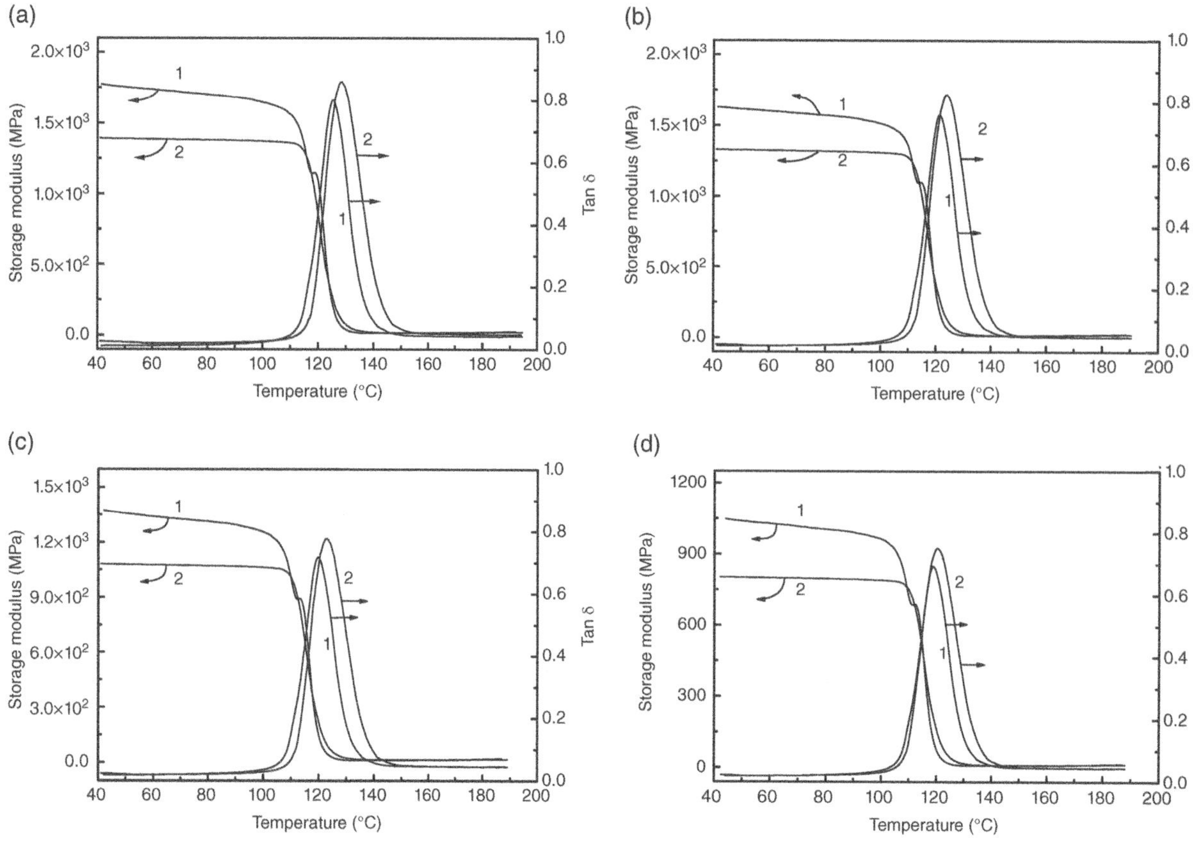

Figure 6.27 DMA spectra of the cured blends of DGFA and FGE: (a) DF91; (b) DF82; (c) DF73; and (d) DF64. Curve 1: as-manufactured sample. Curve 2: as-manufactured sample treated at 140 °C for 20 minutes, and then quenched to room temperature. *Source:* Reprinted from Tian et al. [39]. Copyright 2010, with permission from Elsevier.

for 20 minutes, the samples only show a single transition on the curves of storage moduli versus temperature, despite the fact that the room temperature storage moduli are reduced probably due to plasticization effect of the disconnected furan and maleimide moieties. Evidently, the two-stage variation in storage moduli of the as-manufactured samples reflects retro-DA reaction and glass transition, respectively. The retro-DA reaction happens below the glass transition temperature of the materials, but there is still certain overlap between the two processes.

When the blends are heated above the glass transition temperature, all DA bonds are broken. Thus, the storage moduli in the rubbery region are only related to the cross-link density variations of epoxy networks. According to the equation of state for rubber elasticity, the molecular weight between cross-links of the cured epoxy, M_c, is calculated from [40]:

$$E = 3\frac{\rho}{M_c}RT\left(1-\frac{2}{\phi}\right) \qquad (6.5)$$

where E' stands for storage modulus at rubbery plateau zone, ρ is density, R is the gas constant, T is the absolute temperature, and ϕ is the functionality of epoxy. The plots in Figure 6.28 demonstrate that after the retro-DA reaction, M_c is lowered in all the materials. In other words, the cross-link density of epoxy networks is increased accordingly. The results support the above analysis of reduced enthalpy of retro-DA reaction after cyclic heating and cooling (Table 6.7), and also explain the higher glass transition temperature of the specimens after retro-DA reaction in comparison with the as-manufactured specimens (Figures 6.27 and 6.28).

With respect to the effect of blending with FGE, the higher content of FGE, the higher M_c, either before or after retro-DA reaction. This implies that the thermal reversible DA bonds in the blends are more mobile in the case of high FGE fraction due to the surrounding looser epoxy networks.

6.2.2 Crack Remendability of Cured DGFA/FGE Blends

To assess thermal remendability of the blends, DCDC tests were carried out. Table 6.8 lists healing efficiencies of cured DF10 (i.e. DGFA) measured at different crack

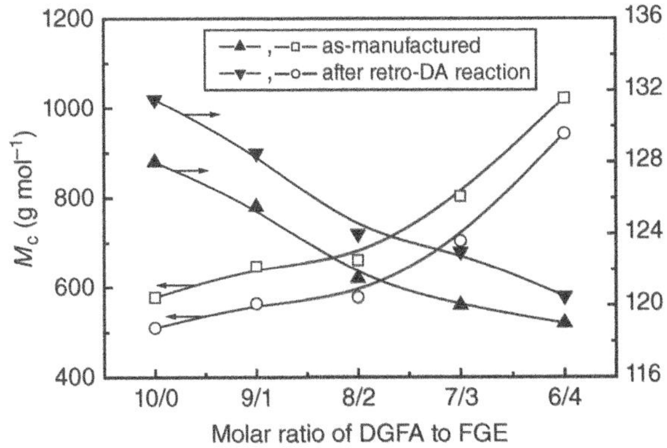

Figure 6.28 Molecular weight between cross-links of the cured epoxy, M_c, and glass transition temperature, T_g, of the cured blends of DGFA and FGE as a function of molar ratio of DGFA to FGE. The data are derived from the corresponding DMA spectra, and M_c values are estimated using storage moduli at 160 °C according to Eq. (6.5). *Source:* Reprinted from Tian et al. [39]. Copyright 2010, with permission from Elsevier.

lengths, and also the average healing efficiencies. Owing to the lower furan fraction in DF10 than in DF01 (i.e. FGE, see Table 6.5 for its values of healing efficiency determined by DCDC tests), the former has to exhibit remarkably lower healing efficiency.

Healing efficiencies of DF91, DF82, DF73, and DF64 are given in Tables 6.9–6.12, respectively. The temperatures for obtaining retro-DA specimens were so selected that they are near the peak retro-DA temperature and below the glass transition temperature. This would guarantee disconnection of the DA bonds as much as possible, while the entire process of crack removal is conducted in glass state, which is critical for keeping profiles of the materials in practical applications. The results show that the addition of FGE greatly increases remendability of the system. The average healing efficiency is improved from 65.9% (for DF10) to 81.1% (for DF64). The increased amount of DA bonds in the blends must have taken effect.

For more molecular information of crack healing in the current system, surface Raman spectra of the specimens were collected. The peak at 1618 cm^{-1} is attributed to the stretching mode of benzene, while those at 1385 and 1448 cm^{-1} are interpreted as breathing of furan rings. The peaks at 794 and 837 cm^{-1} are due to ether bonds resulting from ring opening of epoxide groups. Therefore, when the area of the peak at 1618 cm^{-1} acts as the reference, the concentration of furan rings can be characterized by areas of the peak at 1385 and 1448 cm^{-1}, and that of the cross-linked bonds of epoxy can be described by areas of the peaks at 794 and 837 cm^{-1}. Figure 6.29 indicates that the content of furan groups characterized by peak area ratio, $A_{furan}/A_{benzene}$, is indeed increased with a rise in the content of FGE. Therefore, the blends with higher dosage of FGE contain higher concentration of DA bonds and possess higher healing efficiency.

Figure 6.29 also reveals that the contents of furan groups on fractured surfaces are higher than those on as-manufactured surfaces. Comparatively, the contents of cross-linked bonds of epoxy are almost the same on either fractured or as-manufactured surface. The furan and epoxide groups are homogeneously distributed in the materials on molecular scale, so the higher contents of furan groups on fractured surfaces imply that cracks have propagated preferentially along DA bonds. This agrees with the conclusion by Chen et al. [9] that the energy to break DA adducts is much lower than that to break all the other covalent bonds. Our estimation based on quantum chemistry also indicates that the length of the DA bond is 1.59 Å, which is 0.05 Å longer than the conventional C—C bond length, so that the energy of the DA bond has to be reduced. Owing to this feature of DA bonds, the polymers studied in the present work exhibit much higher healing efficiency than the estimation based on the true

Table 6.8 Healing efficiency of the DF10 specimen.

	$l/R = 1.00$		$l/R = 1.26$		$l/R = 2.14$		
Crack length	σ_{crit} **(MPa)**	η **(%)**	σ_{crit} **(MPa)**	η **(%)**	σ_{crit} **(MPa)**	η **(%)**	η_{avg} **(%)**
Virgin[a]	22.5	—	24.4	—	29.1	—	—
Healed[b]	14.7	65.0	16.1	65.8	19.6	67.0	65.9

[a] As-manufactured specimen.
[b] Cracked virgin specimen was treated at 126 °C for 20 minutes, and then at 80 °C for 72 hours. The treatment temperature 126 °C is close to the peak retro-DA temperature (Table 6.7) and lower than the glass transition temperature (Figure 6.28). When the specimens were treated at 80 °C for reconnection of the furan and maleic groups via DA reaction, the effect of lateral pressure (~0.3 MPa) that helped to further shorten the distance between the cracked faces was evaluated. It was found that there was not a big difference in the measured healing efficiencies whether the lateral pressure was applied to the specimens or not. Therefore, all the specimens of DGFA/FGE blends were healed without the aid of lateral pressure.

Table 6.9 Healing efficiency of the DF91 specimen.

	l/*R* = 1.03		*l*/*R* = 1.65		*l*/*R* = 2.19		
Crack length	σ_{crit} **(MPa)**	η **(%)**	σ_{crit} **(MPa)**	η **(%)**	σ_{crit} **(MPa)**	η **(%)**	η_{avg} **(%)**
Virgin[a]	21.2	—	27.1	—	27.4	—	—
Healed[b]	14.4	68.0	18.6	68.6	18.6	67.9	68.2

[a] As-manufactured specimen.
[b] Cracked virgin specimen was treated at 121 °C for 20 minutes, and then at 80 °C for 72 hours. The treatment temperature 121 °C is close to the peak retro-DA temperature (Table 6.7) and lower than the glass transition temperature (Figure 6.28).

Table 6.10 Healing efficiency of the DF82 specimen.

	l/*R* = 1.00		*l*/*R* = 1.26		*l*/*R* = 2.14		
Crack length	σ_{crit} **(MPa)**	η **(%)**	σ_{crit} **(MPa)**	η **(%)**	σ_{crit} **(MPa)**	η **(%)**	η_{avg} **(%)**
Virgin[a]	18.9	—	20.5	—	24.5	—	—
Healed[b]	13.94	73.8	15.21	74.2	18.13	74	74.0

[a] As-manufactured specimen.
[b] Cracked virgin specimen was treated at 117 °C for 20 minutes, and then at 80 °C for 72 hours. The treatment temperature 117 °C is close to the peak retro-DA temperature (Table 6.7) and lower than the glass transition temperature (Figure 6.28).

Table 6.11 Healing efficiency of of DF73 specimen.

	l/*R* = 1.1		*l*/*R* = 1.53		*l*/*R* = 2.18		
Crack length	σ_{crit} **(MPa)**	η **(%)**	σ_{crit} **(MPa)**	η **(%)**	σ_{crit} **(MPa)**	η **(%)**	η_{avg} **(%)**
Virgin[a]	16.52	—	18.67	—	20.9	—	—
Healed[b]	12.64	76.5	14.33	76.7	16.2	77.5	76.9

[a] As-manufactured specimen.
[b] Cracked virgin specimen was treated at 116 °C for 20 minutes, and then at 80 °C for 72 hours. The treatment temperature 116 °C is close to the peak retro-DA temperature (Table 6.7) and lower than the glass transition temperature (Figure 6.28).

Table 6.12 Healing efficiency of the DF64 specimen.

	l/*R* = 0.89		*l*/*R* = 1.38		*l*/*R* = 2.02		
Crack length	σ_{crit} **(MPa)**	η **(%)**	σ_{crit} **(MPa)**	η **(%)**	σ_{crit} **(MPa)**	η **(%)**	η_{avg} **(%)**
Virgin[a]	11.6	—	13.65	—	15.6	—	—
Healed[b]	9.4	81.0	11.1	81.3	12.61	81.0	81.1

[a] As-manufactured specimen.
[b] Cracked virgin specimen was treated at 114 °C for 20 minutes, and then at 80 °C for 72 hours. The treatment temperature 114 °C is close to the peak retro-DA temperature (Table 6.7) and lower than the glass transition temperature (Figure 6.28).

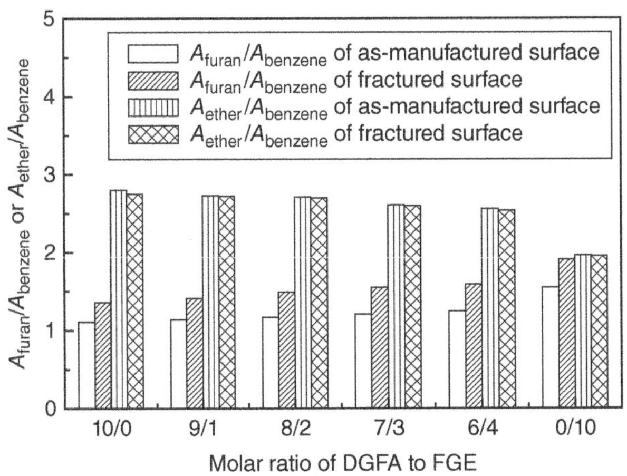

Figure 6.29 Characteristic surface Raman peak area ratios of the cured blends of DGFA and FGE as a function of molar ratio of DGFA to FGE. *Source:* Reprinted from Tian et al. [39]. Copyright 2010, with permission from Elsevier.

concentration of furan groups. For DF64, for example, the amount of furan groups only makes up 40% of the total amount of functional groups (i.e. the sum of furan and epoxide groups), but the healing efficiency is 81.1%.

Besides crack remendability, the mechanical performance of the blends should also be characterized. As shown in Table 6.13, although tensile and flexural properties of the blends decrease with increasing concentration of FGE due to its lower cross-link density, most blends have comparable load bearing capability like the cured commercial bisphenol-A epoxy. The moduli of DF91–DF73 are slightly higher than that of bisphenol-A epoxy, while their strengths are moderately lower. On the whole, the properties data are satisfactory, and the blends might replace traditional epoxy resin in future applications.

6.3 Concluding Remarks

Epoxy with intrinsic self-healing capability can be prepared by first integrating furan and epoxide groups in one molecule. Then, the monomer is cured by bismaleimide and anhydride. The resultant cross-linked polymer contains two types of intermonomer linkage: thermally reversible DA bonds from the reaction between furan and maleimide groups; and thermally irreversible bonds from the reaction between epoxide and anhydride groups. As a result of successive retro-DA and DA reactions, crack healing of the cured epoxy can be conducted in a controlled manner through chain reconnection. In the meantime, the cross-linked epoxide networks play the role of fixing phase and provide the polymer with adequate strength. Crack healing can thus be completed below the glass transition temperature of the polymer. This is critical for the possible application of the self-healing epoxy to avoid or reduce the possibility of material distortion in the course of crack healing.

To improve thermal remendability and mechanical performance of the epoxy, two types of the aforementioned bifunctional epoxy monomer are synthesized and then blended. One of the monomers possesses more furan groups than epoxide groups, while the other has more epoxide groups. Owing to the inherent structural similarity between the monomers, molecular miscibility of the blended components is guaranteed. Having been cured by maleimide and anhydride, the blends exhibit balanced healability and mechanical properties. They can also be healed in the glass state.

The proposed technical routes show the feasibility of constructing intrinsic self-healing polymer through both intramolecular and intermolecular designs.

Table 6.13 Mechanical properties of the cured blends of DGFA and FGE.

Properties	Bisphenol-A epoxy cured by MHHPA [35]	DF10	DF01	DF91	DF82	DF73	DF64
Young's modulus (GPa)	1.8	2.5	0.9	2.3	2.1	1.77	1.35
Tensile strength (MPa)	65	53	19	48.2	44.5	37.6	28.6
Elongation at break (%)	—	1.5	0.5	1.4	1.5	1.6	1.4
Flexural modulus (GPa)	2.6	4.6	1.6	4.2	3.8	3.3	2.4
Flexural strength (MPa)	128	110	32	100	92.4	78	58

References

1 Syrett, J.A., Becer, C.R., and Haddleton, D.M. (2010). Self-healing and self-mendable polymers. *Polymer Chemistry* 1: https://doi.org/10.1039/c0py00104j.

2 Prins, L.J., Reinhoudt, D.N., and Timmerman, P. (2001). Noncovalent synthesis using hydrogen bonding. *Angewandte Chemie International Edition 40*: 2382–2426.

3 Bergman, S.D. and Wudl, F. (2008). Mendable polymers. *Journal of Materials Chemistry 18*: 41–62.

4 Craven, J.M. (1969). Cross-linked thermally reversible polymers produced from condensation polymers with pendant furan groups cross-linked with maleimides. US Patent 3435003.

5 Piermattei, A., Karthikeyan, S., and Sijbesma, R.P. (2009). Activating catalysts with mechanical force. *Nature Chemistry 1*: 133–137.

6 Paulusse, J.M.J. and Sijbesma, R.P. (2004). Reversible mechanochemistry of a PdII coordination polymer. *Angewandte Chemie International Edition 43*: 4460–4462.

7 Davis, D.A., Hamilton, A., Yang, J.L. et al. (2009). Force-induced activation of covalent bonds in mechanoresponsive polymeric materials. *Nature 459*: 68–72.

8 Chen, X.X., Dam, M.A., Ono, K. et al. (2002). A thermally re-mendable cross-linked polymeric material. *Science 295*: 1698–1702.

9 Chen, X.X., Wudl, F., Mal, A.K. et al. (2003). New thermally remendable highly cross-linked polymeric materials. *Macromolecules 36*: 1802–1807.

10 Murphym, E.B., Bolanos, E., Shaffner-Hamann, C. et al. (2008). Synthesis and characterization of a single-component thermally remendable polymer network: Staudinger and Stille revisited. *Macromolecules 41*: 5203–5209.

11 Park, J.S., Takahashi, K., Guo, Z. et al. (2008). Towards development of a self-healing composite using a mendable polymer and resistive heating. *Journal of Composite Materials 42*: 2869–2881.

12 Murphy, E.B. and Wudl, F. (2010). The world of smart healable materials. *Progress in Polymer Science 35*: 223–251.

13 Kersey, F.R., Loveless, D.M., and Craig, S.L. (2007). A hybrid polymer gel with controlled rates of cross-link rupture and self-repair. *Journal of the Royal Society Interface 4*: 373–380.

14 Lee, H. and Neville, K. (1972). *Handbook of Epoxy Resins*. New York: McGraw-Hill.

15 Goodman, S.H. (1998). *Handbook of Thermoset Plastics*, 2e. Noyes Publications.

16 Yang, S., Chen, J.-S., Körner, H. et al. (1998). Reworkable epoxies: thermosets with thermally cleavable groups for controlled network breakdown. *Chemistry of Materials 10*: 1475–1482.

17 Chen, J.-S., Ober, C.K., and Poliks, M.D. (2002). Characterization of thermally reworkable thermosets: materials for environmentally friendly processing and reuse. *Polymer 43*: 131–139.

18 Small, J.H., Loy, D.A., Wheeler, D.R. et al. (2001). Method of making thermally removable polymeric encapsulants. US Patent 6271335.

19 Liu, Y.-L. and Hsieh, C.-Y. (2006). Crosslinked epoxy materials exhibiting thermal remendablility and removability from multifunctional maleimide and furan compounds. *Journal of Polymer Science, Part A: Polymer Chemistry 44*: 905–913.

20 Liu, Y.-L. and Chen, Y.-W. (2007). Thermally reverible crosslinked polyamides with high toughness and self-repairing ability from maleimide- and furan-functionalized aromatic polyamides. *Macromolecular Chemistry and Physics 208*: 224–232.

21 Tian, Q., Yuan, Y.C., Rong, M.Z., and Zhang, M.Q. (2009). A thermally remendable epoxy resin. *Journal of Materials Chemistry 19*: 1289–1296.

22 Tian, Q., Rong, M.Z., Zhang, M.Q., and Yuan, Y.C. (2010). Synthesis and characterization of epoxy with improved thermal remendability based on Diels-Alder reaction. *Polymer International 59*: 1339–1345.

23 Utracki, L.A. (1990). *Polymer Alloys and Blends*. Munich: Hanser Gardner Publications.

24 Isayev, A.I. (2010). *Encyclopedia of Polymer Blends: Volume 1: Fundamentals*. New York: Wiley-VCH.

25 Xu, W.B., Bao, S.P., Shen, S.J. et al. (2003). Differential scanning calorimetric study on the curing behavior of epoxy resin/diethylenetriamine/organic montmorillonite nanocomposite. *Journal of Polymer Science Part B: Polymer Physics 41*: 378–386.

26 Boey, F.Y.C. and Qiang, W. (2000). Experimental modeling of the cure kinetics of an epoxy-hexaanhydro-4-methylphthalicanhydride (MHHPA) system. *Polymer 41*: 2081–2094.

27 Chen, X.X. (2003). Novel polymers with thermally controlled covalent cross-linking. PhD thesis. University of California, Los Angeles.

28 Imai, Y., Itoh, H., Naka, K., and Chujo, Y. (2000). Thermally reversible IPN organic-inorganic polymer hybrids utilizing the Diels-Alder reaction. *Macromolecules 33*: 4343–4346.

29 Gousse, C. and Gandini, A. (1999). Diels-Alder polymerization of difurans with bismaleimides. *Polymer International 48*: 723–731.

30 Watanabe, M. and Yoshie, N. (2006). Synthesis and properties of readily recyclable polymers from bisfuranic terminated poly(ethylene adipate) and multi-maleimide linkers. *Polymer 47*: 4946–4952.

31 Dewar, M.J.S. and Pierini, A.B. (1984). Mechanism of the Diels-Alder reaction. Studies of the addition of maleic anhydride to furan and methylfurans. *Journal of the American Chemical Society 106*: 203–208.

32 Isaacs, N.S. and Keating, N. (1992). The rates of a Diels–Alder reaction in liquid and supercritical carbon dioxide. *Journal of the Chemical Society, Chemical Communications 12*: 876–877.

33 Goiti, E., Heatley, F., Huglin, M.B., and Rego, J.M. (2004). Kinetic aspects of the Diels–Alder reaction between poly(styrene-co-furfuryl methacrylate) and bismaleimide. *European Polymer Journal 40*: 1451–1460.

34 Liu, Y.L., Hsieh, C.Y., and Chen, Y.W. (2006). Thermally reversible cross-linked polyamides and thermo-responsive gels by means of Diels–Alder reaction. *Polymer 47*: 2581–2586.

35 Tao, Z.Q., Yang, S.Y., Ge, Z.Y. et al. (2007). Synthesis and properties of novel fluorinated epoxy resins based on 1,1-bis(4-glycidylesterphenyl)-1-(3′-trifluoromethylphenyl)-2,2,2-trifluoroethane. *European Polymer Journal 43*: 550–560.

36 Tian, Q. (2009). Thermally remendable epoxy resin. PhD thesis. Zhongshan University, Guangzhou.

37 Plaisted, T.A. and Nemat-Nasser, S. (2007). Quantitative evaluation of fracture, healing and re-healing of a reversibly cross-linked polymer. *Acta Materialia 55*: 5684–5696.

38 Brown, E.N., Sottos, N.R., and White, S.R. (2002). Fracture testing of a self-healing polymer composite. *Experimental Mechanics 42*: 372–379.

39 Tian, Q., Rong, M.Z., Zhang, M.Q., and Yuan, Y.C. (2010). Optimization of thermal remendability of epoxy via blending. *Polymer 51*: 1779–1785.

40 He, M.J. (1990). *Polymer Physics*, 334. Shanghai: Press of Fudan University (in Chinese).

7

Intrinsic Self-Healing Via Synchronous Fission/Radical Recombination of the C—ON Bond

CHAPTER MENU

Intrinsic self-healing via thermally reversible Diels-Alder (DA) intermonomer linkages has received extensive research attention, as demonstrated in Chapters 1 and 6. Accordingly, cracks in cross-linked polymer made from cycloaddition between multi-furan and multi-maleimide can be repeatedly re-bonded. In a typical run of healing, however, the polymer has to be heated up to retro-DA temperature (120 °C) for disconnecting DA bonds (i.e. de-cross-linking), which is generally higher than the glass transition temperature (T_g), and then cooled down for re-connection of the DA bonds (DA reaction ≈ 80 °C). This means that the material would inevitably lose its shape as a result of (i) molecular cleavage during the first stage of healing (i.e. retro-DA reaction/de-cross-linking) and (ii) drastically increased deformability due to glass transition [1]. It is worth noting that on the one hand, breakage of reversible bonds and improved mobility of molecules are necessary for crack healing [2], while on the other, distortion of end-use products is not allowed for structural applications where mechanical strength properties are critical even if self-healing is proceeding. The requirements seem to be contradictory and are hard to be satisfied simultaneously in conventional polymers.

In fact, natural healing in living bodies does not affect their operations. The circulatory system keeps on working, for instance, despite the fact that blood clotting is occurring in an injured vessel. Evidently, mimicking healing process of organisms to a higher degree would further improve performance of artificial self-healing materials. Innovative ideas about molecular structure design and a reversible mechanism of molecular fracture and combination should be proposed accordingly.

Unlike the healing based on general reversible covalent chemistry, which has to be conducted in a two-step fashion, self-healing driven by dynamic reversible covalent chemistry proceeds in a one-step fashion. Here we show the feasibility of using the dynamically reversible C—ON bond as the healing chemistry, which combines the aforementioned covalent bond breakage and re-connection into one step. The possible material deformation in the course of crack remending would thus be avoided even when the healing is carried out above its T_g. As illustrated in Scheme 7.1, the C—ON bond in an alkoxyamine moiety

Extrinsic and Intrinsic Approaches to Self-Healing Polymers and Polymer Composites, First Edition. Ming Qiu Zhang and Min Zhi Rong.

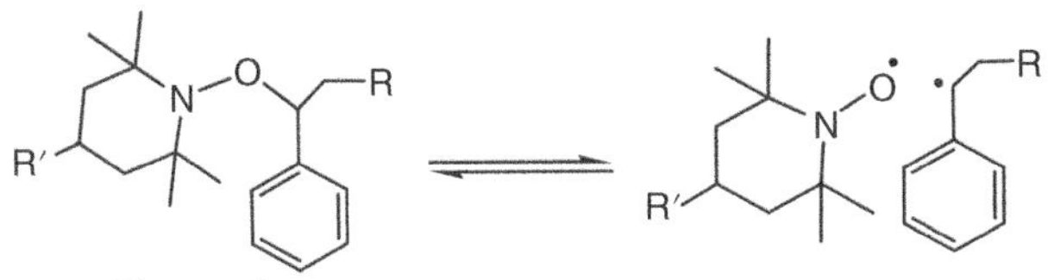

Alkoxyamine

Scheme 7.1 Dissociation–association of alkoxyamine derivatives.

homolytically cleaves upon heating and produces transient carbon-centered and persistent nitroxide radicals in equal amounts [3–8] with very high frequency factor (e.g. $2.4\times1014\,s^{-1}$ in solution [7]). Meanwhile, the carbon-centered radicals would be rapidly trapped by the nitroxide radicals [4]. As a result, under certain homolysis temperature the C—ON bonds in alkoxyamines frequently cleave but immediately recombine. Moreover, the recombination of radicals involves crossover or exchange reaction between radicals that belong to different alkoxyamine moieties before the chain breakage [9]. Taking advantage of this dynamic equilibrium, nitroxide-mediated radical polymerization has become one of the most useful methods for living radical polymerization [7, 8, 10, 11] and various well-defined and complex molecular architectures, many of which are difficult to be prepared via traditional methods, have been synthesized [8, 12–15]. For example, a reversible sol–gel was obtained from linear poly(methacrylic ester)s containing alkoxyamine units on the side chain [16]. Upon heating, the cross-linked gel in solvent (10 wt% anisole solution of polymer) was formed through radical exchange reactions of alkoxyamine moieties, releasing alkoxyamine molecules. The retro-reaction, i.e. de-cross-linking, occurred in the presence of an excess amount of alkoxyamine molecules. Although this kind of sol–gel conversion took place in solvent rather than in solid, it theoretically proved that the reversible reaction between alkoxyamine moieties may be useful to construct a reversible cross-linked structure. Nevertheless, in a hard polymer solid, molecular mobility is greatly restricted. As a result, the release and incorporation of alkoxyamine molecules are impossible.

7.1 Thermal Reversibility of Alkoxyamine in Polymer Solids

To gain a deeper understanding of solid state radical reactions of C—ON bonds, two polymers with the same skeleton (i.e. poly(4-vinylbenzyl chloride), PCMS) but different alkoxyamine moieties as pendant groups and their blends are synthesized and studied [17, 18]. When the blends of polymers **5** and **7** are heated to homolysis temperature, as shown in Scheme 7.2, two types of oxygen-centered nitroxide radical (i.e. macromolecule **9** and small molecule **1**) and their counterpart carbon-centered radicals (i.e. macromolecule **8** and small molecule **10**) are produced. Consequently, a series of crossover or exchange reactions would occur in addition to the recombination of **9–10** and **8–1** pairs. For example, **9** and **8** are combined offering cross-linked polymer **12**, while **1** reacts with **10** giving rise to alkoxyamine **11**. Moreover, termination of carbon-centered radicals **8** and **10** would also be possible, leaving nitroxide radicals **9** and **1** in the system. It is thus known that the polymer blends of **5** and **7** can serve as a qualified specimen for revealing reactions between various radicals resulting from C—ON bond fission, which would help to develop new self-healing polymers and improve healing efficiency of existing materials.

PCMS (M_n = 55 000) was synthesized by free radical polymerization of 4-vinylbenzyl chloride with initiator azobisisobutyronitrile (AIBN). As for preparation of 2,2,6,6-tetramethylpiperidine 1-oxyl (TEMPO) modified PCMS (PCMS-TEMPO) (Scheme 7.3), TEMPO was first reduced to 1-hydroxy-2,2,6,6-tetramethylpiperidine (TEMPO-H) by sodium ascorbate (step 1), which was then deprotonated with NaH in tetrahydrofuran (THF) to obtain TEMPO-Na (step 2). TEMPO-Na reacted with PCMS in THF to afford PCMS-TEMPO in the presence of 18-crown-6 (step 3). The results of elemental analysis indicate that PCMS-TEMPO bearing 1.89 mmol N atoms per gram can be easily achieved and 39% of the chloride has been substituted. Like PCMS-TEMPO, 1-(1-phenyl-ethoxy)-2,2,6,6-tetramethylpiperidin-4-ol (HO-TEMPO-EOPh) modified PCMS (PCMS-TEMPO-EOPh, i.e. polymer **7**) bearing 2.04 mmol N atoms per gram was obtained with 53% displacement of chloride (Scheme 7.4).

To verify crossover reaction between carbon-centered radicals and oxygen-centered nitroxide radicals produced by homolytic cleavage of C—ON bonds in the blends of PCMS-TEMPO and PCMS-TEMPO-EOPh, solutions of the polymers were treated at 125 °C for up to 20 minutes. Here the heat treatment temperature of 125 °C was chosen in accordance with the homolysis temperature of alkoxyamine [8]. As shown in Figure 7.1, the solutions of PCMS-TEMPO-EOPh and PCMS-TEMPO remain unchanged after the heat treatment, while that of PCMS-TEMPO/PCMS-TEMPO-EOPh blends has turned to a gel. Since both PCMS-TEMPO and PCMS-TEMPO-EOPh are linear polymers, the presence of gel means that crossover reaction between macromolecule radicals depicted in Scheme 7.2 must have taken place. The analysis agrees with the conclusions of Otsuka et al. [13, 19], who used similar gels to prepare cross-linked polymers.

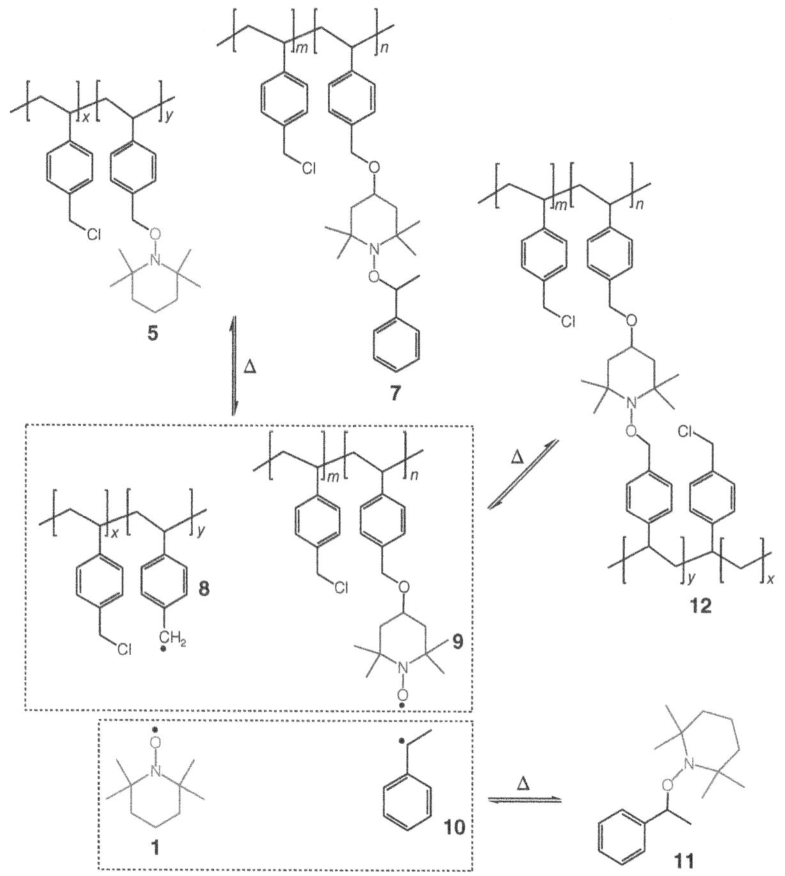

Scheme 7.2 Dynamic crossover or exchange reactions in blended PCMS polymers with alkoxyamine moieties as side groups. **1,** 2,2,6,6-tetramethylpiperidine 1-oxyl (TEMPO); **5,** PCMS-TEMPO; **7,** PCMS-TEMPO-EOPh; **8,** macromolecular carbon-centered radical; **9,** macromolecular nitroxide radical; **10,** ethyl phenyl free radical; **11,** 1-(1-phenyl-ethoxy)-2,2,6,6-tetramethylpiperidine; and **12,** cross-linked polymer.

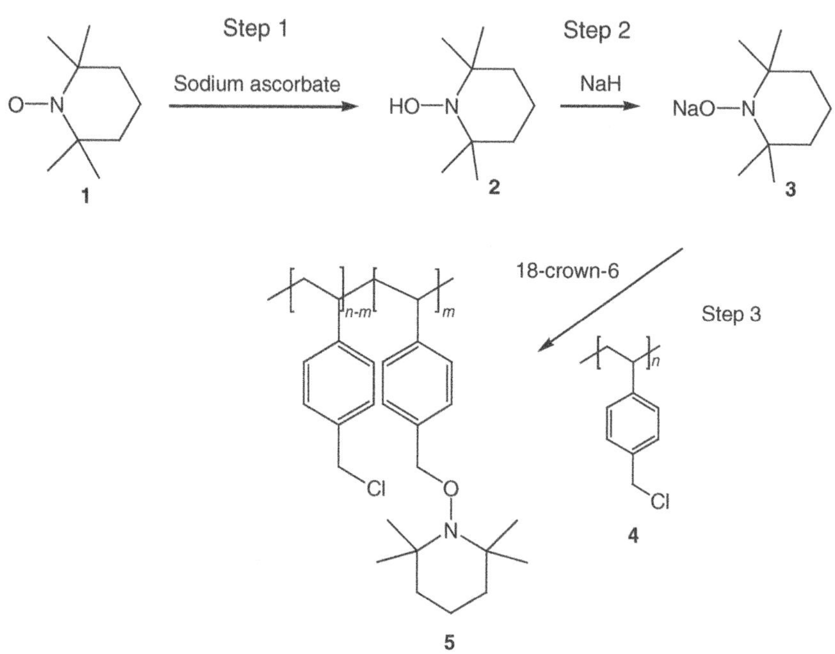

Scheme 7.3 Synthesis of PCMS-TEMPO. **2,** 1-hydroxy-2,2,6,6-tetramethylpiperidine (TEMPO-H); **3,** TEMPO-Na; and **4,** PCMS.

Scheme 7.4 Synthesis of PCMS-TEMPO-EOPh. **6**, 1-(1-phenyl-ethoxy)-2,2,6,6-tetramethylpiperidin-4-ol (HO-TEMPO-EOPh).

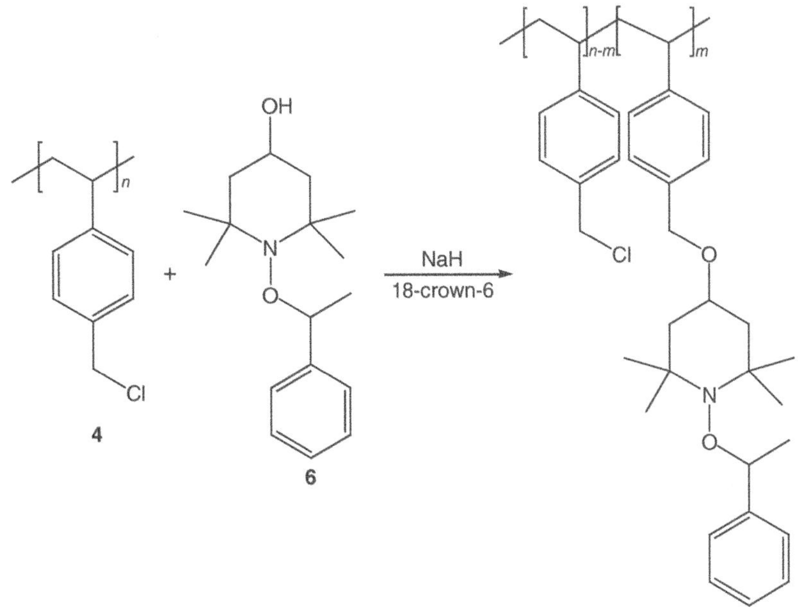

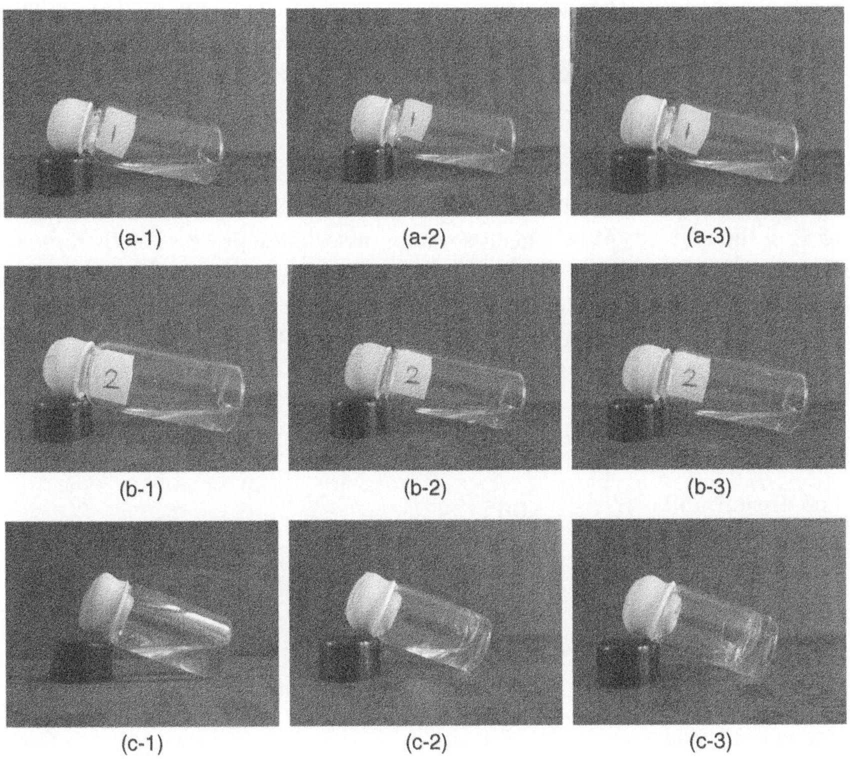

Figure 7.1 Gel formation due to crossover reaction of macromolecular radicals at 125 °C. Time for heat treatment at 125 °C: (1) 0 minutes, (2) 10 minutes, and (3) 20 minutes. (a) PCMS-TEMPO-EOPh, (b) PCMS-TEMPO, and (c) mixture of PCMS-TEMPO and PCMS-TEMPO-EOPh. *Source:* Wang et al. [18], Figure 1, p. 4/with permission of Royal Society of Chemistry.

Additionally, formation of cross-linked structures is also perceived in solidified PCMS-TEMPO/PCMS-TEMPO-EOPh blends. Figure 7.2 gives the differential scanning calorimetry (DSC) heating curves of the polymers. Curves a and b demonstrate that glass transitions of PCMS-TEMPO and PCMS-TEMPO-EOPh occur at 117 and 85 °C, respectively. Curve c indicates that the as-manufactured PCMS-TEMPO/PCMS-TEMPO-EOPh blends exhibit two T_g values of 116 and 85 °C, corresponding to those of PCMS-TEMPO and PCMS-TEMPO-EOPh, respectively. Evidently,

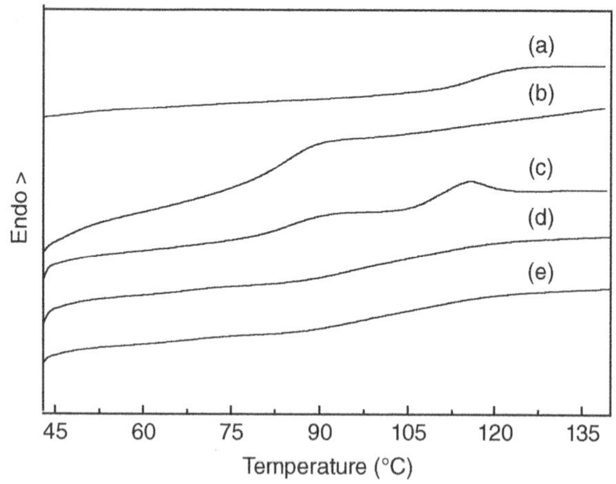

Figure 7.2 DSC heating traces of PCMS-TEMPO, PCMS-TEMPO-EOPh and PCMS-TEMPO/PCMS-TEMPO-EOPh blends (heating rate: 10 °C min^{-1}). (a) As-manufactured PCMS-TEMPO. (b) As-manufactured PCMS-TEMPO-EOPh. (c) As-manufactured blends of PCMS-TEMPO/PCMS-TEMPO-EOPh. (d) Blends of PCMS-TEMPO/PCMS-TEMPO-EOPh treated at 125 °C for 15 minutes and then cooled to room temperature. (e) Blends of PCMS-TEMPO/PCMS-TEMPO-EOPh treated at 125 °C for 30 minutes and then cooled to room temperature.

microphase separation appears in the blends. It is interesting to find that when the blends were treated at 125 °C, the two individual glass transition processes merge forming a single broad glass transition zone with T_g of 98 °C (see curves d and e in Figure 7.2). The value of T_g of the thermally treated blends is independent of the treatment time from 15 to 30 minutes. On the basis of these results, it is known that cross-linkages have been established like the above case, which leads to miscibility of the two components of the polymer blends. The crossover reaction proceeds so fast that 15 minutes is sufficient for building up cross-linked networks in the solid state.

Coinciding with the above C—ON bond fission and radical recombination characteristics, the PCMS-TEMPO/PCMS-TEMPO-EOPh blends should have acquired self-healing ability. To verify this issue, a cross-shaped razor cut was made on a filmy specimen of the blends. Then, the specimen was heated to 125 °C for a period in argon atmosphere, and the crack gradually disappeared as expected. Afterwards, the second cut was made across the first one, and disappeared again as a result of the same repair procedure. The recovery of the intersection of successive cuts represents multi-remendability of the blends.

In contrast, the control tests demonstrate that the cutting damage on PCMS, PCMS-TEMPO and PCMS-TEMPO-EOPh alone cannot be healed under the same condition. The importance of formation of a cross-linking structure among polymer chains to crack healing is thus highlighted.

On the other hand, a careful survey of PCMS, PCMS-TEMPO and PCMS-TEMPO-EOPh indicates that all the cracks on the polymers narrow after heat treatment at 125 °C (which is higher than the T_g of the polymers, see Figure 7.2), because of the relatively larger volumetric expansion in the rubbery state. This means that the improved mobility of polymer chain segments at healing temperature must facilitate surface approach, one of the key stages of crack healing [2].

The above discussion has shown that the C—ON bond is the key factor for successful crack healing, but detailed information about its reversibility in the solid state is not yet available. As many radical reactions are simultaneously involved in the blends of PCMS-based polymers at homolytic temperature (Scheme 7.2) and the concentration of C—ON bonds is low in the polymers, small molecule mimetics are used for the convenience of analysis of the reversible reactions.

HO-TEMPO-EOPh (alkoxyamine **6**), the mimic of polymer **7** (PCMS-TEMPO-EOPh), contains one type of C—ON bond. Scheme 7.5 shows its thermally reversible reaction. When alkoxyamine **6** is cleft into carbon-centered radicals and oxygen-centered nitroxide radicals, the radicals have to be localized owing to the restriction imposed by the solid microstructure. Furthermore, the nitroxide radical **13** possesses high free radical capturing capacity. Nearly no secondary reaction between carbon-centered radicals takes place. The dissociation/association process is fully reversible as a result, which is shown by repeated electron spin resonance (ESR) scans between 20 °C and 125 °C. As can be seen in Figure 7.3, the ESR spectrum at 125 °C is

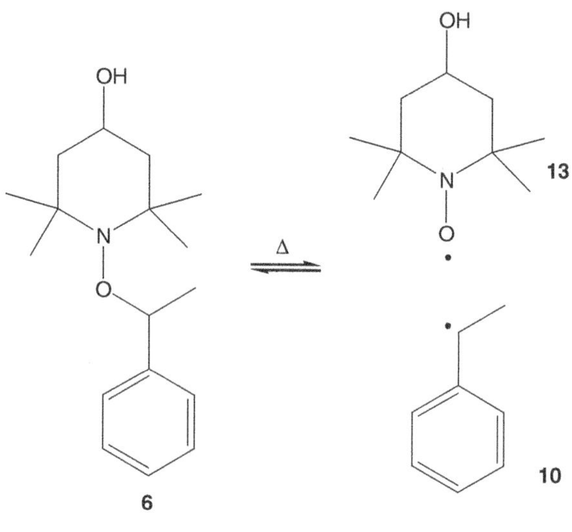

Scheme 7.5 Dissociation–association of alkoxyamine **6** (HO-TEMPO-EOPh). **13,** 4-hydroxy-2,2,6,6-Tetramethylpiperidine 1-oxyl (4-OH-TEMPO).

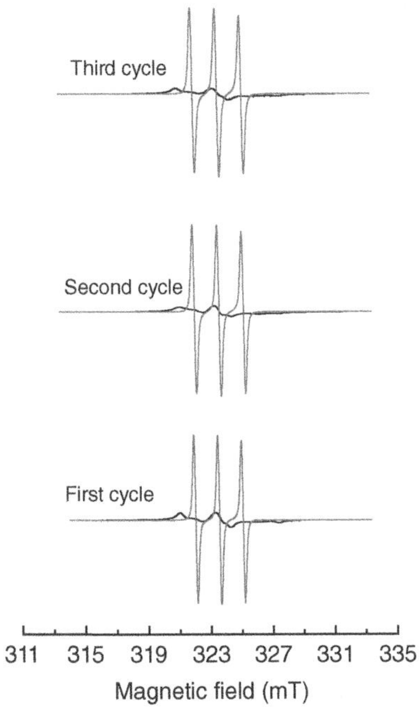

Figure 7.3 Repeated ESR spectra of alkoxyamine **6** (HO-TEMPO-EOPh) recorded during thermal cycling between 20 °C and 125 °C. Black line: 20 °C; and gray line: 125 °C.

characterized by sharp peaks of free radicals, while no signals are detected at 20 °C. The results are identical during three thermal cycles. Furthermore, when the ESR spectra of HO-TEMPO-EOPh were collected at different temperatures, no free radicals were generated below 80 °C, but concentration of free radicals linearly increased with a rise in temperature above 80 °C (Figure 7.4). It is known that the homolysis temperature of HO-TEMPO-EOPh should be higher than 80 °C.

Here the reversibility revealed by small molecule alkoxyamine **6** is different from that of macromolecules carrying the single type of C—ON bond studied in another work of ours (Section 7.2.1, [20]). In the latter case, unpaired radicals are present, as mentioned at the start of the chapter. In consideration of the difference in spatial confinement and mobility between small molecules and macromolecules, we may deduce that the aforementioned incomplete recombination of cleft C—ON bonds should result from poor diffusibility of the macromolecular radicals left after crossover reactions between carbon-centered radicals and nitroxide radicals offered by different alkoxyamine moieties. For the radicals cleft by small molecule alkoxyamines, which are strictly confined by the solid, the space is much smaller than that of the radicals from alkoxyamines attached to macromolecules that are capable of segmental motion. The probability of contact between former radicals is ensured even the homolysis takes place in the solid state, so that the C—ON bonds exhibit full reversibility. With respect to macromolecular chains, scrambling reactions are allowed due to segmental motion but the limited molecular diffusion brings about a few incomplete recombinations of C—ON bonds of the residual radicals. Nevertheless, the degree of irreversible combination of C—ON bonds in cross-linked polystyrene (PS) is not that severe and its influence on thermal reversibility of the material is marginal, as reflected by the repeated remendability [20].

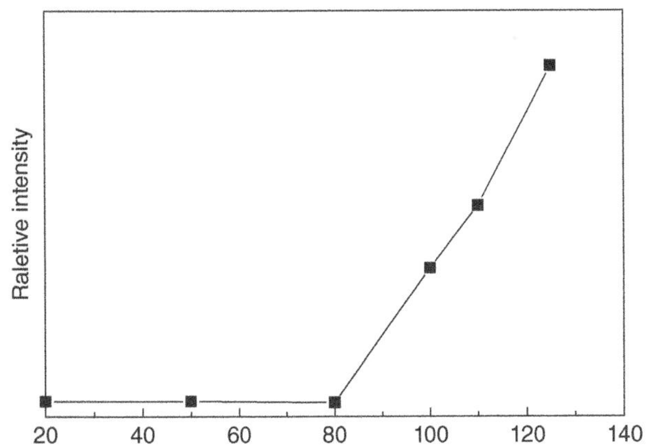

Figure 7.4 Temperature dependence of relative ESR signal intensities.

On the basis of the above investigation, for purposes of studying reversible reactions in the blends of PCMS-TEMPO and PCMS-TEMPO-EOPh, a model mixture of alkoxyamines **6** (HO-TEMPO-EOPh) and **14** (1-isopropoxy -2,2,6,6-tetramethylpiperidin-4-ol) at identical molar ratio was prepared and characterized. The latter acts as the mimic of polymer **5** (PCMS-TEMPO). Its reversible behavior resembles that of HO-TEMPO-EOPh so that the relevant ESR spectra studies are not quoted herein.

From the ESR spectra of the mixture of alkoxyamines **6** and **14** measured at different temperatures during heating, it is found that no signal is perceived at 20 °C and 50 °C. With a gradual increase in temperature, the ESR peak height becomes increasingly stronger like alkoxyamine **6**. Comparatively, the peak of free radicals of the mixture of alkoxyamines **6** and **14** is rather obvious at 80 °C. Clearly, the free radicals should originate from alkoxyamine **14** and the homolysis temperature of the C—ON bond in alkoxyamine **14** is lower than that in alkoxyamine **6**. By plotting ESR signal intensities calculated from the integral area of the spectra against temperature (Figure 7.5), it is seen that free radical content starts to increase from 50 °C and a drastic increase appears above 80 °C. Unlike the linear relationship of alkoxyamine **6** in Figure 7.4, the temperature dependence of free radical concentration of the mixture of alkoxyamines **6** and **14** is parabolically shaped. Different homolysis temperatures of C—ON bonds in alkoxyamines **6** and **14** should account for this phenomenon. In addition, fission of C—ON bonds leads to two types of carbon-centered radical and one type of nitroxide radical in the system. Under conditions of homogeneous mixing of alkoxyamines **6** and **14**, the probability of termination of the carbon-centered radicals would be relatively high (i.e. the reaction between **10** and **15**), so that excessive nitroxide radicals have to be present (Scheme 7.6).

When the system is cooled down, ESR signal intensities abnormally increase with temperature and reach a maximum at 80 °C (Figure 7.6). This might be due to the fact that the homolytic products of alkoxyamine **6** no longer follow fission/radical recombination equilibrium, and a few nitroxide radicals remain unreacted as a result of the combination of partial carbon-centered radicals. Meanwhile, bond fission and radical recombination is synchronously retained among alkoxyamine **14** moieties. The superimposed effects lead to the specific temperature dependence. Even when temperature is cooled to 50 °C, however, the ESR peak heights are almost the same as those at 80 °C. This suggests that alkoxyamine **14** also loses the dynamic homolysis equilibrium because of termination of carbon-centered radicals. At 20 °C, ESR peaks are greatly weakened and broadened, while still keeping a certain intensity, meaning that quite a few nitroxide radicals exist in the system. As compared with the fully reversible results of repeated ESR scans of alkoxyamine **6** during thermal cycles (Figure 7.3), thermal reversibility of the mixture of alkoxyamines **6** and **14** is obviously reduced. These experimental results indicate that so long as two types of carbon-centered radicals coexist in a solid mixture

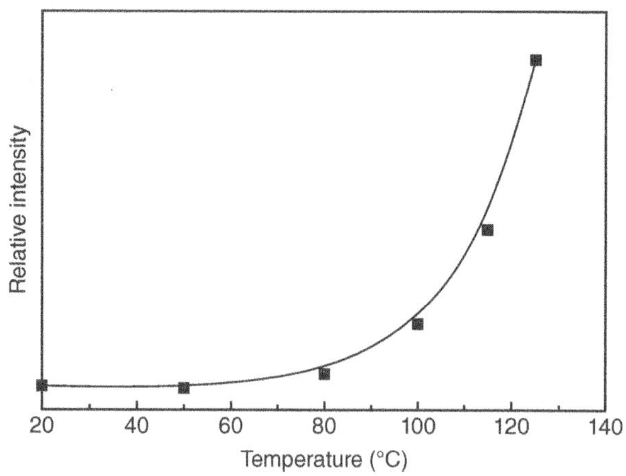

Figure 7.5 Temperature dependence of relative ESR signal intensities.

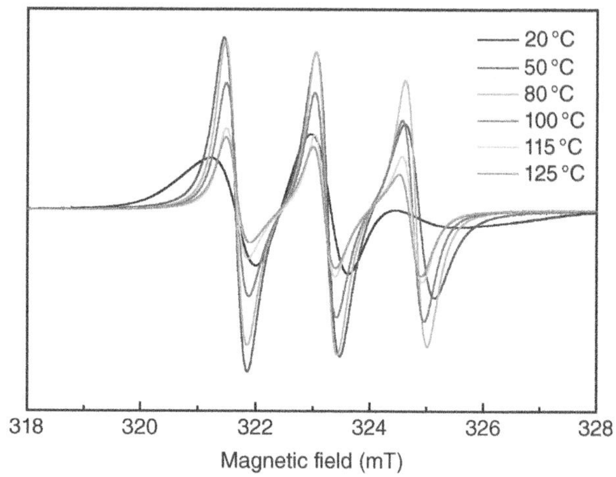

Figure 7.6 ESR spectra of an equimolar mixture of alkoxyamines **6** (HO-TEMPO-EOPh) and **14** (1-isopropoxy-2,2,6,6-tetramethylpiperidin-4-ol) recorded during cooling at different temperatures. *Source:* Reprinted from Wang et al. [18]. Copyright 2012, with permission from the Royal Society of Chemistry.

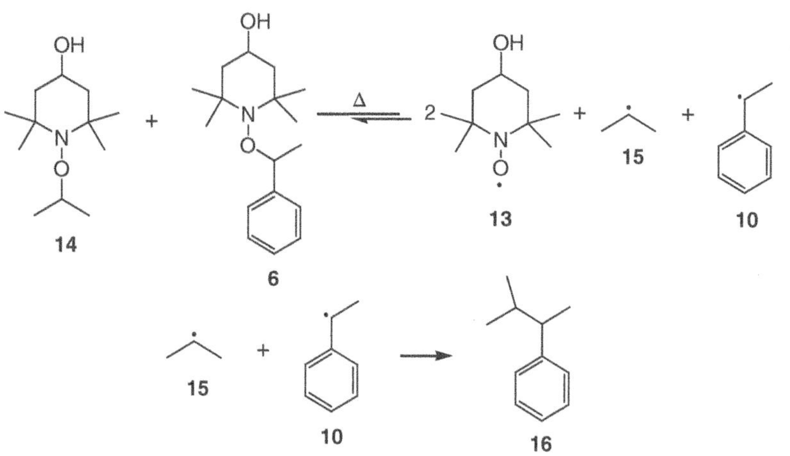

Scheme 7.6 Dissociation–association of a mixture of alkoxyamines **6** and **14**. **14**, 1-isopropoxy-2,2,6,6-tetramethylpiperidin-4-ol; **15**, isopropyl free radical; and **16**, 1-(3-methylbutan-2-yl) benzene.

of alkoxyamines they would react with each other, leading to nitroxide radicals remaining.

With reference to the outcomes of small molecule mimetics, variation in content of free radicals of PCMS-TEMPO/PCMS-TEMPO-EOPh blends as a function of temperature is studied hereinafter. The sample sealed in a NMR tube full of argon was heated up to a certain temperature, then was kept for two minutes allowing temperature equilibrium to be attained, and then the ESR spectrum was recorded. Since several types of radicals are generated in the polymer blends during C—ON bond fission/radical recombination, and these reactions take place in solid polymer that restricts molecular diffusion and collision, it is hard for the reversible reactions to attain equilibrium during the ESR measurements. Nevertheless, the relative ESR intensities collected in the case of a retention time of two minutes are still capable of being compared to examine the effect of temperature on C—ON bond fission, because relative ESR intensities of the polymer blends barely change within two minutes of measurement. Subsequently, the sample was cooled to a certain temperature and then kept for two minutes to record the ESR spectrum. Typical ESR spectra during heating and cooling are shown in Figure 7.7. With increasing temperature, the ESR peaks are significantly intensified due to the increase in radical concentration (Figure 7.7a). Comparatively, the increase in ESR peak intensity is marginal below 100 °C, meaning the thermal energy is inadequate to effectively cleave the C—ON bonds. In the cooling process shown in Figure 7.7b, weakening of ESR peaks is observed as temperature decreases, representing the decrease in radical concentration. Nevertheless, quite a few radicals remain even at 20 °C. According to the above analysis of the mixture of alkoxyamines **6** and **14**, nitroxide radicals should be left due to termination of carbon-centered radicals. In addition, the profile of the ESR spectrum at 20 °C is almost the same as that at 80 °C (except for the obvious right shift of the first peak). Its extents of weakening and widening with decreasing temperature are much smaller as compared with the curve changes of a mixture of alkoxyamines **6** and **14** within the same temperature range (Figure 7.6). This means that there might also be carbon-centered radicals in the remaining radicals.

Furthermore, the intensities of ESR signals in Figure 7.7b continuously decline with temperature, which is quite different from the cooling curves in Figure 7.6. The reaction between carbon-centered radicals in polymers must be remarkably restricted due to the immobilization effect of the surrounding polymer chains on radical diffusion. The probability of termination of carbon-centered radicals has to be lowered accordingly. On the other hand, the cooling process of the polymer blends recorded by ESR

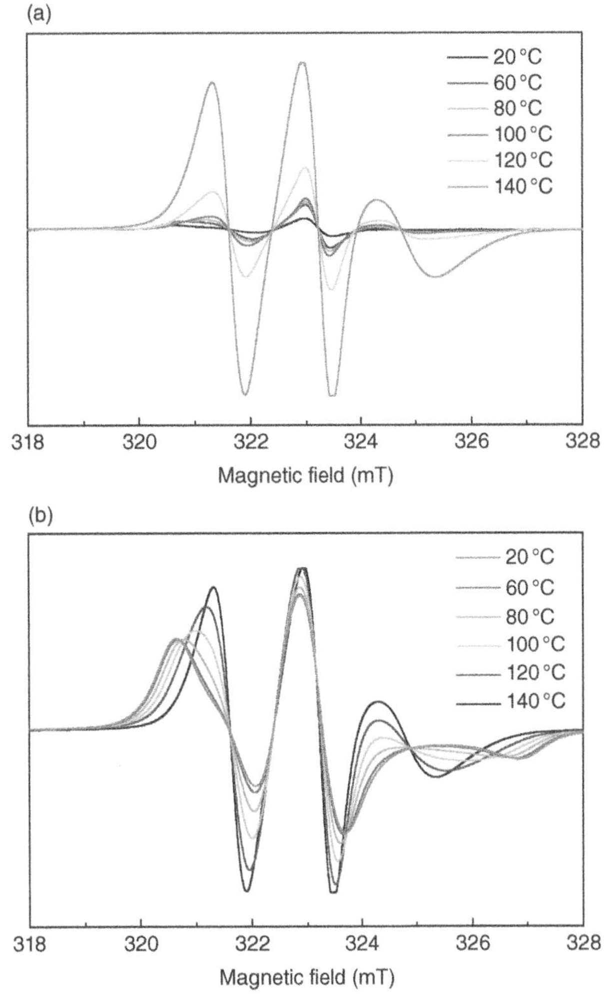

Figure 7.7 ESR spectra of PCMS-TEMPO/PCMS-TEMPO-EOPh blends recorded at different temperatures during (a) heating and (b) cooling. *Source:* Reprinted from Wang et al. [18]. Copyright 2012, with permission from the Royal Society of Chemistry.

spectroscopy is not the exact reverse of their heating process, which resembles the mixture of small molecule alkoxyamines. Much higher radical concentration at the same temperature detected in the course of cooling than heating should result from the remaining nitroxide or carbon radicals generated due to reaction between carbon-centered radicals or incomplete recombination of cleft C—ON bonds. This phenomenon represents the hysteresis effect caused by the kinetic contribution during C—ON bond fission/radical recombination.

To further uncover the cleavage/recombination regulation of C—ON bonds, ESR spectra of the polymer blends were collected during six heating–cooling cycles between 20 °C and 125 °C (Figure 7.8a). Figure 7.8b shows that the relative ESR signal intensities at both 20 and 125 °C increase with the thermal cycle, reflecting the hysteresis effect of polymer solid on C—ON bond fission/radical recombination. During cooling, incomplete radical recombination occurs, strengthening the ESR intensities at 20 °C. It is worth noting that the difference in ESR intensity between 20 °C and 125 °C gradually increases with the heating–cooling cycles. This means that the radicals left during the last heating–cooling cycle may have taken part in C—ON bond fission/radical recombination reactions in the next cycle until equilibrium is reached after about five cycles. Although the hysteresis cannot be eliminated, reversibility of the blends is still measurable and approaches equilibrium. This ensures thermal remendability of the material in practice.

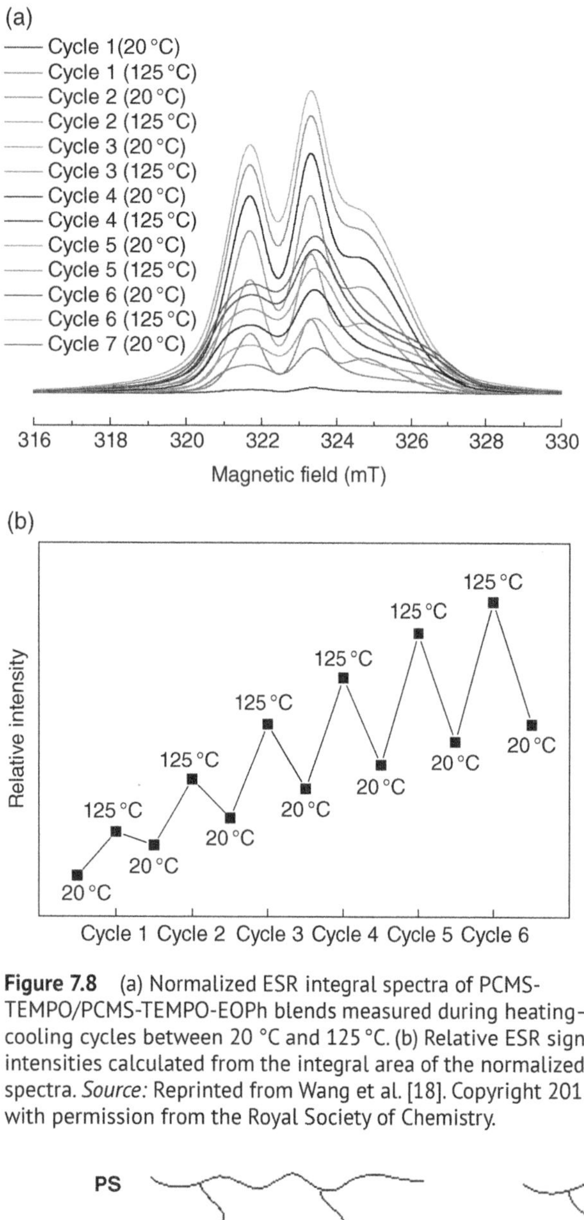

Figure 7.8 (a) Normalized ESR integral spectra of PCMS-TEMPO/PCMS-TEMPO-EOPh blends measured during heating–cooling cycles between 20 °C and 125 °C. (b) Relative ESR signal intensities calculated from the integral area of the normalized spectra. *Source:* Reprinted from Wang et al. [18]. Copyright 2012, with permission from the Royal Society of Chemistry.

On the basis of above investigation, it can thus be concluded that polymer solid hinders diffusion of carbon-centered radicals so that the unwanted secondary reaction is depressed even though complete recombination of the cleft C—ON bond is also somewhat blocked. Nevertheless, polymer solid does more good than harm as viewed from the satisfactory reversibility and multiple self-healing behavior.

7.2 Self-Healing Cross-linked Polystyrene

To verify the ability of the reversible C—ON bond for intrinsic self-healing, cross-linked PS in which the alkoxyamine moiety acts as the reversible cross-linker (Scheme 7.7) is prepared. First, the reaction between diol and methacryloyl chloride produces dimethacrylic ester (Scheme 7.8). Then, cross-linked PS is yielded via bulk polymerization of styrene with dimethacrylic ester as cross-linking agent and benzoyl peroxide (BPO) as initiator.

7.2.1 Synthesis

The diol (4-hydroxy-1-((2′-hydroxy-1′-phenylethyl)oxy)-2,2,6,6-tetramethylpiperidine) is a derivative of TEMPO, which was synthesized and purified as previously reported [21]. Besides, the dimethacrylic ester (4-methacryloyloxy-1-((2′-methacryloyloxy-1′-phenylethyl)oxy)-2,2,6,6-tetramet-hylpiperidine) was prepared following the published procedures [13]. Typically, triethylamine (5.1 g, 50 mmol)

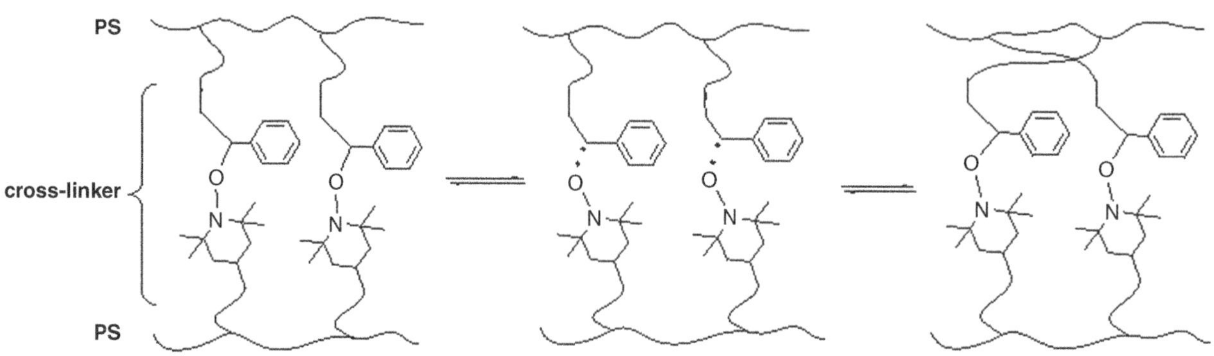

Scheme 7.7 Dynamic de-cross-linking, crossover cross-linking, or exchange reactions of PS.

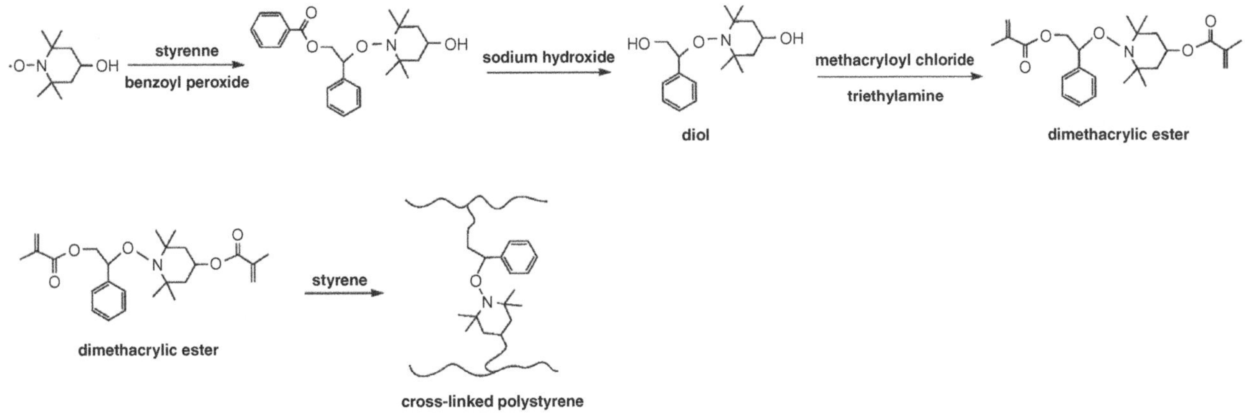

Scheme 7.8 Synthetic route of alkoxyamine-based cross-linked PS.

was slowly injected into the solution of methacryloyl chloride (7.8 g, 75 mmol) and diol (7.3 g, 25 mmol) in anhydrous THF (50 ml). The reaction was carried out with stirring under argon for 10 hours at 25 °C, and then evaporated to dryness. The residue was partitioned between saturated sodium chloride solution and dichloromethane in which the aqueous layer was extracted with dichloromethane several times. The collected organic layers were dried with anhydrous sodium sulfate and evaporated to dryness. The crude product was purified by flash chromatography eluting with 1:20 ethyl acetate/hexane (v/v) to give dimethacrylic ester as a pale yellow oil (9.4 g, 88% yield).

The reversibly cross-linked PS was prepared as follows: BPO (58.1 mg, 0.24 mmol) and a certain amount of cross-linker dimethacrylic ester were dissolved in styrene (7.8 g, 75 mmol) and then charged into a tube. The tube was evacuated and backfilled with argon three times. The mixture was cured for 24 hours at 80 °C and post-cured for 1 hour at 100 °C. Three molar feeding ratios of styrene to cross-linker dimethacrylic ester were used (5:1, 7.5:1, and 9.4:1). The resultant cross-linked PS possesses similar mechanical properties as conventional cross-linked PS with ethyleneglycol dimethacrylate (EGDMA) as cross-linking agent. When the molar feeding ratio of styrene to cross-linker is fixed at 10:1, for example, flexural strength and modulus of the former are 99.4 MPa and 5.0 GPa, respectively, whereas those of the latter are 93.0 MPa and 5.2 GPa, respectively. The control material, irreversibly cross-linked PS excluding alkoxyamine moiety, was made following similar procedures by using EGDMA as cross-linker.

7.2.2 Characterization

ESR spectroscopy, a technique for studying free radicals, is carried out at different temperatures to examine fission behavior of the cross-linker dimethacrylic ester in cross-linked PS. As shown in Figure 7.9, there is no evident

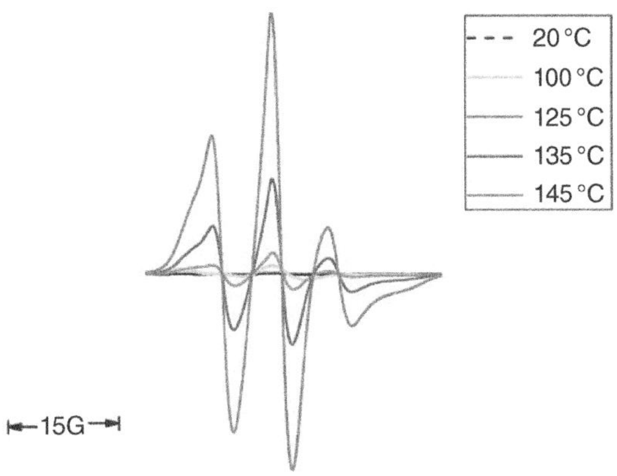

Figure 7.9 Typical ESR spectra of reversibly cross-linked PS (with molar feeding ratio of styrene to cross-linker dimethacrylic ester of 7.5:1) measured at various temperatures.

absorption until 100 °C, and then the relative ESR signal intensity drastically increases with a rise in temperature, indicating that homolytic cleavage of the alkoxyamine moiety (C—ON bond) starts at about 125 °C. It is noted that at each temperature the signal intensity is kept at a certain level within a fixed testing time. Therefore, the ESR spectra can be used to describe the radical concentration in the material, as suggested by Sylvain et al. [22]. To highlight reversibility of the C—ON bonds, Figure 7.10a gives the normalized absorptions recorded by cyclic ESR measurements between 20 °C and 130 °C. Accordingly, the relative intensity of the radical signal estimated from the area under each curve in Figure 7.10a is plotted in Figure 7.10b. It is seen that when temperature alternates between 20 °C and 130 °C, the amount of free radicals fluctuates as a result of scission of the alkoxyamine moiety and recombination of the radicals. These processes are fully reversible, despite the fact that both peak and valley values of the relative intensity slightly increase with the thermal cycles. The latter phenomenon might be attributed to permanent dissociation of a few alkoxyamines caused by irreversible

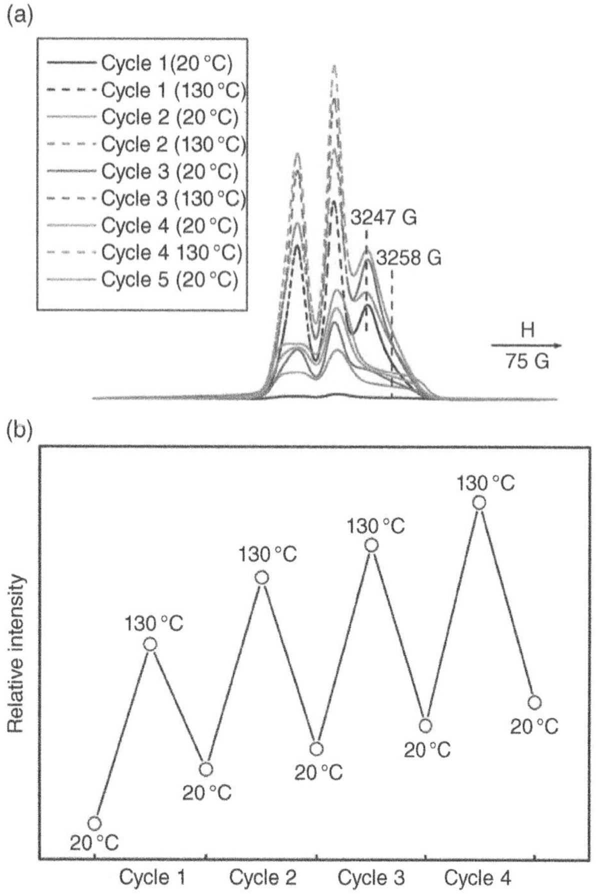

Figure 7.10 (a) Normalized absorption lines obtained from ESR data of reversibly cross-linked molar PS (with molar feeding ratio of styrene to cross-linker dimethacrylic ester of 7.5:1), measured during heating–cooling cycles between 20 °C and 130 °C. (b) Relative ESR signal intensities calculated from the integral area of the normalized absorption lines. *Source:* Reprinted from Yuan et al. [20]. Copyright 2011, with permission from American Chemical Society.

combination of methylene radicals, which results in nitroxide radicals remaining. This can be detected by the temperature dependence of the absorption at around 3247 G (Figure 7.10a). It peaks at 130 °C, but becomes a wide shoulder centered at 3258 G under 20 °C, demonstrating that the radicals at elevated temperature consist of two types of radicals (i.e. methylene and nitroxide radicals) but only nitroxide radicals remain at room temperature (Scheme 7.7). Although radicals are detected at 125 °C and the amount of radicals increases with increasing temperature, the temperature of homolytic cleavage cannot be unlimitedly raised. At ~250 °C thermal degradation of linear PS derived from scission of alkoxyamine moieties is triggered (Figure 7.11). Nevertheless, the reversibly cross-linked PS keeps stable below 200 °C.

To further ascertain the dynamic features of the reversibly cross-linked PS, we compared the temperature dependence of the loss factor, measured by dynamic mechanical analysis (DMA), with that of irreversibly cross-linked PS (with EGDMA as cross-linker), see Figure 7.12a. For the latter polymers, their T_gs determined from tan δ peak positions, significantly increase with the cross-linker content as usual, because of the increase in cross-link density. On the contrary, the reversibly cross-linked PS shows similar T_g at around 125 °C regardless of dimethacrylic ester content. This abnormality reflects the dual role of the reversible cross-linkage in the polymer. On the one hand, the cross-links are in a dynamic equilibrium state of random dissociation/association resulting from the synchronous cleavage and recombination of the C—ON bonds. The relative molecular weight between cross-links has to rapidly change so that its effect on T_g is not statistically significant enough compared with the irreversibly cross-linked PS. On the other hand, the cross-links still take the responsibility

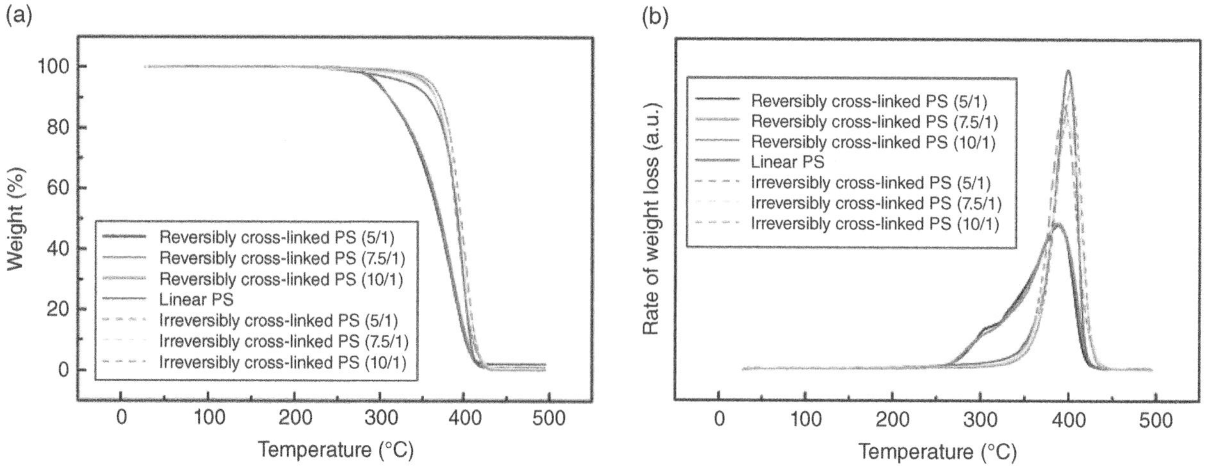

Figure 7.11 (a, b) Pyrolytic behaviors of linear PS, reversibly cross-linked PS, and irreversibly cross-linked PS. The numerals in parentheses represent the molar feeding ratio of styrene to cross-linker.

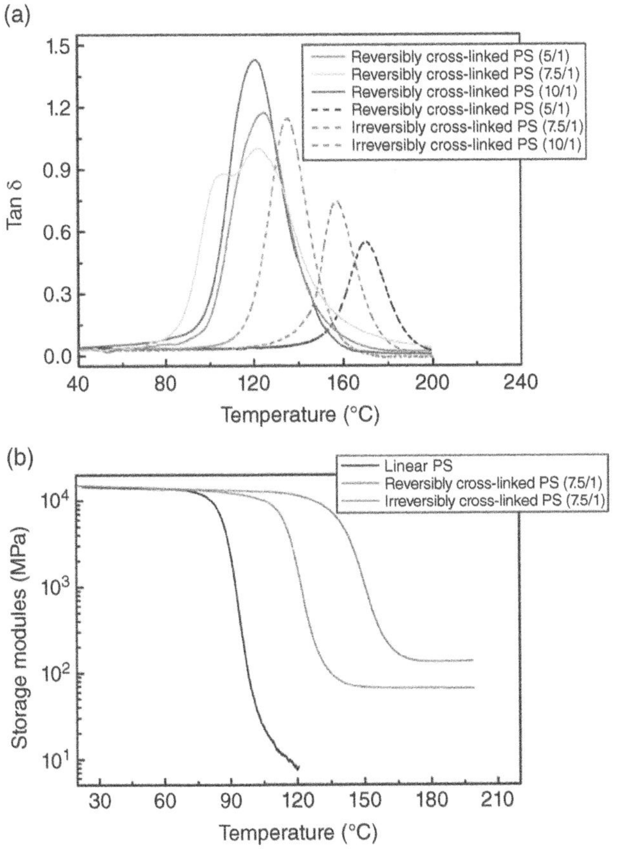

Figure 7.12 Temperature dependences of (a) loss factor of reversibly cross-linked PS and irreversibly cross-linked PS, and (b) storage modulus of linear PS, reversibly cross-linked PS, and irreversibly crosslinked PS. The numerals in parentheses represent the molar feeding ratio of styrene to cross-linker. *Source:* Reprinted from Yuan et al. [20]. Copyright 2011, with permission from American Chemical Society.

of networking PS chains. Otherwise, complete cleavage of cross-linked sites should result in transformation from cross-linked PS to the linear version that has a glass transition at ~100 °C. This argument is supported by the dependence of the area under the tan δ peak on cross-linker content of the materials. For either irreversibly or reversibly cross-linked PS, the area under the tan δ peak decreases with a rise in cross-linker content because higher cross-link density leads to less entanglement and friction between PS chains. Additionally, the storage modulus curve of the reversibly cross-linked specimen shows an elastic plateau above the T_g similar to irreversibly cross-linked PS (Figure 7.12b), which strongly suggests the existence of a cross-linked network even at a temperature as high as 200 °C. However, the lower storage modulus at temperatures above the T_g of reversibly cross-linked PS in comparison with irreversibly cross-linked PS reveals that cross-link density of the former should be lower than that of the latter. This behavior is consistent with the dynamic equilibrium feature of reversible cross-linkage.

A careful study of Figure 7.12a indicates that the PS reversibly cross-linked by dimethacrylic ester with molar feeding ratio of styrene to cross-linker of 7.5:1 exhibits two maxima during glass transition. In addition to the maximum corresponding to T_g at ~125 °C as mentioned above, another one appears at about 105 °C, which is quite close to the T_g of linear PS. It might be related to the heterogeneous network structure derived from dynamic cross-linking/de-cross-linking processes. That is, the maximum at 105 °C should originate from segment motion of free linear PS chains generated by random disconnection of the reversible cross-links. When the polymer involves more reversible cross-linker (e.g. molar feeding ratio of styrene to cross-linker = 5:1), the lengths of the disconnected chains produced by fission of alkoxyamine moieties have to be rather short, which is not enough to form a new damping peak. Accordingly, only a single peak is observed on the loss factor spectrum. As for the polymer with less reversible cross-linker (e.g. molar feeding ratio of styrene to cross-linker = 10:1), the cleft chains are long (or flexible) enough to be immediately reconnected, so that the tan δ peak is still unimodal.

The dynamic cross-linking behavior, or synchronous reversibility, has become an intrinsic performance of the dimethacrylic ester cross-linked PS, as demonstrated by the multiple cyclic DMA scans within the temperature range between 20 °C and 200 °C (Figure 7.13). The marginal increase of T_g with testing cycles is probably due to the appearance of a small amount of irreversible connection between methylene radicals. Nevertheless, most of the reversible links must still remain otherwise their T_g should be significantly increased to the values of irreversibly cross-linked PS. The subsequent multiple remendability tests also support the argument. According to the results of Figure 7.9, it is known that crack healing of PS can be conducted at a temperature higher than 125 °C, allowing synchronous fission of the cross-linkages and radical recombination. Therefore, we set the healing temperature at 130 °C for quantitative characterization of the ability of mechanical property restoration in terms of double cleavage drilled compression (DCDC) tests [1].

Figure 7.14 depicts typical crack length dependences of stress of virgin and healed specimens measured by multiple tests under fixed healing time of 2.5 hours. At molar feeding ratio of styrene to cross-linker of 7.5:1, the specimens can be repeatedly healed, and the efficiency of the first healing event is 75.9%. The mild decline in healing efficiency with healing/refracture cycles might originate from heterogeneity of dynamic recombination of alkoxyamine moieties at the fractured surfaces or reduction of

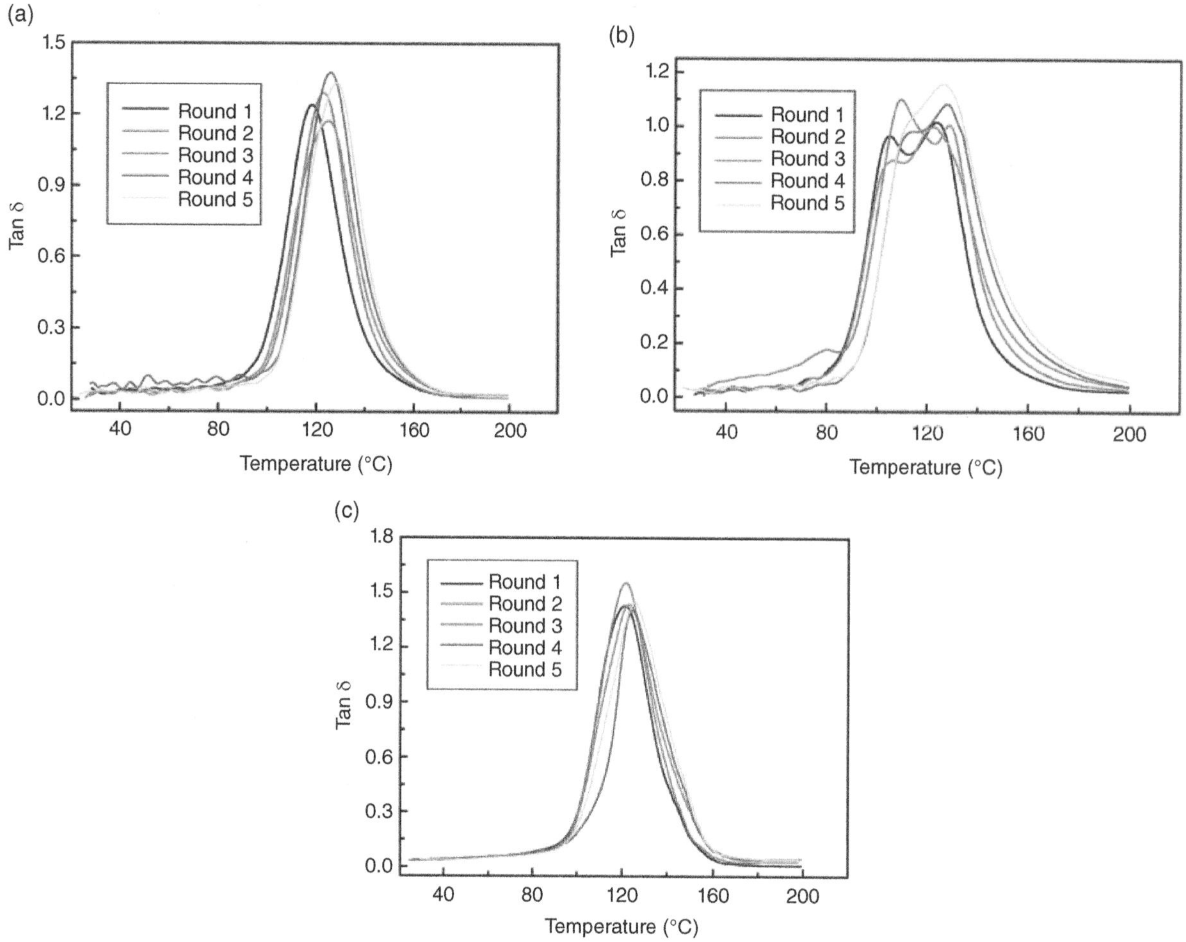

Figure 7.13 DMA spectra of reversibly cross-linked PS measured by repeated scans between 20 °C and 200 °C. Molar feeding ratio of styrene to cross-linker dimethacrylic ester: (a) 5:1; (b) 7.5:1; and (c) 10:1. *Source:* Reprinted from Yuan et al. [20]. Copyright 2011, with permission from American Chemical Society.

reversible alkoxyamine moieties due to permanent connection of methylene radicals. When the healing time is prolonged to six and nine hours, the healing efficiencies are 75.3 and 75.7%, respectively. This means that 2.5 hours is sufficient for crack healing under 130 °C, as thermal equilibrium between dissociation and association of alkoxyamine moieties is reached [22].

It is worth noting that the crack healing temperature (130 °C) is higher than the T_g of the polymer, but the specimens do not show creep distortion during healing. Thermal mechanical analysis (TMA) and DMA data further demonstrate that at the rubbery state the reversibly cross-linked PS has similar deformation resistance as irreversibly cross-linked PS (Figures 7.15 and 7.12b). Simple load bearing tests also confirm the material can stand against pressure at elevated temperature. This suggests that the concept of synchronous covalent bond fission/radical recombination indeed works in developing self-healing polymer, which is robust enough even when macromolecular chains are temporarily broken.

For the reversibly cross-linked PS with molar feeding ratio of styrene to cross-linker of 5:1, low healing efficiency of 26.4% is obtained, which seems to be contrary to expectation because the polymer contains more alkoxyamine moieties. This can be explained by the fact that the higher cross-link density of the polymer reduces the range of motion of radicals. Consequently, exchange reactions (scrambling) of alkoxyamine moieties (Scheme 7.8) across the crack interfaces are limited. In fact, the healing efficiency of the polymer can be increased to 70.1% when the healing temperature is raised to 145 °C. The increased radical concentration and reduced restraint on range of motion of radicals account for the improvement.

With respect to the reversibly cross-linked PS with molar feeding ratio of styrene to cross-linker of 10:1, the

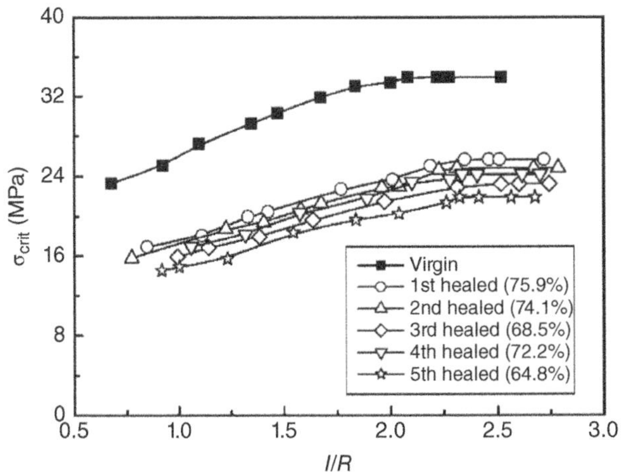

Figure 7.14 Typical critical stress, σ_{crit} versus crack length, *l/R*, of reversibly cross-linked PS specimen with molar feeding ratio of styrene to cross-linker dimethacrylic ester of 7.5:1. The numerals in parentheses represent healing efficiency. Crack healing temperature and time: 130 °C and 2.5 hours, respectively. *Source:* Reprinted from Yuan et al. [20]. Copyright 2011, with permission from American Chemical Society.

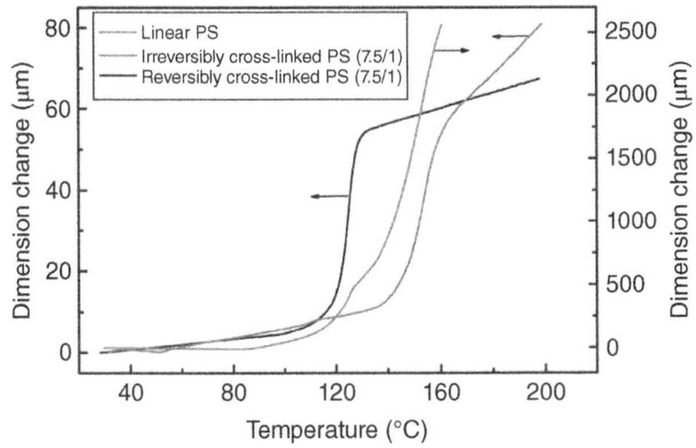

Figure 7.15 Temperature dependence of dimensional change for linear PS, reversibly cross-linked PS, and irreversibly cross-linked PS. The numerals in parentheses represent molar feeding ratio of styrene to cross-linker. The curves were measured by TMA.

measured healing efficiency (32.6%) is also low. This should be attributed to the insufficient amount of radicals despite the range of motion of the radicals being large enough as a result of lower cross-link density.

Considering that thermal treatment itself might contribute to the healing effect, control DCDC tests were performed on irreversibly cross-linked PS. No healing was detected when the specimens with cracks were heated at a temperature 20 °C higher than the T_g of the polymer. It is thus confirmed that chemical reconnection of fractured bonds on crack surfaces dominates the crack healing.

7.3 Self-Healing Epoxy

Epoxy is an important polymeric material widely applied in surface coatings, structural adhesives, printed circuit boards, insulation materials for electronic devices, and advanced composites, because of its mechanical properties, chemical resistance, thermal stability, ease of handling and ease of processing over a wide temperature range. With the development of self-healing technology in recent years, the possibility of imparting intrinsic crack remendability to epoxy has been explored.

Here, reversible covalent bonding through the dynamic exchange reaction of alkoxyamines is introduced to make self-healing epoxy [23]. Under a certain homolysis temperature, C—ON bonds in alkoxyamines undergo reversible radical reaction, producing transient carbon-centered and persistent nitroxide radicals in equal amounts [3–8]. There exists a fast dynamic equilibrium between dissociation and association with a very high equilibrium constant of 1.7×10^{15} [M^{-1}] for bulk living polymerization of styrene at 135 °C [4]. Our early work demonstrated that cross-linked PS containing C—ON bonds as cross-linkers can indeed rebind fractured surfaces repeatedly (Section 7.2). Since covalent bond homolysis and radical recombination synchronously take place in this case, the aforementioned two-step healing strategy based on DA bonds is simplified into a single-step one, and possible material deformation due to collapse of cross-linked networks in the course of crack remending would thus be avoided even when the healing is carried out above the T_g.

For purposes of obtaining alkoxyamine-containing epoxy (diEP, see Scheme 7.9), a new alkoxyamine-based diol is designed and synthesized, which should possess lower homolysis–recombination temperature in comparison with the alkoxyamine-based diol used in our previous work (Section 7.2.1). Meanwhile, Si—O bonds are introduced through reaction between the diol and chlorosilane to improve mobility of molecular chains connected with alkoxyamine groups and facilitate the subsequent healing reaction across crack faces. By hydrosilylation between the silicon hydrogen bond and allyl glycidyl ether, the alkoxyamine-containing epoxy (diEP) is yielded. Considering that the cured version of the alkoxyamine-containing epoxy might be too soft as a result of its flexible long segment structure, traditional bisphenol A diglycidyl ether is incorporated as a rigid component to strengthen the ultimate material. It is hoped that after curing, an epoxy with balanced performance would be yielded (Scheme 7.10), in which the C—ON bonds help to heal cracks via their reversible homolysis–recombination habit.

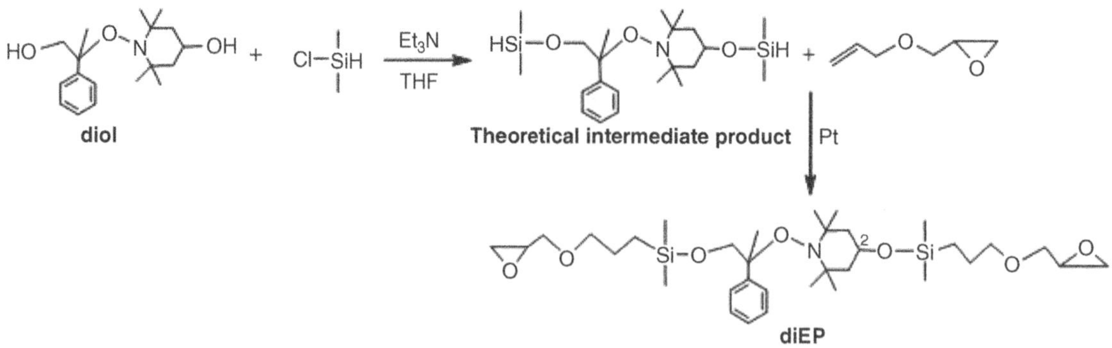

Scheme 7.9 Synthesis of alkoxyamine-containing epoxy (diEP).

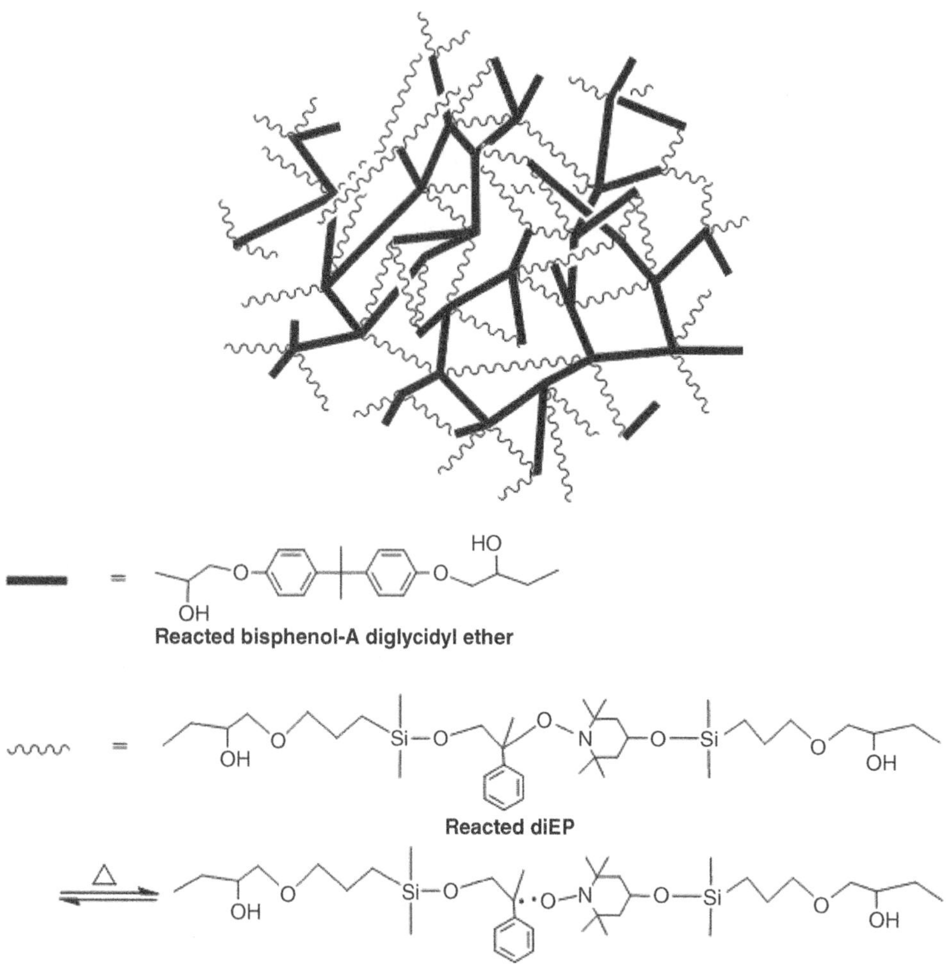

Scheme 7.10 Thermally remendable epoxy containing alkoxyamines.

7.3.1 Synthesis

The alkoxyamine-based diol (4-hydroxy-1-(2′-hydroxy-1′-phenyl-1′-methyl) ethyl-TEMPO) was synthesized as follows: BPO (14.52 g, 0.06 mol) and 4-OH-TEMPO (20.64 g, 0.12 mol) were dissolved in α-methyl styrene (250 ml), heated at 100 °C under argon for three hours, and then cooled to room temperature. The excess α-methyl styrene was removed under reduced pressure. Afterwards, ethanol (300 ml) and KOH aqueous solution (3 N, 120 ml) were added into the system and the mixture was stirred at room temperature for 12 hours, followed by evaporation to dryness and partitioning between dichloromethane and deionized water. The aqueous phase was extracted with dichloromethane three times and the organic phases were collected and dried with anhydrous sodium sulfate for 24 hours. After the solvent was removed, the crude product was purified by column chromatography using ethyl acetate and petroleum ether as the eluent, offering the diol as a pale yellow solid (yield = 34.7%).

The synthesis of diEP referred to the route reported in ref. [24] (Scheme 7.9). The aforementioned alkoxyamine-based diol (3.07 g, 0.01 mol) and triethylamine (3.03 g, 0.03 mol) were dissolved in THF (100 ml), and then chlorodimethylsilane (2.835 g, 0.03 mol) was added dropwise into the system under argon gas flow to ensure an anhydrous reaction environment, followed by stirring at room temperature for five hours. After the generated triethylamine hydrochloride was removed, the mixture was evaporated to dryness under reduced pressure and partitioned between toluene and deionized water. The organic phase was washed with water three times and dried with anhydrous sodium sulfate for 24 hours. Afterwards, Karstedt's catalyst (0.018 g, 7.8×10^{-5} mol) and allyl glycidyl ether (2.394 g, 0.021 mol) were added into the obtained toluene solution of intermediate product, and the compound was heated at 60 °C under an argon atmosphere for 2.5 hours. The solvent and excess allyl glycidyl ether were removed by vacuum distillation to give the product as a pale yellow oil (yield = 63.2%). The obtained alkoxyamine derivative with an epoxy group at both ends is referred to as diEP.

The synthesized diEP was blended with bisphenol A epoxy resin (EPON 828) at a molar ratio of 1:1.5, and then the hardener diethylenetriamine (DETA) was added according to the stoichiometric ratio of the reactive epoxide group of each ingredient. The epoxy specimens (called 828/diEP) used for further characterization were cured in a silicone rubber mold under an argon atmosphere step by step at 30, 80, and 100 °C for 12, 5, and 1 hours, respectively.

To study the relationship between homolysis of alkoxyamine moieties and thermal stability of the material when heated, diEP specimen in the absence of bisphenol A epoxy resin EPON 828 cured by DETA (labeled as N-epoxy) was also prepared. Additionally, another control specimen made from the compounds of bisphenol A epoxy resin EPON 828 and diethylene glycol diglycidyl ether (DGDE) (denoted by 828/DGDE), in which DEDE replaced diEP in 828/diEP, was designed to reveal the effect of entanglement of soft molecular segments on self-healing performance. In this case, the molar ratio of 828/DGDE was still 1.5:1 and DETA served as the hardener again.

7.3.2 Characterization

Since the C—ON bonds in alkoxyamines should play the leading role in self-healing and homolysis–rebonding of C—ON bonds is related to free radicals, ESR was employed to understand the thermally reversible behavior of the materials. Figure 7.16a shows typical ESR spectra of N-epoxy measured at various temperatures. In addition, Figure 7.16b gives the normalized absorption lines corresponding to the ESR curves in Figure 7.16a, from which integrated areas of the relative signal intensities are calculated and plotted as a function of temperature (see Figure 7.16c). Clearly, the radical signal is detectable at 20 °C and its intensity increases with a rise in temperature. The phenomenon resembles the case of cross-linked PS containing alkoxyamine (Section 7.2.2), but the initial temperature for the appearance of the radical signal is much lower than that of the latter (~100 °C). This is because tertiary butyl groups rather than ethyl groups link up with the N—O bonds in the diol (Scheme 7.9), which produce larger steric hindrance, reducing the bond energy of the C—ON bond and lowering its homolysis temperature. On the other hand, the results in Figure 7.16 and Section 7.2.2 confirm that alkoxyamine derivatives attached to different polymers exhibit similar thermal reversibility, which provides ample room to develop different thermally reversible polymers for different requirements.

To further investigate the reversibility of the C—ON bond in cured epoxy containing alkoxyamines, its ESR spectra were collected during repeated heating–cooling cycles between 20 °C and 90 °C (Figure 7.17), as the change in radical relative intensity reflects the homolysis–recombination habit of the C—ON bond. Moreover, comparison between the relative signal intensity after thermocycling and that of the original state at 20 °C would help to evaluate thermal reversibility of the epoxy. It is seen from Figure 7.17 that when temperature alternates between 20 °C and 90 °C, the amount of free radicals fluctuates as a result of scission of alkoxyamine moieties and recombination of the radicals. These processes are fully reversible, but the valley values of the relative intensity slightly increase with the thermal cycles especially in the case of 828/diEP. The latter behavior might be attributed to permanent dissociation of a few alkoxyamines as the combination of tertiary butyl radicals, and the unpaired nitroxide radicals have to be left over [17]. The relatively higher thermal reversibility of N-epoxy than 828/diEP should originate from the larger quantity of soft diEP in the former. On the one hand, flexibility of the abundant soft diEP chains increases the probability of collision between radicals. On the other hand, the molecular chain of diEP (651 g mol^{-1}) is longer than that of EPON 828 (330 g mol^{-1}), thus making

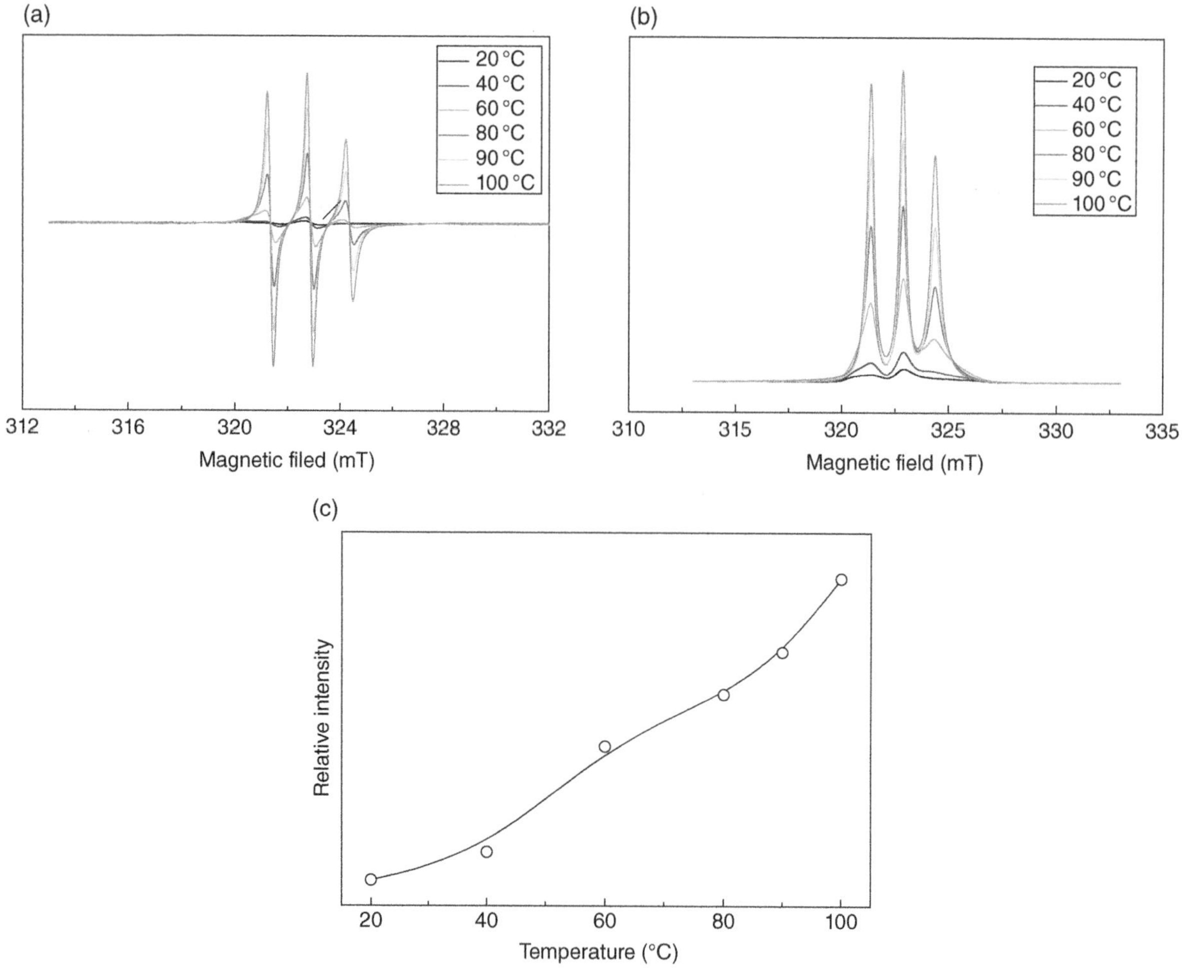

Figure 7.16 (a) Typical ESR spectra of N-epoxy measured at various temperatures. (b) Normalized absorption lines obtained from the data of (a). (c) Temperature dependence of relative ESR signal intensities calculated from integral areas of the normalized absorption curves in (b). *Source:* Reprinted from Yuan et al. [23]. Copyright 2014, with permission from the Royal Society of Chemistry.

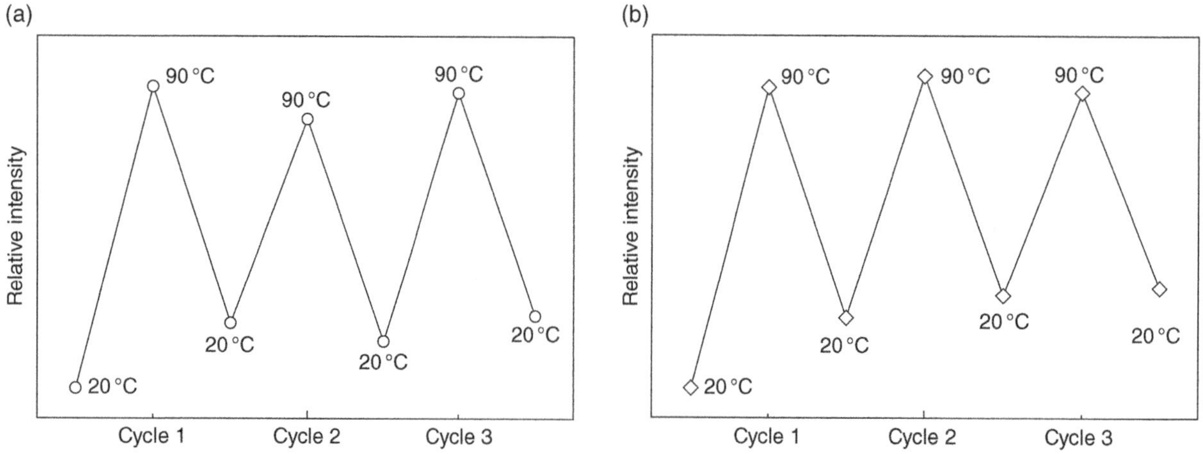

Figure 7.17 Relative ESR signal intensities of (a) N-epoxy and (b) 828/diEP measured during heating–cooling cycles between 20 °C and 90 °C. *Source:* Reprinted from Yuan et al. [23]. Copyright 2014, with permission from the Royal Society of Chemistry.

a lower cross-link density for N-epoxy. Nevertheless, 828/diEP possesses much better reversible performance than the cross-linked PS with alkoxyamines (Section 7.2.2). Apparently the higher stability of tertiary butyl radicals together with soft segments in epoxy are responsible, which can hinder the combination between carbon radicals but benefit reconnection of the carbon and nitroxide radicals attached to the macromolecular chains.

Considering the fact that the C—ON bonds in alkoxyamines frequently cleave but immediately recombine under a certain homolysis temperature, thermal stability of the epoxy materials containing alkoxyamines must be affected accordingly. The thermogravimetric analysis (TGA) curves in Figure 7.18a show that the presence of alkoxyamines significantly decreases the onset temperature of epoxy, which is in proportion to the concentration of alkoxyamines. The homolysis of C—ON bonds and the resultant radicals should account for this. In addition, the onset temperature of 828/DGDE is 30 °C lower than that of EPON 828 (see the inset in Figure 7.18a), implying not only alkoxyamines but also soft segments in the polymer network lower the thermal stability of epoxy.

When looking at the curves of rate of weight loss (Figure 7.18b), two peaks are found for N-epoxy. By reference to the peak position of EPON 828 that does not contain any alkoxyamines, it is known that the peak at higher temperature results from destruction of irreversible covalent bonds, and the other reflects the weight loss caused by cleavage of C—ON bonds.

In addition to TGA that provides temperature-dependent structural information under quiescent condition, DMA enables the issue to be studied in a dynamic way. As shown in Figure 7.19, the N-epoxy specimen can be repeatedly tested, demonstrating the existence of reversible equilibrium between C—ON bond homolysis and recombination; otherwise, even the second test could not be conducted if

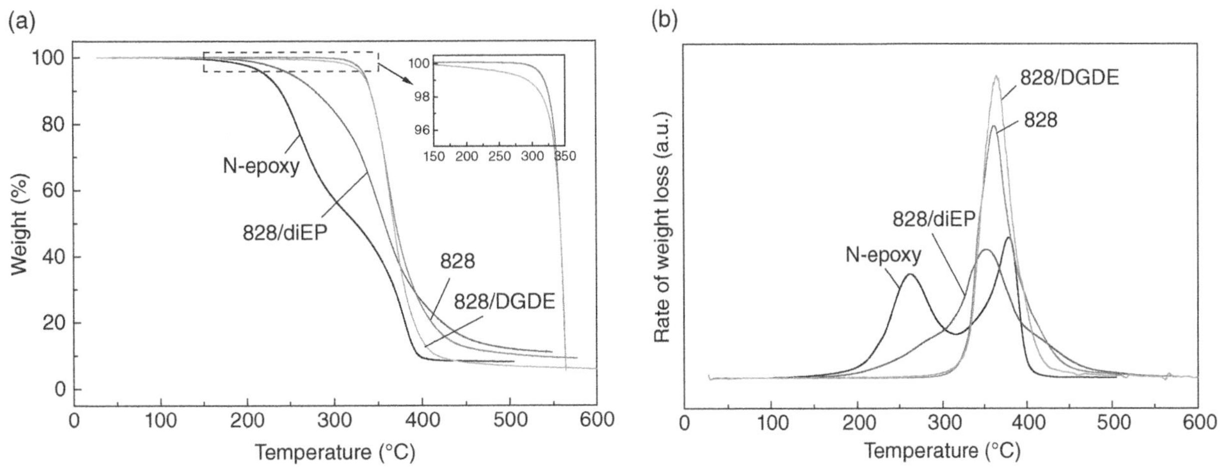

Figure 7.18 (a, b) Pyrolytic behaviors of cured epoxies.

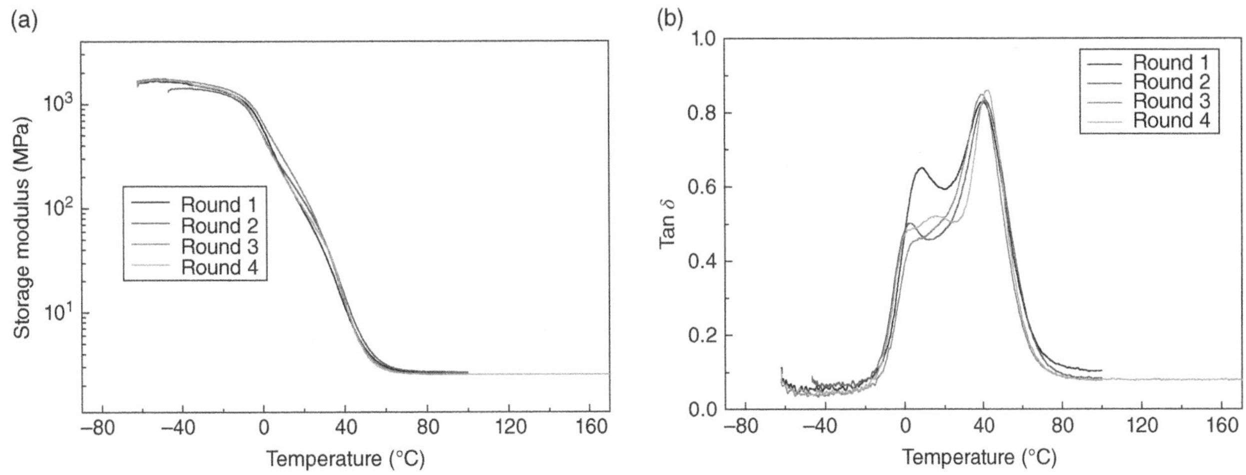

Figure 7.19 Temperature dependence of (a) storage modulus and (b) tan δ of N-epoxy measured by repeated DMA scans.

the radicals failed to be immediately re-bonded and the material behaved like a thermoplastic due to the desctruction of the networks. During the repeated heating–cooling cycles, the temperature dependence of the storage modulus of N-epoxy keeps almost the same, but the profiles of its tan δ peaks are obviously changed. In fact, there are two peaks on the tan δ curves of N-epoxy. Compared with the higher temperature one, both position and height of the lower temperature one change much more remarkably with repeated scans. This should be the result of the dynamic equilibrium feature of the reversible C—ON bonding. That is, the lower temperature peak on the tan δ curve of N-epoxy is contributed by segment motion of the dangling chains generated by disconnection of C—ON bonds, while the higher temperature one represents the undamaged networks. Owing to the random nature of the homolysis–recombination behavior of the C—ON bond, the dangling chain lengths change from time to time so that the position and height of the lower temperature peak are different for different tests. In contrast, the higher temperature peak is basically unchanged due to the relatively definite molecular weight between cross-links of the undamaged networks.

With the incorporation of EPON 828 that contains no reversible C—ON bonds and increases the cross-link density of the system, the tan δ curves of 828/diEP become unimodal as the dangling chains lengths of the disconnected chains produced by homolysis of alkoxyamine moieties have to be much shorter due to the existence of irreversible networks, which cannot arouse a new damping peak (Figure 7.20). Nevertheless, it is worth noting that peak temperature and intensity on the tan δ curves decrease after each test cycle. This should be interpreted as the appearance of a few permanently dissociated alkoxyamines originating from the difficulty of recombination of tertiary butyl radicals and nitroxide radicals in the solid (Figure 7.17b). Accordingly, cross-link density of the epoxy network is reduced with the test cycles (Table 7.1). The analysis is supported by the temperature dependence of the storage modulus, which moves downward during repeated DMA tests (Figure 7.20a). The slight decline in the stiffness should be attributed to the permanently destructed C—ON bonds that could no longer bear load. The decay is marginal on the whole, which is also proved by ESR data, meaning the thermal reversibility of cured 828/diEP would effectively work for crack healing. On the other hand, a comparison of Figures 7.19 and 7.20 indicates that the peak temperature of tan δ of 828/diEP is about 20 °C higher than that of N-epoxy. That is due to the reduced concentration of flexible —Si—O— bonds in 828/diEP.

Figure 7.21 shows three successive macro-repairing processes of an 828/diEP specimen evaluated by impact test. The virgin specimen was first impacted apart, and then the

Table 7.1 Molecular weight between crosslinks, M_c, of the cured epoxies estimated from repeated DMA scans in Figures 7.19 and 7.20 with data of the epoxies without C—ON bonds as references.

	M_c (g mol^{-1})			
Materials	**Cycle 1**	**Cycle 2**	**Cycle 3**	**Cycle 4**
N-epoxy	1685	1664	1729	1790
828/diEP	350	371	384	—
828/DGDE	558	—	—	—
828	161	—	—	—

Source: Reprinted from Yuan et al. [23]. Copyright 2014, with permission from the Royal Society of Chemistry.

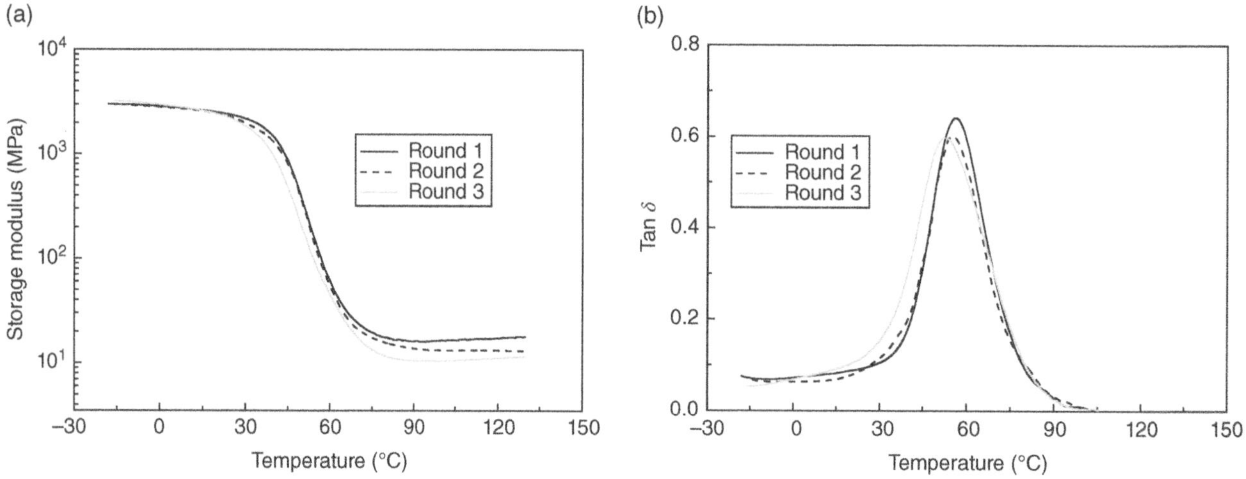

Figure 7.20 Temperature dependence of (a) storage modulus and (b) tan δ of 828/diEP measured by repeated DMA scans.

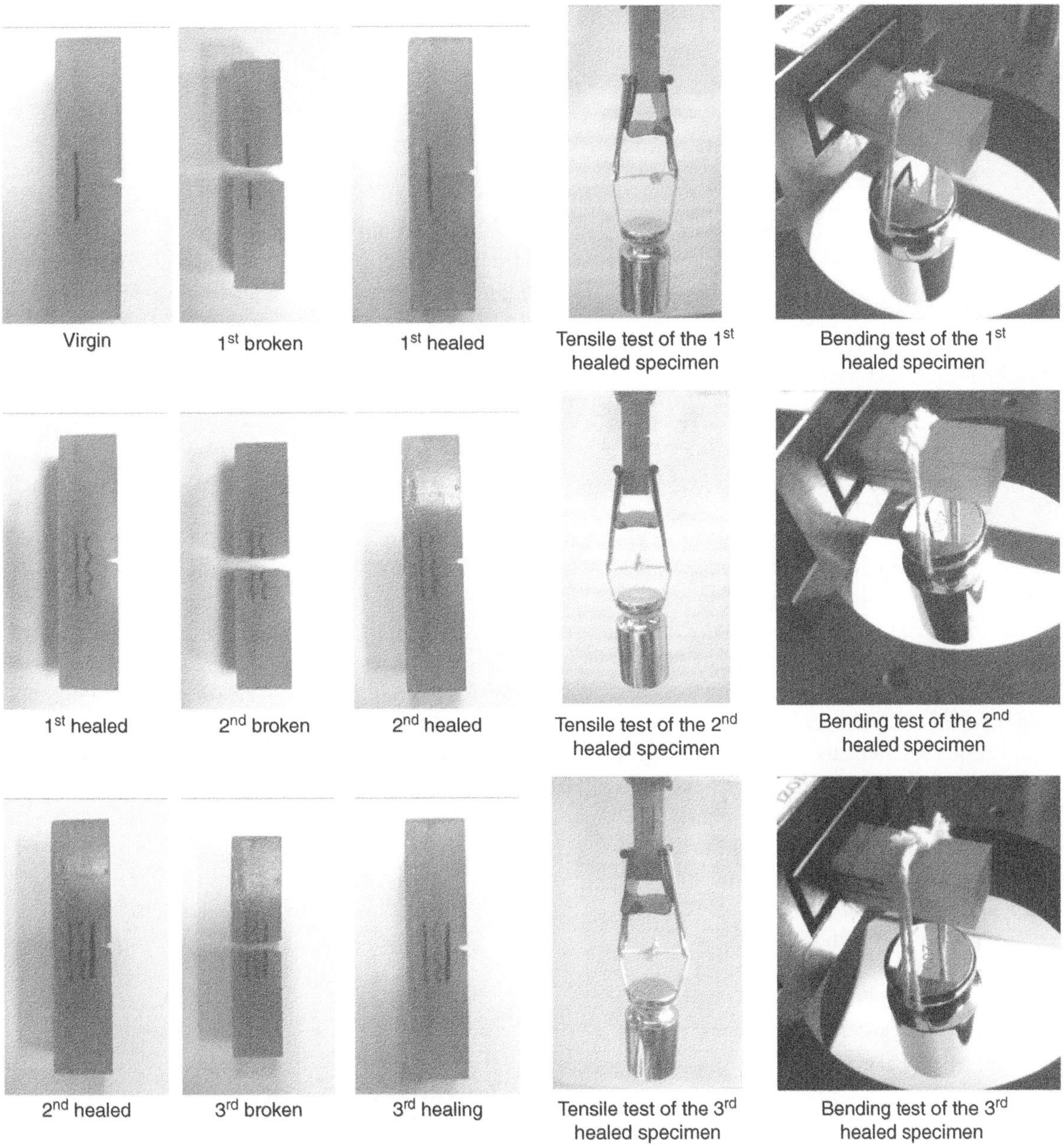

Figure 7.21 Repeated impact damaging-healing of an 828/diEP specimen. *Source:* Yuan et al. [23], Figure 7, p. 7/with permission of Royal Society of Chemistry.

broken parts were recombined. Having been healed in argon at 90 °C for 1.5 hours, no shape change was observed for the specimen and the crack line was absent under visual inspection. It can bear a weight of 200 g (i.e. 15.8 kPa under tensile condition). Afterwards, the healed specimen was subjected to impact loading again, and repaired under the same conditions. The healed version still possessed load bearing ability, and the cracking line was invisible. The third impact–healing tests also gave the same results as before. Clearly, the epoxy has acquired repeated remendability as expected. It is worth noting that the healing temperature is higher than the T_g of the material, but the specimen still keeps its original shape after rehabilitation, which evidently reflects one of the advantages of employing the dynamic reversible C—ON bond as the healing group.

In this work, 90 °C is chosen as the healing temperature for the following reasons. First, the materials should be heated above T_g to ensure higher molecular chain mobobility for the occurrence of the reversible reaction across the crack interface. Secondly, higher homolysis rate or radical concentration at elevated temperature is important for effective healing.

Table 7.2 Healing efficiencies of 828/diEP measured during three successive impact damaging–healing tests with the data of control specimens as references.

Materials	Healing efficiency (%)		
	1st repaired	2nd repaired	3rd repaired
828/diEP	62.0 ± 3.5	59.5 ± 5.7	55.1 ± 3.2
828/DGDE	6.8 ± 4.2	7.2 ± 2.5	6.4 ± 2.7
828/diEP (healed in air)	6.9 ± 3.8	5.4 ± 3.6	6.0 ± 2.9

Source: Reprinted from Yuan et al. [23]. Copyright 2014, with permission from the Royal Society of Chemistry.

Table 7.2 lists that the average healing efficiency of the first healing is 62.0%, while those of the second and third healing are 59.5 and 55.1%, respectively. The decline in healing efficiency with healing–re-fracture cycles might result from incomplete alignment of the specimens and/or permanently irreversible C—ON bond caused by radical deactivation at the fractured surfaces. Compared with the cross-linked PS system [20], the healing efficiency of 828/diEP decreases much more slowly. This is mainly because the long soft segments of diEP in epoxy not only increase the collision probability of radicals to reform covalent C—ON bonds, but also benefit interdiffusion of the dangling chains across the fractured surfaces. Figure 7.22 depicts the healing efficiency of 828/diEP as a function of healing time. With a rise in healing time, the healing efficiency increases and then levels off when the time exceeds two hours. This implies that no more reversible C—ON bonds are involved at the interface with prolonged time. It should be pointed out that although the C—ON healing chemical equilibrium is fairly fast, two hours is still needed to reach steady-state healing. This reflects the importance of interdiffusion of macromolecules and the collision probability of radicals between the fractured surfaces, which act as the control step.

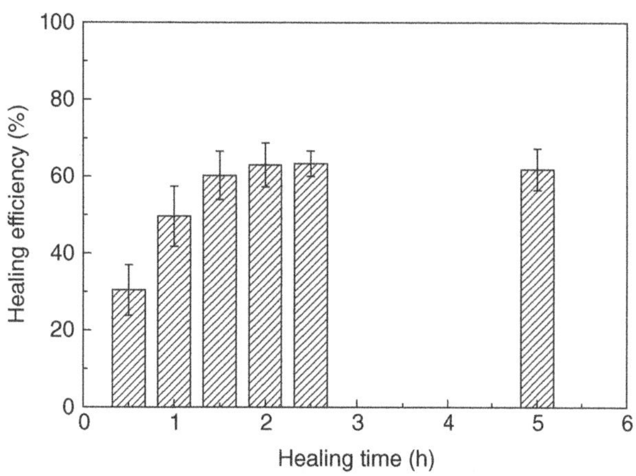

Figure 7.22 Effect of healing time on healing efficiency of 828/diEP. Healing was conducted in argon at 90 °C.

To investigate the effects of C—ON bonds and long soft segments on healing performance of epoxy, controlled experiments were carried out as in the following. First, 828/DGDE was prepared by replacing diEP with DGDE that does not contain any reversible bond but has a long soft structure like diEP. It is found that only ~7% of impact strength can be restored under the same conditions (Table 7.2), suggesting that the reversible C—ON bonding rather than soft chain entanglement or reaction of latent functional groups at the interface makes the decisive contribution to the crack healing. Secondly, the impact ruptured specimens of 828/diEP were exposed to air at 90 °C for 1.5 hours, and then they were cooled down to room temperature, recombined together and healed in air at 90 °C for 1.5 hours. Due to the oxygen-induced deactivation of radicals, the reversible reaction of C—ON bonds cannot proceed in air. As a result, ~6% healing efficiency is detected (Table 7.2). The result proves the importance of C—ON bonding once again.

It is worth noting that the self-healing 828/diEP has not yet achieved full recovery of impact strength owing to the presence of the unmendable 828 structure that acts as the stiff backbone of the networks. To further improve healing efficiency of the epoxy while keeping high strength, novel epoxy monomer should be synthesized instead of blending. The current work provides the knowledge of the structure–property relationship for molecular design.

7.4 Self-Healing Polymers Containing Alkoxyamine with Oxygen Insensitivity and Reduced Homolysis Temperature

The above explorations have proved the feasibility of one-step intrinsic self-healing based on reversible C—ON bonds, but there are still challenges. The relatively higher cleavage temperature of the C—ON bonds (e.g. ~125 °C for the cross-linked PS, see Section 7.2.2) means the healing has to be conducted under external heating. Besides, the oxygen sensitivity of the carbon-centered radicals derived from homolysis of the C—ON bond requires an inert environment for healing. In this context, the synthesis of alkoxyamine with greatly reduced homolysis temperature and oxygen resistance becomes critical [25].

Accordingly, an alkoxyamine derivative, 5-hydroxy-2-(4-hydroxy-2,2,6,6-tetramethylpiperidin-1-yloxy)-2-methyl pentanenitrile (CTPO), is synthesized as a dynamic reversible monomer, which includes a nitrile group (CN) linked with the carbon atom of the C—ON bond (Scheme 7.11a). A few functional moieties like nitrile might have a strong capability of adsorbing electrons and stabilize the carbon radical [26], so it is expected that the C—ON bond in CTPO would become active at lower temperature. The stability of both nitroxide and the specially designed carbon radicals would lead to an oxytolerant reversible chemistry (i.e. healing action) at ambient temperature.

In addition to the introduction of low temperature reversible bonds into macromolecules to realize autonomous self-healing of stiff polymer, the molecular chains on the fractured surface should be mobile enough at room temperature to ensure intimate contact and the subsequent interchange reaction of reversible groups. However, the

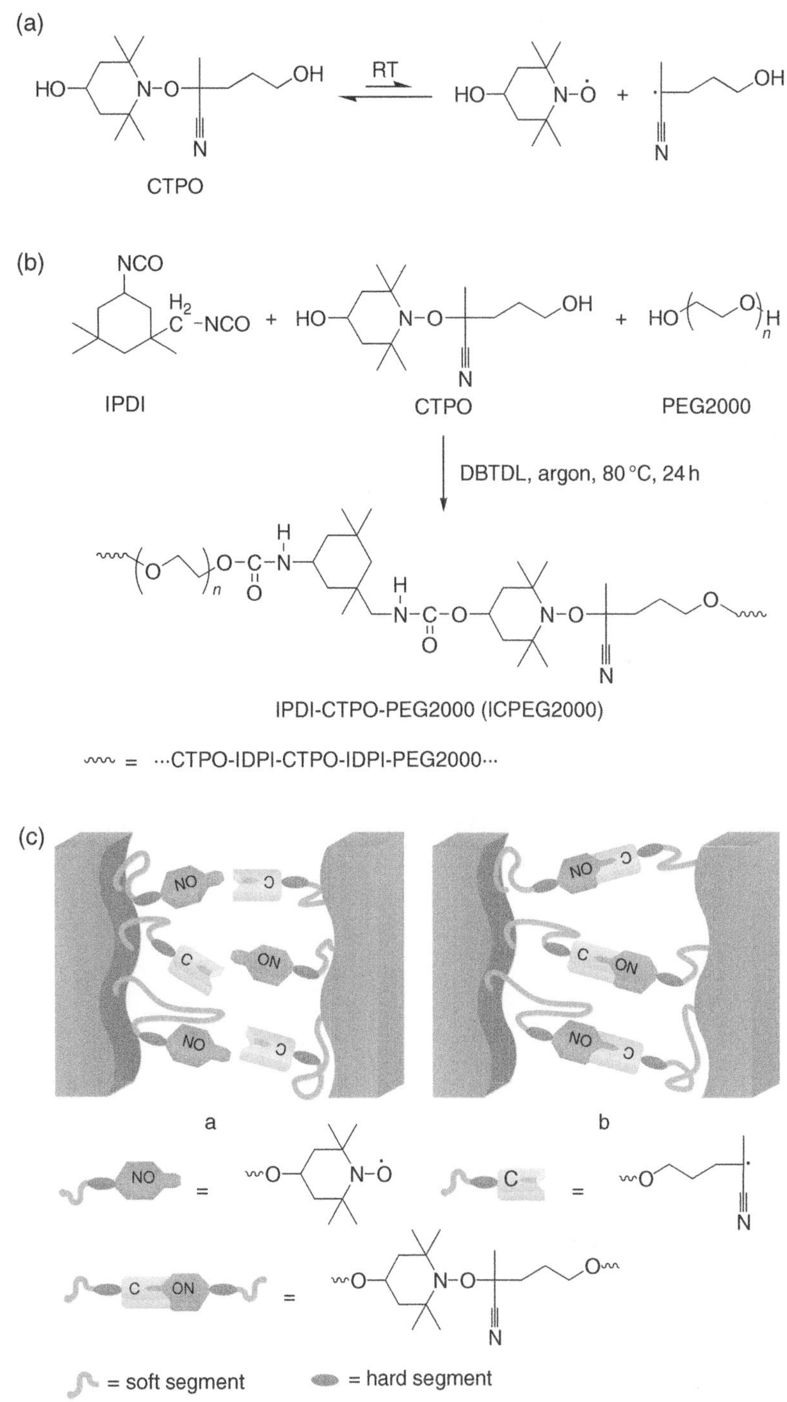

Scheme 7.11 (a) Dissociation–association of the designed alkoxyamine derivative CTPO. (b) Synthesis of the reversible polyurethane ICPEG2000. (c) Schematic diagram of the healing reaction.

polymers would lose a great part of their strength and are no longer stiff materials. Here we copolymerize CTPO with 3-isocyanatomethyl-3,5,5-trimethylcyclohexylisocyanate (IPDI) and polyethylene glycol (PEG2000), forming a polyurethane (PU) in which PEG acts as the soft segment, while IPDI and CTPO act as the hard segment (Scheme 7.11b). PEG possesses a T_g lower than room temperature and can crystallize to strengthen the resultant PU. As a result, the requirements of both strengthening of the polymer and improvement of chain mobility can thus be satisfied at the same time.

7.4.1 Synthesis

A typical process of synthesis of 5-hydroxy-2-(4-hydroxy-2,2,6,6-tetramethylpiperidin-1-yloxy)-2-methyl pentanenitrile (also called 4-cyano-4-TEMPOL-pentanol, i.e. CTPO monomer) is described in the following. 4,4′-Azobis(4-cyano-1-pentanol) (ACPO) (2.520 g, 0.01 mol) and 4-hydroxy-2,2,6,6-tetramethylpiperidinyloxy (1.152 g, 0.0067 mol) were completely dissolved in 150 ml benzene. Then, the mixture was heated to 85 °C and stirred in argon for three hours. Afterwards, the colorless solution was concentrated under vacuum and purified by column chromatography (silica gel) eluting with 5:3 hexane–ethyl acetate (v/v). The combined fractions were recrystallized from hexane to obtain a primrose yellow liquid with a yield of 37.4%.

Preparation of PU ICPEG2000, the model material of the present work, was conducted through bulk polymerization in a two-neck flask. CTPO monomer (2.84 g, 0.01 mol) and PEG2000 (20 g, 0.01 mol) were mixed and heated to 60 °C until PEG2000 was completely melted. Then, IPDI (4.58 g, 0.02 mol, NCO/OH = 1.03, CTPO/PEG2000 = 1, IPDI/CTPO/PEG2000 = 2.06/1/1) and two drops of dibutyltindilaurate (DBTDL) were immediately added to the mixture, which was heated to 80 °C and stirred using a magnetic stirrer in argon for three hours. Afterwards, the mixture was poured into a silicone mold to continue the reaction for 21 hours and then cooled down to obtain the product.

The control material that excludes alkoxyamine moiety, IBPEG2000, was made following similar procedures by using 1,4-butanediol (BDO) instead of CTPO monomer. When PEG2000 diol is replaced by PCL2000 diol, ICPCL2000 can be prepared by a similar process.

7.4.2 Characterization

In the course of synthesis, CTPO monomer is yielded from a simple reaction between 4-hydroxy-2,2,6,6-tetramethyl-piperidinyloxy(TEMPOL)and4,4′-azobis(4-cyano-1-pentanol) (ACPO, a general free radical initiator). The radicals resulting from ACPO initiator are captured by TEMPOL in addition to the formation of CPO as a side reaction (Scheme 7.12). Then, crystalline linear PU ICPEG2000 (i.e. IPDI-CTPO-PEG2000, see Scheme 7.11b) is obtained via bulk polymerization of CTPO with IPDI and PEG2000 under feeding ratios of NCO/OH = 1.03 and CTPO/PEG2000 = 1. The molecular structures of CTPO and the resultant PU ICPEG2000 are confirmed by infrared, Raman and nuclear magnetic resonance spectroscopy.

ICPEG2000 (M_n = 22 000, polydispersity = 1.23) is a stiff solid, ascribed to its higher crystallinity of 59.0% (melting point = 37.2 °C, see Table 7.3), Vickers microhardness of 2.2 GPa, flexural strength of 2.7 MPa and flexural modulus of 0.29 GPa. Besides, ICPEG2000 has a subambient T_g in the range of −60 to −40 °C, determined from tan δ peak position by DMA. This implies that the ICPEG2000 chains should possess high mobility at room temperature as expected, which guarantees the reversible healing reactions among the broken chain fragments on the crack faces (Scheme 7.11c).

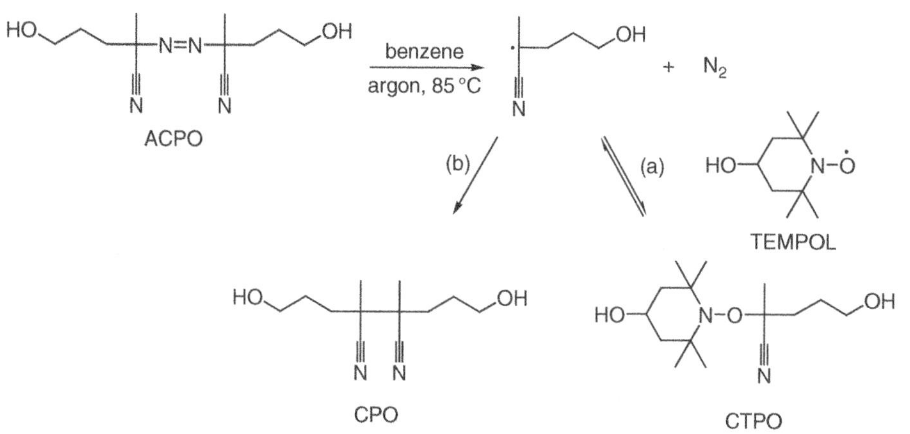

Scheme 7.12 (a, b) Synthesis of CTPO.

Table 7.3 DSC characterization of the polyurethanes.

Polymer	Onset melting temperature (°C)	Peak melting temperature (°C)	Melting enthalpy ($J\,g^{-1}$)	f (%)	X (%)
ICPEG2000	28.7	37.2	85.0	72.9	59.4
IBPEG2000	38.1	45.7	114.9	78.5	74.5
ICPCL2000	42.4	50.0	53.9	72.9	54.8

ICPEG2000 = IPDI-CTPO-PEG2000. IBPEG2000 = IPDI-BDO-PEG2000. ICPCL2000 = IPDI-CTPO-PCL2000. f = weight percentage of PEG2000 or PCL2000 in the polyurethane. X = crystallinity, calculated from: $X = H_f/f\Delta H_f^*$ (where ΔH_f^* is the melting enthalpy of 100% crystallized PEG2000 [196.4 $J\,g^{-1}$] and PCL2000 [135 $J\,g^{-1}$], respectively).

To reveal whether the alkoxyamine units in ICPEG2000 exist in a state of equilibrium of cleavage and recombination at or below room temperature, which is critical for the healing performance, ESR measurements were conducted (Figure 7.23). Since the signal at each temperature is kept at a certain level within a fixed test time, the ESR intensity can be used to describe the radical concentration in the material [22]. For the ESR spectra of CTPO and ICPEG2000 collected below 0 °C (Figure 7.23a1 and b1), it is seen that the low-field peak intensity gradually decreases with a rise in temperature while the peak shape keeps unchanged. This means that the type of free radical within the system is identical, but its life becomes shortened, which agrees with the known habit of nitroxide

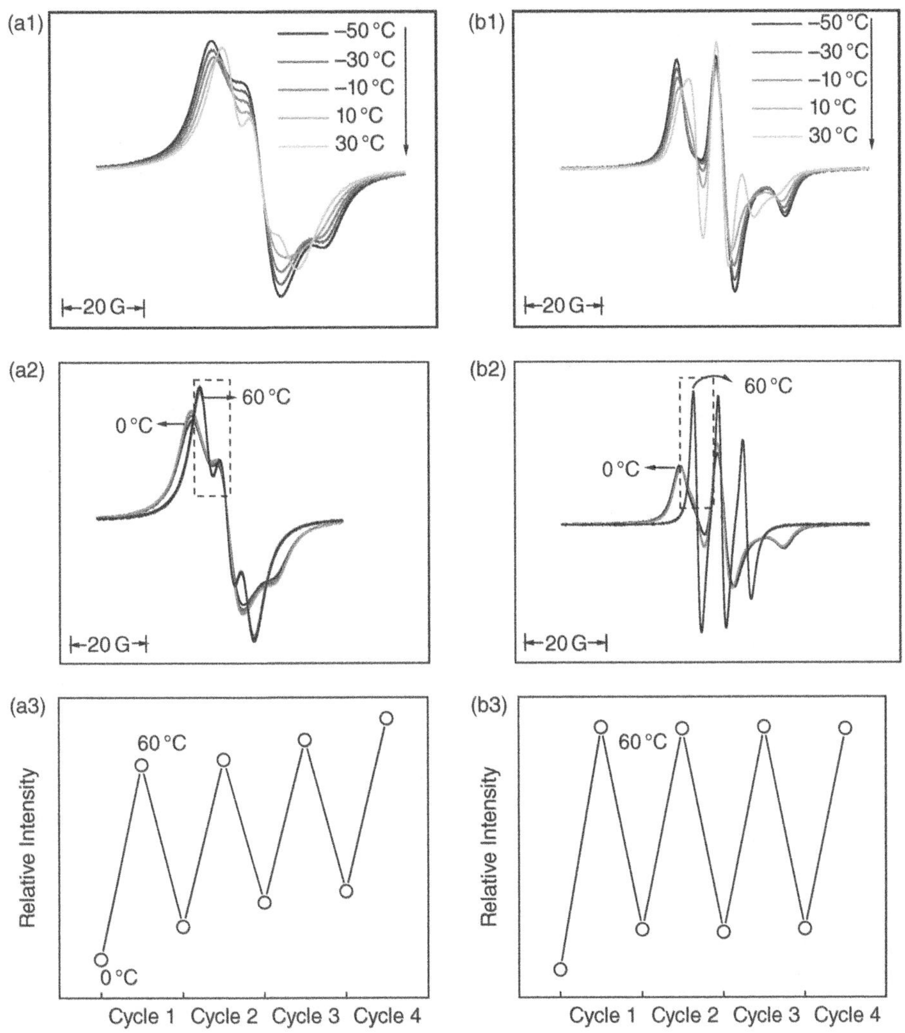

Figure 7.23 Typical ESR spectra of (a1) CTPO and (b1) ICPEG2000 measured at various temperatures. ESR curves of (a2) CTPO and (b2) ICPEG2000 are measured during repeated heating and cooling cycles between 0 °C and 60 °C. Relative ESR signal intensities of (a3) CTPO and (b3) ICPEG2000 are calculated from the low-field peak intensities in (a2) and (b2), respectively. *Source:* Reprinted from Zhang et al. [25]. Copyright 2013, with permission from the Royal Society of Chemistry.

radicals (TEMPOL). The result also reflects that a few nitroxide radicals remain in both CTPO and ICPEG2000. When the temperature exceeds 10 °C, evident changes in low-field peak shape and position are detected due to dissociation and association of C—ON bonds. Furthermore, the higher peak intensity at 30 °C represents an increase of free radical concentration accompanying cleavage of C—ON bonds, which offsets the influence of the shortened life. In this context, it is concluded that the homolysis temperature of CTPO and ICPEG2000 should be higher than 10 °C, and C—ON bonds are bound to be cleft into carbon-centered radicals and oxygen-centered nitroxide radicals at ambient temperature. It is interesting to note that Wang and Wu disclosed an alkoxyamine with a structure similar to CTPO for living radical polymerization [27], but it operated at 130 °C.

On the basis of the above analysis, free radical concentrations of TEMPO and CTPO at different temperatures (−30 to 30 °C) are determined by titration, which quantifies fission behavior of CTPO. During the measurements, nitroxide and carbon radicals are reduced by ascorbic acid, while ascorbic acid is oxidized. For stable nitroxide radicals of TEMPO, the measured radical concentration keeps constant at both −30 and 30 °C. In contrast, the amount of free radicals of CTPO at −30 °C is only about 30% of that measured at 30 °C, demonstrating that fission of C—ON bonds indeed happens at ambient temperature.

Figure 7.23a2 and b2 shows cyclic ESR measurements of CTPO and ICPEG2000 between 0 °C and 60 °C and reveals reversibility of the C—ON bonds. Accordingly, relative intensities of the radical signals estimated from low-field peak intensities are plotted in Figure 7.23a3 and b3, respectively. When temperature alternates between 0 °C and 60 °C, the amount of free radicals fluctuates as a result of scission of the alkoxyamine moiety and recombination of the radicals. These processes are fully reversible. Especially for ICPEG2000, both the peak and valley values of the relative intensity keep identical.

On the basis of the above investigation of room temperature reversibility, the approach proposed by Jones et al. [28] is employed to evaluate self-healing ability of the materials at 15 and 25 °C, which are below the melting point of ICPEG2000 (37.2 °C, see Table 7.3). Efficiency of healing is defined as the ratio of notched Izod impact strengths of healed and virgin material. The reasons why we did not choose tapered double cantilever beam (TDCB) or tensile tests are: (i) the relatively large size of the TDCB specimen requires a larger quantity of polymer to be used, which exceeds our current productive capacity; and (ii) tiny fragments would be split off the tensile ruptured surface due to brittle failure, which hinders intimate contact of the broken specimens during the subsequent healing.

Figure 7.24 depicts the measured results of virgin and repeatedly healed ICPEG2000 specimens. Healing is completed in argon atmosphere and air. Clearly, impact strength of the specimens can be greatly restored after mending at 15 °C (Figure 7.24a). The average healing efficiency of the first healing is 94.6%, while the second and third healings attain 91.7 and 89.8% (Figure 7.24b), respectively. The mild decline in healing efficiency with healing–re-fracture cycles might originate from incomplete alignment of the specimens and/or heterogeneity of dynamic recombination of alkoxyamine moieties at the fractured surfaces. When healing temperature is raised to 25 °C, healing efficiencies of the successive three healing events are 93.4, 92.0, and 91.5% (Figure 7.24b), respectively. The inconspicuous attenuation with healing number can be explained by the increase of radical concentration originating from movement of dissociation/association equilibrium of C—ON bonds to the right (Scheme 7.11a). Meanwhile, the control material IBPEG2000 (i.e. IPDI-BDO-PEG2000, see Table 7.3) is prepared following similar procedures as for ICPEG2000 except that CTPO is replaced by BDO that does not contain any reversible bond. It is found that ruptured specimens of IBPEG2000 cannot be re-connected at either 15 or 25 °C, proving the indispensable role of reversible C—ON bonds in imparting self-remendability to ICPEG2000.

It is worth noting when the broken specimens are repaired in air instead of argon, the polymer still exhibits repeated self-healing behavior with similar healing efficiency (Figure 7.24), and the crack lines are nearly invisible under visual inspection (Figure 7.25). The result coincides with the fact that the reversible reaction of C—ON bonds is not sensitive to oxygen. The detailed mechanism needs further exploration, but the C—ON bond chemistry proves to be feasible for development of self-healing polymer capable of working in air at ambient temperature. The stable nitroxide and carbon radicals linked with CN groups accounts for the manner. In general, the relative stability of a radical center with an attached substituent is described by radical stabilization energy (RSE), calculated from isodesmic reaction [29, 30]. The RSE value of the nitrile group is +5.34 kcal mol^{-1}, suggesting that the cyano-methyl radical is more stable than the methyl radical [30].

Wide angle X-ray diffraction (WXRD) spectra of both ICPEG2000 and IBPEG2000 were collected as a function of temperature. All the curves have similar diffraction peaks with the same position, shape and intensity, demonstrating that the crystalline phase in the two polymers retains unchanged until 25 °C. The result supports the above

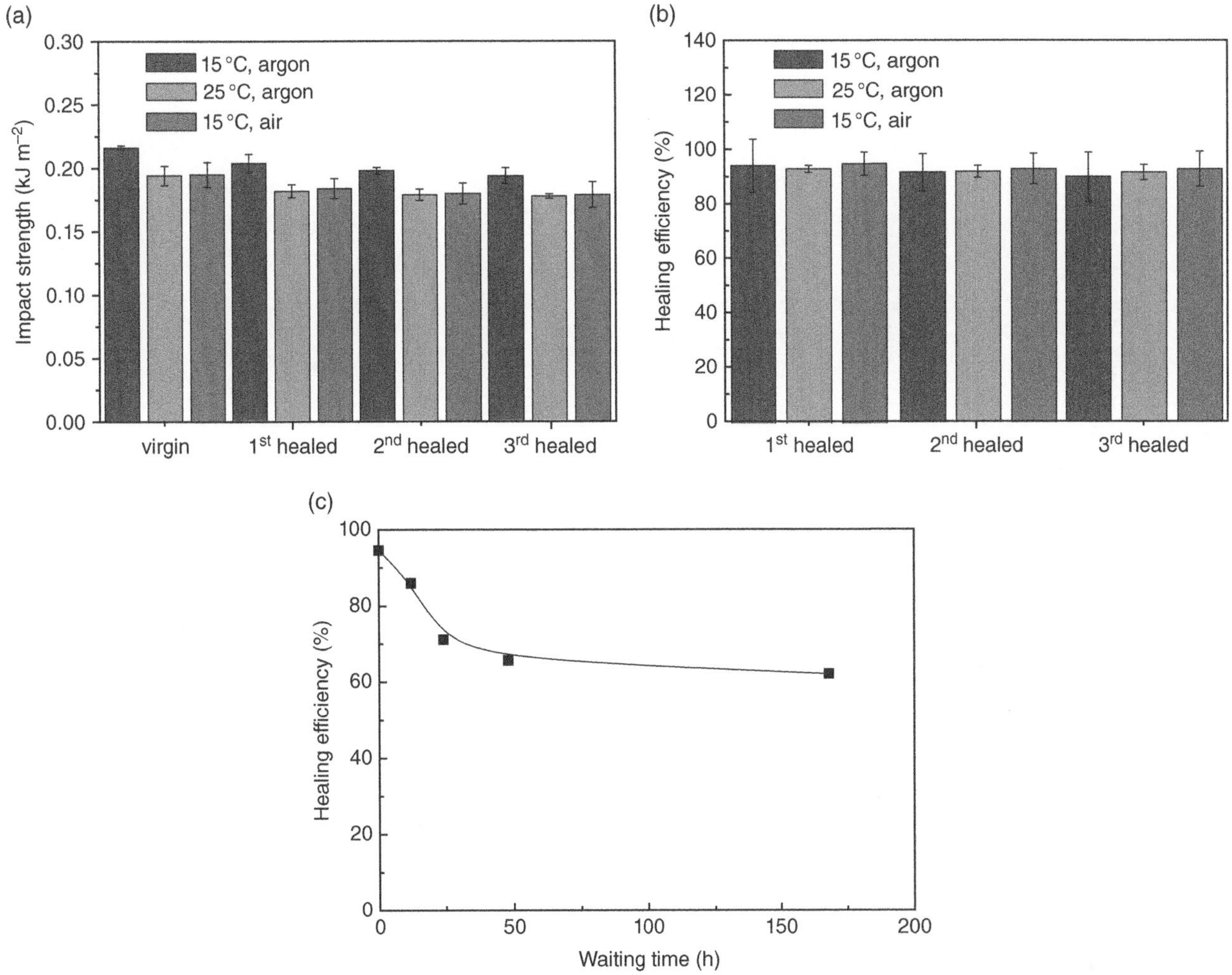

Figure 7.24 (a) Impact strengths of virgin ICPEG2000 specimens and the repeatedly healed versions under different circumstances. (b) Healing efficiencies calculated from (a). (c) Healing efficiency as a function of waiting time (which is the time for the broken specimens to wait for being reconnected in argon at 15 °C). *Source:* Reprinted from Zhang et al. [25]. Copyright 2013, with permission from the Royal Society of Chemistry.

conclusion from another angle, i.e. the remending behavior of ICPEG2000 has nothing to do with crystalline melting.

Wool et al. [2] considered that mutual diffusion of macromolecules across the crack interface is the main factor that controls recovery of mechanical property. This perhaps has the same importance as the reversible reactivity of the C—ON bonds. To check whether interdiffusion of chains takes place between contacted crack surfaces of ICPEG2000, a model system was fabricated as in the following. First, a piece of ICPEG2000 film (~50 μm thick) is jointed to an ICPEG2000 block via fusion bonding. Then, the film is peeled off the block to create a cracked surface carrying broken chain fragments. In the same way, IBPEG2000 film with cracked surface was also produced. Eventually, the two cracked surfaces of the films are brought into contact under gentle pressure of 0.2 MPa at 15 °C for 48 hours (Figure 7.26a), simulating the healing condition of bulk ICPEG2000. By using depth profile analysis of confocal Raman microscopy [31], the characteristic absorption of the CN group at 2227 cm^{-1} along the thickness of the bonded films is recorded (Figure 7.26b,c). Scanning depth dependence of theCN peak intensity shows that the content of CN group gradually decreases from the ICPEG2000 to IBPEG2000 side. Since ICPEG2000 rather than IBPEG2000 contains the CN group, the appearance of the CN signal in IBPEG2000 (i.e. when the scanning depth is >50 μm) must originate from mutual diffusion of ICPEG2000 and IBPEG2000 chains. As ICPEG2000 and IBPEG2000 are chemically similar, the model experiment indicates the possibility of chain diffusion across fractured surfaces of broken ICPEG2000 specimen during healing,

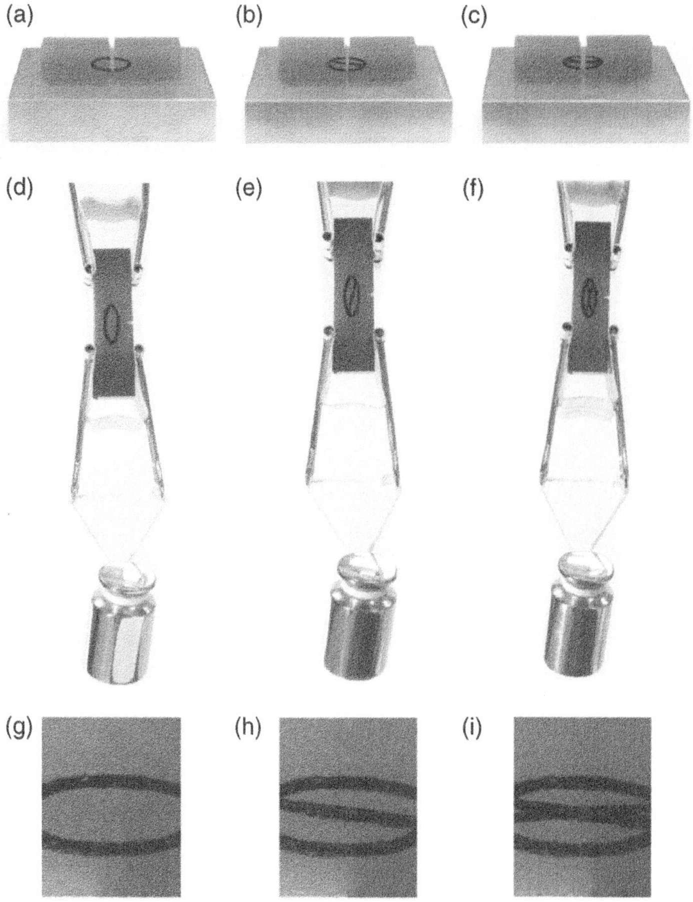

Figure 7.25 Repeated macro-repairing of impacted ICPEG2000 specimen (temperature of impact test = 15 °C). The virgin specimen is first impacted to failure (a), and then the broken parts are recombined. Having been healed in air at 15 °C for 48 hours, the specimen looks like the original without any shape change. It can bear a weight of 100 g (d) and the crack is absent under visual inspection (g). Afterwards, the healed specimen is subjected to impact loading again (b) and repaired under the same conditions. It still possesses load bearing ability (e), and the cracking line is invisible (h). As for the specimen impacted for the third time (c), the healed versions (f) and (i) behave like the former two.

which is a prerequisite for the reversible reaction of C—ON bonds. Nevertheless, it should be noted that the fractured stiff polymers cannot be repaired only by chain diffusion as shown by the above control test of IBPEG2000.

On the other hand, it must be emphasized that the fractured surfaces of the specimen, instead of the molded surfaces, are cohesible. This impressive property comes from the dynamic reversible feature of the C—ON bonds characterized by high equilibrium constant (e.g. 1.7×10^{15} [M^{-1}] for bulk living polymerization of styrene at 135 °C [4]). Moreover, owing to the good compatibility between soft and hard segments that leads to insignificant phase separation, the amorphous PEG phase may help to promote movement of the hard segments containing C—ON bonds and the corresponding reversible reactions.

In fact, the recombination of radicals might take place at the same side of the cracked surface, which partially explains the decay of healing efficiency with repeated rehabilitation. In this case, nitrile-stabilized radicals are coupled with each other, while nitroxide radicals are left over. The overall effect of this side reaction would be represented by a single, irreversible chain cleavage event and an increase in persistent radical concentration. To check its influence in detail, healing efficiency versus waiting time prior to reconnection of the broken specimens is plotted in Figure 7.24c. It is seen that the declining tendency levels off at 50 hours and remains at around 60%, indicating that the side reaction has a limit because of steric hindrance of the broken chains. The side reaction mentioned above might also exist in the bulk, which eventually results in termination of the reversibility. However, the dependence of radical concentration on storage time of CTPO and ICPEG2000 demonstrates that this is not the case (Figure 7.23a2 and b2). The reversible equilibrium of

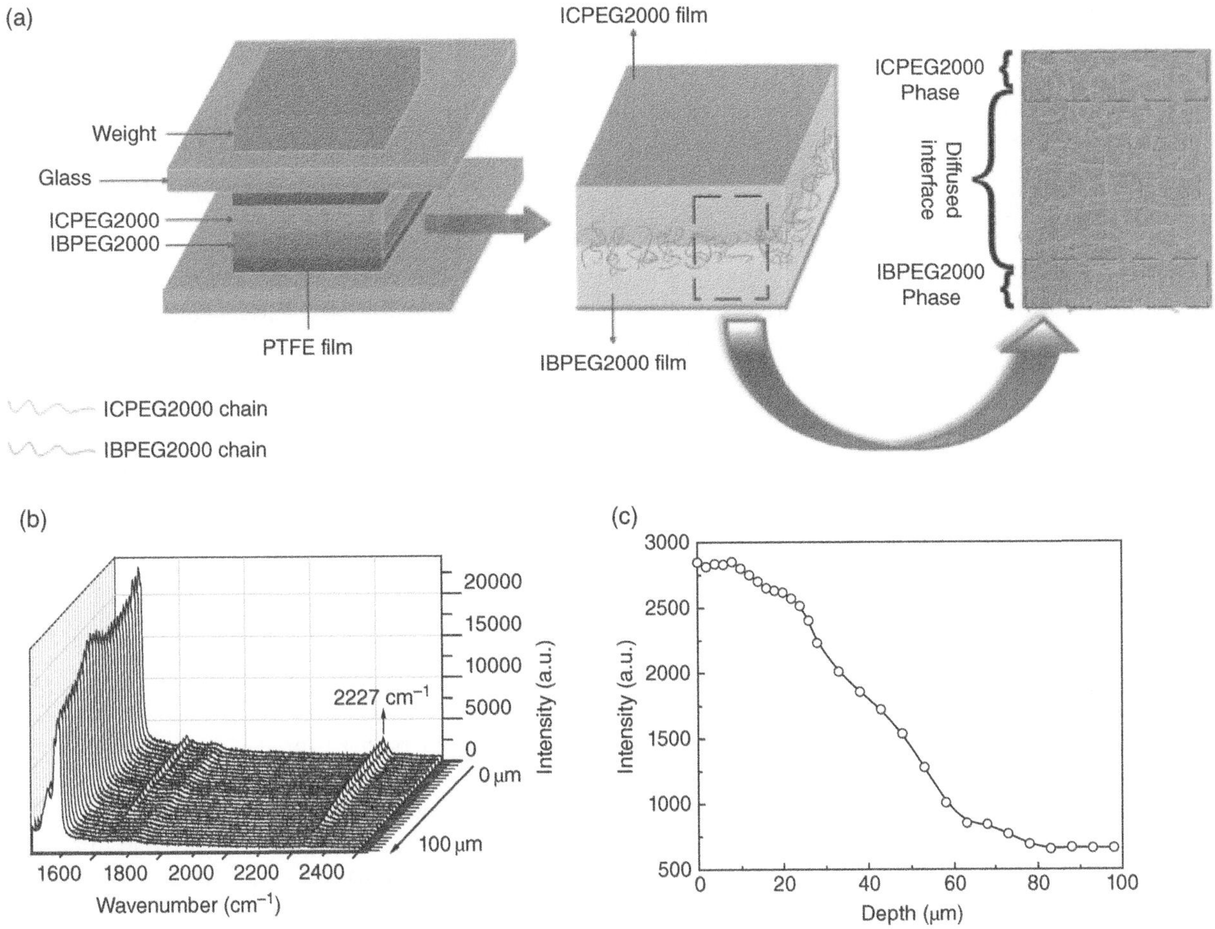

Figure 7.26 (a) Schematic drawing of the sample for depth profile analysis of the interdiffusion phenomenon by using confocal Raman microscopy. (b) Raman spectra collected at the scanning depth from ICPEG2000 (50 μm thick) to IBPEG2000 (50 μm thick) film. (c) Scanning depth dependence of CN peak intensity at 2227 cm^{-1} summarized from (b). *Source:* Reprinted from Zhang et al. [25]. Copyright 2013, with permission from the Royal Society of Chemistry.

C—ON bonds in either CTPO or ICPEG2000 is quite stable and oxytolerant for six months.

7.5 Reversible Shape Memory Polyurethane Network with Intrinsic Self-Healability of Wider Crack

Shape memory polymers (SMPs) have an ability to memorize one or more deformations and recover to permanent shapes from temporary shapes in response to external stimulus. As a result, SMPs can be used as smart materials for advanced applications and have attracted more and more research interest. SMPs are categorized as one-way and two-way versions, depending on whether the deformation is reversible. For one-way SMPs, the transition from temporary shape to permanent shape cannot be repeated by simply reversing the stimulus. A new programming step is necessary to achieve the temporary shape again after recovery.

With respect to two-way (i.e. reversible) SMPs, they can be further classified into two groups: quasi two-way; and two-way [32]. The former takes effect under constant external stress, while the latter does not need the assistance of applied force. There have been an increasing number of reports about external stress-free two-way SMPs probably owing to their convenience.

It is worth noting that in recent years reversible bonds have started to serve as reversible segments in one-way SMPs to replace the melting or glass transitions to fix/release the temporary shape by switching on/off the molecular mobility. The approach broadened the range of base material selection, increased the modes of triggering and brought in additional functionalities. Nevertheless, only very few two-way SMPs take advantage of reversible bonds (i.e. transesterification [33, 34], host–guest inclusion [35], and cinnamates [34]), which mostly act as cross-linkages and have demonstrated attractive performance unavailable for irreversible bonds [33–35].

Previous research in this area indicated that internal stress providers (such as elastomeric network [36, 37], higher melting temperature (T_m) crystallites [38–49], cross-linked elastomer spheres [50], and hydrogen bond network [51]) are necessary for manufacturing free-standing semi-crystalline two-way SMPs. In this work, we propose a novel and simple method to prepare two-way SMP by introducing a new internal stress provider based on reversible dynamic covalent bonds [52]. The key issue lies in coupling alkoxyamine moieties into the main chains of conventional cross-linked PU containing semi-crystalline poly(ε-caprolactone) (PCL) (Schemes 7.13 and 7.14).

It has been known that at certain homolysis temperature the C—ON bonds in alkoxyamines would frequently cleave to create carbon-centered and nitroxide radicals, and the radicals would be immediately recombined. Since recombination of radicals would lead to chain rearrangement [9], significant stress relaxation of the polymer must occur in the meantime as proved by earlier study [53]. In this context, the dynamic reversible C—ON bonds in the cross-linked PU may help to relax the tensile stress applied during programming by making use of the aforementioned dynamic equilibrium. In the case of a suitable homolytic temperature, the amount of the C—ON bonds involved in synchronous bond fission–radical recombination can be limited within an appropriate range. Consequently, the stress imposed on the molecular chains with the C—ON bonds whose dynamicity is initiated can be greatly relaxed, while that on the molecular chains with the C—ON bonds whose dynamicity is not initiated keeps nearly unchanged within the time of interest (Figure 7.27). The greatly relaxed portions tend to maintain the extended status, and the nearly unrelaxed portions tend to recover to the original undeformed shape. The unbalanced internal stress between the partially relaxed and nearly unrelaxed portions becomes the source of the driving force (i.e. internal stress provider), which can cooperate with the reversible melting–crystallization of the crystalline phases and realize two-way shape memory effect (Figure 7.27). Moreover, the dynamic reversibility of the C—ON bonds should also impart shape memory effect aided intrinsic self-healability, reprocessability and reprogrammability to the cross-linked PU as a result of network rearrangement. These smart stimuli-responsive properties are interrelated in nature.

7.5.1 Synthesis

First, hydroxyl-terminated alkoxyamine, 2-(4-hydroxy-2, 2, 6, 6-tetramethylpiperidin-1-yloxy)-*N*-(2-hydroxyethyl)-methylpropanamide, was prepared. Azobis(2-methy-*N*-(2-hydroxyethyl) propionamide) (AMHEPA, 2.88 g, 0.01 mol) and TEMPO (1.72 g, 0.01 mol) were dissolved in dimethyl formamide (DMF, 100 ml). Then, the mixture was heated to 120 °C under stirring in argon for two hours, concentrated in vacuum, and purified by column chromatography

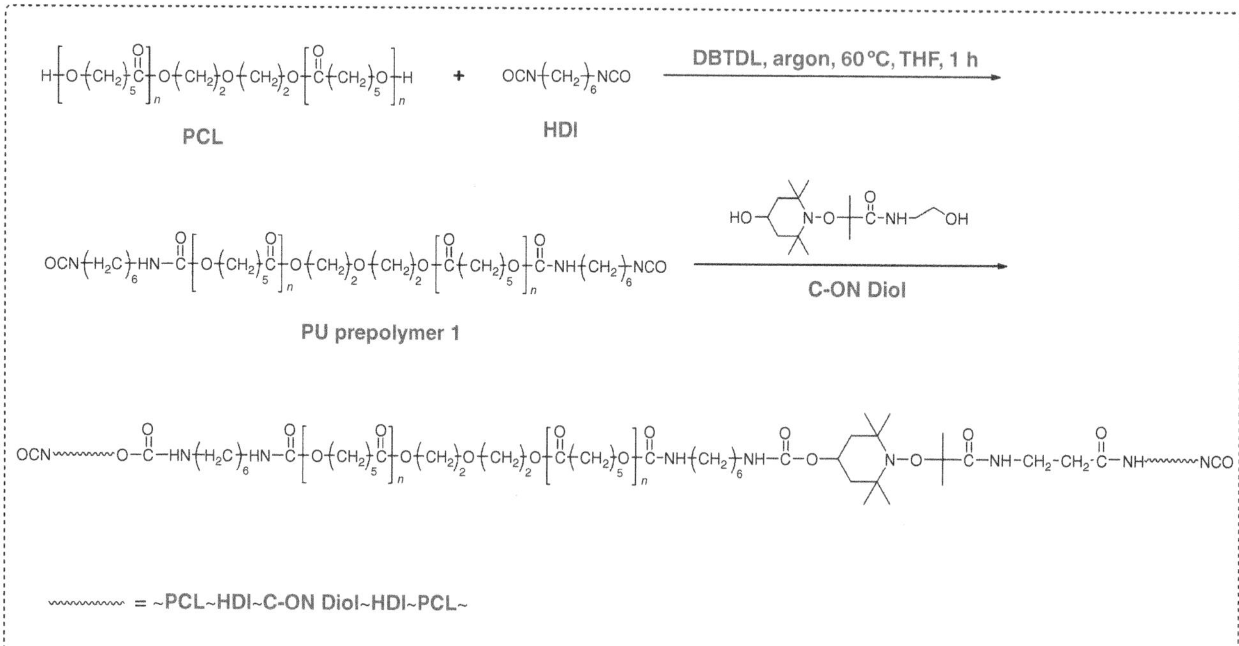

Scheme 7.13 Synthesis of PU prepolymers.

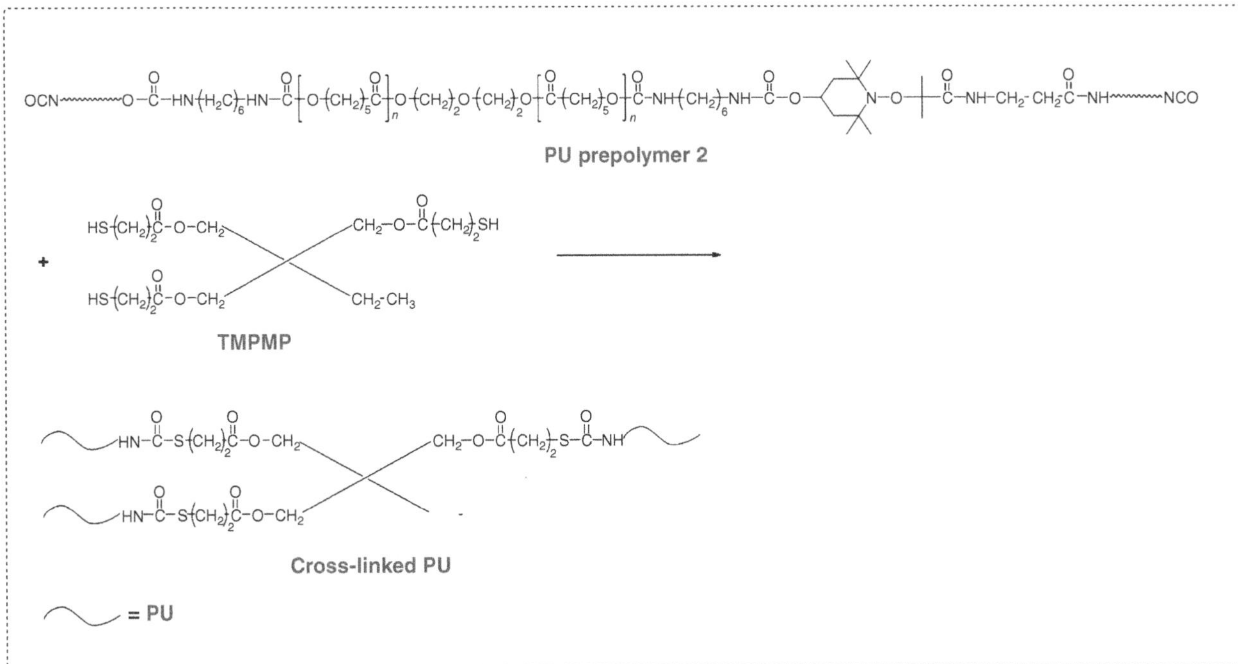

Scheme 7.14 Synthesis of PU prepolymers.

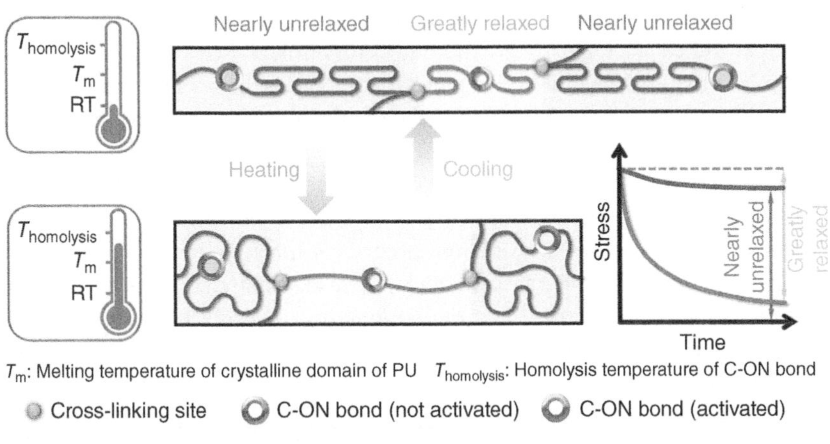

Figure 7.27 Operation principle of the two-way shape memory effect of the programmed cross-linked PU. The symbols representing the C—ON bond (not activated) and C—ON bond (activated) mean whether the dynamicity of the C—ON bond was triggered during programming. When the programmed cross-linked PU is repeatedly heated and cooled within the range between room temperature and onset homolysis temperature of C—ON bond, $T_{homolysis}$ (which is higher than the melting temperature of the crystalline domains), two-way shape memory effect comes into play, and no C—ON bond is triggered. As a result, the two-way shape memory effect is rather stable.

(silica gel) gradient eluted with 5:3 hexane–ethyl acetate (v/v). The resultant was recrystallized from hexane to obtain a primrose yellow liquid with a yield of 74.28%.

Afterwards, PCL (30.00 g, 0.01 mol), hexamethylene diisocyanate (HDI, 3.70 g, 0.022 mol) and two drops of DBTDL were first dissolved in THF. Then, the mixture was heated to 60 °C and stirred in argon for one hour yielding PU prepolymer 1 (Scheme 7.13). The hydroxyl-terminated alkoxyamine (1.45 g, 0.0048 mol) was incorporated as chain extender under stirring for one hour to obtain PU prepolymer 2 (Scheme 7.13) and the cross-linking agent TMPMP (1.91 g, 0.0048 mol) was added for one hour (Scheme 7.14). Eventually, the mixture was poured into a polytetrafluoroethylene (PTFE) mold to remove the solvent at room temperature for 48 hours and the cross-linked PU plate was available.

To highlight the critical role of alkoxyamine, a control sample excluding C—ON bonds was prepared following the above steps, except that hydroxyl-terminated alkoxyamine was replaced by BDO (0.43 g, 0.0048 mol).

To obtain the two-way shape memory effect, the as-synthesized PU specimen was trained following the procedure illustrated in Figure 7.29. First, the dumbbell-shaped specimen of PU was heated to 70 °C for 10 minutes and stretched to 10 times the length of the original one. Next, the specimen was kept at 70 °C for nine days under the constant strain condition in N_2 atmosphere. Afterwards, the specimen was cooled to 25 °C and the applied force was removed. Finally, the specimen was heated to 60 °C for 10 minutes and cooled to 25 °C again.

7.5.2 Characterization

The dynamic reversible bonding manner of the polymer should be identified at the beginning as it determines the expected smart functionalities. Considering the fact that radicals are involved in the reversible fission and recombination of the core units, C—ON bonds, ESR spectra of the material were collected (Figure 7.29). As can be seen, the cross-linked PU has inherited the habit of hydroxyl terminated alkoxyamine. The radical intensity starts to increase after 60 °C due to cleavage of C—ON bonds (Figure 7.29), meaning that bonding and de-bonding of the cross-linked macromolecules become more effective with a rise in temperature. Furthermore, the dynamic bonding is thermally reversible as reflected by the cyclic change in the radical concentration resulting from successive scission of alkoxyamine moieties and recombination of the radicals during thermal cycle between 20 °C and 130 °C (Figure 7.29d). The slight increase of the ESR signal intensity at 20 °C after each cycle suggests that a few alkoxyamines are permanently dissociated due to the combination of carbon free radicals [54]. Nevertheless, the difference between the peak and valley values of the relative intensity keeps almost unchanged.

The data in Figure 7.29c suggest that relatively fewer C—ON bonds take part in the reversible homolysis and recombination when the temperature is 70 °C. Accordingly, the degree and speed of stress relaxation of PU at 70 °C have to be much lower than those at elevated temperature, e.g. 110 °C (Figure 7.30a). Considering the fact that the specimen should be stretched at a temperature > onset temperature of homolysis of C—ON bonds > T_m of PU (i.e. PCL phase) for constructing the internal stress provider, 70 °C seems to be a reasonable value. Otherwise, too many C—ON bonds would turn to be dynamic at higher temperature, leading to obviously weaker mechanical strength of the cross-linked PU and hence raising the difficulty of stretching. This also explains why alkoxyamine does not act as cross-linkage of the PU network but appears in the main chain. Relatively speaking, the disconection–reconnection of alkoxyamine in the main chain would result in slower stress relaxation under identical circumstances.

For reference, Figure 7.30b shows the stress relaxation behaviors of the control material, which proceed in a much milder way than the case including reversible C—ON bonds at the same testing temperature (Figure 7.30a). A characteristic relaxation time at 110 °C of the control material, for example, is 616.9 minutes, much longer than 92.1 minutes of the cross-linked PU with C—ON bonds. The data highlight the effect of accelerated relaxation of reversible C—ON bonds. On the other hand, it should be noted that stress relaxation is evenly conducted for the polymer excluding C—ON bonds, so that the uneven internal stress relaxation in the PU caused by partial activated C—ON bonds would not take place. That is, the planned creation of difference in internal stress between the differently relaxed regions would not be available in the control.

According to the conceptual design described at the start of this chapter and the above experimental results, the programming procedures of the cross-linked PU are worked out for acquiring two-way shape memory effect. The variation in the structure–property relationship in the course of programming is analyzed in the following. It is seen from Figure 7.28 that the cross-linked PU specimen can be favorably elongated at 70 °C (i.e. steps ① and ②). In this case the cross-linking network in the specimen gradually changed from isotropic to highly oriented because of the synergistic effect of the melted PCL phase and fission of C—ON bonds [55]. In step ③, the elongated specimen was kept at 70 °C for nine days in N_2 atmosphere, and the applied force, which was used to stretch the specimen to 10 times the original length, sluggishly relaxed as only a few C—ON bonds are engaged in reversible homolysis and

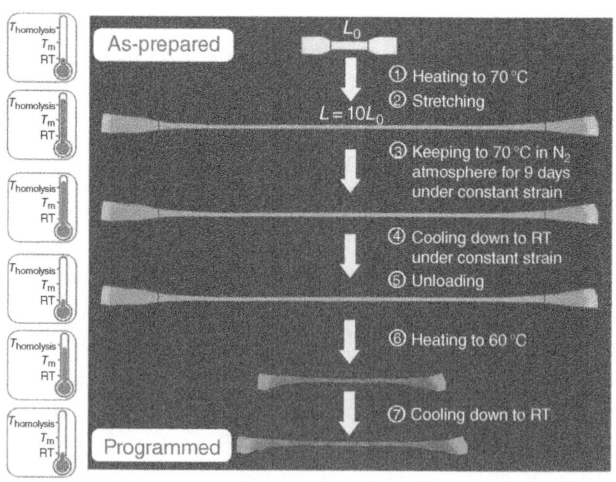

Figure 7.28 Programming of the cross-linked PU toward two-way shape-memory effect.

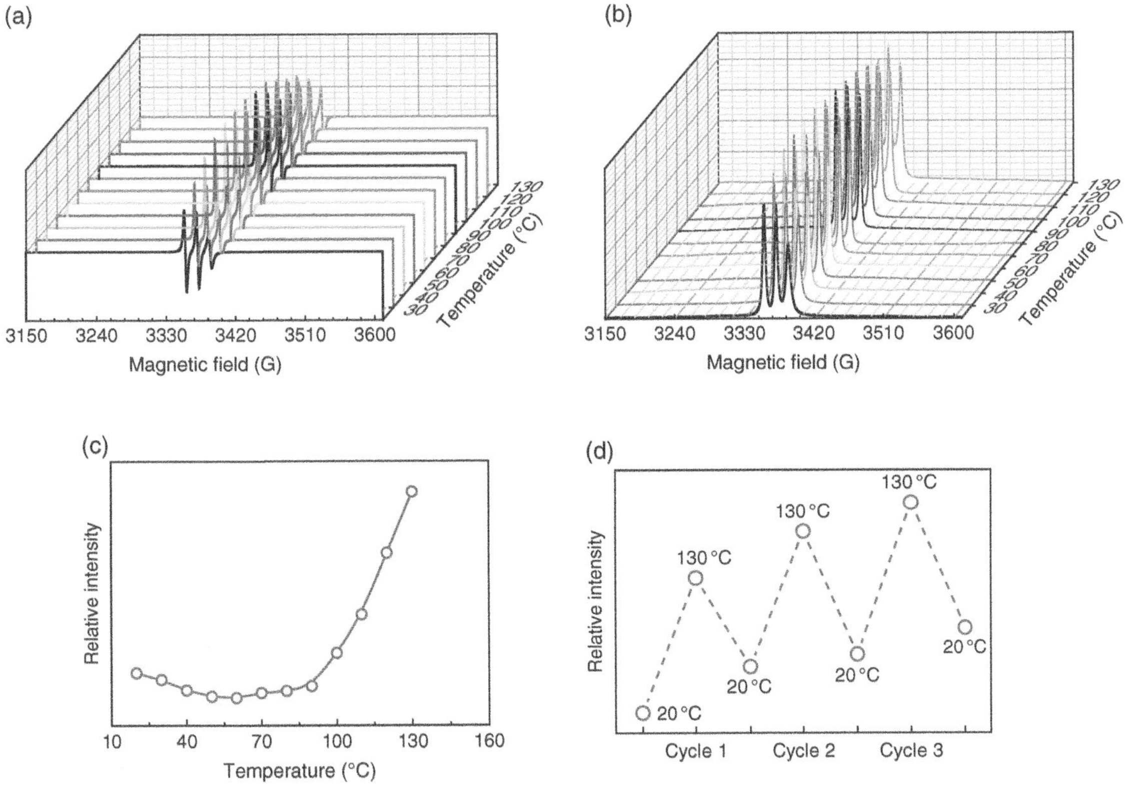

Figure 7.29 (a) Typical ESR spectra of the cross-linked PU collected at various temperatures. (b) Normalized absorption lines converted from the data of (a). (c) Temperature dependence of relative ESR signal intensities calculated from integral areas of the normalized absorption curves in (b). (d) Relative ESR signal intensities measured during heating–cooling cycles between 20 °C and 130 °C.

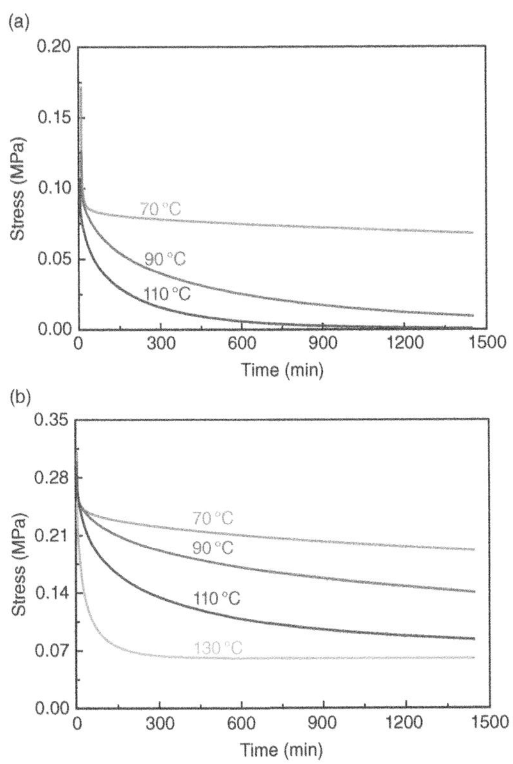

Figure 7.30 Typical stress relaxation curves of (a) cross-linked PU and (b) the control measured at different temperatures.

recombination (Figure 7.29). After the nine-day relaxation, the PU network consisted of two parts (greatly relaxed and nearly unrelaxed) depending on where the dynamic C—ON bonds were randomly triggered. The subsequent cooling at the fixed strain (i.e. step ④) led to oriented crystalline phases from the melted PCL and freeze of the dynamic C—ON bonds. When the applied force was removed at room temperature (step ⑤), the temporary shape of the specimen remained unchanged because of insufficient mobility of the macromolecules. After the second heating to 60 °C (i.e. step ⑥), the oriented crystals melted and transformed into entropically favorable random coils, so that the specimen length became shortened as a result of entropy increase induced contraction. Meanwhile, the nearly unrelaxed part of the PU network tended to recover to the original size of the specimen (i.e. L_0), and the greatly relaxed part of the PU network tended to maintain the metastable status at a strain of 1000% (i.e. $10L_0$) owing to its memory of the temporary shape caused by stress relaxation. The competition between the two opposite tendencies plus the contraction due to melting of PCL eventually resulted in the ultimate specimen length being reduced, but was still longer than the original one. It is clear that the internal stress provided by the greatly relaxed part of the PU network is higher than that by the nearly unrelaxed part, and the net internal stress is tension headed in the same direction of the former. Lastly, the specimen was cooled to room temperature once more offering the programmed version (i.e. step ⑦), and oriented crystalline PCL was recreated under the internal tension. Naturally, the orientation degree of the crystallites was lower than the case subjected to external stress after steps ④ and ⑤, as the internal tension was somewhat offset by the expansion originating from the recrystallization.

Having been trained, the cross-linked PU is able to show two-way shape memory effect free of external stress. As shown in Figure 7.31, an increase of temperature of the specimen (higher than 49.4 °C [T_m of PCL phase] but lower than ~60 °C [onset temperature of homolysis of C—ON

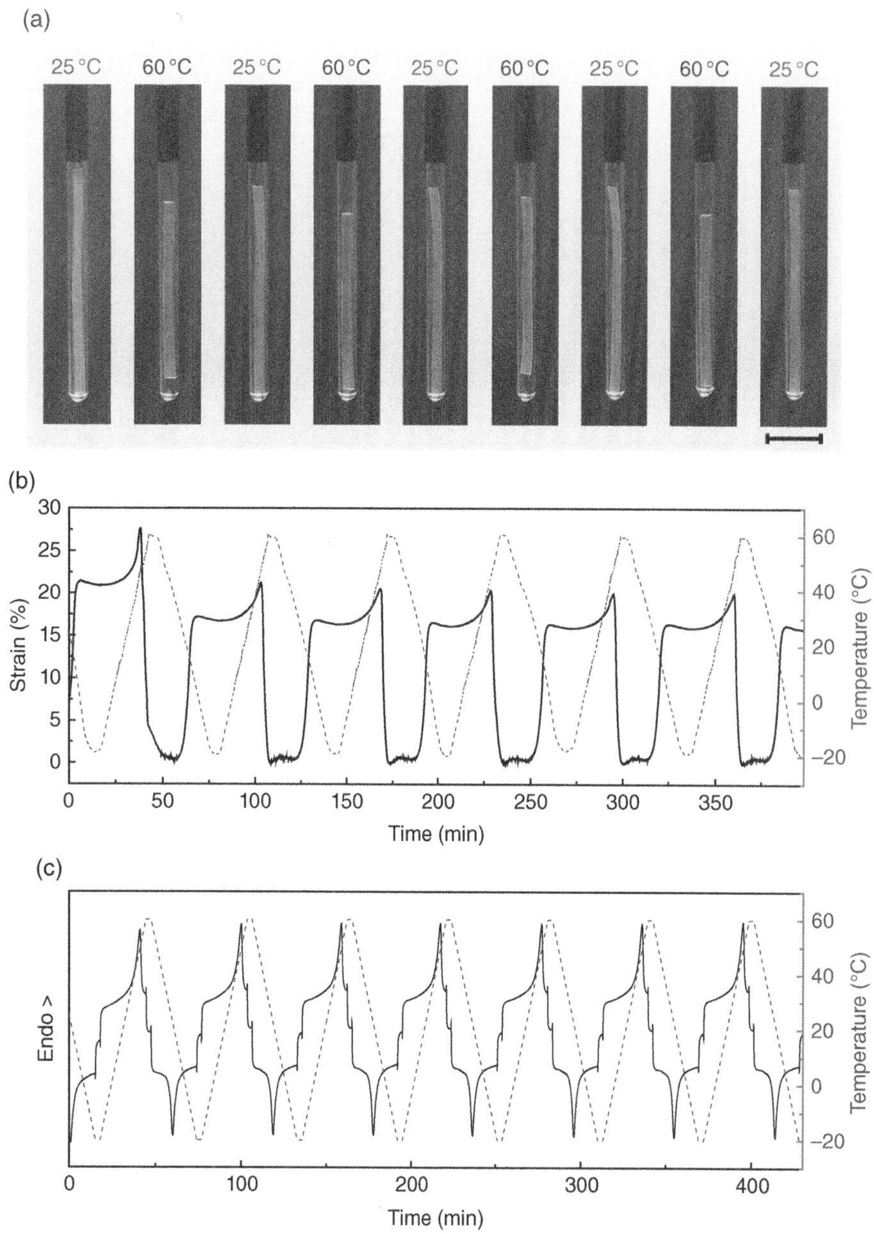

Figure 7.31 (a) Length variation of the programmed cross-linked PU during heating–cooling cycles between 25 °C and 60 °C. The scale bar represents 2 cm in length. (b) Thermomechanical behavior (measured by DMA) and (c) thermogram (measured by DSC) of the programmed PU during repeated heating and cooling cycles between −20 °C and 60 °C (heating rate: 3 °C min^{-1}). The first cycle that was used to remove the thermal history is not displayed for clarity. *Source:* (a) Fan et al. [52], Figure 7, p. 8/with permission of Royal Society of Chemistry. (b, c) Reprinted from Fan et al. [52]. Copyright 2018, with permission from Royal Society of Chemistry.

bonds]) led to melting of the oriented crystalline domains, so that macroscopic shrinkage was detected, which is accompanied by a rise in the internal tension due to the effect of reaction. When the specimen was cooled to room temperature, the oriented crystalline domains were regenerated under the guidance of the internal stress provider, and the latter became less tense. The cyclic melting–recrystallization of the crystalline component works together with the variation in the tension state (less tight and tighter tension) of the internal force provider, ensuring repeated shape transformation of the PU in the absence of external force (Figure 7.27). Therefore, the length of the specimen with two free ends varied vertically up and down (Figure 7.31a). The average reversible strain based on the first three heating and cooling cycles is found to be 23.0 ± 3.64% (Figure 7.31b). An additional test further shows that the orientation degree of the crystalline phases after seven heating–cooling cycles is the same as that of the as-programmed one, which reflects the robustness of the two-way shape memory effect and of the internal stress provider as well.

In addition, the internal stress that helps to implement the two-way shape memory effect can be quantified by elimination of the programing history through heat treatment (Figure 7.32). When the programmed cross-linked PU specimen was heated, its strain negatively increased due to the shrinkage of the two-way shape memory effect and a short plateau appeared at about 60 °C. This means that melting of the crystalline PCL is completed in this case, and the internal stress (tension) maximizes, equilibrating with the contraction force offered by the crystallite melting. By further raising temperature, the dynamicity of the C—ON bonds was triggered and the internal stress stored in the programmed PU network was released. Accordingly, the strain of the specimen kept on decreasing.

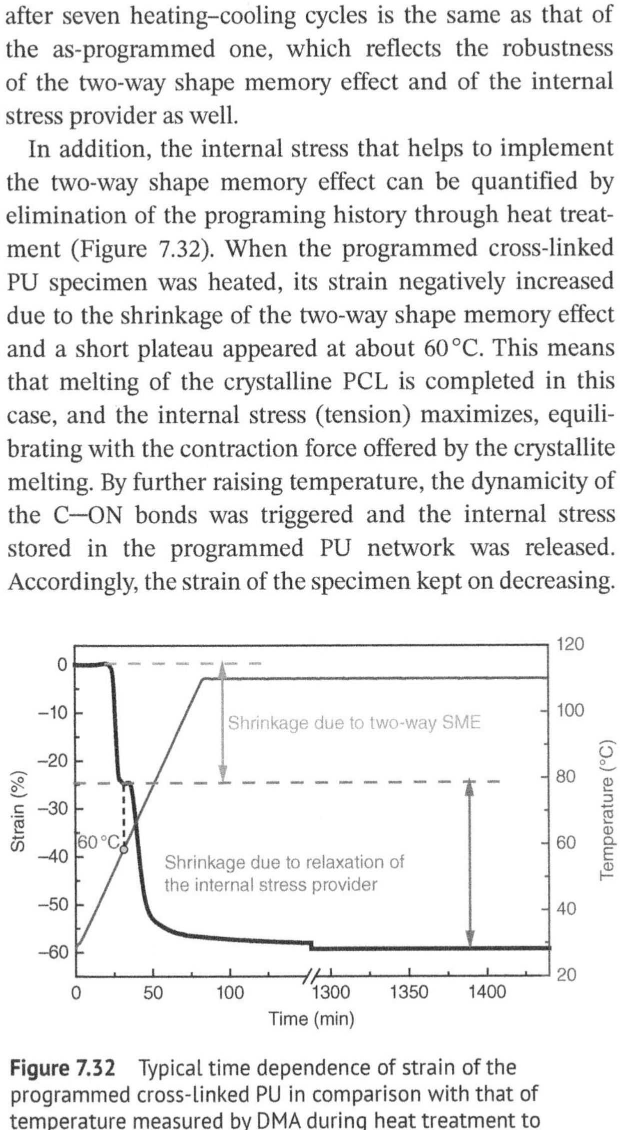

Figure 7.32 Typical time dependence of strain of the programmed cross-linked PU in comparison with that of temperature measured by DMA during heat treatment to eliminate the training history (heating rate: 3 °C min^{-1}).

As the stress relaxation tests indicate that the cross-linked PU can completely relax the applied stress within 24 hours at 110 °C (Figure 7.30a), the heat treatment for elimination of the programing history was also maintained at 110 °C finally. The strain at 24 hours is thus regarded as the smallest value that can be reached by the de-programmed specimen (i.e. the specimen length returned to original one, L_o). In other words, the strain (−24.82% − (−59.32 %) = 34.50%) should result from the relaxation of the internal stress provider. By reviewing the tensile stress–strain curve of the cross-linked PU measured at 60 °C, a stress of 0.26 MPa is found to be required for generating the strain of 34.50%. Therefore, the highest tension offered by the internal stress provider in the programed cross-linked PU is determined to be 0.26 MPa.

Our recent study demonstrated that two-way shape memory effect can be imparted to cross-linked crystalline PU by simply converting the inherent hydrogen bond network into the internal stress provider [51]. The key steps are (i) the stretching at 70 °C, which is higher than T_m of PU and the dissociation temperature of hydrogen bonds (~60 °C), and (ii) the subsequent cooling. Owing to the fact that (i) the cross-linked PU of the present work has similar chemical structure to the former [51] except alkoxyamine moieties are included in the present version, and (ii) the programming procedures and conditions are also similar in the two works, the hydrogen bond network must partly serve as internal stress provider here. In this context, the control PU in which BDO has replaced hydroxyl-terminated alkoxyamine was trained following the same steps described in Figure 7.28 and then de-programmed using the approach suggested in ref. [51]. It is thus known that the internal stress generated by the inherent hydrogen bond network is 0.10 MPa, and the net contribution of the alkoxyamine-induced internal stress equals 0.26–0.10 MPa = 0.16 MPa. The higher internal stress of the latter may result from the higher bond energy of C—ON bonds than that of hydrogen bonds. Besides, compared with the cross-linked PU excluding alkoxyamine moieties, whose crystallinity and average reversible strain due to two-way shape memory effect are 32.9 and 19.9 ± 0.25%, respectively, the current PU has a crystallinity and average reversible strain of 24.3 and 23.0 ± 3.64%, respectively. Clearly, internal stress, rather than crystallinity, is more important for producing high reversible strain.

In fact, the time required for stabilizing the large strain at the beginning of programming (training time for short, refer to step ③ of Figure 7.28) greatly influences the two-way shape memory effect. With a rise in the training time, the average reversible strain first increases and then decreases. The peak value appears at nine days. As the amount of hydrogen bonds would not change with the training time at a constant temperature below the maximum dissociation temperature (~150 °C [51]), the variation must be related to the stress relaxation led by dynamic C—ON bonds. That is, the training time equates with relaxation time. In the case of shorter training time (relaxation time) at 70 °C, only very few macromolecular chains containing C—ON bonds are involved in the relaxation and the net internal tension is not high enough to match the driving force given by the reversible segments, i.e. the crystalline PCL phases. Therefore, the average reversible strain increases with increasing the training time and the fraction of the greatly relaxed portions. When the training time (relaxation time) at 70 °C is excessively long, the net internal tension is much higher than the contraction force of the melted crystalline phases. The average reversible strain has to decrease with further increase in the training time.

Although the programmed cross-linked PU can be de-programmed by heating the specimen to 110 °C (Figure 7.32), the two-way shape memory effect can be readily regained after re-programming following the same procedures (Figure 7.28). This is because the mechanism of shape memory effect is predominated by the dynamic reversibility of the C—ON bonds in the cross-linked PU, as discussed above.

Figure 7.29 indicates that many more C—ON bonds become dynamically reversible above 100 °C, meaning that the polymer should be able to be self-repaired through crossover reactions between radicals that belong to different chains across the interface. To verify this, fractured surfaces of a fully cut dog-bone shaped specimen of the cross-linked PU were manually brought into contact for rehabilitation at 100 °C for a certain time. Then, the healed specimen was measured by tensile test. As shown in Figure 7.33a, the damaged cross-linked PU specimen can indeed regain its tensile strength. The healing efficiency characterized in terms of strength restoration increases with increasing healing time, and reaches the highest value of 92.3% at the healing time of 24 hours.

Furthermore, the specimen can be self-healed multiple times (Figure 7.33b), which is typical for intrinsic self-healing. Comparatively, the control specimen without alkoxyamine possesses very low healing efficiency (24.0%) contributed by entanglements of dangling chains and hydrogen bonds. The results demonstrate the exclusive role of C—ON bonds in the self-healing.

Returning to Figure 7.32, it is seen that when temperature is raised from room temperature to 110 °C, the programmed cross-linked PU specimen undergoes shrinkage due to (i) two-way shape memory effect (up to 60 °C) and (ii) internal stress relaxation (from 60 °C to 110 °C). Such a large contraction strain of 59.32% should be sufficient for closing a wider crack if the material is damaged. More interestingly, since the PU has proved to be self-healable at 100 °C (Figure 7.33), intrinsic self-healing of a wider crack should occur upon closure of the crack as a result of the shape memory effect. Otherwise, no interaction across the cracked surfaces would take place in the absence of intimate contact [2]. Figure 7.34 shows that for a specimen with a wider crack (~250 μm wide and ~0.5 mm deep), when it was heated to 100 °C, the crack was closed with contraction of the specimen. After heating at 100 °C for 24 hours, intrinsic self-healing took place by taking advantage of the intimate contact of cracked surfaces that created conditions for the C—ON bonds on the opposites sides to react with one another. Almost full recovery of tensile strength was perceived. The same results can be obtained when cracks with a similar size were repeatedly created at the same position on the same specimen after healing and re-programming. It means that wider cracks can indeed be narrowed enough owing to the shape memory effect induced contration, accommodating reversible reactions among C—ON bonds from opposite surfaces.

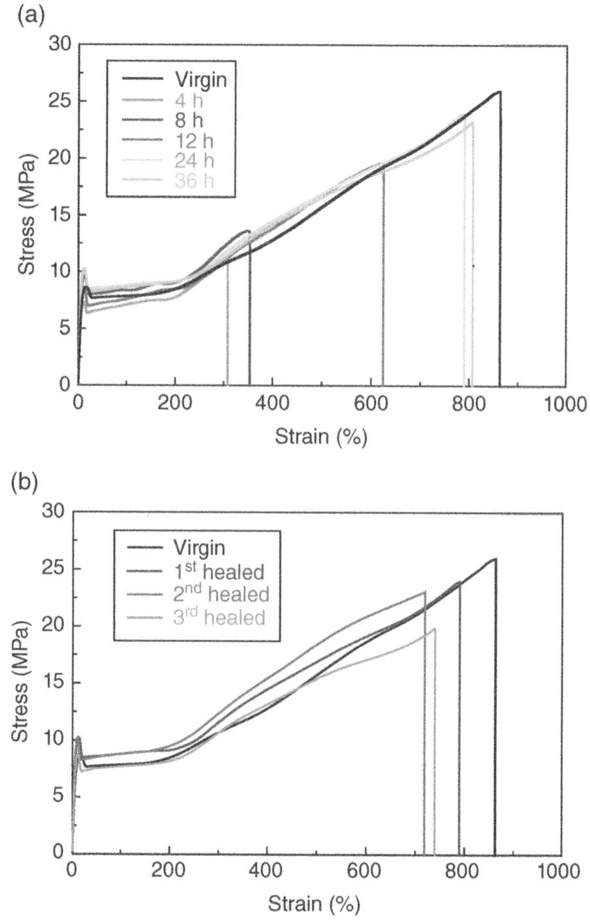

Figure 7.33 Typical tensile stress–strain curves of the cross-linked PU healed (a) at different healing times and (b) for multiple times. Healing temperature: 100 °C; healing time for (b): 24 hours. *Source:* Reprinted from Fan et al. [52]. Copyright 2018, with permission from the Royal Society of Chemistry.

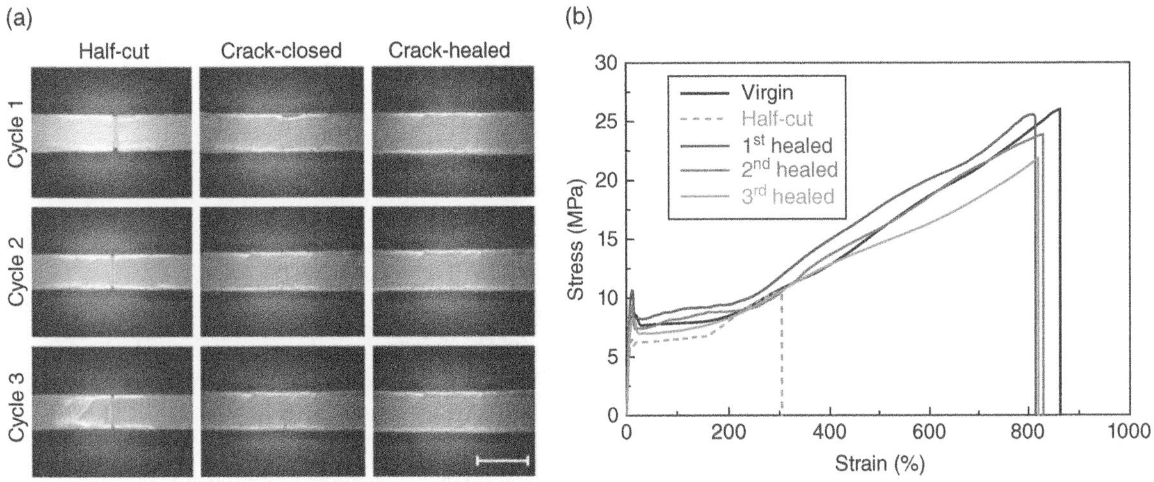

Figure 7.34 Shape memory effect assisted intrinsic self-healing of a wider crack. (a) Photos showing repeated cycles of damaging/crack closure/crack healed. The scale bar represents 5 mm in length. (b) Typical tensile stress–strain curves of the cross-linked PU repeatedly healed with the assistance of shape memory effect (crack size: ~250 μm wide and ~0.5 mm deep). *Source:* (a) Fan et al. [52], Figure 10, p. 10/with permission of Royal Society of Chemistry. (b) Reprinted from Fan et al. [52]. Copyright 2018, with permission from the Royal Society of Chemistry.

In contrast, the unprogrammed specimen with similar half-cut damage can neither autonomously close the crack at 100 °C nor restore the mechanical properties to a high extent. The marginal healing efficiency of 26.5% may result from the healing of the V-shaped crack tip via hydrogen bonding. The experiments further demonstrate the important role of C—ON bonds in allowing shape memory effect and re-bonding of a wider crack.

7.6 Concluding Remarks

Because C—ON bonds in alkoxyamines frequently cleave but immediately recombine at certain homolysis temperature, the two-step healing strategy via DA bonds can be simplified to a single-step one in the case of self-healing based on C—ON bonds. It is conducive to maintain structural stability of products during service. The experimental results have revealed that the polymers carrying alkoxyamine moieties as intra- or intermacromolecular linkages are allowed for self-healing. In addition, self-healing at either low temperature (15 °C and 25 °C) [25] or elevated temperature (130 °C) [20] offers high healing efficiencies.

In general, the factors (i.e. electronic effects, conjugative effect, and steric effect) that are able to reduce the bond energy of C—ON bonds would increase the stability of carbon-centered radicals and decrease the homolysis temperature of C—ON bonds. For example, the homolysis temperature of the C—ON bond has been successfully manipulated through the electron-withdrawing effect of neighboring substituents [25, 54, 56]. The results indicated that with enhancement of the electron-withdrawing effect (e.g. from amido to cyano), the homolysis temperature of the C—ON bond was decreased.

Neveletheless, continuous efforts are needed to hinder the reversible coupling between carbon free radicals generated by homolysis of C—ON bonds, which reduces reversibility of the C—ON bonds.

References

1 Plaisted, T.A. and Nemat-Nasser, S. (2007). Quantitative evaluation of fracture, healing and re-healing of a reversibly cross-linked polymer. *Acta Materialia* 55: 5684–5696.
2 Wool, R.P. and O'Connor, K.M. (1981). A theory of crack healing in polymers. *Journal of Applied Physics* 52: 5953–5963.
3 Veregin, R.P.N., Georges, M.K., Kazmaier, P.M., and Hamer, G.K. (1993). Free radical polymerizations for narrow polydispersity resins: electron spin resonance studies of the kinetics and mechanism. *Macromolecules* 26: 5316–5320.
4 Veregin, R.P.N., Georges, M.K., Hamer, G.K., and Kazmaier, P.M. (1995). Mechanism of living free radical polymerizations with narrow polydispersity: electron spin resonance and kinetic studies. *Macromolecules* 28: 4391–4398.
5 Rowan, S.J., Cantrill, S.J., Cousin, G.R.L. et al. (2002). Dynamische kovalente Chemie. *Angewandte Chemie International Edition* 114: 938–993. (in German).

6 Rowan, S.J., Cantrill, S.J., Cousin, G.R.L. et al. (2002). Dynamic covalent chemistry. *Angewandte Chemie International Edition* 41: 898–952.
7 Mannan, A., Ichikawa, A., and Miura, Y. (2007). Living radical polymerization of styrene mediated by a piperidinyl-N-oxyl radical having very bulky substituents. *Polymer* 48: 743–749.
8 Hawker, C.J., Bosman, A.W., and Harth, E. (2001). New polymer synthesis by nitroxide mediated living radical polymerizations. *Chemical Reviews* 101: 3661–3688.
9 Otsuka, H., Aotani, K., Higaki, Y., and Takahara, A. (2003). Polymer scrambling: macromolecular radical crossover reaction between the main chains of alkoxyamine-based dynamic covalent polymers. *Journal of the American Chemical Society* 125: 4064–4065.
10 Georges, M.K., Veregin, R.P.N., Kazmaier, P.M., and Hamer, G.K. (1993). Narrow molecular weight resins by a free-radical polymerization process. *Macromolecules* 26: 2987–2988.
11 Miura, Y., Nakamura, N., and Taniguchi, L. (2001). Low-temperature “living” radical polymerization of styrene in the presence of nitroxides with spiro structures. *Macromolecules* 34: 447–455.
12 Wu, D.Y., Meure, S., and Solomon, D. (2008). Self-healing polymeric materials: a review of recent development. *Progress in Polymer Science* 33: 479–522.
13 Higaki, Y., Otsuka, H., and Takahara, A. (2004). Dynamic formation of graft polymers via radical crossover reaction of alkoxyamines. *Macromolecules* 37: 1696–1701.
14 Amamoto, Y., Higaki, Y., Matsuda, Y. et al. (2007). Programmed thermodynamic formation and structure analysis of star-like nanogels with core cross-linked by thermally exchangeable dynamic covalent bonds. *Journal of the American Chemical Society* 129: 13298–13304.
15 Yamaguchi, G., Higaki, Y., Otsuka, H., and Takahara, A. (2005). Reversible radical ring-crossover polymerization of an alkoxyamine-containing dynamic covalent macrocycle. *Macromolecules* 38: 6316–6320.
16 Higaki, Y., Otsuka, H., and Takahara, A. (2006). A thermodynamic polymer cross-linking system based on radically exchangeable covalent bonds. *Macromolecules* 39: 2121–2125.
17 Wang, F. (2012). Mechanisms and application of thermally reversible self-healing polymers based on TEMPO derivatives. PhD thesis. Sun Yat-sen University, Guangzhou.
18 Wang, F., Rong, M.Z., and Zhang, M.Q. (2012). Reversibility of solid state radical reactions in thermally remendable polymers with C–ON bonds. *Journal of Materials Chemistry* 22: 13076–13084.
19 Otsuka, H., Aotani, K., Higaki, Y. et al. (2007). Thermal reorganization and molecular weight control of dynamic covalent polymers containing alkoxyamines in their main chains. *Macromolecules* 40: 1429–1434.
20 Yuan, C.E., Rong, M.Z., Zhang, M.Q. et al. (2011). Self-healing of polymers via synchronous covalent bond fission/radical recombination. *Chemistry of Materials* 23: 5076–5081.
21 Hawker, C.J., Barclay, G.G., and Dao, J. (1996). Radical crossover in nitroxide mediated “living” free radical polymerizations. *Journal of the American Chemical Society* 118: 11467–11471.
22 Sylvain, M., Christophe, L.M., Paul, T., and Hanns, F. (2000). Factors influencing the C–O–bond homolysis of trialkylhydroxylamines. *Macromolecules* 33: 4403–4410.
23 Yuan, C.E., Zhang, M.Q., and Rong, M.Z. (2014). Application of alkoxyamine in self-healing of epoxy. *Journal of Materials Chemistry A* 2: 6558–6566.
24 Rossi, N.A.A., Wang, Q., Amine, K., and West, R. (2010). Silicon-containing carbonates-synthesis, characterization, and additive effects for silicon-based polymer electrolytes. *Silicon* 2: 201–208.
25 Zhang, Z.P., Rong, M.Z., Zhang, M.Q., and Yuan, C.E. (2013). Alkoxyamine with reduced homolysis temperature and its application in repeated autonomous self-healing of stiff polymers. *Polymer Chemistry* 4: 4648–4654.
26 Kramer, A. and Nesvadba, P. (2002). 1-Alkoxy-polyallkyl-piperidine derivatives and their use as polymerization regulators. US Patent 6353107 B1.
27 Wang, D.K. and Wu, Z. (1998). Facile synthesis of new unimolecular initiators for living radical polymerizations. *Macromolecules* 31: 6727–6729.
28 Hayes, S.A., Zhang, W., and Jones, F.R. (2007). Self-healing of damage in fibre-reinforced polymer-matrix composites. *Journal of the Royal Society Interface* 4: 381–387.
29 Pasto, D.J. (1988). Radical stabilization energies of disubstituted methyl radicals. A detailed theoretical analysis of the captodative effect. *Journal of the American Chemical Society* 110: 8164–8175.
30 Pasto, D.J., Krasnansky, R., and Zercher, C. (1987). Stabilization energies and structures of substituted methyl radicals. *The Journal of Organic Chemistry* 52: 3062–3072.
31 Hu, C.L., Chen, X.D., Chen, J. et al. (2012). Observation of mutual diffusion of macromolecules in PS/PMMA binary films by confocal Raman microscopy. *Soft Matter* 8: 4780–4788.
32 Huang, M.M., Dong, X., Liu, W.L. et al. (2017). Recent progress in two-way shape memory crystalline polymer and its composites. *Acta Polymerica Sinica* 4: 563–579.
33 Pei, Z., Yang, Y., Chen, Q. et al. (2014). Mouldable liquid-crystalline elastomer actuators with exchangeable covalent bonds. *Nature Materials* 13: 36–41.

34 Jin, B., Song, H., Jiang, R. et al. (2018). Programming a crystalline shape memory polymer network with thermo- and photo-reversible bonds toward a single-component soft robot. *Science Advances* 4: 3865–3870.
35 Miyamae, K., Nakahata, M., Takashima, Y., and Harada, A. (2015). Self-healing, expansion–contraction, and shape-memory properties of a preorganized supramolecular hydrogel through host–guest interactions. *Angewandte Chemie International Edition* 54: 8984–8987.
36 Wu, Y., Hu, J., Han, J. et al. (2014). Two-way shape memory polymer with "switch–spring" composition by interpenetrating polymer network. *Journal of Materials Chemistry A* 2: 18816–18822.
37 Meng, Y., Jiang, J., and Anthamatten, M. (2015). Shape actuation via internal stress-induced crystallization of dual-cure networks. *ACS Macro Letters* 4: 115–118.
38 Behl, M., Kratz, K., Zotzmann, J. et al. (2013). Reversible bidirectional shape-memory polymers. *Advanced Materials* 25: 4466–4469.
39 Wang, K., Jia, Y.G., and Zhu, X.X. (2017). Two-way reversible shape memory polymers made of cross-linked cocrystallizable random copolymers with tunable actuation temperatures. *Macromolecules* 50: 8570–8579.
40 Gao, Y., Liu, W.F., and Zhu, S. (2017). Polyolefin thermoplastics for multiple shape and reversible shape memory. *ACS Applied Materials & Interfaces* 9: 4882–4889.
41 Bothe, M. and Pretsch, T. (2013). Bidirectional actuation of a thermoplastic polyurethane elastomer. *Journal of Materials Chemistry A* 1: 14491–14497.
42 Behl, M., Kratz, K., Noechel, U. et al. (2013). Temperature-memory polymer actuators. *Proceedings of the National Academy of Sciences of the United States of America* 110: 12555–12559.
43 Zhou, J., Turner, S.A., Brosnan, S.M. et al. (2014). Shapeshifting: reversible shape memory in semicrystalline elastomers. *Macromolecules* 47: 1768–1776.
44 Turner, S.A., Zhou, J., Sheiko, S.S., and Ashby, V.S. (2014). Switchable micropatterned surface topographies mediated by reversible shape memory. *ACS Applied Materials & Interfaces* 6: 8017–8021.
45 Gong, T., Zhao, K., Wang, W.X. et al. (2014). Thermally activated reversible shape switch of polymer particles. *Journal of Materials Chemistry B* 2: 6855–6866.
46 Saatchi, M., Behl, M., Nochel, U., and Lendlein, A. (2015). Copolymer networks from oligo(ε-caprolactone) and n-butyl acrylate enable a reversible bidirectional shape-memory effect at human body temperature. *Macromolecular Rapid Communications* 36: 880–884.
47 Li, Q.X., Zhou, J., Vatankhah-Varnoosfaderani, M. et al. (2016). Advancing reversible shape memory by tuning the polymer network architecture. *Macromolecules* 49: 1383–1391.
48 Zharinova, E., Heuchel, M., Weigel, T. et al. (2016). Water-blown polyurethane foams showing a reversible shape-memory effect. *Polymers* 8: 412/1–412/22.
49 Yang, G., Liu, X., Tok, A.Y., and Lipik, V. (2017). Body temperature-responsive two-way and moisture-responsive one-way shape memory behaviors of poly(ethylene glycol)-based networks. *Polymer Chemistry* 8: 3833–3840.
50 Fan, L.F., Rong, M.Z., Zhang, M.Q., and Chen, X.D. (2017). A very simple strategy for preparing external stress-free two-way shape memory polymers by making use of hydrogen bonds. *Macromolecular Rapid Communications* 38: 1700124/1–1700124/7.
51 Fan, L.F., Rong, M.Z., Zhang, M.Q., and Chen, X.D. (2018). A very simple strategy for preparing external stress-free two-way shape memory polymers by making use of hydrogen bonds. *Macromolecular Rapid Communications* 39: 1700714/1–1700714/7.
52 Fan, L.F., Rong, M.Z., Zhang, M.Q., and Chen, X.D. (2018). Dynamic reversible bonds enable external stress-free two-way shape memory effect of a polymer network and the interrelated intrinsic selfhealability of wider crack and recyclability. *Journal of Materials Chemistry A* 6: 16053–16063.
53 Scott, T.F., Draughon, R.B., and Bowman, C.N. (2006). Actuation in crosslinked polymers via photoinduced stress relaxation. *Advanced Materials* 18: 2128–2132.
54 Zhang, Z.P., Lu, Y., Rong, M.Z., and Zhang, M.Q. (2016). A thermally remendable and reprocessable crosslinked methyl methacrylate polymer based on oxygen insensitive dynamic reversible C–ON bonds. *RSC Advances* 6: 6350–6357.
55 Zhang, Z.P., Rong, M.Z., and Zhang, M.Q. (2018). Mechanically robust, self-healable, and highly stretchable "living" crosslinked polyurethane based on a reversible C-C bond. *Advanced Functional Materials* 28: 1706050/1–1706050/8.
56 Zhang, Z.P., Rong, M.Z., and Zhang, M.Q. (2014). Room temperature self-healable epoxyelastomer with reversible alkoxyamines as crosslinkages. *Polymer* 55: 3936–3943.

8

Intrinsic Self-Healing Via Exchange Reaction of the Disulfide Bond

CHAPTER MENU

The disulfide bond is one of the most widely used dynamic linkages in self-healing polymers, which combines both stimulus diversity (e.g. heat, ultraviolet [UV] light, and sunshine) and the flexibility of reaction mechanism (e.g. base-catalyzed disulfide exchange, metal ion-catalyzed disulfide exchange, thiol–disulfide exchange, and radical-mediated exchange). Like the C—ON bond discussed in Chapter 7, the reversible exchange reaction of disulfide bonds also proceeds in a one-step fashion, which benefits self-healing polymers in practice.

In the pioneering work by Klumperman and co-workers, a commercially available epoxy polymer containing disulfide groups with a tetrafunctional thiol was cured [1]. They found that the resulting material was able to fully restore its tensile failure strain at 60 °C. They attributed this recovery to the renewal of cross-links across damaged surfaces due to exchange reactions of disulfide groups. In a later paper by the same group [2], the authors further investigated the healing process and concluded that the dominant mechanism is not the disulfide–disulfide exchange reaction but rather the thiol–disulfide exchange reaction. This work also determined that the thiol groups are readily oxidized, leading to fewer possibilities for thiol–disulfide exchange to occur. Later, Lafont et al. synthesized similar elastomers and demonstrated the effects of thiol cross-linkers on the cohesive and adhesive self-healing ability of these materials at 65 and 100 °C [3]. Again, full restoration of the properties was observed. Rekondo et al. replaced the aliphatic disulfide with an aromatic disulfide to obtain a cross-linked polyureaurethane [4] and found that the resulting thermoset elastomer was capable of self-healing at room temperature, as demonstrated by a healing efficiency of 97% (51% of which was due to the contribution of hydrogen bonding).

On the basis of the earlier work in this field, we carried out a series of studies accordingly. For example, we developed polymers containing disulfide bonds capable of self-healing with high healing efficiency at room temperature [5]. Our goal was to mitigate many of the deficiencies of the existing approaches, including the thiol–disulfide exchange that is oxygen-sensitive and requires heating, and the relatively low healing efficiencies associated with aromatic disulfide exchange.

8.1 Room-Temperature Self-Healable and Remoldable Cross-Linked Polysulfide

The disulfide metathesis of low-molecular-weight compounds at room temperature has been extensively studied. Caraballo et al. [6] employed alkyl and aryl phosphines as catalysts to accelerate the disulfide metathesis of model aliphatic and aromatic disulfides at ambient temperature. Compared with triphenylphosphine, tricyclohexylphosphine

Extrinsic and Intrinsic Approaches to Self-Healing Polymers and Polymer Composites, First Edition. Ming Qiu Zhang and Min Zhi Rong.

exhibited better performance but poor oxidation resistance. The mechanism in this process involves the formation of a sulfur anion that attacks the disulfide bond under alkaline conditions, resulting in metathesis [7]. Arisawa and Yamaguchi proposed a rhodium catalyst for disulfide metathesis [8] and found that the reaction proceeded rapidly in the presence of a phosphine ($RhH(PPh_3)_4$) and trifluoromethanesulfonic acid. Phosphines with electron-donating substituents (such as triarylphosphines) proved to be particularly effective.

Considering that (i) phosphines are more cost effective than rhodium catalysts, (ii) aromatic groups are used to retard disulfide metathesis in the case of phosphine-mediated disulfide metathesis, and (iii) tricyclohexylphosphine is very sensitive to oxidation, we decided to use an aliphatic phosphine with reduced steric hindrance, tri-*n*-butylphosphine (TBP), as the catalyst in the present work to achieve our goal. TBP has previously been employed as a reducing agent to cleave disulfide bonds in aqueous media [9–12], and has also been used to catalyze cycloaddition reactions [13–15]. Here we investigate its performance in both low-molecular-weight and macromolecular systems.

The polymer that serves as a model during our investigations was synthesized from polysulfide diglycidyl ether, using diethylenetriamine (DETA) as the hardener. We anticipated that, if the proposed disulfide metathesis proceeded, the fractured bulk cross-linked polymer would re-bond not only through the reconnection of cleaved S–S bonds, owing to the dynamic exchange nature of the reaction. Furthermore, such polymers might be capable of both self-healing and reprocessing by means of network reshuffling.

The model disulfide metathesis experiments were conducted as in the following. Diethyl disulfide (DEDS, 0.31 g, 2.5 mmol), 2,2′-hydroxy ethyldisulfide (HEDS, 0.39 g, 2.5 mmol), TBP (0.0021 g, 0.01 mmol), 1,8-diazabicyclo [5,4,0]undec-7-ene (DBU, 0.007 g, 0.046 mmol) and acetonitrile (10 ml) were mixed under argon by stirring at 25 °C. Samples were extracted by syringe after specific reaction times to estimate the time dependence of the disulfide metathesis. To assess the effect of the solvent, disulfide metathesis was also conducted in the absence of acetonitrile following the same procedure as above. Scheme 8.1 shows the structures of the substances used in this study that contained disulfide bonds.

To produce modified polysulfides, a quantity of either T131 or G44 (polysulfide T131, M_n = 5000–7000 g mol^{-1}; G44, M_n < 1100 g mol^{-1}) was added to hydroxypropyl acrylate to give an n(—SH) to n(C=C) ratio of 1:1.2, after which 4-dimethylaminopropyridine (DMAP, 0.1% relative to the total mass) was added. After mixing to a uniform state under vacuum at 60 °C for 48 hours to ensure complete reaction between the thiols and the double bonds, the mixture was assessed by Fourier transform infrared (FTIR)

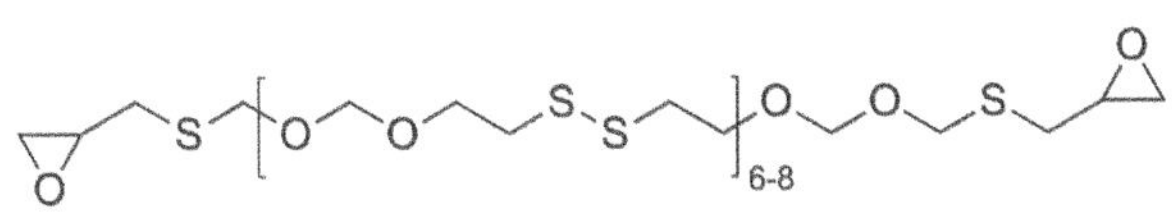

EPS25

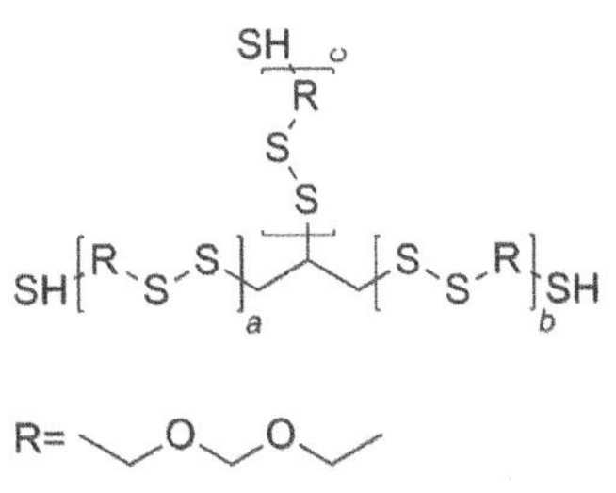

G44 ($a + b + c = n < 7$, cross-linking agent: 0.5 mol% trichloropropane)
T131 ($a + b + c = n = 30-38$, cross-linking agent: 0.5 mol% trichloropropan e)

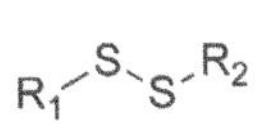

DBDS ($R_1 = R_2 = C_4H_9$), HEDS ($R_1 = R_2 = C_2H_5O$), DEDS ($R_1 = R_2 = C_2H_5$)

Scheme 8.1 Structures of the disulfide compounds.

spectroscopy to identify any remaining thiol groups. In addition, the stability of the molecular chains was assessed by determining molecular weights and the polydispersity index (PDI) by gel permeation chromatography (GPC). The modified T131 (M-T131) and G44 (M-G44), in which all the thiols had been consumed (Scheme 8.2), were subsequently used to study the disulfide exchange reaction.

A typical model exchange trial involving the polysulfides was performed by mixing M-T131 (1g), M-G44 (0.228g), DBU (0.0122g), TBP (0.00368g), and tetrahydrofuran (THF, 10ml) with stirring under argon. Following one hour of reaction at 60 °C, samples were extracted by syringe for the subsequent characterizations. A reaction temperature of 60 °C was applied to allow rapid screening of suitable catalysts.

To ascertain the feasibility of using TBP as a catalyst for disulfide metathesis at room temperature, a model reaction between low-molecular-weight disulfides was initially examined. Figure 8.1 shows high-performance liquid chromatography (HPLC) chromatograms obtained for the system containing DEDS and HEDS in the presence of TBP and the strong base DBU. The latter was added because phosphines are known to be most effective under alkaline conditions. It is evident that, after one hour at room temperature, the metathesis product hydroxyethyl ethyl disulfide (HEED), which is derived from DEDS and HEDS, is detected. With increasing reaction time, the molar fractions of DEDS and HEDS continuously decrease while the fraction of HEED concurrently increases. After three hours, no further changes are seen in the relative intensities of the three peaks in the chromatogram and the proportion of HEED plateaus at about 50%. These results indicate that the exchange reaction has reached equilibrium and demonstrate that TBP can indeed catalyze room temperature disulfide metathesis. Quantification of the reaction reveals that a linear fit to the data obtained at 25 °C gives a rate constant with regression coefficients of 0.9944 and 0.9542 for the first- and second-order plots, suggesting that the reaction likely proceeds via first-order kinetics. Similarly, the reaction at 60 °C also exhibits a higher regression coefficient for the first-order plot than for the second-order one. The activation energy derived from the first-order kinetics is 23.1 kJ mol^{-1}.

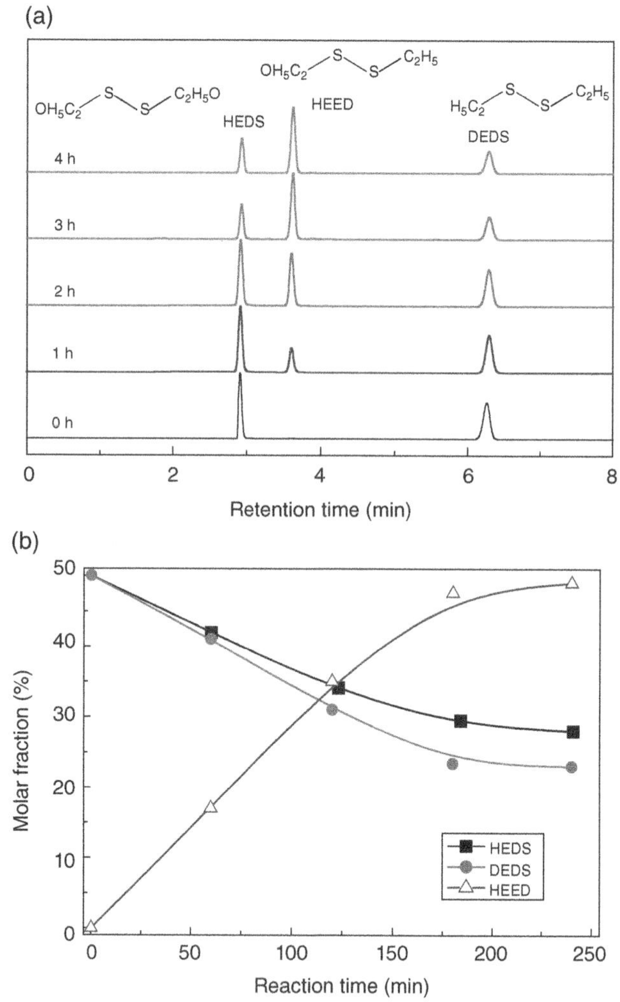

Figure 8.1 (a) HLPC chromatograms of mixtures containing equivalent amounts of DEDS/HEDS mixture as a function of reaction time (25 °C; 1 wt% DBU, 0.3 wt% TBP). The structure of the metathesis product HEED is confirmed by analysis of the mass spectrum. (b) Molar fractions of alkyl disulfides as a function of reaction time, as determined from the data in (a). *Source:* Reprinted from Lei et al. [5]. Copyright 2014, with permission from American Chemical Society.

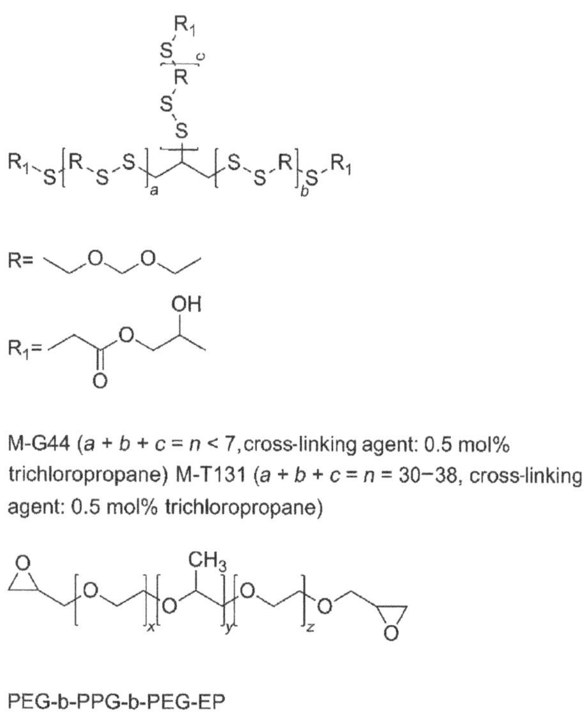

Scheme 8.2 Macromolecules resembling the target polymer.

In addition, the TBP-catalyzed disulfide metathesis between DEDS and DBDS was performed in the absence of a solvent. A similar reaction equilibrium generating a 50% yield of ethyl butyl disulfide (EBDS) was determined by gas chromatography–mass spectrometry (GC–MS) analysis, demonstrating that the solvent (acetonitrile in this case) has no effect on the disulfide metathesis process.

When DBU was replaced by the weak base DMAP, however, no metathesis product was obtained over the equivalent time span, even when the reaction temperature was raised to 80 °C. This further highlights the dependence of the disulfide metathesis on an alkaline environment.

Caraballo et al. [6] suggested that polar solvents may facilitate disulfide metathesis owing to the ion-pair nature of the reaction intermediate. Based on this theory, the metathesis reaction pathways between low-molecular-weight disulfides with catalysis by TBP are depicted in Scheme 8.3. Initially, transient states are formed by reaction between the phosphine salt and the disulfides in the alkaline environment. These then dissociate into a thiolate anion and the cationic intermediate of the phosphine salt. Eventually, cross nucleophilic attack of the thiolate anion at another sulfur atom of the cationic intermediate of the phosphine salt generates both the metathesis product and TBP. When water is present, however, the above reaction proceeds with difficulty because of the presence of phosphine oxide and thiol derivatives. For this reason, it is critical to exclude water from the reaction.

On the basis of the above model reactions, another series of model experiments was performed using polysulfides M-T131 and M-G44 as the main reactants. These materials contain disulfide bonds in the polymeric backbone and have much higher molecular weights than the small molecular compounds, such that they are more like our target material.

As shown in Table 8.1, the molecular weights of M-T131/DBDS when blended with DBU and TBP are significantly lower than those of the original M-T131 or M-T131/DBDS, while the PDI values also show the same trend. These results agree with the trends in Figure 8.1 and suggest that disulfide metathesis can occur in polysulfides in response to catalysis by TBP. As embedding the strong base DBU in the bulk polymer could possibly induce hydrolysis of the polymeric matrix, the weak base DMAP was also examined in the above formulations. Similar reductions in molecular weight and PDI were observed when using DMAP, likely because the S—S bond energy of the polysulfides is lower than that of the low-molecular-weight disulfides. In other words, disulfide metathesis proceeds more readily in macromolecules containing disulfide bonds than in low-molecular-weight disulfides. Based on these results, DMAP was selected for use in conjunction with TBP in the subsequent manufacturing of self-healing cross-linked polymer from the polysulfide diglycidyl ether (EPS25) monomer.

In contrast, because DETA is typically used as a hardener when making cross-linked polymers, its effect on disulfide metathesis should be different. It can be seen from Table 8.1 that neither the molecular weight nor the PDI of M-T131/M-G44/DMAP/DETA exhibit obvious changes as compared with those of the M-T131/M-G44. This situation is analogous to that of M-T131/DBDS in the absence of any other chemicals. Nevertheless, the molecular weights and PDIs of both M-131/M-G44/DETA/TBP and M-T131/DBDS/DETA/TBP are changed following the reaction. Clearly, therefore, DETA can also function as a weak base and thus enhances the catalytic effect of the TBP. Despite this effect, because DETA is consumed when the cross-linked polysulfide is cured, it is present in the cured material at only very low levels and thus the addition of DMAP is still necessary.

As the model experiments have confirmed that disulfide metathesis will take place in macromolecules with the aid of TBP and DMAP, we proceeded to cure EPS25 with DETA in the presence of TBP and DMAP, producing the cross-linked polymer EPS-sh. Scheme 8.4 shows the envisaged disulfide metathesis of the polysulfide network as

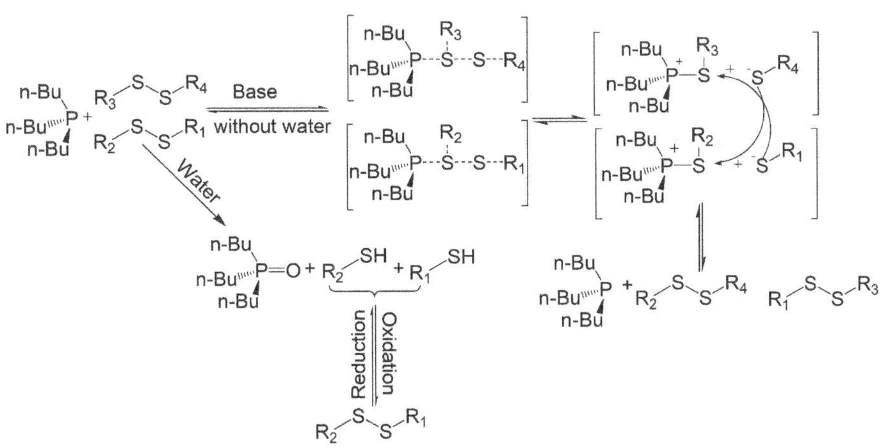

Scheme 8.3 TBP-catalyzed metathesis between low-molecular-weight disulfides.

Table 8.1 Disulfide exchanges of linear polymers containing disulfide bonds.

Materials	DBU	DMAP	DETA	TBP	M_n (×10³)	M_w (×10³)	PDI
M-T131	0	0	0	0	8.0	17.4	2.2
M-G44	0	0	0	0	1.7	2.5	1.5
M-T131/M-G44[a]	0	0	0	0	4.0	12.6	3.2
M-T131/DBDS[a]	0	0	0	0	9.4	14.9	1.9
M-T131/DBDS	1 wt %	0	0	0.3 wt%	5.0	6.2	1.3
M-T131/DBDS	0	1 wt%	0	0.3 wt%	5.2	6.4	1.2
M-T131/DBDS	0	0	1 wt%	0.3 wt%	3.7	8.7	2.4
M-T131/M-G44	1 wt %	0	0	0.3 wt%	2.8	6.2	2.2
M-T131/M-G44	0	1 wt%	0	0.3 wt%	2.8	6.6	2.3
M-T131/M-G44	0	1 wt%	1 wt%	0	3.8	11.8	3.1
M-T131/M-G44	0	0	1 wt%	0.3 wt%	3.7	9.4	2.5

[a] Both M-T131/M-G44 and M-T131/DBDS are equimolar.
Source: Reprinted from Lei et al. [5]. Copyright 2014, with permission from American Chemical Society.

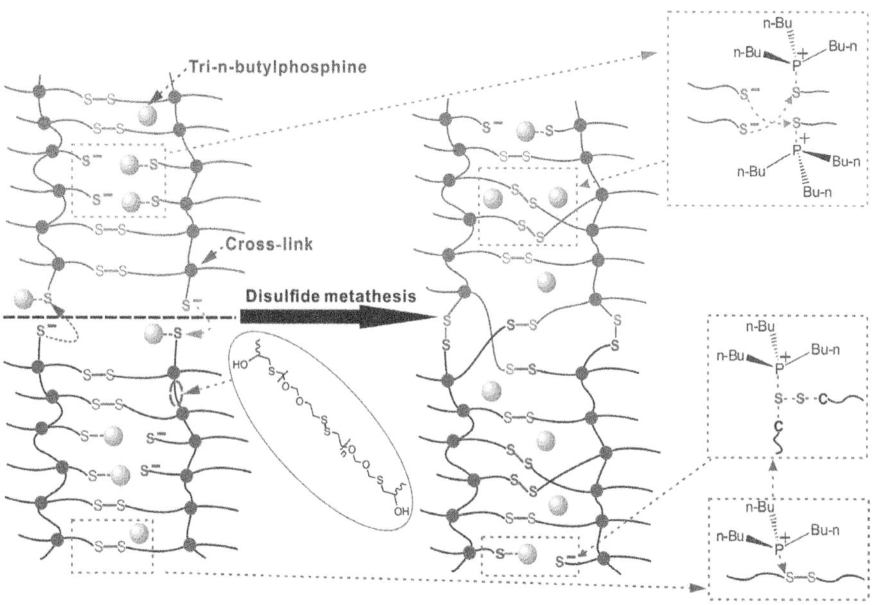

Scheme 8.4 TBP-catalyzed disulfide metathesis within a polysulfide network.

catalyzed by TBP. Based on the action of the organophosphorus compound, transient states are formed between the phosphine salts and disulfide bonds in the polymer. Subsequently ion-pair intermediates consisting of thiolate anions and phosphine salt cations are generated on the macromolecular chains. As a result of the movement of chain segments at room temperature, cross nucleophilic attack of thiolate anions on another sulfur atom by phosphine salt cationic intermediates occurs, leading to exchange of networked chains. Meanwhile, the catalyst returns to its original state. As this disulfide metathesis in the polymer networks repeats itself, rearrangement of the network structure and constant reshuffling of the molecular chains should take place. As a result, the cross-linked polymer should exhibit not only self-healing but also reprocessability based on the dynamic exchanges of disulfide bonds in the presence of the phosphine salt catalyst.

Prior to any discussion of the self-healing performance of the polysulfides, it was vital to obtain some fundamental structural information concerning the polymers.

Infrared spectroscopy was used to examine the effect of curing with DETA on the oxirane rings of EPS25. Figure 8.2a shows that the stretching band of the oxirane rings of EPS25 (at 820–850 cm^{-1}) disappears in the spectra of EPS-sh and control-1, indicating that the system is fully cured. This conclusion is also supported by the data resulting from differential scanning calorimetry (DSC) and DMA tests. As shown in Figure 8.2b, no exothermic events are observed on repeated DSC scans while the glass transition temperature (T_g) of EPS-sh remains unchanged. Figure 8.2c also reveals that only one tan δ peak appears, and its position, a measure of T_g, does not change during multiple measurements.

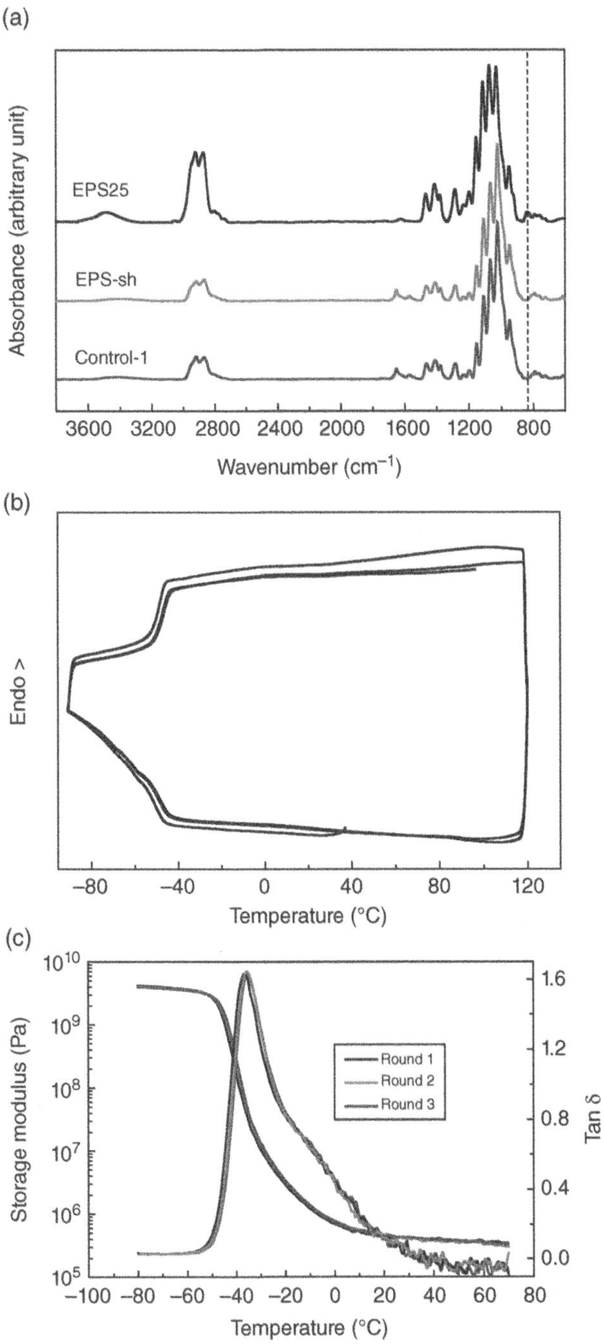

Figure 8.2 (a) FTIR spectra of EPS-sh and control-1 compared with that of EPS25. (b) Repeated DSC heating–cooling scans of EPS-sh. (c) Storage modulus and tan δ of EPS-sh as a function of temperature as obtained by repeated dynamic mechanical analysis (DMA) scans. *Source:* Reprinted from Lei et al. [5]. Copyright 2014, with permission from American Chemical Society.

Based on the DMA results and using Eq. (6.5), the molecular weight between the crosslinks of EPS-sh, M_c, is estimated to be 4.1×10^3, which is much higher than that of control-1 (1.6×10^3). This implies that the degree of cross-linking of the former is lower than that of the latter. This occurs because disulfide metathesis in EPS-sh leads to a dynamic equilibrium between random disconnections and reconnections of S—S bonds. During the DMA tests, a certain fraction of S—S bonds are ruptured and consequently the relative molecular weight between cross-links of EPS-sh appears to be higher than that of control-1 in which disulfide metathesis does not occur because of the absence of TBP. This analysis is confirmed by the higher T_g but lower tan δ peak height of control-1, as the macromolecules are less mobile because of the restraint induced by the higher cross-link density. In addition, the tan δ peak of control-1 is roughly symmetrical, while that of EPS-sh has a shoulder on the higher temperature side, which may originate from dissociation–recombination of S—S bonds.

Nevertheless, EPS-sh chains are still sufficiently cross-linked. Following immersion in the good solvent THF for seven days at room temperature, for example, the material is not dissolved but still maintains its structural integrity, exhibiting only volume expansion.

To gain further insight into the dynamic exchange behavior of disulfide bonds in the polymer EPS-sh, rheological measurements as a function of angular frequency were conducted (Figure 8.3). Figure 8.3a shows that the storage shear modulus, G', decreases with decreasing frequency, while the loss shear modulus, G'', initially decreases with decreasing frequency and then increases. The two curves intersect at a specific frequency, meaning that the rheological behavior of the specimen is no longer dominated by elastic-like behavior ($G' > G''$) but by viscous-like behavior ($G' < G''$) at low frequency. Disulfide metathesis can account for this variation; at certain low frequencies (for a given temperature) the disulfide exchange will begin to have a greater effect on rheological performance because the disconnected network needs time to be reconstructed during metathesis.

A careful examination of the curves in Figure 8.3a shows that the crossover frequency between G' and G'' increases with increasing temperature. This frequency is related to the dissociation rate constant of the reversible network, so the results of Figure 8.3a imply that the disulfide exchange is accelerated with a rise in temperature. This is understandable and agrees with the general behavior of reversible reactions. Owing to the restricted frequency range of the rheometer, however, measurements were not made at frequencies lower than 10^{-2} Hz and therefore do not show that the exchange reaction is not allowed at lower temperatures.

For reference purposes, the rheological behaviors of materials control-1 and control-2 are given in Figure 8.3b,c, respectively. It is interesting to see that the G' and G'' curves intersect at 100 °C (Figure 8.3b), meaning that disulfide metathesis might be enabled even without TBP catalysis. Comparing Figure 8.3a,b, the crossover frequency of EPS-sh at 100 °C is obviously higher than that of control-1, which illustrates once again the important role of TBP. In the case of control-2, which does not contain any S—S bonds, the material behaves like other polymers with permanent covalent bonds. Its G' is independent of frequency ($G' > G''$) over the entire frequency range.

Disulfide metathesis has been found to be possible in EPS-sh and thus should enable self-healing of the cross-linked polysulfide (Scheme 8.4). Figure 8.4a shows that this is indeed the case; the cut EPS-sh regains its tensile strength after the two fracture surfaces are brought together. The healing efficiency increases with increasing time; after 8 hours, about 50% of the original strength is restored. Equilibrium state maximum healing (91%) is observed after 24 hours, indicating almost full recovery. Further extending the healing time to 48 hours does not lead to any increased strength in the healed specimen. To our knowledge, a room temperature healing efficiency of ~91% can be considered as quite a remarkable result for an intrinsic self-healing thermoset material.

Figure 8.4b indicates that EPS-sh can be repeatedly healed. The tensile strength of the virgin specimen cannot be completely restored, so the reconnected interface becomes the weakest part and thus tends to break again during the subsequent tensile tests. This ensures that the second and third tensile failures occur at the same healed portion. The slight decline of efficiency of the second and the third healing can be explained by misalignment of the fractured surfaces. This multiple healing ability is exclusive for intrinsic self-healing materials and represents a valuable feature that distinguishes these materials from extrinsic self-healing versions.

When the healing temperature was raised to 60 °C, a similar healing effect was observed, although only two hours was sufficient to achieve equilibrium healing. This finding is in good agreement with the results of the above rheology study. That is, disulfide exchange becomes faster at higher temperatures.

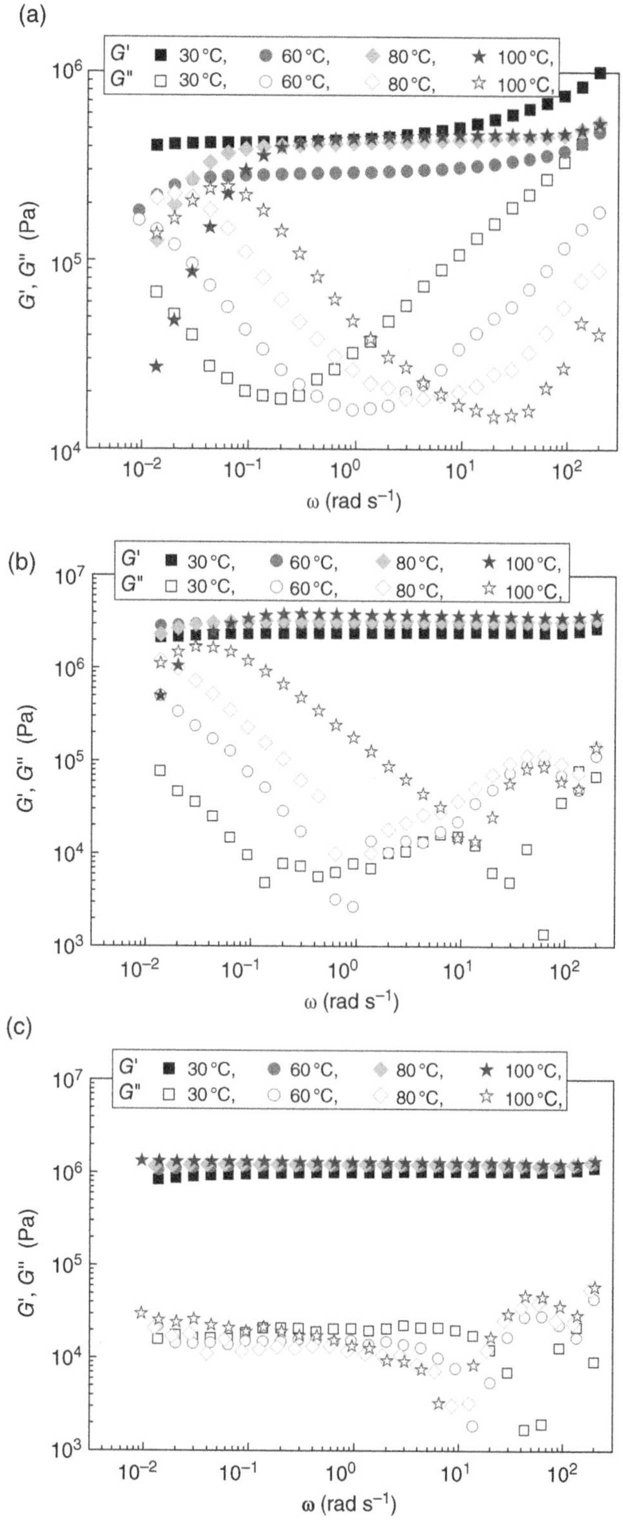

Figure 8.3 Storage shear modulus, G', and loss shear modulus, G'', as functions of oscillatory frequency, ω, for (a) EPS-sh, (b) control-1, and (c) control-2.

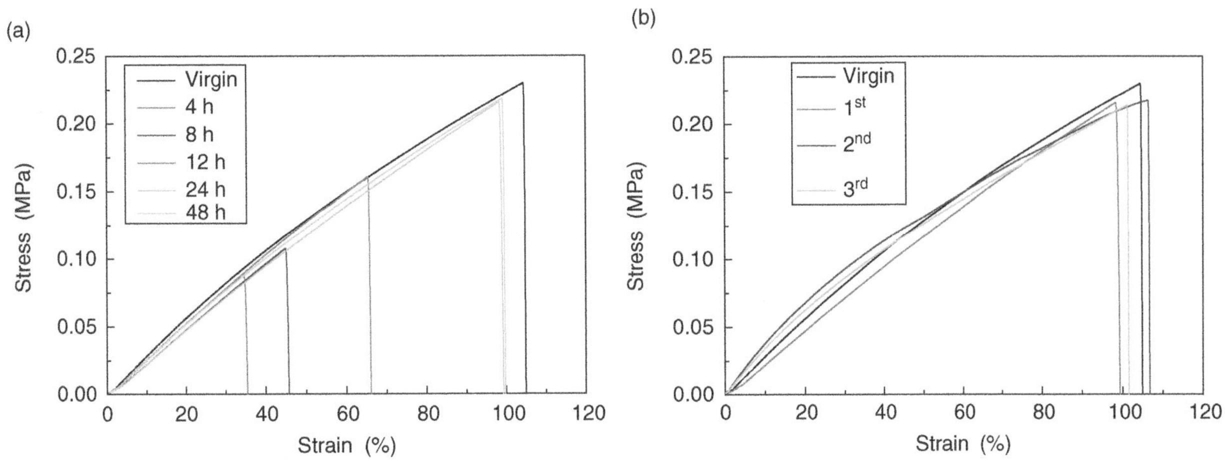

Figure 8.4 Typical tensile stress–strain curves of virgin and healed EPS-sh. (a) Effect of healing time. (b) Effect of repeated repair (healing time = 24 hours). Healing temperature: 25 °C. *Source:* Reprinted from Lei et al. [5]. Copyright 2014, with permission from American Chemical Society.

Tensile tests of the control materials were also performed. In the case of control-1, a marginal healing efficiency of about 13% was observed during the first healing, which is primarily attributed to both dangling chains and hydrogen bonding. The dangling chains represent strands of the fractured networks of the cross-linked polymer that are attached to the undamaged network at one end. These strands have relatively high mobility and help to establish hydrogen bonding among the interdiffused macromolecules across the intimately contacted fractured surfaces. In the case of control-2, which does not contain disulfide bonds, essentially no healing effect was detected. From the above discussion, we know that the disulfide metathesis reaction plays the primary role in the self-healing process.

The effect of exposure of the broken specimens to air was also investigated by intentionally separating the two halves of cut specimens for a time before recombination for healing. Tensile tests showed that high healing efficiency was still obtained even after an isolation time of 72 hours. This behavior differs from that of intrinsic self-healing polymers, which function via hydrogen bonding, because these materials tend to lose their healing ability after long-term separation of the broken surfaces due to self-combination of neighboring free hydrogen bonds. In contrast, metathesis of the disulfide bonds is always in dynamic equilibrium, thus the number of disulfide bonds available for healing does not decrease whether the cut surfaces are in contact with each other or not.

The air-insensitivity of EPS-sh was also established by another group of tests in which samples were evaluated one month after manufacture and were found to repeatedly self-heal to the same extent as the fresh samples (Figure 8.4). Theoretically, TBP may be converted into phosphine oxide and thiol derivatives as a side reaction of disulfide metathesis when water and oxygen are simultaneously present [7]. Because EPS-sh is hydrophobic, however, this side reaction tends not to proceed inside the polymer. Even though the surfaces of the polymer are in contact with the moisture in ambient air, the butyl groups of the TBP would tend to shield the phosphorus atoms from any negative effects. Consequently, the polymer is not sensitive to air exposure.

Classical thermoset materials contain permanent covalent cross-links and thus cannot be reprocessed and reshaped in the same manner as thermoplastics. In such cases, decomposition of the polymer takes place before the flowing temperature is reached. Since the development of covalent adaptable networks [16], however, the idea of introducing dynamic connections that can undergo reversible exchange to form cross-linked materials and thus impart thermal reprocessability to these materials has been reported by several research groups [17–22]. When considering the disulfide metathesis system developed in this work, we anticipated that the cross-linked epoxidized polysulfide should be capable of remolding without heating. To date, there are no reports in the literature of thermosets that exhibit room temperature reprocessability in this manner.

Figure 8.5a,b shows that a cylinder-shaped specimen of EPS-sh could indeed be reshaped into a sheet morphology after compression at room temperature for 24 hours. The disulfide metathesis therefore must have resulted in rearrangement of the cross-linked networks inside the specimen. To further verify the recyclability of the material, the round sheet specimen of EPS-sh was cut into small pieces (Figure 8.5c) and then remolded into a square shape (Figure 8.5d) at room temperature. In comparison, the control-1 and control-2 materials could not be reprocessed under the same conditions because of a lack of disulfide metathesis in these substances.

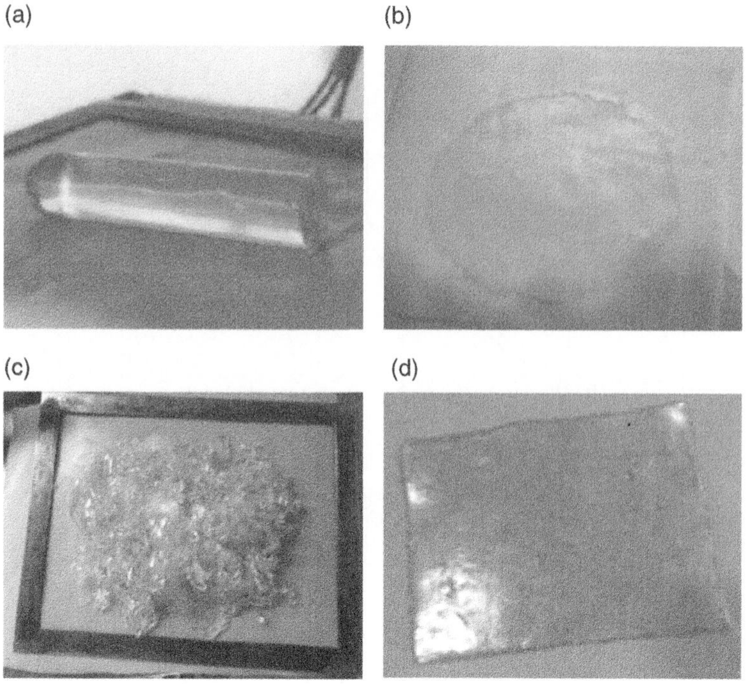

Figure 8.5 (a, b) Reshaping and (c, d) reprocessing of EPS-sh through compression molding at 25 °C for 24 hours. Pressure: 4.5 MPa. The chopped EPS-sh shown in (c) was exposed to air for 48 hours at 25 °C before remolding and the results thus demonstrate the air-insensitivity of the polymer. *Source:* Fan et al. [5], Figure 5, p. 8/with permission of American Chemical Society.

Unlike some reversible reaction-assisted intrinsic self-healing polymers (polyurethane carrying alkoxyamine, for example), in which re-establishment of intramolecular interactions only takes place among disconnected bonds, reconstruction of a new cross-linked network in terms of disulfide metathesis does not necessarily take place at the sites of ruptured S—S bonds (Scheme 8.4). The reaction depends on intermediate-induced rearrangement of networked chains and proceeds so long as the polymers are in close contact with each other. By making use of this feature, reprocessing of the cross-linked polymer is possible. In other words, not every reversible reaction that is capable of giving a self-healing polymer will lead to reprocessability, while reprocessability might serve as a criterion for metathesis.

As disulfide metathesis is responsible for the self-healing and reprocessing effects, the molded surfaces of EPS-sh specimens are not sticky to the touch, which will be a useful feature with regard to handling during practical applications.

In principle, the recycled cross-linked polymer should also be self-healable through disulfide metathesis, just as the original specimen was. As shown in Figure 8.6, the virgin recycled specimen had a tensile strength of 0.23 MPa and failure strain of 104%, both of which are equal to the values measured prior to recycling of the specimen (Figure 8.4). After 24 hours of healing at 25 °C, the tensile strength of the recycled specimen after its first

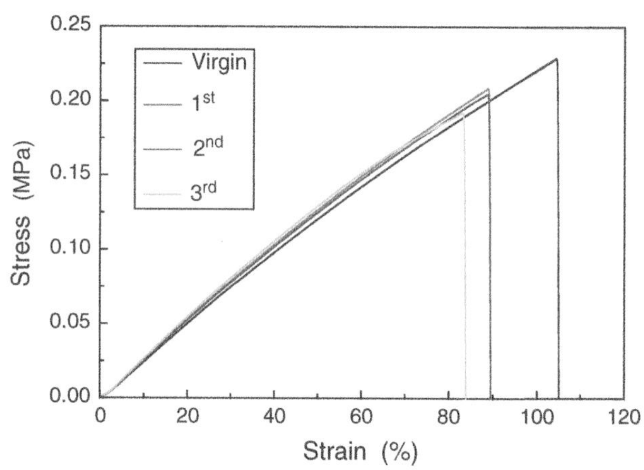

Figure 8.6 Typical tensile stress–strain curves of recycled EPS-sh subjected to successive breakage and healing events. Healing temperature: 25 °C; and healing time: 24 hours.

repair was 0.21 MPa, equaling about 91% of the initial value. The subsequent healing data demonstrate that the multiple self-healing ability of EPS-sh is not affected by reprocessing.

To gain a better understanding of this self-healing polymer, creep tests were carried out at room temperature (Figure 8.7). In general, a creep strain versus time curve obtained with plastics materials consists of four stages: (i) initial rapid elongation, which is independent of time and

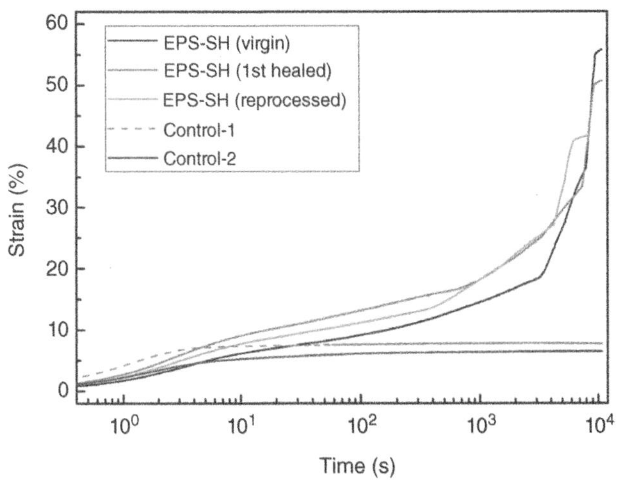

Figure 8.7 Typical tensile creep strain–time curves for EPS-sh and the control specimens, measured at 25 °C.

arises from the elastic and plastic deformations of the polymer; (ii) primary creep, during which the creep rate starts at a relatively high value and then decreases with time; (iii) secondary creep; and (iv) tertiary creep, in which the creep rate increases rapidly and eventually results in creep fracture. Here the virgin self-healing specimen EPS-sh exhibited an insignificant initial rapid elongation stage within the creep curve. Subsequently, the above-noted primary creep was observed, while the transition to the secondary creep was somewhat inconspicuous. The tertiary creep stage lasted for a particularly long time. Both the healed and reprocessed EPS-sh had similar creep behaviors. In contrast, the two control specimens only showed the first three stages of creep within the time span tested and no tertiary creep was observed. Moreover, their strains due to secondary creep were nearly independent of time and were lower than those of the self-healing specimens.

Although the detailed deformation mechanisms involved require further investigation, the differing molecular structures of these materials are likely responsible for the varying creep properties. The self-healing specimen EPS-sh is capable of disulfide metathesis at room temperature and thus rearrangement of the network structure as a result of bond exchange would certainly result in creep or stress relaxation. As for the control materials, because they lack the capability for network rearrangement, their cold flow is restricted to a much greater extent under the same conditions, so that both creep rate and strain are lower than those of EPS-sh. The reduced creep resistance of EPS-sh is a disadvantage of the material with regard to eventual practical applications and products made from this type of polymer would have to avoid working under persistent relatively high stress.

8.2 Sunlight Driven Self-Healing Cross-Linked Polyurethane Containing the Disulfide Bond

Among the reported stimuli for intrinsic self-healing, light is undoubtedly attractive because it is clean, cheap, and readily available [23]. Many studies in this respect used to be focused on the application of UV light. In fact, however, sunlight would be an ideal alternative, which also comes from natural resources. Products made of polymers capable of sunlight stimulated self-healing maintain their structural and property stability in indoor environments or in the shade, and can be unconsciously healed in the outdoors. The negative influence of the photoreversible reaction is thus minimized.

Nevertheless, UV light induced self-healing polymers are not bound to be remendable under sunlight. Coumarin cross-linked polyurethane, for example, exhibited a healing efficiency of 70.2% under UV exposure, but the value fell to 33.5% when illuminated by sunlight [24]. So far, to the best of the authors' knowledge, there are few reports dealing with sunlight self-healing of polymers toward mechanical strength recovery [24, 25]. The primary causes of this are believed to be (i) the UV intensity of sunlight is rather low (0.5–2.5 $mW\,cm^{-2}$ in the daytime in Guangzhou, China, for example) [26] and (ii) the UV-reactivity of the photosensitive moieties in polymers is not high enough.

Earlier work by Otsuka et al. has revealed that disulfide metathesis reactions driven by UV irradiation can be carried out in an aliphatic disulfide-containing polyester [27]. Subsequently, Michal et al. showed that semi-crystalline polydisulfide networks were self-healed through UV triggered disulfide exchange [28] despite the fact that the power intensity of the light source was as high as 2000 $mW\,cm^{-2}$ for heating the sample up to its melting temperature and freeing the frozen disulfide bonds. Clearly, owing to the low UV concentration of sunlight, the key challenge to realization of reversible reactions of disulfide under sunlight exposure lies in the enhancement of its UV sensitivity.

8.2.1 Cross-Linked Polyurethane

8.2.1.1 Bulk Polymer

In the present work, transparent amorphous polyurethane is designed to act as the target polymer. It favors not only light transmission, but also easy adjustment of the T_g and mechanical properties by changing the proportions of the soft and hard segments [29]. In this context, the aliphatic disulfide bond is chosen as the intermonomer photoreversible linkage and all aliphatic monomers are employed for the polymerization, which would impart flexibility and resistance-to-yellowing to the resultant macromolecular backbone. The mobility of the disulfide linkages is thus improved. Furthermore, the possibility of color change due to sunlight exposure that lowers the intensity of incoming light is also reduced. Besides, the electron-donating group is attached to the disulfide bond to weaken its bond energy [29]. Meanwhile, appropriate hydrogen bonding between disulfide chains is established promoting contact among disulfide units. Lastly, cross-linking is introduced to strengthen the polymer [30].

Based on the plan described above, the photoresponsive polyurethane (PU-HEDS-400, see Scheme 8.5), i.e. our target material, was synthesized through the reaction between 2,2′-hydroxy ethyldisulfide and isophorone diisocyanate (IPDI) with polyethylene glycol (PEG400, $M_w = 400$) as chain extender. During the polymerization, the percentages of HEDS and IPDI were so tuned that there were excessive —NCO groups in the system. Consequently, prepolymer with —NCO terminal groups was produced, which was then cross-linked by triethanolamine (TEA). Here, the electron-donating amido groups were connected to the disulfide via ethyl as designed, which would further form intermolecular hydrogen bonds with the counterparts on the neighboring chains.

In a typical experiment, dried IPDI (5.0 g), HEDS (1.6 g), and PEG400 (4.0 g) were added to dichloromethane (100 ml), and the mixture was stirred at 60 °C for 12 hours. Then, TEA (0.4 g) was incorporated into the solution and the system was stirred for six hours to complete the reaction. Finally, the resultant was poured into a plane mold. After volatilization of dichloromethane at room temperature and vacuum drying at 60 °C for 24 hours, the cross-linked polyurethane film (PU-HEDS-400) containing aliphatic disulfide bonds was obtained.

For the convenience of mechanism analysis, a control polyurethane excluding disulfide bonds (PU-BDO-400, Scheme 8.5) was synthesized following the same procedures as described above, except that HEDS (1.6 g) was replaced with 1,4-butanediol (0.9 g). The molecular weight between cross-links = 1568.

Although exchange of disulfide bonds has been found to occur upon irradiation with UV light [27], it is still unclear whether the same reaction is enabled under sunlight. Therefore, a group of model experiments with small molecules (DEDS and HEDS) was carried out first. Both exhibit strong absorption at 250 nm, but the absorption becomes rather weak in the region of sunlight (>290 nm), implying that the exchange reaction should be much more difficult under sunlight as compared with pure UV light.

Figure 8.8a shows that when the equimolar mixture of DEDS and HEDS in acetonitrile is exposed to xenon lamp light, a new peak is detected on the HPLC trace, and the peak area increases with a rise in time. In the meantime, the peak areas of DEDS and HEDS decrease. The equilibrium molar proportion of DEDS, HEDS and the reaction product is 0.33:0.33:0.34 (Figure 8.8b). According to the assignment of the mass spectrum, it is known that the reaction product is ethyl propyl disulfide (EPDS). Clearly, exchange reaction of disulfides can still be activated by sunlight in the absence of catalyst. The conclusion is further supported by the HPLC traces given in Figure 8.8c, where natural sunlight serves as the light source. Additionally, the intensity of the xenon lamp is adjusted from 60 to 100 mW cm^{-2} imitating sunlight at different times of the day. Figure 8.8d shows that the disulfide exchange can proceed even at a light intensity of 60 mW cm^{-2}.

For comparison, the same mixture of DEDS and HEDS in acetonitrile is heated to 60 °C in the dark (which simulates the thermal effect of sunshine), and illuminated by visible light (λ > 400 nm). No exchange reaction product is

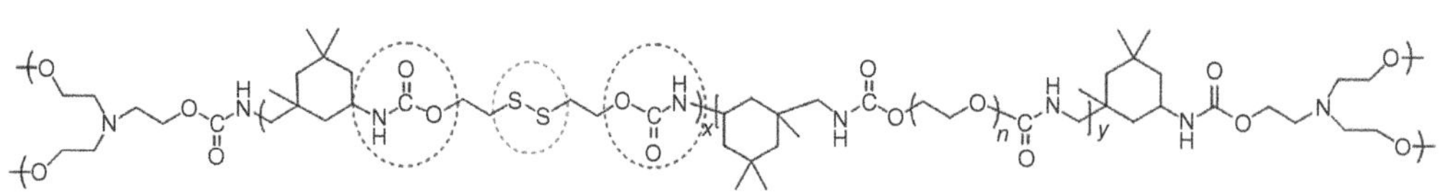

PU-HEDS-400

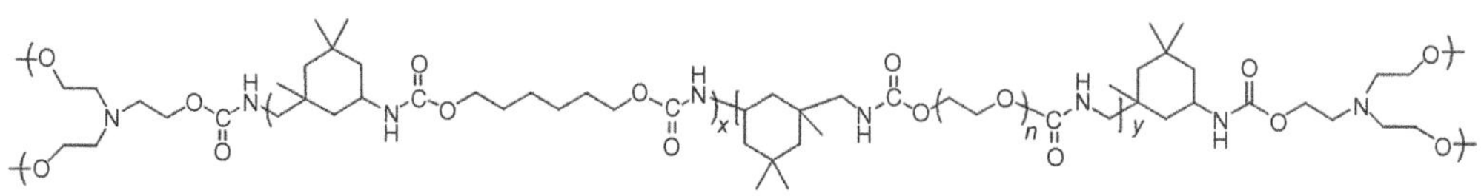

PU-BDO-400

Scheme 8.5 Structures of PU-HEDS-400 and its control PU-BDO-400.

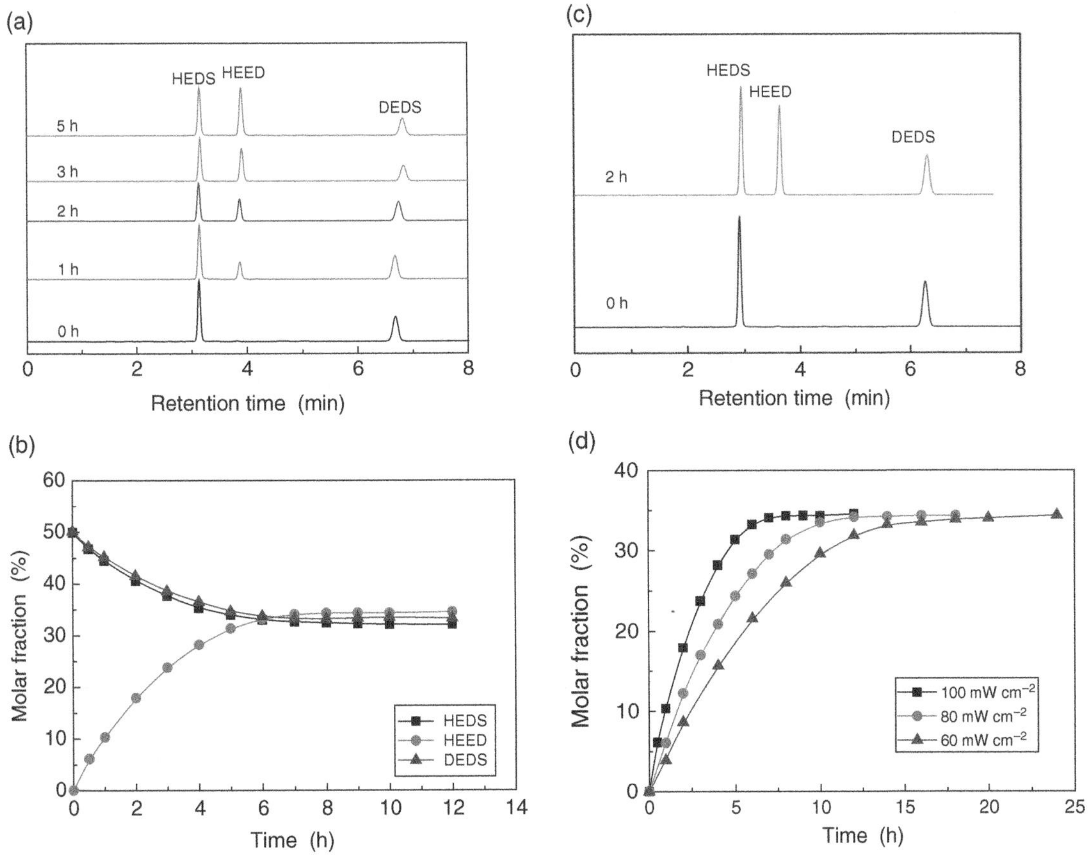

Figure 8.8 (a) HPLC analysis of an equimolar mixture of DEDS and HEDS in acetonitrile exposed to xenon lamp light (intensity = $100\,mW\,cm^{-2}$) as a function of time. (b) Time dependences of molar fractions of the disulfides in an equimolar mixture of DEDS and HEDS in acetonitrile exposed to xenon lamp light (intensity = $100\,mW\,cm^{-2}$). (c) HPLC analysis of an equimolar mixture of DEDS and HEDS in acetonitrile exposed to natural sunlight (noon in July, intensity ≈ $90\,mW\,cm^{-2}$). (d) Time dependence of molar fractions of HEED, the product of the exchange reaction between equimolar DEDS and HEDS in acetonitrile exposed to xenon lamp light with different intensities.

generated in both cases. This confirms that the small amount of UV component of sunlight (3–5% [26]), rather than visible light, can induce photochemical exchange of disulfides. In other words, photochemical exchange of disulfides proves to be rather sensitive to sunlight, while solar heating itself could not activate exchange of disulfides.

The above work confirms that sunlight irradiation can effectively initiate the exchange reaction between small molecular disulfides. Accordingly, we start to check whether the habit is inherited by the target material PU-HEDS-400.

As shown in Figure 8.9, the polymer has strong absorption of UV light, especially below 300 nm, which resembles DEDS and HEDS, and guarantees the proposed response of the material to sunlight. Meanwhile, the polymer is highly transparent and possesses excellent resistance to yellowing, as characterized by the transmittance of over 90% in the visible light range before and after UV aging. The color difference due to aging, ΔE, is measured to be 0.28, which falls into the highest "grade 0" according to the Chinese National Standard GB/T 1766-2008. This means that there is nearly no color change. The results show that the long-term stability of its optical property can be maintained under sunlight irradiation. The absence of phenyl rings and unsaturated bonds in PU-HEDS-400 (Scheme 8.5) are responsible [31, 32].

Solar healable performance of PU-HEDS-400 is first assessed in terms of visual inspection (Figure 8.10). The sample film (0.8 mm thick) with man-made scratches (1–2 μm in width, 3–5 μm in depth) are exposed to sun at noon in July. Just after five minutes, the scars are no longer perceived. In particular, because the second scratch is created across the first one (that has been healed), the disappearance of the intersection of two successive scratches reveals the repeated healability of the polymer.

To further understand the ability of reconnecting broken chains, the dumbbell-shaped film specimen of

PU-HEDS-400 is cut into two pieces. Then, the fracture surfaces are brought together for healing under sunlight from 10:00 a.m. to 4:00 p.m. in July. Tensile test of the healed specimen indicates that 96% of original strength can be restored (Figure 8.11a). This demonstrates that the split PU-HEDS-400 can be chemically re-bonded with the aid of

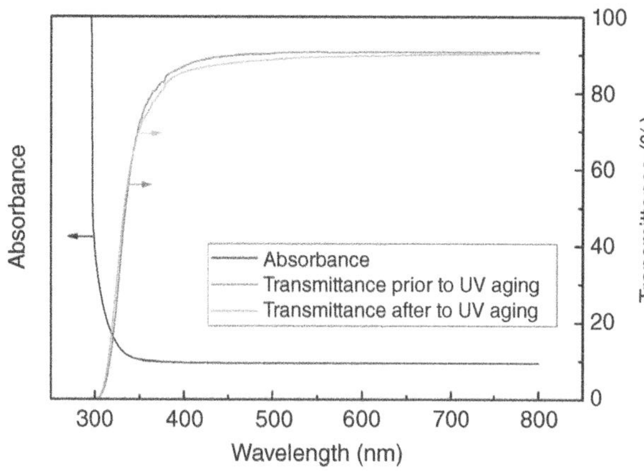

Figure 8.9 Absorbance and transmittance of PU-HEDS-400 (0.8 mm thick) as a function of wavelength. The conditions of UV aging are described in the following. First, the target polymer was exposed to constant UV radiation (340 nm, 0.068 mW cm^{-2}) at 60 °C for 168 hours according to the Chinese National Standard GB/T 23987-2009. Before and after the artificial UV aging, color values of the sample were measured using a 3nh SC-10 colorimeter by the method described in GB/T 11186-1989. Then, color difference, ΔE, was estimated.

sunlight irradiation. When healing of the same polymer is conducted at 50 °C for 12 hours in the dark, however, the healing efficiency is reduced to 43.1%. Similarly, healing efficiency of the control material (PU-BDO-400, which does not contain disulfide bonds) measured under the same illumination conditions as PU-HEDS-400 is 43.4% (Figure 8.11b). By comparing the results of Figures 8.10 and 8.11, it is known that hydrogen bonds make a considerable contribution to the crack healing. The prophase healing should be mainly afforded by hydrogen bonding, so that fast coalescence of fracture is observed (Figure 8.10). Such an interlocking of fractured surface at the early stage of healing is valuable for the subsequent reversible reaction at the interface. In the case of rigid material, the lack of this type of interaction would lead to loose contacts of the broken parts even they are macroscopically brought together, and hence low healing efficiency [29]. The hydrogen bonding built up between amido groups, especially the ones distributed on both sides of the disulfide bond (Scheme 8.6), helps to get the disulfide bonds of different macromolecular chains closer in favor of the disulfide exchange reaction. Much more recovery of strength is detected as a result.

The combined effect of electron-donating groups and hydrogen bonding is reflected by the UV–visible absorption behavior of the dichloromethane solution of prepolymer of PU-HEDS-400, which shows a shoulder peak between 250 nm and 325 nm. In comparison with the absorption of HEDS, the core building block of PU-HEDS-400, the responsiveness to sunlight of PU-HEDS-400 is obviously increased.

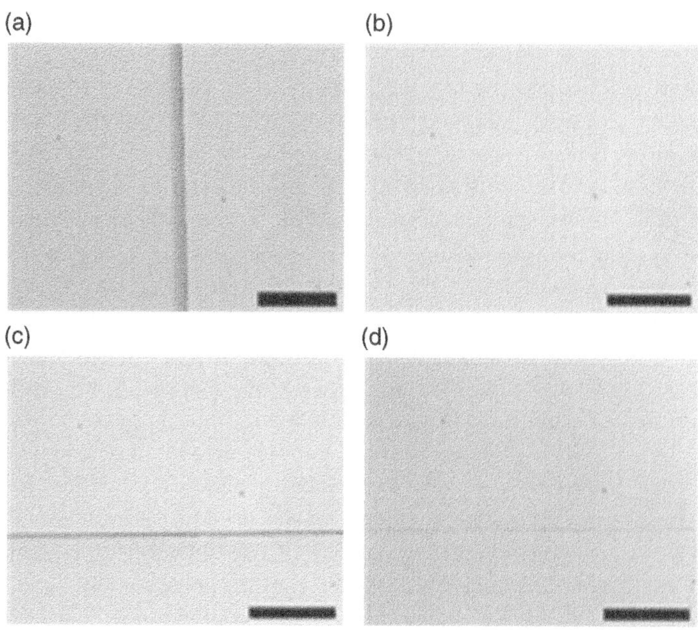

Figure 8.10 Images showing repeated healing of scratched PU-HEDS-400 film (0.8 mm thick). (a) The first scratch; (b) effect of the first repair; (c) the second scratch; and (d) effect of the second repair. Healing conditions: exposure to sun at noon in July for five minutes (intensity ≈ 90 mW cm^{-2}). The scale bars represent 200 μm in length. The specimen is transparent, and the color comes from the substrate. *Source:* Xu et al. [30], Figure 3, p. 6/with permission of Royal Society of Chemistry.

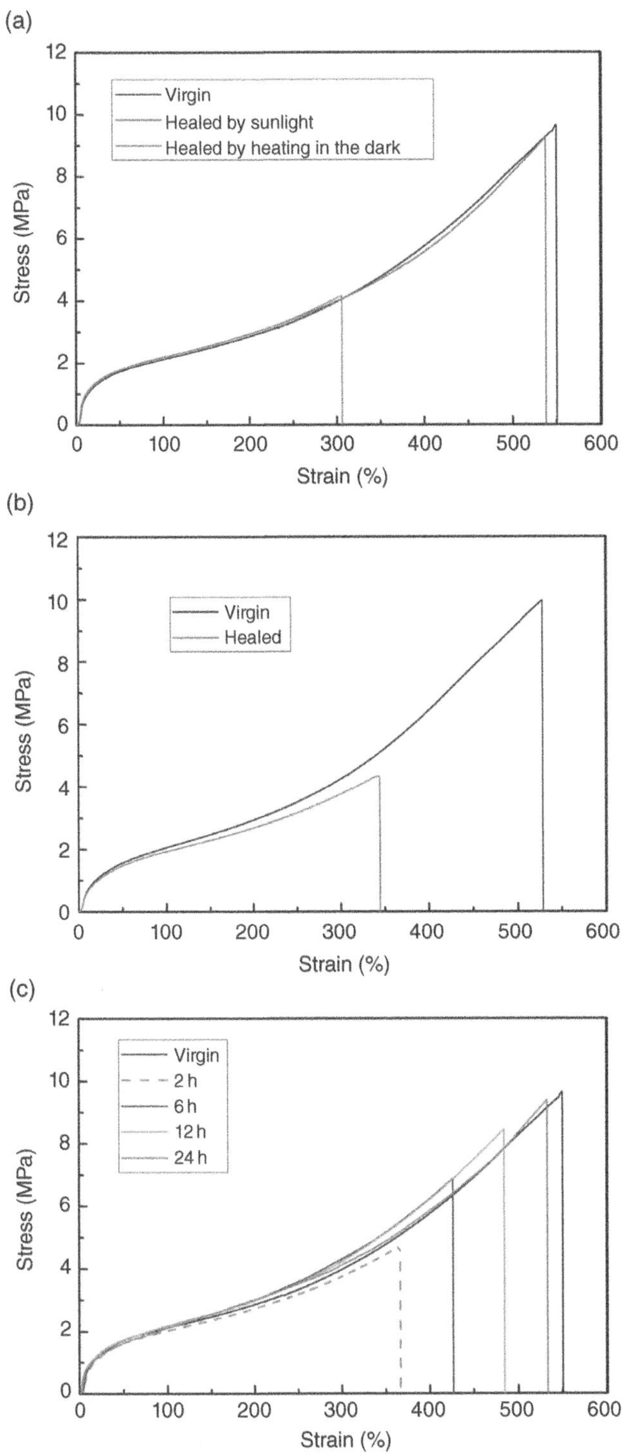

Figure 8.11 (a) Tensile stress–strain curves of virgin and healed PU-HEDS-400 specimens. Sunlight-aided healing is carried out from 10:00 a.m. to 4:00 p.m. in July, while thermal healing is completed at 50 °C for 12 hours in the dark. (b) Tensile stress–strain curves of virgin and healed PU-BDO-400 specimens. Healing is carried out under sunlight from 10:00 a.m. to 4:00 p.m. in July. (c) Tensile stress–strain curves of virgin and healed PU-HEDS-400 specimens. Healing is conducted under xenon lamp light (intensity = 100 mW cm^{-2}) for different times. *Source:* Reprinted from Xu et al. [30]. Copyright 2016, with permission from the Royal Society of Chemistry.

Additionally, the specific structure of PU-HEDS-400 is another important factor for crack healing. PU-HEDS-400 has a T_g of 29.8 °C. The polymer is in the glassy state at room temperature but enters elastomeric state upon illumination by sunlight due to its thermal effect (surface temperature of the specimen is raised to 50 °C by sunshine). Under the circumstances, the flexible polyethylene glycol segments would effectively promote movement of the dangling chains on the fracture surface. Interdiffusion and entanglement of the dangling chains are accelerated across the crack interface [33] creating more opportunities of collision and exchange reaction for the adjacent disulfide bonds as well as forming hydrogen bonds among the macromolecules.

In general, healing efficiency of hydrogen-bonded polymers would quickly decay due to self-combination of the neighboring hydrogen bonds on the same side if the damaged surfaces are left apart for some time, but the disulfide bond does not have this habit. In a verification experiment, the fractured surfaces of PU-HEDS-400 are intentionally separated for four days prior to recombination and the healing efficiency is found to be 79.9%. This means that sunlight-triggered disulfide exchange can still operate across the interface in this case, accounting for the crack healing.

Figure 8.11c shows a more accurate description of the time-dependent healing behavior by using xenon lamp light. The healing efficiency increases with increasing time of exposure, following the conventional rule of healing reaction. Owing to the fact that the intensity of UVA + UVB of our xenon lamp is only one-third of that of sunlight, a longer time has to be taken to completely heal the damaged specimen. The results also suggest that rearrangement of the polymer networks via disulfide exchange is progressively conducted, which eventually offers higher degree of strength restoration than hydrogen bonds. The analysis is supported by the effect of xenon lamp light intensity on healing efficiency. The higher light intensity results in higher healing efficiency at the same exposure time, which agrees with the trend revealed by the model disulfide metathesis experiments (Figure 8.8d).

Since the above result demonstrates that the broken PU-HEDS-400 can be self-healed under sunshine, we naturally associate this manner with the possibility of remolding of the polymer powders. That is, when the polymer is shredded and then placed in the sun, disulfide exchange should also take place among the particle surfaces and reconnect the particles like self-healing. This factually offers a facile way of recycling cross-linked polymer without the need of depolymerization. To examine the feasibility of the above idea, PU-HEDS-400 grains (~2 μm, Figure 8.12a) are compression molded into a blank (~1 mm thick, Figure 8.12b) under 10 MPa for five minutes, which is then clamped between two quartz plates (Figure 8.12c). After successive exposure of each side of the sheet to sunlight for 12 hours,

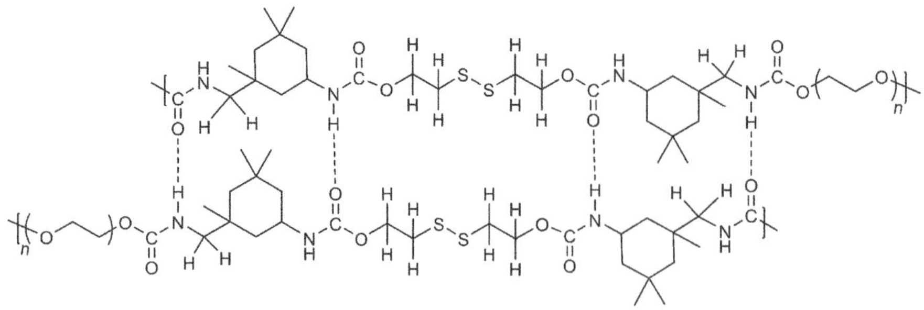

Scheme 8.6 Hydrogen bonds formed between PU-HEDS-400 macromolecules.

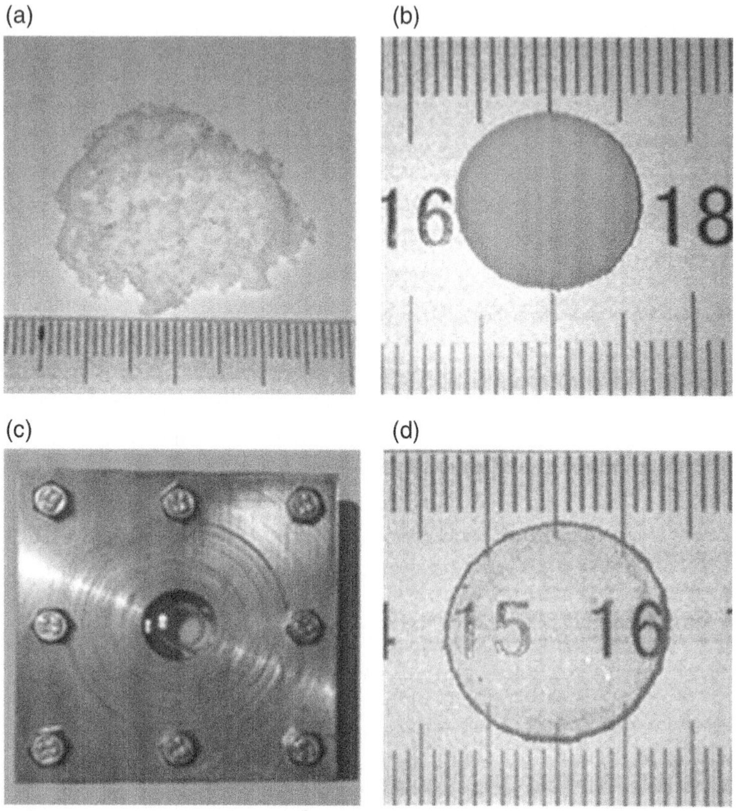

Figure 8.12 Sunlight-aided recycling. (a) Pulverized PU-HEDS-400. (b) Molded blank. (c) The blank clamped between two pieces of quartz plate for sunlight exposure. Each side of the sample is exposed to the sun for two days from 10:00 a.m. to 4:00 p.m. in July so that the accumulated time is 12 hours for each side. (d) The recycled sheet. *Source:* Xu et al. [30], Figure 5, p. 8/with permission of Royal Society of Chemistry.

it is recovered to a transparent bulk again (Figure 8.12d). This can be considered as restoration of transparency. Mechanical testing indicates that after the first run, the recycled polymer regains 92.2% of the original tensile strength (Figure 8.13a). Moreover, the same sample can be repeatedly recycled with similar degrees of restoration (i.e. 89.7 and 92.0% for the second the third runs, respectively), coinciding with the expectation.

In the meantime, the control PU-BDO-400 is reprocessed following the same procedures. Only hydrogen bonding is allowed to be established among the particles, so the recovered sheet regains ~36% of the original strength (Figure 8.13b), resembling the result of the measurement of self-healing capability (Figure 8.11b).

8.2.1.2 Composites with Silver Nanowires as Strain Sensor

By using the above transparent cross-linked polyurethane carrying disulfide in the main chain as the matrix and silver nanowires (AgNWs) as the conductive fillers, we further produce a novel transparent resistive-type strain sensor with high sensitivity and durability [34]. In the meantime, a sandwich-like structure is formed to suppress dislocation of the AgNW assemblies in the interlayer

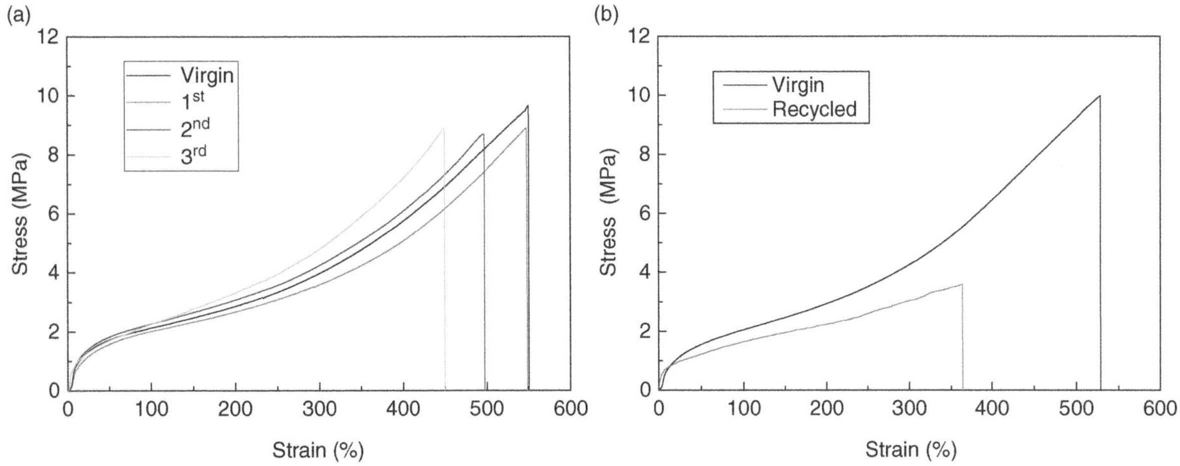

Figure 8.13 Tensile stress–strain curves of (a) virgin and solar recycled PU-HEDS-400 and (b) virgin and solar recycled PU-BDO-400. For the recycling procedures see the caption of Figure 8.12.

during deformation, and the AgNW assemblies are made by a network of welded long AgNWs (AgNWs used to be shorter than 60 μm in length as documented in the literature [35–38] but we were able to prepare a longer version). These measures are believed to benefit improvement of stability and durability of the sensor under large cyclic strains. In addition, density and thickness of the AgNW interlayer is reduced and its homogeneity is increased so as to ensure high sensitivity. On the other hand, sunlight self-healability of the polyurethane is triggered by means of the exchange reaction of the included disulfide bonds, which helps to recover the damage created in the course of service and extends the working life of the sensor.

AgNWs were synthesized by a one-step polyol method. Typically, silver nitrate (4.7×10^{-2} mol) and polyvinylpyrrolidone (PVP k-90, 7.2×10^{-2} mol, M_w = 130 000) were dissolved in 400 ml ethylene glycol, and then transferred into a 1000 ml polyvinylidene fluoride (PVDF) digestion tank. Then 0.05 ml concentrated hydrochloric acid was diluted with 100 ml ethylene glycol to prepare a hydrochloric acid in ethylene glycol solution with molarity of 5×10^{-3} mol l^{-1}, which was poured into the digestion tank and well mixed. The digestion tank was sealed and heated to 140 °C for 15 hours. Then the solution was quenched in a room temperature water bath to stop the reaction. When the AgNW solution was cooled, it was centrifuged at 5000 rpm for 10 minutes and washed three times with ethanol to remove the excess PVP and ethylene glycol. Afterwards, long AgNWs were separated by a stainless standard sieve (200 mesh) and diluted with dichloromethane to a certain concentration for further experiments. The diameter and length of the synthesized AgNWs were estimated to be 100–150 and 100–150 μm, respectively, corresponding to aspect ratios ranging from 667 to 1500. Compared with previous works offering AgNWs with length to diameter ratios from hundreds to one thousand [35–38], ours are at a higher level.

Fabrication of the strain sensor is illustrated in Figure 8.14. The solution of the pre-cross-linked polyurethane

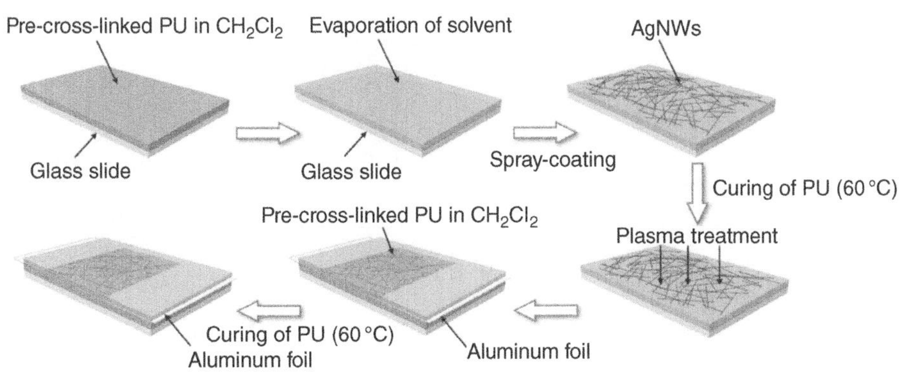

Figure 8.14 Schematic illustration of fabrication of the sandwich structured PU/AgNWs/PU composite strain sensor. *Source:* Reprinted from Song et al. [34]. Copyright 2019, with permission from the Royal Society of Chemistry.

in dichloromethane was poured onto a clean glass slide. When the solvent was evaporated at 25 °C for 12 hours, the AgNW dispersion in dichloromethane of different concentrations was spray-coated on the surface of PU for different times. Next, the film was heated at 60 °C for 24 hours to completely cross-link the PU. Afterwards, the sample surface was treated by argon plasma for 20 minutes at room temperature using a PT-2S plasma surface treatment apparatus (150 W and 40 kHz) to remove the insulating PVP surfactant and to weld the junctions of AgNWs. For electrical tests, both ends of the AgNW networks were attached to aluminum foils. Finally, the solution of the pre-cross-linked polyurethane in dichloromethane was poured onto the surface of AgNWs forming a protective layer, and the sandwich-like PU/AgNWs/PU composite film was heated at 60 °C for 24 hours again and peeled off from the glass slide for the subsequent tests.

For reference, control specimens, AgNWs/PU, were also prepared following the same procedures as described above except that the PU protective layer was not applied.

The objective of the present work is the development of self-healable transparent strain sensor with high sensitivity and durability. Except for the self-healing performance, the rest functionalities are mainly determined by the AgNW network, PU matrix and the interfacial interaction between the AgNWs and PU. Therefore, the influence of the AgNW network on transparency and electrical conductivity is investigated first by using AgNWs/PU as the model material for the convenience of discussion.

Figure 8.15a indicates that with a rise in the concentration of the dispersion of AgNWs in dichloromethane from 3 to 12 $mg\,ml^{-1}$ for spray-coating, the transmittance at 550 nm of the resultant AgNWs/PU composite films significantly decreases. Nevertheless, the conductivity of the composites increases as reflected by the decrease in sheet resistance (Figure 8.15b). Similarly, when the spray-coating times increase at a constant AgNW concentration of 6 $mg\,ml^{-1}$, the same variation tendency of transparency and conductivity is perceived (Figure 8.15a,b). The increased density and thickness of the AgNW network is responsible. Relatively speaking, the composite films containing a AgNW layer made from the dispersion with concentrations of 3 and 6 $mg\,ml^{-1}$ exhibit similar transmittance at 550 nm but different electrical resistance. The latter possesses much higher conductivity. In contrast, spray-coating times greater than twice significantly reduce the transparency, but do not result in a remarkable change in resistance. According to these findings, the composites prepared by single-pass spray-coating of 3 and 6 $mg\,ml^{-1}$ AgNW dispersions were selected for the next round of study.

In general, the relative resistance change and strain is better to be linearly dependent for practical usage. Nonlinearity would cause difficulty for the calibration. The plots in Figure 8.16 demonstrate that only the sandwiched PU/AgNWs/PU composite film at a AgNW concentration of 6 $mg\,ml^{-1}$ has linear electrical response to tensile strain within the strain range from 0 to 25%, and its gauge factor is estimated to be as high as 11.2. The other composite films exhibit obvious nonlinearity, and the gauge factor (slope of the curves) increases with the applied strain. The difference reveals the different robustness of the conductive paths. It is clear that in the case of fewer AgNWs (i.e. AgNW concentration = 3 $mg\,ml^{-1}$), stretching of the composite would easily lead to

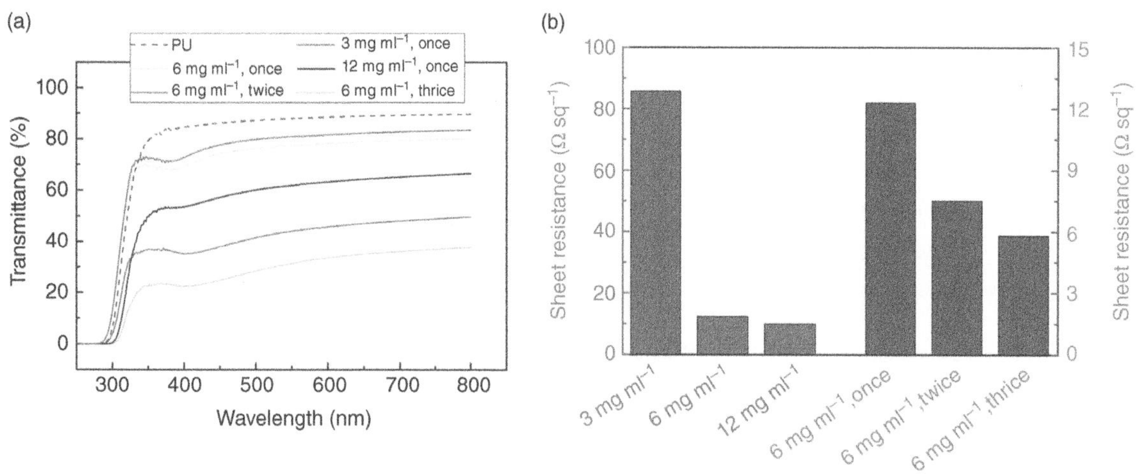

Figure 8.15 (a) UV–visible transmittance spectra of PU and AgNWs/PU composite films, in which the AgNW network was obtained by spray-coating of the dispersion of AgNWs in dichloromethane with different concentrations of AgNWs. (b) Sheet resistance of the AgNWs/PU composite films as a function of concentrations of AgNW dispersion and spray-coating times. The AgNWs in all the composite films were welded by plasma treatment, unless otherwise specified. *Source:* Reprinted from Song et al. [34]. Copyright 2019, with permission from the Royal Society of Chemistry.

disconnection of the sparse nanowire network even if the AgNWs are welded and restrained by the protective PU layer. In this case, there is a lack of AgNWs as load bearers to share the applied force, and the conducting AgNW network has poor deformability. Only when the amount of AgNWs is high enough (i.e. AgNW concentration = 6 mg ml^{-1}) and the AgNWs are welded and covered by the protective layer, the unstable variation in conductive networks is eliminated, while sensitivity of the composite keeps at a high value, because the conduction pathways become compact enough to prevent permanent destruction induced by the limited deformation.

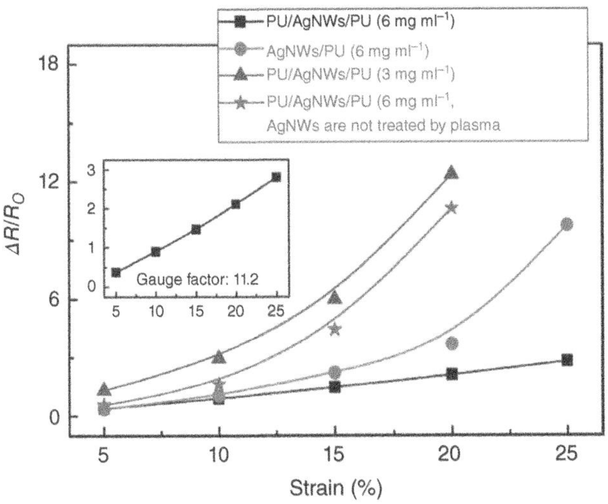

Figure 8.16 Relative resistance change, $\Delta R/R_o$, as a function of tensile strain of different composite films prepared by single-pass spray coating. The inset shows a magnified view of the plot of the PU/AgNWs/PU composite film (AgNW concentration = 6 mg ml^{-1}).

Since the sandwiched PU/AgNWs/PU composite film (AgNW concentration = 6 mg ml^{-1}) turns out to be a candidate for a stable strain sensor, its structure–property relationship is explored in the following. When the PU solution was poured onto the AgNW network to produce the protective layer, the former penetrated into the interconnected pores of the AgNW network due to the strong affinity between the imino groups of PU and the surface of the AgNWs [39]. The cross-sectional scanning electron microscopy (SEM) photo in Figure 8.17a indicates that the resultant mutilayer organization is well constructed and the AgNW network interlayer is evenly embedded between the PU layers. Moreover, the embedded AgNW conductive network is percolated in PU without evident defects (Figure 8.17b), proving that a robust sandwiched structure is successfully fabricated. Compared with the AgNWs/PU film prepared with the same AgNW concentration, the

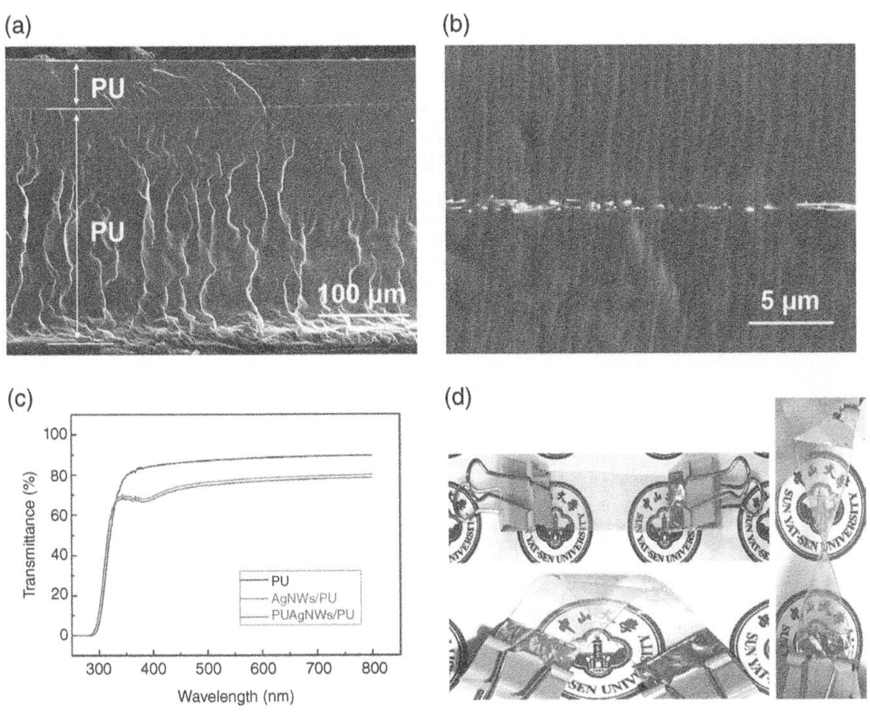

Figure 8.17 (a) SEM photo of a cross-section of the sandwiched PU/AgNWs/PU composite film. (b) Close-up of the embedded AgNW network interlayer of (a). (c) UV-visible transmittance spectra of PU, PU/AgNWs/PU and AgNWs/PU composite films. (d) Photographs showing bending and twisting of sandwiched PU/AgNWs/PU composite film. The AgNW concentration and spray-coating time for preparing the composite films: 6 mg ml^{-1} and single pass, respectively. *Source:* (a, b, d) Song et al. [34], Figure 9, p. 6/with permission of Royal Society of Chemistry. (c) Reprinted from Song et al. [34]. Copyright 2019, with permission from the Royal Society of Chemistry.

transmittance of the sandwiched film at 550 nm is only 1.8% lower (Figure 8.17c). The higher length to diameter ratio of the AgNWs should account for the high transparency, as it results in longer percolation paths and reduced inter-nanowire junctions according to the excluded volume theory [40, 41]. In the meantime, the sheet resistance of PU/AgNWs/PU (12.5 Ω sq.$^{-1}$) is almost identical to that of AgNWs/PU (12.3 Ω sq.$^{-1}$) at the same AgNW concentration of 6 mg ml^{-1}. The slight changes in transmittance and conductivity demonstrate that the addition of the protective thin PU layer has little negative effect. In fact, the transparent sandwiched composite film also has excellent stretchability with elongation at break of more than 500%. It can be bent or twisted by a large angle without any damage (Figure 8.17d). This is important as the composite film is designed to act as a flexible strain sensor, which has to be mounted on or wrapped around curvilinear and biological surfaces.

The above results have shown that the sandwich structured composite film possesses high transparency, excellent stretchability, and flexibility. Therefore, further experiments should be done to understand whether the composite film can be utilized as a high-performance strain sensor. First, the transient relative resistance variations under different strains during fewer cycles of loading–unloading were measured. The results (Figure 8.18a) reveal that the resistance of the composite film increases with increasing tensile strain and then goes down with unloading. The baselines are almost unchanged within the strain range of interest, signifying that the original resistance of the composite film can be fully recovered after releasing the material from the stressed state. In comparison, the baselines of the AgNWs/PU composite film gradually rise during the same tests (Figure 8.18c). Especially for the first cycle, the resistance after unloading becomes much higher than the initial one.

The underlying mechanism involved in these different behaviors can be explained by the AgNWs–PU interaction. On account of the much higher Young's modulus of AgNWs (81–176 GPa) [42, 43] than that of PU (~6.5 MPa), AgNWs can be regarded as rigid fillers during the loading–unloading cycles. When the composite is subjected to tension, the distance between AgNWs has to be enlarged due to the change in their positions and orientations. Besides, the different extensionalities of PU and AgNWs may produce voids at the interface between the ends of AgNWs and

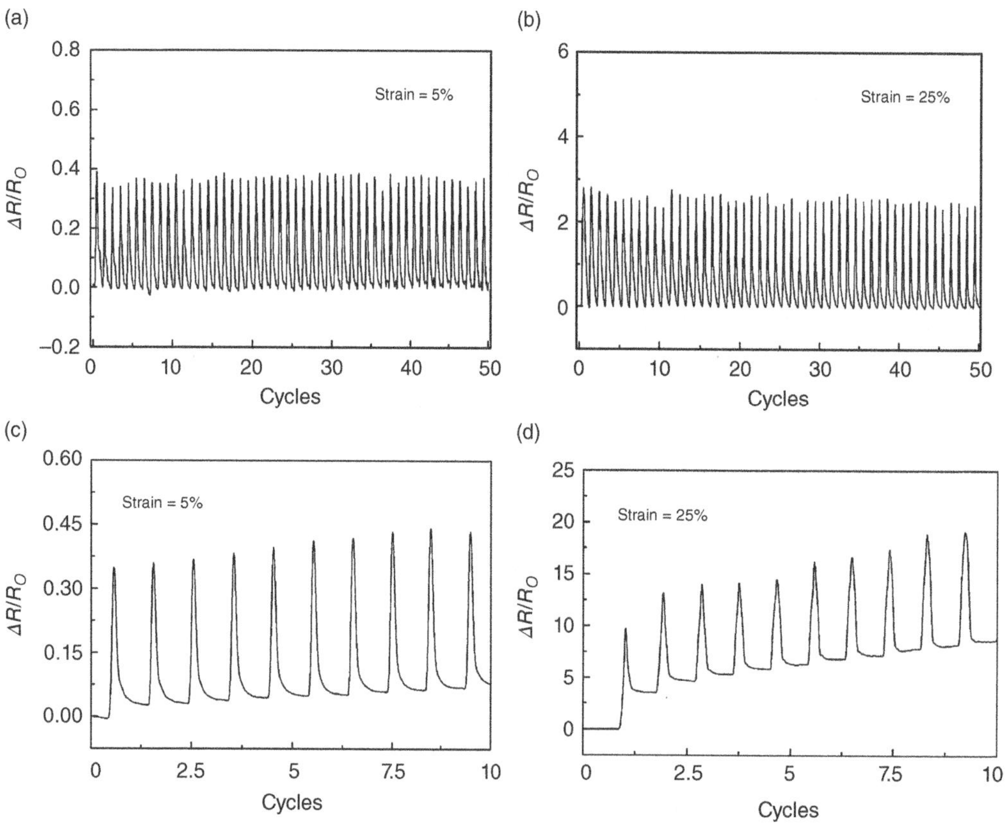

Figure 8.18 Typical relative resistance changes, $\Delta R/R_o$, of (a, b) sandwiched PU/AgNWs/PU and (c, d) AgNWs/PU composite films measured under repeated loading–unloading cycles at different tensile strains. The AgNW concentration and spray-coating time for preparing the specimens: 6 mg ml^{-1} and single pass, respectively.

elongated PU. In the course of unloading, the AgNWs are easy to buckle out of the plane due to the different recovery rates of PU and AgNWs. Buckling and even the subsequent fracture of AgNWs would destroy the electrical pathways to a certain extent and result in irreversible increase of electrical resistance, as characterized by the increased baselines of the AgNWs/PU composite film. Nevertheless, in the case of the sandwiched PU/AgNWs/PU composite film, the integrated structure reduces the probability of interfacial damage and forces the deformed AgNWs to follow their defined paths back without buckling. The conduction network of AgNWs just reversibly changes within the loading–unloading cycles and therefore the relative resistance change can be steadily repeated without changing the baselines.

Durability represents whether the strain sensor can be used steadily for a long time. In addition to the inspection of baseline wander, variation in electrical resistance under longer term cyclic strain should be studied. Figure 8.19a shows that the sandwiched PU/AgNWs/PU composite film is able to stably operate for at least 5000 cycles without failure under a tensile strain of 4%. The maximum cycle of 5816 is recorded when fatigue crack growth starts and the AgNW network becomes extremely unstable as reflected by the drastic increase in resistance.

For comparison, the composite film without the protective AgNWs/PU layer and the sandwiched PU/AgNWs/PU composite film in which the AgNWs are not welded by plasma treatment are tested in the same way. It is seen from Figure 8.19b that in the absence of the protective PU layer, the irreversible buckling and fracture of AgNWs occurs, resulting in not only the upward drifting of the baseline, but also the continuous increase in resistance from the beginning. Figure 8.19c further reveals the critical role of plasma treatment. When the AgNWs in the sandwiched composite films are not welded in advance, the AgNWs are only lapped together. The conductive network is liable to be damaged upon stretching and the relative variation in resistance is raised with time. Therefore, both the protective layer and welding of the AgNWs are critical for enabling the long-term stability of sensing performance.

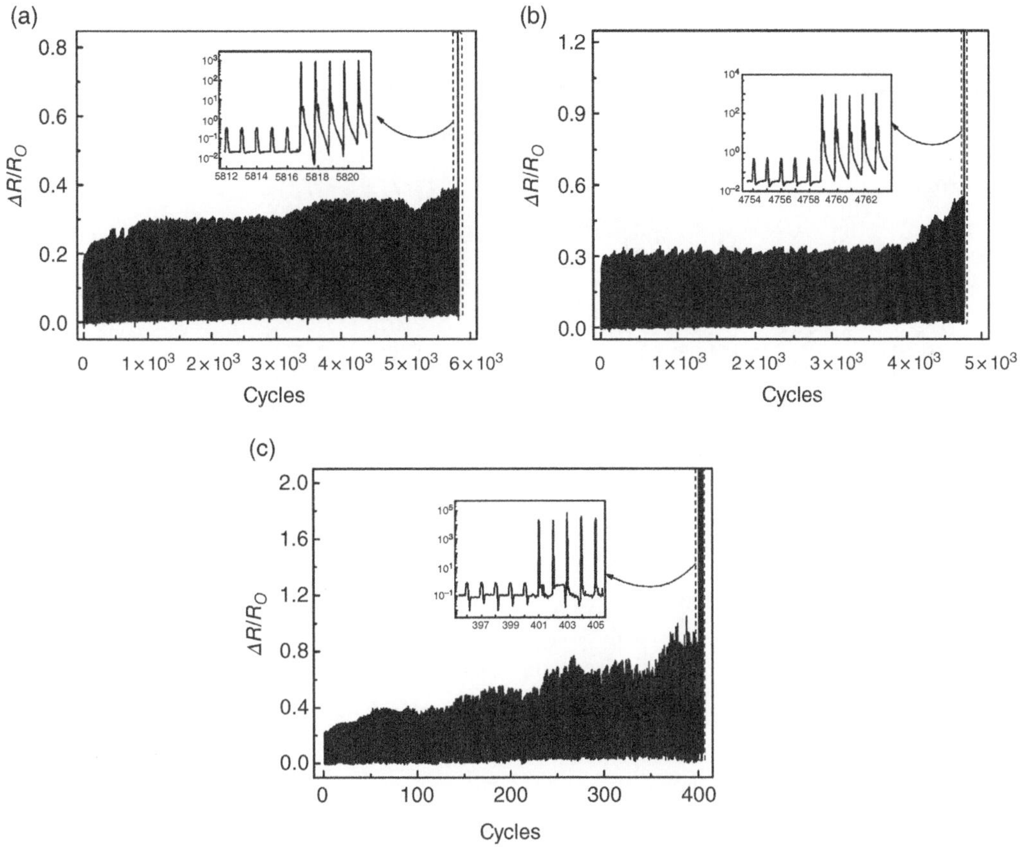

Figure 8.19 Typical relative resistance changes, $\Delta R/R_o$, of (a) sandwiched PU/AgNWs/PU, (b) AgNWs/PU, and (c) sandwiched PU/AgNWs/PU (in which the AgNWs are not treated by plasma) composite films measured under repeated loading–unloading cycles. The AgNW concentration and spray-coating time for preparing the specimens: 6 mg ml^{-1} and single pass, respectively; and tensile strain: 4%. *Source:* Reprinted from Song et al. [34]. Copyright 2019, with permission from the Royal Society of Chemistry.

Self-healing is an effective method for improving durability of material by elimination of the microcracking-induced deterioration. The polyurethane used in the present work proved to be self-healable via the exchange reaction of the included disulfide bonds triggered by the low concentration UV component of sunlight. In the following, research is carried out to check whether this interesting performance is inherited by the PU-based composite films.

For purposes of evaluation, the target materials were scratched to create manual damage in both PU and the AgNW network and then exposed to natural sunlight for rehabilitation. First, AgNWs/PU is discussed. As shown by Figure 8.20a,b, after the composite film is repaired under sunlight, the scratch (~7 μm in width) disappears. Meanwhile, the conductivity is approaching to its initial value with a rise in healing time, meaning that the AgNWs are gradually re-spliced (Figure 8.20c). It is evident that the composite film can be self-healed for recovering its electrical conductivity by means of sunlight irradiation.

AgNWs cannot be reconnected at room temperature after rupture. The restoration of the electrical conductivity must be related to the self-healing ability of the PU substrate. That is, the broken AgNWs at the interface are brought into intimate contact once again by the contraction of PU resulting from crack healing. Careful study of Figure 8.20c shows that the sheet resistance drops slowly at the beginning of healing and then quickly with increase in time. This is because the crack is first closed by hydrogen bonds on the opposite surfaces, followed by the disulfide bond exchange under sunlight across the interface. The covalent bonding makes a greater contribution to the crack healing and conductive path recovery.

The structural characteristics of the conductive AgNW network determines the electrical healing efficiency of the composite film. The denser and thicker AgNW network suggests higher re-connection probability of the broken AgNWs and hence higher electrical healing efficiency (Figure 8.20d). In consideration of the transparency requirement of our sensor (Figure 8.15a), a single pass

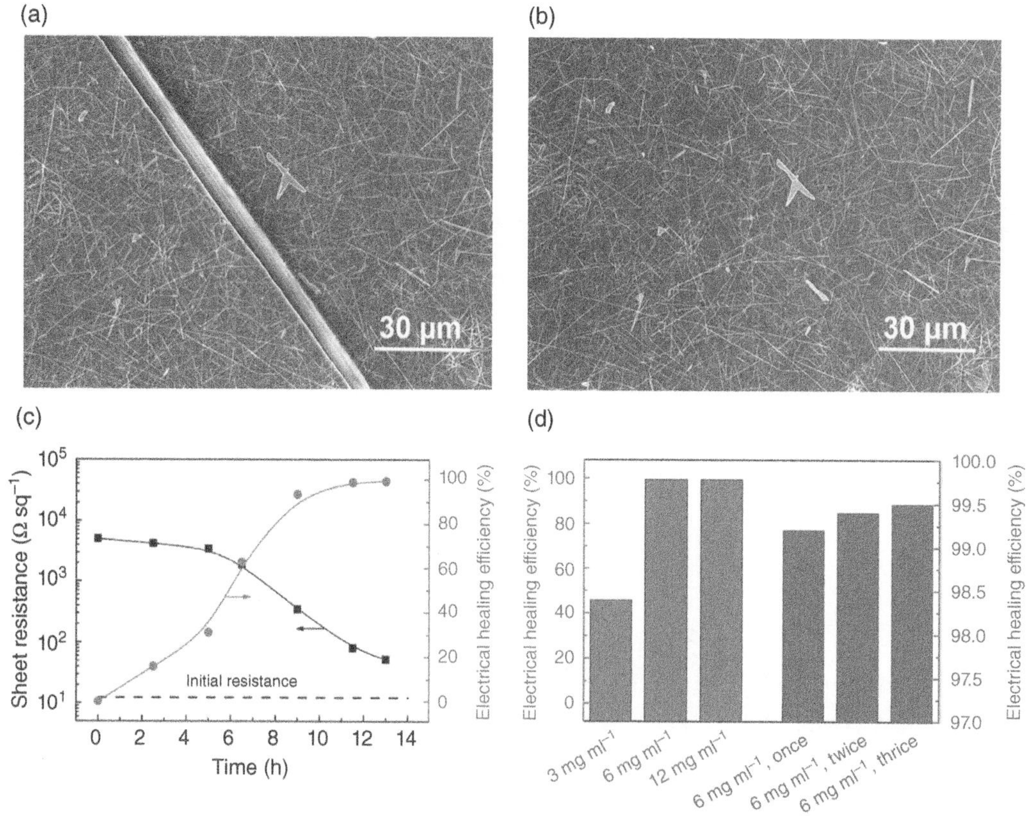

Figure 8.20 SEM images of (a) the scratched and (b) the sunlight healed surface of AgNWs/PU composite film. (c) Sheet resistance and the corresponding electrical healing efficiency of AgNWs/PU composite film as a function of sunlight healing time. (d) Effects of concentrations of AgNW dispersion and spray-coating times on the electrical healing efficiency of AgNWs/PU composite film. The AgNW concentration and spray-coating time for preparing the specimens of (a–c): 6 mg ml^{-1} and single pass, respectively. *Source:* (a, b) Song et al. [34], Figure 12, p. 8/with permission of Royal Society of Chemistry. (c, d) Reprinted from Song et al. [34]. Copyright 2019, with permission from the Royal Society of Chemistry.

spray-coating of 6 mg ml^{-1} AgNW dispersion is still one of the optimal manufacturing conditions. On the other hand, the composite film of AgNWs/PU can also be repeatedly self-healed under sunlight like its PU substrate [30]. Visual inspection shows that the AgNWs/PU composite film is subjected to multiple scratching/healing and all the scars can be recovered with high electrical healing efficiencies (i.e. 99.2, 98.8, and 98.3% corresponding to the successive three healing processes). In particular, the latter scratch is intentionally created across the former one (that had been healed). The disappearance of the intersection of two successive scratches is indicative of the repeated healability of the composite film.

The above study of AgNWs/PU implies that the sandwiched PU/AgNWs/PU must be able to be self-healed under sunlight illumination. Figure 8.21a proves this is the case. The electrical healing efficiencies of the repeated restorations of the scratched specimen are 99.2, 99.2, and 99.3%, respectively (Figure 8.21a). By taking advantage of the recovered constraint effect of the protective layer, the deviation of the healing efficiencies is slightly lower than that of AgNWs/PU. More importantly, the fatigue failed specimen (that had been subjected to cyclic loading–unloading for 5816 cycles, see Figure 8.19a) can behave like the original version after sunlight driven healing (Figure 8.21b), exhibiting long-term stability of resistive sensing.

For reference, the failed specimen was placed in the dark for 24 hours and then tested under the same conditions. Due to the inability of the PU protective layer to self-repair in this case, the constraining effect at the fatigue-induced damaged part has to be weakened. The temporarily recovered conduction paths as a result of reconnection of AgNWs along with unloading are thus easily broken when the test resumes. Consequently, the relative resistance of the specimen keeps on increasing from the very beginning of the cycling and the specimen fails again only after about 400 cycles (Figure 8.21c). The result highlights the importance of self-healing from another angle.

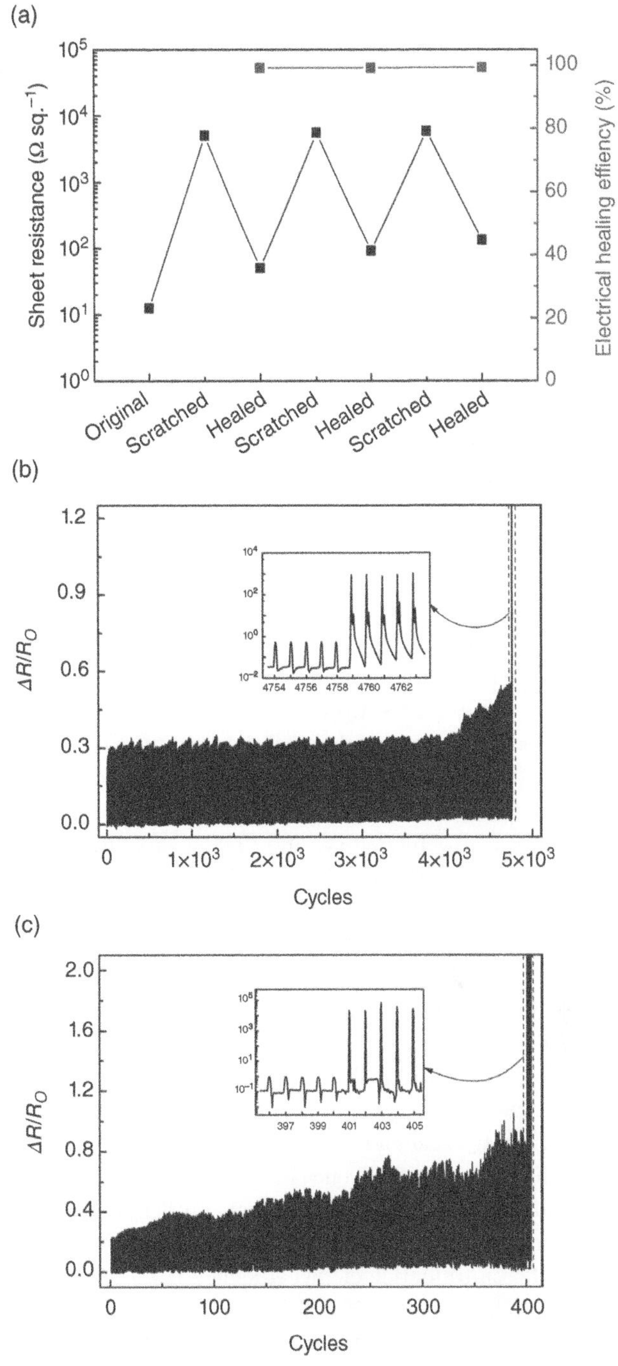

Figure 8.21 (a) Quantification of repeated sunlight driven self-healing of electrical conductivity of the scratched sandwiched PU/AgNWs/PU composite film. (b) Typical relative resistance changes, $\Delta R/R_0$, of the sandwiched PU/AgNWs/PU composite film (that had been failed after 5816 cycles of loading-unloading test [see Figure 8.19a] and then healed under sunlight) measured under repeated loading–unloading cycles. Tensile strain: 4%. (c) Typical relative resistance changes, $\Delta R/R_0$, of the sandwiched PU/AgNWs/PU composite film (that had been failed after 5816 cycles of loading–unloading test [see Figure 8.19a] and then placed in the dark for 24 hours) measured under repeated loading–unloading cycles. Tensile strain: 4%). The AgNW concentration and spray-coating time for preparing the specimens: 6 mg ml^{-1} and single pass, respectively. *Source:* Reprinted from Song et al. [34]. Copyright 2019, with permission from the Royal Society of Chemistry.

8.2.2 Commercial Silicone Elastomer

Silicone elastomer is known for its high chemical stability and good environmental adaptivity over a wide temperature range [44]. Accordingly, it has been found in a wide variety of products, such as automobiles, electronics, medical implants, housewares, sportswear, shoes, and sealants. It would be meaningful if silicone elastomer possesses self-healability to autonomously repair the internal damage created during service, as well as recyclability to renew the waste. The life circle of the material would thus be greatly extended.

So far, there have been some attempts to provide intrinsic self-healing ability to silicone elastomer besides extrinsic self-healability [45]. For example, a living silicone rubber was synthesized [21] and the living reactive anionic species produced during ring-opening polymerization of cyclosiloxane maintains reactivity to siloxane equilibration among different network isomers and cyclic oligomers, which can restore the original strength of broken samples. Furthermore, reversible Diels–Alder (DA) bonds were incorporated to cross-link polysiloxane elastomer, so that the latter was coupled with thermal remendability and remolding performance at elevated temperature [46]. Boroxine is another dynamic covalent bond suitable for self-healing. The polydimethyl siloxane (PDMS) cross-linked by boroxine units can be healed by the reversible dehydration and hydrolysis of boroxine [47]. On the other hand, coordination complexes between the ligands in PDMS and Fe(III) centers were established, leading to self-healing of a silicone elastomer with properties close to biological muscles [48]. Similarly, dynamic complexation between thiol groups in modified silicone-oils and silver nanoparticles was applied for self-healing of fragmented silicone elastomer [49]. As for non-covalent interaction, a tris-urea motif giving six available hydrogen bonding sites was introduced into PDMS chains, so that the elastomer can self-repair the damaged surface via hydrogen bonds [50]. In addition, because the carboxyl end groups in linear carboxyl-terminated PDMS oligomers can react with diethylenetriamine and urea, a supramolecular silicone elastomer with self-healing capacity was obtained [51]. Diffusion of polymer chains can also be utilized for self-healing of PDMS [52].

It is worth noting that silicone elastomers have to be cross-linked in practice by radical polymerization, hydrosilylation, or condensation to turn them into solids with enough strength [53]. The irreversible covalent cross-linked networks do not allow for either self-healing or reprocessing [54–56]. Accordingly, the state-of-the-art recycling technology for the end-of-life irreversible silicones lies in the environmentally unfriendly conversion into low-molecular-weight commodities or monomers with the aid of depolymerization agents, such as boron trifluoride diethyl etherate [55, 57], and zinc salts [56]. Therefore, specific silicones containing reversible bonds as mentioned above had to be designed and prepared to acquire intrinsic self-healability and reprocessing properties. Evidently, the method is not suitable for commercial silicone elastomers, which are used daily and where it is hard to change their molecular structure.

For solving the problem, we propose a facile route to prepare reversibly cured silicone elastomer from α,ω-dihydroxyl polydimethylsiloxane through condensation reaction with disulfide bond-containing silane coupling agent as cross-linker [57], which is similar to the industrial curing process of dealcoholized silicone rubber [53]. The disulfide bond belongs to dynamic reversible covalent bonds, and metathesis of disulfide bonds has successfully realized self-healing and recycling of cured epoxy [1, 3, 5, 58] and polyurethane [4, 30, 59] via network rearrangement. It is necessary is to find out the conditions under which the exchange reaction of the disulfide bonds included in the cured silicone elastomer can effectively proceed. Compared with existing works, the key innovative part of the present investigation is that both the PDMS and silane curing agent are not tailor-made in a laboratory but are market available, which is conducive to the industrialization of the outcomes.

Meanwhile, the cured PDMS elastomers for practical usage have to be strengthened by other polymer resins or inorganic fillers. Since the introduction of inorganic fillers greatly reduces transparency of the materials, addition of polymer resins is a common measure. The mostly used resins are MQ and MTQ resins derived from hydrolysis–condensation between different silanes [53], which consist of end-capping agent (M unit, mono-functionality) and 3D cross-linking segment (T unit, bis-tri-functionality, or Q unit, quadri-functionality). To improve both mechanical strength and reversibility of the PDMS elastomer, a new MTQ resin is synthesized here, in which the silane part of the T unit is substituted by the disulfide bond-containing silane coupling agent used for curing PDMS. The SS-MTQ resin is expected to behave like commercial MTQ resin for the case of reinforcing PDMS elastomer and enhance self-healing and recycling abilities of the elastomer due to the increased amount of reversible disulfide bonds.

After many trials, we found that the silane cross-linked silicone elastomer exhibits dynamicity under sunlight. Unconscious self-healing of cracks and reprocessing of silicone elastomer with the aid of sunshine is naturally attractive because solar energy is green and limitless. In the following, the smart stimulus–response performance and the underlying mechanism are analyzed, with the objective of forming a solid knowledge basis for the development of a technology having practical value.

First, disulfide-containing silicone resin (SS-MTQ) was prepared. Typically, hexamethyldisiloxane (MM, 40 g, 0.25 mol), ethyl alcohol (50 g), deionized water (90 g), and hydrochloric acid (8 g) were successively added into a 500 ml three-neck round-bottom flask. Next, the mixture was stirred at 25 °C for 30 minutes with the protection of nitrogen. Afterwards, the mixed solution of bis(triethoxysilylpropyl) disulfide (Si75, 75 g, 0.16 mol, made by Guangzhou or Aldrich for different tests) and tetraethyl orthosilicate (TEOS, 40 g, 0.19 mol) was dropwise added to the flask, and the reaction mixtures were further stirred at 65 °C for four hours. When the system was cooled to room temperature, the organic layer was separated by a separatory funnel and rewashed with 75% ethyl alcohol three times. A viscous light yellow liquid was obtained as the residual water and ethanol were removed by vacuum distillation at 130 °C for 1.5 hours. The end groups in SS-MTQ are mainly methyl and hydroxyl groups. The latter would interact with PDMS as reinforcement when they are compounded together. By using the back titration method, the hydroxyl content in SS-MTQ was found to be about 1.56 wt%.

The model silicone elastomer (SR-SH) of this work was fabricated according to the conventional preparation procedure of condensed type silicone elastomer in industry (Schemes 8.7 and 8.8). First, HO-PDMS-OH (100.00 phr, viscosity = 3000 cSt unless otherwise specified) was compounded with the silicone resin (SS-MTQ, 5 phr, made by industrial grade Si75 from Guangzhou). Then, the cross-linking agent, bis(triethoxysilylpropyl) disulfide (Si75, 10 phr, industrial grade made by Guangzhou), was added and mixed. After that, the catalyst, dibutyltin dilaurate (DBTDL, 3 phr), was incorporated into the mixture. Having been homogenized, the mixture was cast into an open polytetrafluoroethylene (PTFE) mold, and then degassed under vacuum for five minutes. The elastomer was surface dried within three hours at 25 °C, while the curing was allowed to proceed for five days for more sufficient cross-linking. Its mechanical properties are comparable with those of the commercial version [53].

Besides, a reference material (SR-ref) without disulfide bonds was prepared in the absence of SS-MTQ following the same procedures as that applied for preparing SR-SH, and the cross-linking agent Si75 was replaced by TEOS (5 phr).

Disulfide metathesis in polar solvent (acetonitrile) has been effectively promoted by sunlight [30], but silicone elastomer is a weak polar polymer and disulfide metathesis may be influenced by polarity effect [5]. To ensure whether the catalytic activity of sunlight to disulfide metathesis in nonpolar environment is still valid, model disulfide metathesis experiments of low-molecular-weight disulfides in nonpolar solvent (*n*-hexane) were carried out.

The gas chromatograms of the model disulfide exchange reaction between DEDS and dipropyl disulfide (DPDS) show that an additional peak emerges and intensifies with increasing reaction time, while the peaks of both DEDS and DPDS gradually weaken (Figure 8.22a). By analyzing the mass spectrum of the new substance (Figure 8.22b), it

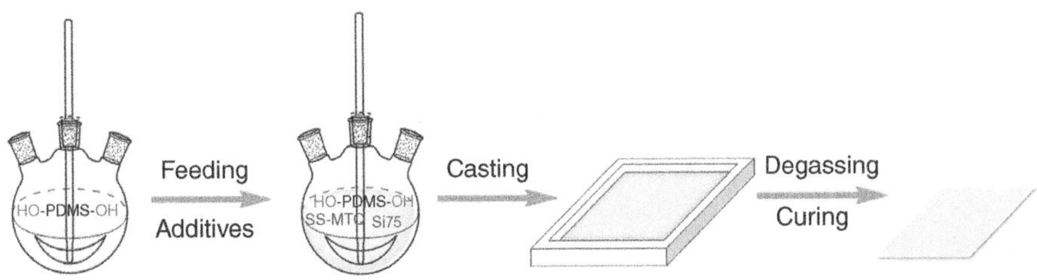

Scheme 8.7 Flow chart showing preparation procedures of the cross-linked silicone elastomer.

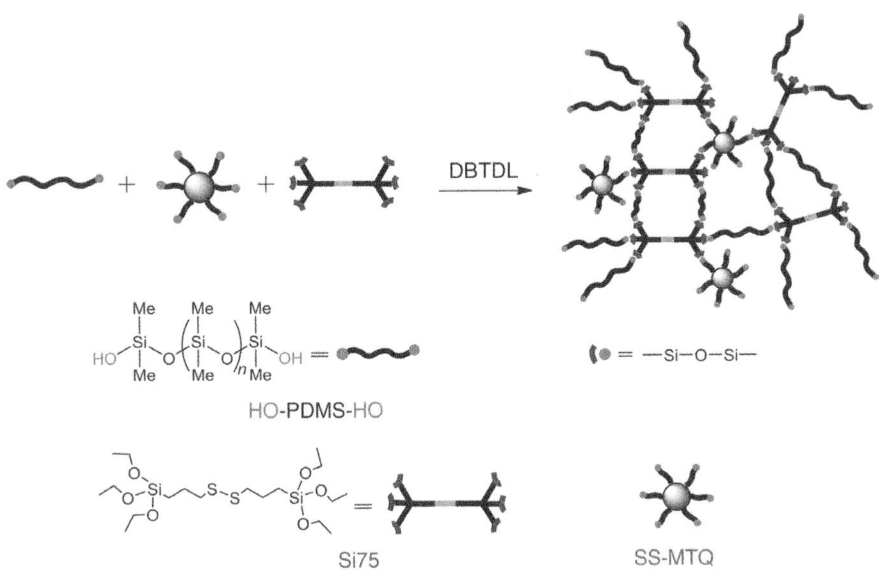

Scheme 8.8 Synthesis of the cross-linked silicone elastomer.

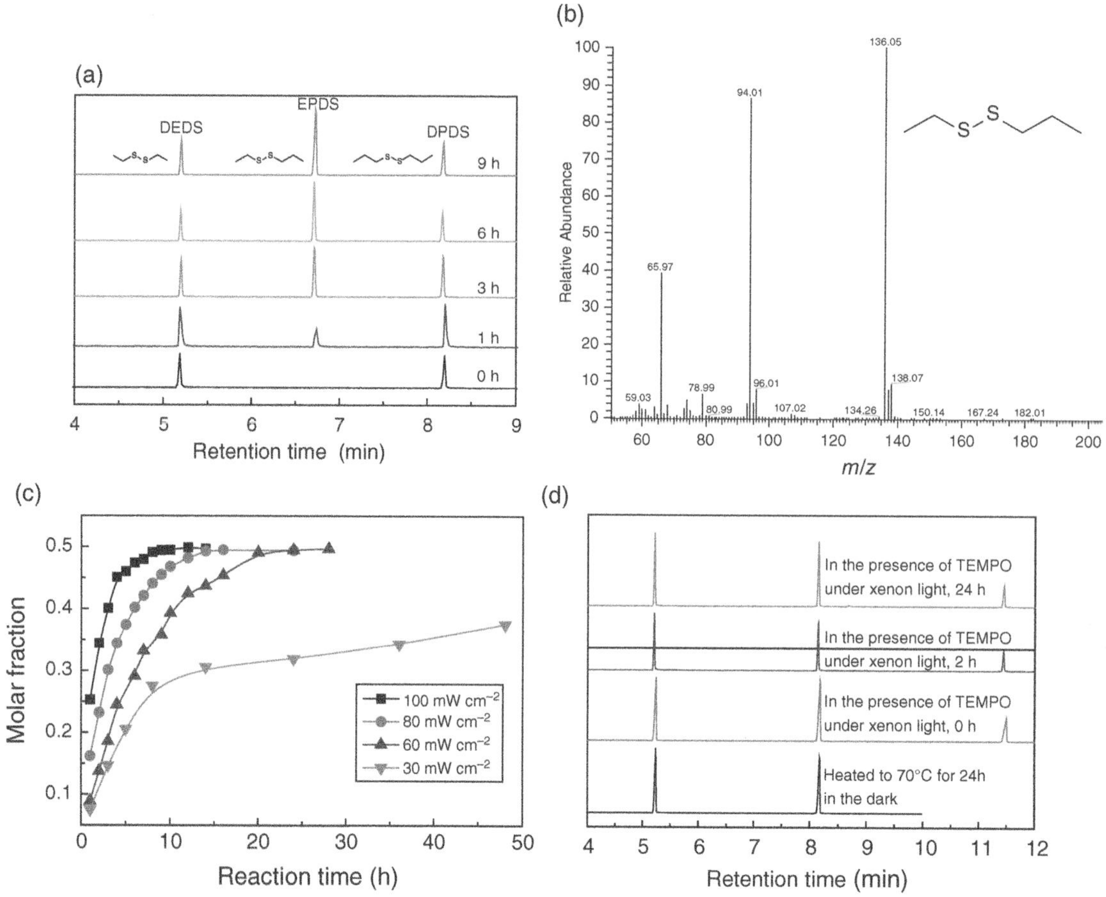

Figure 8.22 (a) GC chromatograms of disulfide metathesis between equimolar DEDS and DPDS in *n*-hexane exposed to xenon lamp light (intensity = 100 mW cm^{-2}). (b) Mass spectrum of the product EPDS of disulfide metathesis between DEDS and DPDS. The signal peak (*m/z*, 136) is a molecular ion peak, which is just the sum of half of DEDS (*m/z*, 122) and DPDS (*m/z*, 150). The ion peaks at *m/z* ≈ 94 and 66 are the fragments of —S—S—C_2H_5 and —S—S—, respectively. (c) Molar fractions of EPDS as a function of reaction time and xenon lamp light intensity. (d) GC chromatograms of disulfide metathesis between equimolar of DEDS and DPDS in *n*-hexane at 70 °C in the dark, or under xenon lamp light (intensity = 100 mW cm^{-2}) in the presence of TEMPO (with molar concentration identical to that of disulfide bonds) without additional heating. *Source:* Reprinted from Xiang et al. [57]. Copyright 2017, with permission from Elsevier.

is identified as the exchange product between DEDS and DPDS, i.e. EPDS. More importantly, the model disulfide metathesis attains the equilibrium state after nine hours, and the ultimate molar ratio of DEDS, DPDS, and EPDS (1:1.02:2.1) is very close to the theoretical value. Accordingly, the reaction equilibrium constant is estimated to be 3.86. It can thus be concluded that under xenon light radiation disulfide metathesis in a nonpolar environment can readily proceed without side reactions. Moreover, disulfide metathesis is still allowed when the light intensity is as low as 40 mW cm^{-2} (Figure 8.22c), which reveals robustness of the reaction despite the fact that sunlight is variable all through the day. When xenon light is absent, however, the disulfide metathesis becomes disabled even at 70 °C (Figure 8.22d). Clearly, sunlight is indispensable for the reaction.

As for the mechanism of light-induced disulfide metathesis, Fairbanks et al. [60] have demonstrated that as exposed to 365 nm UV light (10 mW cm^{-2}) the photogenerated phosphinate radical would attack the disulfide bond, and generate the thiyl radical that initiated the disulfide exchange through chain transfer. Such a radical mechanism was evidenced by Otsuka et al. [27] in the dark, who found that the exchange reaction of disulfide occurred in the presence of benzoyl peroxide (a common initiator of free radical polymerization) but stopped after removing benzoyl peroxide. In accordance with these earlier explorations, radical scavenger (2,2,6,6-tetramethyl-1-piperidinylo xy, TEMPO) was incorporated into the above model disulfide exchange reaction system. As exhibited in Figure 8.22d, there is not any detectable new peak on the gas chromatograms after 24 hours under xenon lamp light.

Since the metathesis product should appear only after 1 hour in the system (Figure 8.22a), the exchange reaction must have been completely hindered by TEMPO, which trapped the radicals necessary for the disulfide metathesis. In this context, it is known that the sunlight-enabled disulfide metathesis also obeys the radical mechanism as in the case of UV illumination [27, 60]. In other words, the UV component in the sunlight must be the key stimulus for the exchange reaction, and the radicals may originate from disulfide scission due to UV radiation [61].

Figure 8.23 shows the transmittance curves of the target material of this study, SR-SH, and its control, SR-ref. Compared with the high transmittance of SR-ref excluding disulfide bonds, the transmittance of SR-SH sharply decreases with the introduction of disulfide bonds as a result of the differences in polarity and solubility parameter between sulfur and silicon atoms. In the UV region, especially below 300 nm, SR-SH has strong absorption, which ensures high responsivity to the low concentration UV component in sunlight.

For purposes of further understanding whether disulfide exchange reaction can take place in cross-linked silicone elastomer, disulfide bond-containing silicone resin (SS-MTQ) is used as the proxy to quantify the chemical structural variation in response to sunlight, because the 3D molecular structure of SS-MTQ resembles the cross-linked structure of silicone elastomer. Here Si75 acts as the T unit (bis-tri-functionality), TEOS as the Q unit (quadri-functionality), and hexamethyldisiloxane as the M unit (mono-functionality). As illustrated in Figure 8.24a, polydispersity of the silicone resin is increased after xenon light irradiation for 12 and 24 hours, while the change in molecular weight is marginal. The variation trends prove that disulfide metathesis has been triggered in the silicone resin as is the case under UV radiation [27], but the total number

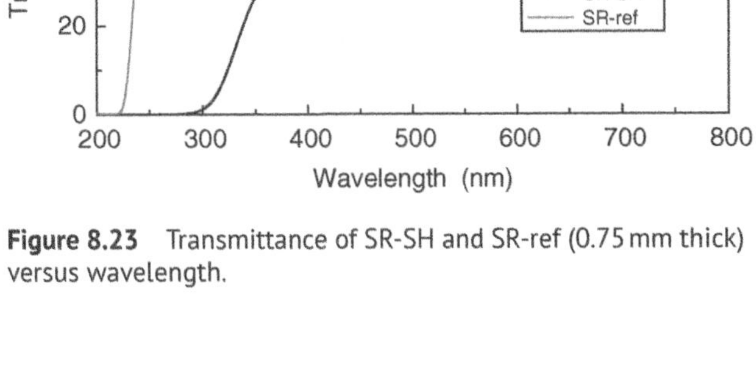

Figure 8.23 Transmittance of SR-SH and SR-ref (0.75 mm thick) versus wavelength.

of cross-linkages would be mostly preserved [62]. When the silicone resin is mixed with Si75 (10 wt%) and then exposed to xenon light, however, a new peak is observed in the low molecular regime of the GPC curve. This means that a significant rearrangement of the silicone resin chains happens due to disulfide metathesis between the macromolecules and small molecular disulfides (Si75). Figure 8.24a also indicates that simple heating without xenon light cannot lead to change in the molecular weight of the silicone resin, as disulfide metathesis is unavailable under the circumstances (Figure 8.22d). This highlights the importance of sunlight-activated disulfide metathesis from another angle.

On the basis of the results of SS-MTQ silicone resin, we start to check the disulfide metathesis in the cross-linked silicone elastomer using the degree of swelling in response to sunlight. The working principle [63, 64] is that (i) the cross-linked networks would expand in good solvent via chain disentanglement enabled by the exchange reaction of the dynamic reversible covalent bonds, and (ii) the cross-linked networks would shrink in poor solvent due to curling up of the chains aided by the exchange reaction of the dynamic reversible covalent bonds. As seen from the first data point in Figure 8.24b, the silicone elastomer SR-SH was first swollen by trichloromethane under xenon light, dried, and then swollen by toluene. Since both trichloromethane and toluene are good solvents of SR-SH, the networks expanded by trichloromethane swelling remain after the xenon lamp is turned off and exhibit higher degree of swelling in toluene. Afterwards, the specimen of SR-SH was first swollen by methanol under xenon light, dried, and then swollen by toluene (see the second data point in Figure 8.24b). Methanol is a poor solvent of SR-SH, so the contracted networks formed during methanol swelling can only reach a rather low degree of swelling even in toluene (good solvent) without the aid of xenon light. The situation is changed so long as the specimen is immersed in trichloromethane again (see the third data point in Figure 8.24b). The repeated fluctuation of degree of swelling of SR-SH demonstrates the effect of sunlight-triggered disulfide metathesis in the cross-linked silicone elastomer. In contrast, the degree of swelling of the control sample, SR-ref, almost does not change after the same treatment due to lack of reversible disulfide bonds.

Considering that disulfide metathesis in the cross-linked silicone elastomer is enabled by sunlight illumination, the material must be able to self-heal by taking advantage of this habit. To quantitatively evaluate such capability, the dumbbell-shaped specimens are bisected, recombined for healing under sunlight radiation, and tested under tension to fail again. As shown in Figure 8.25a, the healed specimen can indeed recover mechanical strength and the

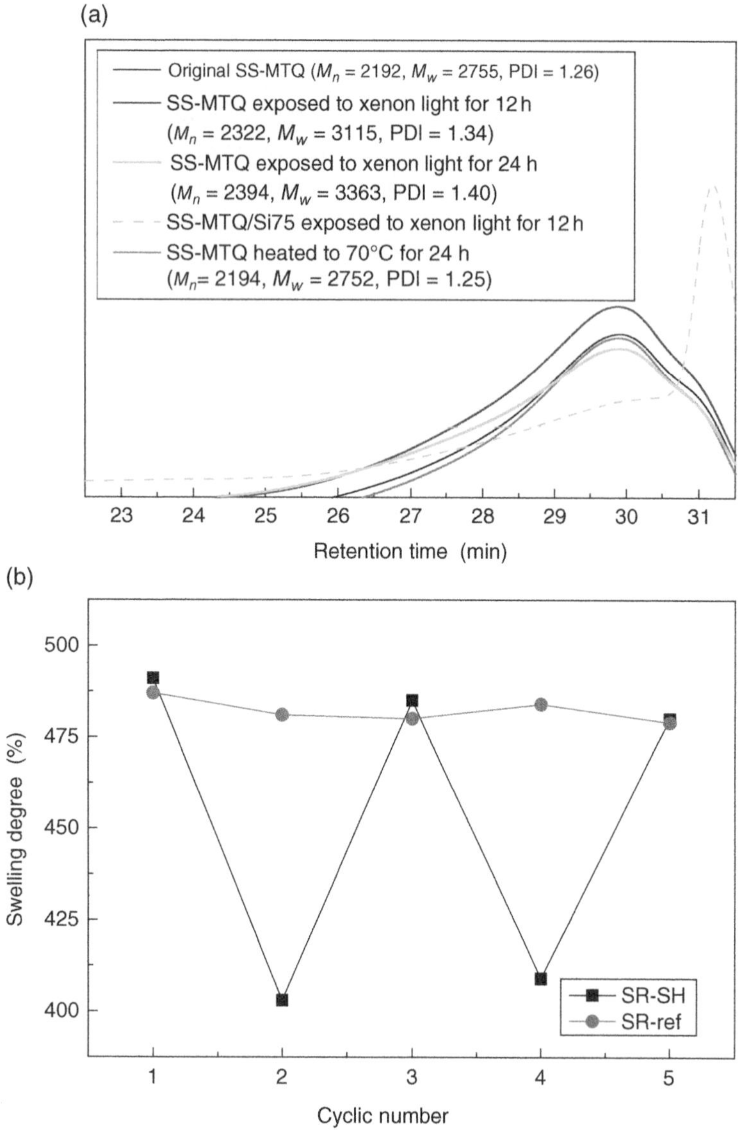

Figure 8.24 (a) GPC curves of SS-MTQ before and after xenon light exposure in comparison with that treated by heating. (b) Swelling degrees of SR-SH and SR-ref in toluene in the dark. Prior to each measurement (data point), the specimen had been swollen by trichloromethane or methanol for 24 hours under xenon light illumination and then dried. *Source:* Reprinted from Xiang et al. [57]. Copyright 2017, with permission from Elsevier.

healing efficiency increases with increasing xenon light intensity at a constant healing time. Obviously, the disulfide metathesis is accelerated with a rise in the light intensity, although the rate of the exchange reaction in the elastomer has to be slower than that in small molecular disulfides (Scheme 8.8) due to the decline in molecular mobility. For achieving better healing effect, the light intensity of the xenon lamp is chosen as $100\,\mathrm{mW\,cm^{-2}}$ for the subsequent study.

Figure 8.25b depicts that the healing efficiency is markedly improved with increasing healing time, especially from 12 to 24 hours. The healing efficiency reaches the maximum at 48 hours. Moreover, the elastomer can be repeatedly healed (Figure 8.25c) owing to the reversibility of disulfide metathesis. The gradual decrease of recovery of mechanical properties should originate from the misalignment between the fractured surfaces [65]. The control specimen, SR-ref, cannot be self-healed under the same circumstances (Figure 8.25d), which demonstrates the critical role of disulfide bonds.

The influence of cross-link density on healing efficiency is given in Figure 8.25e,f. With a rise in the viscosity of the matrix α,ω-dihydroxyl polydimethylsiloxane (i.e. with decreasing matrix molecular weight), the cross-link densities of the resultant silicone elastomer, evaluated by equilibrium swelling, are 1.11×10^{-5}, 0.66×10^{-5} and 0.61×10^{-5} mol cm^{-3}, respectively. Meantime, the corresponding healing efficiencies are 75, 84 and 50%, respectively, which do not monotonously vary with the cross-link density (Figure 8.25e). The result coincides with Figure 8.25f, in which the healing efficiencies are measured to be 70, 84 and 51%, respectively, with increasing

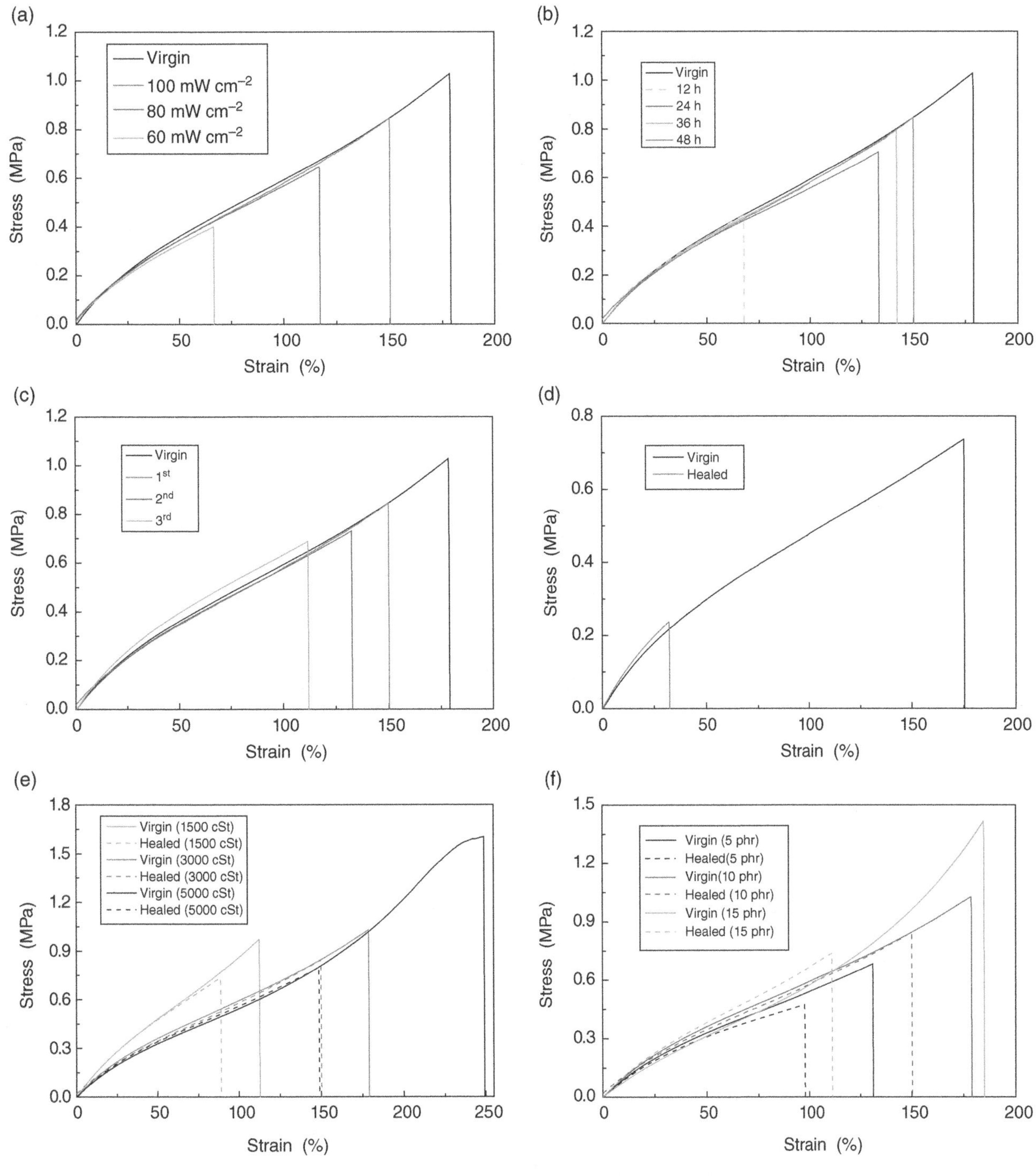

Figure 8.25 Typical tensile stress–strain curves of virgin and healed (a–c, e, f) SR-SH and (d) SR-ref specimens (thickness: 0.75 mm) with the aid of xenon lamp light. (a) Effect of intensity of xenon lamp light (healing time: 48 hours). (b) Effect of healing time (intensity of xenon lamp light: 100 mW cm^{-2}). (c) Effect of repeated healing (intensity of xenon lamp light: 100 mW cm^{-2}; and healing time: 48 hours). (d) Effect of disulfide bonds (intensity of xenon lamp light: 100 mW cm^{-2}; and healing time: 48 h). (e) Effect of matrix viscosity (intensity of xenon lamp light: 100 mW cm^{-2}; and healing time: 48 hours). (f) Effect of dosage of cross-linker (intensity of xenon lamp light: 100 mW cm^{-2}; and healing time: 48 hours). *Source:* Reprinted from Xiang et al. [57]. Copyright 2017, with permission from Elsevier.

dosage of cross-link agent from 5 to 15 phr. The cross-link densities are 0.52×10^{-5}, 0.66×10^{-5} and 0.93×10^{-5} mol cm^{-3}, respectively. In general, the higher the cross-link density means the lower mobility of the dangling chains on the fracture surface [24, 66, 67], so that wetting and diffusion of the macromolecules at the interface are hindered leading to poorer healing effect eventually. Since the dynamic reversible disulfide bonds are brought into the elastomer by the curing agent, however, the higher cross-link density represents a higher amount of disulfide bonds, which should help to offer higher healing efficiency. Therefore, the conflicting effects result in the fact that the highest healing efficiency is observed in the case of moderate cross-link density.

The above self-healing experiments prove that the cross-linked silicone elastomer can re-establish bonding between broken surfaces via disulfide metathesis, which implies that the material could be recycled in the solid state [68–70]. To verify this hypothesis, sheeted SR-SH specimens are pulverized and sandwiched between two quartz plates (Figure 8.26). Having been exposed to xenon light radiation for 48 hours, a new sheet is reformed. As reflected in Figure 8.27a, the recycled SR-SH is able to recover mechanical properties of the original version to a certain extent. The smaller size of the pulverized material corresponds to the higher degree of the strength restoration. This is due to the larger specific surface and more rebinding sites of the smaller particles [30]. On the other hand, because disulfide metathesis is dynamically reversible, the elastomer can be repeatedly recycled (Figure 8.27b).

The above tests are conducted using a xenon lamp as the stimulant of the sun. In the following, we examine the healing and recycling performance under irradiation of natural sunlight. Figure 8.28a indicates that both the

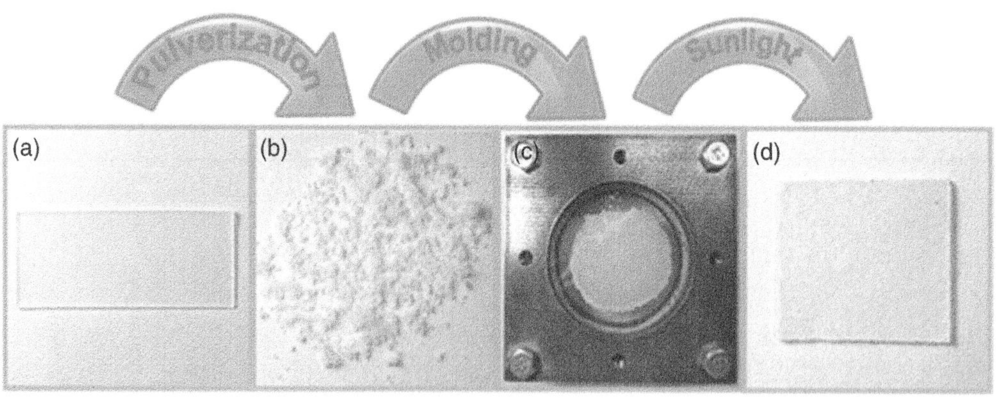

Figure 8.26 Recycling of sheeted SR-SH under 100 mW cm^{-2} xenon light exposure for 48 hours. (a) Original sheet. (b) Pulverized version of (a). (c) Pulverized SR-SH is sandwiched between two quartz plates. (d) Renewed sheet of SR-SH from (c) after xenon light radiation for 48 hours. *Source:* Song et al. [57], Figure 7, p. 7/with permission of Elsevier.

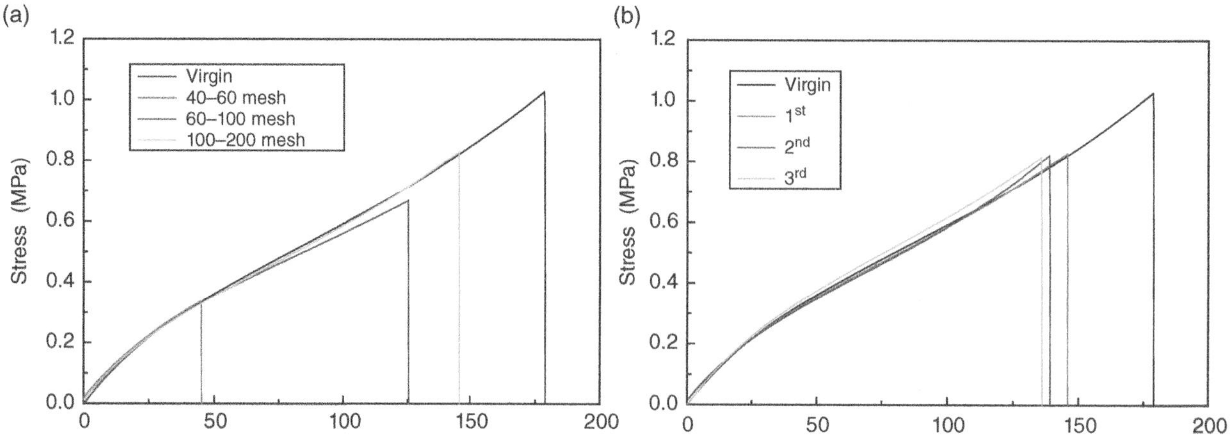

Figure 8.27 Typical tensile stress–strain curves of recycled SR-SH: (a) effect of particle size and (b) effect of repeated recycling (particle size: 100–200 mesh). Recycling conditions: intensity of xenon light: 100 mW cm^{-2}; and compression time: 48 hours.

healed and recycled specimens can restore ~80% mechanical strength of the virgin material, which is similar to the cases exposed to xenon light (Figure 8.25a). Also, the recovery ability of the control specimen of SR-ref is negligible (Figure 8.28b), though its transmittance is greatly superior to SR-SH (Figure 8.23). To exclude the possible contribution of hydrogen bonds to the self-healing, the bisected specimens are intentionally placed in air for 24 hours before recombination allowing for self-combination of the neighboring free hydrogen bonds (if any). By comparing Figure 8.28c with Figure 8.28a, it is known that hydrogen-bond interaction in SR-SH is very weak and the strength recovery is basically a result of disulfide metathesis. Figure 8.28c also illustrates the effect of heating in the dark. The healing efficiency is only ~20%, similar to the values of SR-ref (Figures 8.25d and 8.28b), owing to the physical interaction among the interfacial dangling chains.

8.3 Self-Healing and Reclaiming of Vulcanized Rubber

So far, most intrinsic self-healing polymers are made by introducing reversible bonds into tailor-made macromolecules, which means there is a long way to go before the materials are commercialized. Therefore, imparting intrinsic self-healability to existing polymers would be more feasible for achieving practical application.

Vulcanized rubber developed in the nineteenth century is one of the most used polymers in the world. It has been applied to various industrial products such as tires, conveyor belts, shoe soles, hoses, seals, and gaskets. The large rubber consumption, however, leads to a heavy burden for the environment because of the limited disposal of scraps, which contain nonreversible permanent cross-linked networks. To solve the problem, waste rubbers used to be reclaimed by physical, chemical and biological degradation

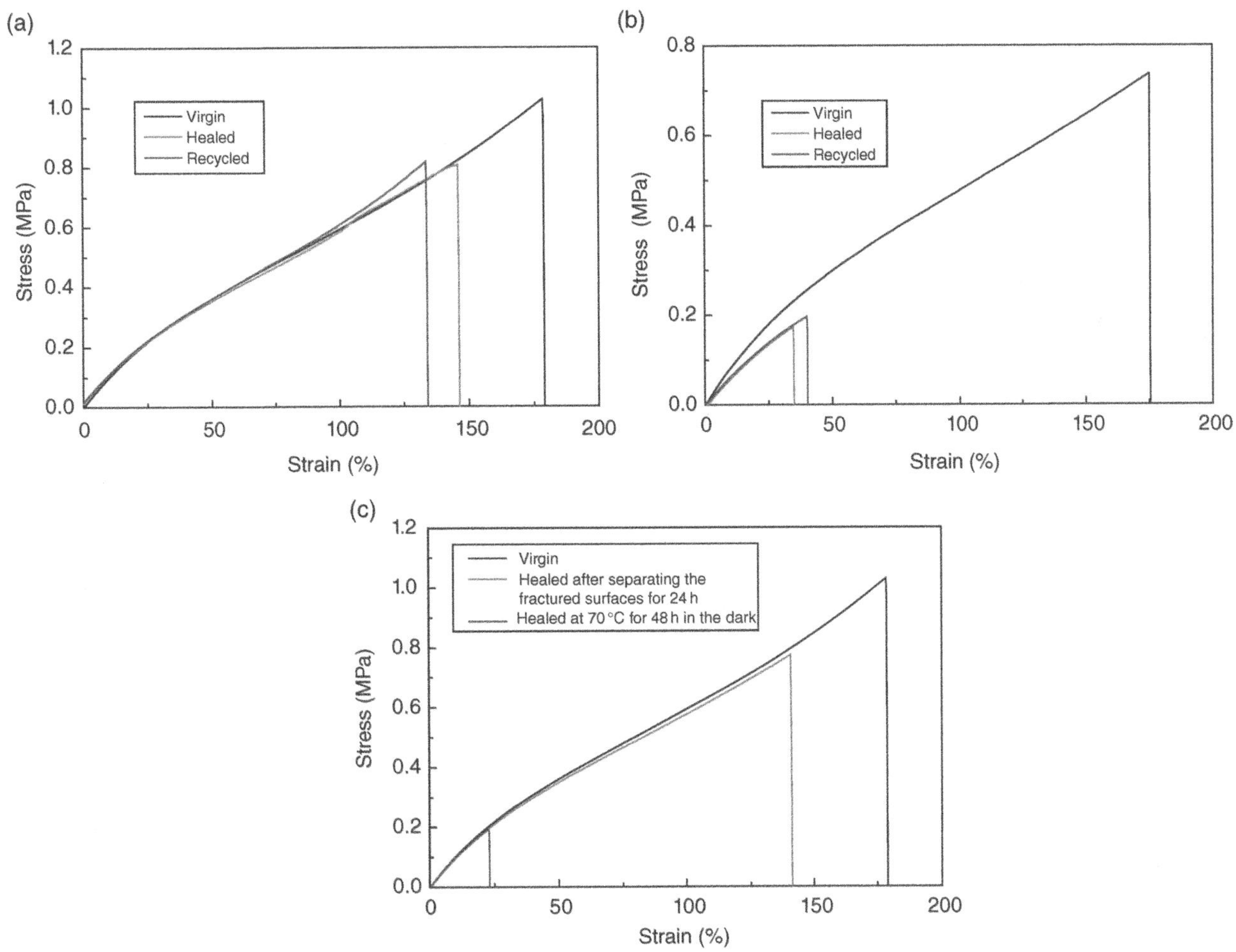

Figure 8.28 Typical tensile stress–strain curves of the virgin, healed and recycled (a, c) SR-SH and (b) SR-ref. Conditions of healing and recycling (unless otherwise specified): exposure to natural sunlight from 9:00 a.m. to 5:00 p.m. for four successive days in June. *Source:* Reprinted from Xiang et al. [57]. Copyright 2017, with permission from Elsevier.

to covert the insoluble and infusible thermoset into small molecular weight fragments with commercial value, or simply burned and buried [71, 72].

Considering that vulcanized rubber made of natural rubber or synthetic rubbers possesses optimized structures and properties after years of development, addition of healing microcapsules into the material or introduction of specific reversible bonds into rubber molecules would not be the most suitable way to couple it with self-healability. In fact, rubber is vulcanized mostly by sulfur or sulfur-donor, resulting in cross-linked networks with monosulfide, disulfide and polysulfide linkages. (According to the accelerator to sulfur ratio, commercial sulfur vulcanization systems are usually classified into conventional (0.1–0.6), semi-efficient (0.7–2.5), and efficient (2.5–12) [73, 74]. The monosulfide, disulfide and polysulfide cross-links produced in the total number of cross-links amount to 0–5, 30–40, and 60–70% for conventional vulcanization, 0–20, 50–70, and 10–30% for semi-efficient vulcanization, and 40–50, 35–50, and 5–20% for efficient vulcanization, respectively [73–79]) Disulfide bonds themselves have enabled self-healing of polymers through exchange or metathesis reaction by using their dynamic reversibility [1–5, 58, 80]. If disulfide exchange or metathesis among different disulfide and polysulfide linkages can also be triggered in sulfur cross-linked networks of vulcanized rubber by a tiny amount of catalyst, the rubber would acquire self-healability accordingly, while keeping its intrinsic quality and original production process. Moreover, reshuffling of macromolecular networks due to dynamic exchange of disulfide and polysulfide bonds would allow vulcanized rubber to be reprocessed without the need of the traditional practice of de-cross-linking [72]. The key issue lies in finding the proper trigger mechanism and catalyst.

To activate the desired reversible reaction of disulfide and polysulfide linkages in vulcanized rubber, heating, rather than light and pH should be used as a means of stimulus in view of practicality and work efficiency. Additionally, it is better if the matching catalyst is not only thermosensitive but also thermotolerant for the following reasons. (i) Rubber vulcanization belongs to a typical thermal processing, which requires all the ingredients to be robust enough to survive the manufacturing. (ii) Healing (or reclaiming) temperature higher than the operating temperature of rubber products is necessary for ensuring stability of structure and properties of the latter during service. Room temperature reshuffling of cross-linked networks that leads to obvious creep of the material at ambient condition is not encouraged in the present case. According to these criteria, thiol, phosphines, and rhodium, which have successfully assisted or catalyzed thermal exchange or metathesis of disulfide, are no longer qualified. Thiol is easy to be oxidized in air. For example, the thiol–disulfide exchange reaction was hindered at temperature above 70 °C as a result of accelerated oxidation of free thiol groups [58]. Phosphine catalyst is vulnerable to oxygen especially at elevated temperature [5, 6]. As for rhodium catalyst, the high cost restricts its large-scale application.

Here, we report $CuCl_2$ as a heat-resisting, oxygen-insensitive and cost-effective catalyst to trigger disulfide metathesis in vulcanized rubber [81]. It was discovered by contrastive analysis of the periodic table of elements and probe trial. $CuCl_2$ used to serve as catalyst for radical addition reaction, chlorination and oxidation, additive for electroplating, photoengraving and feed, and colorant for glass, ceramics, etc. To verify the feasibility of our design, $CuCl_2$-loaded polybutadiene rubber is prepared and characterized in the following.

Since this work is a trial toward development of a technology for possible future application, the basic formula of the model material of this study simulates that for tires (Table 8.2), which contain not only polybutadiene rubber as the base material but also other components such as carbon black, nano-silica, sulfur, stearic acid, and anti-ager. Nevertheless, fractions of the components of our model material differ from those of the industrial product. To increase concentrations of disulfide and polysulfide bonds and mobility of $CuCl_2$ in the sulfur cross-linked network, bis(3-triethoxysilylpropyl) tetrasulfide, a polysulfide organosilane coupling agent and cross-linking agent for the rubber industry, is included. It is expected (i) to take part in the cross-linking reaction of polybutadiene rubber and (ii) to form complex with $CuCl_2$ in the vulcanized rubber. For achieving the targets, part of the organosilane is grafted onto nano-silica in advance, while the rest is directly compounded with all the ingredients. Although incorporation of the organosilane might somewhat weaken the mechanical strength of the model rubber, we prefer to conduct a proof-of-concept study first using the current recipe before working out a solution for real application.

Scheme 8.9 shows the structures of the substances involved in this study that contain disulfide and polysulfide bonds. Polybutadiene rubber acts as the model material. According to the recipe listed in Table 8.2, SiO_2 nanoparticles (100 g) were first mixed with toluene (1000 ml) under ultrasonication. Then, Si-69 (20 ml) was added to the SiO_2 sol with mechanical stirring and reflux for 5 hours. Finally, the SiO_2 grafted by Si-69 (SiO_2-g-Si-69) was centrifuged, ultrasonically washed with cyclohexane three times, and freeze dried for 24 hours. The percentage of grafting was about 10%.

The vulcanized rubber of this work was fabricated as in the following. First, polybutadiene rubber was masticated in a two-roll mill at room temperature. Next, nano-silica (or

Table 8.2 Compositions of the rubbers studied in this work.

Rubbers	Polybutadiene rubber	Sulfur	$CuCl_2$	SiO_2	Si-69[a]	SiO_2-g-Si-69[b]	Carbon black	ZnO	Stearic acid	RD[c]	168[d]	DPTT[e]	DCP[f]
VR-SH[g]	100	1	0.1	0	10	20	5	1	0.5	1	1	1	0
Control 1[h]	100	1	0	0	10	20	5	1	0.5	1	1	1	0
Control 2[i]	100	0	0.1	20	0	0	5	0	0	1	1	0	1.5

[a] Trade name of bis(3-triethoxysilylpropyl)tetrasulfide.
[b] SiO_2 nanoparticles grafted by Si-69.
[c] Trade name of anti-ager poly(1,2-dihydro-2,2,4-trimethyl-quinoline).
[d] Trade name of antioxidant tris(2,4-di-tert-butylphenyl)phosphate.
[e] Dipentamethylenethiuram tetrasulfide serving as accelerator.
[f] Dicumyl peroxide.
[g] Name of the model material of this study.
[h] Reference material containing the same components as VR-SH except that $CuCl_2$ is excluded.
[i] Reference material containing the same components as VR-SH except that peroxide DCP, rather than sulfur, serves as the cross-linking agent. Control 1 and control 2 are used to confirm that both reversible sulfur cross-links and $CuCl_2$ are necessary for healing and reclaiming of vulcanized rubber. Control 1 contains reversible sulfur cross-links without $CuCl_2$, while control 2 contains $CuCl_2$ without reversible sulfur crosslinks.
Source: Reprinted from Xiang et al. [81]. Copyright 2015, with permission from the Royal Society of Chemistry.

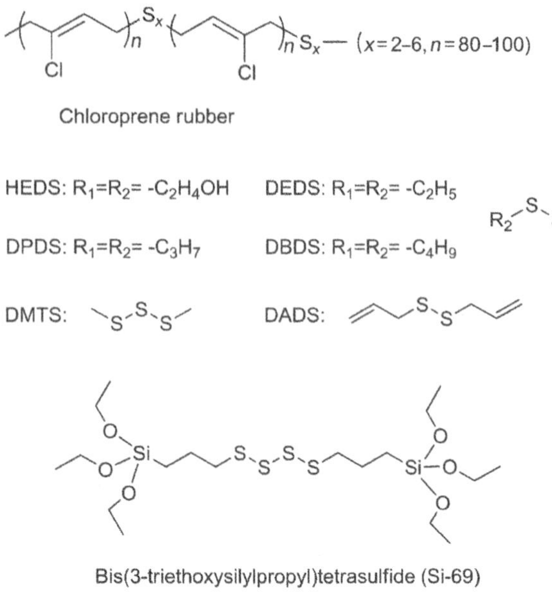

Scheme 8.9 Structures of the disulfide- and polysulfide-containing compounds used in this study.

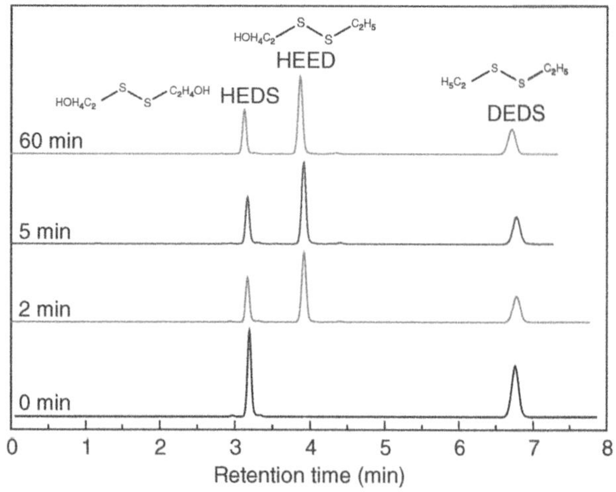

Figure 8.29 HLPC analysis of an equimolar mixture of HEDS and DEDS as a function of time (25 °C, acetonitrile, 0.5 mol% $CuCl_2$). *Source:* Reprinted from Xiang et al. [81]. Copyright 2015, with permission from the Royal Society of Chemistry.

Si-69 grafted nano-silica) and carbon black were mixed with the matrix rubber. Afterwards, Si-69, $CuCl_2$, ZnO, stearic acid, anti-ager RD, and antioxidant 168 were added. Finally, sulfur and accelerator DPTT were incorporated into the premixed compounds. All the raw materials were poured evenly along the nip with alternating cuts. After uniformly compounding, 10 end-roll passes were made before sheeting off. The whole process took about 10 minutes

For comparison, references control 1 and control 2 (Table 8.2) were prepared by the same procedures, except that the curing time of control 1 was 9.70 minutes and that of control 2 was 1.95 minutes as suggested by their t_{90} (the optimum vulcanization time required for reaching a specific torque value which is the sum of minimum torque and 90% of the differentials between maximum and minimum torque).

So far, there is no report on the ability of $CuCl_2$ to catalyze disulfide metathesis. We have to conduct a series of model tests with small molecules first. The HPLC spectra (Figure 8.29) indicate that after the addition of $CuCl_2$ into the equimolar mixtures of two disulfide compounds (HEDS and DEDS) in acetonitrile, a new peak appears, whose area increases with time and reaches the equilibrium within 60 minutes. Meanwhile, the peak areas of HEDS and DEDS decrease. The steady molar proportions of HEDS, DEDS and the reaction product are 24:26:50, which coincide with the theoretical values supposing the reaction product is HEED according to the assignment of the mass spectrum. Evidently, the disulfide exchange reaction can be effectively catalyzed by $CuCl_2$ without side reaction. When the polar acetonitrile is replaced by nonpolar *n*-heptane as solvent, moreover, disulfide metathesis between DPDS and DEDS also takes place in the presence of $CuCl_2$, despite the fact that the reaction temperature has to be higher than 30 °C and a longer time is needed to attain equilibrium. This means that the mechanism of ionic exchange catalyzed by phosphine, which has apparent solvent effect, can be excluded from disulfide metathesis catalyzed by $CuCl_2$. What is more, we also inspected the case in the absence of any catalyst and found that disulfide metathesis did not occur between aliphatic disulfide compounds after a long time regardless of solvent polarity and temperature.

The effect of $CuCl_2$ dosage is examined. It is known that there is a critical threshold for the catalyst. In the case of 0.05 mol% $CuCl_2$, for example, the reaction does not equilibrate within four hours and the conversion is only ~10%. When the dosage of $CuCl_2$ is raised to 0.25 mol%, however, the reaction manages to reach equilibrium after two hours. A much faster equilibrium can be achieved due to further increase of $CuCl_2$ dosage to 0.5 mol%. Therefore, the minimum $CuCl_2$ dosage used for the rubber system should be higher than 0.25 mol%. We have studied the effect of initial molar ratio of reactants on the speed of disulfide metathesis. As the amount of $CuCl_2$ has exceeded the above lower limit, all the reactions are allowed to reach equilibrium within a short time.

Quantification of the above reactions provides the equilibrium constant, K_{eq}, which only varies with reaction temperature and is not affected by initial molar ratio of reactants and catalyst dose. It is thus known that the disulfide metathesis catalyzed by $CuCl_2$ is a dynamic reversible reaction. Besides, the activation energy of disulfide metathesis in acetonitrile is calculated to be 22.3 kJ mol^{-1}, and that in *n*-heptane is 172.3 kJ mol^{-1}. Clearly, more energy needs to be input for the reaction in a nonpolar environment.

To further understand the effect of structures of disulfide molecules on disulfide metathesis, equimolar mixtures of DPDS/diallyl disulfide (DADS) and dibutyl disulfide (DBDS)/dimethyl trisulfide (DMTS) pairs as well as DMTS alone were tested with $CuCl_2$ as catalyst. The allyl disulfide bridge is a common constituent of sulfur cross-linkages in vulcanized rubber, while DMTS simulates polysulfide bonds. The experiments are expected to get close to the actual situation in vulcanized rubber by choosing these reagents. The mass spectra of the metathesis products indicate that allyl propyl disulfide is generated by DPDS/DADS, and methyl butyl disulfide and methyl butyl trisulfide are created by DBDS/DMTS. Moreover, there are also metathesis products (i.e. dimethyl disulfide and dimethyl tetrasulfide) in the system of DMTS and $CuCl_2$, which excludes any disulfide at the beginning. Clearly, disulfide metathesis is allowed to proceed among not only disulfides but also polysulfide molecules, probably because of the similar bonding energies of S—S and $-S_x-$ ($x \geq 3$) bonds (273 and 256 kJ mol^{-1}, respectively). These outcomes suggest that $CuCl_2$ should be applicable for sulfur cross-links of vulcanized rubber that contain both disulfide and polysulfide bonds.

In fact, $CuCl_2$ can ceaselessly catalyze disulfide metathesis. When the third disulfide DBDS was added into the equilibrated DEDS/DPDS system with EPDS as the metathesis product, two new products (i.e. EBDS and propyl butyl disulfide [PBDS]) are yielded and a new equilibrium is reached again. Additionally, the high catalytic activity of $CuCl_2$ is also reflected by the control test with triphenylphosphine (PPh3). It is found that disulfide metathesis is allowed to proceed with the assistance of PPh_3 at 100 °C, but the conversion efficiency is too low to be useful.

In the formula of the rubber to be studied (Table 8.2), there are quite a few chemical additives having different functions. They might deactivate $CuCl_2$ and disable disulfide metathesis. We checked the influence of two representative additives, anti-ager and antioxidant, on disulfide metathesis catalyzed by $CuCl_2$. The dosages of both anti-ager and antioxidant were intentionally much higher than that of $CuCl_2$, while different pairs of disulfides were employed to investigate the possible structural effect of disulfides. Anyhow, the conversion percentage of the reaction approached 45% within two hours for the tests. Hence, the catalytic activity of $CuCl_2$ should be valid for disulfide metathesis in vulcanized rubbers.

In addition to the low-molecular-weight disulfides, a linear macromolecule carrying disulfide and polysulfide bonds on the backbone represented by chloroprene rubber (Scheme 8.9) was also employed as a model compound to examine the catalytic ability of $CuCl_2$ for disulfide metathesis. The molecular weight variation of chloroprene rubber in THF solution with $CuCl_2$ shows that both number average and weight average molecular weights decrease with time, while the molecular weight distribution increases (Figure 8.30). This means that disulfide metathesis has proceeded among the main chains of chloroprene rubber. The variation tendency resembles that reported in reference [27], reflecting the random nature of the metathesis in macromolecules with multiple reversible bonds. As it turns out, disulfide metathesis could occur in disulfide- and polysulfide-containing macromolecules with the assistance of $CuCl_2$.

The above results and discussion have revealed that disulfide metathesis is enabled by $CuCl_2$, but the underlying mechanism is still unknown. The UV–visible spectrum in Figure 8.31a shows that $CuCl_2$ in acetonitrile has two absorption bands. The strong one at 309 nm results from the absorption of $CuCl_2$, while the weak one at 460 nm is ascribed to the complexation effect between $CuCl_2$ and acetonitrile, i.e. $Cu(MeCN)Cl_2$ [82, 83]. Once disulfides are added, the strong peak at 309 nm disappears, and the peak height at 460 nm obviously decreases. In contrast, another strong peak due to absorption of disulfide bonds is observed at 248 nm, which is intensified with a rise in $CuCl_2$ concentration. Similarly, ESR spectra indicate that the g value of $CuCl_2$ in acetonitrile changes from 2.12786 (317 mT) to 2.0616 (327 mT) when equimolar HEDS and DEDS are incorporated (Figure 8.31b), implying variation in the nature of the related chemical bonds [84]. On the other hand, an earlier study of fluorescence of disulfides demonstrated that the formation of complex between Cu^{2+} and disulfides led to quenching effect [85, 86]. The conclusion agrees with our own observation. Figure 8.31c shows that the fluorescence intensity of disulfides decreases with the concentration of $CuCl_2$. The fluorescence is completely quenched when the concentration of $CuCl_2$ is higher than 1 mol%. All these results demonstrate that there is a complex effect between $CuCl_2$ and disulfides.

Amamoto et al. suggested that free radicals are involved in reshuffling of thiuram disulfide moieties under visible light [61]. To see whether the present disulfide metathesis is driven by a similar mechanism, TEMPO was included to capture the possible radicals. The conversion of the disulfide metathesis increases with time like the case without TEMPO (Figure 8.29), but the reaction rate is slowed down after the addition of TEMPO. It might thus be thought that radicals would have been created, which were caught by TEMPO leading to retardation of the reaction. However, the signal intensity of the TEMPO radicals in the reaction system is the same as that in the system without any disulfide under the same conditions. That is,

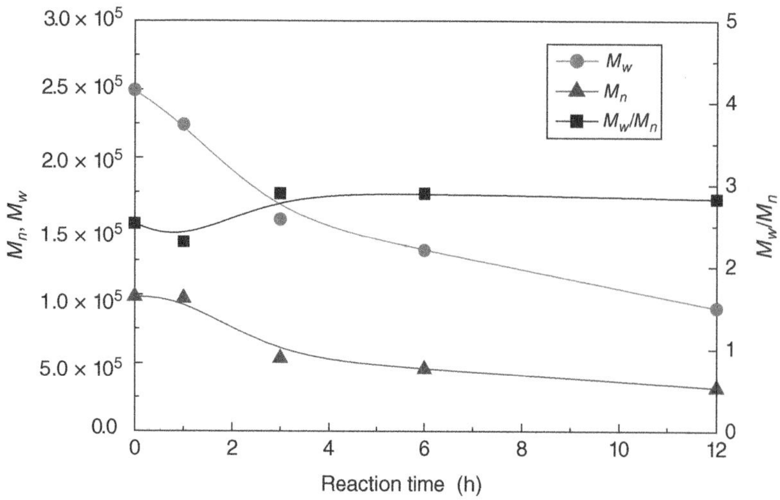

Figure 8.30 Time dependence of number average molecular weight, M_n, weight average molecular weight, M_w, and molecular weight distribution, M_w/M_n, of chloroprene rubber in a solution of THF (3 mg ml^{-1}) at 25 °C ($CuCl_2$: 0.5 mol%). *Source:* Reprinted from Xiang et al. [81]. Copyright 2015, with permission from the Royal Society of Chemistry.

no TEMPO is consumed during the disulfide metathesis. The reduction in reaction rate should result from decay of catalytic activity of $CuCl_2$. No metathesis product is detected only because $CuCl_2$ and TEMPO are mixed for a while in advance. Clearly, the complex effect between $CuCl_2$ and TEMPO restricts the complexation between $CuCl_2$ and disulfide. Besides, if radicals were present in the $CuCl_2$-catalyzed disulfide metathesis, the S-based radicals should have response signals at 350 and 370 mT ($g = 1.925$ and 2.004) in acetonitrile [61]. This is not true for our case as there are no homologous peaks in Figure 8.31b.

Another possible mechanism is related to the formation of ionic intermediates for phosphine-catalyzed disulfide metathesis [5, 6]. However, the above experiment has shown that $CuCl_2$ works in a nonpolar solvent, which disagrees with the solvent effect of phosphines [6]. Additionally, we conducted a group of model tests by introducing diverse inhibitors (i.e. benzoquinone, ethyl acetate, HCl and H_2O) to the disulfide metathesis with $CuCl_2$. The conversion rates of the systems containing inhibitors after 30 minutes are almost the same as that of the system without any inhibitor. Supposing there were mercaptides (RS^-) or sulfenium cations (RS^+) in the reaction process, the inhibitors would interact with the ions and the conversion rate would be decreased. The truth is the opposite, meaning that the reaction has nothing to do with ions.

On the basis of above results and discussion, it is deduced that neither radicals nor ions are involved in the disulfide metathesis catalyzed by $CuCl_2$, which differs from the cases with light catalysis and phosphine catalysis [5, 6, 27, 61].

We found that CuCl can also catalyze disulfide metathesis in the case of higher dosage and reaction temperature, and longer reaction time. Despite the poorer catalytic activity than $CuCl_2$, CuCl takes effect depending on its reducibility, while $CuCl_2$ is often used as oxidant due to the high valence state. The conflicting information demonstrates that the $CuCl_2$ catalyzed disulfide metathesis should not be a redox process.

Figure 8.32a shows the quantum chemical study of $CuCl_2$-catalyzed disulfide metathesis in acetonitrile. All structures are optimized using the B3LYP method with the 6-31G (d) basis set for C, H, O, and S atoms and the SDD basis set for elemental Cu. The PCM solvation model in acetonitrile is performed on the basis of optimized equilibrium geometries by Gaussian Package. Vibrational analysis is performed at the same level of theory. The results indicate that there is complex effect between disulfides and $CuCl_2$, which brings forth the reaction intermediates, IM1. Then, IM1 reaches the transient state, TS1, when the potential energy overcomes the energy barrier, which is the rate-determining step [6]. In the course of transformation, bond lengths of S—S and Cu—Cl also change. Finally, two different sulfur atoms of disulfides in the transient state are split and recombined into a metathesis product and $CuCl_2$-disulfide complex.

According to the results of both experiments and theoretical prediction, the mechanism involved in disulfide metathesis utilizing $CuCl_2$ as catalyst is depicted in Figure 8.32b. First, disulfides and $CuCl_2$ interact with each other as soon as they are mixed together due to the lone pair electrons of sulfur atoms in disulfides and the vacant 3d orbit of Cu(II) [82]. This leads to the formation of an

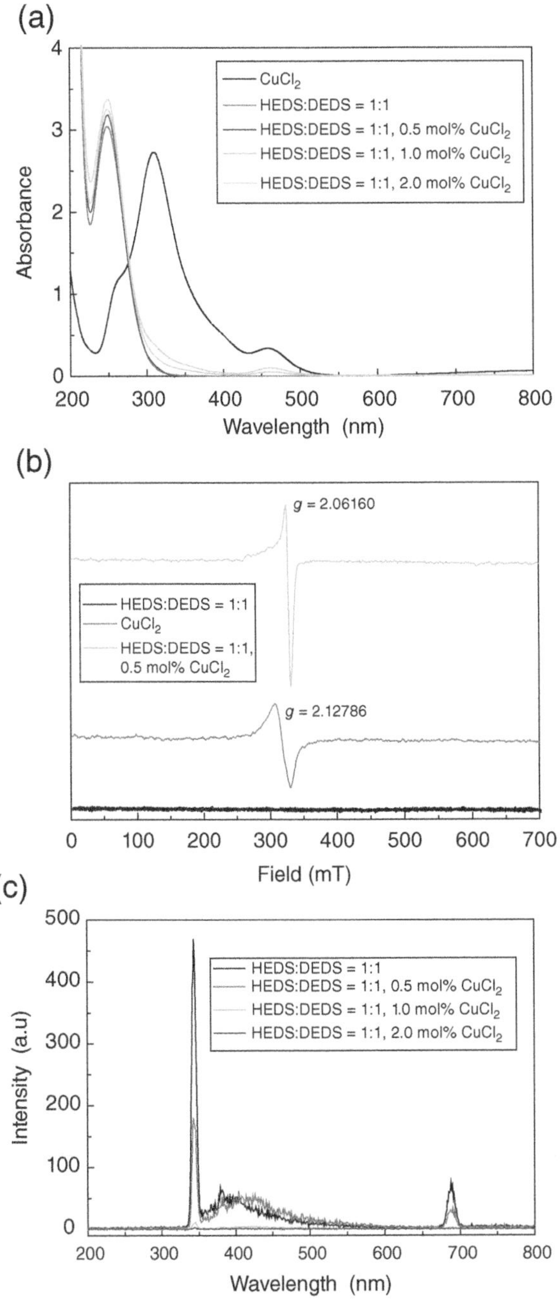

Figure 8.31 (a) UV–visible, (b) electron spin resonance (ESR) and (c) fluorescence spectra (excitation wavelength: 343 nm) of disulfides and $CuCl_2$ in acetonitrile at 25 °C.

intermediate. Then, spatial conformation of the complex intermediate is optimized through adjusting bond angle and bond length between the disulfide and $CuCl_2$. That is, the system enters the transient state. At last, rearrangement and localization of electrons of disulfide bonds make the latter fracture. Recombination of the two different sulfur atoms offers the metathesis product, while $CuCl_2$-disulfide complex remains. The newly generated $CuCl_2$-disulfide complex would rapidly interact with another disulfide, repeating the above processes.

As the model experiments and mechanism investigation have revealed the capability of $CuCl_2$, we start with the work of the model material of this study, VR-SH, whose compositions and curing procedures are carefully described in Table 8.2. The $CuCl_2$ dosage in VR-SH was determined in consideration of speed of disulfide metathesis and aging resistance of the rubber material. The above model experiments have shown that disulfide metathesis can quickly equilibrate with 0.5 mol% $CuCl_2$. However, if the $CuCl_2$ dosage exceeded 1 mol%, the thermal aging property of VR-SH would significantly deteriorate. Therefore, the ideal $CuCl_2$ content in VR-SH was chosen as 0.8 mol% (i.e. 0.1 phr as shown in Table 8.2) to have balanced performance. First, some fundamental properties of the materials are characterized. The percentages of disulfide and polysulfide cross-links in the total number of cross-links in VR-SH are estimated to be 62 and 31%, respectively, while the same parameters of the reference material, control 1, are 54 and 42%, respectively. This means that sufficient amount of reversible sulfur cross-linkages are present in the rubbers like the abovementioned commercial versions. The thermal degradation behaviors show that the initial weight loss temperatures and 50% weight loss temperatures of VR-SH and the references (i.e. control 1 and control 2) are close to each other. Only control 2 possesses slightly higher onset temperature because of the higher C—C bond energy than S—S. It is thus known that the rubbers have acquired good thermostability, which is not affected by the addition of $CuCl_2$. Besides, because (i) the dosage of $CuCl_2$ is low, (ii) the complexation between $CuCl_2$ and disulfide bonds weakens the ability of $CuCl_2$ to induce aging of the rubbers, and (iii) anti-ager is added, thermal aging tests at 110 °C indicated that the aging resistance of the rubbers did not deteriorate in the presence of $CuCl_2$.

Figure 8.33a plots stress relaxation of VR-SH measured at various temperatures. When temperature is lower than 90 °C, VR-SH behaves like conventional polymers. The significant reduction in stress at the beginning stage due to the viscoelastic feature is gradually replaced by a steady plateau. In the case of higher temperature, however, the scenario is quite different. At 110 °C, for example, the stress continues to decrease at slower rate for a long period of time after the initial drastic decline and eventually approaches zero (Figure 8.33a). This means that multiple stress relaxation mechanisms are involved. In contrast, the curves of control 1 and control 2 (Figure 8.33b), in which disulfide metathesis is disabled (Table 8.2), resemble that of VR-SH recorded at 30 °C. Naturally, this relaxation manner of VR-SH is associated with disulfide metathesis as it would lead to rearrangement of the sulfur

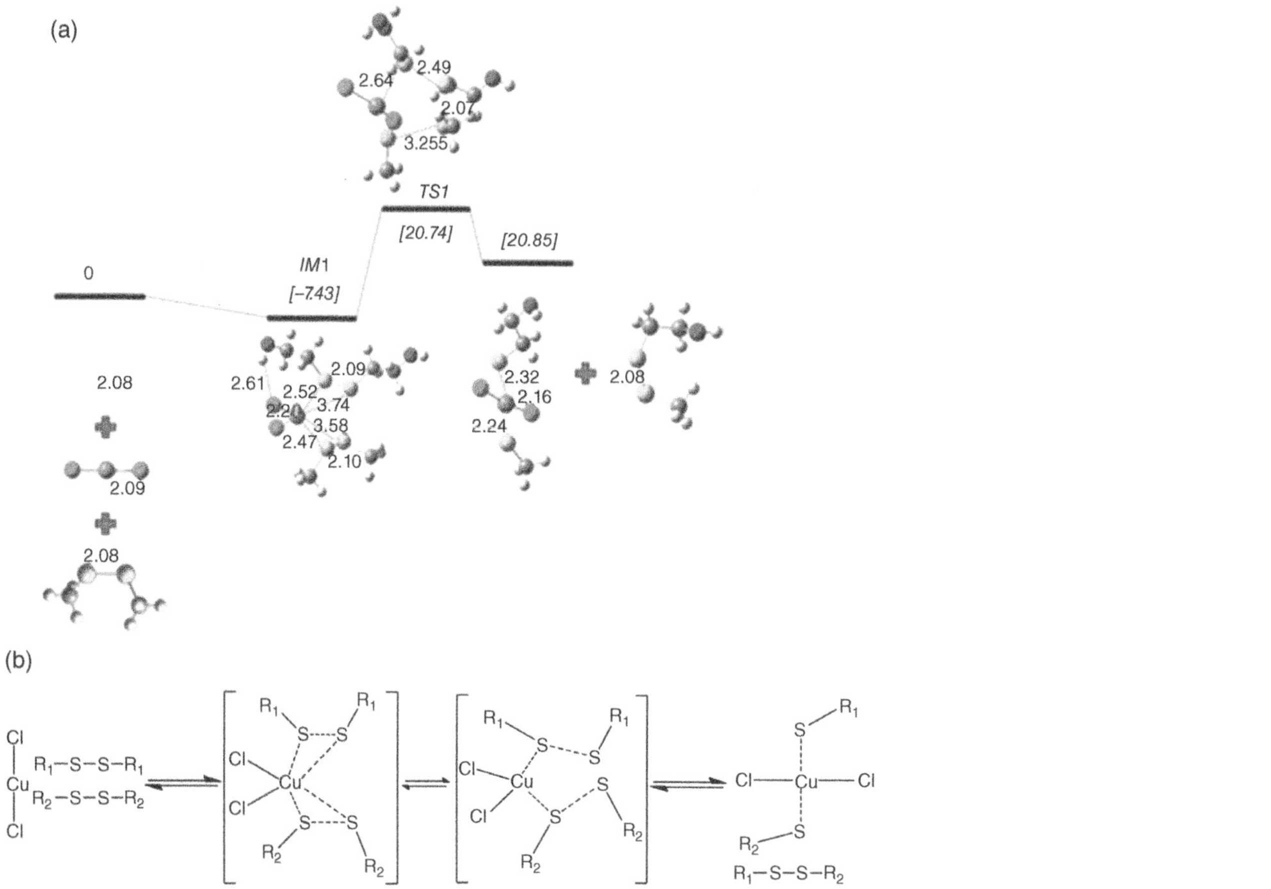

Figure 8.32 (a) Quantum chemical study of $CuCl_2$-catalyzed disulfide metathesis in acetonitrile. Energies are given in kcal mol^{-1}. (b) Mechanism of $CuCl_2$-catalyzed metathesis between low-molecular-weight disulfides.

cross-linked networks once triggered, during which stress relief occurs. Our previous study shows that the activation energy of $CuCl_2$-catalyzed disulfide metathesis of small molecules in nonpolar solvent is higher than that in polar solvent. Here the main component of VR-SH, polybutadiene rubber, is also nonpolar, so that disulfide metathesis takes effect at elevated temperature instead of room temperature.

From a practical perspective, high temperature reshuffling of cross-linked networks of vulcanized rubber is favorable. In this way the crack healing process would not interfere with service stability of rubber products at relatively lower temperature. That is, thermal stability of the material structure is preserved while allowing for reversibility. In normal driving conditions, for example, tire temperature ranges from 30 to 100 °C [87, 88]. Compared with this, $CuCl_2$ seems to be quite suitable for vulcanized rubber as disulfide metathesis induced stress relaxation is obvious only above 110 °C (Figure 8.33a). In this context, 110 °C is selected as the healing (or reclaiming) temperature here. Such a criterion based on accelerated exchange reaction of reversible covalent bonds was also adopted in recycling of polyimine film [89].

Figure 8.34a shows typical tensile stress–strain curves of virgin and healed VR-SH specimens. Healing can indeed proceed as characterized by the fact that the healing efficiency increases with increasing healing time. About 75% tensile strength can be restored after 12 hours. Furthermore, healing of the same place can be repeated (Figure 8.34b). To check the contribution of hydrogen bonding, surfaces of tensile-fractured specimens were not immediately recombined but exposed to air for 12 hours prior to recombination to eliminate the possible hydrogen bond interaction. Accordingly, the healing efficiency approaches 72% for healing time of 12 hours (Figure 8.34c). As a matter of fact, the nonpolar polybutadiene rubber matrix should not possess any hydrogen bonds. Figure 8.34c demonstrates the oxygen insensitivity of the catalytic activity of $CuCl_2$. Comparatively, only ~12% healing efficiency is measured for control 1 and control 2 under the same conditions (Figure 8.34d) as a result of intermolecular entanglements of dangling chains across the fractured surfaces. It can thus

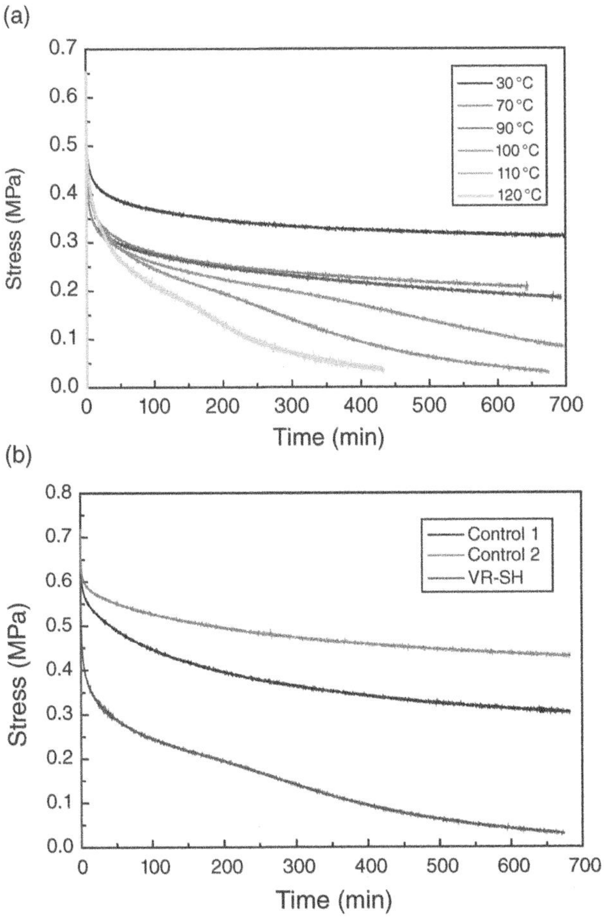

Figure 8.33 Stress relaxation curves of (a) VR-SH measured at various temperatures and (b) VR-SH, control 1 and control 2 measured at 110 °C. *Source:* Reprinted from Xiang et al. [81]. Copyright 2015, with permission from the Royal Society of Chemistry.

be concluded that disulfide metathesis works in healing the vulcanized rubber by triggering rearrangement of inherent sulfur cross-linked networks under catalysis by $CuCl_2$.

During the high temperature vulcanization of the rubber, the catalyst $CuCl_2$ might be turned into Cu_2S, CuS, Cu_2O, and CuO as a result of reaction with the surrounding elements. If so, $CuCl_2$ would lose its ability to catalyze disulfide metathesis, as no metathesis product is generated in the presence of these chemicals. The results confirm that $CuCl_2$ must have kept intact in the vulcanized rubber, otherwise no healing behavior would be observed.

It is worth noting that the mechanical property of VR-SH cannot be fully restored (Figure 8.34a), and the healing efficiency declines with number of healing (Figure 8.34b). Misalignment of the fractured surfaces should be partly responsible for the phenomenon as it leads to pseudo contacts during healing. The non-contact zones make up 6–10% of total area of the fractured surface. This means that the measured healing efficiency should have been higher, if the non-contact regions were intimately contacted with one another leading to additional disulfide metathesis across the interface. Another reason for the suboptimal healing effect lies in the distribution of $CuCl_2$. Although the catalyst particles are well dispersed in the matrix, it is hard to distribute $CuCl_2$ on a molecular level near each disulfide (or polysulfide) bond, so that no disulfide metathesis can be triggered in the regions short of $CuCl_2$ during healing. Although fluidic polysulfide coupling agent Si-69 is introduced to greatly improve mobility (i.e. dispersion) of $CuCl_2$ through the complexation between Si-69 and $CuCl_2$ in the course of compounding, which helps to raise the healing efficiency, $CuCl_2$ is not homogeneously distributed in the matrix. There are always disulfide and polysulfide bonds that fail to take part in the metathesis reaction due to lack of catalyst to form a complex with them, a prerequisite of the reaction (Figure 8.32).

Despite all this, VR-SH proves to acquire remendability due to reconstruction of sulfur cross-linked networks in terms of disulfide metathesis in damaged zones. Taking advantage of the same mechanism, reclaiming of the rubber should also be possible. Accordingly, VR-SH was frozen by liquid nitrogen, ground into small particles and then hot pressed at 110 °C for three hours (Figure 8.35a). Figure 8.35b shows typical tensile stress–strain curves of the reclaimed VR-SH in comparison with that of virgin VR-SH. Evidently, the ground rubber powders can stick to each other under the reclaiming conditions, reforming a compact bulk material with load bearing capability the same as the original rubber before being ground. Careful study of Figure 8.35b indicates that tensile strength and elongation to break of the reclaimed VR-SH increase with decreasing rubber particle size. In general, smaller rubber particles mean larger specific surface area and higher surface concentration of exposed disulfide and polysulfide bonds. Therefore, when smaller rubber particles are closely packed, more disulfide and polysulfide bonds are involved in metathesis. The reconstructed sulfur cross-linked networks become stronger. Although the rubber particle size cannot be further reduced due to limit of the grinding machine in our laboratory, the trend revealed by Figure 8.35b is straightforward. Additionally, Figure 8.35c shows that VR-SH can be repeatedly reclaimed owing to the dynamic reversible characteristics of disulfide metathesis. On the other hand, the abovementioned uneven distribution of $CuCl_2$ catalyst should also partly account for the poorer mechanical properties of the reclaimed VR-SH than the original version.

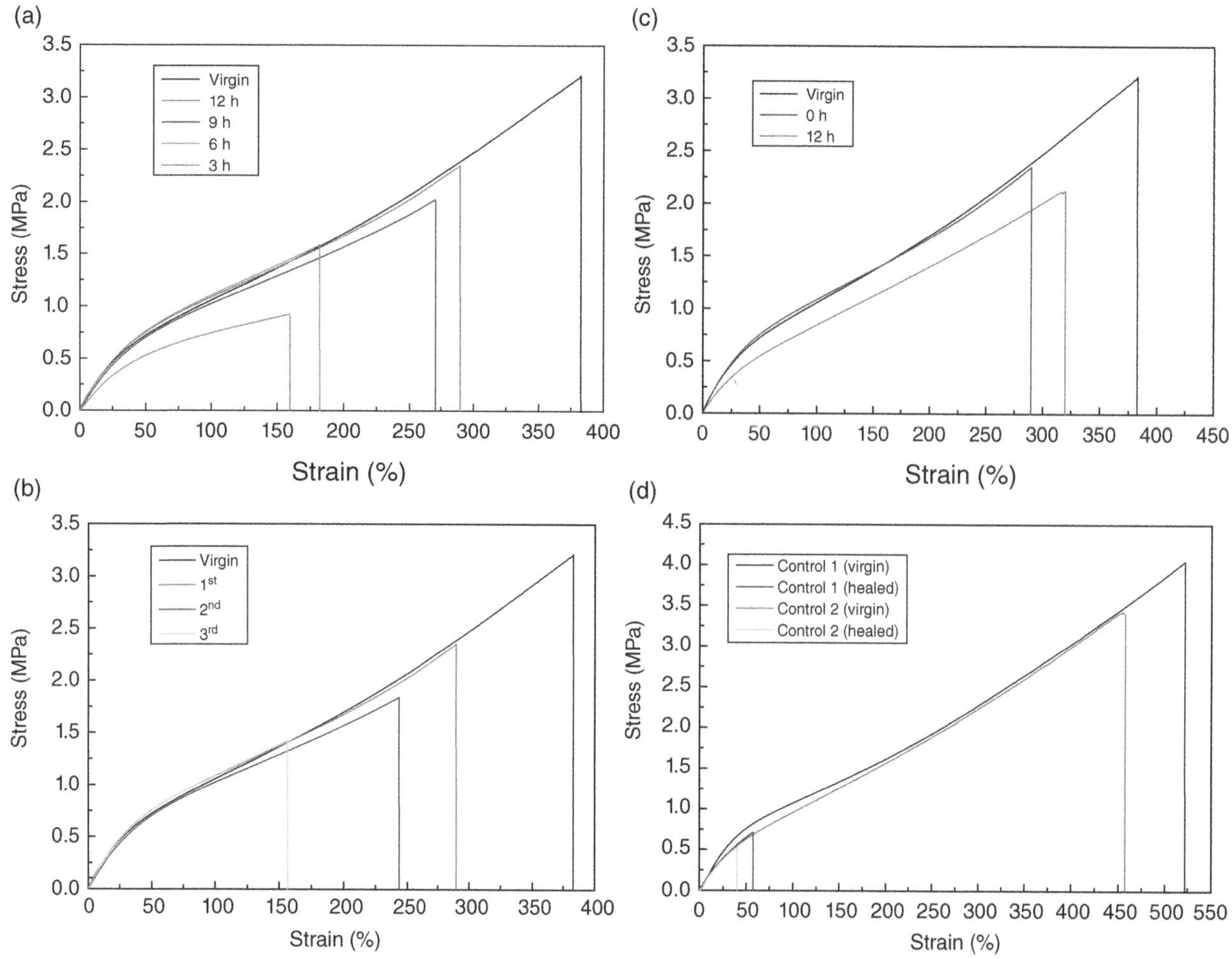

Figure 8.34 (a–c) Typical tensile stress–strain curves of virgin and healed VR-SH specimens. (a) Effect of healing time. (b) Effect of repeated healing (healing time: 12 hours). The tensile strength of the virgin specimen cannot be fully restored, so the reconnected interface became the weakest part, and had to break again during the subsequent tensile tests. This ensured that the second and third tensile failures happened to the same healed portion. (c) Effect of waiting time (which is the time for the broken specimens to wait to be reconnected in air at 25 °C; healing time: 12 hours). (d) Typical tensile stress–strain curves of virgin and healed control 1 and control 2 specimens (healing time: 12 hours). Healing temperature for all the healed specimens: 110 °C. *Source:* Reprinted from Xiang et al. [81]. Copyright 2015, with permission from the Royal Society of Chemistry.

In the meantime, control 1 and control 2 were also ground and tested under the same conditions. On account of the lack of disulfide metathesis, the reclaimed materials possess very poor mechanical strength (Figure 8.35d) like the cases in Figure 8.34d.

Unlike devulcanization, a conventional technique of rubber recycling that cleaves sulfur cross-links of vulcanized rubber, here the reclaiming is based on reshuffling of the cross-linked network through disulfide metathesis. It does not affect the cross-link density and can maintain strength of the rubber in principle. For example, the percentages of disulfide and polysulfide cross-links in the total number of cross-links in reclaimed VR-SH (from the particles of 60 mesh) are 60 and 28%, respectively, almost identical to the data of the abovementioned virgin material.

There were also attempts to recycle vulcanized rubber particles by simple thermal molding without using additional catalyst [90, 91], but the operating principle is related to irreversible scission of sulfur cross-linkages and re-formation of irreversible monosulfide cross-linkages. The reclaiming cannot be repeated multiple times accordingly.

8.4 Concluding Remarks

The exchange reaction of disulfide bonds in polymers can be triggered by heating or the low concentration UV component of sunlight. By taking advantage of this, cross-linked polymers can be repeatedly self-healed, recycled,

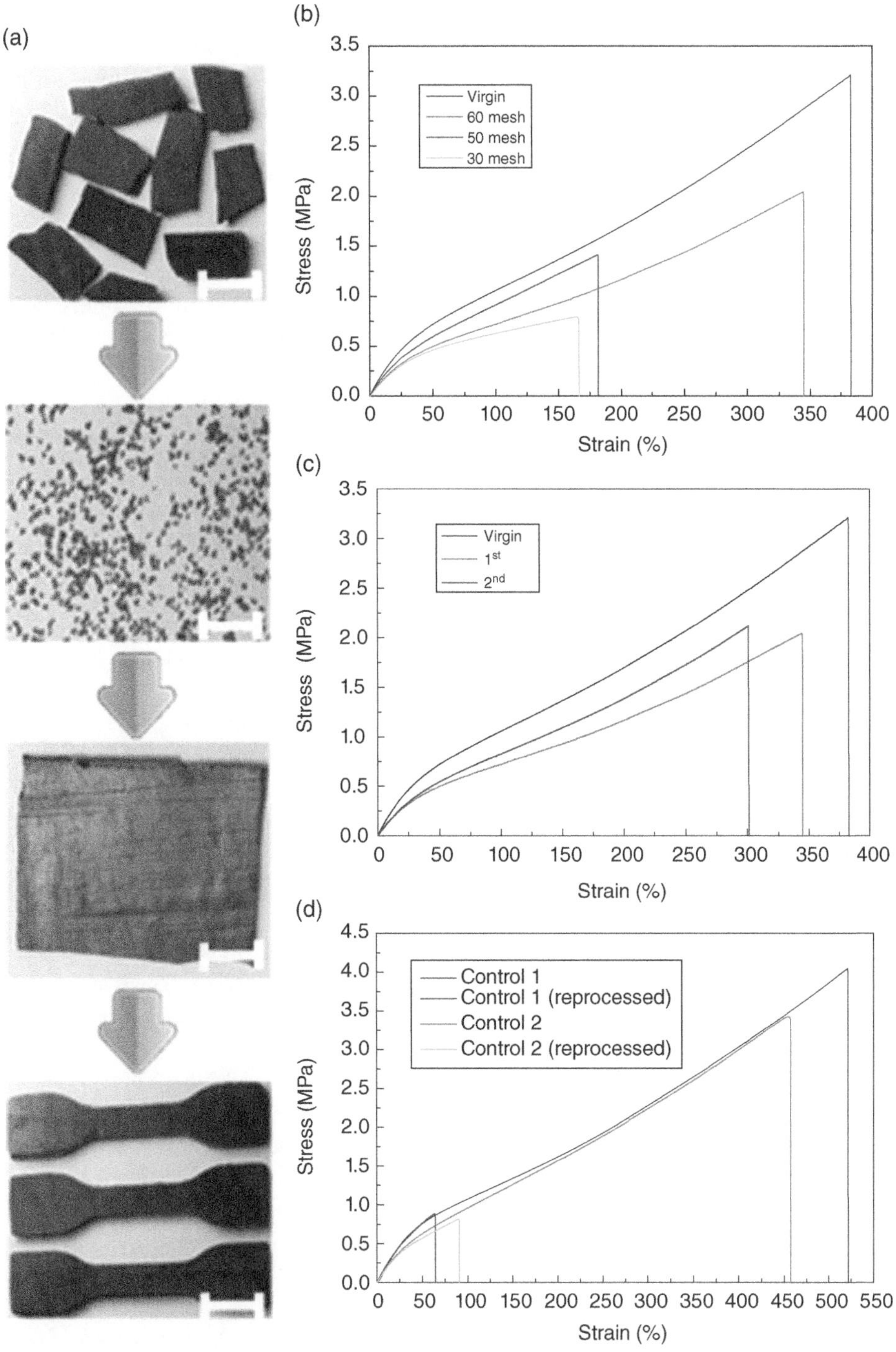

Figure 8.35 (a) Reclaiming of VR-SH. Sheeted rubber was cut, cryogenically ground and compression molded. The scale bars represent 10 mm in length. (b–d) Typical tensile stress–strain curves of virgin and reclaimed rubbers. (b) Effect of particle size of VR-SH. (c) Effect of repeated reclaiming of VR-SH (particle size: 60 mesh). (d) Control 1 and control 2 (particle size: 60 mesh). To categorize the size of the powdered rubber for reclaiming, the latter was run through sieves with different mesh sizes. The size of the screened particles is characterized by the corresponding mesh size of the sieve. The larger the mesh number, the smaller the particle size of the powder. *Source:* (a) Xiang et al. [81], Figure 7, p. 10/with permission of Royal Society of Chemistry. (b–d) Reprinted from Xiang et al. [81]. Copyright 2015, with permission from the Royal Society of Chemistry.

and reshaped. In particular, TBP has proven to be an effective catalyst for promoting air-insensitive metathesis in disulfide-containing polymers. Cross-linked epoxidized polysulfides can be self-healed under its catalysis at room temperature without manual intervention as demonstrated by repeated restoration of tensile strength. In addition, sunlight driven self-healing of cross-linked polyurethane and commercial cross-linked silicone elastomer can be conducted in the absence of any catalyst. The technique may be easier to be developed into an eco-friendly technology for practical application.

Interestingly, our preliminary exploration suggests that metathesis of disulfide bonds of the inherent sulfur cross-links of the vulcanized rubbers can be initiated by the catalyst $CuCl_2$. As a result, the cross-linked networks are allowed to be dynamically rearranged and reshuffled at 110–120 °C. This means that the sulfur cross-links are as stable as the version excluding the catalysts under conventional circumstances without fear of losing stability during service at lower temperature (typically below 110 °C). More importantly, the vulcanized rubbers can be repeatedly self-healed, reshaped, and recycled in solid phase at 110–120 °C without environmentally unfriendly de-cross-linking. The availability of self-healing, reshaping, and reprocessing of vulcanized rubber offers new possibilities of cyclic utilization. Different circulation loops can be selected by various combinations of the smart functionalities according to actual demand. The life cycle of rubbers would thus be extended, while waste of resources reduced.

References

1 Canadell, J., Goossens, H., and Klumperman, B. (2011). Self-healing materials based on disulfide links. *Macromolecules* 44: 2536–2541.

2 Pepels, M., Filot, I., Klumperman, B., and Goossens, H. (2013). Self-healing systems based on disulfide-thiol exchange reactions. *Polymer Chemistry* 4: 4955–4965.

3 Lafont, U., van Zeijl, H., and van der Zwaag, S. (2012). Influence of cross-linkers on the cohesive and adhesive self-healing ability of polysulfide-based thermosets. *ACS Applied Materials & Interfaces* 4: 6280–6288.

4 Rekondo, A., Martin, R., Ruiz de Luzuriaga, A. et al. (2014). Catalyst-free room-temperature self-healing elastomers based on aromatic disulfide metathesis. *Materials Horizons* 1: 237–240.

5 Lei, Z.Q., Xiang, H.P., Yuan, Y.J. et al. (2014). Room-temperature self-healable and remodable cross-linked polymer based on the dynamic exchange of disulfide bonds. *Chemistry of Materials* 26: 2038–2046.

6 Caraballo, R., Rahm, M., Vongvilai, P. et al. (2008). Phosphine-catalyzed disulfide metathesis. *Chemical Communications* 48: 6603–6605.

7 Caraballo, R. (2010). Dynamic sulfur chemistry: screening, evaluation and catalysis. PhD thesis, Royal Institute of Technology, Sweden.

8 Arisawa, M. and Yamaguchi, M. (2003). Rhodium-catalyzed disulfide exchange reaction. *Journal of the American Chemical Society* 125: 6624–6625.

9 Humphrey, R.E. and Potter, J.L. (1965). Reduction of disulfides with tributylphosphine. *Analytical Chemistry* 37: 164–165.

10 Bapat, A.P., Ray, J.G., Savin, D.A., and Sumerlin, B.S. (2013). Redox-responsive dynamic-covalent assemblies: stars and miktoarm stars. *Macromolecules* 46: 2188–2198.

11 Tesoro, G.C. and Sastri, V. (1990). Reversible crosslinking in epoxy resins. I. Feasibility studies. *Journal of Applied Polymer Science* 39: 1425–1437.

12 Sastri, V.R. and Tesoro, G.C. (1990). Reversible crosslinking in epoxy resins. II. New approaches. *Journal of Applied Polymer Science* 39: 1439–1457.

13 Li, E., Jia, P., Liang, L., and Huang, Y. (2014). Phosphine-catalyzed sequential [2 +3] and [3 + 2] annulation domino reaction of γ-benzyl-substituted allenoates with α,β-unsaturated ketimines to construct aza-bicyclo[3,3,0]octane derivatives. *ACS Catalysis* 4: 600–603.

14 Denmark, S.E. and Beutner, G.L. (2008). Lewis base catalysis in organic synthesis. *Angewandte Chemie International Edition* 47: 1560–1638.

15 Fan, R.-H. and Hou, X.-L. (2002). Synthesis of new β-substituted meso-tetraphenylporphyrins via 1,3-dipolar cycloaddition reactions. 1. *The Journal of Organic Chemistry* 68: 726–732.

16 Kloxin, C.J., Scott, T.F., Adzima, B.J., and Bowman, C.N. (2010). Covalent adaptable networks (CANs): a unique paradigm in cross-linked polymers. *Macromolecules* 43: 2643.

17 Watanabe, M. and Yoshie, N. (2006). Synthesis and properties of readily recyclable polymers from bisfuranic terminated poly(ethylene adipate) and multi-maleimide linkers. *Polymer* 47: 4946–4952.

18 Zhang, Y., Broekhuis, A.A., and Picchioni, F. (2009). Thermally self-healing polymeric materials: the next step to recycling thermoset polymers? *Macromolecules* 42: 1906–1912.

19 Sheridan, R.J. and Bowman, C.N. (2013). Understanding the process of healing of thermoreversible covalent adaptable networks. *Polymer Chemistry* 4: 4974–4979.

20 Montarnal, D., Capelot, M., Tournilhac, F., and Leibler, L. (2011). Silica-like malleable materials from permanent organic networks. *Science* 334: 965–968.

21 Zheng, P. and McCarthy, T.J. (2012). A surprise from 1954: siloxane equilibration is a simple, robust, and obvious polymer self-healing mechanism. *Journal of the American Chemical Society* 134: 2024–2027.

22 Zhang, J., Niu, Y., Huang, C. et al. (2012). Self-healable and recyclable triple-shape PPDO–PTMEG co-network constructed through thermoreversible Diels–Alder reaction. *Polymer Chemistry* 3: 1390–1393.

23 Bergman, S.D. and Wudl, F. (2008). Mendable polymers. *Journal of Materials Chemistry* 18: 41–62.

24 Ling, J., Rong, M.Z., and Zhang, M.Q. (2011). Coumarin imparts repeated photochemical remendability to polyurethane. *Journal of Materials Chemistry* 21: 18373–18380.

25 Song, Y.K., Jo, Y.H., Lim, Y.J. et al. (2013). Sunlight-induced self-healing of a microcapsule-type protective coating. *ACS Applied Materials & Interfaces* 5: 1378–1384.

26 Balzani, V., Credi, A., and Venturi, M. (2008). Photochemical conversion of solar energy. *ChemSusChem* 1: 26–58.

27 Otsuka, H., Nagano, S., Kobashi, Y. et al. (2010). A dynamic covalent polymer driven by disulfide metathesis under photoirradiation. *Chemical Communications* 46: 1150–1152.

28 Michal, B.T., Jaye, C.A., Spencer, E.J., and Rowan, S.J. (2013). Inherently photohealable and thermal shape-memory polydisulfide networks. *ACS Macro Letters* 2: 694–699.

29 Zhang, Z.P., Rong, M.Z., Zhang, M.Q., and Yuan, C.E. (2013). Alkoxyamine with reduced homolysis temperature and its application in repeated autonomous self-healing of stiff polymers. *Polymer Chemistry* 4: 4648–4654.

30 Xu, W.M., Rong, M.Z., and Zhang, M.Q. (2016). Sunlight driven self-healing, reshaping and recycling of a robust, transparent and yellowing resistant polymer. *Journal of Materials Chemistry A* 4: 10683–10690.

31 Rosu, D., Rosu, L., and Cascaval, C.N. (2009). IR-change and yellowing of polyurethane as a result of UV irradiation. *Polymer Degradation and Stability* 94: 591–596.

32 Boubakri, A., Guermazi, N., Elleuch, K., and Ayedi, H.F. (2010). Study of UV-aging of thermoplastic polyurethane material. *Materials Science and Engineering: A* 527: 1649–1654.

33 Wool, R.P. and O'Connor, K.M. (1981). A theory of crack healing in polymers. *Journal of Applied Physics* 52: 5953–5963.

34 Song, Y.X., Xu, W.M., Rong, M.Z., and Zhang, M.Q. (2019). A sunlight self-healable transparent strain sensor with high sensitivity and durability based on a silver. *Journal of Materials Chemistry A* 7: 2315–2325.

35 Liu, C. and Yu, X. (2011). Silver nanowire-based transparent, flexible, and conductive thin film. *Nanoscale Research Letters* 6: 75/1–75/8.

36 Coskun, S., Aksoy, B., and Unalan, H.E. (2011). Polyol synthesis of silver nanowires: an extensive parametric study. *Crystal Growth & Design* 11: 4963–4969.

37 Bergin, S.M., Chen, Y.H., Rathmell, A.R. et al. (2012). The effect of nanowire length and diameter on the properties of transparent, conducting nanowire films. *Nanoscale* 4: 1996–2004.

38 Gebeyehu, M.B., Chala, T.F., Chang, S.-Y. et al. (2017). Synthesis and highly effective purification of silver nanowires to enhance transmittance at low sheet resistance with simple polyol and scalable selective precipitation method. *RSC Advances* 7: 16139–16148.

39 Prashant, J. and Pradeep, T. (2005). Potential of silver nanoparticle-coated polyurethane foam as an antibacterial water filter. *Biotechnology and Bioengineering* 90: 59–63.

40 Balberg, I., Binenbaum, N., and Wagner, N. (1984). Percolation thresholds in the three-dimensional sticks system. *Physical Review Letters* 52: 1465–1468.

41 Dalmas, F., Dendievel, R., Chazeau, L. et al. (2006). Carbon nanotube-filled polymer composites. Numerical simulation of electrical conductivity in three-dimensional entangled fibrous networks. *Acta Materialia* 54: 2923–2931.

42 Keshoju, K. and Sun, L. (2009). Mechanical characterization of magnetic nanowire–polydimethylsiloxane composites. *Journal of Applied Physics* 105: 023515/1–023515/5.

43 Wu, B., Heidelberg, A., Boland, J.J. et al. (2006). Microstructure-hardened silver nanowires. *Nano Letters* 6: 468–472.

44 Brook, M.A. (2000). *Silicon in Organic, Organometallic and Polymer Chemistry*. New York: Wiley.

45 Yue, H.B., Fernández-Blázquez, J.P., Beneitoac, D.F., and Vilatela, J.J. (2014). Real time monitoring of click chemistry self-healing in polymer composites. *Journal of Materials Chemistry A* 2: 3881–3887.

46 Zhao, J., Xu, R., Luo, G. et al. (2016). A self-healing, re-moldable and biocompatible crosslinked polysiloxane elastomer. *Journal of Materials Chemistry B* 4: 982–989.

47 Lai, J.C., Mei, J.F., Jia, X.Y. et al. (2016). A stiff and healable polymer based on dynamic-covalent boroxine bonds. *Advanced Materials* 28: 8277–8282.

48 Li, C.H., Wang, C., Keplinger, C. et al. (2016). A highly stretchable autonomous self-healing elastomer. *Nature Chemistry* 8: 618–624.

49 Martín, R., Rekondo, A., Echeberria, J. et al. (2012). Room temperature self-healing power of silicone elastomers

having silver nanoparticles as crosslinkers. *Chemical Communications* 48: 8255–8257.
50 Roy, N., Buhler, E., and Lehn, J.-M. (2013). The tris-urea motif and its incorporation into polydimethylsiloxane-based supramolecular materials presenting self-healing features. *Chemistry – A European Journal* 19: 8814–8820.
51 Yang, L., Lin, Y., Wang, L., and Zhang, A. (2014). The synthesis and characterization of supramolecular elastomers based on linear carboxyl-terminated polydimethylsiloxane oligomers. *Polymer Chemistry* 5: 153–160.
52 Cui, J., Daniel, D., Grinthal, A. et al. (2015). Dynamic polymer systems with self-regulated secretion for the control of surface properties and material healing. *Nature Materials* 14: 790–795.
53 Lai, G.Q. and Xing, S.M. (2009). *Synthesis Technology and Application of Organic Silicon Product*. Beijing: Chemical Industry Press.
54 Döhlert, P. and Enthaler, S. (2016). Conversion of poly(methylhydrosiloxane) waste to useful commodities. *Catalysis Letters* 146: 345–352.
55 Döhlert, P., Pfrommer, J., and Enthaler, S. (2015). Recycling concept for end-of-life silicones: boron trifluoride diethyl etherate as depolymerization reagent to produce difluorodimethylsilane as useful commodity. *ACS Sustainable Chemistry & Engineering* 3: 163–169.
56 Enthaler, S. (2014). Zinc-catalyzed depolymerization of end-of-life polysiloxanes. *Angewandte Chemie International Edition* 53: 2716–2721.
57 Xiang, H.P., Rong, M.Z., and Zhang, M.Q. (2017). A facile method for imparting sunlight driven catalyst-free self-healability and recyclability to commercial silicone elastomer. *Polymer* 108: 339–347.
58 AbdolahZadeh, M., Esteves, A.C.C., van der Zwaag, S., and Garcia, S.J. (2014). Healable dual organic–inorganic crosslinked sol–gel based polymers: crosslinking density and tetrasulfide content effect. *Journal of Polymer Science Part A: Polymer Chemistry* 52: 1953–1961.
59 Yu, K., Taynton, P., Zhang, W. et al. (2014). Reprocessing and recycling of thermosetting polymers based on bond exchange reactions. *RSC Advances* 4: 10108–10117.
60 Fairbanks, B.D., Singh, S.P., Bowman, C.N., and Anseth, K.S. (2011). Photodegradable, photoadaptable hydrogels via radical-mediated disulfide fragmentation reaction. *Macromolecules* 44: 2444–2450.
61 Amamoto, Y., Otsuka, H., Takahara, A., and Matyjaszewski, K. (2012). Self-healing of covalently cross-linked polymers by reshuffling thiuram disulfide moieties in air under visible light. *Advanced Materials* 24: 3975–3980.
62 Hayes, W. and Greenland, B.W. (2013). *Healable Polymer Systems*. Cambridge: Royal Society of Chemistry.
63 Chen, M., Zhong, M., and Johnson, J.A. (2016). Light-controlled radical polymerization: mechanisms, methods, and applications. *Chemical Reviews* 116: 10167–10211.
64 Amamoto, Y., Otsuka, H., Takahara, A., and Matyjaszewski, K. (2012). Changes in network structure of chemical gels controlled by solvent quality through photoinduced radical reshuffling reactions of trithiocarbonate units. *ACS Macro Letters* 1: 478–481.
65 Zhang, M.Q. (2016). "Waking sleeping beauty": regeneration of vulcanized rubber by triggering dynamicity of the inherent sulfur network. *Express Polymer Letters* 10: 627–627.
66 Ling, J., Rong, M.Z., and Zhang, M.Q. (2012). Photo-stimulated self-healing polyurethane containing dihydroxyl coumarin derivatives. *Polymer* 53: 2691–2698.
67 Ling, J., Rong, M.Z., and Zhang, M.Q. (2014). Effect of molecular weight of PEG soft segments on photo-stimulated self-healing performance of coumarin functionalized polyurethanes. *Chinese Journal of Polymer Science* 32: 1286–1297.
68 Xia, N.N., Xiong, X.M., Wang, J. et al. (2016). A seawater triggered dynamic coordinate bond and its application for underwater self-healing and reclaiming of lipophilic polymer. *Chemical Science* 7: 2736–2742.
69 Xia, N.N., Rong, M.Z., and Zhang, M.Q. (2016). Stabilization of catechol–boronic ester bonds for underwater self-healing and recycling of lipophilic bulk polymer in wider pH range. *Journal of Materials Chemistry A* 4: 14122–14131.
70 Xu, C., Cao, L., Lin, B. et al. (2016). Design of self-healing supramolecular rubbers by introducing ionic cross-links into natural rubber via a controlled vulcanization. *ACS Applied Materials & Interfaces* 8: 17728–17737.
71 Adhikari, B., De, D., and Maiti, S. (2000). Reclamation and recycling of waste rubber. *Progress in Polymer Science* 25: 909–948.
72 Myhre, M., Saiwari, S., Dierkes, W., and Noordermeer, J. (2012). Rubber recycling: chemistry, processing, and applications. *Rubber Chemistry and Technology* 85: 408–449.
73 Aprem, A.S., Joseph, K., Mathew, T. et al. (2003). Studies on accelerated sulphur vulcanization of natural rubber using 1-phenyl-2, 4-dithiobiuret/tertiary butyl benzothiazole sulphenamide. *European Polymer Journal* 39: 1451–1460.
74 Studebaker, M.L. (1966). Effect of curing systems on selected physical properties of natural rubber vulcanizates. *Rubber Chemistry and Technology* 39: 1359–1381.
75 Posadas, P., Fernández, A., Brasero, J. et al. (2007). Vulcanization of polybutadiene rubber with

dipentamethylene thiuram tetrasulfide. *Journal of Applied Polymer Science* 106: 3481–3487.

76 González, L., Rodríguez, A., Del Campo, A., and Marcos-Fernández, A. (2002). Crosslink reaction of natural rubber with thiuram sulphur donors in the presence of a thiuram monosulfide. *Journal of Applied Polymer Science* 85: 491–499.

77 Valentín, J.L., Rodríguez, A., Marcos-Fernández, A., and González, L. (2004). Crosslinking of natural and polybutadiene rubbers with thiuram sulfur donors in the presence of a thiuram monosulfide. *Journal of Applied Polymer Science* 93: 1756–1761.

78 Fan, R.L., Zhang, Y., Li, F. et al. (2001). Effect of high-temperature curing on the crosslink structures and dynamic mechanical properties of gum and N330-filled natural rubber vulcanizates. *Polymer Testing* 20: 925–936.

79 Hirayama, D. and Saron, C. (2012). Chemical modifications in styrene–butadiene rubber after microwave devulcanization. *Industrial & Engineering Chemistry Research* 51: 3975–3980.

80 Yoon, J.A., Kamada, J., Koynov, K. et al. (2012). Self-healing polymer films based on thiol–disulfide exchange reactions and self-healing kinetics measured using atomic force microscopy. *Macromolecules* 45: 142–149.

81 Xiang, H.P., Qian, H.J., Lu, Z.Y. et al. (2015). Crack healing and reclaiming of vulcanized rubber by triggering the rearrangement of inherent sulfur crosslinked networks. *Green Chemistry* 17: 4315–4325.

82 Zhong, X.H., Xu, S.L., and Lv, Y.Y. (1998). *Inorganic Chemistry*. Beijing: Science Press.

83 Lamberti, C., Prestipino, C., Capello, L. et al. (2001). The $CuCl_2/Al_2O_3$ catalyst investigated in interaction with reagents. *International Journal of Molecular Sciences* 2: 230–245.

84 Jain, R.K. and Mishra, A.P. (2012). Microwave synthesis, spectral, thermal, and antimicrobial activities of some transition metal complexes involving 5-bromosalicylaldehyde moiety. *Current Chemistry Letters* 1: 163–174.

85 Aliaga, M.E., Andrade-Acuña, D., López-Alarcón, C. et al. (2013). Cu(II)–disulfide complexes display simultaneous superoxide dismutase- and catalase-like activities. *Journal of Inorganic Biochemistry* 129: 119–126.

86 Yang, J.S., Lin, Y.H., and Yang, C.S. (2002). Palladium-catalyzed synthesis of trans-4-(N,N-bis(2-pyridyl)amino) stilbene. A new intrinsic fluoroionophore for transition metal ions. *Organic Letters* 4: 777–780.

87 Lin, Y.J. and Hwang, S.J. (2004). Temperature prediction of rolling tires by computer simulation. *Mathematics and Computers in Simulation* 67: 235–249.

88 Ebbott, T.G., Hohman, R.L., Jeusette, J.P., and Kerchman, V. (1999). Tire temperature and rolling resistance prediction with finite element analysis. *Tire Science and Technology* 27: 2–21.

89 Taynton, P., Yu, K., Shoemaker, R.K. et al. (2014). Heat- or water-driven malleability in a highly recyclable covalent network polymer. *Advanced Materials* 26: 3938–3942.

90 Bilgili, E., Dybek, A., Arastoopour, H., and Bernstein, B. (2003). A new recycling technology: compression molding of pulverized rubber waste in the absence of virgin rubber. *Journal of Elastomers & Plastics* 35: 235–256.

91 Gugliemotti, A., Lucignano, C., and Quadrini, F. (2012). Production of rubber parts by tyre recycling without using virgin materials. *Plastics, Rubber and Composites* 41: 40–46.

Index

Extrinsic and Intrinsic Approaches to Self-Healing Polymers and Polymer Composites, First Edition. Ming Qiu Zhang and Min Zhi Rong.

d

r

s

t

u

v

w

x

y

z

Printed and bound by CPI Group (UK) Ltd, Croydon, CR0 4YY